MECHANICAL ENGINEERING

Edited by **Murat Gökçek**

INTECHOPEN.COM

Mechanical Engineering
Edited by Murat Gökçek

CBS, Edition 2016

Published by InTech

Janeza Trdine 9, 51000 Rijeka, Croatia

Publishing Process Manager Danijela Duric

Technical Editor Teodora Smiljanic

Cover Designer InTech Design Team

Additional hard copies can be obtained from orders@intechopen.com

Mechanical Engineering, Edited by Murat Gökçek
p. cm.
ISBN 978-953-51-0505-3

INTECH
open science | open minds

Contents

Preface IX

Part 1 Power Transmission Systems 1

Chapter 1 **Mechanical Transmissions Parameter Modelling 3**
Isad Saric, Nedzad Repcic and Adil Muminovic

Chapter 2 **Gearbox Simulation Models with Gear and Bearing Faults 17**
Endo Hiroaki and Sawalhi Nader

Chapter 3 **Split Torque Gearboxes: Requirements, Performance and Applications 55**
Abraham Segade-Robleda, José-Antonio Vilán-Vilán, Marcos López-Lago and Enrique Casarejos-Ruiz

Chapter 4 **On the Modelling of Spur and Helical Gear Dynamic Behaviour 75**
Velex Philippe

Chapter 5 **The Role of the Gearbox in an Automatic Machine 107**
Hermes Giberti, Simone Cinquemani and Giovanni Legnani

Chapter 6 **Electrical Drives for Crane Application 131**
Nebojsa Mitrovic, Milutin Petronijevic, Vojkan Kostic and Borislav Jeftenic

Part 2 Manufacturing Processes and System Analysis 157

Chapter 7 **Anisotropic Mechanical Properties of ABS Parts Fabricated by Fused Deposition Modelling 159**
Constance Ziemian, Mala Sharma and Sophia Ziemian

Chapter 8 **Design and Evaluation of Self-Expanding Stents Suitable for Diverse Clinical Manifestation Based on Mechanical Engineering 181**
Daisuke Yoshino and Masaaki Sato

Chapter 9 **Spin and Spin Recovery 209**
Dragan Cvetković, Duško Radaković,
Časlav Mitrović and Aleksandar Bengin

Chapter 10 **Surface Welding as a Way of Railway Maintenance 233**
Olivera Popovic and Radica Prokic-Cvetkovic

Chapter 11 **Study on Thixotropic Plastic Forming of Magnesium Matrix Composites 253**
Hong Yan

Chapter 12 **Development of a Winding Mechanism for Amorphous Ribbon Used in Transformer Cores 277**
Marcelo Ruben Pagnola and Rodrigo Ezequiel Katabian

Chapter 13 **Free Vibration Analysis of Centrifugally Stiffened Non Uniform Timoshenko Beams 291**
Diana V. Bambill, Daniel H. Felix,
Raúl E. Rossi and Alejandro R. Ratazzi

Chapter 14 **Vibration-Based Diagnostics of Steam Turbines 315**
Tomasz Gałka

Chapter 15 **On the Mechanical Compliance of Technical Systems 341**
Lena Zentner and Valter Böhm

Part 3 Thermo-Fluid Systems 353

Chapter 16 **Waste Heat Recycling for Fuel Reforming 355**
Rong-Fang Horng and Ming-Pin Lai

Chapter 17 **Steam Turbines Under Abnormal Frequency Conditions in Distributed Generation Systems 381**
Fabrício A. M. Moura, José R. Camacho,
Geraldo C. Guimarães and Marcelo L. R. Chaves

Chapter 18 **Aeronautical Engineering 401**
Časlav Mitrović, Aleksandar Bengin,
Nebojša Petrović and Jovan Janković

Chapter 19 **Numerical Modeling of Wet Steam Flow in Steam Turbine Channel 443**
Hasril Hasini, Mohd. Zamri Yusoff and Norhazwani Abd. Malek

Chapter 20 **Experimental Study on Generation of Single Cavitation Bubble Collapse Behavior by a High Speed Camera Record 463**
Sheng-Hsueh Yang, Shenq-Yuh Jaw and Keh-Chia Yeh

Chapter 21 **Noise Reduction in Butterfly Valve Cavitation by Semicircular Fins and Visualization of Cavitation Flow 483**
Kazuhiko Ogawa

Part 4 Simulations and Computer Applications 501

Chapter 22 **Computer Simulation of Involute Tooth Generation 203**
Cuneyt Fetvaci

Chapter 23 **Applications of Computer Vision in Micro/Nano Observation 527**
Yangjie Wei, Chengdong Wu and Zaili Dong

Chapter 24 **Advanced Free Form Manufacturing by Computer Aided Systems – Cax 555**
Adriano Fagali De Souza and Sabrina Bodziak

Part 5 New Approaches for Mechanical Engineering Education and Organization Systems 587

Chapter 25 **Modern Methods of Education, Research and Design Used in Mechanical Engineering 589**
Borza Sorin-Ioan, Brindasu Paul Dan and Beju Livia Dana

Chapter 26 **Mechanical Engineering Education: Preschool to Graduate School 615**
Emily M. Hunt, Pamela Lockwood-Cooke and Michelle L. Pantoya

Chapter 27 **Use of Discounted Cash Flow Methods for Evaluation of Engineering Projects 631**
Igor Pšunder

Chapter 28 **Configuration Logic of Standard Business Processes for Inter-Company Order Management 647**
Carsten Schmidt and Stefan Cuber

Preface

Mechanical engineering is one of the largest engineering disciplines that present tangible solutions for whole humanity's prosperity and the quality of life. Rapidly increasing demands have been increasing its importance more and more. The many areas within the scope of mechanical engineering include transportation, power generation, energy conversion, machine design, manufacturing and automation, the control of system. The purpose of the *Mechanical Engineering* book is to present to the engineers in industrial areas and to the academic environments the state-of-to-art information on the most important topics of modern mechanical engineering.

This Mechanical Engineering book is organized into the following five parts:

I Power Transmission Systems
II Manufacturing Processes and System Analysis
III Thermo-Fluid Systems
IV Simulations and Computer Applications
V New Approaches in Mechanical Engineering Education and Organization Systems

The first part of this book starts with a collection of articles on the power transmission systems. This section introduces modeling of transmission parameter, the performance and simulation, and dynamics analysis of gearboxes. Section two collects articles about the manufacturing processes and system analysis such as welding, plastic forming, investigation of mechanical properties, and vibration analysis. The third section presents the studies related to thermo-fluid science and it includes topics such as fuel reforming, steam turbines used distributed power production, numerical modeling of wet steam flow, collapse behavior of cavitation bubble, and visualization of cavitation flow. The subsequent fourth part provides a platform to share knowledge about the simulation and computer applications in mechanical engineering. Lastly, section five is a collection of articles that investigate modern education methods and engineering projects in mechanical engineering.

I would like to express my sincere appreciation to all of the authors for their contributions. The successful completion of the book *Mechanical Engineering* has been the result of the cooperation of many people. I would like to thank the Publishing Process Manager Ms. Leda Bace for her support during the publishing process, as well as Mr. Metin Ertufan for inviting me to be the editor of this book.

Dr. Murat Gökçek PhD,
Faculty of Engineering
Department of Mechanical Engineering
Niğde University, Niğde,
Turkey

Part 1

Power Transmission Systems

1

Mechanical Transmissions Parameter Modelling

Isad Saric, Nedzad Repcic and Adil Muminovic
University of Sarajevo, Faculty of Mechanical Engineering,
Department of Mechanical Design,
Bosnia and Herzegovina

1. Introduction

In mechanical technique, transmission means appliance which is used as intermediary mechanism between driving machine (e.g. of engine) and working (consumed) machine. The role of transmission is transmitting of mechanical energy from main shaft of driving machine to main shaft of working machine. The selection of transmission is limited by the price of complete appliance, by working environment, by dimensions of the appliance, technical regulations, etc. In mechanical engineering, so as in technique generally, mechanical transmissions are broadly used. Mechanical transmissions are mechanisms which are used for mechanical energy transmitting with the change of angle speed and appropriate change of forces and rotary torques. According to the type of transmitting, mechanical transmissions could be divided into: transmissions gear (sprocket pair), belt transmissions (belt pulleys and belt), friction transmissions (friction wheels) and chain transmissions (chain pulleys and chain). (Repcic & Muminovic, 2007)

In this chapter, the results of the research of three-dimensional (3D) geometric parameter modelling of the two frequently used types of mechanical transmissions, transmissions gear (different types of standard catalogue gears: spur gears, bevel gears and worms) and belt transmissions (belt pulley with cylindrical external surface, or more exactly, with pulley rim) using CATIA V5 software system (modules: *Sketcher, Part Design, Generative Shape Design, Wireframe and Surface Design* and *Assembly Design*), is shown.

Modelling by computers are based on geometric and perspective transformation which is not more detail examined in the chapter because of their large scope.

It is advisable to make the parameterisation of mechanical transmissions for the purpose of automatization of its designing. Parameter modelling application makes possible the control of created geometry of 3D model through parameters integrated in some relations (formulas, parameter laws, tables and so on). All dimensions, or more precisely, geometric changeable parameter of gear and belt pulley, can be expressed through few characteristic fixed parameters (m, z, z_1, z_2 and N for the selected gear; d, B_k, d_v and s for the selected belt pulley). Geometry of 3D mechanical transmission model is changed by changes of these parameters values. Designer could generate more designing solutions by mechanical transmission parameterisation.

Because AutoCAD does not support parameter modelling, and command system, that it has, does not make possible simple realization of changes on finished model, parameter oriented software systems (CATIA V5, SolidWorks, Mechanical Desktop, and so on) which used analytical expressions for variable connection through parameters are used. CATIA V5 (*Computer-**A**ided **T**hree-dimensional **I**nteractive **A**pplication*) is the product of the highest technological level and represents standard in the scope of designing (Dassault Systemes, IBM, 2011). Currently, it is the most modern integrated CAD/CAM/CAE software system that can be find on the market for commercial use and scientific-research work. The biggest and well-known world companies and their subcontractors use them. It is the most spread in the car industry (Daimler Chrysler, VW, BMW, Audi, Renault, Peugeot, Citroen, etc.), airplane industry (Airbus, Boeing, etc.), and production of machinery and industry of consumer goods. The system has mathematical models and programs for graphical shapes presentation, however users have no input about this process. As a solution, it is written independently from operative computer system and it provides the possibility for program module structuring and their adaptation to a user. In the „heart" of the system is the integrated associational data structure for parameter modelling, which enables the changes on the model to be reflected through all related phases of the product development. Therefore, time needed for manual models remodelling is saved. The system makes possible all geometric objects parametering, including solids, surfaces, wireframe models and constructive elements. (Karam & Kleismit, 2004; Saric et al., 2010) Whole model, or part of model, can parameterise in the view of providing of more flexibility in the development of new variants designing solutions. Intelligent elements interdependence is given to a part or assembly by parameterising. The main characteristic of parameter modelling in CATIA V5 system is the great flexibility, because of the fact that parameters can be, but do not have to be, defined in any moment. Not only changing of parameter value, but their erasing, adding and reconnecting, too, are always possible. (Karam & Kleismit, 2004) Total ***G**raphical **U**ser **I**nterface* (GUI) programmed in C++ program is designed like tools palette and icons that can be find in Windows interface. Although it was primarily written for Windows and Windows 64-bit, the system was written for AIX, AIX 64-bit, HP-UX, IRIX and Solaris operative system. To obtain the maximum during the work with CATIA V5 system, optimized certificated hardware configurations are recommended (Certified hardware configurations for CATIA V5 systems, 2011).

Parameter modelling in CATIA V5 system is based on the concept of knowledge, creating and use of parameter modelled parts and assemblies. (Saric et al., 2009) Creating of 3D parameter solid models is the most frequently realized by combining of the approach based on ***F**eatures **B**ased **D**esign* - FBD and the approach based on Bool's operations (***C**onstructive **S**olid **G**eometry* - CSG). (Amirouche, 2004; Shigley et al. 2004; Spotts at el., 2004) The most frequent parameter types in modelling are: *Real, Integer, String, Length, Angle, Mass*, etc. They are devided into two types:

- internal parameters which are generated during geometry creating and which define its interior features (depth, distance, activity, etc.) and
- user parameters (with one fixed or complex variables) which user specially created and which define additional information on the: *Assembly Level, Part Level* or *Feature Level.*

So, parameter is a variable we use to control geometry of component, we influence its value through set relations. It is possible to do a control of geometry by use of tools palette *Knowledge* in different ways:

- by creating of user parameters set and by their values changing,
- by use of defined formulas and parameter laws that join parameters,
- by joining of parameters in designed tables and by selection of appropriate configured set.

The recommendation is, before components parameterising, to:

1. check the component complexity,
2. notice possible ways of component making,
3. notice dimensions which are going to change and
4. select the best way for component parameterising.

2. Mechanical transmissions parameter modelling

Modelling of selected mechanical transmissions was done in *Sketcher, Part Design* and *Generative Shape Design* modules of CATIA V5 system. As prerequisite for this way of modelling, it is necessary to know modelling methodology in modules *Wireframe and Surface Design* and *Assembly Design* of CATIA V5 system. (Karam & Kleismit, 2004; Dassault Systemes, 2007a, 2007b; Zamani & Weaver, 2007)

After finished modelling procedure, mechanical transmissions can be independently used in assemblies in complex way.

Parameter marks and conventional formulas (Table 1. and 5.) used in mechanical transmissions modelling can be found in references (Repcic & et al., 1998; Repcic & Muminovic, 2007, pgs. 139, 154-155, 160-161). Clear explanations for transmissions gear and belt transmissions can be found in references (Repcic & et al., 1998, pgs. 54-106, 118-151).

2.1 Transmissions gear parameter modelling

Next paragraph is shows 3D geometric parameter modelling of characteristic standard catalogue gears: spur gears, bevel gears and worms.

Gears were selected as characteristic example, either because of their frequency as mechanical elements or because exceptionally complex geometry of cog side for modelling.

Every user of software system for designing is interested in creation of complex plane curve *Spline* which defined geometry of cog side profile.

The control of 3D parameterised model geometry is done by created parameters, formulas and parameter laws shown in the tree in Fig. 1. (Cozzens, 2006) Parameters review, formulas and parameter laws in the *Part* documents tree activating is done through the main select menu (*Tools* → *Options* → *Part Infrastructure* → *Display*).

Spur gear	Bevel gear	Worm
z	*z*1, *z*2	*z*1=1
m		*m*=1,5
a=20 deg		
p=*m**PI		
r=(*z***m*)/2	*r*=(*z*1**m*)/2	*d*=20 mm
rb=*r**cos(*a*)	*rb*=*rc**cos(*a*)	

Spur gear	Bevel gear	Worm
rf=*r*-1.2**m*		
ra=*r*+*m*		*ra*=*d*/2+*ha*
rr=0.38**m*		
ha=*m*		
hf=1.2**m*		
xd=*rb**(cos(*t***PI*)+sin(*t***PI*)**t***PI*)		*N*=6,5
yd=*rb**(sin(*t***PI*)-cos(*t***PI*)**t***PI*)		*L*=(*N*+1)**p*
0≤*t*≤1		*gama*=atan(*m***z*1/*d*)
	delta=atan(*z*1/*z*2)	*dZ*=-*L*/2
	rc=*r*/cos(*delta*)	
	lc=*rc*/sin(*delta*)	
	tc=-atan(Re*lations**yd.Evaluate*(*a*/180deg)/ Re*lations**xd.Evaluate*(*a*/180deg))	
	b=0.3**rc*	
	ratio=1-*b*/*lc*/cos(*delta*)	
	dZ=0 mm	

Table 1. Parameters and formulas

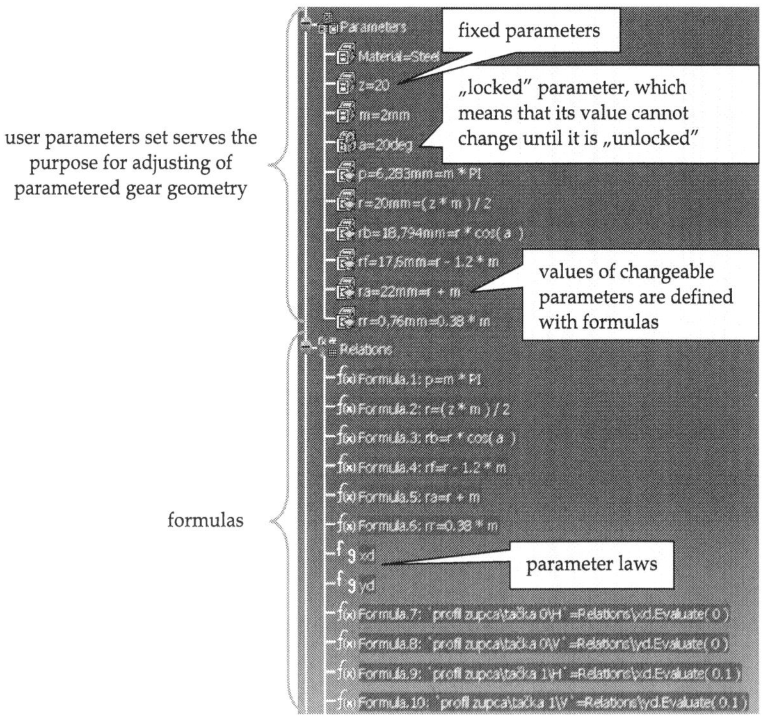

Fig. 1. Gear geometry control

2.1.1 Spur gears parameter modelling

To define fixed parameters (Fig. 2.), we select command *Formula*  from tools palette *Knowledge* or from main select menu. Then, we:

1. choose desired parameter type (*Real, Integer, Length, Angle*) and press the button *New Parameter of type,*
2. type in a new parameter name,
3. assign a parameter value (only in the case if parameter has fixed value) and
4. press the button *Apply* to confirm a new parameter creation.

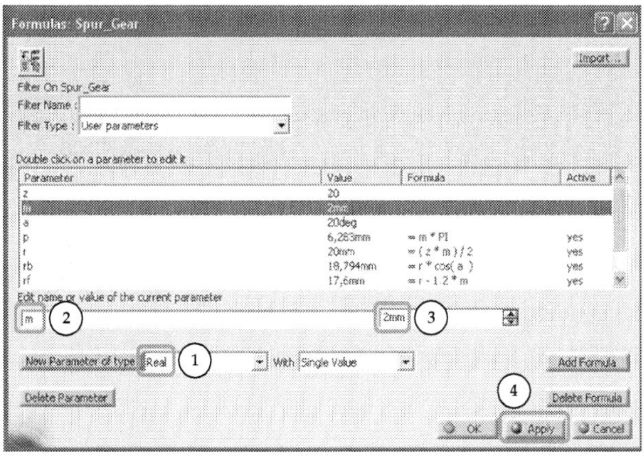

Fig. 2. Fixed parameters defining

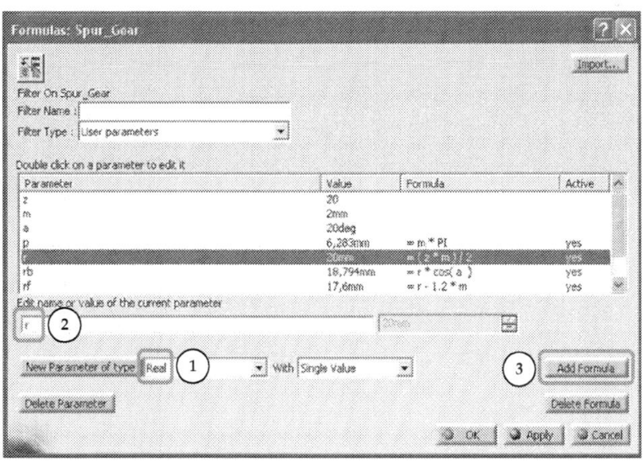

Fig. 3. Changeable parameters defining

Most geometrical gear parameters are changeable and are in the function of fixed parameters m and z (Fig. 3.). We do not need to set values for these parameters, because CATIA V5 system calculates them itself. So, instead of values setting, formulas are defined by choosing the command *Formula* f(x) (Fig. 4.). When formula has been created, it is possible to manipulate with it by the tree, similar as with any other model feature.

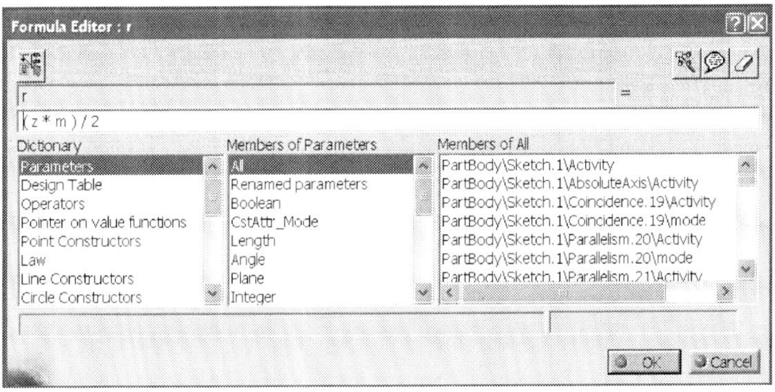

Fig. 4. Formula setting

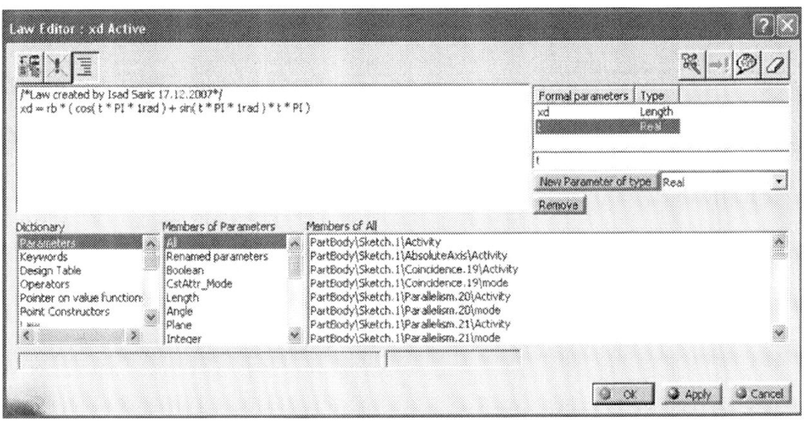

Fig. 5. Setting of parameter laws for calculation of x and y coordinates of involute points

Position of the points on involutes profile of cog side is defined in the form of parameter laws (Fig. 5.). For coordinate points of involute $(x0,y0)$, $(x1,y1)$, ... , $(x4,y4)$ we most frequently define a set of parameters. To create parameter laws, we choose the command *Law* fog from tools palette *Knowledge*. Then, we give two laws in parameter form, which we are going to be used for calculation of x and y coordinate points of involute

$$xd = rb * (\cos(t * PI * 1rad) + \sin(t * PI * 1rad) * t * PI) \tag{1}$$

$$yd = rb * (\sin(t * PI * 1rad) - \cos(t * PI * 1rad) * t * PI) \quad (2)$$

While we use law editor, we have to take into account the following:

- trigonometric functions, specially angles, are not considered as numbers, and because of that angle constants like $1rad$ or $1\deg$ must be used,
- *PI* is the value of the number π .

For the purpose of accuracy checking of previously conducted activities, review of formulas, parameter laws and values of all defined fixed and changeable parameters is activated in the tree of *Part* document (*Tools* → *Options* → *Knowledge*).

The example of spur gear parameter modelling is shown in the next paragraph. All dimensions, or more precisely, geometric changeable parameters of spur gear are in the function of fixed parameters m and z. We can generate any spur gear by changing parameters m and z.

Part Number	m	z	d_g	d	b_z	b_g	b_k	t_k
G2-20	2	20	30	15	20	35	5	2,35
G3-40	3	40	60	25	30	50	8	3,34
G4-60	4	60	120	30	40	60	10	3,34

Table 2. Selected spur gears parameters

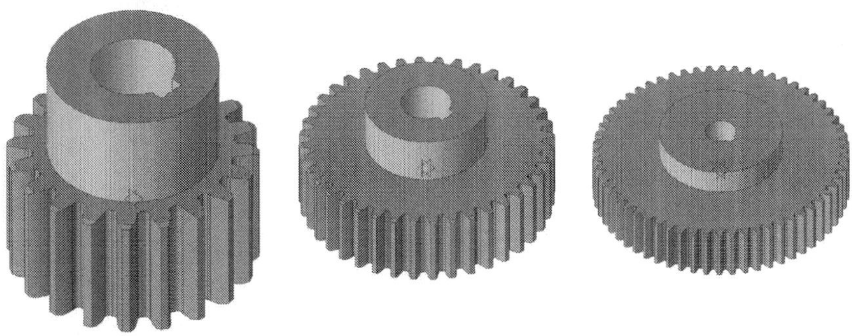

1° G2-20 (m=2 mm, z=20) 2° G3-40 (m=3 mm, z=40) 3° G4-60 (m=4 mm, z=60)

Fig. 6. Different spur gears are the result of parameter modelling

Fig. 6. shows three different standard catalogue spur gears made from the same CATIA V5 file, by changing parameters m and z. (Saric et al., 2009, 2010)

2.1.2 Bevel gears parameter modelling

The example of bevel gear parameter modelling is shown in the next paragraph. All dimensions, or more precisely, geometric changeable parameters of bevel gear are in the function of fixed parameters m, z_1 and z_2. We can generate any bevel gear by changing parameters m, z_1 and z_2.

Part Number	m	z_1	z_2	d_g	d	b_z	b_g	Connection between hub and shaft
B2-25	2	25	25	40	12	10,6	25,52	M5
DB3-15	3	15	30	36	18	17	36,26	M6
FB4-15	4	15	60	48	20	34	59,9	M8

Table 3. Selected bevel gears parameters

Fig. 7. shows three different standard catalogue bevel gears made from the same CATIA V5 file, by changing parameters m, z_1 and z_2. (Saric et al., 2009, 2010)

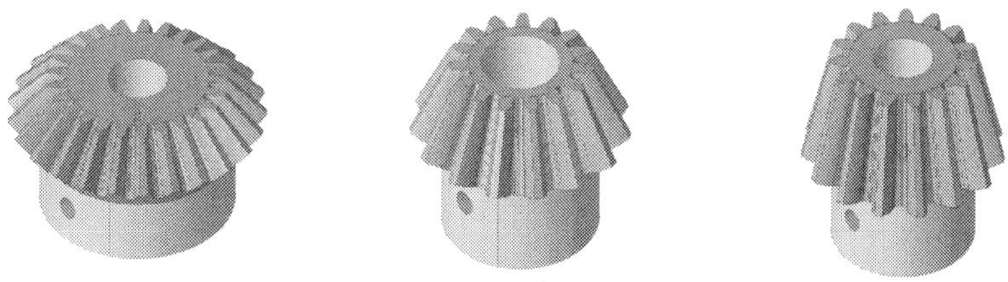

1° B2-25 (m=2 mm, z_1=25, z_2=25) 2° DB3-15 (m=3 mm, z_1=15, z_2=30) 3° FB4-15 (m=4 mm, z_1=15, z_2=60)

Fig. 7. Different bevel gears are the result of parameter modelling

2.1.3 Worms parameter modelling

The example of worm parameter modelling is shown in the next paragraph. All dimensions, or more precisely, geometric changeable parameters of worm are in the function of fixed parameters m, z_1 and N. We can generate any worm by changing parameters m, z_1 and N.

Part Number	m	z_1	d_g	d	L	L_g	Connection between hub and shaft
W1,5-1	1,5	1	23	10	35	45	M5
W2,5-2	2,5	2	35	15	45	60	M6
W3-3	3	3	41	20	55	70	M8

Table 4. Selected worms parameters

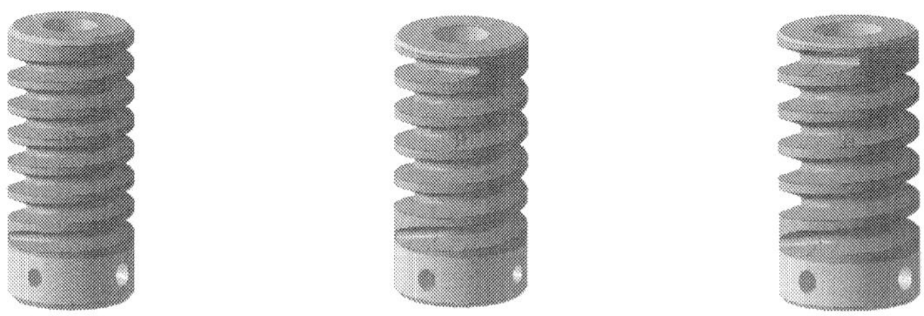

1° W1,5-1 (m=1,5 mm, z_1=1, N=6,5) 2° W2,5-2 (m=2,5 mm, z_1=2, N=5) 3° W3-3 (m=3 mm, z_1=3, N=5)

Fig. 8. Different worms are the result of parameter modelling

Fig. 8. shows three different standard catalogue worms made from the same CATIA V5 file, by changing parameters m, z_1 and N. (Saric et al., 2009, 2010)

2.2 Belt transmissions parameter modelling

This application includes wide area of the industry for the fact that belt transmitting is often required. Generally, belt transmitting designing process consists of needed drive power estimate, choice of belt pulley, length and width of belt, factor of safety, etc. Final design quality can be estimated by efficiency, compactness and possibilities of service. If engineer does not use parameter modelling, he/she must pass through exhausting phase of design, based on learning from the previous done mistakes, in order to have standard parts like belt pulleys and belts, mounted on preferred construction. This process is automatized by parameter modelling. In such process, characteristics that registered distance between belt pulleys, belts length, etc., are also created. Such characteristics, also, register links, belt angle speeds and exit angle speed. The results for given belts length can be obtained by the feasibility study. Few independent feasibility studies for the different belts lengths are compared with demands for compactness. In such a way, several constructions of belt transmitting can be tested, and then it is possible to find the best final construction solution.

The example of belt pulley parameter modelling is shown in the next paragraph. The belt pulley K is shown in the Fig. 9., and it consists of several mutual welded components: hub G, pulley rim V, plate P and twelve side ribs BR. All dimensions, or more precisely, geometric changeable parameters of belt pulley are in function of fixed parameters d, B_k, d_v and s. We can generate any belt pulley with cylindrical external surface by changing parameters d, B_k, d_v and s.

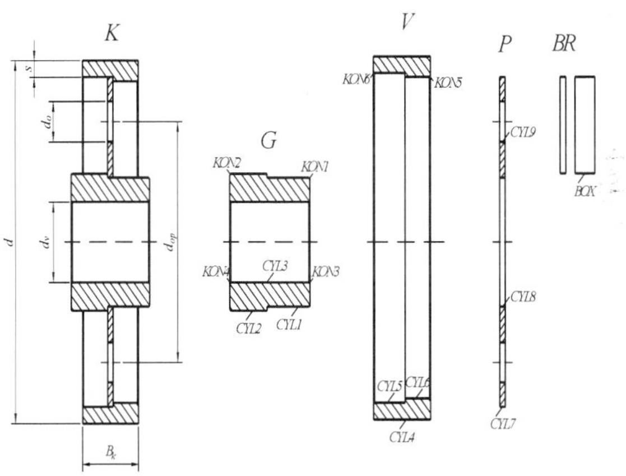

Fig. 9. Modelling of belt pulley parts with cylindrical external surface

Dimensions of hub depends from diameter of shaft d_v, on which hub is set. Shaft diameter is the input value through which the other hub dimension are expressed.

Hub shape can be obtained by adding and subtraction of cylinders and cones shown in the Fig. 9.

$$G = CYL1 + CYL2 - CYL3 - KON1 - KON2 - KON3 - KON4 \tag{3}$$

Pulley rim of belt pulley depends from diameter of belt pulley d, pulley rim width B_k, diameter of shaft d_v and minimal pulley rim thickness s.

$$V = CYL4 - CYL5 - CYL6 - KON5 - KON6 \tag{4}$$

Plate dimensions depend from diameter of belt pulley d, minimal pulley rim thickness s and diameter of shaft d_v.

$$P = CYL7 - CYL8 - 6 \cdot CYL9 \tag{5}$$

Side ribs are side set rectangular plates which can be shown by primitive in the form of prism.

$$BR = BOX \tag{6}$$

Whole belt pulley is obtained by adding of formed forms.

$$K = G + V + P + 6 \cdot BR \tag{7}$$

Belt pulley with cylindrical external surface	
$CYL1$: D=1,6 $\cdot d_v$, H=0,75 $\cdot d_v$	$CYL9$: H= 0,1 $\cdot d_v$
$CYL2$: D=1,7 $\cdot d_v$, H=0,65 $\cdot d_v$+2 mm	$KON1$: D=1,6 $\cdot d_v$, H=1 mm, angle 45°
$CYL3$: D=d_v, H=1,4 $\cdot d_v$+2 mm	$KON2$: D=1,7 $\cdot d_v$, H=1 mm, angle 45°
$CYL4$: D=d, H=B_k	$KON3$: D=d_v, H=1 mm, angle 45°
$CYL5$: D=d-2 $\cdot s$, H=B_k/2+0,05 $\cdot d_v$+1 mm	$KON4$: D=d_v, H=1 mm, angle 45°
$CYL6$: D=d-2 $\cdot s$-0,1 $\cdot d_v$, H=B_k/2-0,05 $\cdot d_v$-1 mm	$KON5$: D=d-2 $\cdot s$-0,1 $\cdot d_v$, H=1 mm, angle 45°
$CYL7$: D=d-2 $\cdot s$, H=0,1 $\cdot d_v$	$KON6$: D=d-2 $\cdot s$, H=1 mm, angle 45°
$CYL8$: D=1,6 $\cdot d_v$, H=0,1 $\cdot d_v$	BOX: A=[(d-2 $\cdot s$)-1,8 $\cdot d_v$]/2, B=0,35 $\cdot B_k$, C=0,1 $\cdot d_v$

Table 5. Parameters and formulas

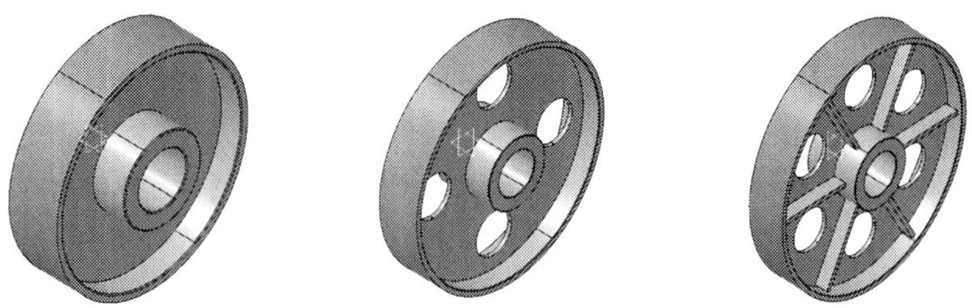

1° K200 (d=200 mm, B_k=50 mm, d_v=50 mm, s=4 mm)

2° K315 (d=315 mm, B_k=63 mm, d_v=60 mm, s=4,5 mm, d_{op}=204 mm, d_o=60 mm)

3° K400 (d=400 mm, B_k=71 mm, d_v=70 mm, s=5 mm, d_{op}=250 mm, d_o=70 mm)

Fig. 10. Different belt pulleys with cylindrical external surface are the results of parameter modelling

Fig. 10. shows three different standard belt pulleys with cylindrical external surface made from the same CATIA V5 file, by changing parameters d, B_k, d_v and s. (Saric et al., 2009)

Use of side ribs that are posed between holes on the plate is recommended during modelling of belt pulleys with longer diameter (Fig. 10.).

Rotary parts of belt pulley shown in the Fig. 9., can be modelled in a much more easier way. More complex contours, instead of their forming by adding and subtraction, they can be formed by rotation. In the first case, computer is loaded with data about points inside primitive which, in total sum, do not belong inside volume of component. In the second case, rotary contour (bolded line in the Fig. 11.) is first defined, and, then, primitive of desired shape is obtained by rotation around rotate axis.

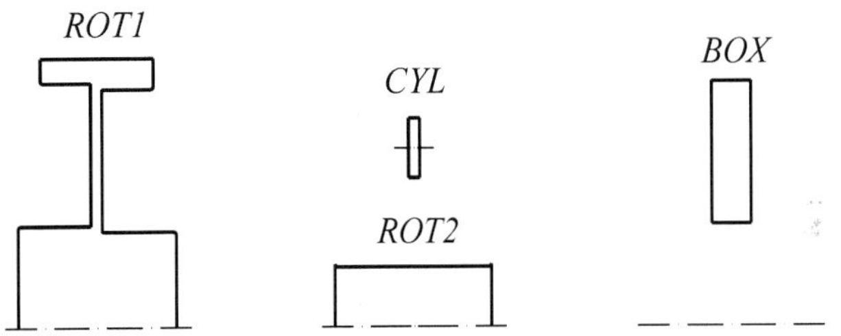

Fig. 11. Modelling of rotary forms

For primitives, shown in the Fig. 11., final form is obtained after the following operations

$$K = \left(ROT1 - ROT2 - 6 \cdot CYL\right) \cup 6 \cdot BOX \tag{8}$$

3. Conclusion

Designer must be significantly engaged into the forming of the component shape. Because of that reason, once formed algorithm for the modelling of the component shape is saved in computer memory and it is used when there is need for the modelling of the same or similar shape with similar dimensions. (Saric et al., 2009)

Parts which are not suitable for interactive modelling are modelled by parameters. In the process of geometric mechanical transmission modelling in CATIA V5 system, we do not have to create shape directly, but, instead of that, we can put parameters integrated in geometric and/or dimensional constraints. Changing of characteristic fixed parameters gives us a 3D solid model of mechanical transmission. This way, designer can generate more alternative designing samples, concentrating his attention on design functional aspects, without special focus on details of elements shape. (Saric et al., 2010)

For the purpose of final goal achieving and faster presentation of the product on the market, time spent for the development of the product is marked as the key factor for more profit gaining. Time spent for process of mechanical transmissions designing can be reduced even by 50% by parameter modelling use with focus on the preparatory phase (Fig. 12.).

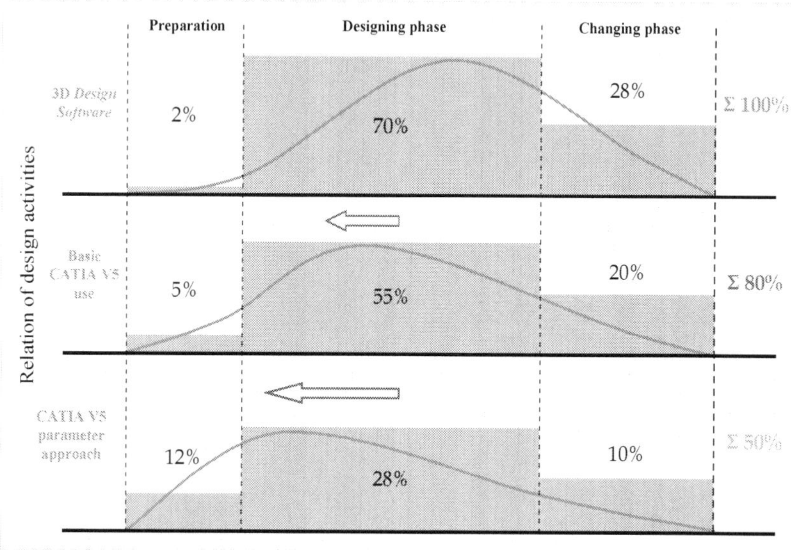

Fig. 12. Relation of design activities and reducing of time spent for design by parameter modelling

These are the advantages of parameter modelling use:

- possibility to make family of parts with the same shape based on one created model,
- forming of libraries basis of standard mechanical elements which take up computer memory, similar to the classic approach of 3D geometric modelling, is not necessary,
- use of parameters enables global modification of whole assembly (automatic reconfiguration),
- development of the product is faster, etc. (Saric et al., 2009)

We can conclude that CATIA V5 system offers possibility of geometric association creation defined by relations established between parameters. Therefore, components parameterisation must obligatory apply in combination with today's traditional geometric modelling approach. Direct financial effects can be seen in production costs reduction, which increases the productivity. Therefore profit is bigger and price of products are lower. (Saric et al., 2010)

Obtained 3D model from CATIA V5 system is used as the base for technical documentation making, analysis of stress and deformation by *Finite Element Method* (FEM), generating of NC/CNC programs for production of the parts on machine (CAM/NC), *Rapid Prototyping* (RP), etc.

4. Acknowledgment

Researches were partially financed by WUS Austria under supervision of Austrian Ministry of foreign affairs as the part of CDP+ project (No. project: 43-SA-04).

5. Nomenclature

A	mm	side rib length
a	°	line of contact angle
B	mm	side rib width
B_k	mm	pulley rim width
BR	-	side rib
b	mm	cog width
b_g	mm	hub width
b_k	mm	keyway width
b_z	mm	gear width
C	mm	side rib thickness
CYL	-	cylinder
D	mm	appropriate diameter of belt pulley components
d	mm	gear pitch circle; diameter of the hole for shaft; diameter of belt pulley
d_g	mm	interior diameter of hub
d_o	mm	diameter of the hole on the plate
d_{op}	mm	diameter on which holes on plate are set
d_v	mm	diameter of shaft
delta	°	a half of an angle of front cone
dZ	mm	translation of geometry over z axis; translation of worm surface over z axis
G	-	hub
gama	°	angle of helix
H	mm	appropriate length of belt pulley components
ha	mm	addendum part of cog height
hf	mm	root part of cog height
K	-	belt pulley
KON	-	cone
L	mm	helix length
L_g	mm	hub length
lc	mm	cone axis length
m	mm	module
N	-	number of helix
P	-	plate
PI	-	value of number π
p	mm	step on pitch circle
ROT	-	rotation
r	mm	pitch radius
ra	mm	addendum radius
ratio	-	factor of scaling exterior to interior cog profile
rb	mm	basic radius
rc	mm	length of generating line of back (additional) cone
rf	mm	root radius
rr	mm	radius of profile root radius
s	mm	minimal pulley rim thickness

t	-	involutes function parameter
t_k	mm	keyway depth
tc	°	cutting angle used for contact point putting in zx plane
V	-	pulley rim
xd	mm	x coordinate of involutes cog profile generated on the base of parameter t
yd	mm	y coordinate of involutes cog profile generated on the base of parameter t
(x,y)	mm	coordinates of involute points
z	-	cog number
z_1	-	cog number of driver gear; number of turn of a worm
z_2	-	cog number of following gear

6. References

Amirouche, F. (2004). *Principles of Computer-Aided Design and Manufacturing* (2nd edition), Prentice Hall, ISBN 0-13-064631-8, Upper Saddle River, New Jersey

Certified hardware configurations for CATIA V5 systems. (May 2011). Available from: <http://www.3ds.com/support/certified-hardware/overview/>

Cozzens, R. (2006). *Advanced CATIA V5 Workbook: Knowledgeware and Workbenches Release 16.* Schroff Development Corporation (SDC Publications), ISBN 978-1-58503-321-8, Southern Utah University

Dassault Systemes. (2007a). *CATIA Solutions Version V5 Release 18 English Documentation*

Dassault Systemes. (2007b). *CATIA Web-based Learning Solutions Version V5 Release 18 Windows*

Dassault Systemes - PLM solutions, 3D CAD and simulation software. (May 2011). Available from: <http://www.3ds.com/home/>

International Business Machines Corp. (IBM). (May 2011). Available from: <http://www.ibm.com/us/en/>

Karam, F. & Kleismit, C. (2004). *Using Catia V5*, Computer library, ISBN 86-7310-307-X, Cacak

Repcic, N. et al. (1998). *Mechanical Elements II Part*, Svjetlost, ISBN 9958-10-157-7 (ISBN 9958-11-075-X), Sarajevo

Repcic, N. & Muminovic, A. (2007). *Mechanical Elements II*, Faculty of Mechanical Engineering, Sarajevo

Saric, I., Repcic, N. & Muminovic, A. (2006). 3D Geometric parameter modelling of belt transmissions and transmissions gear. *Technics Technologies Education Management - TTEM*, Vol. 4, No. 2, (2009), pp. 181-188, ISSN 1840-1503

Saric, I., Repcic, N. & Muminovic, A. (1996). Parameter Modelling of Gears, *Proceedings of the 14th International Research/Expert Conference „Trends in the Development of Machinery and Associated Technology - TMT 2010"*, pp. 557-560, ISSN 1840-4944, Mediterranean Cruise, September 11-18, 2010

Shigley, J.E., Mischke, C.R. & Budynas, R.G. (2004). *Mechanical Engineering Design* (7th edition), McGraw-Hill, ISBN 007-252036-1, New York

Spotts, M.F., Shoup, T.E. & Hornberger, L.E. (2004). *Design of Machine Elements* (8th edition), Prentice Hall, ISBN 0-13-126955-0, Upper Saddle River, New Jersey

Zamani, N.G. & Weaver, J.M. (2007). *Catia V5: Tutorials Mechanism Design & Animation*, Computer library, ISBN 978-86-7310-381-5, Cacak

2

Gearbox Simulation Models with Gear and Bearing Faults

Endo Hiroaki[1] and Sawalhi Nader[2]
[1]*Test devices Inc.,*
[2]*Prince Mohammad Bin Fahd University (PMU), Mechanical Engineering Department, AlKhobar*
[1]*USA*
[2]*Saudi Arabi*

1. Introduction

Simulation is an effective tool for understanding the complex interaction of transmission components in dynamic environment. Vibro-dynamics simulation of faulty gears and rolling element bearings allows the analyst to study the effect of damaged components in controlled manners and gather the data without bearing the cost of actual failures or the expenses associated with an experiment that requires a large number of seeded fault specimens. The fault simulation can be used to provide the data required in training Neural network based diagnostic/prognostic processes.

2. Key elements in gearbox simulation

2.1 Transmission error

Gears, by their inherent nature, cause vibrations due to the large pressure which occurs between the meshing teeth when gears transmit power. Meshing of gears involves changes in the magnitude, the position and the direction of large concentrated loads acting on the contacting gear teeth, which as a result causes vibrations. Extended period of exposure to noise and vibration are the common causes of operational fatigue, communication difficulties and health hazards. Reduction in noise and vibration of operating machines has been an important concern for safer and more efficient machine operations.

Design and development of quieter, more reliable and more efficient gears have been a popular research area for decades in the automotive and aerospace industries. Vibration of gears, which directly relates to noise and vibration of the geared machines, is typically dominated by the effects of the tooth meshing and shaft revolution frequencies, their harmonics and sidebands, caused by low (shaft) frequency modulation of the higher tooth-mesh frequency components. Typically, the contribution from the gear meshing components dominates the overall contents of the measured gearbox vibration spectrum (see Figure-2.1.1).

Transmission Error (TE) is one of the most important and fundamental concepts that forms the basis of understanding vibrations in gears. The name 'Transmission Error' was originally coined by Professor S. L. Harris from Lancaster University, UK and R.G. Munro, his PhD student at the time. They came to the realization that the excitation forces causing the gears to vibrate were dependent on the tooth meshing errors caused by manufacturing and the bending of the teeth under load [1].

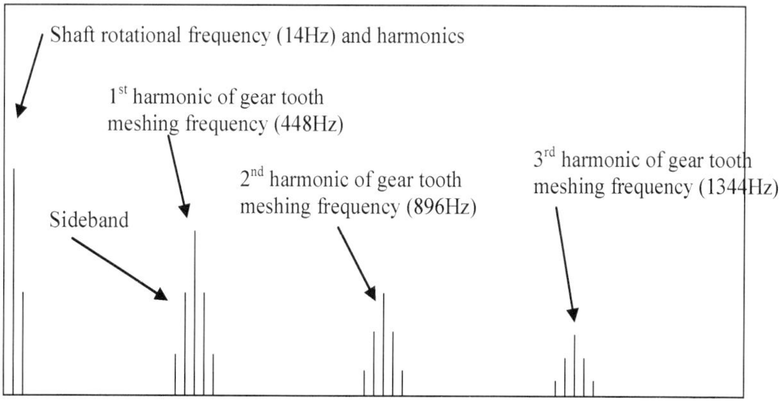

(Number of Teeth = 32)

Fig. 2.1.1. Typical spectrum composition of gear vibration signal.

TE is defined as the deviation of the angular position of the driven gear from its theoretical position calculated from the gearing ratio and the angular position of the pinion (Equation-2.1.1). The concept of TE is illustrated in Figure-2.1.2.

$$TE = \left(\theta_{gear} - \frac{R_{pinion}}{R_{gear}} \theta_{pinion} \right) \tag{2.1.1}$$

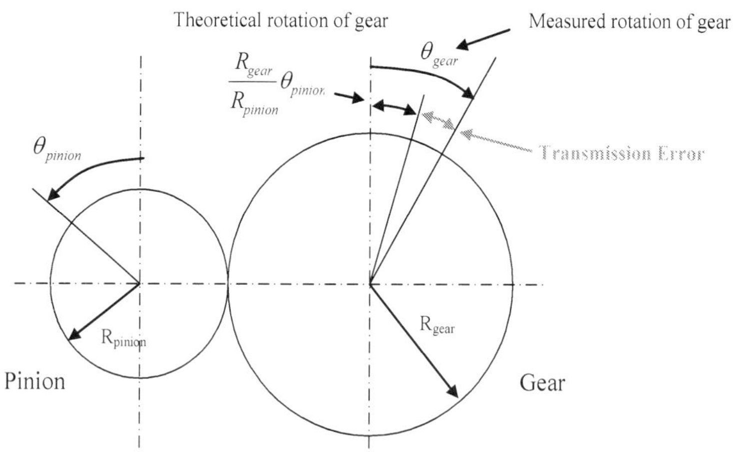

Fig. 2.1.2. Definition of Transmission Error.

What made the TE so interesting for gear engineers and researchers was its strong correlation to the gear noise and the vibrations. TE can be measured by different types of instruments. Some commonly used methods are: Magnetic signal methods, straingauge on the drive shaft, torsional vibration transducers, tachometers, tangential accelerometers and rotary encoders systems. According to Smith [2], TE results from three main sources: 1) Gear geometrical errors, 2) Elastic deformation of the gears and associated components and 3) Errors in mounting. Figure-2.1.3 illustrates the relationship between TE and its sources.

Transmission Error exists in three forms: 1) Geometric, 2) Static and 3) Dynamic. Geometric TE (GTE) is measured at low speeds and in the unloaded state. It is often used to examine the effect of manufacturing errors. Static TE (STE) is also measured under low speed conditions, but in a loaded state. STE includes the effect of elastic deflection of the gears as well as the geometrical errors. Dynamic TE (DTE) includes the effects of inertia on top of all the effects of the errors considered in GTE and in STE. The understanding of the TE and the behaviour of the machine elements in the geared transmission system leads to the development of realistic gear rotor dynamics models.

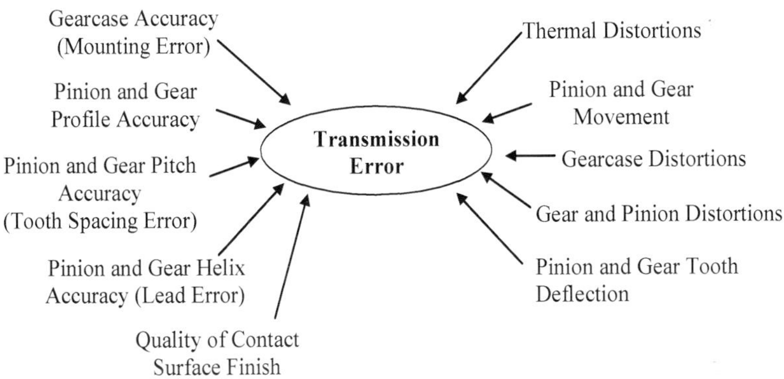

Fig. 2.1.3. Sources of Transmission Error.

2.2 Effect of gear geometric error on transmission error

A typical GTE of a spur gear is shown in Figure-2.2.1. It shows a long periodic wave (gear shaft rotation) and short regular waves occurring at tooth-mesh frequency. The long wave is often known as: Long Term Component: LTC, while the short waves are known as: Short Term Component: STC.

The LTC is typically caused by the eccentricity of the gear about its rotational centre. An example is given in Figure-2.2.2 to illustrate how these eccentricities can be introduced into the gears by manufacturing errors; it shows the error due to a result of the difference between the hobbing and the shaving centres.

The effect of errors associated with gear teeth appears in the STC as a localized event. The parabolic-curve-like effect of tooth tip relief is shown in Figure-2.2.3(a). The STC is caused mainly by gear tooth profile errors and base pitch spacing error between the teeth. The effect of individual tooth profile errors on the GTE is illustrated in Figure-2.2.3(a). The GTE of a meshing tooth pair is obtained by adding their individual profile errors. The STC of gear GTE is synthesised by superposing the tooth pair GTEs separated by tooth base pitch angles (Figure-2.2.3 (b)).

Another common gear geometric error is tooth spacing or pitch errors, shown in Figure-2.2.4. The tooth spacing error appears in GTE as vertical raise or fall in the magnitude of a tooth profile error.

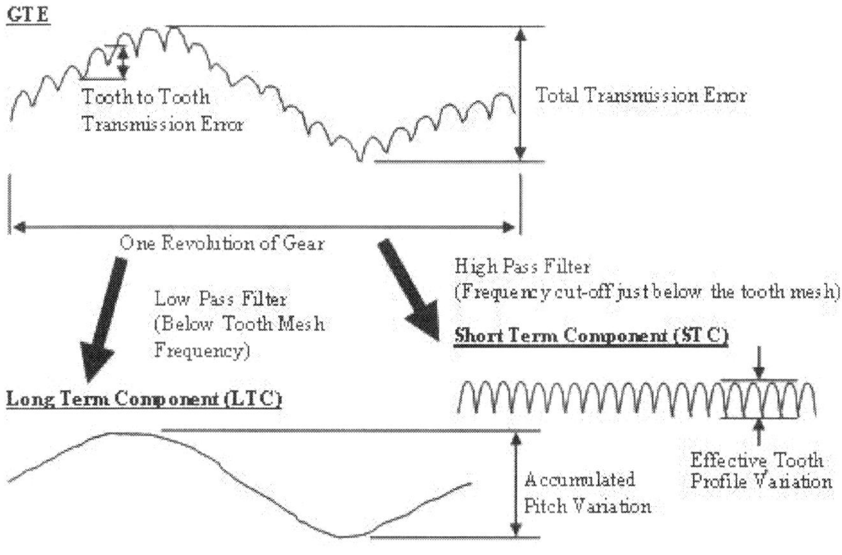

Fig. 2.2.1. A typical Geometrical Transmission Error.

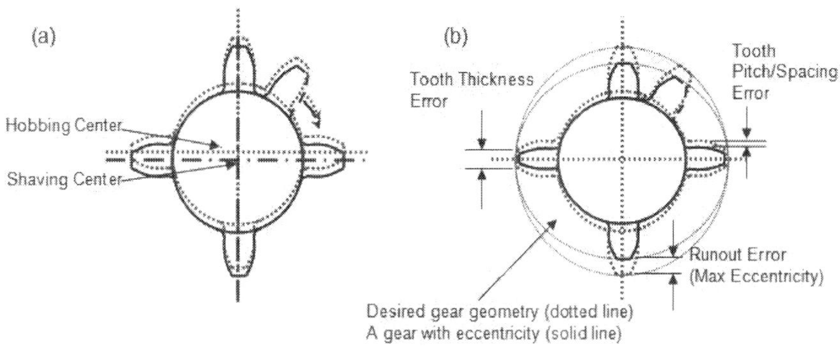

Fig. 2.2.2. (a) Eccentricity in a gear caused by manufacturing errors, (b) Resulting errors in gear geometry.

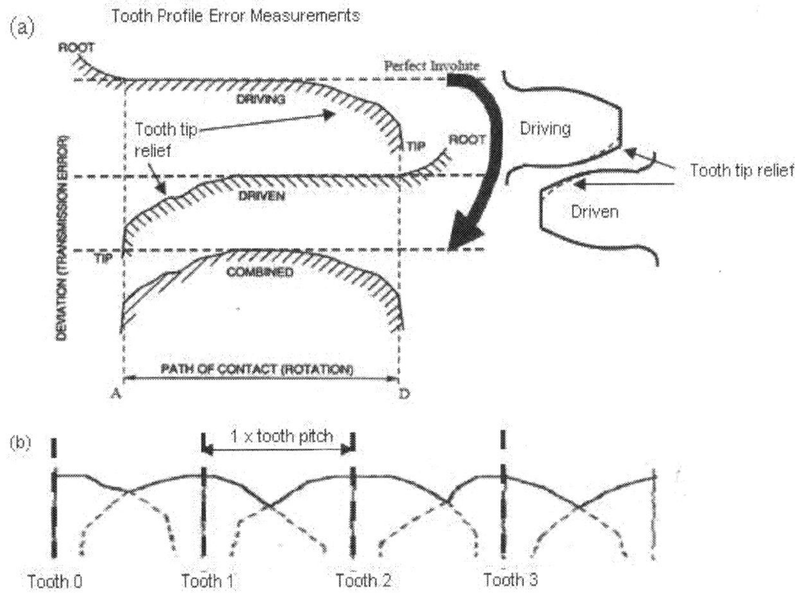

Fig. 2.2.3. (a) GTE of a meshing tooth pair (b) Resulting Short Term Component of GTE.

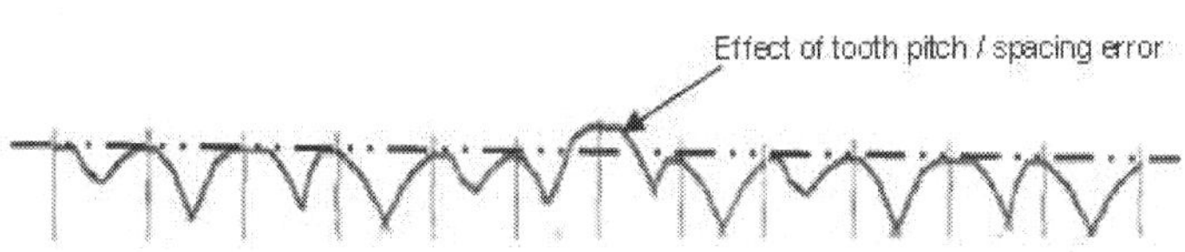

Fig. 2.2.4. Effect of spacing error appearing in short term component of GTE.

2.3 Effect of load on transmission error

Elastic deflections occurring in gears are another cause of TE. Although gears are usually stiff and designed to carry very large loads, their deflection under load is not negligible. Typical deflection of gear teeth occurs in the order of microns (μm). Although it depends on the amount of load gears carry, the effect of the deflection on TE may become more significant than the contribution from the gear geometry.

A useful load-deflection measure is that 14N of load per 1mm of tooth face width results in 1μm of deflection for a steel gear: i.e. stiffness = $14E10^9$ N/m/m for a tooth pair meshing at the pitch line. It is interesting to note here that the stiffness of a tooth pair is independent of its size (or tooth module) [3]. Deflection of gear teeth moves the gear teeth from their theoretical positions and in effect results in a continuous tooth pitch error: see Figure-2.3.1 (a). The effect of the gear deflection appears in the TE (STE) as a shifting of the GTE: Figure-2.3.1 (b).

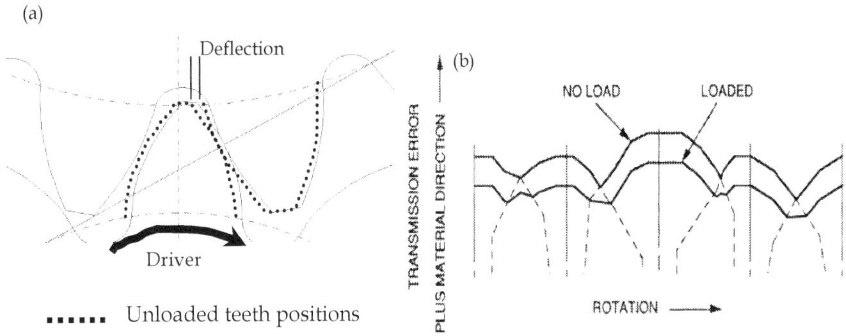

Fig. 2.3.1. (a) Deflection of gear tooth pair under load, (b) Effect of load on transmission error (TE).

Consider the more general situation where the deflection in loaded gears affects the TE significantly. Note that the following discussion uses typical spur gears (contact ratio = 1.5) with little profile modification to illustrate the effect of load on TE. Figure-2.3.2 illustrates the STE caused by the deflection of meshing gear teeth. The tooth profile chart shows a flat

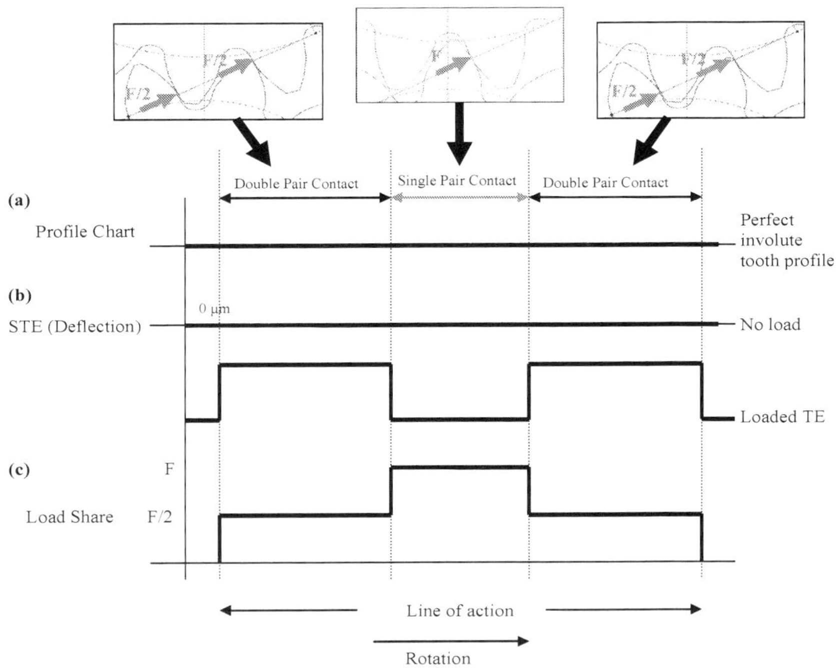

Fig. 2.3.2. Effect of Load on TE, (a) Tooth Profile Chart, (b) Static Transmission Error, (c) Loading acting on a tooth

line indicating the ideal involute profile of the tooth: Figure-2.3.2 (a). The effect of mesh stiffness variation due to the change in the number of meshing tooth pairs appears as steps in the STE plot: Figure-2.3.2 (b). The amount of deflection increases when a single pair of teeth is carrying load and decreases when the load is shared by another pair. The share of force carried by a tooth through the meshing cycle is shown in Figure-2.3.2 (c).

A paper published jointly by S.L. Harris, R. Wylie Gregory and R.G. Munro in 1963 showed how transmission error can be reduced by applying appropriate correction to the involute gear profile [4, 5]. The Harris map in Figure-2.3.3 shows that any gear can be designed to have STE with zero variation (i.e. a flat STE with constant offset value) for a particular load. The basic idea behind this technique is that the profile of gear teeth can be designed to cancel the effect of tooth deflection occurring at the given load.

Additionally, variation of TE can be reduced by increasing the contact ratio of the gear pair. In other words, design the gears so that the load is carried by a greater number of tooth pairs.

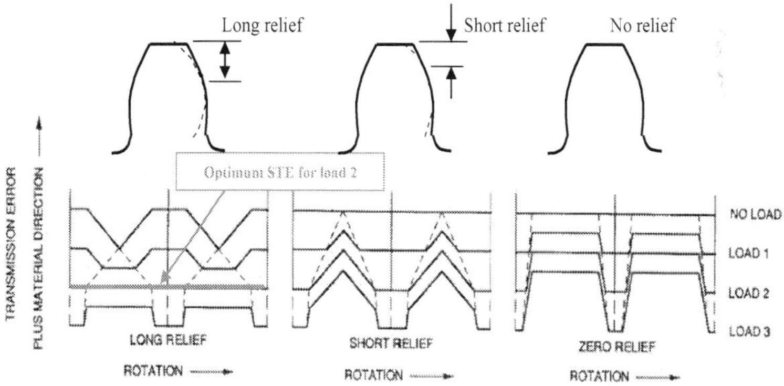

Fig. 2.3.3. Optimum tooth profile modification of a spur gear.

2.4 Modelling gear dynamics

It is a standardized design procedure to perform STE analysis to ensure smoothly meshing gears in the loaded condition. It was explained in this section how the strong correlation between the TE and the gear vibration makes the TE a useful parameter to predict the quietness of the gear drives. However, a more realistic picture of the gear's dynamic properties can not be captured without modelling the dynamics of the assembled gear drive system. Solution of engineering problems often requires mathematical modelling of a physical system. A well validated model facilitates a better understanding of the problem and provides useful information for engineers to make intelligent and well informed decisions.

A comprehensive summary of the history of gear dynamic model development is given by Ozguven and Houser [6]. They have reviewed 188 items of literature related to gear dynamic simulation existing up to 1988. In Table-2.4.1, different types of gear dynamics models were classified into five groups according to their objectives and

Group-1: Simple Dynamic Factor Models Most early models belong to this group. The model was used to study gear dynamic load and to determine the value of dynamic factor that can be used in gear root stress formulae. Empirical, semi-empirical models and dynamic models constructed specifically for determination of dynamic factor are included in this group.
Group-2: Models with Tooth Compliance Models that consider tooth stiffness as the only potential energy storing element in the system. Flexibility of shafts, bearings etc is neglected. Typically, these models are single DOF spring-mass systems. Some of the models from this group are classified in group-1 if they are designed solely for determining the dynamic factor.
Group-3: Models for Gear Dynamics A model that considers tooth compliance and the flexibility of the relevant components. Typically these models include torsional flexibility of shafts and lateral flexibility of bearings and shafts along the line of action.
Group-4: Models for Geared Rotor Dynamics This group of models consider transverse vibrations of gear carrying shafts as well as the lateral component *(NOTE: Transverse: along the Plane of Action, Lateral: Normal to the Plane of Action).* Movement of the gears is considered in two mutually perpendicular directions to simulate, for example, whirling.
Group-5: Models for Torsional Vibrations The models in the third and fourth groups consider the flexibility of the gear teeth by including a constant or time varying mesh stiffness. The models belonging to this group differentiate themselves from the third and fourth groups by having rigid gears mounted on flexible shafts. The flexibility at the gearmesh is neglected. These models are used in studying pure (low frequency) torsional vibration problems.

Table 2.4.1. Classification of Gear Dynamic Models. (Ozguven & Houser [6])

functionality. Traditionally lumped parameter modelling (LPM) has been a common technique that has been used to study the dynamics of gears. Wang [7] introduced a simple LPM to rationalize the dynamic factor calculation by the laws of mechanics. He proposed a model that relates the GTE and the resulting dynamic loading. A large number of gear dynamic models that are being used widely today are based on this work. The result of an additional literature survey on more recently published materials by Bartelmus [8], Lin & Parker [9, 10], Gao & Randall [11, 12], Amabili & Rivola [13], Howard et al [14], Velex & Maatar [15], Blankenship & Singh [17], Kahraman & Blankenship [18] show that the fundamentals of the modelling technique in gear simulations have not changed and the LMP method still serves as an efficient technique to model the wide range of gear dynamics behaviour. More advanced LPM models incorporate extra functions to simulate specialized phenomena. For example, the model presented by P. Velex and M. Maatar [15] uses the individual gear tooth profiles as input and calculates the GTE directly from the gear tooth profile. Using this method they simulated how the change in contact behaviour of meshing gears due to misalignment affects the resulting TE.

FEA has become one of the most powerful simulation techniques applied to broad range of modern Engineering practices today. There have been several groups of researchers who attempted to develop detailed FEA based gear models, but they were troubled by the challenges in efficiently modelling the rolling Hertzian contact on the meshing surfaces of gear teeth. Hertzian contact occurs between the meshing gear teeth which causes large concentrated forces to act in very small area. It requires very fine FE mesh to accurately model this load distribution over the contact area. In a conventional finite element method, a fully representative dynamic model of a gear requires this fine mesh over each gear tooth flank and this makes the size of the FE model prohibitively large.

Researchers from Ohio State University have developed an efficient method to overcome the Hertzian contact problem in the 1990s' [16]. They proposed an elegant solution by modelling the contact by an analytical technique and relating the resulting force distribution to a coarsely meshed FE model. This technique has proven so efficient that they were capable of simulating the dynamics of spur and planetary gears by [19, 20] (see Figure-2.4.1). For more details see the CALYX user's manuals [21, 22].

For the purpose of studies, which require a holistic understanding of gear dynamics, a lumped parameter type model appears to provide the most accessible and computationally economical means to conduct simulation studies.

A simple single stage gear model is used to explain the basic concept of gear dynamic simulation techniques used in this chapter. A symbolic representation of a single stage gear system is illustrated in Figure-2.4.2. A pair of meshing gears is modelled by rigid disks representing their mass/moment of inertia. The discs are linked by line elements that represent the stiffness and the damping (representing the combined effect of friction and fluid film damping) of the gear mesh. Each gear has three translational degrees of freedom (one in a direction parallel to the gear's line of action, defining all interaction between the gears) and three rotational degree of freedoms (DOFs). The stiffness elements attached to the centre of the disks represent the effect of gear shafts and supporting mounts. NOTE: Symbols for the torsional stiffnesses are not shown to avoid congestion.

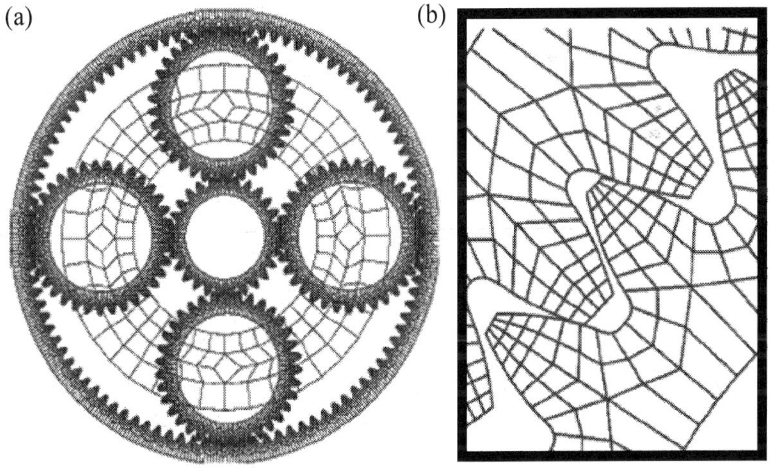

Fig. 2.4.1. (a) Parker's planetary gear model and (b) FE mesh of gear tooth. Contacts at the meshing teeth are treated analytically. It does not require dense FE mesh.
(Courtesy of Parker et al. [20])

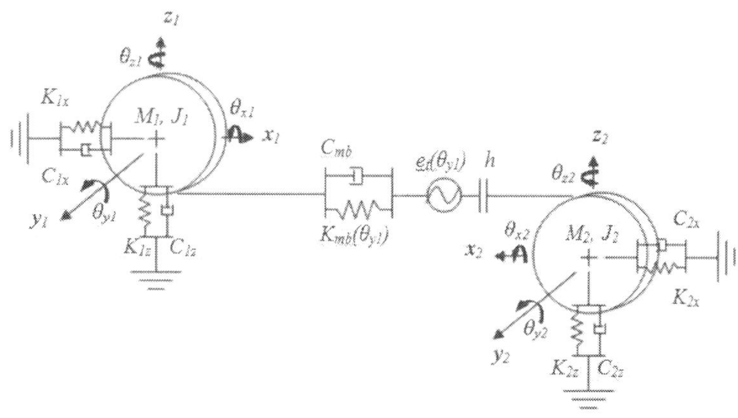

x_i, y_i, z_i Translation at i_{th} Degrees of Freedom.

θ_{xi}, θ_{yi}, θ_{zi} Rotation about a translational axis at i_{th} Degrees of Freedom.

C, C_{mb} Damping matrices. The subscript '*mb*' refers to damping at the gearmesh. Typically for C_{mb}, $\zeta = 3 \sim 7\%$.

K, K_{mb} Linear stiffness elements. The subscript '*mb*' refers to stiffness at the gearmesh.

h An 'on/off' switch governing the contact state of the meshing gear teeth.

$\underline{e_t}$ A vector representing the combined effect of tooth topography deviations and misalignment of the gear pair.

i Index: i=1,2, 3 ...etc.

Fig. 2.4.2. A Typical Lumped Parameter Model of Meshing Gears.

The linear spring elements representing the Rolling Element Bearings (REB) are a reasonable simplification of the system that is well documented in many papers on gear simulation. For the purpose of explaining the core elements of the gear simulation model, the detail of REB as well as the casing was omitted from this section; more comprehensive model of a gearbox, with REB and casing, will be presented later in section 2.5.

Vibration of the gears is simulated in the model as a system responding to the excitation caused by a varying TE, 'e_t' and mesh stiffness 'K_{mb}'. The dominant force exciting the gears is assumed to act in a direction along the plane of action (PoA). The angular position dependent variables 'e_t' and 'K_{mb}' are expressed as functions of the pinion pitch angle (θ_{y1}) and their values are estimated by using static simulation. Examples of similar techniques are given by Gao & Randall [11, 12], Du [23] and Endo and Randall [61].

Equations of motion derived from the LPM are written in matrix format as shown in Equation-2.4.1. The equation is rearranged to the form shown in the Euqation-2.4.2; the effect of TE is expressed as a time varying excitation in the equation source. The dynamic response of the system is simulated by numerically solving the second order term (accelerations) for each step of incremented time. The effect of the mesh stiffness variation is implemented in the model by updating its value for each time increment.

$$M\ddot{\underline{x}} + C\left(\dot{\underline{x}} - \dot{\underline{e}}_t\right) + K\left(\underline{x} - \underline{e}_t(\theta)\right) = F_s \tag{2.4.1}$$

$$M\ddot{\underline{x}} + C\dot{\underline{x}} + K\underline{x} = F_s + hC_{mb}(\theta)\dot{\underline{e}}_t + hK_{mb}(\theta)\underline{e}_t \tag{2.4.2}$$

where,

$\underline{x}$, $\dot{\underline{x}}$, $\ddot{\underline{x}}$ Vectors of translational and rotational displacement, velocity and acceleration.

θ Angular position of pinion.

K, K_{mb} Stiffness matrices (where K includes the contribution from K_{mb}). The subscript '*mb*' refers to stiffness at the gearmesh.

C, C_{mb} Damping matrices (C including contribution from C_{mb}). The subscript '*mb*' refers to damping at the gearmesh.

h An 'on/off' switch governing the contact state of the meshing gear teeth.

F Static force vector.

$\underline{e}_t$, $\dot{\underline{e}}_t$ A vector representing the combined effect of tooth topography deviations.

2.5 Modelling rolling element bearings and gearbox casing

For many practical purposes, simplified models of gear shaft supports (for example, the effect of rolling element bearings (REBs) and casing were modelled as simple springs with constant stiffnesses) can be effective tools. However, fuller representations of these components become essential in the pursuit of more complete and accurate simulation modelling.

For a complete and more realistic modelling of the gearbox system, detailed representations of the REBs and the gearbox casing are necessary to capture the interaction amongst the gears, the REBs and the effects of transfer path and dynamics response of the casing.

Understanding the interaction between the supporting structure and the rotating components of a transmission system has been one of the most challenging areas of designing more detailed gearbox simulation models. The property of the structure supporting REBs and a shaft has significant influence on the dynamic response of the system. Fuller representation of the REBs and gearbox casing also improves the accuracy of the effect transmission path that contorts the diagnostic information originated from the faults in gears and REBs. It is desired in many applications of machine health monitoring that the method is minimally intrusive on the machine operation. This requirement often drives the sensors and/or the transducers to be placed in an easily accessible location on the machine, such as exposed surface of gearbox casing or on the machine skid or on an exposed and readily accessible structural frame which the machine is mounted on.

The capability to accurately model and simulate the effect of transmission path allows more realistic and effective means to train the diagnostic algorithms based on the artificial intelligence.

2.5.1 Modelling rolling element bearings

A number of models of REBs exist in literatures [24, 25, 26, 27] and are widely employed to study the dynamics and the effect of faults in REBs. The authors have adopted the 2 DoF model originally developed by Fukata [27] in to the LPM of the gearbox. Figure-2.5.1 (a) illustrates the main components of the rolling element bearing model and shows the load zone associated with the distribution of radial loads in the REB as it supports the shaft. Figure-2.5.1 (b) explains the essentials of the bearing model as presented by [28]. The two degree-of-freedom REB model captures the load-deflection relationships, while ignoring the effect of mass and the inertia of the rolling elements. The two degrees of freedom (x_s, y_s) are related to the inner race (shaft). Contact forces are summed over each of the rolling elements to give the overall forces on the shaft.

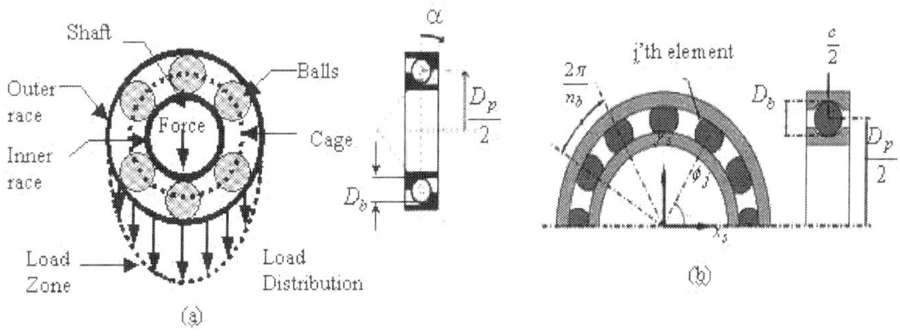

Fig. 2.5.1. (a) Rolling element bearing components and load distribution; (b) Two degree of freedom model. [28]

The overall contact deformation (under compression) for the j'th -rolling element δ_j is a function of the inner race displacement relative to the outer race in the x and y directions $((x_s - x_p), (y_s - y_p))$, the element position ϕ_j (time varying) and the clearance (c). This is given by:

$$\delta_j = (x_s - x_p)\cos\phi_j + (y_s - y_p)\sin\phi_j - c - \beta_j C_d \quad (j = 1,2..) \tag{2.5.1}$$

Accounting for the fact that compression occurs only for positive values of δ_j, γ_j (contact state of δ_j the rolling elements) is introduced as:

$$\gamma_j = \begin{cases} 1, & \text{if } \delta_j > 0 \\ 0, & \text{otherwise} \end{cases} \tag{2.5.2}$$

The angular positions of the rolling elements ϕ_j are functions of time increment dt, the previous cage position ϕ_o and the cage speed ω_c (can be calculated from the REB geometry and the shaft speed ω_s assuming no slippage) are given as:

$$\phi_j = \frac{2\pi(j-1)}{n_b} + \omega_c dt + \phi_o \quad \text{with} \quad \omega_c = (1 - \frac{D_b}{D_p})\frac{\omega_s}{2} \tag{2.5.3}$$

The ball raceway contact force f is calculated by using traditional Hertzian theory (non-linear stiffness) from:

$$f = k_b\,\delta^n \tag{2.5.4}$$

The load deflection factor k_b depends on the geometry of contacting bodies, the elasticity of the material, and exponent n. The value of n=1.5 for ball bearings and n=1.1 for roller bearings. Using Equation-2.5.4 and summing the contact forces in the x and y directions for a ball bearing with n_b balls, the total force exerted by the bearings to the supporting structure can be calculated as follows:

$$f_x = k_b \sum_{j=1}^{n_b} \gamma_j \delta_j^{1.5} \cos\phi_j \quad \text{and} \quad f_y = k_b \sum_{i=1}^{n_b} \gamma_i \delta_j^{1.5} \sin\phi_j \qquad (2.5.5)$$

The stiffness of the given REB model is non-linear, and is time varying as it depends on the positions of the rolling elements that determine the contact condition. The effect of slippage was introduced to the model by adding random jitters of 0.01-0.02 radians to the nominal position of the cage at each step.

2.5.2 Gearbox casing model – Component mode synthesis method

Lumped parameter modelling (LPM) is an efficient means to express the internal dynamics of transmission systems; masses and inertias of key components such as gears, shafts and bearings can be lumped at appropriate locations to construct a model. The advantage of the LPM is that it provides a method to construct an effective dynamic model with relatively small number of degrees-of-freedoms (DOF), which facilitates computationally economical method to study the behaviour of gears and bearings in the presence of nonlinearities and geometrical faults [32, 33, 34, 35].

One of the limitations of the LPM method is that it does not account for the interaction between the shaft and the supporting structure; i.e. casing flexibility, which can be an important consideration in light weight gearboxes, that are common aircraft applications. Not having to include the appropriate effect of transmission path also results in poor comparison between the simulated and measured vibration signals.

Finite Element Analysis (FEA) is an efficient and well accepted technique to characterize a dynamic response of a structure such as gearbox casings. However, the use of FEA results in a large number of DOF, which could cause some challenges when attempt to solve a vibro-dynamic model of a combined casing and the LMP of gearbox internal components. Solving a large number of DOFs is time consuming even with the powerful computers available today and it could cause a number of computational problems, especially when attempting to simulate a dynamic response of gear and bearing faults which involves nonlinearities.

To overcome this shortcoming, a number of reduction techniques [36, 37] have been proposed to reduce the size of mass and stiffness matrix of FEA models. The simplified gearbox casing model derived from the reduction technique is used to capture the key characteristics of dynamic response of the casing structure and can be combined with the LPM models of gears and REBs.

The Craig-Bampton method [37] is a dynamic reduction method for reducing the size of the finite element models. In this method, the motion of the whole structure is represented as a combination of boundary points (so called master degree of freedom) and the modes of the structure, assuming the master degrees of freedom are held fixed. Unlike the Guyan reduction [38], which only deals with the reduction of stiffness matrix, the Craig-Bumpton method accounts for both the mass and the stiffness. Furthermore, it enables defining the frequency range of interest by identifying the modes of interest and including these as a part of the transformation matrix. The decomposition of the model into both physical DOFs (master DOFs) and modal coordinates allows the flexibility of connecting the finite elements to other substructures, while achieving a reasonably good result within a required frequency range. The Craig-Bumpton method is a very convenient method for modelling a geared transmission system as the input (excitation) to the system is not defined as forces, but as geometric mismatches at the connection points (i.e. gear transmission error and bearing geometric error). The following summary of the Craig-Bampton method is given based on the references [39-41].

In the Craig-Bampton reduction method, the equation of motion (dynamic equilibrium) of each superelement (substructure), without considering the effect of damping, can be expressed as in Equation-2.5.6:

$$[M]\{\ddot{u}\}+[k]\{u\}=\{F\} \tag{2.5.6}$$

Where $[M]$ is the mass matrix, $[k]$ is the stiffness matrix, $\{F\}$ is the nodal forces, $\{u\}$ and $\{\ddot{u}\}$ are the nodal displacements and accelerations respectively. The key to reducing the substructure is to split the degrees of freedom into masters $\{u_m\}$ (at the connecting nodes) and slaves $\{u_s\}$ (at the internal nodes). The mass, the stiffness and the force matrices are re-arranged accordingly as follows:

$$\overbrace{\begin{bmatrix} M_{mm} & M_{ms} \\ M_{sm} & M_{ss} \end{bmatrix}}^{M}\begin{Bmatrix} \ddot{u}_m \\ \ddot{u}_s \end{Bmatrix}+\overbrace{\begin{bmatrix} k_{mm} & k_{ms} \\ k_{sm} & k_{ss} \end{bmatrix}}^{k}\begin{Bmatrix} u_m \\ u_s \end{Bmatrix}=\begin{Bmatrix} F_m \\ 0 \end{Bmatrix} \tag{2.5.7}$$

The subscript m denotes master, s denotes slave. Furthermore, the slave degrees of freedom (internals) can be written by using generalized coordinates (modal coordinates by (q) using the fixed interface method, i.e. using the mode shapes of the superelement by fixing the master degrees of freedom nodes (connecting/ boundary nodes). The transformation matrix (T) is the one that achieves the following:

$$\begin{Bmatrix} u_m \\ u_s \end{Bmatrix}=T\begin{Bmatrix} u_m \\ q \end{Bmatrix} \tag{2.5.8}$$

For the fixed interface method, the transformation matrix (T) can be expressed as shown in Equation-2.5.9:

$$T=\begin{bmatrix} I & 0 \\ G_{sm} & \phi_s \end{bmatrix} \tag{2.5.9}$$

where,

$$G_{sm}=-k_{ss}^{-1}k_{sm} \tag{2.5.10}$$

and ϕ_s is the modal matrix of the internal DOF with the interfaces fixed.

By applying this transformation, the number of DOFs of the component will be reduced. The new reduced mass and stiffness matrices can be extracted using Equations 2.5.11 & 2.5.12 respectively:

$$M_{reduced} = T^t M T \tag{2.5.11}$$

and

$$k_{reduced} = T^t k T \tag{2.5.12}$$

Thus Equation-2.5.7 can be re-written in the new reduced form using the reduced mass and stiffness matrices as well as the modal coordinates as follows:

$$\overbrace{\begin{bmatrix} M_{bb} & M_{bq} \\ M_{qb} & M_{qq} \end{bmatrix}}^{M_{reduced}} \begin{Bmatrix} \ddot{u}_m \\ \ddot{q} \end{Bmatrix} + \overbrace{\begin{bmatrix} k_{bb} & 0 \\ 0 & k_{qq} \end{bmatrix}}^{k_{reduced}} \begin{Bmatrix} u_m \\ q \end{Bmatrix} = \overbrace{\begin{Bmatrix} F_m \\ 0 \end{Bmatrix}}^{F_{reduced}} \tag{2.5.13}$$

Where M_{bb} is the boundary mass matrix i.e. total mass properties translated to the boundary points. k_{bb} is the interface stiffness matrix, i.e. stiffness associated with displacing one boundary DOF while the others are held fixed. The M_{bq} is the component matrix (M_{qb} is the transpose of M_{bq}).

If the mode shapes have been mass normalized (typically they are) then:

$$k_{qq} = \begin{bmatrix} \backslash & & 0 \\ & \lambda_i & \\ 0 & & \backslash \end{bmatrix} \tag{2.5.14}$$

where λ_i is the eigenvalues; $\lambda_i = k_i / m_i = \omega^2{}_i$, and,

$$M_{qq} = \begin{bmatrix} \backslash & & 0 \\ & I & \\ 0 & & \backslash \end{bmatrix} \tag{2.5.15}$$

Finally the dynamic equation of motion (including damping) using the Craig-Bampton transform can be written as:

$$\begin{bmatrix} M_{bb} & M_{bq} \\ M_{qb} & I \end{bmatrix} \begin{Bmatrix} \ddot{u}_m \\ \ddot{q} \end{Bmatrix} + \begin{bmatrix} 0 & 0 \\ 0 & 2\zeta\omega \end{bmatrix} \begin{Bmatrix} \dot{u}_m \\ \dot{q} \end{Bmatrix} + \begin{bmatrix} k_{bb} & 0 \\ 0 & \omega^2 \end{bmatrix} \begin{Bmatrix} u_m \\ q \end{Bmatrix} = \begin{Bmatrix} F_m \\ 0 \end{Bmatrix} \tag{2.5.16}$$

where $2\zeta\omega$ = modal damping (ζ = fraction of critical damping)

For more detailed explanation of the techniques related to the modelling of rolling element bearings and the application of component mode synthesis (CMS) techniques, refer to the works by Sawalhi, Deshpande and Randall [41].

2.6 Solving the gear dynamic simulation models

A block diagram summarizing the time integral solution of a typical dynamic model with some time varying parameters is shown in Figure-2.6.1. There are a range of numerical algorithms available today to give solution of the dynamic models: direct time integration, harmonic balancing and shooting techniques, to name some commonly recognized methods.

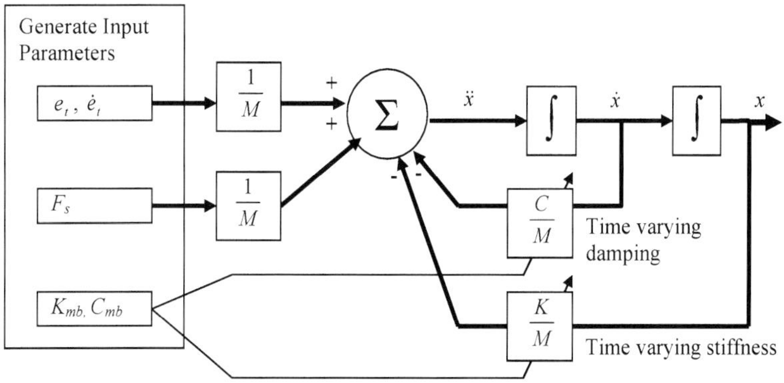

(NOTE: '∫' stands for integration over a single step of incremented time)

Fig. 2.6.1. Dynamic Simulation Process

Sometimes the solution for gears requires simulation of highly non-linear events, for example, rattling and knocking in gears, which involve modelling of the contact loss. The works presented by R. Singh [42], Kahraman & Singh [43], Kahraman & Blankenship [18] and Parker & Lin [9, 10] show some examples of the "stiff" problems involving non-linearity due to contact loss and clearances.

The solution for these Vibro-Impact problems presents difficulties involving ill-conditioning and numerical "stiffness". In [42] Singh explains that ill-conditioning of a numerical solution occurs when there is a component with a large frequency ratio: ratio of gear mesh frequency to the natural frequency of the component.

The numerical stiffness in the gear dynamic simulation becomes a problem when gears lose contact. The relationship between the elastic force, relative deflection and gear mesh stiffness is illustrated in Figure-2.6.2. The gradient of the curve represent the gear mesh stiffness.

Contact loss between the gears occurs when the force between the gears becomes zero. The gears are then unconstrained and free to move within the backlash tolerance. The presence of a discontinuity becomes obvious when the derivative of the curve in Figure-2.6.2 (i.e. mesh stiffness) is plotted against the relative deflection. The discontinuity in the stiffness introduces instability in the numerical prediction.

In more formalized terms the "stiffness" of a problem is defined by local Eigen-values of the Jacobian matrix. Consider an equation of motion expressed in simple first order vector form '$f(x, t)$', (Equation-2.6.1). Typically, the solution of an equation of motion is obtained by linearizing it about an operating point, say 'x_0', (Equation-2.6.2). Usually, most of the higher

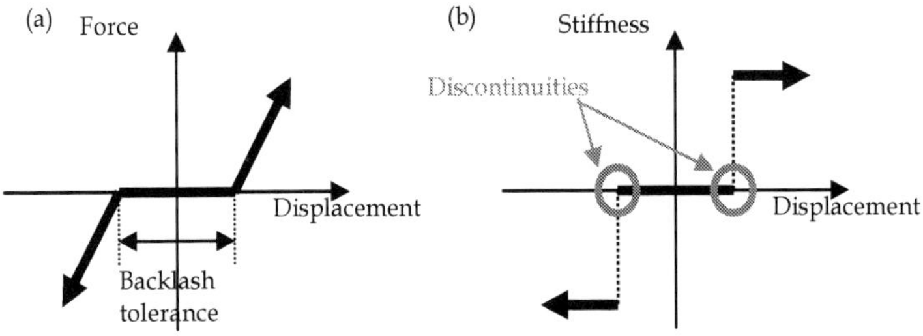

Fig. 2.6.2. Non-linearity due to contact loss in meshing gears; a) Force vs. Displacement, b) derivative of former, i.e. Stiffness vs. Displacement

order terms above the 1st derivative are ignored for linearization, which leaves the following expression (Equation-2.6.3). The differential term '*J*' is called the Jacobian matrix (or Jacobian in short). The problems involving gear contact losses are "Stiff" problems because of the discontinuity in system derivatives (Jacobians). For a more detailed discussion on this topic refer to the work presented by Singh [42].

$$\underline{f}(x,t) = {dx}/{dt} \tag{2.6.1}$$

$$\underline{f}(x,t) \cong \underline{f}(x_0,t) + \left({df}/{dx}\right)_{x_0} (x - x_0) \tag{2.6.2}$$

$$\underline{f}(x,t) \cong \underline{f}(x_0,t) + \underline{J}(x - x_0) \tag{2.6.3}$$

3. Modelling gearbox faults

The study of gear faults has long been an important topic of research for the development of gear diagnostic techniques based on vibration signal analysis. Understanding how different types of gear tooth faults affect the dynamics of gears is useful to characterise and predict the symptoms of the damage appearing in vibration signals [44, 45]. The strong link between the TE and the vibration of the gears was explained earlier. The effect of different types of gear tooth faults on TE can be studied by using the static simulation models. The result of static simulation can be then used to determine how different types of gear faults can be modelled into the dynamic simulation.

Gears can fail for a broad range of reasons. Finding a root cause of damage is an important part of developing a preventative measure to stop the fault from recurring. Analysis of gear failure involves a lot of detective works to link the failed gear and the cause of the damage. Comprehensive guidelines for gear failure analysis can be found in Alban [46], DeLange [47] and DANA [48]. AGMA (American Gear Manufacturers Association) recognizes four types of gear failure mode and a fifth category which includes everything else: Wear, Surface Fatigue, Plastic Flow, Breakage and associated gear failures [49].

The effect of gear tooth fillet cracks (TFC) and spalls on gear transmission error was studied in detail by using a static simulation models (FEA and LTCA (HyGears [50])). A pair of meshing gears were modelled and analysed in step incremented non-linear static environment. Note: the transmission error obtained from the static simulation models are referred to as Motion Errors (ME) here forth by following the HyGears convention.

It was explained earlier that the interaction between two meshing gears can be expressed in the dynamic model as time-varying stiffness, damping and gear tooth topological error elements linking the two lumped mass moments of inertia. The effect of gear tooth faults can be implemented into the dynamic simulation model as changes to these parameters. The understanding gained from the detailed simulation model studies of TFCs and spalls on gear motion has lead to the method of modelling the effect of the faults in dynamic model. The relevance between the types of gear faults to the selected parameters will be explained through subsequent sections.

Further to the simulation of gear tooth faults, this chapter also briefly touches on the modelling of spalls in rolling element bearings (REB), which is also a common type of faults in geared transmission systems.

3.1 Modelling the effect of a fatigue crack in tooth fillet area of a gear

Classical tooth root fillet fatigue fracture is the most common cause of gear tooth breakages (Figure-3.1.1). Stress raisers, such as micro cracks from the heat treatment, hob tears, inclusions and grinding burns are common causes that initiate the cracks. The cracks occurring in the gear tooth fillet region progressively grow until the whole tooth or part of it breaks away. The breakage of a tooth is a serious failure. Not only the broken part fails, but serious damage may occur to the other gears as a result of a broken tooth passing though the transmission [48].

Fig. 3.1.1. A spur gear missing two teeth. They broke away due to the propagation of fatigue cracks. (Courtesy of DANA [48])

The research conducted for NASA by Lewicki [51] sets a clear guideline for predicting the trajectory of cracks occurring at the gear tooth fillet. Lewicki predicted crack propagation paths of spur gears with a variety of gear tooth and rim configurations, including the effect of: rim and web thickness, initial crack locations and gear tooth geometry factors (Diametral pitch, number of teeth, pitch radius and tooth pressure angle). A summary of the results is presented in Figure-3.1.2.

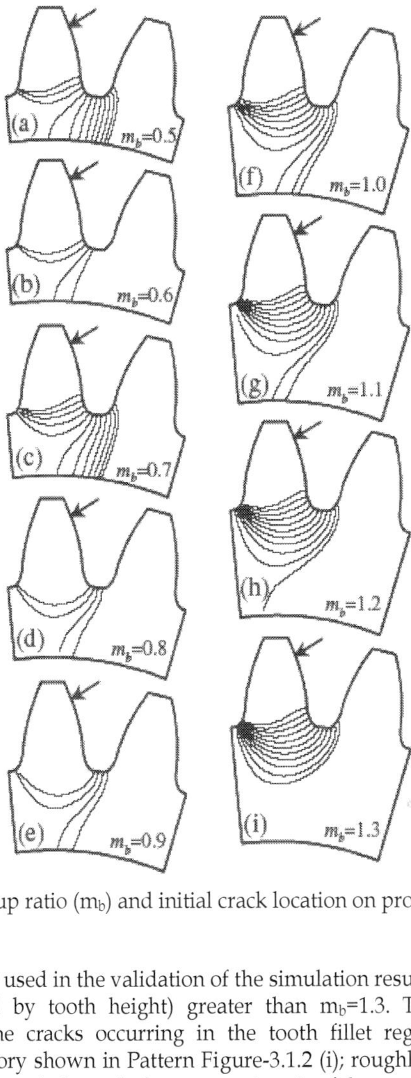

Fig. 3.1.2. Effect of backup ratio (m_b) and initial crack location on propagation path. (Lewicki [51])

A set of spur gears later used in the validation of the simulation result have a "backup ratio" (rim thickness divided by tooth height) greater than m_b=1.3. Therefore based on the Lewicki's prediction the cracks occurring in the tooth fillet region are most likely to propagate in the trajectory shown in Pattern Figure-3.1.2 (i); roughly 30~45° in to the tooth relative to the radial line path through the symmetric axis of the spur gear tooth profile.

Full 3D modelling of a propagating gear tooth crack is one of the actively researched areas. Some examples of simulation studies using the Boundary Element Method (BEM) are given in [52, 53, 54, 55]. The simulation studies using 3D models show complex behaviour crack growth from the small crack seeded at the middle of the gear tooth fillet. An example is shown in Figure-3.1.3 from "Modelling of 3D cracks in split spur gear", by Lewicki [52]. The crack front expands rapidly across the width of the gear tooth as it progresses into the thickness of the tooth. The tooth fillet crack (TFC) model used in this work assumes 2D

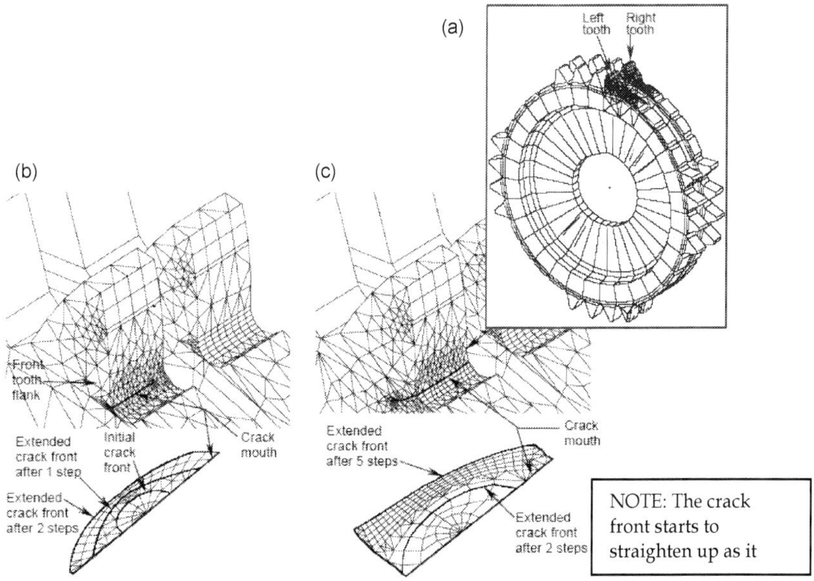

Fig. 3.1.3. 3D Crack propagation model. a) Boundary Element Model of Split Spur Gear, b) Close up of view of the gear teeth and crack section at earlier stage of development and c) more progressed crack. (Lewicki [53])

conditions which approximates the weakening of the cracked gear tooth when the crack is extended across the whole width of the tooth face.

The motion error (ME), as obtained from the finite element (FE) model of a gear pair (32x32 teeth) presented in Figure-3.1.4, is shown in Figure-3.1.5. The MEs of the gears at different

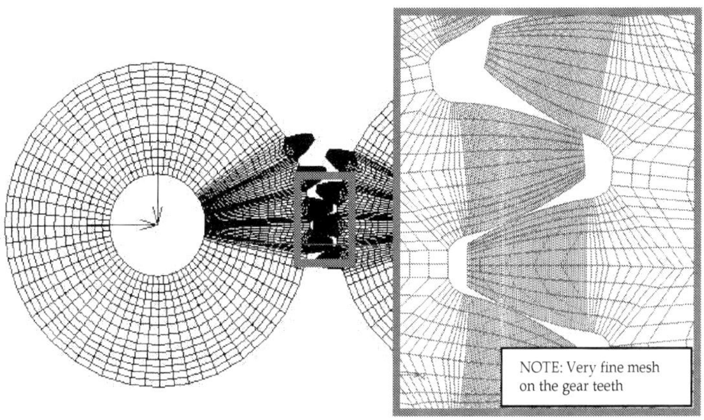

Fig. 3.1.4. Spur Gears (32x32) and its FE mesh (L), detailed view of the mesh around the gear teeth (R). [56]

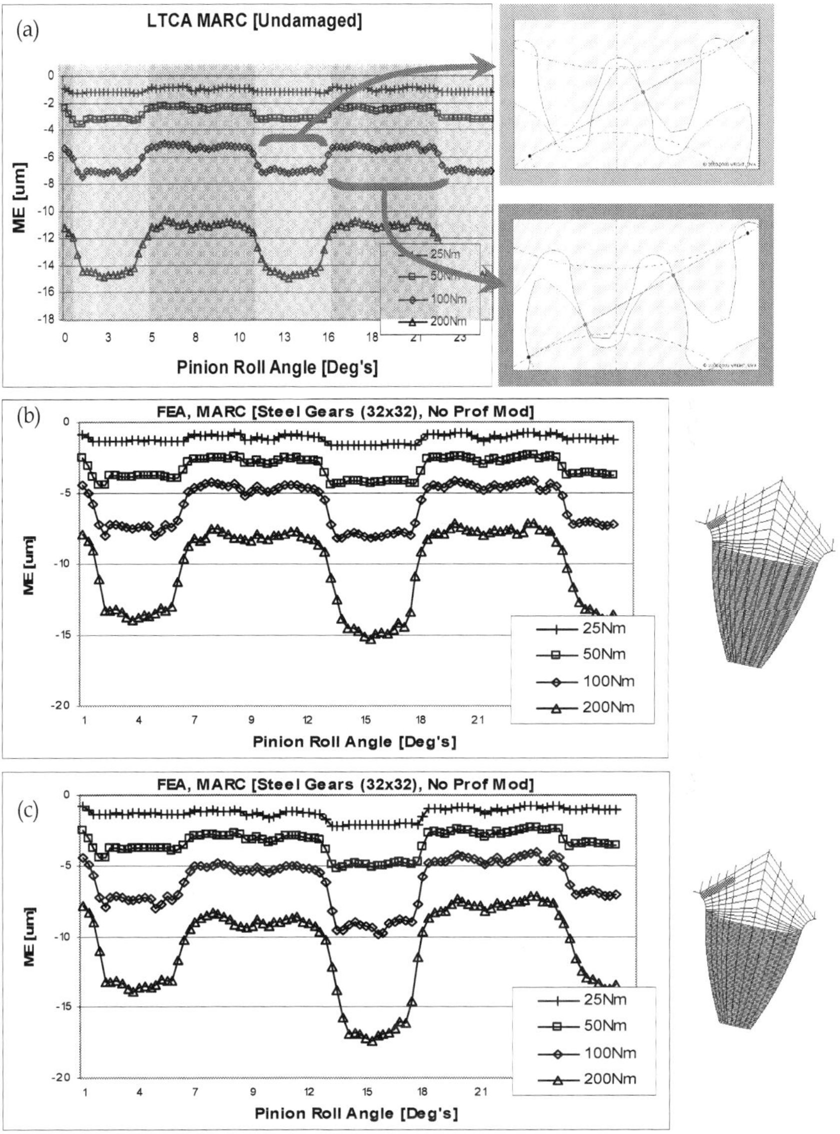

Fig. 3.1.5. Motion Errors of (32x32) teeth gear pairs; (a) Undamaged gear set; (b) TFC (L=1.18mm); (c) TFC (L=2.36mm)

amount of loadings (25, 50, 100 & 200Nm) are compared in the same plot. The magnitude of MEs increases with the larger loads as the deflection of the meshing teeth become greater with the larger loads. The change in the amount of ME is roughly in linear relationship with the load, which reflects the linear elastic behaviour of the gear tooth deflection. Note the square pattern of the ME, which resulted from the time (or angular position) dependent variation of gear mesh stiffness, due to the alternating single and double tooth engagement.

The plots presented in Figures 3.1.5 (b) & (c) show the MEs of the gears with tooth fillet cracks of three different sizes. The localized increase in the amount of ME over a period of ME pattern is a direct consequence of the reduced gear tooth stiffness caused by the TFC.

The plots shown in Figure-3.1.6 (a) ~ (c) are the residual ME (RME) obtained by taking the difference between the MEs of uncracked gears and the ones with TFCs. The RMEs show a "double stepped" pattern that reflects the tooth meshing pattern of the gears, where smaller step with less deflection occurs as the crack tooth enters the mesh and share the load with the adjacent tooth; the larger second step follows when the cracked tooth alone carries the load.

The RMEs of the TFCs show linearly proportional relationship between the amount of loading and the change in RME for a given crack size. The linear relationship between the loading and the amount of tooth deflection on a cracked gear tooth indicates that the effect of TFC can be modelled effectively as a localized change in the gear mesh stiffness.

The plots in Figure-3.1.7 show the transmission errors (TEs) measured from a pair of plastic gears with a root fillet cut (Figure-3.1.7 (a)), which the cut replicates a tooth fillet crack. The TEs of Figure 3.1.7 (c1-c3 and d1-d3) are compared to the simulated patterns of MEs (figure 3.1.7 (b)). Composite TEs (CTEs) and the zoomed view of the CTEs are shown in Figure-3.1.7 (c1 & d1) and (c2 & d2) respectively. The CTE combines the both long and short term components of the TE (LTC and STC) presented earlier in section 2.2 (figure 2.2.1). The resemblance between the simulated MEs and measured the TEs confirms the validity of the simulated effect of TFC. The STCs (c3 & d3) were obtained by high pass filtering the CTE. The pattern in the STCs shows clear resemblance to the simulated TFC effect (Figure3.1.7 (b)).

The simulation model used in this study does not consider the effect of plasticity. This assumption can be justified for a gear tooth with small cracks where localised effect of plasticity at the crack tip has small influence on the overall deflection of the gear tooth, which is most likely the case for the ideal fault detection scenario.

More recent work published by Mark [57, 60] explains that the plasticity can become a significant factor when work hardening effect can cause a permanent deformation of the cracked tooth. In this case, the meshing pattern of the gears changes more definitively by the geometrical error introduced in the gears by the bent tooth. In some cases the bent tooth result in rather complex meshing behaviour that involves tooth impacting. Further explanation on this topic is available from the works published by Mark [57, 60].

Within the limitation of the simulated TFC model discussed above, the approach to model the TFC as a localized variation in the gear mesh stiffness is acceptable for a small crack emerging in the gear tooth fillet area. For the purpose of developing a dynamic simulation model of a geared transmission system with an emerging TFC the model presented here offers a reasonable approach.

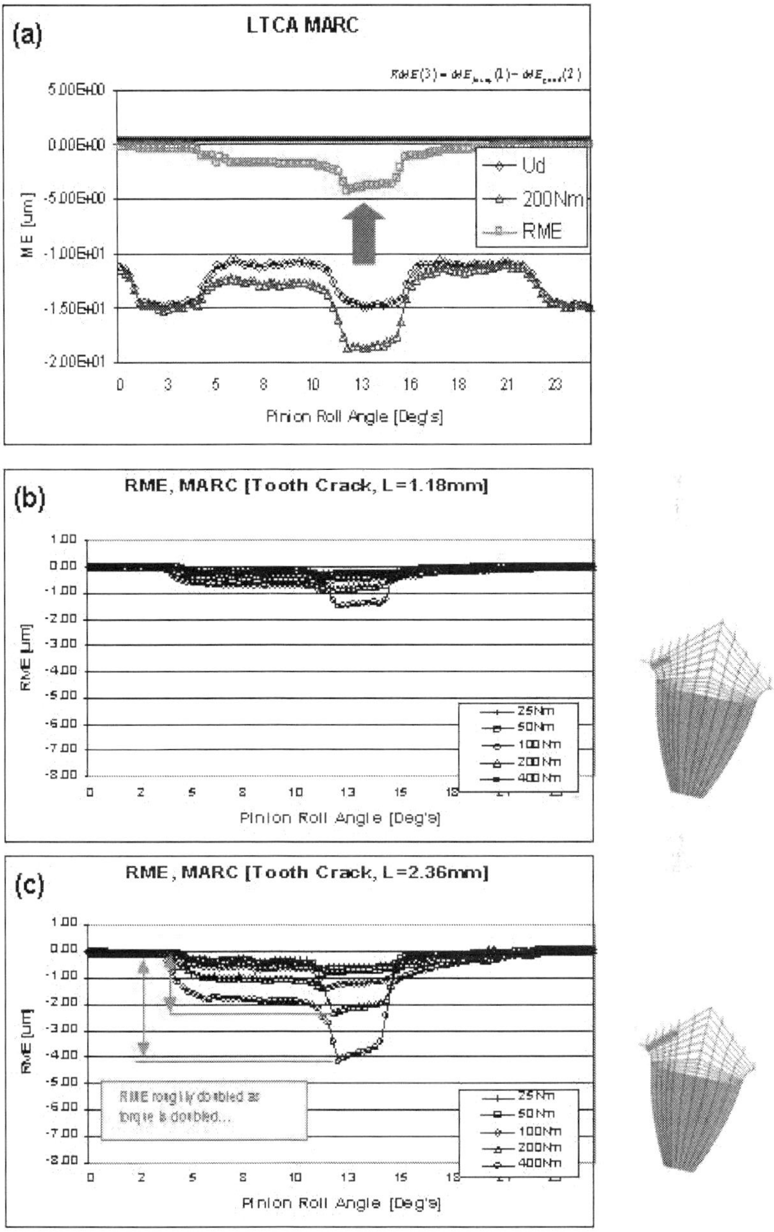

Fig. 3.1.6. Comparison of RMEs of gears with TFCs; (a) Illustrated definition of RME; (b) RMEs of TFC sizes L=1.18mm; (c) RMEs of TFC sizes L=2.36mm.

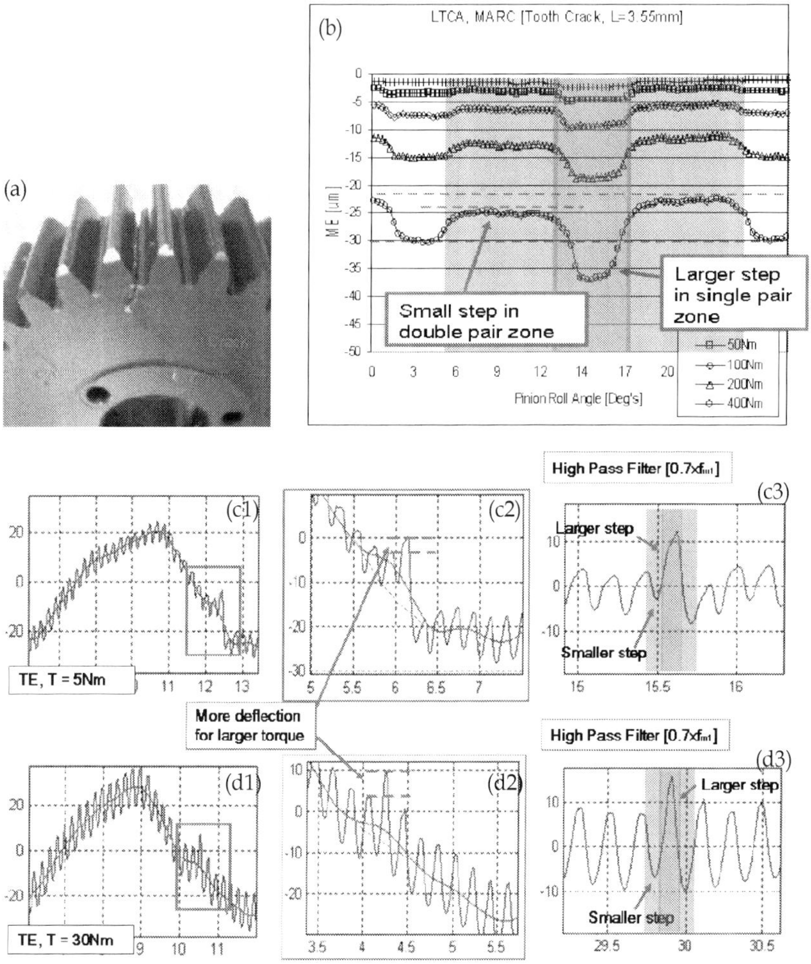

Fig. 3.1.7. Comparison of simulated vs. measured transmission errors; (a) Picture of a gear with a seeded TFC; (b) MEs from a FE model. (c1) ~ (c3) TEs of the gears with TFC loaded with 5Nm; (d1) ~ (d3) TEs of the gear with TFC loaded with 30Nm; CTE (c1, d1), Zoomed views (c2, d2) and STC (c3, d3).

3.2 Modelling the effect of a spall on a tooth face of a gear

Symptoms of surface fatigue vary, but they can generally be noticed by the appearance of cavities and craters formed by removal of surface material. The damage may start small or large and may grow or remain small. In some cases gears cure themselves as they wear off the damage: Initial pitting [47]. The terms "Spalling" and "Pitting" are often used indiscriminately to describe contact fatigue damages. Figure-3.2.1 shows some examples of

(a)  (b) 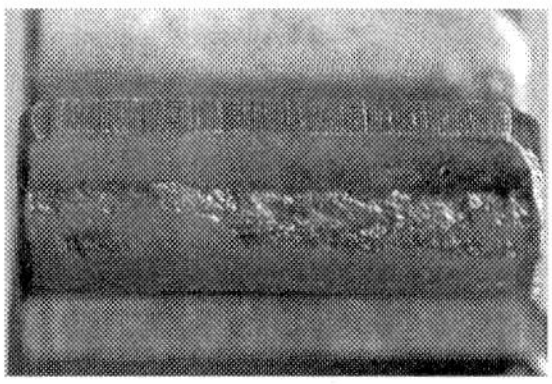

Fig. 3.2.1. Examples of (a) progressive pitting and (b) severe spalling damages on gear teeth. (DANA [48])

spalls and pitting occurring on a gear tooth. This work follows the definition of contact surface damage given by Tallian [58]: Spalling designated only as macro-scale contact fatigue, reserving the term pitting for the formation of pores and craters by processes other than fatigue cracking.

There are three distinctive phases in the development of surface fatigue damage:

1. Initial Phase: Bulk changes in the material structure take place around the highly stressed area under the contact path. Change in hardness, residual stress and some microscopic changes in the grain structure of the metal.
2. A Long Stable Phase: Microscopic flow occurs in the highly stressed area changing the material structure and residual stress conditions at the microscopic level. The change in the structure brought out by the microscopic flow can be observed by eyes in the illuminated etched areas.
3. Macroscopic Cracking: This is the last failure phase instituting the crack growth.

Spalls have a distinctive appearance that is characterised by how they were formed. A fully developed spall typically has its diameter much larger than its depth. The bottom of the spall has a series of serrations caused by propagating fatigue cracks running transverse to the direction of rolling contact. The bottom of the spall parallels the contact surface roughly at the depth of maximum unidirectional shear stress in Hertzian contact.

The sidewalls and the wall at the exiting side of the spall (as in the exiting of rolling contact) are often radially curved as they are formed by material breaking away from the fatigued area. The entrance wall of the spall is characterised by how it was initiated. Tallian [58] explains that shallow angled entry (less than 30° inclination to the contact surface) occurs when the spall is initiated by cracks on the surface. Spalls with steep entry (more than 45°) occur when the spall is initiated by subsurface cracks.

Surface originated spalls are caused by pre-existing surface damage (nicks and scratches). It is also known that lubrication fluid could accelerate the crack propagation when contact occurs in such a way that fluid is trapped in the cracks and squeezed at extremely high pressure as the contacting gear teeth rolls over it.

The subsurface originated spalls are caused by presence of inclusions (hard particles and impurities in the metal) and shearing occurs between the hard and the soft metal layers formed by case hardening. A spall caused by the initial breakage of the gear tooth surface continues to expand by forming subsequent cracks further down the rolling direction. Figure-3.2.2 illustrates the formation and expansion of spall damage by Ding & Rieger [59].

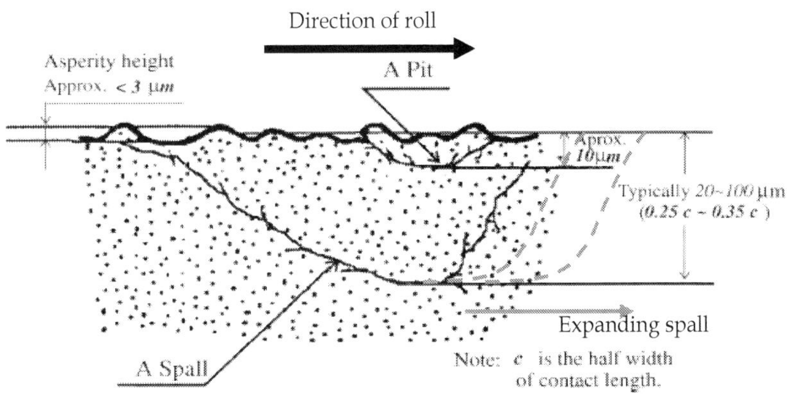

Fig. 3.2.2. Formation and expansion of spalls. (Ding & Rieger [59])

The result of contact stress analysis from HyGears [50] shows the occurrence of high stress concentration at the entrance and exit walls of the spall (Figure-3.2.3). The high stress concentration at these edges implies the likelihood of damage propagation in that direction. The result of the simulation corresponds to Tallian's [58] observation that spalls tend to expand in the direction of rolling contact. In the HyGears simulation the spall was modelled as a rectangular shaped recess on the tooth surface in the middle of the pitch line.

Typical spur gear tooth surfaces are formed by two curvatures: profile and lead curvatures. The 2D models are limited to simulating the effect of the spall crater on the gear tooth profile only. 3D simulation is required to comprehend the complete effect of spalls on the

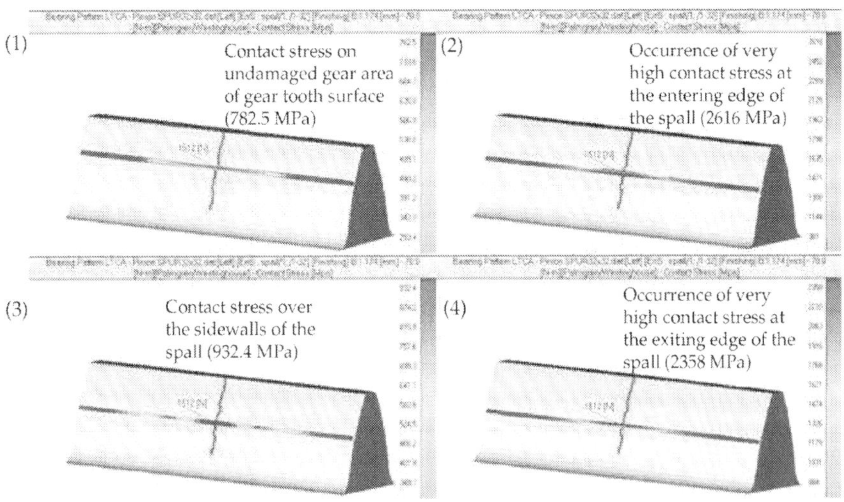

Fig. 3.2.3. HyGears analysis of contact stresses around the spall. [62]

contact surface of the gears. A study was carried out to investigate the effect of spalls on the ME using a 3D gear tooth model [62]. The result of the study showed that the effect of the spalls on the ME is completely dominated by the displacement caused by the topological error of the gear tooth surface caused by the fault.

Formation of spalls is a complicated process and it is one of the active research topics which have been studied for some time. Although there are several items in the literature that describe the properties and process of formation of spalls, no literature was sighted which defines the specific definition of shapes and sizes of spalls occurring on spur gear teeth.

Papers presented by Badaoui et al. [63, 64] and Mahfoudh et al. [65] show a successful example of simulating the effect of a spall on spur gear tooth by modelling the fault as prismatic slot cut into the gear tooth surface. Their results were validated by experiment. The model of the spall used in this work follows the same simplification of the general shape of the spalling fault. The simulation of the spall is bounded by these following assumptions:

1. A spall is most likely to initiate at the centre of the pitchline on a gear tooth where maximum contact stress is expected to occur in the meshing spur gear teeth as the gears carry the load by only one pair of teeth.
2. The spall expand in size in the direction of rolling contact until it reaches the end of the single tooth pair contact zone as shown in Figure-3.2.4.
3. When the spall reaches the end of the single tooth pair contact zone, the spall will then expand across the tooth face following the position of the high contact stresses. The contact stress patterns shown in Figure-3.2.3 strongly support this tendency of spall growth.

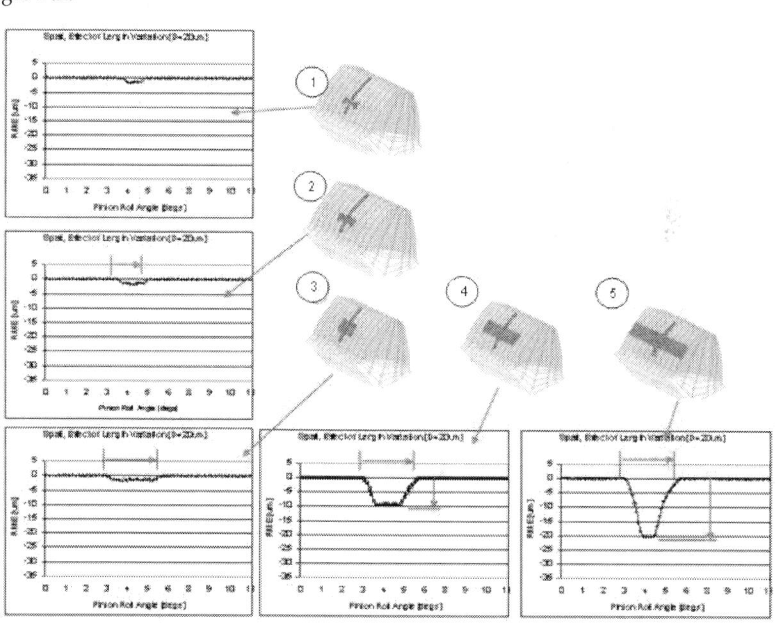

Fig. 3.2.4. A model of spall growth pattern and the resulting RME. [62]

The models of spalls were developed by following the set of assumptions described above. Resulting plots of the RMEs are presented in Figure-3.2.4 along with the shape illustration of spalls on a gear tooth. Note that the RMEs of the spalls form bucket shapes and their length and the depth are determined by the length and width of the spall; i.e. the shape of RME and the progression of the spall is directly related. This information can be used in diagnosis and prognosis of the fault.

The plots in Figure-3.2.5 show TEs measured from a pair of plastic gears with a spall (Figure-3.2.5 (a)). The change in the pattern of the TE due to the spall is comparable to the simulated pattern of the MEs (Figure-3.2.5 (b)). Composite TEs (CTEs) and the zoomed views of the CTEs are shown in Figure-3.2.5 (c1~c3). The resemblance between the simulated MEs and measured TEs confirms the validity of the simulated effect of TFC. The STCs (d1~d3) were obtained by high pass filtering the CTE. The pattern in the STCs shows clear correlation between the simulated and measured patterns of spall motion errors.

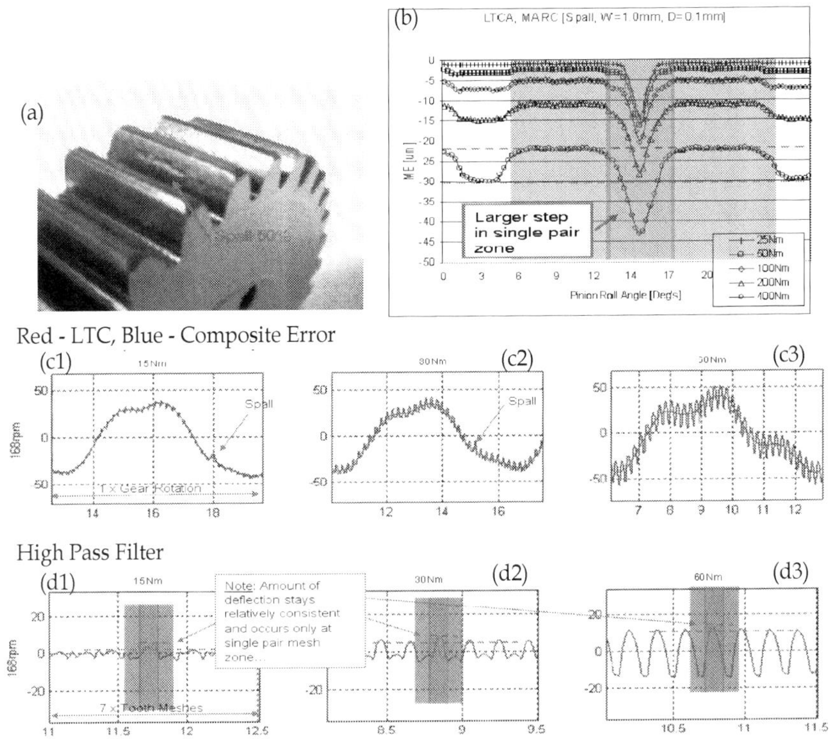

Fig. 3.2.5. Comparison of simulated vs. measured transmission errors; (a) Picture of a gear with a seeded Spall; (b) MEs from a FE model; (c1) ~ (c3) CTEs of the gear with the spall; (d1) ~ (d3) STCs of the gear with the spall; Applied torque: 15Nm (c1 & d1), 30Nm (c2, d2), 60Nm (c3, d3).

3.3 Simulating the effect of TFCs and spalls in a gear dynamics model

The FEA model based study of TFCs and spalls has lead to identifying some useful properties of the faults that can be used to model the effect of the faults in the lamped parameter type gear dynamics models.

The Residual Motion Errors (RME) of TFCs have shown double stepped patterns that were load dependent. The change in the amount of deflection in the gear mesh (i.e. ME) with a cracked tooth is influenced by the size of the crack and also by the amount of loading on the gears. The simulation result showed that the linearly proportional relationship between the torque applied to the gears and the resulting RME value. The bucket shaped RMEs of spalls were not affected by the loading condition but purely driven by the change in the contact path patterns of the meshing teeth, due to the geometrical deviation of the gear tooth surface caused by a spall. It was also understood from the simulation studies that the size and the shape of a spall affect the length and the depth of the bucket.

Based on the observation above, the effect of TFC was modelled as locally reduced tooth meshing stiffness and the spalls as direct displacement due to the topological alteration of the gear tooth surface. In the gear dynamic model is shown in Figure-3.3.1, the effect of a TFC was implemented as a reduction in stiffness "K_m" over one gear mesh cycle. The change in the value of K_m was calculated from the FEA model mapped into an angular position dependent function in the gear dynamic model. A spall was implemented as a localized displacement mapped on the "e_t". An illustration of a TFC and a spall models in a gear rotor dynamic model is shown in Figure-3.3.1 [61].

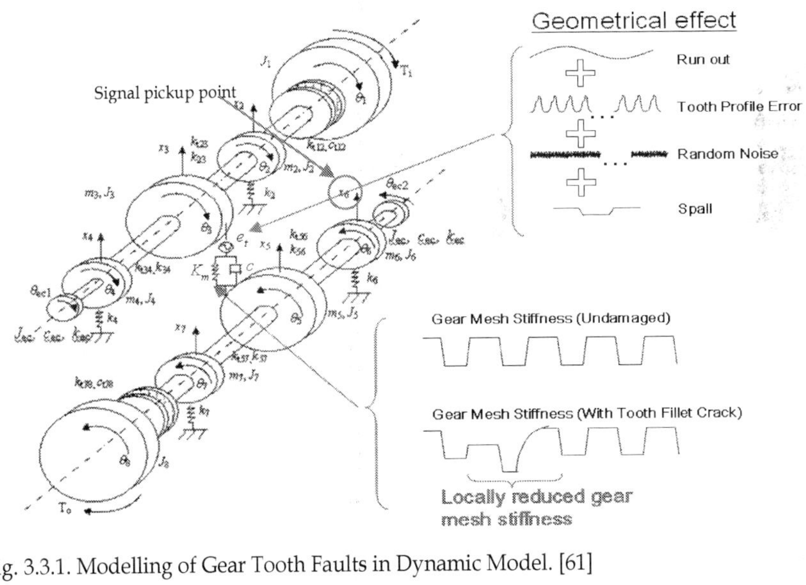

Fig. 3.3.1. Modelling of Gear Tooth Faults in Dynamic Model. [61]

Figures 3.3.2 (a1~a3) and 3.3.3 (a1~a3) show the simulated vibration signal (acceleration) from the LPM shown in the Figure-3.3.1. The signals were measured from the free end of the driven shaft (see Figure-3.3.1). The residual signals shown in (b1~b3) of Figures 3.3.2 and 3.3.3 were obtained by subtracting the simulated vibration of undamaged gearbox from the damaged one. Two identical simulation models, one with a gear tooth fault (TFC or spall) and the other undamaged, were run in parallel and the difference between the two model out puts were taken to separate the effect of the gear faults. The impact like effect of the gear tooth faults is seen in both the gears with a TFC and a spall.

A comparison of the simulated signals shows that the magnitude of the TFC impulses is affected by the amount of torque applied to the gears, while the spall impulses are not. This response is consistent with the observation made in the static simulation of the gears in mesh.

Careful observation of the residual signal also reveals that the fault information is not only buried in the dominant effect of gearmesh, but also somewhat distorted by the effect of transmission path from the gearmesh to the point where the signal was measured. The effect of transmission path appears in the residual signal As the transient "tail" effect convolved over the impulse due to the gear fault (see Equation-3.3.1 and Figure-3.3.4 for illustration).

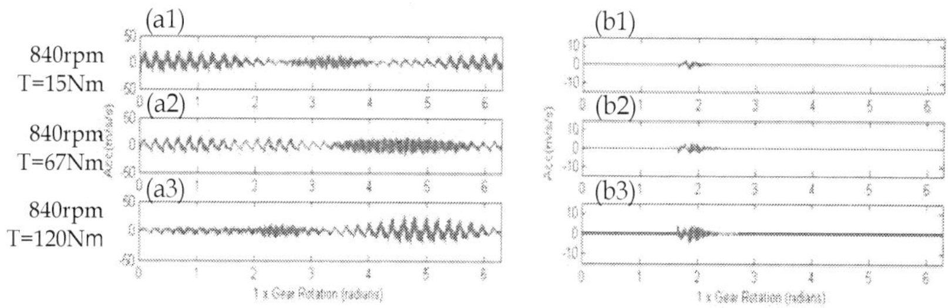

Fig. 3.3.2. Simulated gearbox vibration signal with the effect of a TFC.

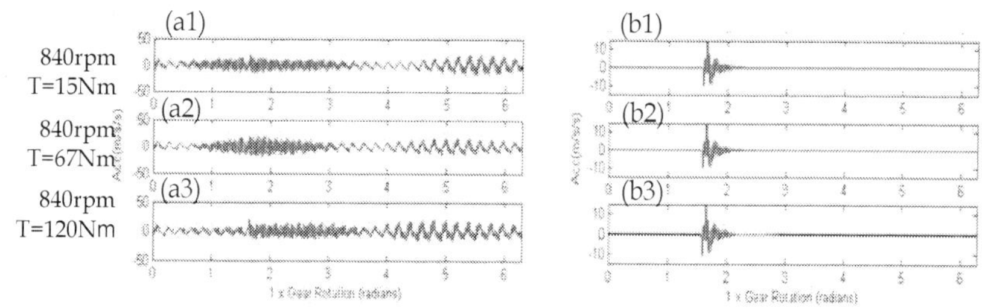

Fig. 3.3.3. Simulated gearbox vibration signal with the effect of a spall.

The residual signal of the TFC and the spall provide a useful means to understand the nature of diagnostic information of the faults. In machine condition monitoring signal processing techniques are often developed to detect and quantify the symptoms of a damage buried in a background noise. By being able to see the "clear" effect of a fault, the most effective signal processing technique can be applied to target and monitor the symptoms of the damage. The idea of using the simulated fault signals to design and improve the fault detection and machine condition monitoring techniques has been put to effective uses by Randall, Sawalhi and Endo [34, 35, 62, 66].

$$y = (e + w + n) * h \text{ (Note: * represents convolution)} \quad (3.3.1)$$

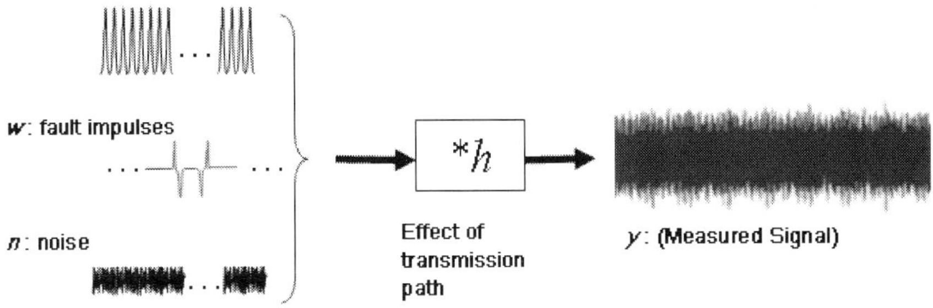

Fig. 3.3.4. Vibration of a Gearbox. [62]

3.4 Modelling a spall in a rolling element bearing

This section discusses a several simple but an effective methods of modelling a spall in the rolling element bearing model introduced previously in section 2.5.1. The ideas behind each modelling approach was discussed by using the example of modelling a spall in the outer race of a bearing. The same method can be easily expanded in to modelling an inner race spall and ball faults. More detailed explanation on this topic is available from the work published by Sawalhi and Randall [29, 30, 31, 34, 35, 66, 67].

The simplest form of a spall model can be implemented to the REB model by assuming instantaneous contact loss between the bearing races and rollig an element(s) as it pass over the spall. So, in reference to the rolling element model described above, the presence of a spall of a depth (C_d) over an angular distance of ($\Delta\phi_d$) can be modelled by using the fault switch β_j which defines the contact state of rolling elements over a defined angular position (ϕ_d). In effect, this mothod models the spall as a step function as shown in Figure-3.4.1 and further illustrated by Figure-3.4.2 (a), in which β_j is defined as follows:

$$\beta_j = \begin{cases} 1, & if\ \phi_d < \phi_j < \phi_d + \Delta\phi_d \\ 0, & otherwise \end{cases} \quad (3.4.1)$$

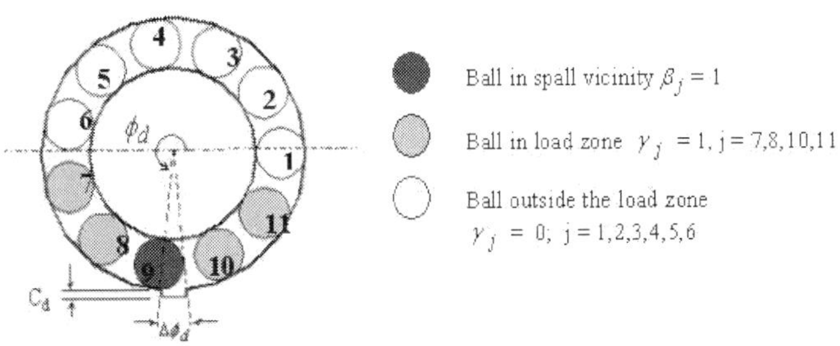

Fig. 3.4.1. Spall definition on the outer race [66]

Fig. 3.4.2. Modified model of a spall based on a more realistic ball trajectory. [66]

The outer race spall is fixed in location between ϕ_d and $\phi_d + \Delta\phi_d$. This normally occurs in the load zone. An inner race spall rotates at the same speed as the rotor, i.e. $\phi_d = \omega_c dt + \phi_{do}$ (ϕ_{do} : initial starting location of the spall).

This model of the spall assumes that the rolling element will lose contact suddenly once it enters the spall region, and will regain contact instantly when exiting from that area (Figure-3.4.2 (a)). The abrupt change in the rolling element positions at the entry and exit of the spall results in very large impulsive forces in the system, which is not quite realistic. An modification on the previous model was introduced in [28] in which the depth of the fault (C_d) was modelled as a function of (ϕ_j), Figure-3.4.2 (b). The improvement on the model is to represent more realistic trajectory of the rolling element movement based on the relative size of the rolling element and the depth of the spall. Although, the profile of the trajectory appears much less abrupt than the earlier version; and apears to have only one position that may result in impulse, it still resulted in two impulses which does not agree with the experimental observation.

Careful observation of the interaction between the rolling element and spall leads to the trajectory is shown in Figure-3.4.3. The entry path of the rolling element has been represented as having a fixed radius of curvature (equal to that of the rolling element); entry of the rolling element in to the spall is therefore somewhat smoother. The smoother change in curvature at the entry would then represent a step in acceleration. On exiting the spall, the centre of the rolling element would have to change the direction suddenly, this representing a step change in velocity or an impulse in acceleration. This has been modelled as a sudden change (i.e. similar to the original model [28]). The resulting acceleration signal

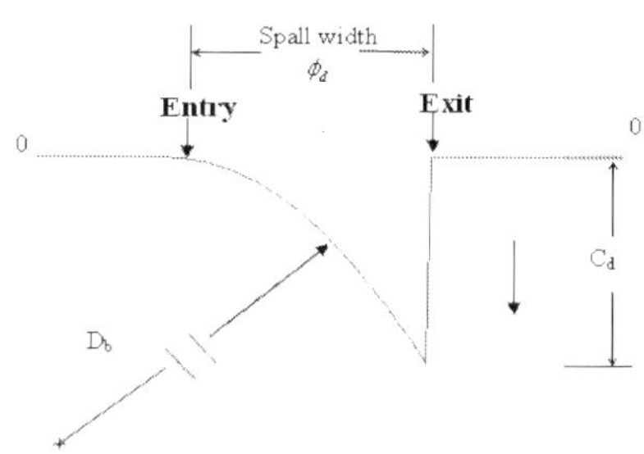

Fig. 3.4.3. A correlated model of a spall based on experimental data.

from this model appears to agree with the experimental observation, however author recommends further validation on this modelling approach and more updates are expected based on the findings.

For more detailed explanations on the modelling of REB fault refer to the works published by Sawalhi & Randall [29, 30, 31, 34, 35, 66, 67].

4. Conclusion

The techniques for modelling the effect of gearbox faults: tooth fillet cracks, tooth face spalls and bearing spalls, were presented and discussed in this chapter. The main purpose of the damage modelling is to simulate the effect of the faults on the dynamics of a geared transmission system that can be used in improving the understanding of the diagnostic information that manifest in the vibration signal mix from a gearbox.

The fault detection and diagnostic techniques based on vibration signal analysis are the ideal non-destructive machine health monitoring method, that can be applied in a minimally intrusive manner; i.e. by attaching an accelerometer on a gearbox casing. However, the dynamic interaction amongst the machine elements of a gearbox is often complex and the vibration signals measured from the gearbox is not easy to interpret. The diagnostic information that directly related to an emerging fault in a gear or a bearing is typically buried in the dominating signal components that are driven by the mechanisms of the transmission system themselves: For example, gear meshing signals.

Traditionally, the researchers worked on development of a signal processing technique for the gearbox diagnosis have embarked on their endeavours by making educated assumptions on the properties of the diagnostic information of the faults. These assumptions are often based on their careful observation of a measured vibration signals. However, the relevance of this approach is often somewhat limited by the simple fact that it's not easy to observe the key details of the fault signals from the signal mix.

It was demonstrated in this chapter how simulation models can be put to effective uses for studying the properties of the fault signals in greater details. A method of isolating the fault signals from the simulated gearbox signal mix was described in the section 3.3. The residual signals obtained from this process showed how the faults manifested in the resulting vibration signals in the "cleaned" state. The observation of the simulated residual signals has led to an improved understanding of the characteristics of impulses caused by the faults and the distorting effect of the transmission path (from the origin of the fault signal to the measurement location). The improved understanding of the fault signal obtained from the simulation studies led to the development of more effective signal processing techniques [34, 35, 62, 66].

The models of the gearbox faults presented in this work require further refinement. Some of the areas of future improvement aforementioned in the main body of the chapter include; improving the understanding of; the effect of plastic deformation in gear TFC, the effect of spall shapes and the effect of non-linear dynamic interaction of the gears and bearings. Improving the correlation between the simulated and the measured signals is a good way to demonstrate the understanding of the effect of faults in a geared transmission system. This knowledge compliments the design and development efforts in vibration signal analysis based machine condition monitoring technologies. In the near future, accurately simulated signals of a faulty gearbox can aid the machine learning process of fault diagnosis algorithms based on neural networks. Performing this task in experiments are time consuming and costly exercise; simulation model based approach appears much more desirable.

The authors hope that the work presented in this chapter will stir the thoughts and the new ideas in readers that will contribute to the advancement of the gear engineering and the technologies in detecting and diagnosing the incipient faults in geared transmission systems.

5. Nomenclature

Unless otherwise stated the following tables defines the symbols and the acronyms used in this chapter.

$\overline{A_1A_2}$	A vector connecting points A_1 and A_2
Ø	Pressure Angle
LoA	Line of Action
PoA	Plane of Action
LoC	Line of Centres
R_b, r_b	Gear Base Radius and Pinion Base Radius
R_{gear}, r_{pinion}	Gear Pitch Radius and Pinion Pitch Radius
R_o and r_o	Gear Outer Radius Pinion Outer Radius
P_b	Base pitch
CR	Contact Ratio
N	Number of teeth on a gear
TE	Transmission Error (Measured experimentally)
GTE	Geometrical Transmission Error
STE	Static Transmission Error
LTC	Long Term Composite of TE
STC	Short Term Composite of TE
ME	Motion Error (Numerically calculated TE)
x_i, y_i, z_i	Translation at i_{th} Degrees of Freedom
θ_{xi}, θ_{yi}, θ_{zi}	Rotation about a translational axis at i_{th} Degrees of Freedom
K, K_{mb}	Linear stiffness elements. The subscript 'mb' refers to stiffness at the gearmesh
C, C_{mb}	Damping matrices. The subscript 'mb' refers to damping at the gearmesh.
H	An 'on/off' switch governing the contact state of the meshing gear teeth.
$\underline{e_t}$	A vector representing the combined effect of tooth topography deviations and misalignment of the gear pair.
T	Torque

$\underline{x}$, $\underline{\dot{x}}$, $\underline{\ddot{x}}$ Vectors of displacement, velocity and acceleration (translation)

σ Stress

ε Strain

6. References

[1] J. D. Smith, and D. B. Welbourn, 2000, "Gearing Research in Cambridge", Cambridge Gearing Publications

[2] R. E. Smith, 1983, "Solving Gear Noise Problems with Single Flank Composite Inspection", The Gleason Works, Rochester, NY, USA

[3] R. G. Munro, D. Houser, 2004, "Transmission Error Concepts", Notes from Gear Noise Intensive Workshop, 19th ~ 21st July 2004, Melbourne, Australia

[4] R. W. Gregory, S. L. Harris and R. G. Munro, 1963a, "Dynamic Behaviour of Spur Gears", Proc IMechE, Vol. 178(1), pp 207-218

[5] R. W. Gregory, S. L. Harris and R. G. Munro, 1963b, "Torsional Motions of a Pair of Spur Gears", Proc IMechE Applied Mechanics Convention, Vol. 178 (3J), pp 166-173

[6] H. N. Ozguven and D. R. Houser, 1988a, "Mathematical Models Used in Gear Dynamics - A Review", Journal of Sound and Vibration, 121(3), pp 383-411

[7] C. C. Wang, 1985, "On Analytical Evaluation of Gear Dynamic Factors Based on Rigid Body Dynamics", Journal of Mechanism, Transmission and Automation in Design, Vol. 107, pp 301-311

[8] W. Bartelmus, 2001, "Mathematical Modelling and Computer Simulations as an Aid to Gearbox Diagnostics", Mechanical Systems and Signal Processing, 15 (5), pp 855-871

[9] J. Lin and R. G. Parker, 2002, "Mesh Stiffness Variation Instabilities in Two-Stage Gear Systems", Transaction of ASME, Jan 2002, Vol. 124, pp 68-76

[10] R. G. Parker and J. Lin, 2001, "Modelling, Modal Properties, and Mesh Stiffness Variation Instabilities of Planetary Gears", NASA, NASA/CR-2001-210939

[11] Y. Gao and R. B, Randall, 1999, "The Effect of Bearing and Gear Faults in Rolling Element Bearing Supported Gear Systems", DSTO Aeronautical and Maritime Research Laboratory, July 1999, CEVA-99-02

[12] Y. Gao and R. B. Randall, 2000, "Simulation of Geometric, Static and Dynamic Gear Transmission Errors", DSTO Aeronautical and Maritime Research Laboratory, Jan 2000, CEVA-2000-01

[13] M. Amabili, and A. Rivola, 1997, "Dynamic Analysis of Spur Gear Pairs: Steady-State Response and Stability of the SDOF Model with Time Varying Meshing Damping", Mechanical System and Signal Processing, 11 (3), pp 375-390

[14] I. Howard, S. Jia and J. Wang, 2001, "The Dynamic Modelling of a Spur Gear in Mesh Including Friction and A Crack", Mechanical Systems and Signal Processing, 15 (5), pp 831-853

[15] P. Velex and M. Maatar, 1996, "A Mathematical Model for Analysing the Influence of Shape Deviations and Mounting Error on Gear Dynamic Behaviour", Journal of Sound and Vibration, 191(5), pp 629-660

[16] S. M. Vijayakar, 1991, "A Combined Surface Integral and Finite Element Solution for a Three-Dimensional Contact Problem", International Journal of Numerical Methods in Engineering, 31, pp 524-546

[17] G. W. Blankenship and R. Singh, 1995, "A New Gear Mesh Interface Dynamic Model to Predict Multi-Dimensional Force Coupling and Excitation", Mechanical, Machine Theory, Vol. 30, No.1, pp43-57

[18] A. Kahraman and G. W. Blankenship, 1996, "Interactions between Commensurate parametric and Forcing Excitations in A System with Clearance", Journal of Sound and Vibration, 194 (3), pp 317-336

[19] R. G. Parker, 2000, "Non-Linear Dynamic Response of a Spur Gear Pair: Modelling and Experimental Comparisons", Journal of Sound and Vibration, 237(3), pp 435-455

[20] R. G. Parker, V. Agashe and S. M. Vijayakar, 2000, "Dynamic Response of a Planetary Gear System Using a Finite Element/Contact Mechanics Model", Transaction of ASME, Sep 2000, Vol. 122, pp 304-310

[21] CALYX, 2002, CALYX USER MANUAL, Advanced Numerical Solutions, Hilliard Ohio, U.S.A

[22] HELICAL, 2002, HELICAL 3D USER's MANUAL, Advanced Numerical Solutions, Hilliard Ohio, U.S.A

[23] S. Du, 1997, "Dynamic Modelling and Simulation of Gear Transmission Error for Gearbox Vibration Analysis", PhD dissertation, University of New South Wales

[24] N. S. Feng, E. J. Hahn, and R. B. Randall, 2002, "Using transient analysis software to simulate vibration signals due to rolling element bearing defects", Proceedings of the 3rd Australian congress on applied Mechanics, 689-694, Sydney.

[25] I. K. Epps and H. McCallion, 1994, "An investigation into the characteristics of vibration excited by discrete faults in rolling element bearings" (extract from PhD thesis of I.K. Epps), Annual Conference of the Vibration Association of New Zealand, Christchurch.

[26] D. Ho, 1999, "Bearing Diagnostics and Self Adaptive Noise Cancellation", PhD dissertation, UNSW.

[27] S. Fukata, E. H. Gad, T. Kondou, T. Ayabe and H. Tamura, 1985, "On the Vibration of Ball Bearings", Bulletin of JSME, 28 (239), pp. 899-904.

[28] A. Liew, N.S. Feng and E. J. Hahn, 2002, "Transient rolling element bearing systems", Trans. ASME Turbines and Power, 124(4), pp. 984-991

[29] N. Sawalhi and R.B. Randall, 2011, Vibration response of spalled rolling element bearings: observations, simulations and signal processing techniques to track the spall size. Mechanical Systems and Processing, 25 (3), (2011)

[30] N. Sawalhi and R.B. Randall, 2010, " Improved simulations for fault size estimation in rolling element bearings", paper presented at the seventh International Conference on Condition Monitoring and Machinery Failure Prevention Technologies, Stratford-upon-Avon, England, 22-24 June (2010).

[31] N. Sawalhi and R.B. Randall, 2008, "Localised fault diagnosis in rolling element bearings in gearboxes", paper presented at the Fifth International Conference on Condition Monitoring and Machinery Failure Prevention Technologies, Edinburgh, 15-18 July (2008).

[32] J. Sopanen & A Mikkola, 2003, "Dynamic model of a deep-groove ball bearing including localized and distributed defects. Prt 1: Theory", Proc of IMechE, vol. 217, Prt K, J of Multi-body Dynamics, 2003, pp 201-211.

[33] T. Tiwari, K. Gupta, O. Prakash, 2000, "Dynamic response of an unbalanced rotor supported on ball bearings", J of Sound & Vibration 238 (2000), pp757-779.

[34] N. Sawalhi & R. B. Randall, 2008, "Simulating gear and bearing interaction in the presence of faults: Prt II - Simulation of the vibrations produced by extended bearing faults", MSSP, Vol. 22, pp1952-1996 (2008).
[35] N. Sawalhi & R. B. Randall, 2008, "Simulating gear and bearing interaction in the presence of faults: Prt I - The combined gear bearing dynamic model and simulation of localized bearing faults", MSSP, Vol. 22, pp1924-1951 (2008).
[36] Z. -Q. Qu, 2004, "Model order reduction technique: with applications in Finite Element Analysis", Springer, 2004.
[37] R. Craig & M. Bumpton, 1968, "Coupling of substructures of dynamic analysis", AIAA Journal, 67 (1968), pp1313-1319.
[38] B. M. Irons, 1967, "Structural eigenvalue problem-elimination of unwanted variables", AIAA Journal, 3(5), pp961-962.
[39] L. Despnade, N.Sawalhi, R.B.Randall, "Improved Gearbox Simulations for Diagnostic and prognostics Purposes Using Finite Element Model Reduction Techniques", paper presented at the 6th Australasian Congress on Applied Mechanics, Perth, Australia, 12-15 December (2010
[40] C. Carmignani, P. Forte & G. Melani, 2009, "Component modal synthesis modelling of a gearbox fro vibration monitoring simulation", The 6th Int'l conf on Condition Monitoring Manhinery Failure Prevention Technology, Dublin, Ireland.
[41] N. Sawalhi, L. Deshpande and R. B. Randall, 2011, "Improved simulation of faults in gearboxes for diagnostic and prognostic purposes using a reduced finite element model of casing", AIAC 14 Fourteenth Australian International Aerospace Congress.
[42] R. Singh, 1989, "Analysis of Automotive Neutral Gear Rattle", Journal of Sound and Vibration, Vol. 131 (2), pp177-196
[43] A. Kahraman and R. Singh, 1991, "Interaction between Time-Varying Mesh Stiffness and Clearance Non-Linearity in Geared System", Journal of Sound and Vibration", 146, pp135-156
[44] D. P. Townsend, 1997, "Gear and Transmission Research at NASA", NASA-TM-107428
[45] A. K. Wong, 2001, "Vibration-Based Helicopter Health Monitoring - An Overview of the Research Program in DSTO", DSTO-HUMS2001
[46] L. E. Alban, 1985, "Systematic Analysis of Gear Failure", American Society of Metals, Metals Park OH, ISBN-0871702002
[47] G. DeLange, 1999, "Analyzing Gear Failures", Hydrocarbon Processing, Features: 4/00, www.hydrocarbonprocessing.com
[48] DANA, 2002, Raodranger, "Understanding Spur Gear Life", EATON and DANA Corporation, TRSM-0913, (www.roadranger.com)
[49] ANSI/AGMA, 1995, "Appearance of Gear Teeth - Terminology of Wear and Failure", American Gear Manufacturers Association, ANSI/AGMA 1010-E95
[50] HyGears, HyGears © V2.0 Gear Design and Analysis Software, Involute Simulation Software Inc
[51] D. G. Lewicki, 2001, "Gear Crack Propagation Path Studies - Guidelines for Ultra-Safe Design", NASA-TM- 2001 211073
[52] T. J. Curtin, R. A. Adey, J. M. W. Bayman and P. Marais, 1998, Computational Mechanics Inc, Billerica, Massachusetts, 01821, USA

[53] D. G. Lewicki, 1998, "Three-Dimensional Gear Crack Propagation Studies", NASA-TM-1998-208827

[54] S. Pehan, T. K. Hellen, J. Flasker and S. Glodes, 1997, "Numerical Methods for Determining Stress Intensity Factors VS Crack Depth in Gear Tooth Roots", Int J of Fatigue, Vol. 19, No. 10, pp677-685

[55] M. Guagliano and L. Vergani, 2001, "Effect of Crack Closure on Gear Crack Propagation", Int J or Fatigue, Vol. 23, pp67-73

[56] H. Endo, C. Gosselin, R. B. Randall, (Sept 2004), "The effects of localized gear tooth damage on the gear dynamics - A comparison of the effect of a gear tooth root crack and a spall on the gear transmission error", IMechE, 8th International Conference on Vibrations in Rotating Machinery, University of Wales, Swansea

[57] W. D. Mark, C. P. Reagor and D. R. McPherson, 2007, "Assessing the role of plastic deformation in gear-health monitoring by precision measurement of failed gears", MSSP, 21: pp177-192 (2007)

[58] T. E. Tallian, 1992, "Failure Atlas fir Hertz Contact Machine Elements", ASME Press, ISBN 0-7918-0008-3

[59] Y. Ding and N. F. Rieger, 2003, "Spalling formation Mechanism for Gears", Wear, 254 2003), pp1307-1317

[60] W. D. Mark and C. P. Reagor, 2007, "Static transmission error vibratory excitation contribution from plastically deformed gear teeth caused by tooth bending fatigue damage", MSSP, 21: pp885-905 (2007)

[61] H. Endo, R. B. Randall and C. Gosselin, 2004, "Differential Diagnosis of Spall vs. Cracks in the Gear Tooth Fillet Region", Journal of Failure Analysis and Prevention, Vol4, Issue 5, Oct 2004

[62] H. Endo, R. B. Randall and C. Gosselin, 2008, "Differential Diagnosis of Spall vs. Cracks in the Gear Tooth Fillet region: Experimental Validation", Journal of Mechanical System and Signal Processing, due to be published on 19/Jan/2009, Vol23 Issue3.

[63] M. EL Badaoui, J. Antoni, F. Guillet and J. Daniere, 2001a, "Use of the Moving Cepstrum Integral to Detect and Localise Tooth Spalls in Gears", Mechanical System and Signal Processing, 2001, 15 (5) , 873-885

[64] M. EL Badaoui, V. Cahouet, F. Guillet, J. Daniere and P. Velex, 2001b, "Modelling and Detection of Localized Tooth Defects in Geared Systems", Transaction of ASME, Sep, 2001, Vol. 123, pp422-430

[65] J. Mahfoudh, C. Bard, M. Alattass and D. Play, 1995, "Simulation of Gearbox Dynamic Behaviour with Gear Faults", I Mech E, C492/045/95

[66] N.Sawalhi, R.B.Randall and H.Endo, "Gear and bearing fault simulation applied to diagnostics and prognostics", paper presented at The 19th International Congress and Exhibition on Condition Monitoring and Diagnostic Engineering Management, Luleå, Sweden, June (2006).

[67] N. Sawalhi, and R.B. Randall, Vibration response of spalled rolling element bearings: observations, simulations and signal processing techniques to track the spall size. Mechanical Systems and Processing, 25 (3), (2011).

3

Split Torque Gearboxes: Requirements, Performance and Applications

Abraham Segade-Robleda, José-Antonio Vilán-Vilán, Marcos López-Lago and Enrique Casarejos-Ruiz
University of Vigo, Spain

1. Introduction

Although the simplest gear systems are those with just one gear engagement area between a pair of gears, alternatives are available for applications where it is necessary to transmit a very high torque in a very small space. One option to increase power density is to use the split torque systems that were mainly developed for the aviation industry. These gear systems are based on a very simple idea: division of the transmission of force between several contact areas, thereby increasing the contact ratio. This gives rise, however, to the problem of meshing four gears (Fig. 1).

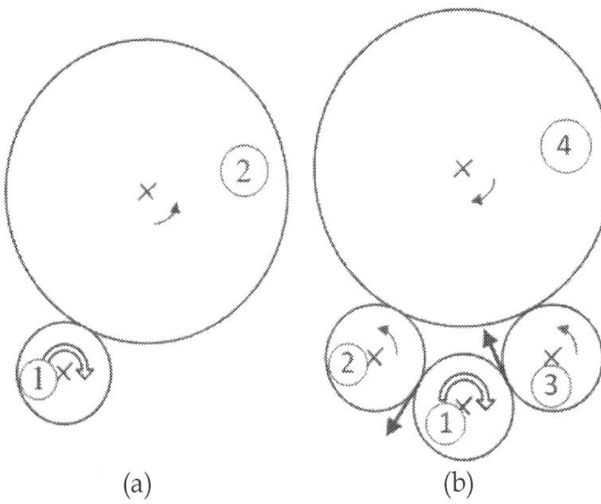

Fig. 1. (a) Standard gearbox assembly; (b) Split torque gearbox assembly

Split torque gearboxes are configurations where a driving pinion (1) meshes with two intermediate idler pinions (2, 3), which simultaneously act on another gear (4). From now on, this assembly will be called four-gear meshing. In this case, the torque split is from gear (1) to gears (2) and (3) which engage gear (4). This gear assembly results in the reduction in gear speed causing an increase in available torque; hence, the split torque transmission means we can use smaller gears.

The greater the number of gears that engage the same crown, the lower the torque exercised by each pinion. Gear assembles can have up to 14 gears engaging a single crown, as happens, for example, in tunnel boring machines.

This chapter explores four-gear meshing in a gear assembly that ensures a 50%torque split for each meshing area. Split torque gears are studied from two perspectives: first, the most common applications of split torque gearboxes in the aeronautical sector and second, the two most restrictive aspects of their application, namely:

- The geometric limitation of the four-gear assembly that requires simultaneous engagement for all four gears. Note that the four gears do not mesh correctly in just any position, although they may seem to do so initially. We will describe the conditions for simultaneous meshing of the four gears in general terms below.
- Torque split between the two gearbox paths must be as balanced as possible to ensure that neither of the paths is overloaded. The technology available to ensure proper torque split between two paths will be discussed below.

2. Applications

Gear transmission requirements for aircraft are very demanding, with a standard gear ratio between engine and rotor of 60:1 (Krantz, 1996). Moreover, the gear transmission system should be safe, reliable, lightweight and vibration-free. One of the most limiting factors is weight and there are three fundamental transmission parameters that greatly affect this factor:

1. The number of transmission stages. The greater the number of stages used to achieve the final gear ratio, the heavier the transmission, given that more common elements such as shafts and bearings are necessary.
2. The number of transmission paths, the basis for split torque gearboxes. Torque is divided between several transmission paths, resulting in a contact force in the smaller gear that means that smaller, and consequently lighter, gears can be used.
3. The final stage transmission ratio. Using a greater transmission ratio in the final stage enables weight to be reduced. This is because torque in previous steps is lower, making it possible to use smaller gears.

In helicopters, planetary gear systems are typically used for the final transmission stage, with planets consisting of between 3 and 18 gears and with planetary gearing transmission ratios between 5:1 and 7:1 (Krantz, 1996; White, 1989).

Using split-path arrangements with fixed shafts in the final transmission stage is a relatively recent development that offers a number of advantages over conventional systems, being several of them based on weight reduction for the overall transmission:

- It allows torque to be transmitted through various paths. This is a major advantage because when torque is split, the contact force between teeth is less and, hence, smaller, lighter gears can be used, therefore reducing the overall weight. Split torque however has the disadvantage that the torque must be shared equally between the paths. The problems associated with split torque are discussed in Section 4.

- It allows transmission path redundancy. Thus, if one transmission path fails during flight, operation can always be assured through another path. In many cases, consequently, gear transmissions are sized so that a single path can handle 100% of engine power.
- It achieves final-stage transmission ratios of around 10:1 to 14:1 (Krantz, 1996; White, 1989). This improvement over the 5:1 to 7:1 ratios for planetary gearboxes (Krantz, 1996; White, 1989) is reflected in a corresponding reduction in the weight of the transmission system.

Several patents for transmission systems that apply split torque have been filed by Sikorky Aircraft Corporation and McDonnell Douglas Helicopters (Gmirya & Kish, 2003; Gmirya, 2005; Craig et al., 1998) that refer either to complete or improved power transmission systems from the rotorcraft or aircraft engine to the rotor or propeller. Other studies that describe various aspects of split torque transmission systems, particularly their use in helicopter gearboxes (White, 1974, 1983, 1989, 1993, 1998), conclude that such gears have a number of advantages over traditional gear systems.

Below we describe two helicopter transmission systems that use multiple path gearboxes. The first is a helicopter gearbox used for laboratory tests of torque divided into two stages, and the second is a commercial helicopter three-stage gearbox that combines bevel, spur and helical gears.

2.1 Helicopter gearbox for laboratory testing

The gear transmission described below was used to perform numerous tests on the operation of split-torque transmissions (Krantz et al., 1992; Krantz, 1994, 1996; Krantz & Delgado, 1996), which can be considered a standard for aeronautical applications. The full assembly is depicted in Fig. 2.

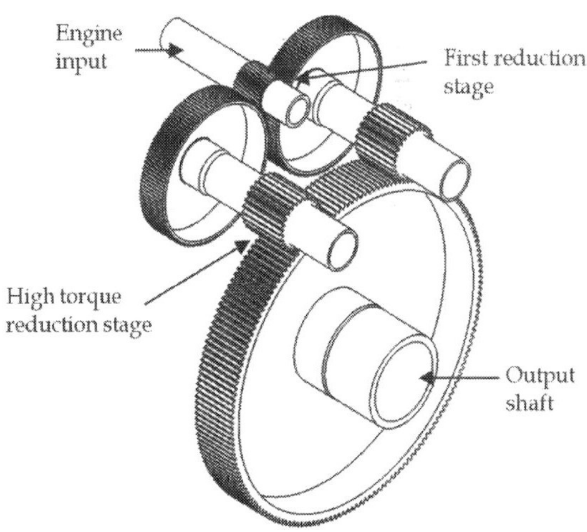

Fig. 2. Helicopter transmission for laboratory testing

The transmission is sized for input of 373 kW at a speed of 8780 rpm. As can be observed in Fig. 2, the transmission has two stages:

- First reduction stage. The first stage is a helical gear with a input pinion with 32 teeth and two output gears with 124 teeth each. The gear ratio is 3.875:1, resulting in an output speed of 2256.806 rpm. This is the stage where torque is split between the input pinion and the two output gears.
- High torque reduction stage. The output shaft is driven by a gear which is driven simultaneously by two spur pinions, each coaxial to the gear in the first reduction stage. The ratio between the gear teeth is 27/176, so the transmission ratio is 6.518:1, resulting in an output shaft speed of 347.6 rpm.

This configuration results in torque of 9017.56 Nm. being transmitted through two paths.

2.2 Commercial helicopter transmission

This gear transmission, studied in depth by White (1998), is sized for two engines, each with a continuous rating of 1200 kW turbine at a nominal speed of 22976 rpm. The main rotor speed is 350 rpm for an overall speed reduction ratio of nearly 66:1. Fig. 3 depicts a plan view of the gear transmission and Fig. 4 is a three-dimensional view showing the gears.

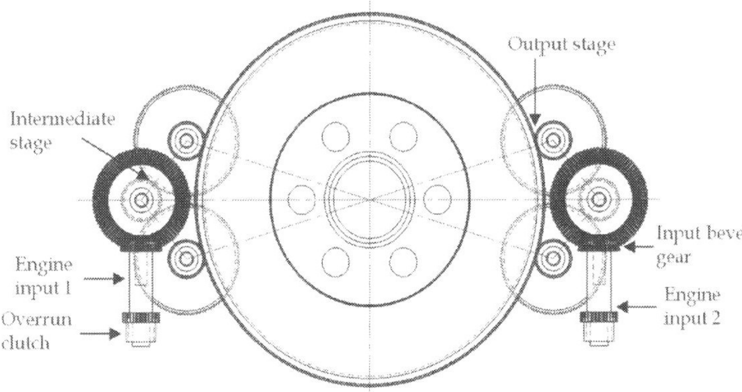

Fig. 3. Arrangement of gear trains between engines.

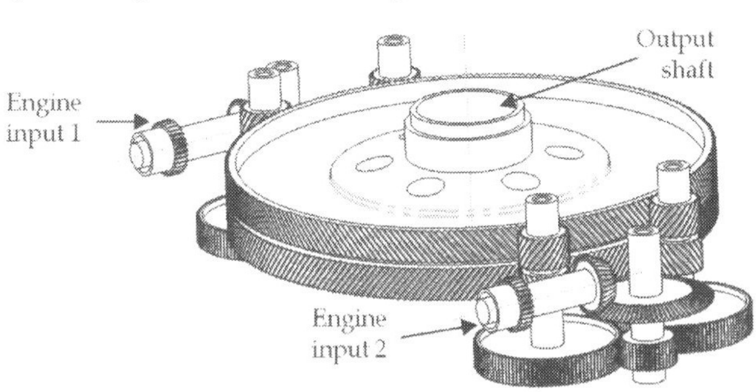

Fig. 4. Three-dimensional view of the gear train arrangement

Total transmission reduction is achieved by three gearing stages, clearly depicted in Fig. 3 and Fig. 4:

- Engine input. Engine torque is accepted by an overrun clutch, mounted with a bevel pinion. This bevel gear, with a between-teeth ratio of 34/84, produces a transmission ratio of 2.470:1. In this stage, the output velocity is 9299 rpm.
- Intermediate stage. Dual offset spur gears are driven by a single pinion. The between-teeth ratio of 41/108 produces a transmission ratio of 2.634:1, resulting in an output shaft speed of 3530 rpm. This meshing results in the first split in torque between the two intermediate gears.
- High-torque output stage. A double-helical gear is driven by a pinion coaxial with each intermediate stage gear. In this stage, the torque is split again between the two helical pinions, with the result that the output shaft simultaneously receives torque from four pinions for each bevel gear. In this transmission it is very convenient to combine torque split with reduction, as greater torque is transmitted in each stage.

The between-teeth gear ratio is 23/232, so the transmission ratio is 10.087:1, resulting in an output shaft speed of 350 rpm.

This configuration uses double-helical gearing at the output stage to drive the output shaft. The helical pinions have opposing angles, which ensures equilibrium between the axial forces. When a double gear operates on the output shaft, the area of support is twice that of a simple gear. This causes a reduction in contact force, which in turn results in a reduction ratio that is twice that of the simple case, with the corresponding reduction in weight and mechanical load.

Overall, this constitutes a transmission ratio of 65.64:1, with the total torque in the output shaft exercised by each engine of 28818Nm, split between the four pinions that engage the output shaft crown. This calculation is based on estimating overall losses, with each input engine operating independently, of 12%.

One of the main problems in split torque transmission is ensuring equal torque split between the paths. To ensure correct torque split, a long, torsionally flexible shaft is used between the intermediate-stage spur gear and the output-stage helical pinions. Section 4 describes the methods most frequently used to ensure correct torque split between paths.

3. Feasible geometric configurations

To ensure simultaneous meshing of four gears (Fig. 1), configuration must comply with certain geometric constraints. A number of studies describe the complexity of simultaneous gearing in split torque gearboxes (Kish, 1993a) and in planetary gear systems (Henriot, 1979, Parker & Lin, 2004); other studies approach the problem generically (Vilán-Vilán et al., 2010), describing possible solutions that ensure the simultaneous meshing of four gears.

For four gears to mesh perfectly, the teeth need to mesh simultaneously at the contact points. The curvilinear quadrilateral and the pitch difference are defined below in order to express the meshing condition. From now on we will use this nomenclature of our own devising -that is, curvilinear quadrilateral - to indicate the zone defined by portions of pitch circles in the meshing area (Fig. 5). The pitch difference is the sum of pitches in the input and output gears minus the sum of pitches in the idler gears at the curvilinear quadrilateral. For perfect engagement between the four gears, the pitch difference must coincide with a whole number of pitches.

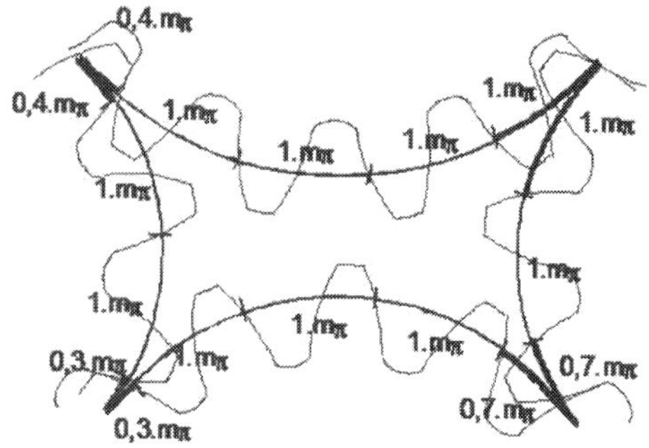

Fig. 5. General conditions for simultaneous meshing of four gears

A relationship is thus established between the position of the gears, as defined by the relative distance between centres, and the number of teeth in each of the gears. Below we explore two possible cases of over-constrained gears:

- CASE 1. Four outside gears.
- CASE 2. Three outside gears and one ring gear.

3.1 Case 1. Four outside gears

For a gearbox with the geometry illustrated in Fig. 6, it is possible to locate the different positions that will produce suitable meshing between gears, in function of the number of teeth in each gear, by defining the value of the angles α, β, δ and γ.

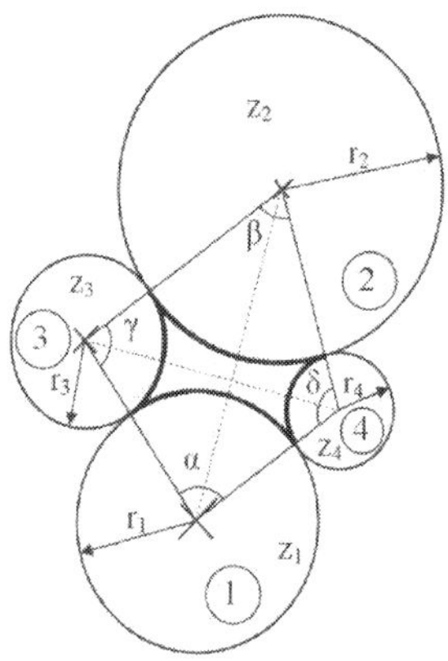

Fig. 6. Nomenclature for the four-gear case

The condition described in the previous section can be mathematically expressed as follows (see Nomenclature):

$$r_1 \cdot \alpha + r_2 \cdot \beta - r_3 \cdot \gamma - r_4 \cdot \delta = n \cdot (m\pi) \quad n \in Z \tag{1}$$

where n is the pitch difference in the curvilinear quadrilateral. As previously mentioned, n must be a whole number to ensure suitable meshing between gears.

We thus obtain an equation with four unknowns (α, β, γ, δ). The three remaining relationships can be obtained from the quadrilateral that joins the centres of the pitch circles (this quadrilateral will be denoted the rectilinear quadrilateral). Finally, we come to a transcendental equation (2) from which α can be obtained according to the number of teeth in the gears.

$$e_1 - f \cdot \cos\left[A_1 \cdot \boldsymbol{\alpha} + B_1 \cdot \arccos\left(\frac{c_1 - a_1 + b_1 \cdot \cos\boldsymbol{\alpha}}{d_1}\right) + C_1\right] = g_1 - \\ -h_1 \cdot \cos\left[A'_1 \cdot \boldsymbol{\alpha} + B'_1 \cdot \arccos\left(\frac{c_1 - a_1 + b_1 . \cos\boldsymbol{\alpha}}{d_1}\right) + C'_1\right] \quad (2)$$

Once the angle α has been determined, we can calculate:

$$\beta = \arccos\left(\frac{c_1 - a_1 + b_1 \cdot \cos\alpha}{d_1}\right) \quad (3)$$

$$\gamma = A_1 \cdot \alpha + B_1 \cdot \beta + C_1 \quad (4)$$

$$\delta = A'_1 \cdot \alpha + B'_1 \cdot \beta + C'_1 \quad (5)$$

a_1, b_1, c_1, d_1, e_1, f_1, g_1, h_1, A_1, B_1, C_1, A_1', B_1' and C_1' are numerical relationships among the teeth number from each wheel that must mesh simultaneously. The value of each coefficient is listed in the Appendix.

The transcendental equation for obtaining α has several solutions, all representing possible assemblies for the starting gears. For example, for four-gear meshing with the next teeth numbers: z_1=30, z_2=50, z_3=20 and z_4=12 (see Nomenclature), all the possible solutions for the gear can be encountered. In this case solutions are n = -12, -11, -3, -2, -1, 0, 1, 2, 3, 4, 7, 29 and 30, where n is the pitch difference between the two sides of the curvilinear quadrilateral (a whole number that ensures suitable meshing). Fig. 7 shows some of the possible meshing solutions.

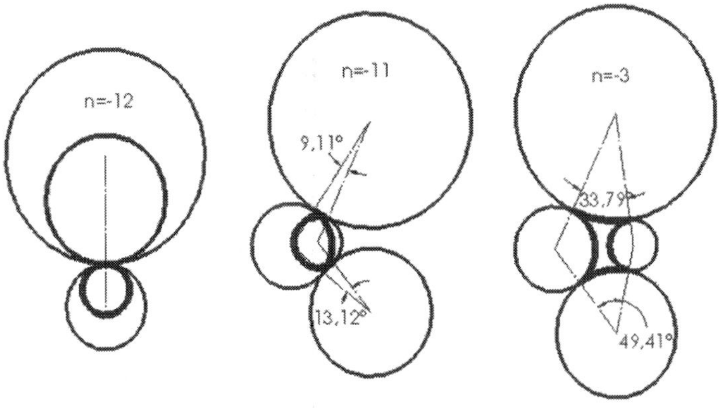

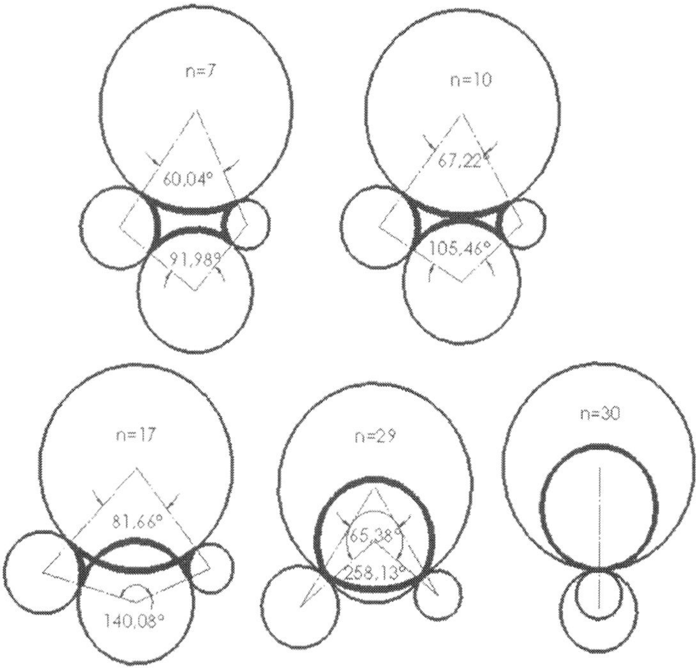

Fig. 7. Feasible solutions for given numbers of teeth

3.2 Case 2. Three outside gears and one ring gear

In this case torque is transmitted from a driving pinion (1) to a ring gear (2) through two idler pinions (3) and (4). Two solutions are available depending on the geometry of the rectilinear quadrilateral that joins the centres of the pitch circles, either crossed (Fig. 8 (a)) or non-crossed (Fig. 8 (b)). The starting equation is different for each of these cases.

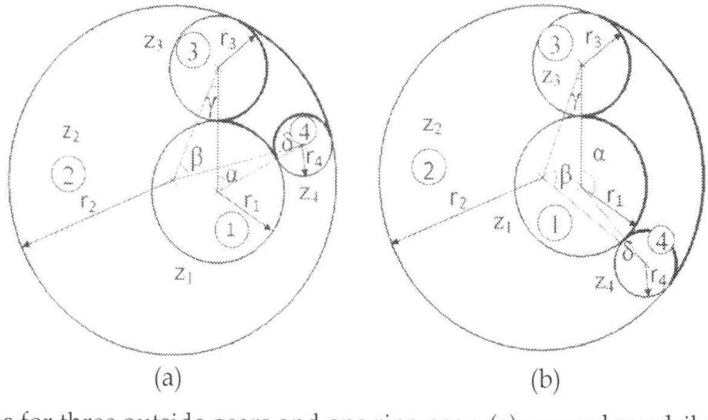

Fig. 8. Solutions for three outside gears and one ring gear: (a) crossed quadrilateral (b) non-crossed quadrilateral

For the crossed quadrilateral configuration, the starting equation is (see Nomenclature):

$$z_1 \cdot \alpha + z_2 \cdot \beta + z_3 \cdot \gamma - z_4 \cdot \delta = \pi \cdot (2 \cdot n + z_3 + z_4) \tag{6}$$

Finally, we come to the same transcendental equation (2), where the coefficients are a_2, b_2, c_2, d_2, e_2, f_2, g_2, h_2, A_2, B_2, C_2, A_2', B_2' and C_2', whose values are listed in the Appendix.

For the non-crossed quadrilateral configuration, the starting equation is:

$$z_1 \cdot \alpha - z_2 \cdot \beta - z_3 \cdot \gamma - z_4 \cdot \delta = \pi \cdot (2 \cdot n - 2 \cdot z_2 + z_3 + z_4) \tag{7}$$

Finally, we come to the same transcendental equation (2), where the coefficients become a_2, b_2, c_2, d_2, e_2, f_2, g_2, h_2, A_3, B_3, C_3, A_3', B_3' and C_3', whose values are listed in the Appendix.

3.3 A particular case: Outside meshing with equal intermediate pinions

A common split torque gear assembly is one with two equally sized idler pinions (Fig. 9).

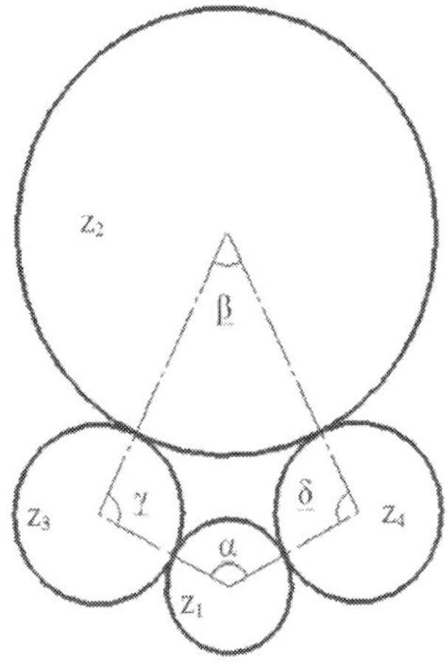

Fig. 9. Idler pinions in an outside gear

The solution is obtained by particularizing the general solution for four outside wheels and imposing the condition $z_3 = z_4$, or $\gamma = \delta$. The following equations are defined for the curvilinear quadrilateral:

$$z_1 \cdot \alpha + z_2 \cdot \beta - 2 \cdot z_3 \cdot \gamma = n \cdot 2\pi \tag{8}$$

$$\alpha + \beta + 2 \cdot \gamma = 2\pi \tag{9}$$

$$(z_1 + z_3) \cdot \sin\left(\frac{\alpha}{2}\right) = (z_2 + z_3) \cdot \sin\left(\frac{\beta}{2}\right) \tag{10}$$

Resolving the system, the following transcendental function in α is obtained:

$$\frac{z_1 + z_3}{z_2 + z_3} \cdot \sin\left(\frac{\alpha}{2}\right) = \sin\left[\frac{z_3 + n}{z_2 + z_3} \cdot \pi - \frac{z_1 + z_3}{z_2 + z_3} \cdot \left(\frac{\alpha}{2}\right)\right] \quad (11)$$

The solutions for the other angles can now be obtained:

$$\beta = 2 \cdot \arcsin\left[\frac{z_1 + z_3}{z_2 + z_3} \cdot \sin\left(\frac{\alpha}{2}\right)\right] \quad (12)$$

$$\gamma = \pi - \frac{\alpha}{2} - \frac{\beta}{2} \quad (13)$$

4. Load sharing

The main problem in the design of split torque gearboxes is to ensure that torque is equally split between different paths. Small deviations in machining can result in one of the paths with 100% of torque and the other path operating entirely freely (Kish & Webb, 1992). This situation causes excessive wear in one of the paths and renders the torque split system ineffective.

Below we describe approaches to ensuring equal torque split between different paths in split torque gear arrangements. The main types are:

1. Geared differential. This differential mechanism, frequently used in the automotive sector, delivers equal torques to the drive gears of a vehicle.
2. Pivoted systems. These use a semi-floating pinion constrained both to pivot normal to the line of action and to seek a position where tooth loads are equal.
3. Quill shafts. A torsion divider with a separate gear and pinion, each supported on its own bearings, are connected through the quill shaft, which allows torsional flexibility.

The use of any of these systems to ensure correct torque split makes the gearbox heavier and assembly and maintenance more complex, which is why a number of authors do not support the use of systems that ensure torque split. Described below are the main systems that ensure correct torque split and discussed also are the proposals of authors who advocate for not using special systems.

4.1 Geared differential

One way to ensure correct torque split between two branches is to use a differential system. The great disadvantage of this system, however, is that resistive torque lost in one branch leads to loss of the full engine torque. Different differential mechanisms can be used, with assemblies very similar to those in vehicles or to the system depicted in Fig. 10. Assembled at the entry point to the gearbox is an input planetary system that acts as a differential that ensures load sharing. This transmission accepts power from three input engines, each of which has a differential system that ensures balanced torque splitting. Power is input from each engine to the sun gear of the differential planetary system. The carrier is the output to a bevel pinion that drives one torque splitting branch while the ring gear drives the other torque splitting branch. As the carrier and the ring gear rotate in opposite directions, the bevel pinions are arranged on opposite sides to ensure correct rotation direction. Each output bevel gear drives one pinion which then combines power into the output gear.

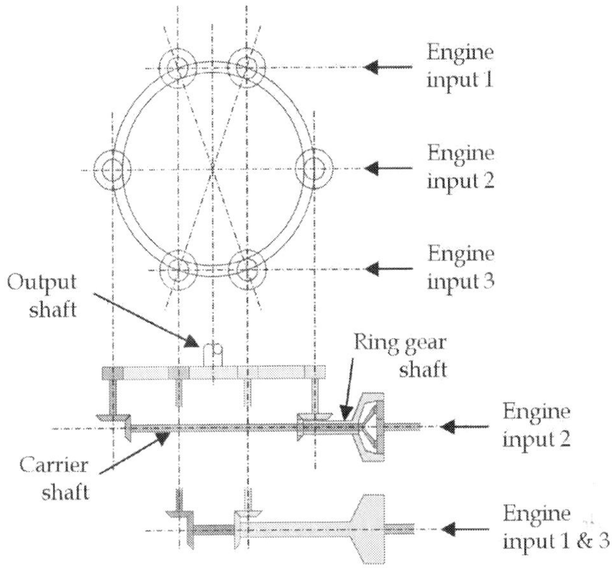

Fig. 10. Schematic view of split torque main transmission

4.2 Pivoted systems

One type of pivoted systems is described in detail in a patent (Gmirya, 2005) for split torque reduction applied to an aerial vehicle propulsion system (Fig. 11). *"The input pinion (64) engages with gears (66) and (68). The input pinion is defined along the gear shaft A_G, the first gear*

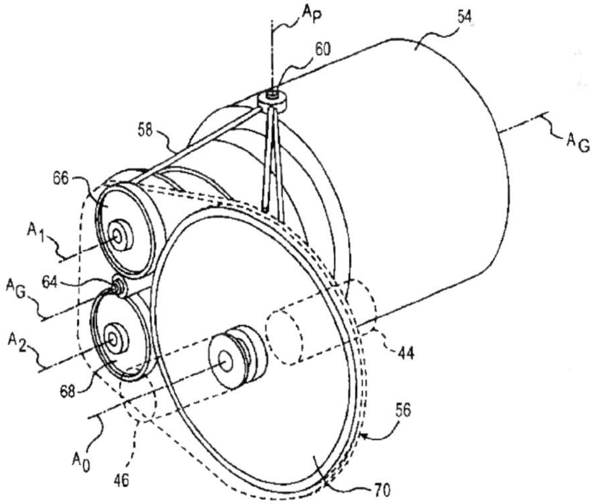

Fig. 11. Perspective view of the split torque gearbox with pivoted engine support (Gmirya, 2005)

(66) defines a first gear rotation shaft A_1 and the second gear (68) defines a second gear rotation shaft A_2. The axes A_G, A_1 and A_2 are preferably located transversally to the pivot axis A_p. The first gear (66) and the second gear (68) engage an output gear (70). The output gear (70) defines an output rotation shaft A_0 and is rotationally connected to the translational driveshaft (44) and the rotor driveshaft (46) to power, respectively, the translational propulsion system and the rotor system".

The assembly transmits torque from the pinion (64), which operates at very high revolutions, to the output shaft (44 -46) via two paths. The pivot system works as follows: since the input pinion (64) meshes with two gears (66) and (68), the pivoted engine arrangement permits the input pinion (64) to float until gear loads between the input gear (64), the first gear (66) and the second gear (68) are balanced. Irrespective of gear teeth errors or gearbox shaft misalignments, the input pinion will float and split torque between the two gears.

4.3 Quill shafts

Below we describe assemblies used in systems that allow some torsion in the split torque shafts (Smirnov, 1990; Cocking, 1986) in order to minimize the difference in torque split between paths. These systems achieve their goal in several ways:

- Conventional systems (Kish, 1993a) assemble intermediate shafts with some torsional flexibility so that angular deviation produced between the input and output pinions adjusts the torque transmitted via the two paths.
- Other systems are based on elastomeric elements in the shaft (Isabelle et al., 1992, Kish & Webb, 1992) or materials with a lower elastic modulus (Southcott, 1999), such as an idler pinion constructed of nylon or a similar material (Southcott, 1999). This solution is not explored here because the torque transmitted is reduced.
- Yet other systems operate on the basis of spring elements (Gmirya & Vinayak, 2004).

The use of such elements in the design adds weight and makes both initial assembly and maintenance more complex, thereby losing to some degree the advantages of split torque gearboxes. Described below are the most representative types of quill shaft.

4.3.1 Conventional quill shafts

Conventional quill shaft design involves assembly on three different shafts (Fig. 12). The

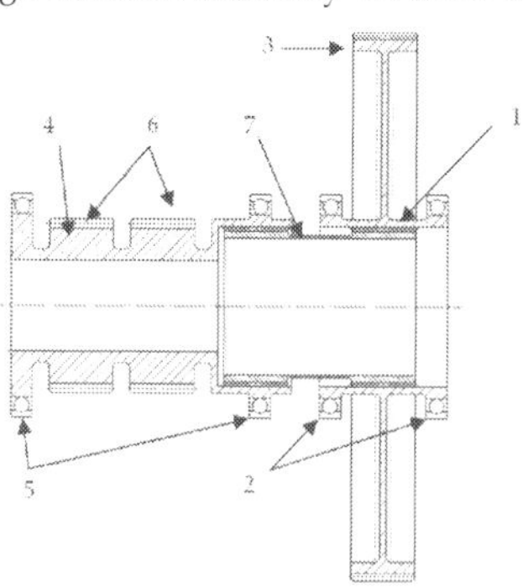

Fig. 12. Conventional assembly of a quill shaft

input shaft (1) is assembled with two separate bearings (2) and the input gear (3).The output shaft (4) is assembled with two separate bearings (5) and, in this case, two output pinions (6). The quill shaft is a third shaft (7) that connects the other two shafts. Due to a lower polar moment of inertia, it admits torsional flexibility, resulting in a small angular deviation between the input and output shafts. The value of the angular deviation is proportional to the transmitted torque; thus, if one path transmits more torque than the other, the angular deviation is greater, allowing the shaft that transmits less torque to increase its load.

4.3.2 Quill shafts based on elastomeric elements

Elastomeric elements are frequently used in quill shafts given their low elastic modulus. For example, one system (Isabelle et al., 1992), based on using elastomers (Fig. 13), consists of *"an annular cylindrical elastomeric bearing (14) and several rectangular elastomeric bearing pads (16). The elastomeric bearing (14) and bearing pads (16) have one or more layers (60); each layer (60) has an elastomer (62) with a metal backing strip (64) secured by conventional means such as vulcanization, bonding or lamination".*

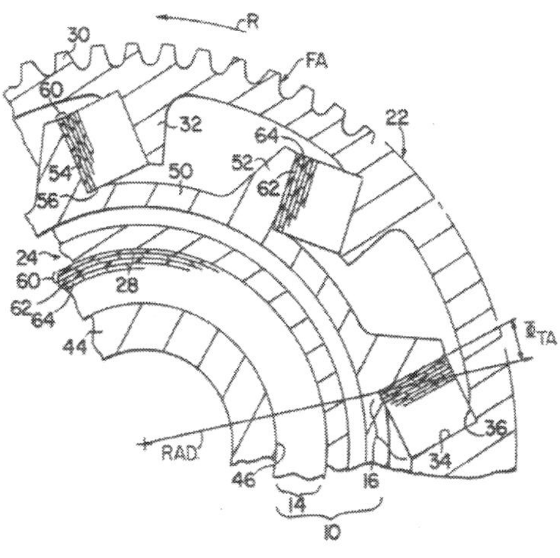

Fig. 13. Elastomeric load sharing device (Isabelle et al., 1991)

The annular cylindrical elastomeric bearing (14) absorbs possible misalignments between shafts resulting from defects in assembly. The rectangular elastomeric bearing pads (16) are responsible for providing torsional flexibility to the shafts of the possible gear paths in order to ensure equal torque transmission.

Another elastomer-based system (Kish & Webb, 1992) (Fig. 14) consists of an assembly with *"a central shaft (21) and a pair of bull pinions (22) and (23). The shaft (21) is supported by the bearings (24) and (25); a gear flange (26) at the end of the shaft has bolt holes (27) and teeth (28) on the outer circumference. A spur gear (29) is held to the flange (26) using upper and lower rims (30) and (31), consisting of flat circular disks (32) with bolt holes (33) and an angled outer wall (34). Gussets (35) between the wall and the disk increase rim stiffness to minimize deflection. One or more elastomer layers (36), bonded to the outer surface (37) of the wall (34), act as an elastomeric torsional isolator".*

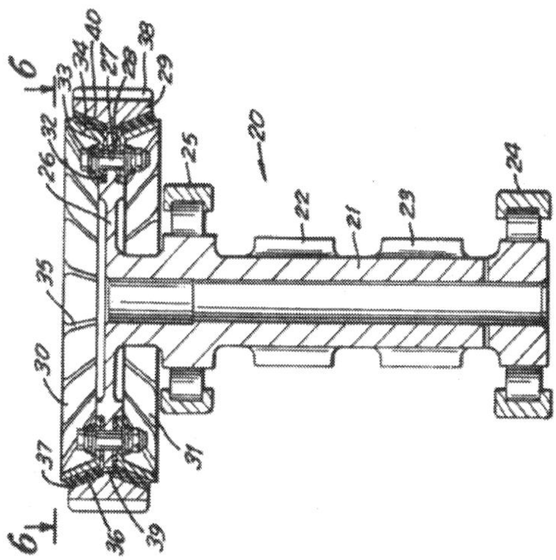

Fig. 14. Gear assembly using an elastomeric torsional isolator (Kish & Webb, 1992)

This assembly was tested in the Advanced Rotorcraft Transmission project (Kish, 1993a), by comparing it with conventional quill shafts. It was concluded that the torque split was excellent and also had other advantages such as lower transmission of force to supports, less vibration and less noise during operation.

The main problem with using elastomers to achieve proper torque split is their degradation over time, especially when used in high-torque gear transmissions where temperatures are high and there is contact with oil. Some authors therefore propose the use of metallic elements to achieve the same effect as the quill shaft.

4.3.3 Quill shafts based on spring elements

Some authors propose the use of metallic elements to achieve the same effect as the quill shaft. One such system (Gmirya & Vinayak, 2004) (Fig. 15) is based on achieving this effect by using *"at least one spring element (30) placed between and structurally connecting the gear shaft (32) and the outer ring of gear teeth (34). The gear shaft (32) has flange elements (36) that project radially outboard of the shaft. The ring of gear teeth (34), similarly, has a flange element (38) that projects radially inward towards the gear shaft"*. In this case, a pair of spring elements (30) is arranged on each side of the gear teeth flange element (38)".

This assembly is designed in such a way that the spring elements absorb torsional deflection between the gears, thereby ensuring proportional torque split between paths.

4.4 No use of special systems

Split torque gearboxes are used in order to reduce the weight of the gear system, so the simplest option is assembly without special systems for regulating torque split. Several authors support this option, for example, Kish (1993a, 1993b), who concluded from tests that acceptable values can be achieved without using any special torque split system, simply by

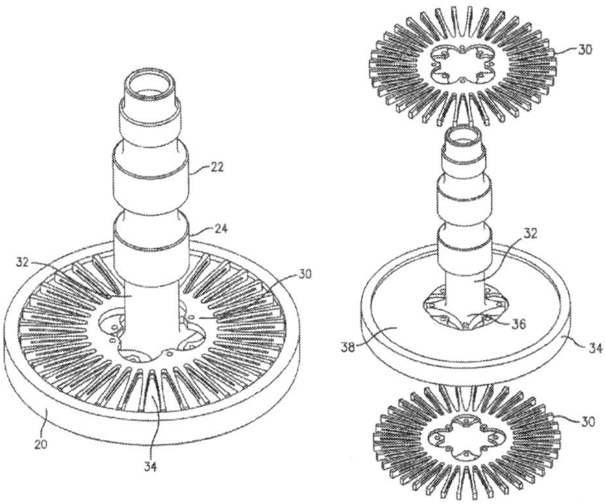

Fig. 15. Load sharing gear in combination with a double helical pinion (Gmirya & Vinayak, 2004)

ensuring manufacturing according to strict tolerances and correct assembly. Krantz (1996) proposed the use of the clocking angle as a design parameter to achieve adequate torque split between paths. This author has studied the effects of gearshaft twisting and bending, and also tooth bending, Hertzian deformations within bearings and the impact of bearing support movement on load sharing.

Krantz (1996) defined the clocking angle as β and described the assembly prepared for measurement (Fig. 16): *"The output gear is fixed from rotating and a nominal counter-clockwise*

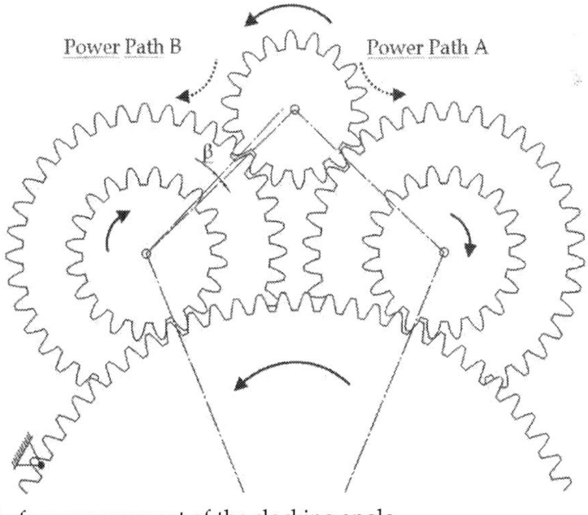

Fig. 16. Assembly for measurement of the clocking angle

torque is applied to the input pinion so that the gear teeth come into contact. When all the gear teeth for both power paths come into contact, then the clocking angle β is, by definition, equal to zero. If the teeth of one power path are not in contact, then the clocking angle β is equal to the angle that the first-stage gear would have to be rotated relative to the second-stage pinion to bring all the teeth into contact".

The tests show that suitable (47 per cent/53 per cent) load sharing can be achieved merely by taking into account the clocking angle and ensuring proper machining and assembly.

This research into the clocking angle has been followed up by subsequent authors (Parker & Lin, 2004) who have studied how contact between different planetary gears is sequenced.

5. Conclusion

Choosing the correct assembly for aircraft power transmission is a key factor in the quest for weight reduction. Although technological advances in mechanical components can help achieve weight reduction in gear systems, their influence is much less than that of choosing the correct gear assembly.

Planetary gear or split torque systems are typically used in helicopter gear transmissions. The fundamental advantage of the split torque systems is that less weight is achieved by equal torque transmission and gear transmission ratios. This advantage is based primarily on arguments as follows:

- In the final transmission stage where the greatest torque is achieved, the use of several paths for the transmission ratio means that, given equal torque and stress levels in the teeth, the ratio between output torque/weight will be better in torque split gear systems than in planetary gear systems.
- In the final transmission stage, transmission ratios of around 5:1 or 7:1 are achieved by planetary gearboxes used with a single stage, compared to 10:1 or 14:1 for split torque gears used in the final stage.
- The possibility of achieving higher transmission ratios in split torque gearboxes makes it possible to use a smaller number of gear stages, resulting in lighter gear systems.
- Split torque gearboxes need fewer gears and bearings that planetary gearboxes, which means lower transmission losses.
- A key factor for aircraft use is that split torque gearboxes improve reliability by using multiple power paths; thus, if one path fails, operation is always assured through another path.
- The main disadvantage of the split torque gearboxes is when torque split between the possible paths is uneven; however, several solutions are available to ensure correct torque split.

These arguments would indicate the advisability of using this type of transmission in aircraft gear systems.

6. Nomenclature

m gear module

r_i radius of the pitch circle of wheel i

z_i	number of teeth in wheel i
α	angle formed by the lines between centres, between wheels 3 4 1
β	angle formed by the lines between centres, between wheels 3 2 4
γ	angle formed by the lines between centres, between wheels 2 3 1
δ	angle formed by the lines between centres, between wheels 1 4 2
n	pitch difference between the two sides of the curvilinear quadrilateral

7. Appendix

The numerical relationships among the teeth number used in the text are listed below. C_1, C_1', C_2, C_2', C_3 and C_3' are functionS of n, a whole number which represents the pitch difference in the curvilinear quadrilateral.

$$a_1 = (z_1 + z_3)^2 + (z_1 + z_4)^2 \tag{14}$$

$$b_1 = 2 \cdot (z_1 + z_3) \cdot (z_1 + z_4) \tag{15}$$

$$c_1 = (z_2 + z_3)^2 + (z_2 + z_4)^2 \tag{16}$$

$$d_1 = 2 \cdot (z_2 + z_3) \cdot (z_2 + z_4) \tag{17}$$

$$e_1 = (z_1 + z_3)^2 + (z_2 + z_3)^2 \tag{18}$$

$$f_1 = 2 \cdot (z_1 + z_3) \cdot (z_2 + z_3) \tag{19}$$

$$g_1 = (z_1 + z_4)^2 + (z_2 + z_4)^2 \tag{20}$$

$$h_1 = 2 \cdot (z_1 + z_4) \cdot (z_2 + z_4) \tag{21}$$

$$A_1 = \frac{z_1 + z_4}{z_3 - z_4} \tag{22}$$

$$B_1 = \frac{z_2 + z_4}{z_3 - z_4} \tag{23}$$

$$C_1 = 2\pi \cdot \frac{z_4 + n}{z_4 - z_3} \tag{24}$$

$$A_1' = \frac{z_1 + z_3}{z_4 - z_3} \tag{25}$$

$$B_1' = \frac{z_2 + z_3}{z_4 - z_3} \tag{26}$$

$$C_1' = 2\pi \cdot \frac{z_3 + n}{z_3 - z_4} \tag{27}$$

$$a_2 = (z_1 + z_3)^2 + (z_1 + z_4)^2 \tag{28}$$

$$b_2 = 2 \cdot (z_1 + z_3) \cdot (z_1 + z_4) \tag{29}$$

$$c_2 = (z_2 - z_3)^2 + (z_2 - z_4)^2 \tag{30}$$

$$d_2 = 2 \cdot (z_2 - z_3) \cdot (z_2 - z_4) \tag{31}$$

$$e_2 = (z_1 + z_3)^2 + (z_2 - z_3)^2 \tag{32}$$

$$f_2 = 2 \cdot (z_1 + z_3) \cdot (z_2 - z_3) \tag{33}$$

$$g_2 = (z_1 + z_4)^2 + (z_2 - z_4)^2 \tag{34}$$

$$h_2 = 2 \cdot (z_1 + z_4) \cdot (z_2 - z_4) \tag{35}$$

$$A_2 = \frac{z_1 + z_4}{z_4 - z_3} \tag{36}$$

$$B_2 = \frac{z_2 - z_4}{z_4 - z_3} \tag{37}$$

$$C_2 = \pi \cdot \frac{2 \cdot n + z_3 + z_4}{z_3 - z_4} \tag{38}$$

$$A_2' = \frac{z_1 + z_3}{z_4 - z_3} \tag{39}$$

$$B_2' = \frac{z_2 - z_3}{z_4 - z_3} \tag{40}$$

$$C_2' = \pi \cdot \frac{2 \cdot n + z_3 + z_4}{z_3 - z_4} \tag{41}$$

$$A_3 = \frac{z_1 + z_4}{z_3 - z_4} \tag{42}$$

$$B_3 = \frac{z_4 - z_2}{z_3 - z_4} \tag{43}$$

$$C_3 = \pi \cdot \frac{2 \cdot n - 2 \cdot z_2 + z_3 + 3 \cdot z_4}{z_4 - z_3} \tag{44}$$

$$A_3' = \frac{z_1 + z_3}{z_4 - z_3} \tag{45}$$

$$B_3' = \frac{z_3 - z_2}{z_4 - z_3} \tag{46}$$

$$C_3' = \pi \cdot \frac{2 \cdot n - 2 \cdot z_2 + 3 \cdot z_3 + z_4}{z_3 - z_4} \tag{47}$$

8. References

Cocking, H. (1986). The Design of an Advanced Engineering Gearbox. *Vertica.* Vol 10, No. 2, Westland Helicopters and Hovercraft PLC, Yeovil, England, pp. 213-215

Craig, G. A.; Heath, G. F. & Sheth, V. J. (1998). Split Torque Proprotor Transmission, McDonnell Douglas Helicopter Co., *U.S. Patent* Number 5,823,470

Gmirya, Y.; Kish, J.G. (2003). Split-Torque Face Gear Transmission, Sikorsky Aircraft Corporation, *U.S. Patent* Number 6,612,195

Gmirya, Y. & Vinayak, H. (2004). Load Sharing Gear for High Torque, Split-Path Transmissions. Sikorsky Aircraft Corporation. *International Patent PCT.* WO 2004/094093

Gmirya, Y. (2005). Split Torque Gearbox With Pivoted Engine Support, Sikorsky Aircraft Corp., *U.S. Patent Number* 6,883,750

Henriot G. (1979). *Traité théorique et pratique des engrenages (I)*, Ed. Dunod, 6th Edition, ISBN 2-04-015607-0, Paris, France, pp. 587-662

Isabelle, C.J; Kish & J.G, Stone, R.A. (1992). Elastomeric Load Sharing Device. United Technologies Corporation. *U.S. Patent* Number 5,113,713

Kish, J.G. & Webb, L.G. (1992). Elastomeric Torsional Isolator. United Technologies Corporation. *U.S. Patent* Number 5,117,704

Kish, J.G. (1993a). *Sikorsky Aircraft Advanced Rotorcraft Transmission (ART) Program* - Final Report. NASA CR-191079, NASA Lewis Research Center, Cleveland, OH

Kish, J.G. (1993b). Comanche Drive System. *Proceedings of the Rotary Wing Propulsion Specialists Meeting,* American Helicopter Society, Williamsburg, VA, pp. 7

Krantz, T.L.; Rashidi, M. & Kish, J.G. (1992). *Split Torque Transmission Load Sharing,* in: Technical Memorandum 105,884, NASA Lewis Research Center, Cleveland, Ohio. Army Research Laboratory Technical Report 92-C-030

Krantz, T.L. (1994). *Dynamics of a Split Torque Helicopter Transmission,* in: Technical Memorandum 106,410, NASA Lewis Research Center, Cleveland, Ohio. Army Research Laboratory Technical Report ARL-TR-291

Krantz, T.L. (1996). *A Method to Analyze and Optimize the Load Sharing of Split Path Transmissions,* in: Technical Memorandum 107,201, NASA Lewis Research Center, Cleveland, Ohio. Army Research Laboratory Technical Report ARL-TR-1066

Krantz, T.L. & Delgado, I.R. (1996). *Experimental Study of Split-Path Transmission Load Sharing,* in: Technical Memorandum 107,202, NASA Lewis Research Center, Cleveland, Ohio. Army Research Laboratory Technical Report ARL-TR-1067

Parker, R.G., Lin, J. (2004). Mesh phasing relationships in planetary and epicyclic gears, *Journal of Mechanical Design*. Vol. 126, Issue 2, pp. 365-370, ISSN 1050-0472, eISSN 1528-9001

Smirnov, G. (1990). *Multiple-Power-Path Nonplanetary Main Gearbox of the Mi-26 Heavy-Lift Transport Helicopter,* in: Vertiflite, Mil Design Bureau, Moscow, Vol. 36, pp. 20-23, ISSN 0042-4455

Southcott, B. (1999). Gear Arrangement. Adelaide Gear Pty. Ltd. *U.S. Patent* Number 5,896,775

Vilán-Vilán, J.A.; Segade-Robleda, A.; López-Lago, M & Casarejos-Ruiz, E. (2010), Feasible Geometrical Configurations for Split Torque Gearboxes With Idler Pinions, *Journal of Mechanical Design*, Vol. 132, Issue 12, pp. 121,011, 8 pages, ISSN 1050-0472, eISSN 1528-9001

White, G. (1974). New Family of High-Ratio Reduction Gear With Multiple Drive Paths, *Proceedings of the Institution of Mechanical Engineers.*, Vol. 188, pp. 281-288, ISSN 020-3483

White, G. (1983). A 2400 kW Lightweight Helicopter Transmission Whit Split Torque Gear Trains, ASME Paper 84 Det-91

White, G. (1989). Split-Torque Helicopter Transmission With Widely Separated Engines. *Proceedings of the Institution of Mechanical Engineers.* Vol 203, Number G1, pp. 53-65, ISSN 0954-4100

White, G. (1993). *3600 HP Split-Torque Helicopter Transmission,* in: Mechanical Systems Technology Branch Research Summary, 1985-1992, NASA Technical Memorandum 106,329

White, G. (1998). Design study of a split-torque helicopter transmission, in: *Proceedings of the Institution of Mechanical Engineers, Part G: Journal of Aerospace Engineering*. Vol. 212, No. 2, pp. 117-123, ISSN 0954-4100, eISSN 2041-3025.

4

On the Modelling of Spur and Helical Gear Dynamic Behaviour

Velex Philippe
University of Lyon, INSA Lyon, LaMCoS UMR CNRS, France

1. Introduction

This chapter is aimed at introducing the fundamentals of spur and helical gear dynamics. Using three-dimensional lumped models and a thin-slice approach for mesh elasticity, the general equations of motion for single-stage spur or helical gears are presented. Some particular cases including the classic one degree-of-freedom model are examined in order to introduce and illustrate the basic phenomena. The interest of the concept of transmission errors is analysed and a number of practical considerations are deduced. Emphasis is deliberately placed on analytical results which, although approximate, allow a clearer understanding of gear dynamics than that provided by extensive numerical simulations. Some extensions towards continuous models are presented.

2. Nomenclature

b : face width

C_m, C_r : pinion, gear torque

$e(M)$, $E_{MAX}(t)$: composite normal deviation at M, maximum of $e(M)$ at time t.

E, E^* : actual and normalized depth of modification at tooth tips

$\mathbf{F}_e(t) = \int_{L(t,\mathbf{q})} k(M)\delta e(M)\mathbf{V}(\mathbf{M})dM$: time-varying, possibly non-linear forcing term associated with tooth shape modifications and errors

$\mathbf{G} = \mathbf{V}_0 \mathbf{V}_0^T$

$H(x)$: unit Heaviside step function ($H(x)=1$ *if* $x>1; H(x)=0$ *otherwise*)

k_m, $k(t,\mathbf{q})$: average and time-varying, non-linear mesh stiffness

$k(t) = k_m(1+\alpha\varphi(t))$, linear time-varying mesh stiffness

k_0 : mesh stiffness per unit of contact length

$k(M)$, mesh stiffness per unit of contact length at M

$k_{\Phi p}$: modal stiffness associated with (ω_p , $\mathbf{\Phi}_p$)

$\left[\mathbf{K}_G(t)\right] = \int_{L(t,\mathbf{q})} k(M)\mathbf{V}(\mathbf{M})\ \mathbf{V}(\mathbf{M})^T dM$: time-varying, possibly non-linear gear mesh stiffness matrix

$L(t,\mathbf{q})$: time-varying, possibly non-linear, contact length

$L_m = \varepsilon_\alpha \dfrac{b}{\cos\beta_b}$: average contact length

$\bar{m} = \dfrac{I_{02} I_{01}}{Rb_1^2 I_{02} + Rb_2^2 I_{01}}$: equivalent mass

$m_{\Phi p}$: modal mass associated with (ω_p , $\mathbf{\Phi_p}$)

$\mathbf{n}_1$: outward unit normal vector with respect to pinion flanks

$NLTE$: no-load transmission error

O_1, O_2 : pinion, gear centre

Pb_a : apparent base pitch

Rb_1, Rb_2 : base radius of pinion, of gear

$(\mathbf{s},\mathbf{t},\mathbf{z})$: coordinate system attached to the pinion-gear centre line, see Figs. 1&2

T_m : mesh period.

TE , TE_S : transmission error, quasi-static transmission error under load

$\mathbf{V}(\mathbf{M}), \mathbf{V}_0$, structural vector, averaged structural vector

$\mathbf{W}$: projection vector for the expression of transmission error, see (44-1)

$(\mathbf{X},\mathbf{Y},\mathbf{z})$: coordinate system associated with the base plane, see Fig. 2

$\mathbf{X}_0 = \bar{\mathbf{K}}^{-1}\mathbf{F}_0$: static solution with averaged mesh stiffness (constant)

$\mathbf{X}_S$, $\mathbf{X}_D$ $\mathbf{X}$: quasi-static, dynamic and total (elastic) displacement vector (time-dependent)

Z_1, Z_2 : tooth number on pinion, on gear

α : small parameter representative of mesh stiffness variations, see (30)

α_p : apparent pressure angle

β_b : base helix angle

$\delta_m = \dfrac{F_S}{k_m} = \mathbf{V}^{\mathrm{T}}\mathbf{X}_0$: static mesh deflection with average mesh stiffness

$\delta e(M) = E_{MAX}(t) - e(M)$: instantaneous initial equivalent normal gap at M

$\Delta(M)$: mesh deflection at point M

ε_α : theoretical profile contact ratio

ε_β : overlap contact ratio

$\Lambda = \dfrac{Cm}{Rb_1 b k_0}$, deflection of reference

$\mathbf{\Phi_p}$: p^{th} eigenvector of the system with constant averaged stiffness matrix

ς_P : damping factor associated with the pth eigenfrequency

Γ : dimensionless extent of profile modification (measured on base plane)

$\tau = \dfrac{t}{T_m}$, dimensionless time

ω_p : p^{th} eigenfrequency of the system with constant averaged stiffness matrix

$\varpi_{pm} = \dfrac{\omega_p}{n\Omega_1}$, dimensionless eigenfrequency

Ω_1, Ω_2 : pinion, gear angular velocity

$\bar{\mathbf{A}}$: vector $\mathbf{A}$ completed by zeros to the total system dimension

$(\bullet)^* = \dfrac{(\bullet)}{\delta_m}$, normalized displacement with respect to the average static mesh deflection

$(\hat{\bullet}) = \dfrac{(\bullet)}{k_m}$, normalized stiffness with respect to the average mesh stiffness

3. Three-dimensional lumped parameter models of spur and helical gears

3.1 Rigid-body rotations – State of reference

It is well-known that the speed ratio for a pinion-gear pair with perfect involute spur or helical teeth is constant as long as deflections can be neglected. However, shape errors are present to some extent in all gears as a result of machining inaccuracy, thermal distortions after heat treatment, etc. Having said this, some shape modifications from ideal tooth flanks are often necessary (profile and/or lead modifications, topping) in order to compensate for elastic or thermal distortions, deflections, misalignments, positioning errors, etc. From a simulation point of view, rigid-body rotations will be considered as the references in the vicinity of which, small elastic displacements can be superimposed. It is therefore crucial to characterise rigid-boy motion transfer between a pinion and a gear with tooth errors and/or shape modifications. In what follows, $e(M)$ represents the equivalent normal deviation at the potential point of contact M (sum of the deviations on the pinion and on the gear) and is conventionally positive for an excess of material and negative when, on the contrary, some material is removed from the ideal geometry. For rigid-body conditions (or alternatively under no-load), contacts will consequently occur at the locations on the contact lines where $e(M)$ is maximum and the velocity transfer from the pinion to the gear is modified compared with ideal gears such that:

$$\left(Rb_1\,\Omega_1 + Rb_2\,\Omega_2\right)\cos\beta_b + \frac{dE_{MAX}(t)}{dt} = 0 \tag{1}$$

where $E_{MAX}(t) = \max_M(e(M))$ with $\max_M()$, maximum over all the potential point of contact at time t

The difference with respect to ideal motion transfer is often related to the notion of no-load transmission error $NLTE$ via:

$$\frac{d}{dt}(NLTE) = Rb_1\,\Omega_1 + Rb_2\,\Omega_2 = -\frac{1}{\cos\beta_b}\frac{dE_{MAX}(t)}{dt} \tag{2}$$

Using the Kinetic Energy Theorem, the rigid-body dynamic behaviour for frictionless gears is controlled by:

$$J_1\,\Omega_1\,\dot{\Omega}_1 + J_2\,\Omega_2\,\dot{\Omega}_2 = C_m\,\Omega_1 + C_r\,\Omega_2 \tag{3}$$

with J_1, J_2: the polar moments of inertia of the pinion shaft line and the gear shaft line respectively. C_m, C_r : pinion and gear torques.

The system with 4 unknowns ($\Omega_1, \Omega_2, C_m, C_r$) is characterised by equations (2) - (3) only, and 2 parameters have to be imposed.

3.2 Deformed state – Principles

Modular models based on the definition of gear elements (pinion-gear pairs), shaft elements and lumped parameter elements (mass, inertia, stiffness) have proved to be effective in the simulation of complex gear units (Küçükay, 1987), (Baud & Velex, 2002). In this section, the theoretical foundations upon which classic gear elements are based are presented and the corresponding elemental stiffness and mass matrices along with the possible elemental forcing term vectors are derived and explicitly given. The simplest and most frequently used 3D representation corresponds to the pinion-gear model shown in Figure 1. Assuming that the geometry is not affected by deflections (small displacements hypothesis) and provided that mesh elasticity (and to a certain extent, gear body elasticity) can be transferred onto the base plane, a rigid-body approach can be employed. The pinion and the gear can therefore be assimilated to two rigid cylinders with 6 degrees of freedom each, which are connected by a stiffness element or a distribution of stiffness elements (the discussion of the issues associated with damping and energy dissipation will be dealt with in section 4.3). From a physical point of view, the 12 degrees of freedom of a pair represent the generalised displacements of i) traction: u_1, u_2 (axial displacements), ii) bending: v_1, w_1, v_2, w_2 (translations in two perpendicular directions of the pinion/gear centre), $\varphi_1, \psi_1, \varphi_2, \psi_2$ (bending rotations which can be assimilated to misalignment angles) and finally, iii) torsion:

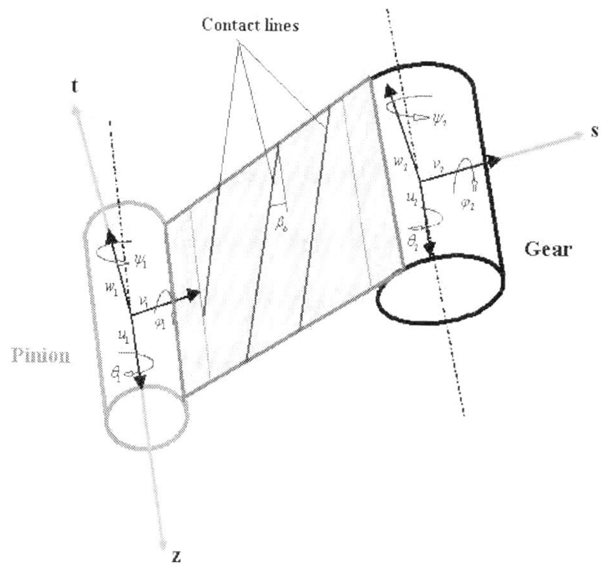

Fig. 1. A 3D lumped parameter model of pinion-gear pair.

θ_1, θ_2 which are small angles associated with deflections superimposed on rigid-body rotations $\Theta_1 = \int_0^t \Omega_1(\sigma)d\sigma$ (pinion) and $\Theta_2 = \int_0^t \Omega_2(\sigma)d\sigma$ (gear). Following Velex and Maatar (1996), screws of infinitesimal displacements are introduced whose co-ordinates for solid k (*conventionally k=1 for the pinion, k=2 for the gear*) can be expressed in two privileged coordinate systems: i) $(\mathbf{s},\mathbf{t},\mathbf{z})$ such that $\mathbf{z}$ is in the shaft axis direction (from the motor to the load machine), $\mathbf{S}$ is in the centre-line direction from the pinion centre to the gear centre and $\mathbf{t} = \mathbf{z} \times \mathbf{s}$ (Fig. 1) or, ii) $(\mathbf{X},\mathbf{Y},\mathbf{z})$ attached to the base plane (Fig. 1):

$$\{S_k\}\begin{Bmatrix} \mathbf{u_k}(\mathbf{O_k}) = v_k\,\mathbf{s} + w_k\,\mathbf{t} + u_k\mathbf{z} \\ \boldsymbol{\omega}_\mathbf{k} == \varphi_k\,\mathbf{s} + \psi_k\,\mathbf{t} + \theta_k\,\mathbf{z} \end{Bmatrix} or \begin{Bmatrix} \mathbf{u_k}(\mathbf{O_k}) = V_k\,\mathbf{X} + W_k\mathbf{Y} + u_k\mathbf{z} \\ \boldsymbol{\omega}_\mathbf{k} == \Phi_k\,\mathbf{X} + \Psi_k\,\mathbf{Y} + \theta_k\,\mathbf{z} \end{Bmatrix} \quad k=1,2 \tag{4}$$

where O_1, O_2 are the pinion and gear centres respectively

3.3 Deflection at a point of contact – Structural vectors for external gears

Depending on the direction of rotation, the direction of the base plane changes as illustrated in Figure 2 where the thicker line corresponds to a positive rotation of the pinion and the finer line to a negative pinion rotation about axis $(O_1, \mathbf{z})$.

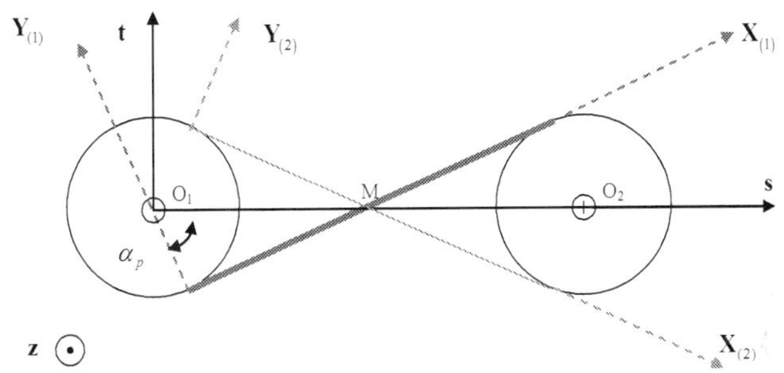

Fig. 2. Directions of rotation and planes (lines) of action. (*the thicker line corresponds to a positive rotation of pinion*)

For a given helical gear, the sign of the helix angle on the base plane depends also on the direction of rotation and, here again; two configurations are possible as shown in Figure 3.

Since a rigid-body mechanics approach is considered, contact deflections correspond to the interpenetrations of the parts which are deduced from the contributions of the degrees-of-freedom and the initial separations both measured in the normal direction with respect to the tooth flanks. Assuming that all the contacts occur in the theoretical base plane (or plane of action), the normal deflection $\Delta(M)$ at any point M, potential point of contact, is therefore expressed as:

$$\Delta(M) = \mathbf{u}_1(\mathbf{M}).\mathbf{n}_1 - \mathbf{u}_2(\mathbf{M}).\mathbf{n}_1 - \delta e(M) \tag{5}$$

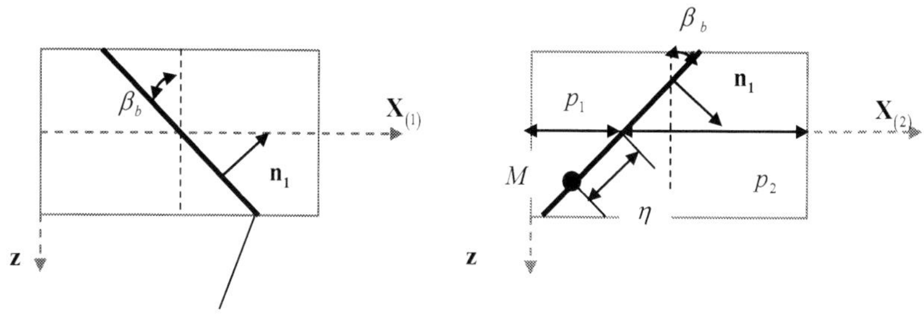

Fig. 3. Helix angles on the base plane.

where $\delta e(M) = \max_M(e(M)) - e(M)$ is the equivalent initial normal gap at M caused by tooth modifications and/or errors for example, $\mathbf{n}_1$ is the outward unit normal vector to pinion tooth flanks (Fig.3)

Using the shifting property of screws, one obtains the expression of $\Delta(M)$ in terms of the screw co-ordinates as:

$$\Delta(M) = \mathbf{u}_1(\mathbf{O}_1).\mathbf{n}_1 + (\boldsymbol{\omega}_1 \times \mathbf{O}_1\mathbf{M}).\mathbf{n}_1 - \mathbf{u}_2(\mathbf{O}_2).\mathbf{n}_1 - (\boldsymbol{\omega}_2 \times \mathbf{O}_2\mathbf{M}).\mathbf{n}_1 - \delta e(M) \tag{6}$$

which is finally expressed as:

$$\Delta(M) = \begin{bmatrix} \mathbf{n}_1 \\ \mathbf{O}_1\mathbf{M} \times \mathbf{n}_1 \\ -\mathbf{n}_1 \\ -\mathbf{O}_2\mathbf{M} \times \mathbf{n}_1 \end{bmatrix}^T \cdot \begin{bmatrix} \mathbf{u}_1(\mathbf{O}_1) \\ \boldsymbol{\omega}_1 \\ \mathbf{u}_2(\mathbf{O}_2) \\ \boldsymbol{\omega}_2 \end{bmatrix} - \delta e(M) \tag{7}$$

or, in a matrix form:

$$\Delta(M) = \mathbf{V}(\mathbf{M})^T \mathbf{q} - \delta e(M) \tag{8}$$

where $\mathbf{V}(\mathbf{M})$ is a structural vector which accounts for gear geometry (Küçükay, 1987) and $\mathbf{q}$ is the vector of the pinion-gear pair degrees of freedom (*superscript T refers to the transpose of vectors and matrices*)

The simplest expression is that derived in the $(\mathbf{X},\mathbf{Y},\mathbf{z})$ coordinate system associated with the base plane leading to:

$$\mathbf{V}(\mathbf{M})^T = \langle \cos\beta_b,\ 0,\ \varepsilon\sin\beta_b,\ -\zeta\,\varepsilon\,Rb_1\sin\beta_b,\ \eta - \varepsilon\,p_1\sin\beta_b,\ \zeta\,Rb_1\cos\beta_b, \\ -\cos\beta_b,\ 0,\ -\varepsilon\sin\beta_b,\ -\zeta\,\varepsilon\,Rb_2\sin\beta_b,\ -[\eta + \varepsilon\,p_2\sin\beta_b],\ \zeta\,Rb_2\cos\beta_b \rangle$$

$$\mathbf{q}^T = \langle V_1\ \ W_1\ \ u_1\ \ \Phi_1\ \ \Psi_1\ \ \theta_1\ \ V_2\ \ W_2\ \ u_2\ \ \Phi_2\ \ \Psi_2\ \ \theta_2 \rangle \tag{9}$$

where Rb_1, Rb_2 are the pinion, gear base radii; β_b is the base helix angle (*always considered as positive in this context*); p_1, p_2, η are defined in Figure 3; $\varepsilon = \pm 1$ depending on the sign of the helix angle; $\zeta = +1$ for a positive rotation of the pinion and $\zeta = -1$ for a negative rotation of the pinion.

An alternative form of interest is obtained when projecting in the $(\mathbf{s}, \mathbf{t}, \mathbf{z})$ frame attached to the pinion-gear centre line:

$$\begin{aligned}\mathbf{V}(\mathbf{M})^T = \langle & \cos\beta_b \sin\alpha_p, \ \zeta\cos\beta_b\cos\alpha_p, \ \varepsilon\sin\beta_b, \ -\zeta\varepsilon Rb_1\sin\beta_b\sin\alpha_p - \zeta(\eta - \varepsilon p_1\sin\beta_b)\cos\alpha_p \\ & \left[-\varepsilon Rb_1\sin\beta_b\cos\alpha_p + (\eta - \varepsilon p_1\sin\beta_b)\sin\alpha_p\right], \ \zeta Rb_1\cos\beta_b, \ -\cos\beta_b\sin\alpha_p, \ -\zeta\cos\beta_b\cos\alpha_p, \\ & -\varepsilon\sin\beta_b, \ -\zeta\varepsilon Rb_2\sin\beta_b\sin\alpha_p + \zeta(\eta + \varepsilon p_2\sin\beta_b)\cos\alpha_p \\ & -\left[\varepsilon Rb_2\sin\beta_b\cos\alpha_p + (\eta + \varepsilon p_2\sin\beta_b)\sin\alpha_p\right], \ \zeta Rb_2\cos\beta_b \ \rangle\end{aligned}$$

$$\mathbf{q}^T = \langle v_1 \ \ w_1 \ \ u_1 \ \ \varphi_1 \ \ \psi_1 \ \ \theta_1 \ \ v_2 \ \ w_2 \ \ u_2 \ \ \varphi_2 \ \ \psi_2 \ \ \theta_2 \rangle \tag{10}$$

3.4 Mesh stiffness matrix and forcing terms for external gears

For a given direction of rotation, the usual contact conditions in gears correspond to single-sided contacts between the mating flanks which do not account for momentary tooth separations which may appear if dynamic displacements are large (of the same order of magnitude as static displacements). A review of the mesh stiffness models is beyond the scope of this chapter but one usually separates the simulations accounting for elastic convection (i.e., the deflection at one point M depends on the entire load distribution on the tooth or all the mating teeth (Seager, 1967)) from the simpler (and classic) thin-slice approach (the deflection at point M depends on the load at the same point only). A discussion of the limits of this theory can be found in Haddad (1991), Ajmi & Velex (2005) but it seems that, for solid gears, it is sufficiently accurate as far as dynamic phenomena such as critical speeds are considered as opposed to exact load or stress distributions in the teeth which are more dependent on local conditions. Neglecting contact damping and friction forces compared with the normal elastic components on tooth flanks, the elemental force transmitted from the pinion onto the gear at one point of contact M reads:

$$\mathbf{dF}_{1/2}(\mathbf{M}) = k(M)\Delta(M)dM\,\mathbf{n}_1 \tag{11}$$

with $k(M)$: mesh stiffness at point M per unit of contact length

The resulting total mesh force and moment at the gear centre O_2 are deduced by integrating over the time-varying and possibly deflection-dependent contact length $L(t,\mathbf{q})$ as:

$$\{F_{1/2}\}\begin{cases}\mathbf{F}_{1/2} = \displaystyle\int_{L(t,\mathbf{q})} k(M)\Delta(M)dM\,\mathbf{n}_1 \\ \mathbf{M}_{1/2}(\mathbf{O}_2) = \displaystyle\int_{L(t,\mathbf{q})} k(M)\Delta(M)\mathbf{O}_2\mathbf{M}\times\mathbf{n}_1\,dM\end{cases} \tag{12-1}$$

Conversely the mesh force wrench at the pinion centre O_1 is:

$$\{F_{2/1}\}\begin{cases} \mathbf{F}_{2/1} = -\int\limits_{L(t,\mathbf{q})} k(M)\Delta(M)dM\,\mathbf{n}_1 \\ \mathbf{M}_{2/1}(\mathbf{O}_1) = -\int\limits_{L(t,\mathbf{q})} k(M)\Delta(M)\mathbf{O}_1\mathbf{M}\times\mathbf{n}_1\,dM \end{cases} \tag{12-2}$$

The mesh inter-force wrench can be deduced in a compact form as:

$$\{F_M\}\begin{Bmatrix} \{F_{2/1}\} \\ \{F_{1/2}\} \end{Bmatrix} = -\int\limits_{L(t,\mathbf{q})} k(M)\Delta(M)\mathbf{V}(\mathbf{M})dM \tag{13}$$

and introducing the contact normal deflection $\Delta(M) = \mathbf{V}(\mathbf{M})^T\mathbf{q} - \delta e(M)$ finally leads to:

$$\{F_M\} = -\left[\mathbf{K}_G(t)\right]\mathbf{q} + \mathbf{F}_e(t) \tag{14}$$

where $\left[\mathbf{K}_G(t)\right] = \int\limits_{L(t,\mathbf{q})} k(M)\mathbf{V}(\mathbf{M})\;\mathbf{V}(\mathbf{M})^T dM$ is the time-varying gear mesh stiffness matrix

$\mathbf{F}_e(t) = \int\limits_{L(t,\mathbf{q})} k(M)\delta e(M)\mathbf{V}(\mathbf{M})dM$ is the excitation vector associated with tooth shape modifications and errors

3.5 Mass matrix of external gear elements–Additional forcing (inertial) terms

For solid k (pinion or gear), the dynamic sum with respect to the inertial frame can be expressed as:

$$\mathbf{\Sigma}_\mathbf{k}^0 = m_k\left[\left(\ddot{v}_k - e_k\dot{\Omega}_k\sin\Theta_k - e_k\Omega_k^2\cos\Theta_k\right)\mathbf{s} + \left(\ddot{w}_k + e_k\dot{\Omega}_k\cos\Theta_k - e_k\Omega_k^2\sin\Theta_k\right)\mathbf{t} + \ddot{u}_k\mathbf{z}\right] \tag{15}$$

where m_k and e_k are respectively the mass and the eccentricity of solid k

A simple expression of the dynamic moment at point O_k can be obtained by assuming that O_k is the centre of inertia of solid k and neglecting gyroscopic components (complementary information can be found in specialised textbooks on rotor dynamics (see for instance (Lalanne & Ferraris, 1998)):

$$\mathbf{\delta}_\mathbf{k}^0(O_k) \cong I_k\ddot{\phi}_k\mathbf{s} + I_k\ddot{\psi}_k\mathbf{t} + I_{0k}\left(\dot{\Omega}_k + \ddot{\theta}_k\right)\mathbf{z} \tag{16}$$

where I_k is the cross section moment of inertia and I_{0k} is the polar moment of solid k

Using the same DOF arrangement as for the stiffness matrices, a mass matrix for the pinion-gear system can be deduced as (note that the same mass matrix is obtained in the $(\mathbf{X},\mathbf{Y},\mathbf{z})$ coordinate system):

$$\left[\mathbf{M}_G\right] = \mathbf{diag}\left(m_1, m_1, m_1, I_1, I_1, I_{01}, m_2, m_2, m_2, I_2, I_2, I_{02}\right) \tag{17-1}$$

along with a forcing term associated with inertial forces (whose expression in $(\mathbf{X},\mathbf{Y},\mathbf{z})$ has the same form on the condition that angles $\Theta_{1,2}$ are measured from $\mathbf{X}$ and $\mathbf{Y}$):

$$\mathrm{F_G}(t)=\langle m_1e_1(\dot{\Omega}_1\sin\Theta_1+\Omega_1^2\cos\Theta_1)\ \ -m_1e_1(\dot{\Omega}_1\cos\Theta_1-\Omega_1^2\sin\Theta_1)\ \ 0\ \ 0\ \ 0\ \ -I_{01}\dot{\Omega}_1$$
$$m_2e_2(\dot{\Omega}_2\sin\Theta_2+\Omega_2^2\cos\Theta_2)\ \ -m_2e_2(\dot{\Omega}_2\cos\Theta_2-\Omega_2^2\sin\Theta_2)\ \ 0\ \ 0\ \ 0\ \ -I_{02}\dot{\Omega}_2\rangle \qquad (17\text{-}2)$$

3.6 Usual simplifications

Examining the components of the structural vectors in (9) and (10), it can be noticed that most of them are independent of the position of the point of contact M with the exception of those related to bending slopes $\Psi_{1,2}$ or $\varphi_{1,2},\psi_{1,2}$. Their influence is usually discarded especially for narrow-faced gears so that the mesh stiffness matrix can be simplified as:

$$\left[\mathbf{K}_G(t)\right]\cong\int_{L(t,\mathbf{q})}k(M)dM\ \mathbf{V}_0\mathbf{V}_0^T=k(t,\mathbf{q})\mathbf{G} \qquad (18)$$

where $\mathbf{V}_0$ represents an average structural vector and $k(t,\mathbf{q})$ is the time-varying, possibly non-linear, mesh stiffness function (scalar) which plays a fundamental role in gear dynamics.

3.6.1 Classic one-DOF torsional model

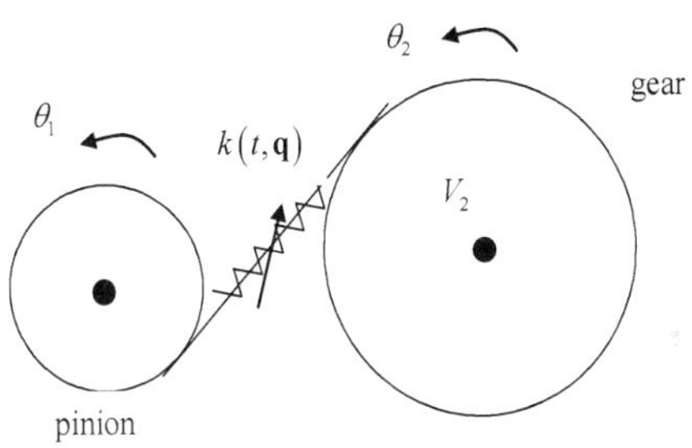

Fig. 4. Basic torsional model.

Considering the torsional degrees-of-freedom only (Figure 4), the structural vector reads (keeping solely the non-zero components):

$$\mathbf{V}(\mathbf{M})=\mathbf{V}_0=\begin{bmatrix}\zeta Rb_1\\ \zeta Rb_2\end{bmatrix}\cos\beta_b \qquad (19)$$

and the following differential system is derived $(\zeta^2=1)$:

$$\begin{bmatrix}I_{01} & 0\\ 0 & I_{02}\end{bmatrix}\begin{bmatrix}\ddot{\theta}_1\\ \ddot{\theta}_2\end{bmatrix}+k(t,\theta_1,\theta_2)\cos^2\beta_b\begin{bmatrix}Rb_1^2 & Rb_1Rb_2\\ Rb_1Rb_2 & Rb_2^2\end{bmatrix}\begin{bmatrix}\theta_1\\ \theta_2\end{bmatrix}=$$
$$=\begin{bmatrix}Cm\\ Cr\end{bmatrix}+\int_{L(t,\mathbf{q})}k(M)\delta e(M)dM\begin{bmatrix}\zeta Rb_1\\ \zeta Rb_2\end{bmatrix}\cos\beta_b-\begin{bmatrix}I_{01}\dot{\Omega}_1\\ I_{02}\dot{\Omega}_2\end{bmatrix} \qquad (20)$$

Note that the determinant of the stiffness matrix is zero which indicates a rigid-body mode (the mass matrix being diagonal). After multiplying the first line in (20) by $Rb_1 I_{02}$, the second line by $Rb_2 I_{01}$, adding the two equations and dividing all the terms by $\left(I_{10} Rb_2^2 + I_{20} Rb_1^2\right)$, the semi-definite system (20) is transformed into the differential equation:

$$\hat{m}\ddot{x} + k(t,x)x = F_t + \zeta \cos\beta_b \int_{L(t,x)} k(M)\delta e(M)dM - \kappa \frac{d^2}{dt^2}(NLTE) \tag{21}$$

With $x = Rb_1\theta_1 + Rb_2\theta_2$, relative apparent displacement

$\hat{m} = \dfrac{I_{02} I_{01}}{Rb_1^2 I_{02} + Rb_2^2 I_{01}}$, equivalent mass

$\kappa = \Omega_1^2 \dfrac{I_{02}}{Rb_2^2}$ when the pinion speed Ω_1 and the output torque C_r are supposed to be constant.

3.6.2 A simple torsional-flexural model for spur gears

The simplest model which accounts for torsion and bending in spur gears is shown in Figure 5. It comprises 4 degrees of freedom, namely: 2 translations in the direction of the line of action V_1, V_2 (at pinion and gear centres respectively) and 2 rotations about the pinion and gear axes of rotation θ_1, θ_2. Because of the introduction of bending DOFs, some supports (bearing/shaft equivalent stiffness elements for instance) must be added.

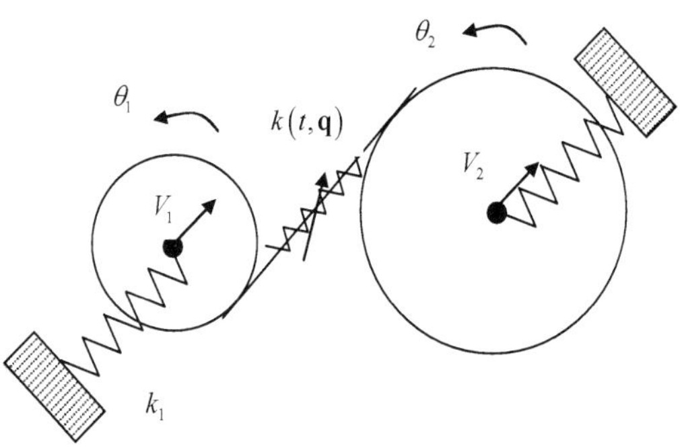

Fig. 5. Simplified torsional-flexural spur gear model.

The general expression of the structural vector $\mathbf{V}(\mathbf{M})$ (9) reduces to:

$$\mathbf{V}_0{}^T = \langle 1 \quad \zeta Rb_1 \quad -1 \quad \zeta Rb_2 \rangle \tag{22}$$

Re-writing the degree of freedom vector as $\mathbf{q}^{*T} = \langle v_1 \quad Rb_1\theta_1 \quad v_2 \quad Rb_2\theta_2 \rangle$, the following parametrically excited differential system is obtained for linear free vibrations:

$$\mathbf{M}\ddot{\mathbf{q}}^* + \mathbf{K}(t)\mathbf{q}^* = 0 \tag{23-1}$$

$$\mathbf{M}=\begin{bmatrix} m_1 & & & \\ & I_{01}/Rb_1^2 & & \\ & & m_2 & \\ & & & I_{02}/Rb_2^2 \end{bmatrix}; \quad \mathbf{K}(t)=\begin{bmatrix} k(t)+k_1 & \zeta k(t) & -k(t) & \zeta k(t) \\ & k(t) & -\zeta k(t) & k(t) \\ & & k(t)+k_2 & -\zeta k(t) \\ & & & k(t) \end{bmatrix} \quad (23\text{-}2)$$

Remark: The system is ill-conditioned since rigid-body rotations are still possible (no unique static solution). In the context of 3D models with many degrees of freedom, it is not interesting to solve for the normal approach $Rb_1\theta_1 + Rb_2\theta_1$ as is done for single DOF models. The problem can be resolved by introducing additional torsional stiffness element(s) which can represent shafts; couplings etc. thus eliminating rigid-body rotations.

4. Mesh stiffness models – Parametric excitations

4.1 Classic thin-slice approaches

From the results in section 2-5, it can be observed that, in the context of gear dynamic simulations, the mesh stiffness function defined as $k(t,\mathbf{q}) = \int_{L(t,\mathbf{q})} k(M)dM$ plays a key role.

This function stems from a 'thin-slice' approach whereby the contact lines between the mating teeth are divided in a number of independent stiffness elements (with the limiting case presented here of an infinite set of non-linear time-varying elemental stiffness elements) as schematically represented in Figure 6.

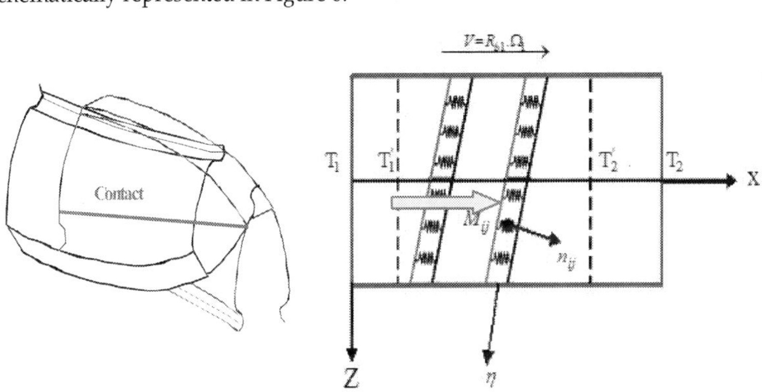

Fig. 6. 'Thin-slice' model for time-varying mesh stiffness.

Since the positions of the teeth (and consequently the contact lines) evolve with time (or angular positions), the profiles slide with respect to each other and the stiffness varies because of the contact length and the individual tooth stiffness evolutions. The definition of mesh stiffness has generated considerable interest but mostly with the objective of calculating accurate static tooth load distributions and stress distributions. It has been shown by Ajmi and Velex (2005) that a classic 'thin-slice' model is sufficient for dynamic calculations as long as local disturbances (especially near the tooth edges) can be ignored. In this context, Weber and Banascheck (1953) proposed a analytical method of calculating tooth deflections of spur gears by superimposing displacements which arise from i) the contact between the teeth, ii) the tooth itself considered as a beam and, iii) the gear body (or foundation) influence. An analytical expression of the contact compliance was obtained using the 2D Hertzian theory for cylinders in contact which is singular as far as the normal approach between the parts (contact deflection) is concerned. The other widely-used formulae for tooth contact deflection comprise the analytical formula of Lundberg (1939), the approximate Hertzian approach originally used at Hamilton Standard (Cornell, 1981) and the semi-empirical formula developed by Palmgren (1959) for rollers. The tooth bending radial and tangential displacements were derived by equating the work produced by one individual force acting on the tooth profile and the strain energy of the tooth assimilated to a cantilever of variable thickness. Extensions and variants of the methodology were introduced by Attia (1964), Cornell (1981) and O'Donnell (1960, 1963) with regard to the foundation effects. Gear body contributions were initially evaluated by approximating them as part of an elastic semi-infinite plane loaded by the reactions at the junction with the tooth. A more accurate expression for this base deflection has been proposed by Sainsot *et al.* (2004) where the gear body is simulated by an elastic annulus instead of a half-plane. Figure 7 shows two examples of mesh stiffness functions (no contact loss) calculated by combining Weber's and Lundberg's results for a spur and a helical gear example. It can be observed that the stiffness fluctuations are stronger in the case of conventional spur gears compared with helical gears for which the contact variations between the teeth are smoother.

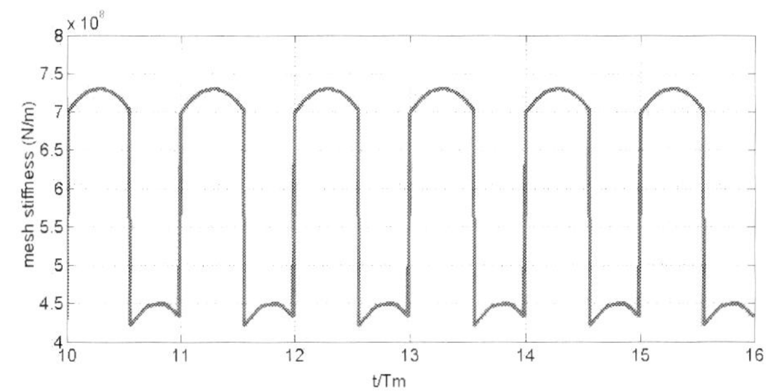

a - Spur gear

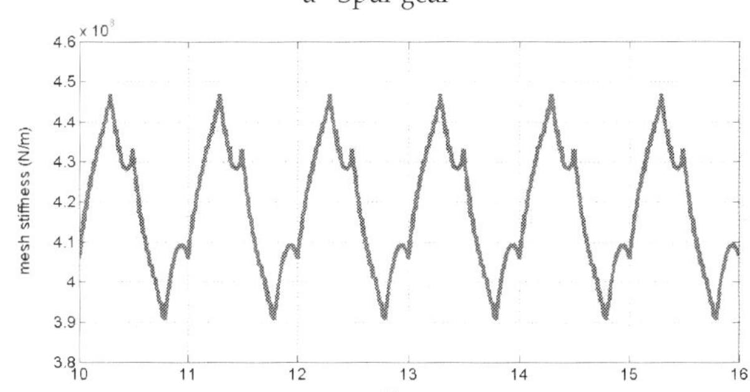

b - Helical gear $(\beta = 30°)$

Fig. 7. Examples of mesh stiffness functions for errorless gears.

Although the results above are based on simplified bi-dimensional approaches, they are still widely used in gear design. For example, the mesh stiffness formulae in the ISO standard 6336 stem from Weber's analytical formulae which were modified to bring the values in closer agreement with the experimental results. Another important simplification brought by the ISO formulae is that the mesh stiffness per unit of contact length k_0 is considered as approximately constant so that the following approximation can be introduced:

$$\int_{L(t,\mathbf{q})} k(M)dM \cong k_0 \int_{L(t,\mathbf{q})} dM = k_0 L(t,\mathbf{q}) \tag{24}$$

where $L(t,\mathbf{q})$ is the time-varying (possibly non-linear) contact length.

4.2 Contact length variations for external spur and helical gears

Considering involute profiles, the contact lines in the base plane are inclined by the base helix angle β_b (Figure 8) which is nil for spur gears. All contact lines are spaced by integer multiples of the apparent base pitch Pb_a and, when the pinion and the gear rotate, they all undergo a translation in the $\mathbf{X}$ direction at a speed equal to $Rb_1 \Omega_1$.

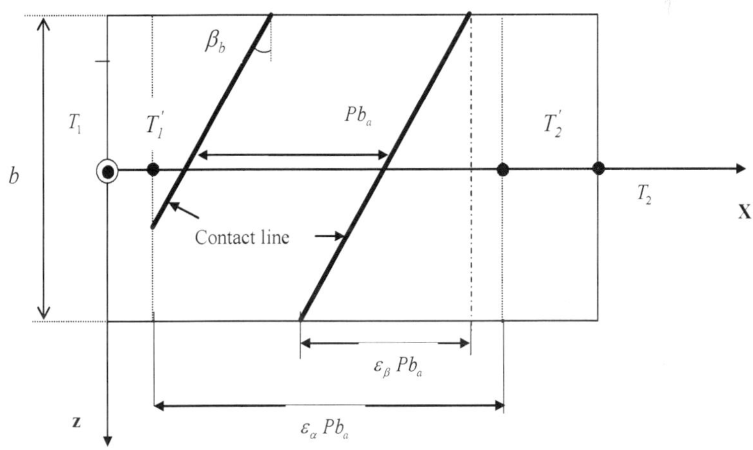

Fig. 8. Base plane and contact lines (b : face width; $\mathbf{Z}$: axial direction (direction of the axes of rotation); T_1, T_2 : points of tangency on pinion and gear base circles and T_1', T_2' : limits of the contact area on base plane).

It transpires from this geometrical representation that the total length of contact between the pinion and the gear is likely to vary with time and, based on the simple stiffness equation (24), that mesh stiffness is time-varying and, consequently, contributes to the system excitation via parametric excitations.

The extent of action on the base plane is an important property measured by the contact ratio ε_α which, in simple terms, represents the 'average number' of tooth pairs in contact (possibly non integer) and is defined by:

$$\varepsilon_\alpha = \frac{T_1'T_2'}{Pb_a} = \frac{\sqrt{Ra_1^2 - Rb_1^2} + \sqrt{Ra_2^2 - Rb_2^2} - E\sin\alpha_p}{\pi m \cos\alpha_p} \qquad (25\text{-}1)$$

with Ra_1, Ra_2: external radius of pinion, of gear; Rb_1, Rb_2: base radius of pinion, of gear; $E = \left\| O_1\vec{O}_2 \right\|$: centre distance

In the case of helical gears, the overlap due to the helix is taken into account by introducing the overlap ratio ε_β defined as:

$$\varepsilon_\beta = \frac{b\tan\beta_b}{Pb_a} = \frac{1}{\pi}\frac{b}{m}\frac{\tan\beta_b}{\cos\alpha_p} \qquad (25\text{-}2)$$

and the sum $\varepsilon = \varepsilon_\alpha + \varepsilon_\beta$ is defined as the total contact ratio.

Introducing the dimensionless time $\tau = \frac{t}{T_m}$ where $T_m = \frac{Pb_a}{Rb_1\Omega_1}$ is the mesh period i.e. the time needed for a contact line to move by a base pitch on the base plane, a closed form expression of the contact length $L(\tau)$ for ideal gears is obtained under the form (Maatar & Velex, 1996), (Velex et al., 2011):

$$\frac{L(\tau)}{L_m} = 1 + 2\sum_{k=1}^{\infty} Sinc(k\varepsilon_\alpha) Sinc(k\varepsilon_\beta) \cos\left(\pi k(\varepsilon_\alpha + \varepsilon_\beta - 2\tau)\right) \qquad (26)$$

with: $L_m = \varepsilon_\alpha \frac{b}{\cos\beta_b}$, average contact length

$Sinc(x) = \frac{\sin(\pi x)}{\pi x}$ is the classic sine cardinal function which is represented in Figure 9.

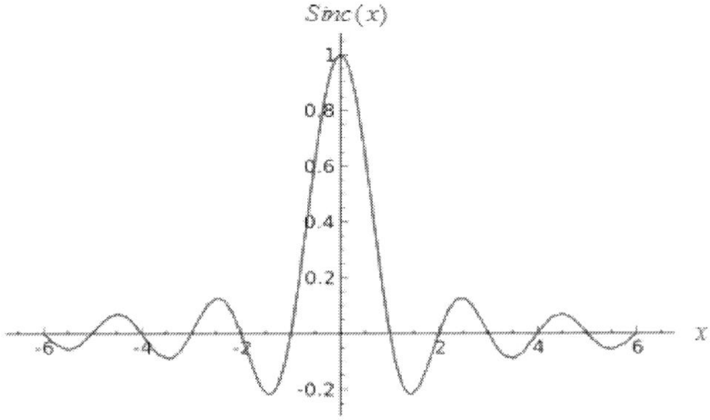

Fig. 9. Evolutions of $Sinc(x) = \frac{\sin(\pi x)}{\pi x}$.

The following conclusions can be drawn:

a. for spur gears, $\varepsilon_\beta = 0$ and $Sinc(k\varepsilon_\beta) = 1$
b. it can observed that the time-varying part of the contact length disappears when either ε_α or ε_β is an integer
c. harmonic analysis is possible by setting $k = 1,2,...$ in (27) and it is possible to represent the contact length variations for all possible values of profile and overlap contact ratios on a unique diagram. Figure 10 represents the RMS of contact length variations for a realistic range of contact and overlap ratios. It shows that:
   - contact length variations are significant when ε_α is below 2 and ε_β below 1
   - contact length is constant when $\varepsilon_\alpha = 2$ ($\varepsilon_\alpha = 1$ has to avoided for a continuous motion transfer) and / or $\varepsilon_\beta = 1$
   - for overlap ratios ε_β above 1, contact length variations are very limited.

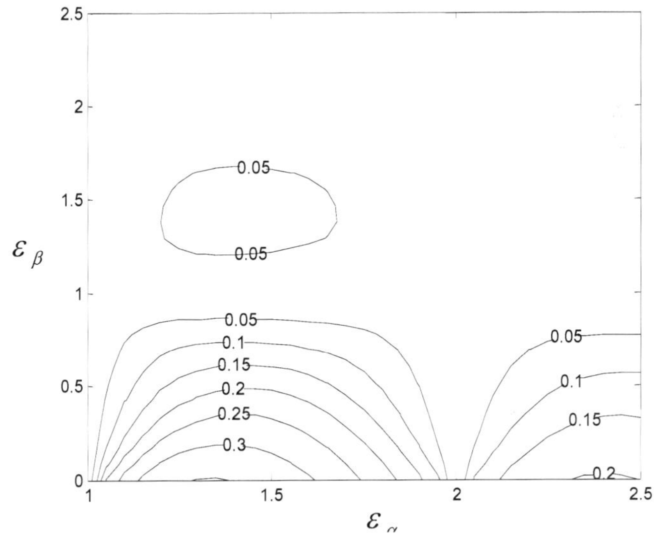

Fig. 10. Contour plot of the R.M.S. of $L(\tau)/L_m$ for a range of profile and transverse contact ratios.

4.3 Approximate expressions – Orders of magnitude

Mesh stiffness can be determined using the Finite Elements Method but it is interesting to have orders of magnitude or approximate values at the design stage. For solid gears made of steel, an order of magnitude of the mesh stiffness per unit of contact length k_0 is $\cong 1.3 \; 10^{10}$ N/m². More accurate expressions can be derived from the ISO 6336 standard which, for solid gears, gives:

$$k_0 \cong \cos\beta \frac{0.8}{q} \tag{27}$$

Focusing on dynamic tooth loads, it is interesting to introduce the percentage of modal strain energy stored in the gear mesh which, for a given pair (ω_p , $\mathbf{\Phi}_\mathrm{P}$), is defined as:

$$\rho_p = k_m \frac{\mathbf{\Phi}_\mathrm{P}^T \bar{\mathbf{V}}_0 \bar{\mathbf{V}}_0^T \mathbf{\Phi}_\mathrm{P}}{\mathbf{\Phi}_\mathrm{P}^T \left[[\mathbf{K}_0] + k_m \bar{\mathbf{V}}_0 \bar{\mathbf{V}}_0^T \right] \mathbf{\Phi}_\mathrm{P}} = v_{\Phi p}^2 \frac{k_m}{k_{\Phi p}} \tag{37}$$

with $\quad v_{\Phi p} = \mathbf{\Phi}_\mathrm{P}^T \bar{\mathbf{V}} = \bar{\mathbf{V}}^T \mathbf{\Phi}_\mathrm{P}$

$k_m, k_{\Phi p}$: average mesh stiffness and modal stiffness associated with (ω_p , $\mathbf{\Phi}_\mathrm{P}$)

It as been shown (Velex & Berthe., 1989) that ρ_p is a reliable indicator of the severity of one frequency with regard to the pinion-gear mesh and it can be used to identify the potentially critical speeds ω_p for tooth loading which are those with the largest percentages of modal strain energy in the tooth mesh. If the only excitations are those generated by the meshing (the mesh frequency is $Z_1\Omega_1$), the tooth critical speeds can be expressed in terms of pinion speed as:

$$\Omega_1 = \omega_p / kZ_1 \qquad k = 1,2,... \tag{38}$$

Based on the contact length variations and on the transmission error spectrum, the relative severity of the excitations can be anticipated.

Remark: The critical frequencies are supposed to be constant over the speed range (gyroscopic effects are neglected). Note that some variations can appear with the evolution of the torque versus speed (a change in the torque or load can modify the average mesh stiffness especially for modified teeth).

For the one DOF tosional model in Figure 4, there is a single critical frequency $\omega = \sqrt{k_m / \hat{m}}$ whose expression can be developed for solid gears of identical face width leading to:

$$\Omega_1 \cong \frac{\Lambda}{k} \frac{\cos \alpha_p}{MZ_1^2} \sqrt{\frac{b}{B}} \sqrt{\cos \beta_b} \sqrt{\varepsilon_\alpha} \sqrt{1+u^2} \tag{39}$$

where $k = 1,2,...$ represents the harmonic order; $\Lambda = \sqrt{\dfrac{8k_0}{\pi\rho}}$ (ρ is the density), for steel gears $\Lambda \cong 210^3\ ms^{-1}$; M is the module (in meter); B is the pinion or gear thickness (supposed identical); b is the effective contact width (which can be shorter than B because of chamfers for example); $u = \dfrac{Z_1}{Z_2}$, speed ratio.

5.3 Dynamic response

5.3.1 The problem of damping

Energy dissipation is present in all geared systems and the amount of damping largely controls the amplification at critical speeds. Unfortunately, the prediction of damping is still a challenge and, most of the time; it is adjusted in order to fit with experimental evidence. Two classical procedures are frequently employed:

a. the assumption of proportional damping (Rayleigh's damping) which, in this case, leads to:

$$[\mathbf{C}] = a[\mathbf{M}] + b\left[[\mathbf{K}_0] + k_m \bar{\mathbf{V}}_0 \bar{\mathbf{V}}_0^T\right] \tag{40}$$

with: a,b, two constants to be adjusted from experimental results

b. the use of (a limited number of) modal damping factors ς_p :

The damping matrix is supposed to be orthogonal with respect to the mode-shapes of the undamped system with the averaged stiffness matrix such that:

$$\mathbf{\Phi}_p^T[\mathbf{C}]\mathbf{\Phi}_p = 2\varsigma_p\sqrt{k_{\Phi p}\, m_{\Phi p}} \tag{41-1}$$

$$\mathbf{\Phi}_p^T[\mathbf{C}]\mathbf{\Phi}_q = 0 \tag{41-2}$$

with: ς_p : modal damping factor associated with mode p

$k_{\Phi p}, m_{\Phi p}$: modal stiffness and mass associated with mode p

or introducing the modal damping matrix $[\mathbf{C}_\Phi]$:

$$[\mathbf{C}_\Phi] = diag\left(2\varsigma_p\sqrt{k_{\Phi p}\, m_{\Phi p}}\right),\ p = 1, N\mathrm{mod} \tag{41-3}$$

Following Graig (1981), the damping matrix can be deduced by a truncated summation on a limited number of modes Nr leading to the formula:

$$[\mathbf{C}] = \sum_{p=1}^{Nr} \frac{2\varsigma_p\,\omega_p}{m_{\Phi p}}\left([\mathbf{M}]\mathbf{\Phi}_p\right)\left([\mathbf{M}]\mathbf{\Phi}_p\right)^T \tag{42}$$

with: $\omega_p = \sqrt{\dfrac{k_{\Phi p}}{m_{\Phi p}}}$

Regardless of the technique employed, it should be stressed that both (41) and (42) depend on estimated or measured modal damping factors ς_p for which the data in the literature is rather sparse. It seems that $0.02 \le \varsigma_p \le 0.1$ corresponds to the range of variation for modes with significant percentages of strain energy in the meshing teeth. The methods also rely on the assumption of orthogonal mode shapes which is realistic when the modal density (number of modes per frequency range) is moderate so that inter-modal couplings can be neglected.

5.3.2 Linear response

Based on the previous developments, the linear response of gears to mesh parametric excitations can be qualitatively assessed. Response peaks are to be expected at all tooth critical speeds and every sub-harmonic of these critical speeds because mesh stiffness time variations may exhibit several harmonics with significant amplitudes. Figure 11, taken from Cai and Hayashi (1994), is a clear example of such typical dynamic response curves when the gear dynamic behaviour is dominated by one major tooth frequency ω_n (and can be simulated by using the classic one DOF model). The amplifications associated with each peak depends on i) the excitation amplitude (Eq. (27) can provide some information on the amplitude associated with each mesh frequency harmonic) and ii) the level of damping for this frequency. For more complex gear sets, interactions between several frequencies can happen but, as far as the author is aware, the number of frequencies exhibiting a significant percentage of modal strain energy in the tooth mesh seems very limited (frequently less than 5) thus making it possible to anticipate the potential dangerous frequency coincidences for tooth durability.

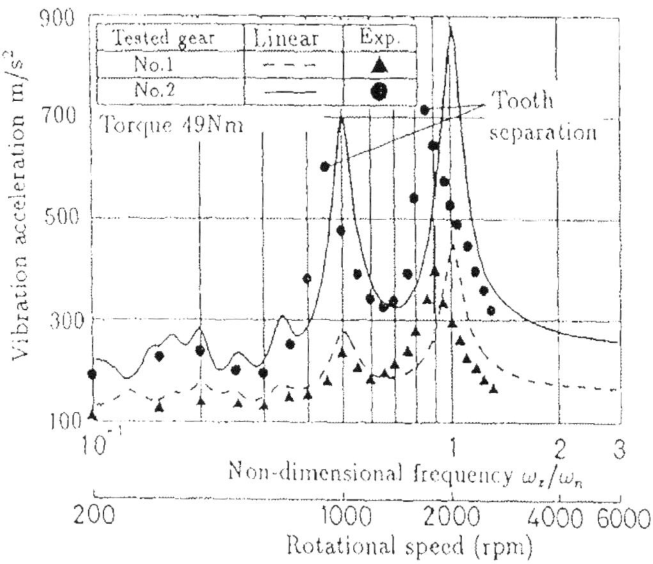

Fig. 11. Examples of dynamic response curves (Cay & Hayashi, 1994).

5.3.3 Contact condition – Contact losses and shocks

Only compressive contact forces can exist on tooth flanks and using (11), this imposes the following unilateral condition in case of contact at point M:

$$\mathbf{dF}_{1/2}(\mathbf{M}) \cdot \mathbf{n}_1 = k(M)\Delta(M)dM > 0 \tag{43}$$

or, more simply, a positive mesh deflection $\Delta(M)$.

If $\Delta(M) \leq 0$, the contact at M is lost (permanently or temporarily) and the associated contact force is nil. These constraints can be incorporated in the contact force expression by introducing the unit Heaviside step function $H(x)$ such that $H(x)=1$ if $x>0$ and $H(x)=0$ otherwise. Finally, one obtains:

$$\mathbf{dF}_{1/2}(\mathbf{M}) = k(M)\Delta(M)H(\Delta(M))dM\,\mathbf{n}_1 \tag{44}$$

It can observed that contact losses are related to the sign of $\Delta(M) \cong \mathbf{V}_0^T\mathbf{X} - \delta e(M)$, from which, it can be deduced that two cases have to be considered:

a. $\delta e(M)$ is larger than the normal approach $\mathbf{V}_0^T\mathbf{X}$ which, typically, corresponds to large amplitudes of tooth modifications reducing the actual contact patterns, to spalls on the flanks (pitting) where contact can be lost, etc.
b. the amplitude of the dynamic displacement $\mathbf{X}$ is sufficiently large so that the teeth can separate ($\mathbf{X}$ is periodic and can become negative in some part of the cycle).

Momentary contact losses can therefore occur when vibration amplitudes are sufficiently large; they are followed by a sequence of free flight within the tooth clearance until the teeth collide either on the driving flanks or on the back of the teeth (back strike). Such shocks are particularly noisy (rattle noise) and should be avoided whenever possible. Analytical investigations are possible using harmonic balance methods and approximations of $H(x)$ (Singh et al., 1989), (Comparin & Singh, 1989), (Kahraman & Singh, 1990), (Kahraman & Singh, 1991), and numerical integrations can be performed by time-step schemes (Runge-Kutta, Newmark, etc.). The most important conclusions are:

a. contact losses move the tooth critical frequencies towards the lower speeds (softening effect) which means that predictions based on a purely linear approach might be irrelevant. The phenomenon can be observed in Fig. 11 where the experimental peaks are at lower speed than those predicted by the linear theory.

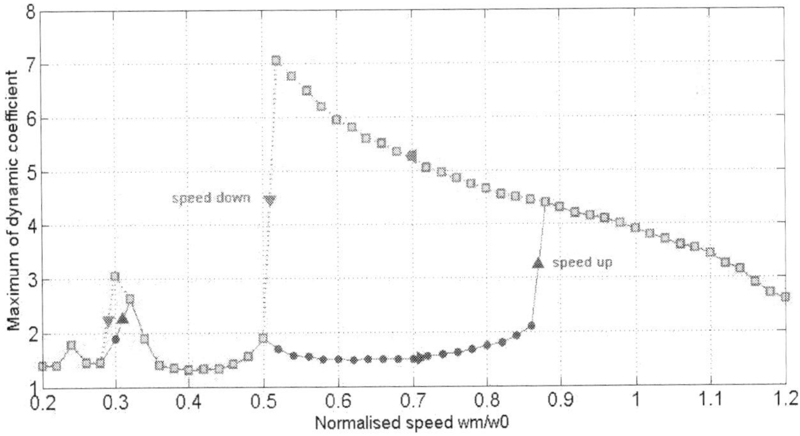

Fig. 12. Dynamic response curves by numerical simulations - Amplitude jumps - Influence of the initial conditions (speed up vs speed down), *2% of the critical damping, Spur gears* $Z_1 = 30$, $Z_2 = 45$, $M = 2mm$, *standard tooth proportions.*

b. When contact losses occur, response curves exhibit amplitude jumps (sudden amplitude variations for a small speed variation),
c. Because of a possibly strong sensitivity to initial conditions, several solutions may exist depending on the kinematic conditions i.e., speed is either increased or decreased
d. damping reduces the importance of the frequency shift and the magnification at critical tooth frequency.

These phenomena are illustrated in the response curves in Figure 12.

6. Transmission errors

6.1 Definitions

The concept of transmission error (*TE*) was first introduced by Harris (1958) in relation to the study of gear dynamic tooth forces. He realised that, for high speed applications, the problem was one of continuous vibrations rather than a series of impacts as had been thought before. Harris showed that the measure of departure from perfect motion transfer between two gears (which is the definition of *TE*) was strongly correlated with excitations and dynamic responses. *TE* is classically defined as the deviation in the position of the driven gear (for any given position of the driving gear), relative to the position that the driven gear would occupy if both gears were geometrically perfect and rigid.

NB: *The concept embodies both rigid-body and elastic displacements which can sometimes be confusing.*

Figure 13 illustrates the concept of transmission error which (either at no-load or under load) can be expressed as angular deviations usually measured (calculated) on the driven member (gear) or as distances on the base plane.

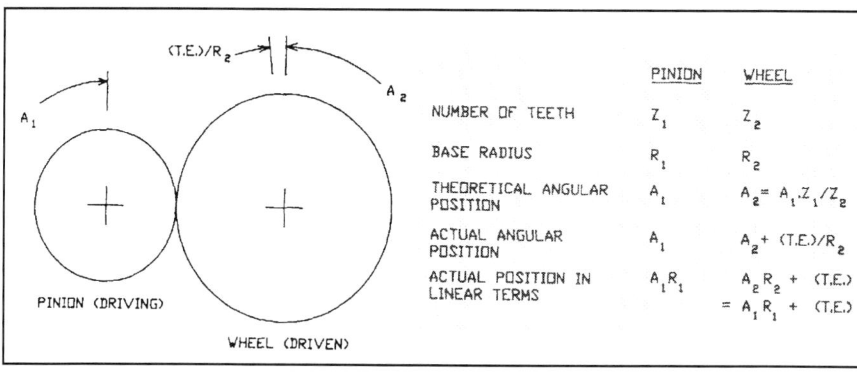

Fig. 13. Concept of transmission error and possible expressions (after Munro, (1989)).

Figures 14 and 15 show typical quasi-static *T.E.* traces for spur and helical gears respectively. The dominant features are a cyclic variation at tooth frequency (mesh frequency) and higher harmonics combined with a longer term error repeating over one revolution of one or both gears.

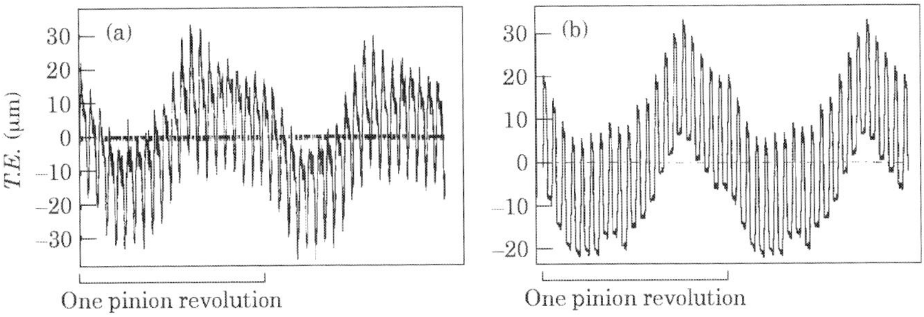

Fig. 14. Examples of quasi-static T.E. measurements and simulations - Spur gear (Velex and Maatar, 1996).

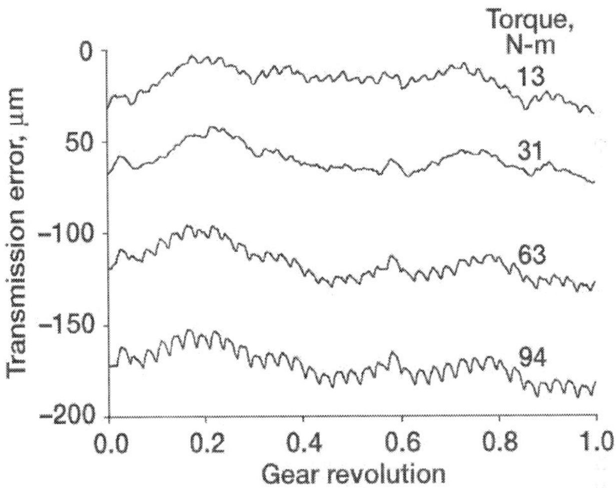

Fig. 15. T.E. measurements at various loads - Helical gear example.
NASA measurements from www.grc.nasa.gov/WWW/RT2001/5000/5950oswald1.html.

6.2 No-load transmission error (NLTE)

No-load *T.E.* (*NLTE*) has already been introduced in (2); it can be linked to the results of gear testing equipment (single flank gear tester) and is representative of geometrical deviations. From a mathematical point of view, *NLTE* is derived by integrating (2) and is expressed as:

$$NLTE = -\frac{E_{MAX}(t)}{\cos \beta_b} \tag{45}$$

6.3 Transmission errors under load

The concept of transmission error under load (TE) is clear when using the classic single degree of freedom torsional model (as Harris did) since it directly relies on the angles of torsion of the pinion and the gear. For other models (even purely torsional ones), the definition of TE is ambiguous or at least not intrinsic because it depends on the chosen cross-sections (or nodes) of reference for measuring or calculating deviations between actual and perfect rotation transfers from the pinion to the gear. Following Velex and Ajmi (2006), transmission error can be defined by extrapolating the usual experimental practice based on encoders or accelerometers, i.e., from the actual total angles of rotation, either measured or calculated at one section of reference on the pinion shaft (subscript I) and on the gear shaft (subscript II). TE as a displacement on the base plane reads therefore:

$$TE = Rb_1\left[\int_0^t \Omega_1\, d\xi + \theta_I\right] + Rb_2\left[\int_0^t \Omega_2\, d\xi + \theta_{II}\right] = Rb_1\,\theta_I + Rb_2\,\theta_{II} + NLTE \qquad (46)$$

with ξ, a dummy integration variable and θ_I, θ_{II}, the torsional perturbations with respect to rigid-body rotations (degrees of freedom) at node I on the pinion shaft and at node II on the gear shaft.

Introducing a projection vector $\mathbf{W}$ of components Rb_1 and Rb_2 at the positions corresponding to the torsional degrees of freedom at nodes I and II and with zeros elsewhere, transmission error under load can finally be expressed as:

$$TE = \mathbf{W}^T\,\mathbf{X} + NLTE \qquad (47\text{-}1)$$

which, for the one DOF model, reduces to:

$$TE = x + NLTE \qquad (47\text{-}2)$$

6.4 Equations of motion in terms of transmission errors

For the sake of clarity the developments are conducted on the one-DOF torsional model. Assuming that the dynamic contact conditions are the same as those at very low speed, one obtains from (21) the following equation for quasi-static conditions (i.e., when Ω_1 shrinks to zero):

$$k(t,x)x_S = F_t + \zeta\cos\beta_b \int_{L(t,x)} k(M)\delta e(M)dM \qquad (48)$$

which, re-injected in the dynamic equation (21), gives:

$$\tilde{m}\ddot{x} + k(t,x)x = k(t,x)x_S - \kappa\frac{d^2}{dt^2}(NLTE) \qquad (49)$$

From (47-2), quasi-static transmission error under load can be introduced such that $x_S = TE_S - NLTE$ and the equation of motion is transformed into:

$$\tilde{m}\ddot{x} + k(t,x)x = k(t,x)\left[TE_S - NLTE\right] - \kappa\frac{d^2}{dt^2}(NLTE) \qquad (50)$$

An alternative form of interest can be derived by introducing the dynamic displacement x_D defined by $x = x_S + x_D$ as:

$$\tilde{m}\ddot{x}_D + k(t,x)x_D = -\tilde{m}\frac{d^2}{dt^2}(TE_S) + (\tilde{m} - \kappa)\frac{d^2}{dt^2}(NLTE) \tag{51}$$

The theory for 3D models is more complicated mainly because there is no one to one correspondence between transmission error and the degree of freedom vector. It can be demonstrated (Velex and Ajmi, 2006) that, under the same conditions as for the one DOF model, the corresponding differential system is:

$$[\mathbf{M}]\ddot{\mathbf{X}}_D + [\mathbf{K}(t,\mathbf{X})]\mathbf{X}_D \cong -[\mathbf{M}]\hat{\mathbf{D}}\frac{d^2}{dt^2}(TE_S) + \left[\frac{1}{Rb_2}\mathbf{I}_P + [\mathbf{M}]\hat{\mathbf{D}}\right]\frac{d^2}{dt^2}(NLTE) \tag{52}$$

where $\hat{\mathbf{D}} = k_m \cos\beta_b \left[\bar{\mathbf{K}}\right]^{-1}\bar{\mathbf{V}}$, $\mathbf{X}_D = \mathbf{X} - \mathbf{X}_S$, dynamic displacement vector

6.5 Practical consequences

From (51) and (52), it appears that the excitations in geared systems are mainly controlled by the fluctuations of the quasi-static transmission error and those of the no-load transmission error as long as the contact conditions on the teeth are close to the quasi-static conditions (this hypothesis is not verified in the presence of amplitude jumps and shocks). The typical frequency contents of *NLTE* mostly comprise low-frequency component associated with run-out, eccentricities whose contributions to the second-order time-derivative of NLTE can be neglected. It can therefore be postulated that the mesh excitations are dominated by $\frac{d^2}{dt^2}(TE_S)$. This point has a considerable practical importance as it shows that reducing the dynamic response amplitudes is, to a certain extent, equivalent to reducing the fluctuations of TE_S. Profile and lead modifications are one way to reach this objective. Equation (50) stresses the fact that, when total displacements have to be determined, the forcing terms are proportional to the product of the mesh stiffness and the difference between TE_S and $NLTE$ (and not TE_S !). It has been demonstrated by Velex et al. (2011) that a unique dimensionless equation for quasi-static transmission error independent of the number of degrees of freedom can be derived under the form:

$$\cos\beta_b\, \hat{k}(t,\mathbf{X}_S)\, TE_S^*(t) = 1 - \int_{L(t,\mathbf{X}_S)} \hat{k}(M) e^*(M)\, dM \tag{53}$$

with $\hat{A} = \frac{A}{k_m}$, $A^* = \frac{A}{\delta_m}$, for any generic variable A (normalization with respect to the average mesh stiffness and the average static deflection).

Assuming that the mesh stiffness per unit of contact length is approximately constant (see section 2-5), analytical expressions for symmetric profile modifications (identical on pinion and gear tooth tips as defined in Fig. 16) rendering $TE_S(t)$ constant (hence cancelling most of the excitations in the gear system) valid for spur and helical gears with $\varepsilon_\alpha \leq 2$ can be found under the form:

$$E = \frac{\Gamma\Lambda}{2\Gamma - 1 + \frac{1}{\varepsilon_\alpha}} \tag{54}$$

submitted to the condition $\Gamma \geq \frac{\varepsilon_\alpha - 1}{2\varepsilon_\alpha}$

with E: tip relief amplitude; Γ: dimensionless extent of modification (such that the length of modification on the base plane is $\Gamma \varepsilon_\alpha Pb_a$) and $\Lambda = \frac{Cm}{Rb_1 b\, k_0}$: deflection of reference.

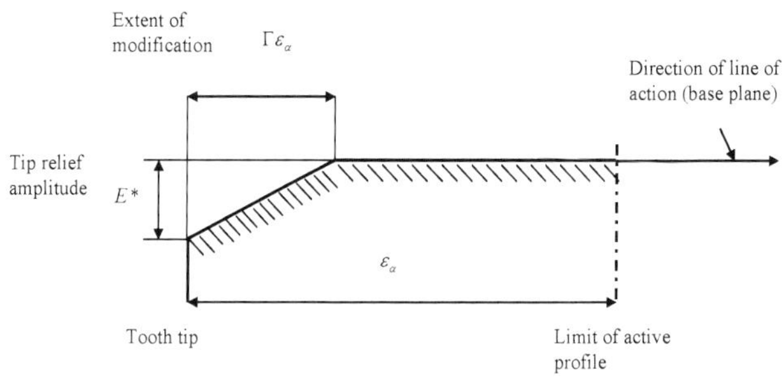

Fig. 16. Definition of profile relief parameters

Based on these theoretical results, it can be shown that quasi-static transmission error fluctuations for ideal gears with profile relief depend on a very limited number of parameters: i) the profile and lead contact ratios which account for gear geometry and ii) the normalised depth and extent of modification. These findings, even though approximate, suggest that rather general performance diagrams can be constructed which all exhibit a zone of minimum TE variations defined by (54) as illustrated in Figure 17 (Velex et al., 2011).It is to be noticed that similar results have been obtained by a number of authors using very different models (Velex & Maatar, 1996), (Sundaresan et al., 1991), (Komori et al., 2003), etc.

The dynamic factor defined as the maximum dynamic tooth load to the maximum static tooth load ratio is another important factor in terms of stress and reliability. Here again, an approximate expression can be derived from (51-52) by using the same asymptotic expansion as in (34) and keeping first-order terms only (Velex & Ajmi, 2007). Assuming that TE_S and $NLTE$ are periodic functions of a period equal to one pinion revolution; all forcing terms can be decomposed into a Fourier series of the form:

$$-[\mathbf{M}]\hat{\mathbf{D}}\frac{d^2}{dt^2}(TE_S)+\left[\frac{1}{Rb_2}\mathbf{I}_P+[\mathbf{M}]\hat{\mathbf{D}}\right]\frac{d^2}{dt^2}(NLTE)=-\Omega_1^2\sum_{n\geq 1} n^2\left[A^*_n \sin n\Omega_1 t + B^*_n \cos n\Omega_1 t\right] \tag{55}$$

and an approximate expression of the dimensionless dynamic tooth load can be derived under the form:

$$r(t)=\frac{F_D(t)}{F_S}\cong 1+\sum_p \sqrt{\rho_p \hat{k}_{\Phi p}}\,\Upsilon_{pn}(t) \tag{56}$$

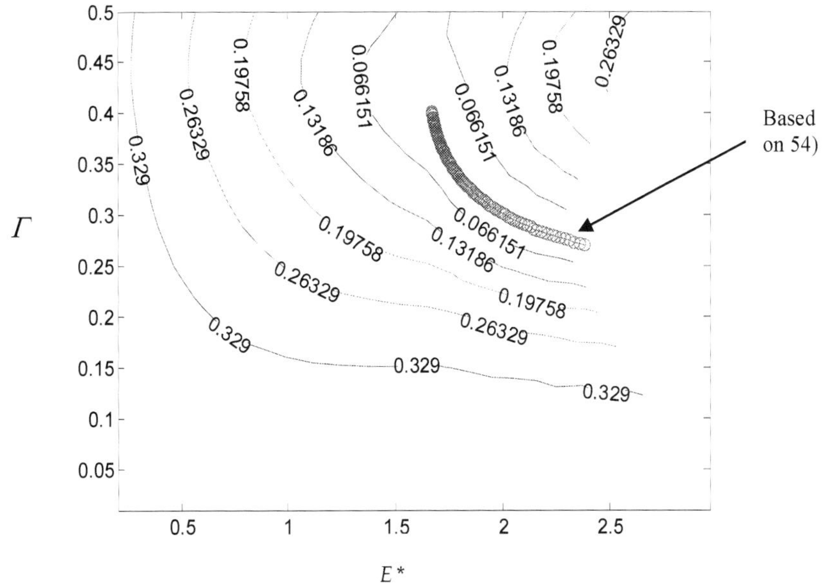

Fig. 17. Example of performance diagram: contour lines of the RMS of quasi-static transmission error under load - Spur gear $\varepsilon_\alpha = 1.67$.

with

$$\mathrm{Y}_{pn}(t)=\sum_{n\geq 1}\frac{\bar{A}^*{}_n\left[\left(\varpi_{pn}\right)^2-1\right]+2\bar{B}^*{}_n\,\varsigma_p\left(\varpi_{pn}\right)}{\left[\left(\varpi_{pn}\right)^2-1\right]^2+4\varsigma_p^2\left(\varpi_{pn}\right)^2}\sin n\Omega_1 t+\frac{\bar{B}^*{}_n\left[\left(\varpi_{pn}\right)^2-1\right]-2\bar{A}^*{}_n\,\varsigma_p\left(\varpi_{pn}\right)}{\left[\left(\varpi_{pn}\right)^2-1\right]^2+4\varsigma_p^2\left(\varpi_{pn}\right)^2}\cos n\Omega_1 t$$

$$\varpi_{pn}=\frac{\omega_p}{n\Omega_1}$$

Equation (56) makes it possible to estimate dynamic tooth loads with minimum computational effort provided that the modal properties of the system with averaged stiffness matrix and the spectrum of TE_S (predominantly) are known. One can notice that the individual contribution of a given mode is directly related to its percentage of strain energy in the meshing teeth and to the ratio of its modal stiffness to the average mesh stiffness. These properties can be used for identifying the usually limited number of critical mode shapes and frequencies with respect to tooth contact loads. They may also serve to test the structural modifications aimed at avoiding critical loading conditions over a range of speeds. It is worth noting that, since α is supposed to be a small parameter, the proposed methodology is more suited for helical gears.

7. Towards continuous models

7.1 Pinion, gear distortions

In the case of wide-faced gears, gear body deflections (especially those of the pinion) cannot be neglected and the torsion/bending distortions must be modelled since they can strongly affect the contact conditions between the teeth. For solid gears, one of the simplest approaches consists in modelling gear bodies by two node shaft finite elements in bending, torsion and traction as described in Ajmi and Velex (2005) which are connected to the same mesh interface model as that described in section 3 and Fig. 6. Assuming that any transverse section of the pinion or gear body originally plane remains plane after deformation (a fundamental hypothesis in Strength of Materials), gear bodies can then be sliced into elemental discs and infinitesimal gear elements using the same principles as those presented in section 2. The degrees of freedom of every infinitesimal gear element are expressed by using the shape functions of the two-node, six DOFs per node shaft element. By so doing, all the auxiliary DOFs attributed at every infinitesimal pinion and gear are condensed in terms of the degrees of freedom of the shaft nodes leading to a (global) gear element with 24 DOFs.

7.2 Thin-rimmed applications

The approach in 6.1 is valid for solid gears but is irrelevant for deformable structures such as thin-rimmed gears in aeronautical applications for example where the displacement field cannot be approximated by simple polynomial functions as is the case for shafts. Most of the attempts rely on the Finite Element Method applied to 2D cases (Parker et al., 2000), (Kahraman et al., 2003) but actual 3D dynamic calculations are still challenging and do not lend themselves to extensive parameter analyses often required at the design stage. An

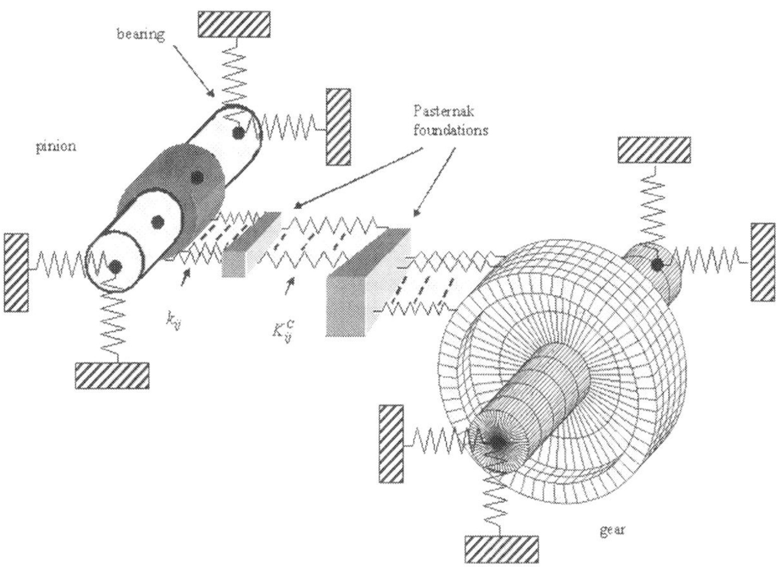

Fig. 18. Example of hybrid model used in gear dynamics (Bettaieb et al., 2007).

alternative to these time-consuming methods is to use hybrid FE/lumped models as described by Bettaieb et al, (2007). Figure 18 shows an example of such a model which combines i) shaft elements for the pinion shaft and pinion body, ii) lumped parameter elements for the bearings and finally iii) a FE model of the gear + shaft assembly which is sub-structured and connected to the pinion by a time-varying, non-linear Pasternak foundation model for the mesh stiffness. The computational time is reduced but the modelling issues at the interfaces between the various sub-models are not simple.

8. Conclusion

A systematic formulation has been presented which leads to the definition of gear elements with all 6 rigid-body degrees-of-freedom and time-varying, possibly non-linear, mesh stiffness functions. Based on some simplifications, a number of original analytical results have been derived which illustrate the basic phenomena encountered in gear dynamics. Such results provide approximate quantitative information on tooth critical frequencies and mesh excitations held to be useful at the design stage.

Gear vibration analysis may be said to have started in the late 50's and covers a broad range of research topics and applications which cannot all be dealt with in this chapter: multi-mesh gears, power losses and friction, bearing-shaft-gear interactions, etc. to name but a few. Gearing forms part of traditional mechanics and one obvious drawback of this long standing presence is a definite sense of déjà vu and the consequent temptation to construe that, from a research perspective, gear behaviour is perfectly understood and no longer worthy of study (Velex & Singh, 2010). At the same time, there is general agreement that although gears have been around for centuries, they will undoubtedly survive long into the 21st century in all kinds of machinery and vehicles.

Looking into the future of gear dynamics, the characterisation of damping in geared sets is a priority since this controls the dynamic load and stress amplitudes to a considerable extent. Interestingly, the urgent need for a better understanding and modelling of damping in gears was the final conclusion of the classic paper by Gregory et al. (1963-64). Almost half a century later, new findings in this area are very limited with the exception of the results of Li & Kahraman (2011) and this point certainly remains topical. A plethora of dynamic models can be found in the literature often relying on widely different hypotheses. In contrast, experimental results are rather sparse and there is certainly an urgent need for validated models beyond the classic results of Munro (1962), Gregory et al. (1963), Kubo (1978), Küçükay (1984 &87), Choy et al. (1989), Cai & Hayashi (1994), Kahraman & Blankenship (1997), Baud & Velex (2002), Kubur et al. (2004), etc. especially for complex multi-mesh systems. Finally, the study of gear dynamics and noise requires multi-scale, multi-disciplinary approaches embracing non-linear vibrations, tribology, fluid dynamics etc. The implications of this are clear; far greater flexibility will be needed, thus breaking down the traditional boundaries separating mechanical engineering, the science of materials and chemistry.

9. References

Ajmi, M. & Velex, P. (2005). A model for simulating the dynamic behaviour of solid wide-faced spur and helical gear, *Mechanisms and Machine Theory*, vol. 40, n°2, (February 2005), pp. 173-190. ISSN: 0094 -114X.

However, a more realistic model of the transmission has to take into account the inevitable loss of power to evaluate how it affects the correct sizing of the motor-reducer coupling and the resulting performance of the machine.

In general, transmissions are very complex, as the factors responsible for the losses of power. In this chapter they are taken into account just considering the transmission mechanical efficiency η defined as the ratio between the power outgoing from the transmission (W_{out}) and the incoming one (W_{in}), or through the extensive factors (T_{in}, T_{out}) and the intensive ones (ω_{in}, ω_{out}) of the power itself:

$$\eta = \frac{W_{out}}{W_{in}} = \frac{T_{out}}{T_{in}}\frac{\omega_{out}}{\omega_{in}} = \mu\tau \leq 1 \tag{4}$$

The loss of power within the transmission leads to a reduction of available torque downstream of the transmission. Indeed, if the coefficient of multiplication of forces for an ideal transmission is $\mu = \tau^{-1}$, when $\eta \neq 1$ it becomes $\mu' < \mu$, thus leading to a consequent reduction of the transmitted torque ($T'_{out} < T_{out}$).

Let's define W_M and W_L respectively as the power upstream and downstream of the transmission. When the power flows from the motor to the load the machine is said to work with *direct power flow*, otherwise, the functioning is said to be *backward*. Depending on the machine functioning mode, the transmission power losses are described by two different mechanical efficiency values η_d and η_r, where:

$$\eta_d = \frac{W_L}{W_M} \quad \text{(direct power flow)} \qquad \eta_r = \frac{W_M}{W_L} \quad \text{(backward power flow)}. \tag{5}$$

3. A single d.o.f. machine

An automatic machine is a system, usually complex, able to fulfill a particular task. Regardless of the type of machine, it may be divided into simpler subsystems, each able to operate only one degree of freedom and summarized by the three key elements: motor, transmission and load (Fig. 2).

3.1 The load and the servo-motor

The power supplied by the motor depends on the external load applied T_L and on the inertia acting on the system $J_L\dot{\omega}_L$. Since different patterns of speed ω_L and acceleration $\dot{\omega}_L$ generate different loads, the choice of a proper law of motion is the first project parameter that should be taken into account when sizing the motor-reducer unit. For this purpose, specific texts are recommended (Ruggieri et al. (1986), Melchiorri (2000)).

Once the law of motion has been defined, all the characteristics of the load are known.

Electric motor, and among these brushless motors, are the most widespread electrical actuators in the automation field, and their working range can be approximately subdivided into a *continuous* working zone (delimited by the motor rated torque) and a *dynamic* zone (delimited by the maximum motor torque $T_{M,max}$). A typical shape of the working zones is displayed in Fig.3. Usually the motor rated torque decreases slowly with the motor speed ω_M from T_{M,N_c} to $T_{M,N}$. To simplify the rated torque trend and to take a cautious approach, it's usually considered constant and equal to $T_{M,N}$ up to the maximum allowed motor speed

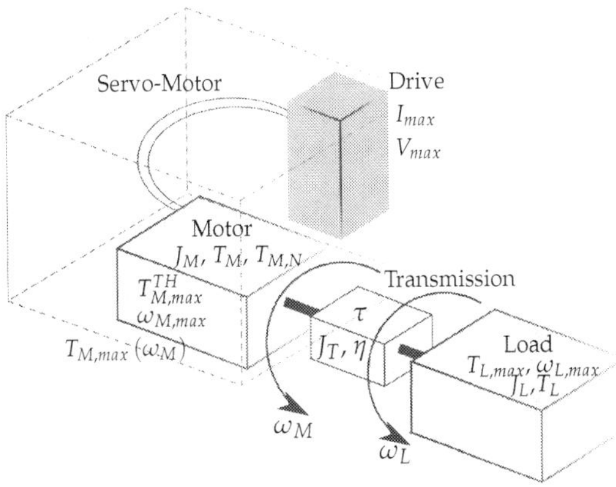

Fig. 2. Scheme of a generic machine actuated by an electric motor

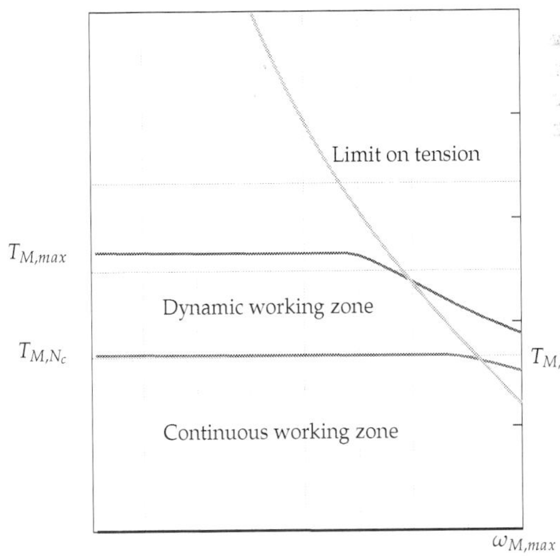

Fig. 3. An example of a speed/torque curve of a common brushless motor

$\omega_{M,max}$, whereas $T_{M,max}$ decreases from a certain value of ω_M. The nominal motor torque $T_{M,N}$, which is specified by the manufacturer in the catalogs, is defined as the torque that can be supplied by the motor for an infinite time, without overheating. Conversely, the trend of the maximum torque $T_{M,max}$ is very complex and depends on many factors. For this reason it is difficult to express it with an equation.

3.2 Conditions to the right coupling between motor and load

Frequently in industrial applications, the machine task is cyclical with period t_a much smaller than the motor thermal time constant. Once the task has been defined, a motor-task

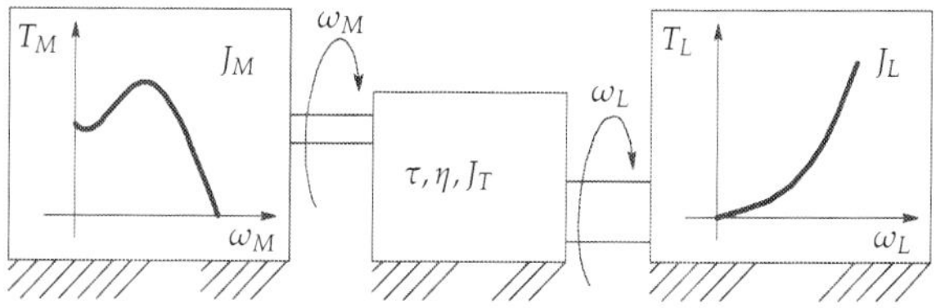

Fig. 4. Scheme of a generic single degree of freedom machine

combination is feasible if there exists a transmission ratio such that:

1. the maximum speed required from the motor is smaller than the maximum speed achievable ($\omega_{M,max}$);

$$\omega_M \leq \omega_{M,max} \tag{6}$$

2. a certain p-norm [1] of the motor torque $\|T_M(t)\|_p$ is smaller than a corresponding motor specific limit T_p. The norms that have most physical meaning for motor torques are Van de Straete et al. (1998):

$$\|T_M(t)\|_2 = T_{rms} = \left(\frac{1}{t_a}\int_0^{t_a} T_M^2(t)dt\right)^{\frac{1}{2}} \tag{7}$$

$$\|T_M(t,\omega)\|_\infty = T_{Max}(\omega) = \max|T_M(t,\omega)| \quad 0 \leq t \leq t_a \tag{8}$$

Since the motor torque is proportional to the current, 2-norm is a measure of mean square current through the windings and it should be limited by the nominal torque $T_{M,N}$ to avoid overheat. The ∞-norm is a measure of peak current and its limit is the maximum torque that can be exerted $T_{M,max}$. These limits translate into:

$$T_{rms} \leq T_{M,N} \tag{9}$$

$$T_{Max}(\omega) \leq T_{M,max}(\omega) \qquad \forall\omega \tag{10}$$

where $T_{M,max}$ depends on ω as shown in Fig. 3.

While the terms on the right of inequalities (6), (9), (10) are characteristics of each motor, ω_M, T_{rms} and T_{Max} depend on the load and on the mechanical features of the motor and the speed reducer.

3.3 Mathematical model of a machine

The machine functioning can be described as an instantaneous balance of power. In the case of direct power flow it can be expressed as:

$$\eta_d\left(\frac{T_M}{\tau} - \frac{J_M\dot{\omega}_L}{\tau^2}\right) = T_L + (J_L + J_T)\,\dot{\omega}_L \tag{11}$$

[1] In general, the p-norm $\|\cdot\|_p$ of a time function $f(t)$ over the period t_a is defined as:

$$\|f(t)\|_p = \left(\frac{1}{t_a}\int_0^{t_a} f^p(t)dt\right)^{\frac{1}{p}}$$

while in backward power flow it is:

$$\left(\frac{T_M}{\tau} - \frac{J_M \dot{\omega}_L}{\tau^2}\right) = \eta_r \left[T_L + (J_L + J_T)\,\dot{\omega}_L\right] \tag{12}$$

Equations (11), (12) can be written as:

$$T_{M,d} = \frac{\tau T_L^*}{\eta_d} + J_M \frac{\dot{\omega}_L}{\tau} \tag{13}$$

$$T_{M,r} = \tau T_L^* \eta_r + J_M \frac{\dot{\omega}_L}{\tau} \tag{14}$$

where:

$$T_L^* = T_L + (J_L + J_T)\,\dot{\omega}_L \tag{15}$$

is the total torque applied to the outgoing transmission shaft.

To unify these different operating conditions, a general mechanical efficiency function is introduced by Legnani et al. (2002). It is defined as:

$$\eta = \begin{cases} \eta_d & \text{(direct power flow)} \\ 1/\eta_r & \text{(backward power flow)} \end{cases} \tag{16}$$

where η, η_d and η_r are constants.

In the case of backward power flow it results $\eta > 1$. Note that this does not correspond to a power gain, but it is simply an expedient for unifying the equations of the two working conditions (direct/backward power flow). In fact, in this case, the effective efficiency is $\eta_d = 1/\eta < 1$.

4. Selecting the transmission

Conditions (6), (9), (10) can be expressed, for each motor, as constraints on acceptable transmission ratios. However, each transmission is characterized not only by its reduction ratio, but also by its mechanical efficiency and its moment of inertia. This chapter analyzes how these factors affect the conditions (6), (9), (10) and how they reduce the range of suitable transmission ratios for a given motor.

4.1 Limit on the maximum achievable speed

Since each motor has a maximum achievable speed $\omega_{M,max}$, a *limit transmission ratio* τ_{kin} can be defined as the minimum transmission ratio below which the system cannot reach the requested speed:

$$\tau_{kin} = \frac{\omega_{L,max}}{\omega_{M,max}} \tag{17}$$

The condition of maximum speed imposed by the system can be rewritten in terms of minimum gear ratio τ_{kin} to guarantee the required performance. Eq.(6) becomes:

$$\tau \geq \tau_{kin} \tag{18}$$

4.2 Limit on the root mean square torque

When the direction of power flow during a working cycle is mainly direct or mainly backward, using the notation introduced in eq. (16), the root mean square torque can be expressed as:

$$T_{rms}^2 = \int_0^{t_a} \frac{T_M^2}{t_a} dt = \int_0^{t_a} \frac{1}{t_a} \left(\frac{\tau T_L^*}{\eta} + J_M \frac{\dot{\omega}_L}{\tau} \right)^2 dt \tag{19}$$

Developing the term in brackets and using the properties of the sum of integrals, it is possible to split the previous equation into the following terms:

$$\begin{aligned} \int_0^{t_a} \frac{1}{t_a} \left(\frac{\tau T_L^*}{\eta} \right)^2 dt &= \frac{T_{L,rms}^{*2} \tau^2}{\eta^2} \\ \int_0^{t_a} \frac{1}{t_a} \left(J_M \frac{\dot{\omega}_L}{\tau} \right)^2 dt &= \frac{J_M^2}{\tau^2} \dot{\omega}_{L,rms}^2 \\ \int_0^{t_a} \frac{1}{t_a} \left(\frac{2 T_L^* J_M \dot{\omega}_L}{\eta} \right)^2 dt &= 2 \frac{J_M}{\eta} \left(T_L^* \dot{\omega}_L \right)_{mean} \end{aligned} \tag{20}$$

In cases in which the power flow changes during the cycle it is not possible to choose the proper value of η and eq.(19) is no longer valid. In this circumstance, any individual working cycle must be analyzed to check if one of the two conditions (eq.(11) or eq.(12)) can be reasonably adopted. For example, in the case of purely inertial load ($T_L = 0$) it is[2]:

$$|T_{M,d}| \geq |T_{M,r}| \tag{21}$$

and the equation for direct power flow can be prudentially adopted. Same considerations can be done when the the load is mainly resistant (limited inertia). For all these cases, the condition of eq.(9) becomes:

$$\frac{T_{M,N}^2}{J_M} \geq \frac{\tau^2}{J_M} \frac{T^{*2}_{L,rms}}{\eta^2} + J_M \frac{\dot{\omega}_{L,rms}^2}{\tau^2} + 2 \frac{\left(T_L^* \dot{\omega}_L \right)_{mean}}{\eta} \tag{22}$$

4.2.1 The accelerating factor and the load factor

Considering eq.(22) two terms can be introduced: the *accelerating factor* Legnani et al. (2002)

$$\alpha = \frac{T_{M,N}^2}{J_M} \tag{23}$$

[2] For purely inertial load eq.(15) can be written as:

$$T_L^* = (J_L + J_T)\, \dot{\omega}_L$$

Then eq.(21) becomes:

$$\left(\frac{\tau (J_L + J_T)}{\eta_d} + \frac{J_M}{\tau} \right) |\dot{\omega}_L| \geq \left(\tau (J_L + J_T)\, \eta_r + \frac{J_M}{\tau} \right) |\dot{\omega}_L|$$

and thus:

$$\frac{1}{\eta_d} \geq \eta_r$$

which is always true since both η_d and η_r are defined as positive and smaller than 1.

that characterizes the performance of a motor Giberti et al. (2010b), and the *load factor*:

$$\beta = 2\left[\dot{\omega}_{L,rms}T^*_{L,rms} + (\dot{\omega}_L T^*_L)_{mean}\right] \tag{24}$$

defining the performance required by the task. The unit of measurement of both factors is *W/s*.

Coefficient α is exclusively defined by parameters related to the motor and therefore it does not depend on the machine task. It can be calculated for each motor using the information collected in the manufacturers' catalogs. On the other hand, coefficient β depends only on the working conditions (applied load and law of motion) and it is a term that defines the power rate required by the system.

4.2.2 Range of suitable transmission ratios

Introducing α and β, equation (22) becomes:

$$\alpha \geq \frac{\beta}{\eta} + \left[\frac{T^*_{L,rms}}{\eta}\left(\frac{\tau}{\sqrt{J_M}}\right) - \dot{\omega}_{L,rms}\left(\frac{\sqrt{J_M}}{\tau}\right)\right]^2 \tag{25}$$

Since the term in brackets is always positive, or null, the load factor β represents the minimum value of the right hand side of equation (25). It means that the motor accelerating factor α must be sufficiently greater than the load factor β/η, so that inequality (22) is verified. This check is a first criterion for discarding some motors. If $\alpha > \beta/\eta$, a range of useful transmission ratio exists and can be obtained by solving the biquadratic inequality:

$$\left(\frac{T^{*\,2}_{L,rms}}{\eta^2 J_M}\right)\tau^4 + \left(\frac{\beta}{\eta} - \alpha - 2\frac{T^*_{L,rms}}{\eta}\dot{\omega}_{L,rms}\right)\tau^2 + J_M\dot{\omega}^2_{L,rms} \leq 0 \tag{26}$$

Inequality (26) has 4 different real solutions for τ. As the direction of the rotation is not of interest, only the positive values of τ are considered. A range of suitable transmission ratios is included between a minimum τ_{min} and a maximum gear ratio τ_{max} for which the condition in equation (9) is verified:

$$\tau_{min}, \tau_{max} = \eta\sqrt{J_M}\frac{\sqrt{\alpha - \frac{\beta}{\eta} + \frac{4\dot{\omega}_{L,rms}T^*_{L,rms}}{\eta}} \pm \sqrt{\alpha - \frac{\beta}{\eta}}}{2T^*_{L,rms}} \tag{27}$$

$$\tau_{min} \leq \tau \leq \tau_{max} \tag{28}$$

>From equation 27 it is evident that a solution exists only if $\alpha \geq \frac{\beta}{\eta}$.

4.2.3 The optimum transmission ratio

The constraint imposed by equation (25) becomes less onerous when a suitable transmission is selected, with a transmission ratio τ that annuls the terms in brackets. This value of τ is called *optimum transmission ratio*. Considering an ideal transmission ($\eta = 1$, $J_T = 0\ kgm^2$) one gets:

$$\tau = \tau_{opt} = \sqrt{\frac{J_M\dot{\omega}_{L,rms}}{T^*_{L,rms}}} \tag{29}$$

that, for a purely inertial load ($T_L = 0$), coincides with the value introduced in Pasch et al. (1984):

$$\tau' = \sqrt{\frac{J_M}{J_L}} \tag{30}$$

The choice of the optimum transmission ratio allows system acceleration to be maximized (supplying the same motor torque) or to minimize the torque supplied by the motor (at the same acceleration).

For real transmissions, eq.(29) takes the general form:

$$\tau = \tau_{opt,\eta} = \sqrt{\frac{J_M \dot{\omega}_{L,rms}}{T^*_{L,rms}}\eta} = \tau_{opt}\sqrt{\eta} \tag{31}$$

showing the dependence of the optimum transmission ratios on the mechanical efficiency η. Eq. (31) shows how the optimization of the performance of the motor-reducer unit through the concept of inertia matching is considerably affected by the mechanical efficiency. In the following (par. 6.3) this effect will be graphically shown.

4.3 Limit on the motor maximum torque

As shown in Fig.3, each motor can supply a maximum torque $T_{M,max}(\omega_M)$ that depends on the speed ω_M. However this relationship cannot be easily described by a simple equation. As a result, it is difficult to express condition (10) as a range of suitable transmission ratios.

Moreover the maximum torque that can be exerted depends on the maximum current supplied by the drive system. For this reason, these conditions will be checked only once the motor and the transmission have been chosen. It has to be:

$$T_{M,max}(\omega_M) \geq \max\left|\frac{\tau T_L}{\eta} + \left(\frac{J_M + J_T}{\tau} + \frac{J_L \tau}{\eta}\right)\dot{\omega}_L\right| \qquad \forall\omega. \tag{32}$$

This test can be easily performed by superimposing the motor torque $T_M(\omega_M)$ on the motor torque/speed curve.

4.4 Checks

Once the transmission has been chosen, it is important to check the operating conditions imposed by the machine task satisfy the limits imposed by the manufacturers. Main limits concern:

- the maximum achievable speed;
- the maximum torque applicable on the outcoming shaft;
- the nominal torque applicable on the outcoming shaft;
- the maximum permissible acceleration torque during cyclic operation (over 1000/h) using the load factor.

There is no a standard procedure for checking the transmission. Each manufacturer, according to his experience and to the type of transmission produced, generally proposes an empirical procedure to check the right functioning of his products.

These verifications can be carried out by collecting information from catalogs.

5. Ideal transmission

If both the inertia of the transmission and the power losses are negligible, the gearbox can be considered as ideal ($J_T = 0$ kgm^2, $\eta = 1$). In this cases the problem is easier and the equations describing the dynamical behavior of the machine and the corresponding operating conditions can be simplified.

The limit on the maximum achievable speed remains the same (6), since it arises from kinematic relationships, while limits on the root mean square torque (7) and maximum torque (8) are simplified. Inequality (22) can be written as:

$$\frac{T_{M,N}^2}{J_M} \geq \tau^2 \frac{T^{*2}_{L,rms}}{J_M} + J_M \frac{\dot{\omega}^2_{L,rms}}{\tau^2} + 2(T_L^* \dot{\omega}_L)_{mean}. \tag{33}$$

whose solutions are between

$$\tau_{min}, \tau_{max} = \frac{\sqrt{J_M}}{2T^*_{L,rms}} \left[\sqrt{\alpha - \beta + 4\dot{\omega}_{L,rms} T^*_{L,rms}} \pm \sqrt{\alpha - \beta} \right]. \tag{34}$$

Accordingly, the condition on the maximum torque can be expressed as:

$$T_{M,max}(\omega_M) \geq \max \left| \tau T_L + \left(\frac{J_M}{\tau} + J_L \tau \right) \dot{\omega}_L \right| \qquad \forall \omega. \tag{35}$$

5.1 Selection of gearbox and motor

The main steps to select the gear-motor are:

STEP 1: Creation of a database containing all the commercially available motors and reducers useful for the application. For each motor the accelerating factor (α_i) must be calculated. Once the database has been completed it can be re-used and updated each time a new motor-reducer unit selection is needed.

STEP 2: Calculation of the load factor β, on the basis of the features of the load (T_L^*).

STEP 3: Preliminary choice of useful motors: all the available motors can be shown on a graph where the accelerating factors of available motors are compared with the load factor. All the motors for which $\alpha < \beta$ can be immediately rejected because they cannot supply sufficient torque, while the others are admitted to the next selection phase. Figure 7 is related to the industrial example discussed in par.5.2 and displays with circles the acceleration factors α_i of the analyzed motors, while the horizontal line represents the load factor β.

STEP 4: Identification of the ranges of useful transmission ratios for each motor preliminarily selected in step 3. For these motors a new graph can be produced displaying for each motor the value of the transmission ratios τ_{max}, τ_{min}, τ_{opt} and $\tau_{M,lim}$. The graph is generally drawn using a logarithmic scale for the y-axis, so τ_{opt} is always the midpoint of the adoptable transmission ratios range. In fact:

$$\tau_{opt}^2 = \tau_{min}\tau_{max} \quad \Leftrightarrow \quad \log \tau_{opt} = \frac{\log \tau_{min} + \log \tau_{max}}{2} \tag{36}$$

A motor is acceptable if there is at least a transmission ratio τ for which equations (18, 28) are verified. These motors are highlighted by a vertical line on the graph. Figure 8 is related to the same industrial example and shows the useful transmission ratios for each motor preliminarily selected.

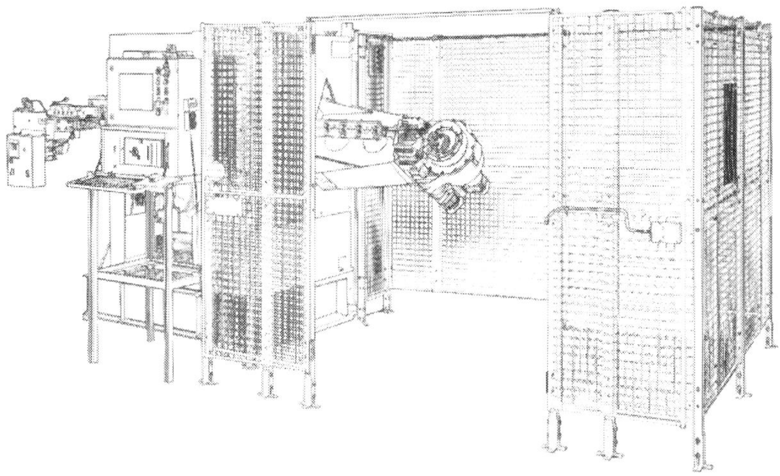

Fig. 5. CNC wire bending machine

STEP 5: Identification of the useful commercial speed reducers: the speed reducers available are represented by horizontal lines. If one of them intersects the vertical line of a motor, this indicates that the motor can supply the required torque if that specific speed reducer is selected. Table 2 sums up the acceptable combinations of motors and speed reducers for the case shown in figure 8. These motors and reducers are admitted to the final selection phase.

STEP 6: Optimization of the selected alternatives: the selection can be completed using different criteria such as economy, overall dimensions, space availability or any other depending on the specific needs.

STEP 7: Checks (see Sec.4.4).

5.2 Example

Fig. 5 shows a CNC wire bending machine. The system automatically performs the task of bending in the plane, or three-dimensionally, a wire (or tape) giving it the desired geometry. The machine operation is simple: semi-finished material is stored in a hank and is gradually unrolled by the unwinding unit. The straightened wire is guided along a conduit to the machine's bending unit, which consists of a rotating arm on which one or more bending heads are mounted. Each head is positioned in space by a rotation of the arm around the axis along which the wire is guided in order to shape it in all directions 6.

The production capacity of the machine is related to the functionality of the heads, while bending productivity depends strongly on arm speed, which allows the heads to reach the position required for bending. The design of the system actuating the rotating arm (selection of motor and speed reducer) is therefore one of the keys to obtaining high performance.

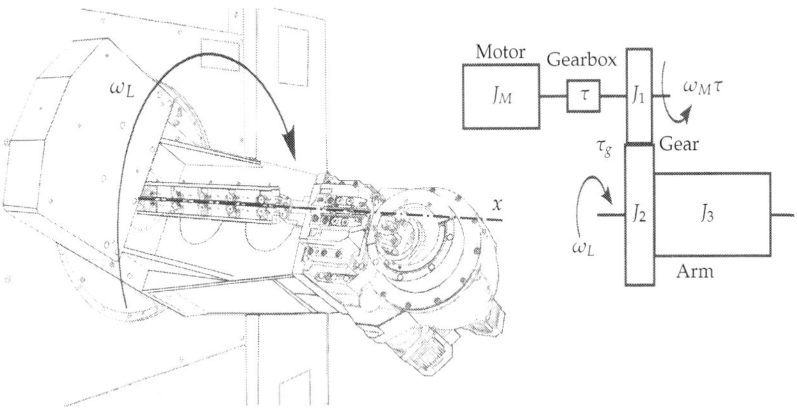

Fig. 6. The moving arm and mechanical system layout

Consider now only the bending unit: the motor, with its moment of inertia J_M, is connected through a planetary reducer with transmission ratio τ to a pair of gear wheels that transmits the rotation to the arm.

The pair of gear wheels has a ratio $\tau_2 = 1/5$ which is dictated by the overall dimension of the gearbox and cannot be modified. Putting $J_1 = 0.0076$ [kgm^2], $J_2 = 1.9700$ [kgm^2] and $J_3 = 26.5$ [kgm^2] respectively as the moment of inertia referred to the axes of rotation of the two wheels and of the arm, the comprehensive moment of inertia referred to the output shaft of the planetary gear is:

$$J_L = J_1 + (J_2 + J_3)\tau_2^2 = 1.1464 \text{ [kgm}^2\text{]}$$

Since the load is purely inertial[3], the load factor can easily be calculated as:

$$\beta = 4 J_L \dot{\omega}_{L,rms}^2 \tag{37}$$

where $\dot{\omega}_{L,rms}$ is a function of the law of motion used. The choice of the law of motion depends on the kind of operation requested and, in the most extreme case, it consists of a rotation of $h = 180°$ in $t_a = 0.6$ [s]. After this, a stop of $t_s = 0.2$ [s] before the next rotation is normally scheduled.

The value of $\dot{\omega}_{L,rms}$ can be expressed through the mean square acceleration coefficient ($c_{a,rms}$) using the equation:

$$\dot{\omega}_{L,rms} = c_{a,rms} \frac{h}{t_a^2} \frac{1}{\tau_2} \sqrt{\frac{t_a}{t_a + t_s}}$$

As is known (Van de Straete et al., 1999), the minimum mean square acceleration law of motion is the cubic equation whose coefficient is $c_{rms} = 2\sqrt{3}$. Moreover, this law of motion has the advantage of higher accelerations, and therefore high inertial torques, corresponding to low velocities. Substituting numerical values in eq.(37) one gets: $\beta = 7.8573 \cdot 10^4$ [W/s].

[3] In the selecting phase frictions are not considered.

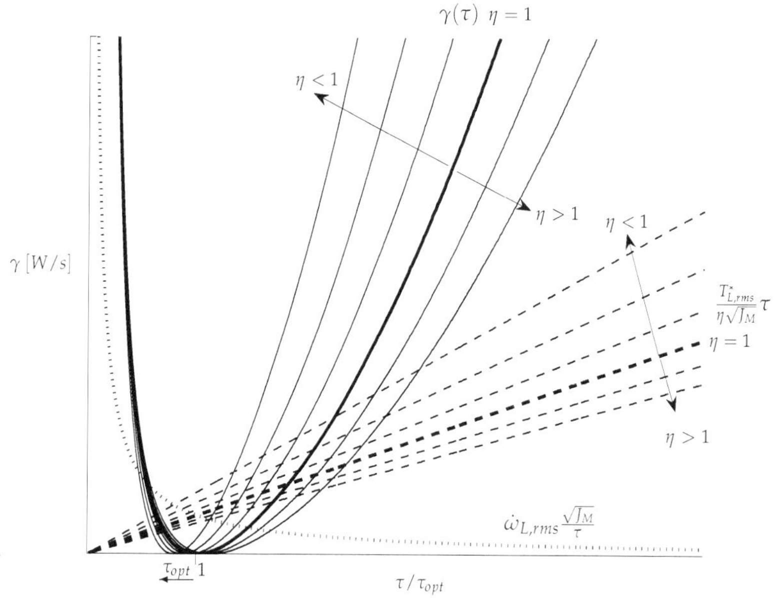

Fig. 11. Variation of the extra-power rate factor as a function of the transmission ratio and mechanical efficiency.

7. Guidelines for the motor-reducer selection

The theoretical steps presented can be summarized by a series of graphs for evaluating the effect of the transmission on the choice of the motor that make the selection process easy to use.

Firstly, to ensure the condition on the root mean square torque it should be verified that the accelerating factor α of each available motor is greater than the limit β/η. However, since the transmission has not yet been selected and thus its efficiency and inertia are still unknown, it is possible to perform only a first selection of acceptable motors, eliminating those for which $\alpha < \beta$.

For each selectable motor a $\eta\tau$ graph can be plotted (Fig. 14), with the limits on the transmission ratio defined by equations (18), (28). Since these functions depend on the transmission moment of inertia, such limits are plotted for each available transmission.

Available transmissions can be inserted in the $\eta\tau$ graph (there is an example in Fig. 14) using their coordinates (η_{di}, τ_i) and (η_{ri}, τ_i) which can easily be found in manufacturers' catalogs. Each speed reducer appears twice: to the left of the dashed line for direct power flow, to the right for backward power flow.

Remember that for tasks characterized by mainly backward power flow, only the transmissions on the right half plane should be considered ($\eta > 1$). For all other cases only the transmissions on the left half plane ($\eta < 1$) should be taken into account.

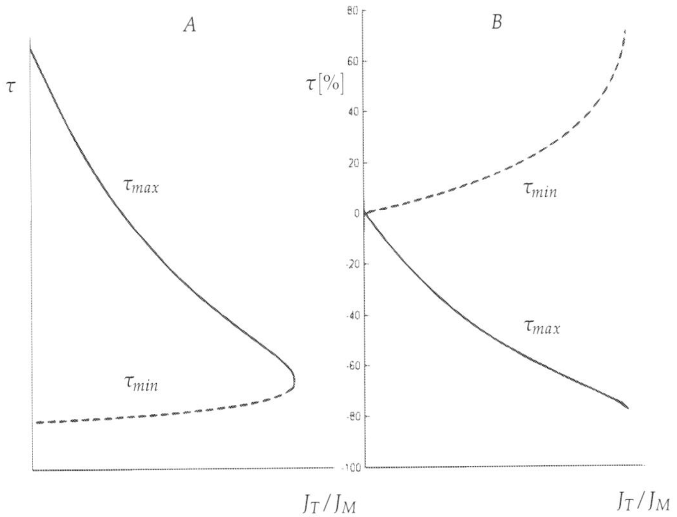

Fig. 12. A - reduction of range of useful transmission ratio as function of J_T/J_M (for T_L = cost); B - percentage change of maximum and minimum transmission ratios as function of J_T/J_M (for T_L = cost).

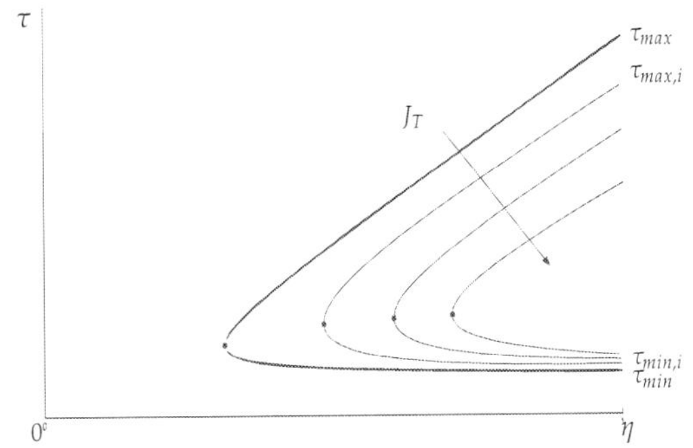

Fig. 13. Effect of the transmission moment of inertia on the $\eta\tau$ graph (for T_L = cost).

The ith speed reducer is acceptable if it lies inside the area limited by the limit transmission ratio (τ_{kin}) and the corresponding maximum and minimum ratios ($\tau_{max,i}$, $\tau_{min,i}$). These reducers are highlighted on the graph with the symbol ⊙, unacceptable ones with ⊗

Table 4 resumes all the alternatives both for the direct and backward power flow modes.

Note that, for the direct power flow mode, transmissions which would be acceptable with normal selection procedures, are now discarded (e.g. transmission T_2 because of insufficient efficiency, reducer T_4 because of its excessively high moment of inertia).

Moreover it is evident that, for the motor considered, there are no acceptable transmissions with a ratio equal to the optimum. More generally it could happen that a motor, while meeting

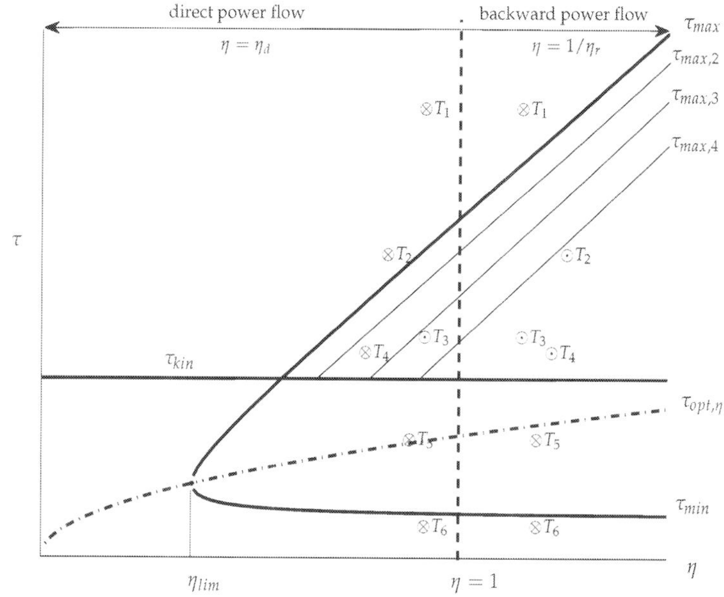

Fig. 14. Graph $\eta\tau$ for the selection of the speed reducer for a specific motor: ($\odot$ acceptable transmission, $\otimes$ unacceptable transmission)

Transmission	Direct P.F.	Cause	Backward P.F.	Cause
T_1	unaccept.	$\tau_1 > \tau_{max}$	unaccept.	$\tau_1 > \tau_{max}$
T_2	unaccept.	$\tau_2 > \tau_{max}$	accept.	—
T_3	accept.	—	accept.	—
T_4	unaccept.	$\tau_4 > \tau_{max,4}$	accept.	—
T_5	unaccept.	$\tau_5 < \tau_{kin}$	unaccept.	$\tau_5 < \tau_{kin}$
T_6	unaccept.	$\tau_6 < \tau_{kin} \wedge \tau_6 < \tau_{min}$	unaccept.	$\tau_6 < \tau_{kin} \wedge \tau_6 < \tau_{min}$

Table 4. Overview of acceptable and unacceptable transmissions for the example in Fig. 14

all the constraints mentioned, may have a range of acceptable gear ratios within which there is no reducer commercially available. For this reason it cannot be chosen.

The choice of the motor-reducer unit is made easy by comparing the $\eta\tau$ graphs for each selectable motor.

The resulting graphs give an overview of all the possible pairings of motors and transmissions that satisfy the original conditions. They allow the best solution to be selected from the available alternatives, in terms of cost, weight and dimensions, or other criteria considered important according to the application.

Once the motor and the transmission have been selected and all their mechanical properties are known, the final checks can be performed.

8. Conclusions

The correct choice of the motor-reducer unit is a key factor in automation applications. Such a selection has to be made taking into account the mechanical constraints of the components, in particular the operating ranges of the drive system and the mechanical features of the transmission. The paper investigates the effects of these constraints on the correct choice of the motor-reducer unit at the theoretical level and illustrated a method for its selection that allows the best available combination to be chosen using a practical approach to the problem.

It identifies the influence of the transmission's mechanical efficiency and inertia on the coupling between motor and reducer itself, showing how they affect the optimum solution. The procedure, based on the production of a chart containing all the information needed for the correct sizing of the system, sums up all the possible solutions and allows them to be quickly compared to find the best one.

9. Nomenclature

Symbol	Description
T_M	motor torque
J_M	motor moment of inertia
$T_{M,N}$	motor nominal torque
$T_{M,max}$	motor maximum torque
$\omega_M, \dot{\omega}_M$	motor angular speed and acceleration
T_L	load torque
J_L	load moment of inertia
T_L^*	generalized load torque
$T_{L,rms}^*$	generalized load root mean square torque
$T_{L,max}$	load maximum torque
$\omega_L, \dot{\omega}_L$	load angular speed and acceleration
$\dot{\omega}_{L,rms}$	load root mean square acceleration
t_a	cycle time
$\tau = \omega_L/\omega_M$	transmission ratio
τ_{opt}	optimal transmission ratio
η	transmission mechanical efficiency
η_d	transmission mechanical efficiency (direct power flow)
η_r	transmission mechanical efficiency (backward power flow)
J_T	transmission moment of inertia
α	accelerating factor
β	load factor
γ	extra-power rate factor
τ_{min}, τ_{max}	minimum and maximum acceptable transmission ratio
τ_{kin}	minimum kinematic transmission ratio
$\omega_{M,max}$	maximum speed achievable by the motor
$\omega_{L,max}$	maximum speed achieved by the load
W_M	motor side power
W_L	load side power

10. References

Pasch, K.A., Seering, W.P., "On the Drive Systems for High-Performance Machines", *Transactions of ASME*, 106, 102-108 (1984).

Cusimano, G., "A Procedure for a Suitable Selection of Laws of Motion and Electric Drive Systems Under Inertial Loads", *Mechanism and Machine Theory* 38, 519-533 (2003)

Chen, D.Z., Tsai, L.W., "The Generalized Principle of Inertia Match for Geared Robotic Mechanism", *Proc. IEEE International Conference on Robotics and Automation* , 1282-1286 (1991)

Van de Straete, H.J, Degezelle, P., De Shutter, J., Belmans, R., "Servo Motor Selection Criterion for Mechatronic Application", *IEEE/ASME Transaction on Mechatronics* 3, 43-50 (1998).

Van de Straete, H.J, De Shutter, J., Belmans, R., "An Efficient Procedure for Checking Performance Limits in Servo Drive Selection and Optimization", *IEEE/ASME Transaction on Mechatronics* 4, 378-386 (1999)

Van de Straete, H.J, De Shutter, J., Leuven, K.U., "Optimal Variable Transmission Ratio and Trajectory for an Inertial Load With Respect to Servo Motor Size", *Transaction of the ASME* 121, 544-551 (1999)

Giberti, H, Cinquemani, S., Legnani, G., "A Practical Approach for the Selection of the Motor-Reducer Unit in Electric Drive Systems" *Mechanics Based Design of Structures and Machines*, Vol. 39, Issue 3, 303-319 (2011)

Giberti, H, Cinquemani, S., Legnani, G., "Effects of transmission mechanical characteristics on the choice of a motor-reducer", *Mechatronics* Vol. 20, Issue 5, 604-610 (2010)

Giberti, H, Cinquemani, "On Brushless Motors Continuous Duty Power Rate", *ASME 2010 10th Biennial Conference on Engineering Systems Design and Analysis, ESDA 2010*, Istanbul, Turkey, 12-14 July (2010).

Legnani, G., Tiboni, M., Adamini, R., "Meccanica degli Azionamenti", Ed. Esculapio, Italy, (2002)

Roos, F., Johansson, H., Wikander, J., "Optimal Selection of Motor and Gearhead in Mechatronic Application", *Mechatronics* 16, 63-72 (2006)

Magnani P.L., Ruggieri G. Meccanismi per Macchine Automatiche, UTET, Italy (1986)

Melchiorri C., Traiettorie per azionamenti elettrici, Esculapio, Italy (2000)

6

Electrical Drives for Crane Application

Nebojsa Mitrovic[1], Milutin Petronijevic[1],
Vojkan Kostic[1] and Borislav Jeftenic[2]
[1]*University of Nis, Faculty of Electronic Engineering,*
[2]*University of Belgrade, Faculty of Electrical Engineering,*
Serbia

1. Introduction

A crane is the type of machine mainly used for handling heavy loads in different industry branches: metallurgy, paper and cement industry. By the construction, cranes are divided into the overhead and gantry cranes. An overhead crane, also known as a bridge crane, is a type of crane where the hook and line mechanism runs along a horizontal beam that itself travels on the two widely separated rails. Often it is in a factory building and runs along rails mounted on the two long walls. A gantry crane is similar to an overhead crane designed so that the bridge carrying the trolley is rigidly supported on two or more legs moving on fixed rails embedded in the floor. Stationary or mobile units can be installed outdoors or indoors. Some industries, for example port containers application or open storage bins, require wide span gantry cranes. In outdoor applications, the influence of the wind on the behavior of the drive may be considerable (Busschots, 1991). Wind and skew can significantly influence a safe operation of the crane. This will certainly dispose the type design of the crane (lattice or box type design) from a mechanical aspect as well as the selection, size and control of crane electrical drives.

Electrical technology for crane control has undergone a significant change during the last few decades. The shift from Ward Leonard system to DC drive technology and the advent of powerful Insulated Gate Bipolar Transistors (IGBTs) during the 1990s enabled the introduction of the AC drive (Backstrand, 1992; Paul et al., 2008). Conventional AC operated crane drives use slip ring induction motor whose rotor windings are connected to power resistance in 4 to 5 steps by power contactors. Reversing is done by changing the phase sequence of the stator supply through line contactors. Braking is achieved by plugging. The main disadvantage is that the actual speed depends on the load. An electronic control system has recently been added to continuously control rotor resistor value. Nowadays, these systems are replaced by frequency converters supplied squirrel-cage induction motors for all types of motion (Paul et al., 2008). Control concept based on application of Programmable Logic Controllers (PLC) and industrial communication networks (Field-buses) are a standard solution which is used in complex applications (Slutej et al.,1999).

An overhead and gantry cranes are typically used for moving containers, loading trucks or material storage. This crane type usually consists of three separate motions for transporting material. The first motion is the hoist, which raises and lowers the material. The second is the trolley (cross travel), which allows the hoist to be positioned directly above the material for placement. The third is the gantry or bridge motion (long travel), which allows the entire crane to be moved along the working area. Very often, in industrial applications additional drives as auxiliary hoist, power cable reel and conveyer belt are needed. Therefore, generally, a crane is complex machinery.

Depending on the crane capacity each of the mentioned drives, can be realized as multi-motor. The term multi-motor drive is used to describe all the drives in a technological process. If the controlled operation of the drives is required by the process based on the controlled speed of the individual drives, the expression controlled multi-motor drives is adequate. For many of such drives, the mechanical coupling on the load side is typical (Jeftenic et al., 2006; Rockwell, 2000). In applications with cranes, coupling of the individual motors is realized by the mechanical transmition device, and it is usually technologically unbreakable.

2. Possible load sharing configurations overview

Controlled drives are usually fed from the power converter, which is also true for controlled multi-motor drives. The kind, the type and the number of converters used depend on the type of motors, their power ratings, and of the kind of the multi-motor drive. The control and regulation also depend on the type of the multi-motor drive, but also on the type of the converter selected, therefore the selection of the converter and the controller for these drives must be analyzed together. Regarding the power supply of the motor, the following cases are possible (Jeftenic et al., 2006):

- multiple motors fed by a single converter (multiple motors - single converter),
- motors controlled by separate converters (multiple motors - multiple converters).

In crane applications multi-motor drives are used very often and a proportional share of power between motors is required. Load sharing is a term used to describe a system where multiple converters and motors are coupled and used to run one mechanical load (Rockwell, 2000). In the strictest sense, load-sharing means that the amount of torque applied to the load from each motor is prescribed and carried out by each converter and motor set. Therefore, multiple motors and converters powering the same process must contribute its proportional share of power to the driven load.

Multiple motors that are run from a single converter do not load share because torque control of individual motors is not possible. The load distribution, in that case, is influenced only by the correct selection of the torque-speed mechanical characteristic. For the squirrel-cage induction motors, there is no economical method for the adjustment of the mechanical characteristic of the ready-made motors, but this has to be done during the selection. For the slip-ring induction motor, the mechanical characteristic can be adjusted afterwards, with the inclusion of the rotor resistors. Motors that are controlled by separate converters without any interconnection also do not share the load. The lack of interconnection defeats any possible comparison and error signal generation that is required to compensate for the differences in the load that is applied to any single drive and motor set.

Control topologies for load sharing consider the presence of interconnection, i.e. information knowledge about load (motor current or torque). There are three categories of load sharing techniques: common speed reference, torque follower and speed trim follower.

The common speed reference is the simplest the least precise and the less flexible form of load sharing to set up, Fig. 1a). The precision of this control depends on the drives control algorithm, the motor characteristics and the type of load to be controlled.

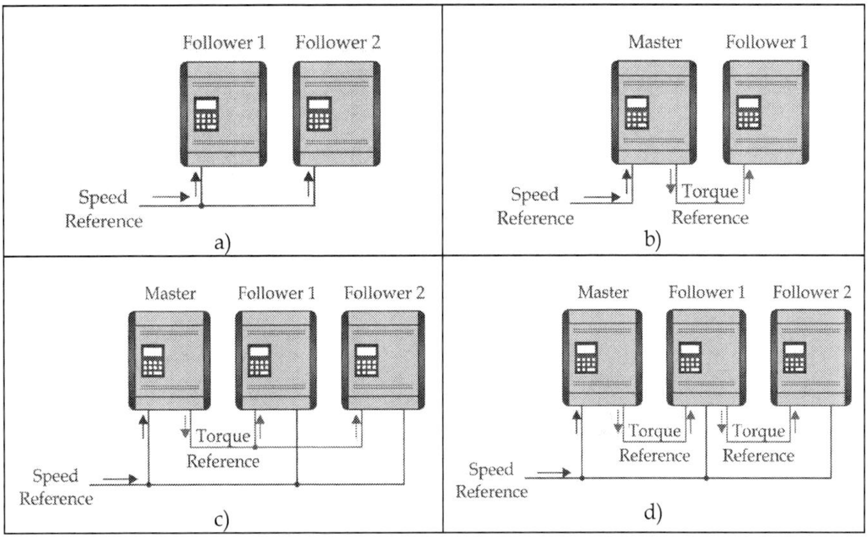

Fig. 1. Load sharing configuration a) Common speed reference, b) Torque follower, c) and d) Speed trim follower.

The torque follower type of load sharing requires the frequency converter to have the capability of operation in "torque mode", Fig. 1b). If speed regulation is required, one of the converters ("master") may be in "speed mode". In speed mode controller provides a torque command at output which can be distributed to the other converters ("slaves" or "torque followers"). The second converter operates in torque regulation mode with the torque reference of the master as command. This torque signal may be scaled to divide load sharing in any desired ratio.

In speed trim follower configuration, Fig. 1.c) and d), all converters are operated in speed regulation mode and receive the same speed reference. The torque reference of the master is sent to the follower converters. Each follower converter compares its own torque reference with that of the master, Fig. 1c). The output of the comparator is an error signal that trims the speed of the follower. Alternative configuration cascades the torque reference comparison, Fig. 1d). The first follower compares the master to its internal value. The second follower compares the foregoing follower to its internal value etc.

2.1 Torque and power requirements for crane drives

Speed control is an essential feature in crane drives. It is required for allowing soft starting and stopping of the travel motions for enabling its correct positioning of load. For the lifting drive the speed control in a wide speed range, from zero to nominal values, is required.

Because of the precision when raising and lowering load, the possibility of working at a very low speed and hold a load in the standstill is required, without using the mechanical brakes.

The torque and power that have to be delivered by the drive may be obtained from the torque versus speed characteristic from the load (so-called mechanical characteristics) and the differential equation of motion, (Belmans et al., 1993; Mitrovic et al., 2011).

The differential equation of motion, describing the behavior of the drive is:

$$J\frac{d\omega}{dt} = T_e - T_l \quad (1)$$

where T_e is the electromagnetic torque of the motor, T_l is the torque of the load, J is the inertia of the drive. If $T_e > T_l$, the system accelerates ($d\omega/dt > 0$), if $T_l/T_e < 0$ it decelerates ($d\omega/dt>0$). The stedy state operation is reached if $T_l = T_e$ and $\omega = const.$

Multiplying equation (1) by the rotating speed ω, yields the power:

$$P_e = \omega T_e = \omega T_l + \omega J\frac{d\omega}{dt} = P_m + P_d \quad (2)$$

This equation shows that the mechanical power P_e=ωT_e, obtained after the electromechanical conversion in the motor, is equal to the power absorbed by the load P_m=ωT_l only when the speed does not change. Otherwise, the amount corresponding to change in kinetic energy must be added (if the speed increases) or subtracted (if the speed decreases):

$$P_d = \frac{dE_{kin}}{dt} = \omega\frac{Jd\omega}{dt} \quad (3)$$

In the following, the travel and hoist motion of the crane drives will be analyzed.

The mechanical characteristic is given in Fig.2a) for travel motion. Apart from the zone around zero, the torque is constant. The available torque is used for accelerating the system. For a travel in one direction, braking and reversing to full speed in other direction, the speed reference signal is given by top curve of Fig.2b). The torque reference signal is generated by converter (second curve), leading to the machine actual speed. Multiplying the actual and torque reference, yields the actual power (third curve). The peak power is found at the end of the acceleration period. If wind forces are taken into consideration, the torque versus speed curve is shifted horizontally as shown in Fig.2c). The torque and speed reference remain the same, as well as the actual speed. However, the torque reference and the actual power differ, as shown on Fig.2d).

During acceleration, kinetic energy is stored in the system. To stop the crane, this energy must be absorbed by the drive. In the indoor situation, this energy is well known and only present for a short period of time. For outdoor applications, the wind forces may become very important. When travelling in the same direction as the wind, the wind drives the crane and a situation may occur, where a continuous electrical braking is required. The drive must be capable of handling this inverse power direction either by consuming the power in a resistor or preferably by feeding it back to the supply.

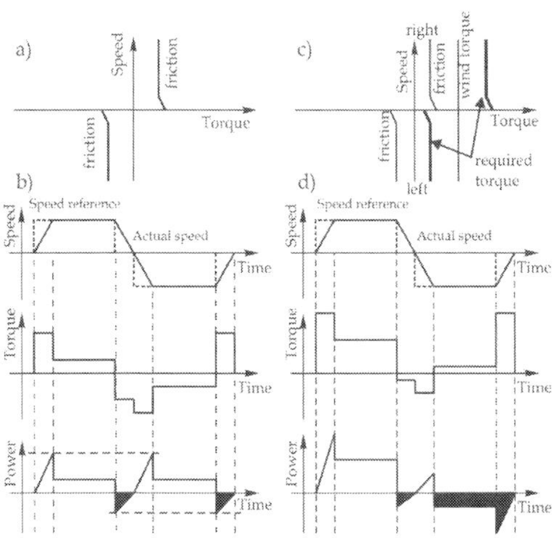

Fig. 2. Power and torque requirements for travel motion, a) and b) without wind influence, c) and d) with wind influence.

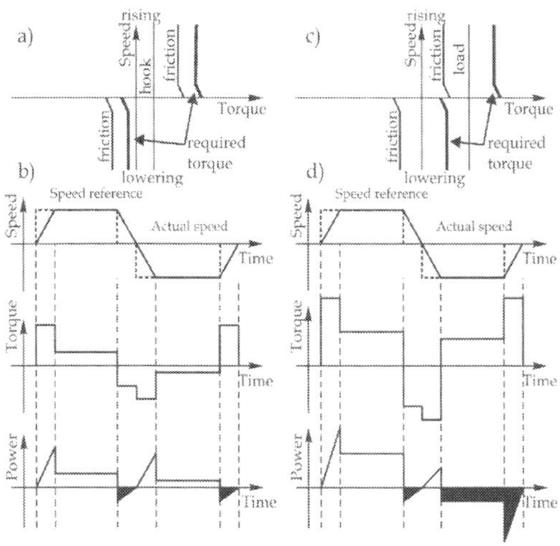

Fig. 3. Power and torque requirements for hoist a) and b) without load, c) and d) with load.

The hoist torque-speed characteristic is shown in Fig.3a) for an unloaded hook. The characteristic resembles the one for the travel motion. However, it is always asymmetric with respect to the vertical axis, due to the gravitation force. This asymmetry becomes more pronounced when the hook is loaded (Fig.3c). For both unloaded and loaded situation, the speed, torque and power are given in Fig.3b) and Fig.4d). Again the amount of braking power is indicated. The worst braking case with a hoist motion, is when sinking a loaded hook. It should be noted that the weight of the hook may be considerable. The hook may be simple, or may consist of several parts to handle the load. For the hoist motion, the speed control is very important in the low speed range: avoiding damage to the load when putting it down and minimizing the stress on the mechanical brakes.

2.2 Frequency converters for induction motor drives

Today's adjustable speed drives (ASD) in the low and mid power range are normally based on the concept of variable voltage, variable frequency (VVVF). Fig.4 shows the basic concept of a single variable speed drive. The three-phase AC supply network is rectified. The DC capacitor, which links the supply rectifier to the inverter, assures that the inverter sees a constant DC voltage from which it generates the required supply voltage and frequency to the motor.

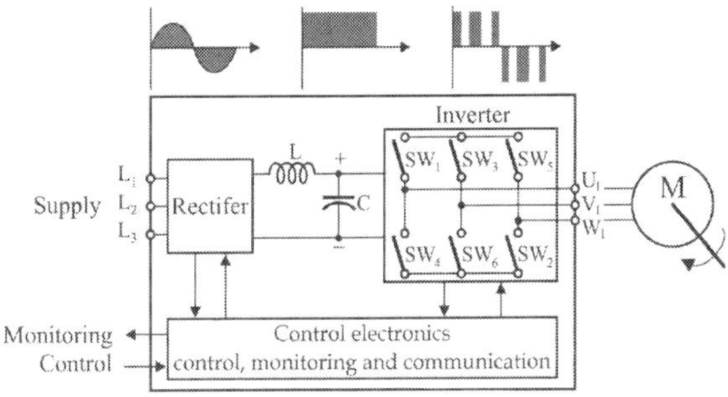

Fig. 4. Basic concept of a variable speed drive.

General classification divides induction motor control schemes into scalar and vector-based methods (Petronijevic et al., 2011). Opposite to scalar control, which allows control of only output voltage magnitude and frequency, the vector-based control methods enable control of instantaneous voltage, current and flux vectors. In numerous industrial applications, such as HVAC (heating, ventilation and air conditioning), fan or pump applications, good dynamic performances are not usually the main control objective. Fig. 5a) illustrates a V/Hz open loop control scheme, where pulse-width modulation (PWM) is realised applying the space vector technique (SVPWM) and output voltage fundamental component amplitude is modified with a voltage drop compensation U_{s0} at low output frequencies. Basic control structure of rotor field oriented control (RFO) is illustrated in Fig. 5b). Two inner PI-controlled current loops for d and q stator current components are shown, as well as synchronous speed estimator (ω_e, based on reference stator currents components). In its basic version, direct torque control (DTC) consists of a three-level hysteresis comparator for torque control and a two-level hysteresis comparator for flux control as shown in Fig.5.c).

Type of the front end converter, regardless of the control schemes, depends on the power and torque requirements of the drive. Adjustable speed drives in industrial applications are usually characterized by a power flow direction from the AC distribution system to the load (Rashid, 2001). This is, for example, the case of an ASD operating in the motoring mode. In this instance, the active power flows from the DC side to the AC side of the inverter. However, there are an important number of applications in which the motor begins to act as a generator and regenerates energy back into the DC bus of the drive. Moreover, this could be a transient condition as well a normal operating condition. This is known as the regenerative operating mode. For example, these regenerative conditions can occur when quickly decelerating a high inertia load and this can be considered as transient condition. The speed control of a load moving vertically downward (hoist) can be considered as normal operating condition.

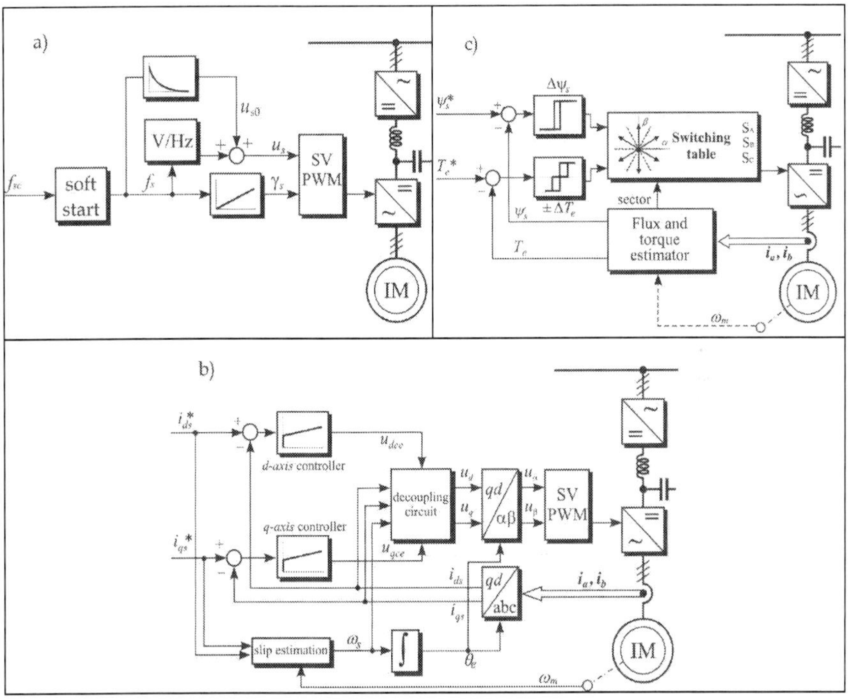

Fig. 5. Control scheme: a) V/Hz, b) vector control, c) DTC control.

Drive applications can be divided into three main categories according to speed and torque. The most common AC drive application is a single quadrant application where speed and torque always have the same direction, i.e. the power flow from inverter to process. The second category is two-quadrant applications where the direction of rotation remains unchanged but the direction of torque can change, i.e. the power flow may be from drive to motor or vice versa. The third category is fully four-quadrant applications where the direction of speed and torque can freely change. These applications are typically elevators, winches and cranes (ABB, 2011).

In order for an AC drive to operate in quadrant II or IV in speed-torque plane, a means must exist to deal with the electrical energy returned to the drive by the motor. The typical pulse width modulated AC drive is not designed for regenerating power back into the three phase supply lines. Electrical energy returned by the motor can cause voltage in the DC link to become excessively high when added to existing supply voltage. Various drive components can be damaged by this excessive voltage. These regenerative conditions can occur when:

- quickly decelerating a high inertia load,
- controlling the speed of a load moving vertically downward (hoist, declining conveyor),
- a sudden drop in load torque occurs,
- the process requires repetitive acceleration and deceleration to a stop,
- controlling the speed of an unwind application.

In standard drives the rectifier is typically a 6-pulse diode rectifier only able to deliver power from the AC network to the DC bus but not vice versa. If the power flow changes as in two or four quadrant applications, the power fed by the process charges the DC capacitors and the DC bus voltage starts to rise. The capacitance is a relatively low value in an AC drive resulting in fast voltage rise, and the components of a frequency converter may only withstand voltage up to a certain specified level.

In order to prevent the DC bus voltage rising excessively, the inverter itself prevents the power flow from process to frequency converter. This is done by limiting the braking torque to keep a constant DC bus voltage level. This operation is called overvoltage control and it is a standard feature of most modern drives. However, this means that the braking profile of the machinery is not done according to the speed ramp specified by the user.

There are two technologies available to prevent the AC drive from reaching the trip level: Dynamic braking and active front end regeneration control. Each technology has its own advantages and disadvantages.

3. Dynamic braking

A dynamic brake consists of a chopper and a dynamic brake resistor. Fig.6 shows a simplified dynamic braking schematic. The chopper is the dynamic braking circuitry that senses rising DC bus voltage and shunts the excess energy to the dynamic brake resistor. A chopper contains three significant power components: The chopper transistor is an IGBT. The chopper transistor is either ON or OFF, connecting the dynamic braking resistor to the DC bus and dissipating power, or isolating the resistor from the DC bus. The current rating of the chopper transistor determines the minimum resistance value used for the dynamic braking resistor. The chopper transistor voltage control regulates the voltage of the DC bus during regeneration. The dynamic braking resistor dissipates the regenerated energy in the form of heat.

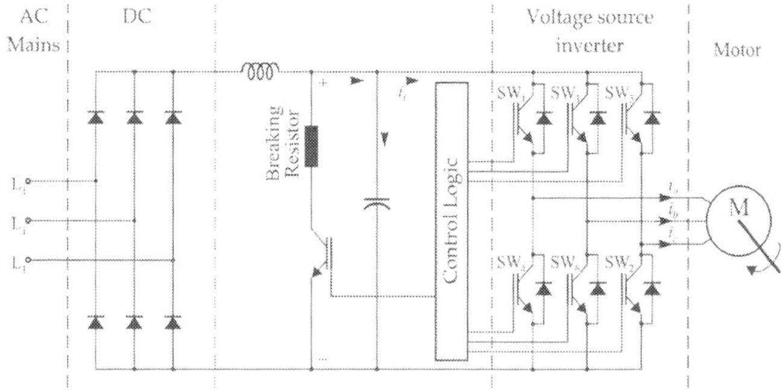

Fig. 6. Voltage source inverter with diode front end rectifier and dynamic brake module.

As a general rule, dynamic braking can be used when the need to dissipate regenerative energy is on an occasional or periodic basis. In general, the motor power rating, speed, torque, and details regarding the regenerative mode of operation will be needed in order to estimate what dynamic braking resistor value is needed. The peak regenerative power of the drive must be calculated in order to determine the maximum resistance value of the dynamic braking resistor.

The peak breaking power required to decelerate the load, according to equation (4) is:

$$P_b = \frac{J\,\omega_b(\omega_b - \omega_0)}{t_b} \tag{4}$$

where t_b represents total time of deceleration, ω_b and ω_0 initial and final speed in the process of braking.

The value of P_b can now be compared to the drive rating to determine if external braking module is needed. If peak braking power is 10% greater than rated drive power external braking module is recommended. Compare the peak braking power to that of the rated motor power, if the peak braking power is greater than 1.5 time that of the motor, then the deceleration time, needs to be increased so that the drive does not go into current limit.

The peak power dynamic brake resistance value can be calculated as:

$$R_{db} = \frac{V_{dc}^2}{P_b} \tag{5}$$

The choice of the dynamic brake resistance value should be less than the value calculated by equation (5). If a dynamic braking resistance value greater than the ones imposed by the choice of the peak regenerative power is made and applied, the drive can trip off due to transient DC bus overvoltage problems. Once the approximate resistance value of the dynamic braking resistor is determined, the necessary power rating of the dynamic braking resistor can be calculated. The power rating of the dynamic braking resistor is estimated by applying what is known about the drive's motoring and regenerating modes of operation.

To calculate the average power dissipation the braking duty cycle must be determined. The percentage of time during an operating cycle (t_c) when braking occurs (t_b) is duty cycle ($\varepsilon = t_b / t_c$). Assuming the deceleration rate is linear, average power is calculated as follows:

$$P_{av} = \frac{t_b}{t_c}\frac{P_b}{2}\frac{\omega_b + \omega_0}{\omega_b} \tag{6}$$

Steady state power dissipation capacity of dynamic brake resistors must be greater than that average. If the dynamic braking resistor has a large thermodynamic heat capacity, then the resistor element will be able to absorb a large amount of energy without the temperature of the resistor element exceeding the operational temperature rating.

Fig.7a) shows the experimental results (DC voltage and chopper current) for the variable frequency drive with braking module in DC link and external braking resistor, under a step change of induction motor load in regenerative regime. Danfoss frequency (series VLT 5000) converter is used in experimental set-up. For the supply voltage of 400 V, DC link voltage is about 540 V. When negative load torque is applied, DC link voltage rises. The chopper transistor voltage control regulates the voltage of the DC bus during regeneration to near 800 V allowing current flow in the resistor. Regenerative energy is then realised into heat. After the end of the regenerative period, DC voltage returns to a value that corresponds to a motor regime. The Fig.7b) shows the line voltage and current at the input of the diode rectifier.

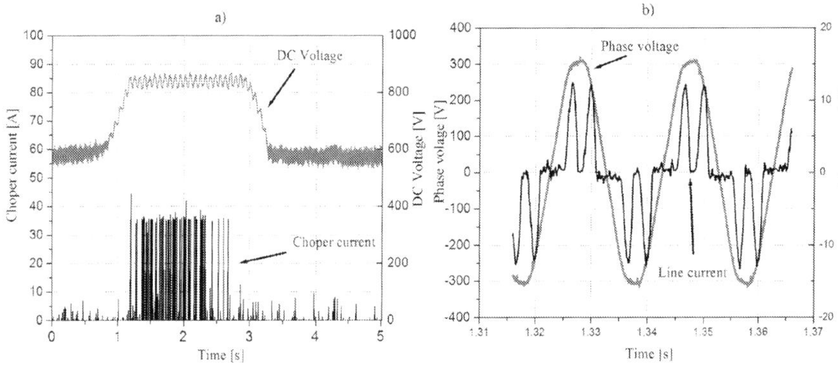

Fig. 7. a) DC voltage and chopper current, b) line voltage and current.

A voltage source PWM inverter with diode front-end rectifier is one of the most common power configurations used in modem variable speed AC drives, (Fig. 6). An uncontrolled diode rectifier has the advantage of being simple, robust, and low cost. However, it allows only unidirectional power flow. Therefore, energy returned from the motor must be dissipated on a power resistor controlled by a chopper connected across the dc link. A further restriction is that the maximum motor output voltage is always less than the supply voltage.

4. Active front end rectifier

Various alternative circuits can be used to recover the load energy and return it to power supply. One such scheme is shown in Fig. 8 and presents the most popular topology used in ASD. The diode rectifier is replaced with PWM voltage source rectifier. This is already an industrially implemented technology and known as most successful active front end (AFE) solution in ASD if regenerative operation is needed (e.g. for lowering the load in crane) and therefore was chosen by most global companies: Siemens, ABB, and others.

The term Active Front End Inverter refers to the power converter system consisting of the line-side converter with active switches such as IGBTs, the DC link capacitor bank, and the load-side inverter. The line-side converter normally functions as a rectifier. But, during regeneration it can also be operated as an inverter, feeding power back to the line. The line-side converter is popularly referred to as a PWM rectifier in the literature. This is due to the fact that, with active switches, the rectifier can be switched using a suitable pulse width modulation technique.

The PWM rectifier basically operates as a boost chopper with AC voltage at the input, but DC voltage at the output. The intermediate DC-link voltage should be higher than the peak of the supply voltage. The required DC-link voltage needs be maintained constant during rectifier as well as inverter operation of the line side converter. The ripple in DC link voltage can be reduced using an appropriately sized capacitor bank. The AFE inverter topology for a motor drive application, as shown in Fig.8, has two three-phase, two-level PWM converters, one on the line side, and another on the load side. The configuration uses 12 controllable switches. The line-side converter is connected to the utility through inductor. The inductor is needed for boost operation of the line-side converter.

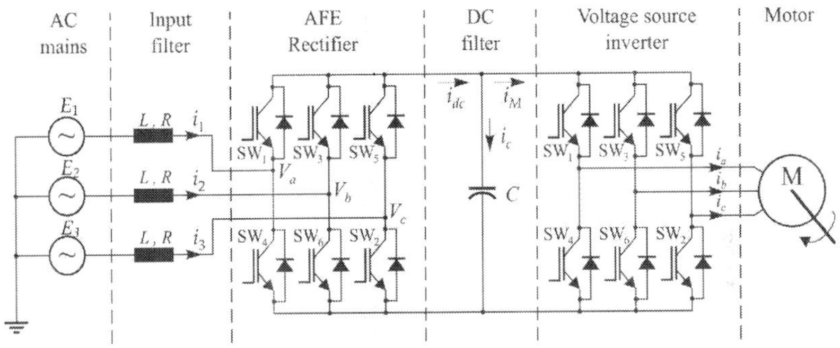

Fig. 8. Active front end inverter topology.

For a constant dc-link voltage, the IGBTs in the line-side converter are switched to produce three-phase PWM voltages at *a*, *b*, and *c* input terminals. The line-side PWM voltages, generated in this way, control the line currents to the desired value. When DC link voltage drops below the reference value, the feed-back diodes carry the capacitor charging currents, and bring the DC-link voltage back to reference value.

The steady state characteristics as well as differential equations describing the dynamics of the front-end rectifier can be obtained independent of an inverter and motor load. This is because the DC-link voltage can be viewed as a voltage source, if V_{dc} is maintained constant for the full operating range. The inverter is thus connected to the voltage source, whose terminal voltage V_{dc}, remains unaffected by any normal inverter and motor operation (Jiuhe et al., 2006).

Furthermore, as shown in Fig.8, the rectifier can also be viewed as connected to the voltage source V_{dc}. Thus, the rectifier is able to control magnitude and phase of PWM voltages V_{abc} irrespective of line voltages E_{123}.

The dynamic equations for each phase can be written as,

$$\begin{bmatrix} E_1 \\ E_2 \\ E_3 \end{bmatrix} = L\frac{d}{dt}\begin{bmatrix} i_1 \\ i_2 \\ i_3 \end{bmatrix} + R\begin{bmatrix} i_1 \\ i_2 \\ i_3 \end{bmatrix} + \begin{bmatrix} V_{a0} \\ V_{b0} \\ V_{c0} \end{bmatrix} \tag{7}$$

In synchronous rotating *d-q* reference frame Equations 8 and 9 represent the dynamic *d-q* model of an active front end inverter in a reference frame rotating at an angular speed of ω.

$$L\frac{di_{qe}}{dt} = E_{qe} - \omega L i_{de} - R i_{qe} - V_{qe} \quad (8)$$

$$L\frac{di_{de}}{dt} = E_{de} + \omega L i_{qe} - R i_{de} - V_{de} \quad (9)$$

The differential equation governing DC link voltage also needs to be added to the above set of system equations to completely define system dynamics:

$$C\frac{dV_{dc}}{dt} = i_{dc} - i_{M} \quad (10)$$

where, i_{dc} is the total DC link current supplied by the rectifier, while i_M is the load-side DC current which is the result of induction motor operation.

In Equations 8 and 9, the terms E_{qe} and E_{de} are computed from source voltages, E_1, E_2, and E_3. Since line voltages are known, the angular frequency ω, can be easily estimated. The PWM voltages V_{qe} and V_{de} are the two inputs to the system which are generated using the sine-triangle PWM controller. L and R represent series impedance.

Equations (8 and 9) shows that *d-q* current is related with both coupling voltages $\omega L i_q$ and $\omega L i_d$, and main voltage E_d nd E_q, besides the influence of PWM voltage V_{qe} and V_{de}. Voltage V_{qe} and V_{de} are the inputs, controlled in such a way as to generate desired currents. Now define new variables V'_{qe} and V'_{de} such that (Hartani & Miloud, 2010):

$$V_{qe} = -V'_{qe} - \omega L i_{qe} + E_{qe} \quad (11)$$

$$V_{de} = -V'_{de} + \omega L i_{de} + E_{de} \quad (12)$$

So that the new system dynamic equations become:

$$L\frac{di_{qe}}{dt} = -i_{qe}R + V'_{qe} \quad (13)$$

$$L\frac{di_{se}}{dt} = -i_{de}R + V'_{de} \quad (14)$$

We can see from equations that the two axis current are totally decoupled because V'_{qe} and V'_{de} are only related with i_{qe} and i_{de} respectively.

The simple proportional-integral (PI) controllers are adopted in the current and voltage regulation, Fig.9. The control scheme of the PWM rectifier is based on a standard cascaded two-loop control scheme implemented in a *d-q* rotating frame: a fast control loop to control the current in the boost inductors and a much slower control loop to maintain constant dc-link voltage. The reference angle for the synchronous rotating *d-q* frame θ, is calculated, based on the three input phase voltages.

For the current control loop *d-q* synchronously rotating reference frame with the fundamental supply voltage frequency is used (Odavic et al., 2005). The line currents (i_1, i_2, i_3) are measured and transformed to the *d-q* reference frame, Fig.10.

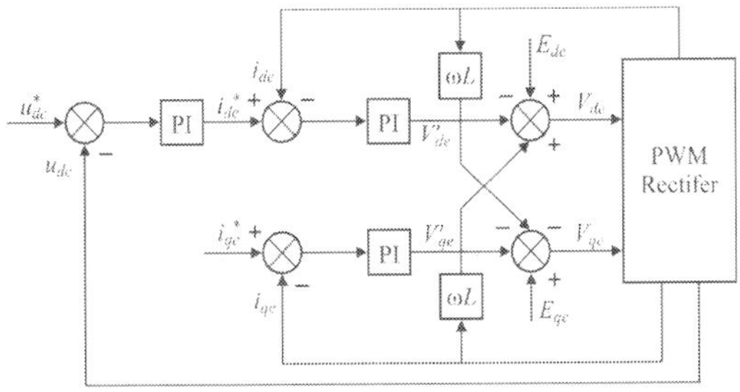

Fig. 9. Decoupled current control of PWM rectifier.

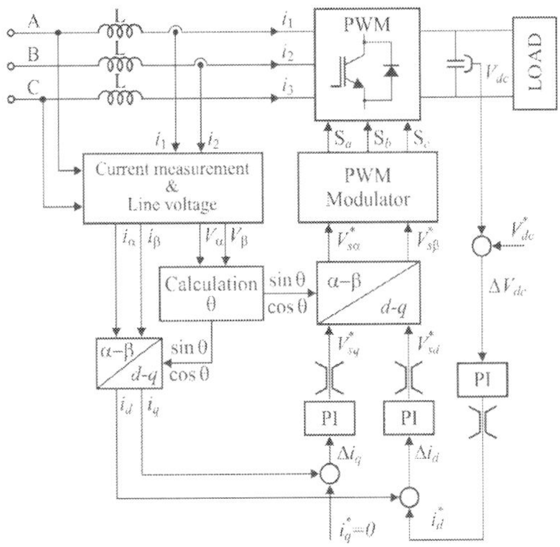

Fig. 10. Simplified block diagram of the AFE.

To get information about the position of the line voltage vector PLL (phase locked loop) is implemented. PI controllers for the *d-q* components of line current are identical and ω*L* terms are included to eliminate the coupling effect among the *d* and *q* components. Outputs of the line current PI controllers present *d* and *q* components of the voltage across the line inductance. Subtracting this voltage from the supply voltage gives the converter voltage from the AC side that is used to get the modulation signal for proper switching of six switching devices.

The main task of the sinusoidal front end is to operate with the sinusoidal line current; so *d* and *q* components of the line current reference are DC values. Using this approach of control it is possible to control the output voltage of converter as well as the power factor of converter in the same time. To achieve unity power factor the reference of *q* current component need to be set on zero.

Based on analysis, the simulation model of the whole is built using Matlab/Simulink to test the performance of the active front end rectifier. On the load side is the field oriented induction motor drive with topology as shown in Fig.5c). The whole system behavior is simulated as a discrete control system. The AC source is an ideal balanced three phase voltage source with frequency 50 Hz. The phase to phase voltage is 400 V. capacitor in DC link is 4700 μF. The line resistor and line inductance of each phase is 0.1 Ω and 3 mH respectively. The induction motor rated power is 30 kW.

The induction machine is initially running at a constant speed reference (100 rad/s) and under a no load regime. From this situation, we apply a rated load torque during a time interval of 0.5 s and then we remove the load. This case corresponds to a step up and a step down torque perturbation.

The following figures summarize the results of the simulation. Fig. 11 and Fig.12 shows the behavior of the rectifier under a step change of the induction motor load. Fig.11 refers to the motor mode of operation (lifting) and Fig.2 to the generator mode (lowering).

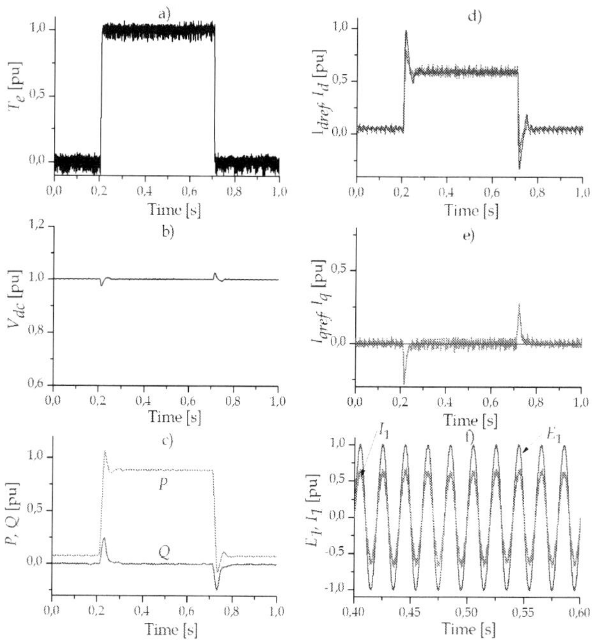

Fig. 11. Behavior of the rectifier under a step change of the induction motor load (motor operation).

Fig. 11b) and Fig. 12b) show the behavior of the DC link voltage of the PWM rectifier in response to a step change in the load. Fig. 11c) and Fig. 12c) presents the active and reactive power on the line side of converter. Since that the q component of the set point current is I_{qref}=0 (Fig.11e and Fig.12e) there is only *d* current component (Fig.11d and Fig.12d), reactive power is zero. Fig.11f) and Fig.12f) show the behavior of steady state voltage and current delivered by the source when the line side converter works in the rectifier and regenerative mode. The input current is highly sinusoidal and keeps in phase with the voltage, reaching a unity power factor.

The simulation results show that the rectifier has excellent dynamic behavior and following advantages:

- Power flow between the motor and the mains supply is possible in both directions, and so this makes the drive more efficient than when a braking resistor is used. As well as recovering energy during deceleration, it is possible to recover energy from overhauling load such as a crane when the load is being lowered.
- Good quality input current waveforms are possible with unity power factor.
- It is possible to boost the DC link voltage to a level that is higher than would be possible with a simple diode rectifier.

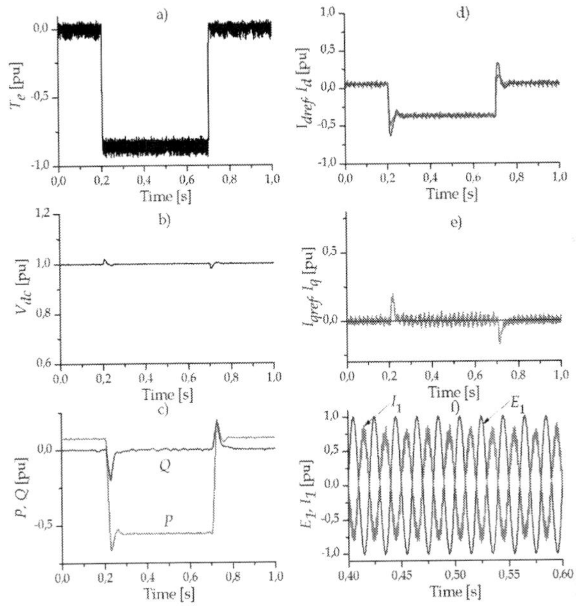

Fig. 12. Behavior of the rectifier under a step change of the induction motor load (generator operation).

5. Case study 1: Derrick crane

The experimental behavior analysis of some drives is considered in a derrick crane, which serves for load handling in many industry branches. The main task in adjustable speed drives design is a safe, multi-axis movement that allows material handling throughout the working area. The derrick crane with following technical details has been taken for experimentation with adjustable frequency drive:

- Main hoist load capacity: 60 t;
- Auxiliary load capacity: 12.5 t;
- Main hoist height: 46 m;
- Auxiliary hoist height: 49 m;

- Jib boom radius: 0-82.5°;
- Length of runway rail path: 350 m
- Working conditions: outdoor.

Using AFE rectifier/regenerative unit on common DC bus, six groups of inverter-motor combinations are supplied:

- hoist motion with 2x55 kW vector control inverter, the motor is a six pole, 2x45 kW,
- auxiliary hoist motion with a 55 kW vector control inverter, supplying a four pole, 45 kW motor,
- jib motion with 2x55 kW vector control inverter, the motor is a six pole, 2x45 kW,
- travel motion with 3x7.5 kW vector control inverter, the motor is six pole, 3x7.5 kW,
- motor driven cable reel with 3 kW vector control inverter supplying 3 kW motor,
- auxiliary drives with 8x1.1 kW and motors (8x1.1 kW).

The rating of the AFE rectifier/regenerative unit output at cosφ=1 and 400 V supply voltage is 177 kW. This is far less than the sum of the ratings of the individual invertors, being 300 kW. In the Fig.13 crane with indicated drives is shown.

Fig. 13. Derrick crane with indicated drives.

Fig. 14 shows the power circuit topology of the derrick crane using AFEs at the input side. The AFE is connected upstream to the standard frequency inverter and consists of three components:

- Active Infeed Converter.
- Line Filter Module (EMC filter, line contactor and charging circuit).
- Line Filter Choke

Power flows from the line through the input transformer and the input reactance into AFE, creating a common DC bus. The inverters take energy from the common DC bus to control the induction motors for the different movements.

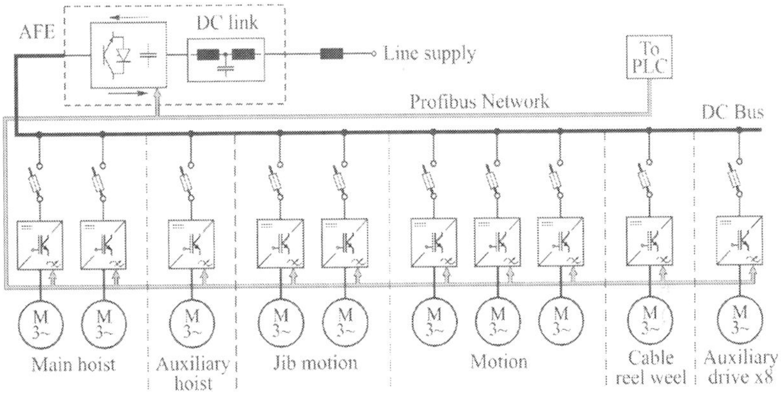

Fig. 14. Single line power circuit topology.

Fig.15a) shows a hoist movement with the 30% of full load and Fig.15b) for auxiliary hoist with an unloaded hook measured in similar conditions. In Fig.15a), curve 1 gives the actual speed signal (reference speed signal is given at 100% from the crane driver joystick command). Curves 2, 3 and 4 show the torque, power and motor current, respectively. After an acceleration period (ending at 5 s), a constant torque is delivered. This transition in torque level coincides with reaching the prescribed speed. At 17.5 s, the speed reference signal is made zero (stop command). The driving torque becomes zero and the system decelerates due to gravity. After 20 s, zero speed is reached, and the drive has to deliver the torque required for holding the load before closing the mechanical brake. After that the same the same measurements were performed during lowering. Due to high friction losses, torque was required to start the decent. After short initial period, only the dynamic friction is present, yielding a small driving torque. After the acceleration, the power flow is reversed and the drive lowers the load at a constant speed. From 43 s on, the system is braked with maximum torque, until standstill.

The same measurements were performed during lowering and hoist movement with the 30% of full load, Fig. 16a). The reference speed was 25, 50, 75 and 100% of rated speed. As on the Fig.15 actual speed, motor torque, power and current are shown. After that for jib-boom motion the same signal was record as shown on Fig.16b). In both figure regenerative periods during the lowering, at any reference speed, can be seen.

From the Fig.15 and Fog.16, can be seen periods when the energy recovery occurs at the point of load lowering. It is very important to point out that the AFE topology allows for fully regenerative operation, which is quite important for crane application.

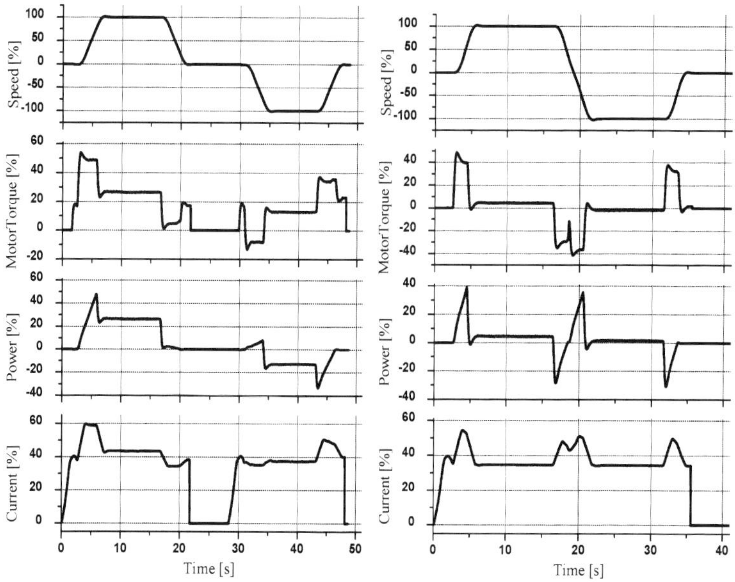

Fig. 15. Measured pattern of the a) hoist motion, b) auxiliary hoist.

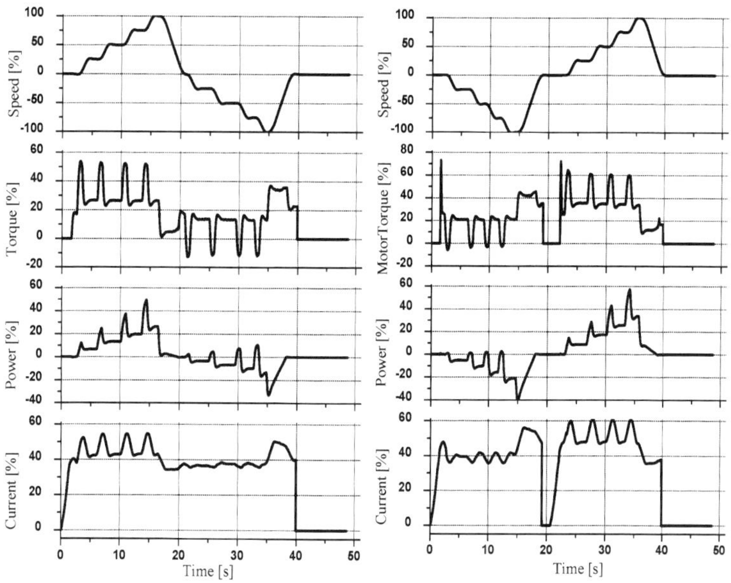

Fig. 16. Measured pattern at 25, 50, 75 and 100% of rated speed a) hoist motion, b) jib motion.

Different tests have been performed on the system to show some of the capabilities in the AFE inverter system. The measurements are done at steady-state operation. During experiments, the DC link voltage is boosted to 650 V. The first test is rectifier system operation when the induction machine operates as motor during lifting of the load, Fig.17a), and second test is regenerative operation during lowering of the load, Fig.17.b). Both figures show the measured line currents, line voltages and DC voltage. It can be observed a high stationary performance both in motor and generator operation. The line current is nearly a sine wave with unity power factor while DC voltage is unchanged.

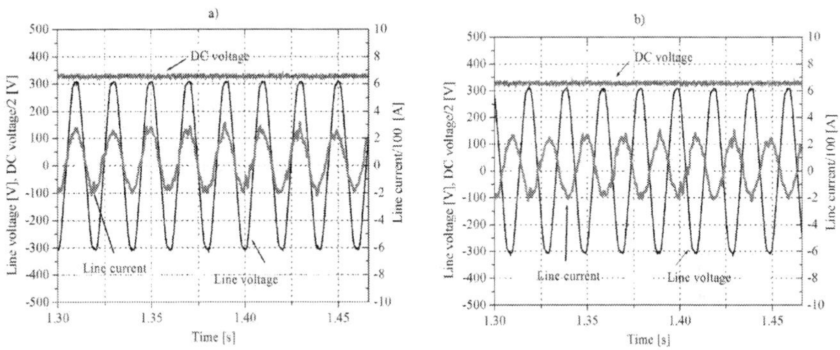

Fig. 17. Waveforms under steady-state operation: Line voltage, line current and dc link voltage a) motor operation, b) generator operation.

6. Case study 2: Wide span gantry crane

The experimental behavior analysis of some drives is considered on the example of crane with wide span, which in sugar factory serves for continuous transport of sugar beet from the reception position to the factory storage.

The crane with following details has been taken for experimentation with adjustable frequency drive:

- Handling capacity: 500 t/h;
- Gantry span: 64.5 m;
- Runway rail path: 300 m;
- Hoist height: 18 m;
- Working conditions: outdoor.

Gantry crane for sugar beet storage is designed from the following functional parts:

1. Gantry drive (16 m/min) with four induction motors of 5.5 kW, two per leg.
2. System conveyor belts (2 m/s) with "battered" (30kW), horizontal (30kW) and "butterfly" conveyor (11kW).
3. Trolley drive (12 m/min) with four motors of 1.1kW.
4. "Butterfly" hoist (3 kW).
5. Motor driven cable reel (1.1 kW).
6. Decentralized crane control system with appropriate PLC, Profibus communication between converters and other intelligent devices (encoders, operator panels etc.).

In the Fig.18 gantry crane with indicated drives is shown. All motors are three phase fed by frequency converters.

Certainly, the most complicated is the gantry drive, from following reasons:

- that is multi motor drive which consists of two motors on each side,
- the span is wide,
- construction is lattice, therefore it is elastic,
- plant is located outdoor so the influence of the wind may be considerable,
- the length of runway rail path is 300 m.

Basic requirements set in front of this drive are: equal load distribution between motors located on the same side, as well as skew elimination between fixed and free gantry leg.

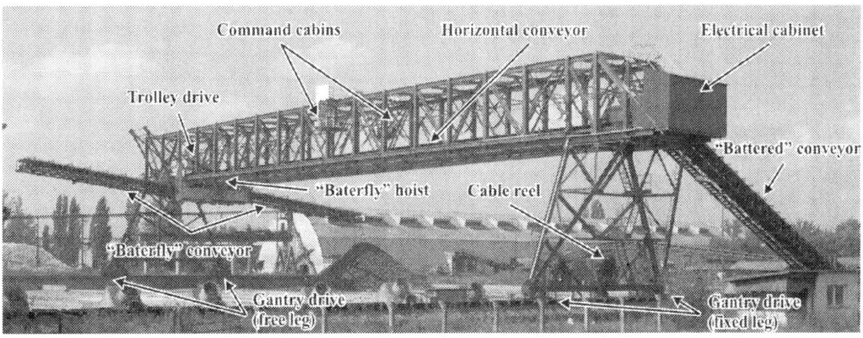

Fig. 18. Gantry crane with indicated drives.

6.1 Load sharing

Although the motors have the same power, there are few necessary reasons to do the load distribution: different wheel diameter, unequal adhesion, geometrical imperfection of the construction, slipping of the pinion wheel due to wet or frozen rails. Load distribution is resolved by using speed trim load sharing configuration, Fig.1c). Load distribution controller is realized by PLC.

In the Fig.19 the principle block scheme for load distribution between two rail coupled induction motors (IM_1 and IM_2) fed by frequency converters (FC1 and FC2) is shown. Starting point at design of load sharing controller is that the less loaded motor should accelerate in order to take over the part of load from the more loaded motor. Information about the load can be obtained in different ways. The easiest one is by motor current. Modern converters used in drives, enable to obtain information about the motor torque in percentage in relation to rated torque.

As we can see in the Fig.19, the speed reference of only one motor (n^*_2) is updated in relation to the main speed reference ($n^*=n^*_1$). Reference correction Δn^* is proportional to the difference of estimated electromagnetic torque ($\Delta T_e = T_{e1} - \Delta T_{e2}$). Proportional gain of load sharing regulators is denoted as K_{LS} .

In order to ensure the stabile operation of the motors during the large external disturbances, especially at low speed when estimation of electromagnetic torque in speed sensorless drives looses on accuracy, it is necessary to limit the correction value Δn^*, as shown in Fig.19.

For the purpose of suggested algorithm verification the trolley load sharing is analyzed. Because of the short distance between left and right side the skew may be neglected. Trolley drive consists of four motors, two on each side (IM_1-IM_2 on left and IM_3-IM_4 on right side). Frequency converters are set on speed sensorless vector control mode. Motors have the common reference speed. In the Fig.20a) motors torque without load distribution is shown. At reference speed, in steady state, we can see that even the motors have the same rated power, load torques are different. Estimated motor torque is not applied in control algorithm. Speed between left and right side is different because it depends of motor characteristics and load, as shown in Fig.20a).

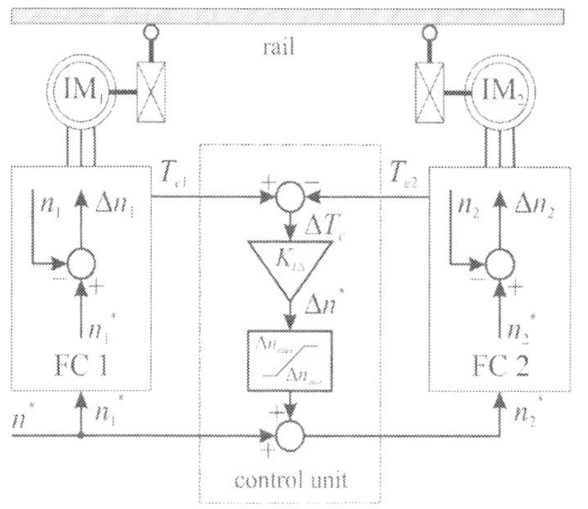

Fig. 19. The principle of load sharing based on estimated torque.

Effect of load sharing is shown in Fig.20b). The approximately equal motors torque on the same leg can be easily seen. Used system enables that speed of every motor is regulated, but also the load difference is controlled. In this way the load difference is being maintained on the desired accuracy.

Depending on the purpose of drives and needed accuracy of maintaining load distribution, load controller can be with only proportional effect, but also with proportional integrated effect. In our case only proportional controller with K_{LS}=1 p.u. is used. Output from the load controller is restricted on only several percentages of maxsimum speed reference (in our example $\Delta n_{min\text{-}max}$=2%). That is quite enough to provide necessary load regulation and not to "break" the drive speed regulation by too big effect on the speed reference. This solution can be applied for all kinds of multi motor drives on cranes.

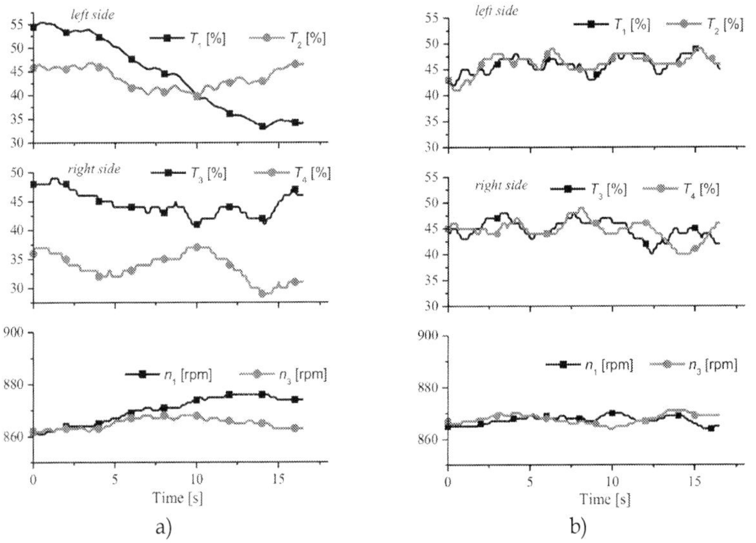

Fig. 20. Motors torque and speed: a) without load sharing; b) with load sharing.

6.2 Skew elimination

On most rails mounted wide span gantry cranes skewing problem is associated with poor rail conditions, uneven wheel wear, wind influence, wheel slippage or unequal load conditions when the trolley is operating at one end of the crane bridge. The skewing of the crane can cause excessive wheel abrasion and stress, especially to the wheel flanges. It can also produce horizontal or lateral forces that can result in unusual stresses to the crane runway beams and building structure. This often results in differing diameters of drive wheels, which subsequently cause the crane to skew.

The crane construction consists of opposed pairs of end truck assemblies (left hand side is named as free leg and right hand side is named as fixed leg). These are movable along a track and a long transverse support member between the end truck assemblies. Each end truck assembly includes two sets of trolleys and an upper load bar laterally interconnects the two sets of trolleys.

The hardware for skew elimination consists of a PLC with Field-bus communication, two absolute multi-turn encoders, two proximity sensors and four frequency converters for motor supply of trolley drives, as shown in Fig.21). On each end truck, one of the converters is master and the other one is slave. The master-slave references distribution is modified according to the load sharing principle as shown in Fig.1c).

The main devices for skew tracking are two absolute encoders (E_1 and E_2) installed on a special, non-tractive wheel (so called free wheel), in order to avoid slipping. Encoders measure the traveled distance, and absolute position is transferred to the anti-skew control subsystem in PLC, as shown in Fig.21). The fixed leg frequency converter (FC_1) is set as a master for gantry drive skew elimination algorithm; while in this case, the speed reference for the frequency converter on the free leg (FC_2) is modified with the anti-skew controller output.

The control scheme for skew elimination between the master and slave motor of gantry drive is shown in detail in Fig.22). As it can be seen, we propose a simple proportional (P) controller acting as an additional, outer correction loop, which supplies speed control loop. The speed reference of one motor (n^*_2) is updated in relation to the main speed reference ($n^*=n^*_1$) with reference correction value Δn^*. If the encoder position difference ΔE related to maximum allowed skew is known, the controller gain K_{SC} can be calculated, (Mitrovic et al., 2010).

In order to ensure the stabile and safe operation of the motors during the large external disturbances and at low speed, when estimation of electromagnetic torque in speed sensor-less drives looses on accuracy, it is necessary to limit the correction value Δn^*.

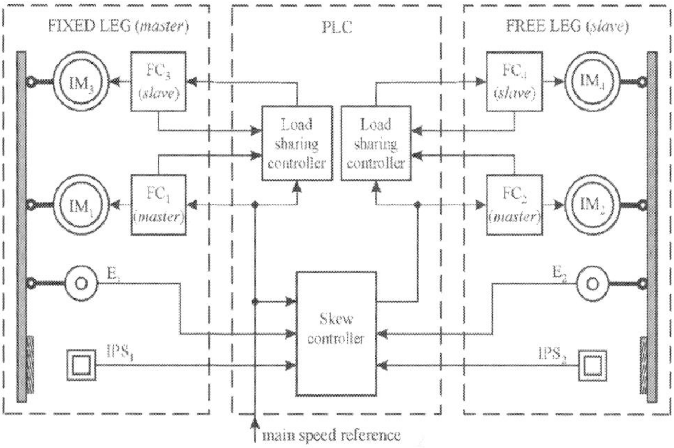

Fig. 21. Block scheme of gantry drive.

A reliable operation (even in terms of key components failure - for example encoders) requests an additional external disturbance compensator (EDC) which includes several pairs of position bars (or markers, M) and inductive proximity sensors (IPS). The EDC takes into account all external influences on the position difference of the two encoders: the free wheels diameter difference and an accidental wheel and encoder joint slipping.

The proximity sensors are fitted on the end truck holders, while the position bars are equidistantly mounted along the rails. During the crane movement, proximity sensors detect the moment when the fixed (or free) leg passes above the markers and so register the crane actual skew. Now when both legs are positioned on the markers, absolute encoders measure the trajectory difference, as shown in Fig.22). In fact, this difference is the real skew (s) of the crane, determined at each crossing over the markers. If the difference is greater than the length of the markers (l_m) that means the crane skew is bigger than allowed. For this reason it is required that the length of markers matches the allowed skew of the crane. The distance between successive markers (l_{ms}) depends on the length of marker and maximum expected liner speed difference between the legs.

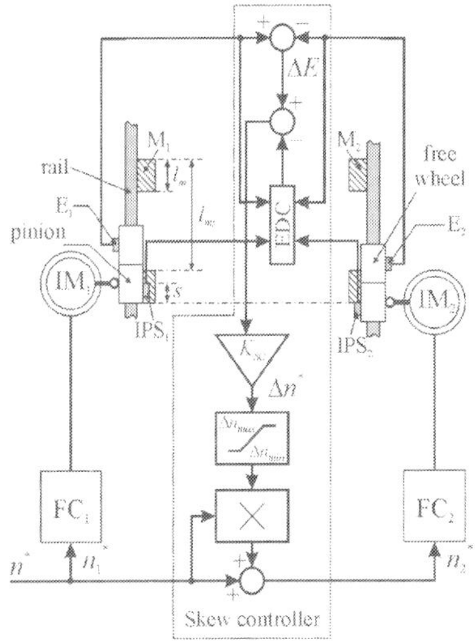

Fig. 22. The principle block diagram of skew controller.

The limited number of the necessary input data for the calculation and design of the skew controller allows quick adjustment of parameters and the choice of EDC components, and the proposed algorithm makes it suitable for industrial applications.

In the analyzed example, the loads of fixed and free legs are different, partly because of asymmetry of gantry, but mostly because of the trolley moving along the gantry. The calculated critical skew of gantry structure is 100 cm, but during normal operation the maximum allowed skew is 50 cm. A preview of gantry drive parameters, controller design and set-up values are taken from reference (Mitrovic et al., 2010).

At the beginning, we analyzed the behavior of gantry drives without a skew controller and the main results are shown in Fig.23a).

In this case, the load-sharing controllers for the fixed and free gantry leg are applied. Three working sections are noticeable: crane acceleration, steady state operation, and crane deceleration. The encoder measures the motor speed, while torque is estimated from the frequency converters. The measured data are collected in PLC SCADA system. The observed variables are master motor (IM_1) speed n_1, speed difference n_1-n_2 between the master motor (IM_1) on the fixed leg and the master motor (IM_2) on the free leg, torque differences between motors on the same leg (IM_1-IM_3, IM_2-IM_4) and the value of actual skew (*s*). In this case, as the skew is not controlled, it can be seen the increase of the value. During the crane skew, motors (IM_1 and IM_3) on the fixed leg are more loaded than the motors (IM_2 and IM_4) on the free leg. In addition to that, the effects of the load-sharing controller can be noticed because the motors on the same leg share loads approximately, i.e. torque difference oscillates about zero value.

The next experiment was performed including the skew controller and under the similar operational regimes as in the previous case: acceleration, steady state operation and deceleration. The experimental results are shown in Fig.23b). During the crane acceleration/deceleration, due to different loads between the fixed and free leg temporarily skew can be observed. The skew controller action eliminates this start-up disturbance in a few seconds. Simultaneously, with the action of a skew regulator, load-sharing controllers provide motor loading in proportion to their rated power. At constant speed operation, the trolley moves between the fixed and free leg, which causes additional differences in loads, but the proposed controller successfully compensates for these disturbances. In the case of crane deceleration, it can be seen that the characteristic case of the free leg stopping is postponed in order to complete the elimination of skew and for the fine position adjustment.

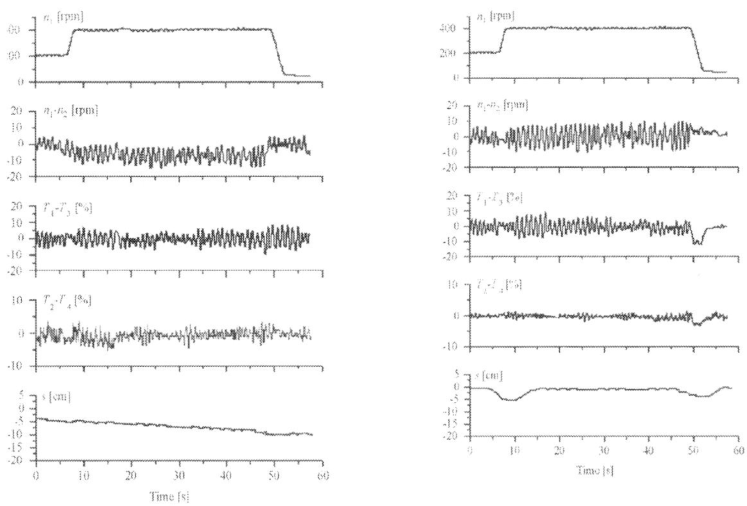

a) b)

Fig. 23. Behavior of gantry drives: a) without skew controller; b) with skew controller.

7. Conclusion

The application of squirrel cage induction motors supplied from the frequency converters (also known as adjustable speed drive) have become the standard solution for the modern crane drives. However, the standard configuration of the inverter can not be used for some drives primarily due to regenerative operation, which in some cases may be intermittent (long travel and cross travel) and continuous (lowering). The power and torque requirements in details are described and analyzed for such drives. From the aspect of the required power crane drives are often implemented as a multi motor. One of the important issue in this case is load distribution between the motor proportional to the motor power rating which can be resolved by applying the modern converters in one of the master-follower configuration.

This chapter describes the solutions that are commonly used in modern crane drives. In case that it is a casual recuperating the dynamic braking is used. If continious regeneretation occur active front end rectifier capable to returning energy into the supply network is used. The following two case studies are selected. Case study 1 is typical because the AFE is used which in addition of power recovery possibility also serves to supply all the drives on the common DC bus. Case study 2 deals with gantry cranes that have a large span. Solutions of two problems that occur in these types of cranes are shown as follows: load distribution between multiple motor and skew problem as a result of a large span.

8. Acknowledgment

This paper is supported by Project Grant III44004 (2011-2014) financed by Ministry of Education and Science, Republic of Serbia.

9. References

ABB, Technical guide No.8, Electrical Braking, 2011.

Backstrand, J.E. (1992). The Application of Adjustable Frequency Drives to Electric Overhead Cranes, *Industry Applications Society Annual Meeting*, 1992, Conf. Rec.1992 IEEE 4-9 Oct. 1992, vol.2, pp.1986 - 1991.

Belmans R., Bisschots F. & Timmer R. (1993). Practical Design Considerations for Braking Problems in Overhead Crane Drives, *Conf. Rec. IEEE-IAS*, Vol.1, pp. 473-479.

Busschots F., Belmans R. & Geysen W. (1991). Application of Field Oriented Control in Crane Drives, *Proc. IEEE-IAS, Annual Meeting, Dearborn, Michigan, USA*, September 28-October 4, 1991, pp. 347-353.

Hartani, K. & Miloud, Y. (2010). Control Strategy for Three Phase Voltage Source PWM Rectifier Based on the Space Vector Modulation, *Advances in Electrical and Computer Eng.*, Vol.10, Issue 3, pp. 61-65.

Jeftenic B., Bebic M. & Statkic S. (2006). Controlled Multi-motor Drives, *Intern. Symp. SPEEDAM* 2006, Taormina (Sicily) - Italy, 23-26 May 2006 , pp. 1392-1398.

Jiuhe, W., Hongren, Y., Jinlong Z & Huade, L. (2006). Study on Power Decoupling Control of Three Phase Voltage Source PWM Rectifiers, *Power Electronics and Motion Control Conference*, 2006.

Mitrovic N., Petronijevic M., Kostic V. & Bankovic B. (2011). Active Front End Converter in Common DC Bus Multidrive Application, XLVI *Proc. of Inter. Conf. ICEST* 2011, Serbia, Nis, Vol.3, pp. 989-992, 2011.

Mitrovic, N., Kostic, V., Petronijevic, M. & Jeftenić, B. (2010). Practical Implementation of Load Sharing and Anti Skew Controllers for Wide Span Gantry Crane Drives", *JME*, Vol. 56, no. 3, pp. 207-216, 2010.

Odavic M., Jakopovic Z. & Kolonic F. (2005). Sinusoidal Active Front End under the Condition of Supply Distortion, *AUTOMATIKA* 46(2005), 3–4, pp.135–141, 2005.

Paul, A. K., Banerje, I., Snatra, B.K. & Neogi, N. (2008). Application of AC Motors and Drives in Steel Industries, XV *Natinal Power System Conference*, Bombay, Dec.2008, pp.159-163.

Petronijevic M., Veselic B., Mitrovic N., Kostic V. & Jeftenic, B. (2011). Comparative Study of Unsymmetrical Voltage Sag Effects on Adjustable Speed Induction Motor Drives, *Electric Power Applications, IET* , vol.5, no.5, pp.432-442, May 2011

Rashid H., *Power Electronics Handbook*, Academic Press, San Diego, 2001.

Rockwell Automation (2000), Load Sharing for the 1336 PLUS II AC Drive, *Publication number* 1336E-WP001A-EN-P, June 2000.

Slutej, A., Kolonic, F. & Jakopovic, Z. (1999). The New Crane Motion Control Concept with Integrated Drive Controller for Engineered Crane Application, ISIE'99, *Proc.of the IEEE International Symposium*, Volume 3, 1999, pp.1458 - 1461.

Part 2

Manufacturing Processes and System Analysis

7

Anisotropic Mechanical Properties of ABS Parts Fabricated by Fused Deposition Modelling

Constance Ziemian[1], Mala Sharma[1] and Sophia Ziemian[2]
[1]*Bucknell University,*
[2]*Duke University,*
USA

1. Introduction

Layered manufacturing (LM) methods have traditionally been used for rapid prototyping (RP) purposes, with the primary intention of fabricating models for visualization, design verification, and kinematic functionality testing of developing assemblies during the product realization process (Caulfield et al., 2007). Without any need for tooling or fixturing, LM allows for the computer-controlled fabrication of parts in a single setup directly from a computerized solid model. These characteristics have proven beneficial in regard to the objective of reducing the time needed to complete the product development cycle (Chua et al., 2005).

There are numerous LM processes available in the market today, including stereolithography (SLA), fused deposition modeling (FDM), selective laser sintering (SLS), and three-dimensional printing (3DP), all of which are additive processes sharing important commonalities (Upcraft & Fletcher, 2003). For each of these processes, the object design is first represented as a solid model within a computer aided design (CAD) software package and then exported into tessellated format as an STL file. This faceted model is then imported into the relevant LM machine software where it is mathematically sliced into a series of parallel cross-sections or layers. The software creates a machine traverse path for each slice, including instructions for the creation of any necessary scaffolding to support overhanging slice portions. The physical part is then fabricated, starting with the bottom-most layer, by incrementally building one model slice on top of the previously built layer. This additive layering process is thus capable of fabricating components with complex geometrical shapes in a single setup without the need for tooling or human intervention or monitoring.

In recent years, layered manufacturing processes have begun to progress from rapid prototyping techniques towards rapid manufacturing methods, where the objective is now to produce finished components for potential end use in a product (Caulfield et al., 2007). LM is especially promising for the fabrication of specific need, low volume products such as replacement parts for larger systems. This trend accentuates the need, however, for a thorough understanding of the associated mechanical properties and the resulting behaviour of parts produced by layered methods. Not only must the base material be durable, but the mechanical properties of the layered components must be sufficient to meet in-service loading and operational requirements, and be reasonably comparable to parts produced by more traditional manufacturing techniques.

Fused deposition modeling (FDM) by Stratasys Inc. is one such layered manufacturing technology that produces parts with complex geometries by the layering of extruded materials, such as durable acrylonitrile butadiene styrene (ABS) plastic (Figure 1). In this process, the build material is initially in the raw form of a flexible filament. The feedstock filament is then partially melted and extruded though a heated nozzle within a temperature controlled build environment. The material is extruded in a thin layer onto the previously built model layer on the build platform in the form of a prescribed two-dimensional (x-y) layer pattern (Sun et al., 2008). The deposited material cools, solidifies, and bonds with adjoining material. After an entire layer is deposited, the build platform moves downward along the z-axis by an increment equal to the filament height (layer thickness) and the next layer is deposited on top of it.

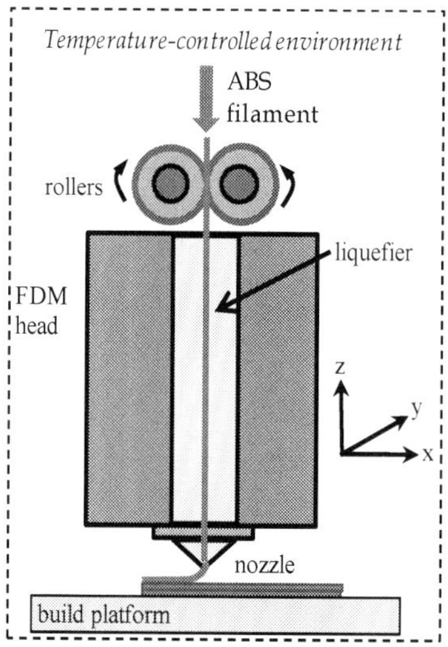

Fig. 1. Schematic of the FDM process

If the model requires structural support for any overhanging geometry, a second nozzle simultaneously extrudes layers of a water soluble support material in this same manner. Once the build process is completed, the support material is dissolved and the FDM part can be viewed as a laminate composite structure with vertically stacked layers of bonded fibers or rasters (Sood et al., 2011). Consequently, the mechanical properties of FDM parts are not solely controlled by the build material of the original filament, but are also significantly influenced by a directionally-dependent production process that fabricates components with anisotropic characteristics associated with the inherent layering.

Several researchers have specifically considered the anisotropic characteristics of FDM parts in recent years. Rodriguez et al. (2001) investigated the tensile strength and elastic modulus of FDM specimens with varying mesostructures in comparison with the properties of the ABS monofilament feedstock. They determined that the tensile strength was the greatest for parts with fibers aligned with the axis of the tension force. Ahn et al. (2002) designed a factorial experiment to quantify the effects of model temperature, bead width, raster orientation, air gap, and ABS color on both tensile and compressive strengths of FDM parts. It was determined that both air gap and raster orientation had significant effects on the resulting tensile strength, while compressive strength was not affected by these factors. Their results include a set of recommended build rules for designing FDM parts. A similar study was completed by Sood et al. (2010), with varying factors of layer thickness, build orientation, raster angle, raster width, and air gap. These researchers implemented a central composite experiment design and analyzed the functional relationship between process parameters and specimen strength using response surface methodology. Their results indicate that the tested factors influence the mesostructural configuration of the built part as well as the bonding and distortion within the part. On the basis of this work, Sood et al. (2011) further examined the effect of the same five process parameters on the subsequent compressive strength of test specimens. Their work provides insight into the complex dependency of compressive stress on these parameters and develops a statistically validated predictive equation. Results display the importance of fiber-to-fiber bond strength and the control of distortion during the build process. Lee et al. (2005) concluded that layer thickness, raster angle, and air gap influence the elastic performance of compliant ABS prototypes manufactured by fused deposition. A study conducted by Es Said et al. (2000) analyzed the effect of raster orientation and the subsequent alignment of polymer molecules along the direction of deposition during fabrication. These researchers considered the issue of volumetric shrinkage and raster orientation with respect to tensile, flexural and impact strengths. Lee et al. (2007) focused on the compressive strength of layered parts as a function of build direction. They determined that the compressive strength is greater for the axial FDM specimens than for the transverse. The foregoing studies reveal the directional dependence or anisotropy of the mechanical properties of FDM parts as a result of mesostructure and fiber-to-fiber bond strength, and provide numerous insights and recommendations regarding significant process parameters and the development of component build rules.

The goal of this project is to quantitatively analyze the potential of fused deposition modeling to fully evolve into a rapid manufacturing tool. The project objective is to develop an understanding of the dependence of the mechanical properties of FDM parts on raster orientation and to assess whether these parts are capable of maintaining their integrity while under service loading. The study utilizes the insights provided by previous researchers and further examines the effect of fiber orientation on a variety of important mechanical properties of ABS components fabricated by fused deposition modeling. This study uses FDM build recommendations provided in previous work, as well as the defined machine default values, in order to focus analysis specifically on the significant issue of fiber or raster orientation, i.e. the direction of the polymer beads (roads) relative to the loading direction of the part. Tensile, compressive, flexural, impact, and fatigue strength properties of FDM specimens are examined, evaluated, and placed in context in comparison with the properties of injection molded ABS parts.

2. Experimental procedure

2.1 Materials

All of the FDM specimens tested and analyzed in this study were acrylonitrile butadiene styrene (ABS). ABS is a carbon chain copolymer belonging to styrene ter-polymer chemical family. It is a common thermoplastic that is formed by dissolving butadiene-styrene copolymer in a mixture of acrylonitrile and styrene monomers, and then polymerizing the monomers with free radial initiators (Odian, 2004). The result is a long chain of polybutadiene crisscrossed with shorter chains of poly(styrene-co-acrylonitrile). The advantage of ABS is that it combines the strength and rigidity of the acrylonitrile and styrene polymers with the toughness of the polybutadiene rubber. The proportions can vary from 15 to 35% acrylonitrile, 5 to 30% butadiene, and 40 to 60% styrene. In this study, the resulting composition was 90-100% acrylonitrile/butadiene/styrene resin, with 0-2% mineral oil, 0-2% tallow, and 0-2% wax.

2.2 Specimen preparation and equipment

This project included five different mechanical tests: tension, compression, flexural (3-point bend), impact, and tension-tension fatigue. Three unique specimen designs were required. The tension, flexural, and fatigue specimens were all thin rectangular slabs (Figure 2a) fabricated to be 190.5 mm long, 12.7 mm wide, and 2.6 mm thick in accordance with ASTM D3039 (ASTM, 1998), ASTM D790 (ASTM, 2007), and ASTM D3479 (ASTM, 2007a) standards respectively. Compression specimens were cylindrical and fabricated with dimensions conforming to the ASTM D695 standard (ASTM, 1996). Each cylinder was 25.4 mm long and 12.7 mm diameter (Figure 2b). Impact specimens were fabricated with dimensions conforming to the ASTM D256 standard (ASTM, 2010). The geometry was a v-notched rectangular block of 63.5 mm long, 25.4 mm wide, and 25.4 mm thick (Figure 2c). The v-notch was modeled within the computer solid model of the specimen and was produced directly on the FDM machine.

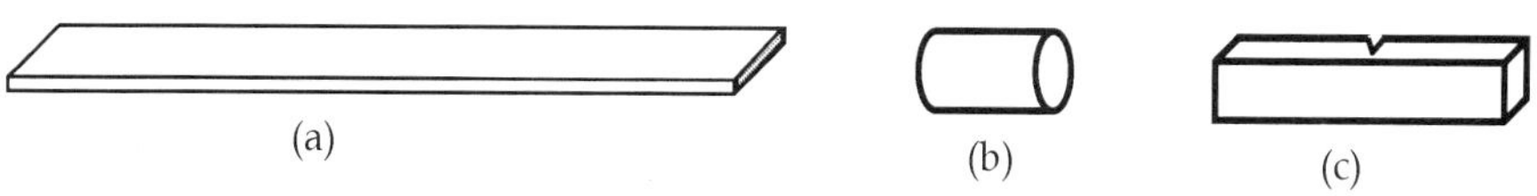

Fig. 2. Specimen geometries associated with each test

All FDM specimens were fabricated with a Stratasys Vantage-*i* machine. Solid models were first created using Pro/Engineer® software, and then tessellated and exported in STL format. Digital models were then sliced using the Vantage machine's Insight software, and layer extrusion tool paths were generated, i.e. raster patterns used to fill interior regions of each layer, to represent the four different fiber orientations studied in each test.

Layered specimens were all fabricated in a build orientation that aligned the minimum part dimension with the z-axis of the machine, i.e. perpendicular to the build platform. In this orientation, five to ten replicate specimens were built with each of four different raster patterns relative to the part loading direction, for each of the five different tests completed. The four raster orientations included: (a) longitudinal or 0°, i.e. rasters aligned with long dimension of the specimen, (b) diagonal or 45°, i.e. rasters at 45° to the long dimension of the specimen, (c) transverse or 90°, i.e. rasters perpendicular to long dimension of the specimen, and (d) default or +45°/-45° criss-cross, i.e. representing the machine's default raster orientation (Figure 3).

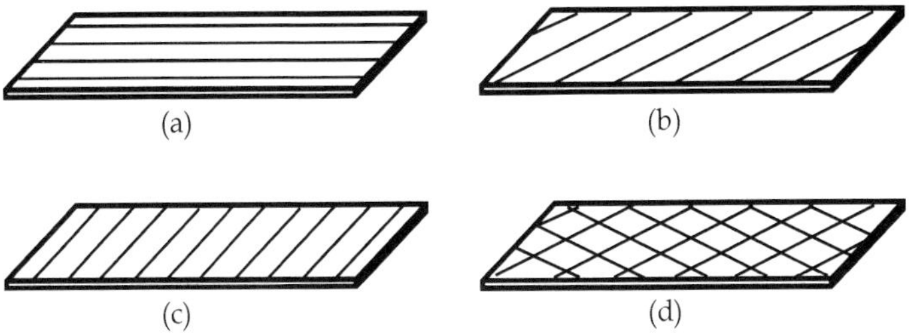

Fig. 3. Four different raster orientations investigated

All FDM specimens were built while holding all other machine process settings at the recommended or default values displayed in Table 1.

Factor	Value/Level
Air gap	0.0 mm
Nozzle	T12
Road width	0.3048 mm
Slice height	0.1778 mm
Part interior fill style	Solid normal
Part fill style	Perimeter/raster
Liquefier temperature	320 °C
Envelope temperature	80 °C

Table 1. Fixed FDM process settings

In order to measure the reference strength and behaviour of the ABS filament material, for comparisons with the layered parts, additional specimens were fabricated by injection molding for the same five tests. Aluminium molds for each of the three previously described geometries (Figure 2) were designed using Pro/Engineer® software, and manufactured on a Haas VF-1 CNC machining center. Mold cavity dimensions were the same as those described for the FDM specimens, with slight increases to compensate for the shrinkage of molded ABS at approximately 0.005 cm/cm. Parting lines and runners were located with an effort to avoid potential stress concentrations or anomalies in the resulting specimens that might affect test results. All molded specimens were fabricated from the same material as the layered models by feeding the FDM-ABS filament into a polymer granulator and cutting it into pellets of 3-5 mm in length. The pellets were then fed into the hopper of a Morgan Press G-100T injection molding machine. Molding parameters were set to the recommended values for ABS plastic, including nozzle temperature of 270 °C, mold preheat temperature of 120° C, clamping force of 71 kN (16,000 lb), and injection pressure of 41 MPa (6000 psi). Ten replicate specimens were molded for each of the five tests.

Tensile, compressive, flexural, and tension-fatigue tests were performed on an Instron model 3366 dual column uniaxial material testing with .057 micron displacement precision and up to 0.001 N force accuracy. The machine has a 10kN load force capacity. Impact strength was studied on a TMI impact tester. Resulting fracture surfaces were subsequently prepared by gold sputtering and analyzed with a JSM 500-type JEOL Scanning Electron Microscope (SEM).

3. Results and discussion

3.1 Tension testing

The tension specimens were thin rectangular slabs made in compliance with ASTM D3039 (ASTM, 1998). Fabrication utilized the FDM T12-nozzle, providing an individual layer or slice height of 0.1778 mm. The specimen thickness of 2.6 mm was subsequently achieved with a total of 15 layers (Figure 4).

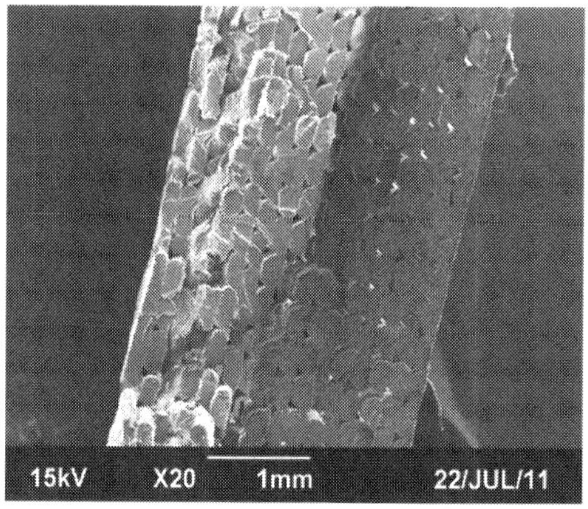

Fig. 4. SEM image displaying 15 layer thickness

A summary of the tension test results for the four raster orientations is displayed in Table 2. The mean ultimate and yield strengths (0.2% offset) were largest for the longitudinal (0°) raster orientation, 25.72 and 25.51 MPa respectively, and weakest for the transverse (90°) raster orientation, 14.56 and 14.35 MPa respectively. The mean ultimate strength of the 90° specimens represented only 56.23% of that of the 0° raster specimens, followed by the 45° specimens at 61.45% and the +45°/-45° specimens at 74.09%.

Raster Orientation	*Mean Yield Strength* (MPa), *Std Dev*	*Mean Ultimate Strength* (MPa), *Std Dev*	*Mean Effective Modulus* (MPa), *Std Dev*
Longitudinal (0°)	25.51, *0.73*	25.72, *0.91*	987.80, *19.98*
Diagonal (45°)	15.68, *0.27*	16.22, *0.27*	741.78, *20.28*
Transverse (90°)	14.35, *0.08*	14.56, *0.05*	738.77, *7.91*
Default (+45°/-45°)	18.90, *0.53*	19.36, *0.39*	768.01, *33.31*

Table 2. Tension test results

A one-way analysis of the variance (ANOVA) was completed in order to consider the equivalence of the population means for the four raster orientations. The results, appearing in Table 3, include a calculated F-test statistic of $F(3,16) = 490.98$ and a p-value of 0.0001, indicating a significant difference between some or all of the mean ultimate strength (UTS) values associated with the four raster orientations at a level of significance of $\alpha = 0.05$. The coefficient of determination associated with this analysis was $R^2 = 0.9886$.

Source	DF	SS	MS	F	P
Raster Angle	3	363.991	121.330	460.98	0.0001
Error	16	4.211	0.263		
Total	19	368.202			

Table 3. ANOVA results comparing mean ultimate tensile strengths of 4 raster orientations

Further analysis in the form of post hoc comparisons was performed to determine which raster orientations differed in mean UTS. Tukey's method (Montgomery, 2009), creating a set of 95% simultaneous confidence intervals for the difference between each pair of means, indicated that the difference was significant for all pairwise comparisons of mean UTS values (Table 4). The difference between the mean UTS of the longitudinal rasters (25.72) and that of the transverse rasters (14.56) was the most significant. These results confirm that raster orientation has a significant effect on the tensile strengths of the FDM specimens. Tensile strength is thus verified to be affected by the directional processing and subsequent directionality of the polymer molecules, signifying an anisotropic property.

Raster Orientation (i)	**Raster Orientation (j)**	**Difference of Mean UTS (i-j)**	**95% Confidence Interval**	
			Lower Bound	**Upper Bound**
Longitudinal	45-Degree	9.50*	8.579	10.437
	Transverse	11.16*	10.235	12.093
	Default	6.36*	5.438	7.296
45-Degree	Transverse	1.66*	0.727	2.585
	Default	-3.14*	-4.07	-2.212
Transverse	Default	-4.8*	-5.726	-3.868

* Mean difference is significant at the 0.05 level

Table 4. Post hoc Tukey HSD multiple comparisons of mean tensile strengths

The quantitative data analysis was followed by detailed physical inspection of the specimens at both macro and microscopic levels. Macroscopically, the fracture patterns of the specimens varied somewhat as a function of the raster orientation of the two-dimensional layers and the resulting weakest path for crack propagation (Figure 5). The 90° specimens failed in the transverse direction and the 45° specimens failed along the 45° line. The 0° specimens failed primarily in the transverse direction, although there was some fiber pullout and delamination intermittently evident as well. The +45°/-45° specimens broke at intersecting fracture paths along ±45°, resulting in a saw-tooth fracture pattern across the specimen width. It is likely that fracture paths controlled by weak interlayer bonding are affected by the residual stresses that result from the volumetric shrinkage of the polymer layers during solidification and cooling. In addition, interlayer porosity and air gaps serve to reduce the actual load-bearing area across the layers, providing an easy fracture path.

In specimens with the longitudinal (0°) raster orientation, the molecules tend to align along the stress axis direction. This produces the strongest individual two-dimensional layers subjected to tension loading. During the testing of these specimens, stress whitening due to

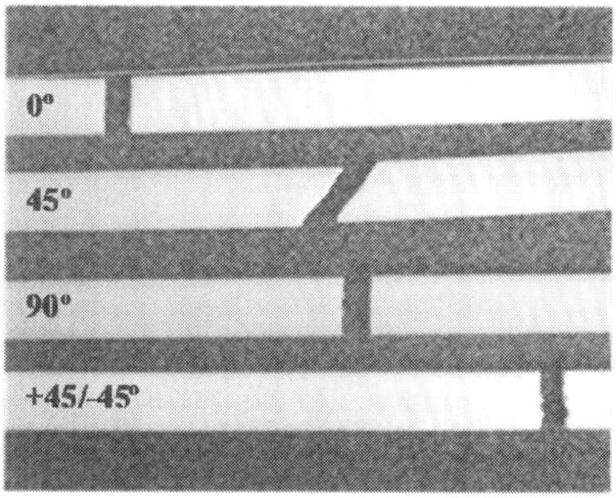

Fig. 5. Failure modes of the specimens with each of the four raster orientations

craze formation and growth was observed to develop prior to reaching the yield stress. Failure occurred at whitened areas from plastic deformation where some evidence of localized fiber delamination was observed. The fracture surfaces were further analyzed with a scanning electron microscope, and displayed failure that was predominantly brittle in nature with localized micro-shearing on each fiber face (Figure 6). The tensile strength of these specimens is thus more heavily dependent on the strength of the ABS monofilament than specimens with fibers running at orientations other than 0° with the stress axis (Rodriguez et al., 2001).

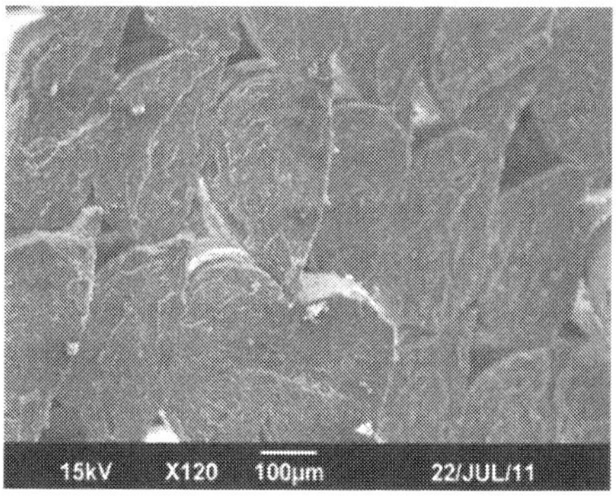

Fig. 6. SEM image of fracture surface of a 0° raster specimen

In contrast, the specimens with transverse (90°) raster orientations did not display obvious crazing during testing, and failure occurred predominantly at the weak interface between layered ABS fibers (Figure 7). These specimens experienced brittle interface fracture. Weak interlayer bonding or some amount of interlayer porosity was evident in the failure of many of the specimens with raster orientations other than 0°, and appeared to be the cause of layer delamination along the fiber orientation during loading. The tensile strength of these specimens depended much more heavily upon the fiber-to-fiber fusion and any air gap resulting between the fibers, as opposed to the strength of the fibers themselves.

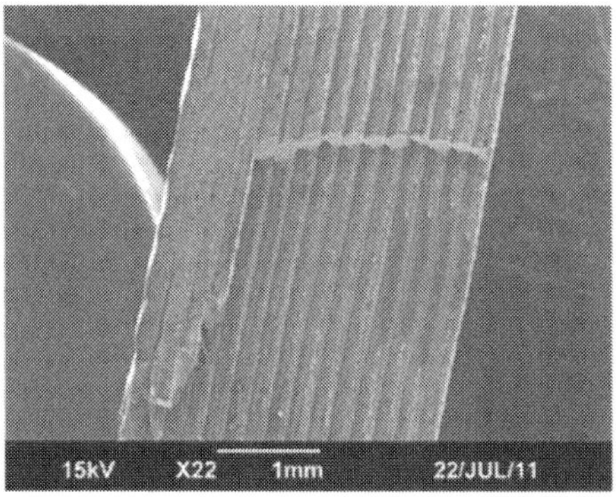

Fig. 7. SEM image of fracture surface of a 90° raster specimen

Under microscopic examination, the +45°/-45° specimens displayed multiple failures of individual raster fibers in both shear and tension (figure 8a). Failure occurred by the pulling and eventual rupturing of individual fibers whereby the material separated at a +45°/-45° angle relative to the tensile load, creating the saw-tooth like appearance evident at the macro-scale (figure 5). Failure of the 45° raster specimens was similar to that of the default +45°/-45° in that it was a brittle shear failure on each of the individual fibers at the microscopic level, as each raster was pulled in tension and failed at 45 degrees relative to the loading axis (figure 8b). Macroscopically, the samples also displayed a characteristic shear failure along the 45° line with the tensile load (figure 5).

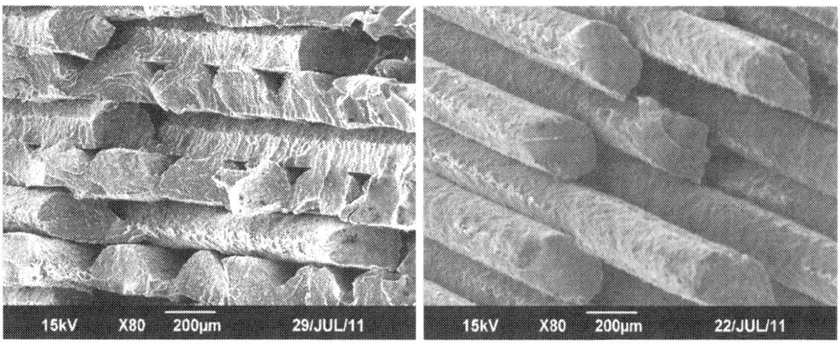

(a) (b)

Fig. 8. SEM image of fracture surfaces of (a) +45°/-45° and (b) 45° raster specimen

The air gap that forms during fabrication and remains present between fibers of the FDM specimens is a significant factor in considering tensile ultimate and yield strengths and comparing these properties to those of injection molded ABS specimens. Although the FDM machine setting indicated a desired air gap of 0.0 mm, the fiber geometry inherently causes the presence of triangular air voids as seen in the SEM image in Figure 9. These voids influence the effective tensile strengths and effective elastic moduli of the FDM parts by decreasing the physical cross-sectional area of material specimens. This is in part why the

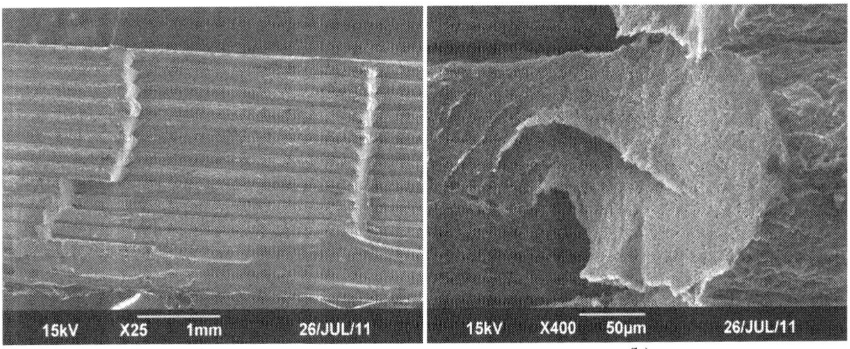

(a) (b)

Fig. 13. SEM image of fractured 90° flexural specimen magnified (a) 25X and (b) 400X

rasters, from which the crack then splits to the right and left and eventually climbs, as seen in the shear failure of individual rasters in layers. Upon closer microscopic examination, it was observed that individual rasters showed shear failure with a clearly defined exaggerated shear lip on the top of each fiber; something that was not observed in the analysis of 90° raster specimens that failed in tension testing (Figure 7). In both the 45° and the 90° raster specimens, there is little localized plastic deformation before failure initiates and fibers begin to break.

3.4 Impact testing

The impact study utilized an Izod test configuration with a notched specimen held as a vertical cantilevered beam as shown in Figure 14. In this position, the material was subjected to a load in the form of an impact blow from a weighted pendulum hammer striking the notched side of the specimen. The test measures the impact energy or notch toughness, and the results are expressed in energy absorbed per unit of thickness at the notch in units of J/cm. The impact energy absorbed by the specimen during failure is measured by calculating the change in the potential energy of the hammer. The change in potential energy is proportional to difference in the height of the hammer from its initial position to the maximum height achieved after impact.

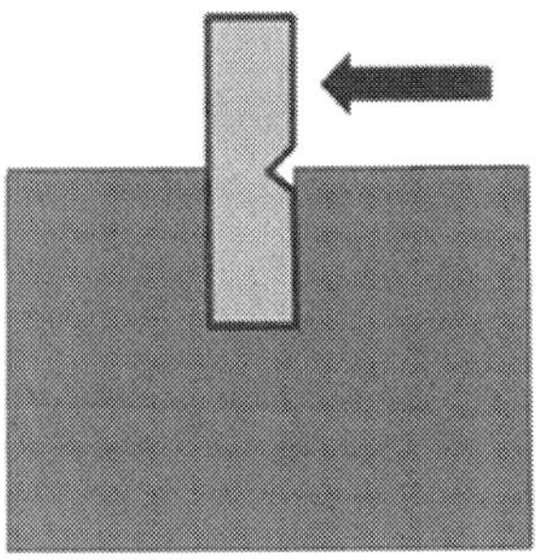

Fig. 14. Izod impact test configuration

Impact tests were completed on 10 specimens with each of the four raster orientations. The mean impact energy results are displayed in Table 9. The absorbed energy was the highest for the longitudinal (0°) fiber orientation (2.989 J/cm) and the lowest for the transverse (90°) orientations (1.599 J/cm). The 45° and +45°/-45° default specimens broke with mean impact resistances between those of the 0° and 90° specimens.

Raster Orientation	Mean Impact Energy (J/cm)	Standard Deviation	Fracture Type
Longitudinal (0°)	2.991	0.103	Hinged
Diagonal (45°)	2.339	0.483	Hinged & Complete
Transverse (90°)	1.599	0.014	Complete
Default (+45°/-45°)	2.514	0.338	Hinged & Complete

Table 9. Impact test results

The relative impact strengths of the four raster orientations correlated well with the tensile strength results. In addition, the variation of the impact strengths was smallest for the transverse orientation (Figure 15), coinciding with the variation of the tensile test results.

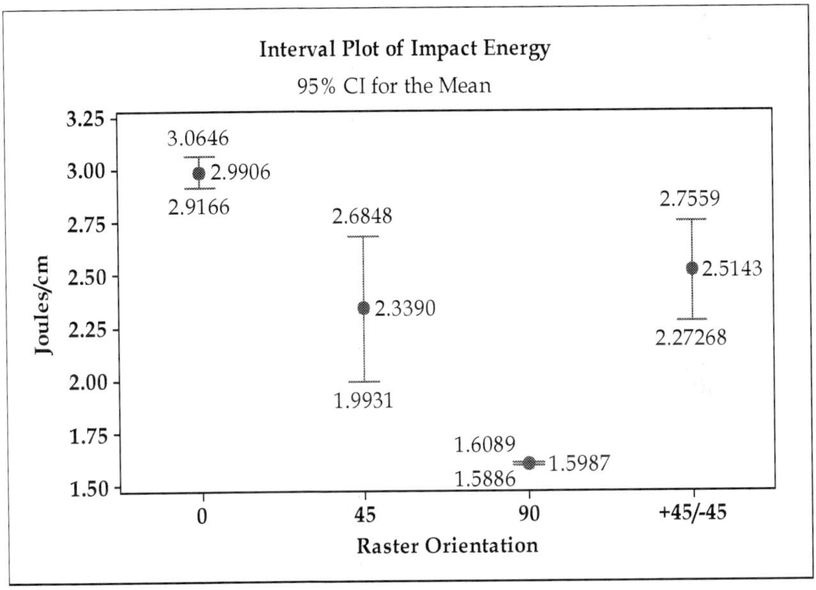

Fig. 15. Interval plot of impact test results

A one-way ANOVA was conducted to compare the effect of raster orientation on mean impact energies in 0°, 45°, 90°, and +45°/-45° conditions. There was a significant effect of raster orientation on impact energies at the $p < 0.05$ level for the four conditions, with $F(3, 36) = 37.23$, $p = 0.0001$. Post hoc comparisons using indicated that the mean impact energy for the 45° diagonal condition (2.339 J/cm) did not significantly differ from that of the +45°/-45° condition (2.514 J/cm), applying a 95% confidence interval (Table 10). All other paired comparisons indicated statistically significant differences in mean impact energies. These results suggest that impact strength has anisotropic characteristics.

Raster Orientation (*i*)	Raster Orientation (*j*)	Difference of Mean Impact Energy (*i-j*)	95% Confidence Interval	
			Lower Bound	Upper Bound
Longitudinal	45-Degree	0.652*	0.291	1.012
	Transverse	1.392*	1.031	1.753
	Default	0.473*	0.116	0.837
45-Degree	Transverse	0.740*	0.379	1.101
	Default	-0.175	-0.536	0.184
Transverse	Default	-0.916*	-1.276	-0.555

* Mean difference is significant at the 0.05 level

Table 10. Post hoc Tukey HSD multiple comparisons of mean impact energies

Inspection indicated that the fracture patterns of the specimens varied as a function of the raster orientations. The longitudinal (0°) and transverse (90°) pieces fractured along a path oriented 90° to the length of the specimen. The fracture surface appeared smooth along the layers of each of the transverse specimens, all of which experienced clean and complete separation of the specimen into two discrete pieces. Weak interfaces parallel to the crack front affected the transverse raster specimens by providing a straightforward path for crack propagation, resulting in the least amount of energy absorption. Failure initiated from the side on the notch where the energy absorption was greatest, thus causing the rasters to fracture longitudinally along the length of the fiber through several layers until the energy adsorption decreased significantly and the sample broken in half (Figure 16a). Upon closer examination, it can be seen that individual fibers plastically deformed and twisted at the ends upon catastrophic failure, whereby halves of the samples separated (Figure 16b).

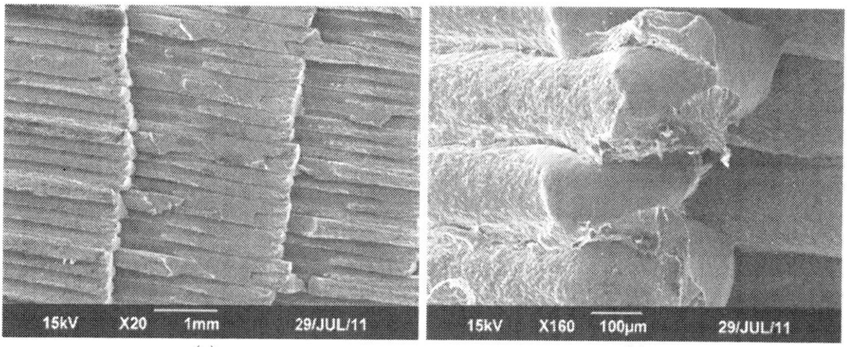

(a) (b)

Fig. 16. SEM image of fractured 90° impact specimen magnified (a) 20X and (b) 160X

The fracture surfaces of the 0° raster specimens were macroscopically rougher, in contrast, displaying slightly jagged edges of individual broken fibers along the transverse fracture plane. Interlayer delamination was also evident and resulted in hinged specimens at failure. Figure 17 displays a 0° specimen where the slightly varied fiber length is evident, along with the formation of a hinge. The longitudinal specimens had the highest mean impact strength, correlating well with the results of three-point bend tests.

Fig. 17. SEM image of impact fractured 0° raster specimen

Weak interfaces running at ±45° to the crack front require relatively high amounts of energy as a result of a mixed fracture mode. As a result, a rougher texture was evident for the fracture surfaces of the default raster specimens, and several hinged as shown on the bottom of Figure 18a. The fracture patter of the 45° specimens, in contrast, fractured more consistently along the 45° maximum shear plane as shown on the top of Figure 18a. The macroscopic roughness of the fracture surface was the result of the varying fiber lengths, evident in the SEM image of Figure 18b.

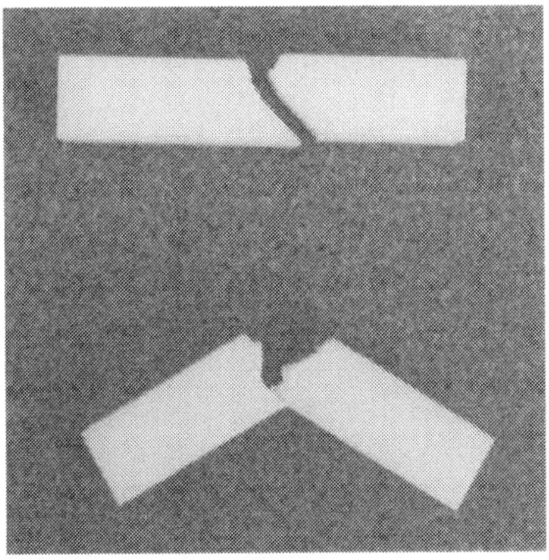

(a)

(b)

Fig. 18. (a) Fracture patterns of 45° & +45°/-45° specimens; (b) SEM image of impact fractured 45° raster specimen

3.5 Tension-tension fatigue testing

In analyzing the feasibility of fused deposition modeling to fully evolve into a rapid manufacturing tool, it is important to assess the fatigue properties of FDM specimens and their dependence upon raster orientation. A comprehensive fatigue study is warranted and is currently underway by the authors. At this time, however, a pilot fatigue study utilizing a tension-tension loading configuration has been completed. The fatigue tests utilized specimens originally with natural undamaged surfaces. The maximum load of the fatigue cycles was set to 70% of the mean failure load as determined in the static tensile tests for each specific raster orientation. The minimum cycle load was set as 1/10 of the maximum load, and the resulting stress ratio R for the fatigue tests was 0.1. The value of 70% of the mean failure load was selected following a preliminary study focused on determining a reasonable compromise between the occurrences of excessively long fatigue lives and being too close to the static strength.

All fatigue tests were performed at room temperature. To eliminate any heating effects due to considerable strains, the fatigue loading was applied at a low frequency, 0.25 Hz. The specimen surface was observed during testing and the number of cycles to final failure was determined. The test results appear in Table 11. The 45° raster orientation fractured with the smallest mean number of cycles to failure (1312), representing only 26.7% of that of the +45°/-45° raster orientation (4916).

Raster Orientation	Mean No. Cycles to Failure	Standard Deviation
Longitudinal (0°)	4557	694
Diagonal (45°)	1312	211
Transverse (90°)	1616	195
Default (+45°/-45°)	4916	150

Table 11. Tension-tension fatigue test results

A one-way ANOVA was completed in order to consider the equivalence of the mean number of cycles to failure for the four raster orientations. The results, appearing in Table 12, provide a calculated F-test statistic of $F(3,16) = 124.95$ and a p-value of 0.0001, indicating a significant difference between some or all of the mean cycle values associated with the four raster orientations at a level of significance of $\alpha = 0.05$. The resulting coefficient of determination associated with this analysis was $R^2 = 0.9591$.

Source	DF	SS	MS	F	P
Raster Angle	3	54093034	18031011	124.95	0.0001
Error	16	2308959	144310		
Total	19	56401993			

Table 12. One-way ANOVA results for tension-tension fatigue testing

Tukey post hoc comparisons indicated that the difference between the mean number of cycles to failure was not significant for the pairwise comparisons of 0° with +45°/-45°, or 45° with 90° raster specimens. All other pairwise comparisons were significantly different.

These results confirm that certain raster orientations have a significant effect on the tension-fatigue properties of the FDM specimens.

The failure modes of the specimens were similar to those for static tension testing (Figure 5), except that several of the 0° raster specimens fractured with a more uneven and almost toothed appearance during fatigue testing. This is shown in Figure 19a where the clusters of rasters have broken at various fiber lengths showing an erratic crack path most probably driven by the areas of weakest fiber bonds and voids between fibers. This SEM image also shows the smooth, brittle, tensile failure on each individual raster face.

The fracture surfaces of the +45°/-45° raster specimens, in contract, showed a mixed mode repeated failure of individual fibers by shearing and tension (Figure 19b). Upon close examination of the individual raster faces, failure initiation sites can be observed at multiple locations. In areas of closely bonded clusters of fibers, "river patterns" can be observed and are believed to occur at large crack growth rates. At the same time, patterns resembling "fish scales" are observed and are often an indication of small crack growth rates. This change of the pattern indicates the existence of a dynamic transition of failure mode.

(a) (b)

Fig. 19. SEM images of fatigue fractured specimens with: (a) 0° rasters (b) +45°/-45° rasters

There was some level of correlation between the tension-tension fatigue results and the static tension test results. While the 0° raster orientation achieved the maximum tensile strength, the +45°/-45° specimens survived the most fatigue cycles to failure on average. However, the mean number of cycles to failure for the 0° raster orientation was not found to be statistically different than the +45°/-45° specimens at a level of significance of $\alpha = 0.05$.

Although these fatigue tests only serve the purpose of a pilot study, the results indicate that the directionality of the polymer molecules and the presence of air gaps and porosity result in anisotropic behaviour of FDM specimens under tension-fatigue loading.

4. Conclusion

The mechanical properties of ABS specimens fabricated by fused deposition modelling display anisotropic behaviour and are significantly influenced by the orientation of the layered rasters and the resulting directionality of the polymer molecules. The presence of air gaps and the quantity of air voids between the rasters or fibers additionally influences the strength and effective moduli in regard to all of the tests completed in this study.

a. Tension tests indicate that the ultimate and yield strengths are the largest for the 0° raster orientation, followed by the +45°/-45°, 45°, and 90° orientations in descending order. The differences between mean ultimate tensile strengths are significant for all pairwise comparisons of different raster orientations. Fracture paths are affected by the directionality of the polymer molecules and the strength of individual layers. The longitudinal specimens benefit from the alignment of molecules along the stress axis.
b. The compression test data indicates that the 45° raster specimens are significantly weaker in compression than the other raster orientations, and they distort prior to failure as a result of shearing along the raster axes. The other three raster orientations have mean yield strengths that are significantly larger than that of the 45° raster specimens, and that are statistically equal to each other at a level of significance of $\alpha = 0.05$.
c. The results of both three-point bend and impact tests correlate well with tension test results, again indicating that the yield strengths are the largest for the 0° raster orientation, followed by the +45°/-45°, 45°, and 90° orientations in descending order. The 0° rasters offer the most resistance to bending due to the largest effective raster lengths. As raster angle increases, the effective length and associated flexural and impact strengths decrease. Mean flexural and impact strengths are significantly affected by raster orientations, with the pairwise comparison of 45° and 90° rasters as the only one with no statistical difference.
d. Preliminary tension-tension fatigue tests indicate anisotropic behaviour on the basis of raster orientations. The difference between the mean number of cycles to failure was statistically significant for all pairwise comparisons other than 0° with +45°/-45°, and 45° with 90° raster specimens. Failure modes are similar to those seen in static tension tests.

The results of this project are useful in defining the most appropriate raster orientation for FDM components on the basis of their expected in-service loading. Results are also useful to benchmark future analytical or computational models of FDM strength or stiffness as a function of void density. Additional research currently in progress includes a thorough fatigue analysis of FDM specimens with varying raster orientations.

5. References

Ahn, S., Montero, M., Odell, D., Roundy, S. & Wright, P. (2002), Anisotropic Material Properties of Fused Deposition Modeling ABS. *Rapid Prototyping Journal*, Vol. 8, No. 4, pp. 248 -257, ISSN 1355-2546

ASTM Standard D256. (2010). Standard Test Methods for Determining the Izod Pendulum Impact Resistance of Plastics. *ASTM International*, West Conshohocken, Pennsylvania, DOI: 10.1520/D0256-10, Available from: <www.astm.org>

ASTM Standard D695. (1996). Standard Test Method for Compressive Properties of Rigid Plastics. *ASTM International*, West Conshohocken, Pennsylvania, DOI: 10.1520/D0695-10, Available from: <www.astm.org>

ASTM Standard D790. (2010). Standard Test Methods for Flexural Properties of Unreinforced and Reinforced Plastics and Electrical Insulating Materials. *ASTM*

International, West Conshohocken, Pennsylvania, DOI: 10.1520/D0790-10, Available from: <www.astm.org>

ASTM Standard D3039/D3039M - 08. (2008). Standard Test Method for Tensile Properties of Polymer Matrix Composite Materials. *ASTM International*, West Conshohocken, Pennsylvania, DOI: 10.1520/D3039_D3039M-08, Available from: <www.astm.org>

ASTM Standard D3479. (2007). Standard Test Method for Tension-Tension Fatigue of Polymer Matrix Composite Materials. *ASTM International*, West Conshohocken, Pennsylvania, DOI: 10.1520/D3479M-96R07, Available from: <www.astm.org>

Caulfield, B., McHugh, P. & Lohfeld, S. (2007). Dependence of mechanical properties of polyamide components on build parameters in the SLS process. *Journal of Materials Processing Technology*, Vol. 182, pp. 477–488, ISSN 0924-0136

Chua, C., Feng, C., Lee, C. & Ang G. (2005). Rapid investment casting: direct and indirect approaches via model maker II. *International Journal of Advanced Manufacturing Technology*, Vol. 25, pp. 11–25, ISSN

Es Said, O., Foyos, J., Noorani, R., Mendelson, M., Marloth, R. & Pregger, B. (2000). Effect of layer orientation on mechanical properties of rapid prototyped samples. *Materials and Manufacturing Processes*, Vol. 15, No. 1, pp. 107–22, ISSN 1532-2475.

Lee, B., Abdullah, J. & Khan, Z. (2005). Optimization of rapid prototyping parameters for production of flexible ABS object. *Journal of Materials Processing Technology*, Vol. 169, pp.54–61, ISSN 0924-0136

Lee, C., Kim, S., Kim, H. & Ahn, S. (2007). Measurement of anisotropic compressive strength of rapid prototyping parts. *Journal of Materials Processing Technology*, Vol. 187–188, pp. 627–630, ISSN 0924-0136

Montgomery, D. (2009). *Design and Analysis of Experiments* (7th Edition), John Wiley & Sons, ISBN 978-0-470-12866-4, Hoboken, New Jersey

Odian, G. (2004). *Principles of Polymerization*(4th Edition), John Wiley & Sons, ISBN 978-0-471-27400-1, Hoboken, New Jersey

Riley, W., Sturges, L. & Morris, D. (2006). *Mechanics of Materials* (6th Edition), John Wiley & Sons, ISBN 978-0-471-70511-6, Hoboken, New Jersey

Rodriguez, J., Thomas, J. & Renaud, J. (2001). Mechanical Behavior of Acrylonitrile Butadiene Styrene (ABS) Fused Deposition Materials. Experimental Investigation. *Rapid Prototyping Journal*, Vol. 7, No. 3, pp. 148-158, ISSN 1355-2546

Rodriguez, J., Thomas, J. & Renaud, J. (2003). Mechanical behavior of acrylonitrile butadiene styrene fused deposition materials modeling. *Rapid Prototyping Journal*, Vol. 9, No. 4, pp. 219-230, ISSN1355-2546

Sood A., Ohdar R. & Mahapatra, S. (2010). Parametric appraisal of mechanical property of fused deposition modelling processed parts. *Materials & Design*, Vol. 31, No. 1, pp. 287–95, ISSN 0261-3069

Sood, A., Ohdar, R. & Mahapatra, S. (2011). Experimental investigation and empirical modeling of FDM process for compressive strength improvement. *Journal of Advanced Research*, DOI:10.1016/j.jare.2011.05.001, ISSN 2090-1232

Sun, Q., Rizvi, G., Bellehumeur, C. & Gu, P. (2008). Effect of processing conditions on the bonding quality of FDM polymer filaments. *Rapid Prototyping Journal*, Vol. 14, No. 2, pp. 72 - 80, ISSN 1355-2546

Upcraft, S. & Fletcher, R. (2003). The rapid prototyping technologies. *Assembly Automation*, Vol.23, No.4, pp. 318–330, ISSN 0144-5154

8

Design and Evaluation of Self-Expanding Stents Suitable for Diverse Clinical Manifestation Based on Mechanical Engineering

Daisuke Yoshino and Masaaki Sato
Department of Biomedical Engineering, Graduate School of Biomedical Engineering,Tohoku University, Japan

1. Introduction

Atherosclerosis is one of the most prominent diseases that induce dysfunction of circulation, and it is a disease of large and medium size arteries. If cholesterol presenting at high concentration in a blood injures an intima, a white corpuscle, i.e. a monocyte, goes into the intima and mutates into a foam cell. Then, smooth muscle cells migrate from the media to the intima, and they grow proliferously there. Based on these phenomena, cholesterol and other lipid materials accumulate in the intima. Atherosclerosis has become a serious problem in the developed countries that are aging. Therefore, countermeasures to the atherosclerosis have become important. Although there are various medical treatments for the atherosclerosis, a stent placement has received much attention as a minimally invasive procedure for vascular stenotic lesion based on the coronary atherosclerosis, the arteriosclerosis obliterans, etc. A stent is a cylindrical tube-shaped medical device that can expand the stenotic lesion in a blood vessel continuously. When considering the expansion method of a stent, two types are available. One is a self-expanding type that can expand by itself when released from the sheath of a catheter. Another is a balloon-expandable type that must be expanded forcibly using a balloon catheter. Because the self-expanding stent continues to expand to the memorized diameter at the stenotic lesion, it has the long-term patency of a vascular wall. In the present study, the main target is the self-expanding type.

Recently, the severe problem of in-stent restenosis has arisen in a blood vessel with a stent placed and left in it. In-stent restenosis results from the neointimal thickening in the blood vessel based on the hyperplasia of smooth muscle cells. The hyperplasia of smooth muscle cells is caused by a mechanical stimulus from the stent to the vascular wall. The drug-eluting stent (DES) containing immunosuppressive agents is already in clinical use to resolve this problem (Morice et al., 2002). It can be said that the DES is more effective in preventing the development of restenosis than a bare metal stent (BMS). However, it can be said that the DES does not help to improve the life prognosis or to prevent myocardial infarction (Babapulle et al., 2004; Kastrati et al., 2007; Lagerqvist et al., 2007). It is also reported that the DES might cause deterioration in the life prognosis, although the BMS does not (Nordmann et al., 2006). Depletion of immunosuppressive agents has been pointed

Step 4. Selection of stent with the lowest flexural rigidity in adequate range: If the stent, which is in the adequate range, is selected, it can expand the stenotic part in the blood vessel sufficiently. The stent sometimes has too high flexural rigidity for the lesion because of selection only in terms of the stent stiffness in radial direction. When selecting a stent, its flexural rigidity should be considered. But we have no basis of the flexural rigidity for selecting a stent. Therefore, it is decided to select the stent that has the lower flexural rigidity than any other stent being in the adequate range. For the assumed symptom, SENDAI (SD8, d_s = 8 mm) and Zilver® (ZIL, d_s = 10 mm) are most suitable.

In this section, we introduced the method to select a stent suitable for the patient's symptom based on mechanical properties of the stent. It is considered that the selection method can help doctors greatly in clinical sites. Commercially available stents are targeted for this selection method. There are limitations to selecting a suitable stent using this method. Therefore, a novel stent has to be designed for providing the stent more suitable for the patient's symptom. The method to design more suitable stent will be described in the following sections.

3. Design support system for self-expanding stents

A design support system for a self-expanding stent using CAD and CAE is introduced in this section. This support system can improve the efficiency of the suitable stent design.

3.1 Design variables of SENDAI stent

Figure 7 shows a two-dimensional diagram of a SENDAI stent. Each wire section is constructed from 12 loosely curved S-shaped wires. The strut section of the stent connects them using three bridge wires. On the two-dimensional shape of the SENDAI stent in Fig. 7, the design variables that might affect the mechanical properties of the stent are set. Here, l_w and l_b are the length of the wire and the bridge wire along the axial direction, θ_w and θ_b are the angle of the wire and the bridge wire to the axial direction, t_w and t_b are line element width of the wire and the bridge wire, and r_i and r_o are the inner radius and the outer radius of the wire end part. Every wire is structured in an arc shape. When the design variables l_w, θ_w, and others are given, the arc shape is determined and wire section is constructed through laying out these arc shapes continuously. Furthermore, the number of wires n_w, the number of bridge wires n_b, and the thickness of the tube material t_s are also design variables.

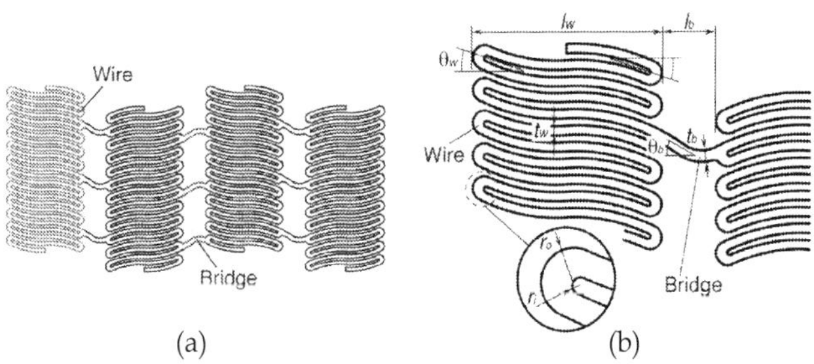

Fig. 7. Two-dimensional diagram of SENDAI stent (a) and main design variables for SENDAI stent (b)

3.2 Framework of design support system

Figure 8 shows the design support system for a self-expanding stent. The left-hand side of the figure shows the production process of the SENDAI stent. It has three production stages: a 'manufacture' stage, during which the NC data are created based on the two-dimensional diagram to manufacture the initial stent shape while the initial stent is manufactured by using laser processing; an 'expansion' stage, during which the initial stent is forcibly expanded in the radial direction by inserting a tapered rod into the stent as it is given shape-memory treatment; and an 'evaluation' stage, during which the performance of the expanded stent is tested. The right-hand side of Fig. 8 shows the flow of the shape design for self-expanding stents being proposed. A three-dimensional model of the initial stent manufactured by using laser processing is created from the two-dimensional shape by using 3D CAD. Then, by dividing into finite elements, the finite element model representing the initial stent is created based on the 3D CAD model, and the expanded stent shape is predicted by applying an expansion analysis using the finite element method for large deformation. Based on this prediction, a rigidity analysis is conducted using a non-linear finite element method. The mechanical properties of the stent are evaluated from the results. This process corresponds to the actual production process of the 'manufacture,' 'expansion,' and 'evaluation' stages.

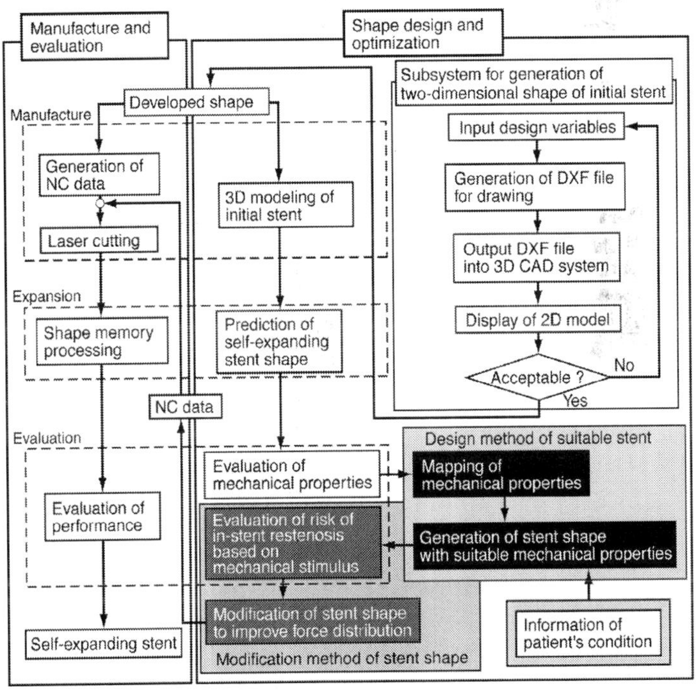

Fig. 8. Design support system for self-expanding stent. The left-hand side shows the production process of the SENDAI stent. The right-hand side shows the flow of the proposed design support system.

This support system has design method of a self-expanding stent suitable for the patient's symptom based on mechanical properties of a stent. This method is available to introduce into the existing design support system described above, and has two stages which are the design and modification methods. In the first stage, a stent shape with mechanical properties suitable for the patient's symptom is determined and designed. In the second stage, to modify the stent shape in consideration of the risk of in-stent restenosis realizes designing the stent shape more suitable for the patient's symptom. The risk of in-stent restenosis is evaluated based on a mechanical stimulus to a vascular wall by insertion of a stent. These two stages of the design method will hereinafter be described in more detail (see in Section 6).

After the two-stage design method, the design support system ends with the generation of the NC data of the designed stent necessary for moving onto the actual production process. For change in a stent shape, a subsystem for generating the two-dimensional shape of the initial stent was introduced, and this is available for changing the two-dimensional shape flexibly. It can also be used to generate a two-dimensional diagram of the stent in the first place.

4. Mechanical properties of stent

It is important to obtain sensitivities of mechanical properties of a stent to design variables when designing the stent suitable for patient's symptom. In this section, important mechanical properties, namely, radial stiffness, flexural rigidity, and shear rigidity, are evaluated and the maps of their sensitivities are made. The expanded stent shape is used for evaluation of mechanical properties. The expanded shape is predicted from the expansion analysis (see in (Yoshino & Inoue, 2010) for details).

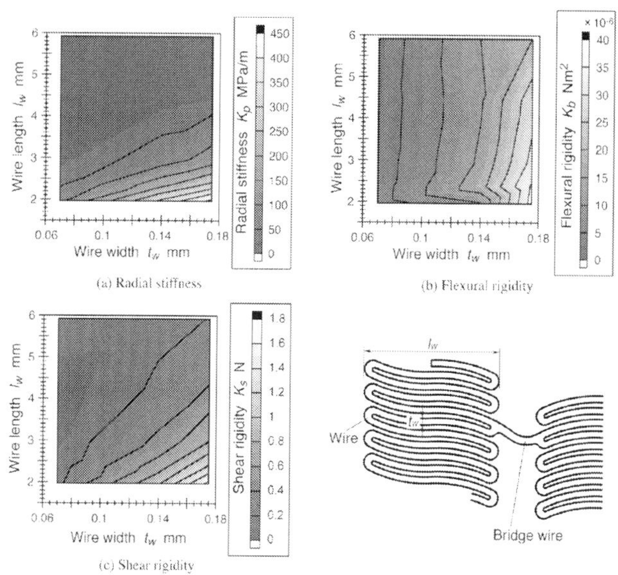

Fig. 9. Sensitivities of the mechanical properties of the SENDAI stent 6 mm diameter to thedesign variables: (a) radial stiffness, (b) flexural rigidity, and (c) shear rigidity

We must know the relationship between the mechanical properties of a stent and the design variables in order to design a stent with specific properties. The mechanical properties of the SENDAI stent, such as radial stiffness, flexural rigidity, and shear rigidity, were evaluated, and their sensitivities to the design variables were also defined, as shown in Fig. 9 (Yoshino & Inoue, 2010). The wire length along the axial direction and the wire width were selected as design variables. Isolines on the maps of mechanical properties are very important for proposing designs. The isoline is plotted onto the maps based on the required mechanical property, and design variables of the proposed design are determined from the isoline. In addition, we assumed the mechanical properties of the stent material as illustrated in Fig. 10: Young's modulus of 28 GPa, and Poisson's ratio of 0.3.

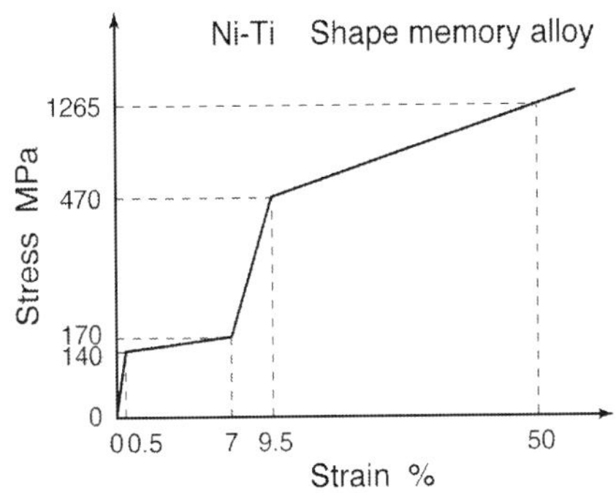

Fig. 10. Assumed stress-strain relationship for the stent material

5. Estimation of force on vascular wall caused by insertion of self-expanding stent

The forces on a blood vessel are classified into two categories: internal pressure caused by the expansion of the stent and the force resulting from the straightening of the blood vessel, which is a phenomenon whereby a curved blood vessel is straightened by the stent. Straightening of blood vessels often occurs in cases involving the use of a closed-cell stent. As a result, a problem occurs in that the straightening easily encourages kinking of the blood vessel at the flexural area distant from the stented lesion (Tamakawa et al., 2008).

In this section, focusing on a method by which to improve the force distribution on the vascular wall according to the symptoms of the patient, we introduce a method to compute the distribution of the contact force between the stent and the blood vessel under the assumption that the stent is inserted into a straight blood vessel. In the method, the stent and the blood vessel are simplified as axisymmetrical models. Then a method for calculating the distribution of the straightening force on the vascular wall is introduced.

5.1 Computation of contact force distribution on vascular wall caused by expansion of stents

The expansion of a stent in a blood vessel induces pressure on the contact surfaces of the stent and the blood vessel. Analysis of pressure or contact force based on the complicated shape of stent is a difficult and time-consuming task. Therefore, a simplified calculation method using the axisymmetrical models is presented as shown in Fig. 11. The stent is modeled by a number of rings, indicated by broken lines in Fig. 11. The wire section is represented by 11 rings. The blood vessel is similarly modeled by rings, where the ring intervals are the same as those of the stent. In addition, it is assumed that the blood vessel is much longer than the stent. It is also assumed that these rings deform axisymmetrically due to the uniformly distributed radial force.

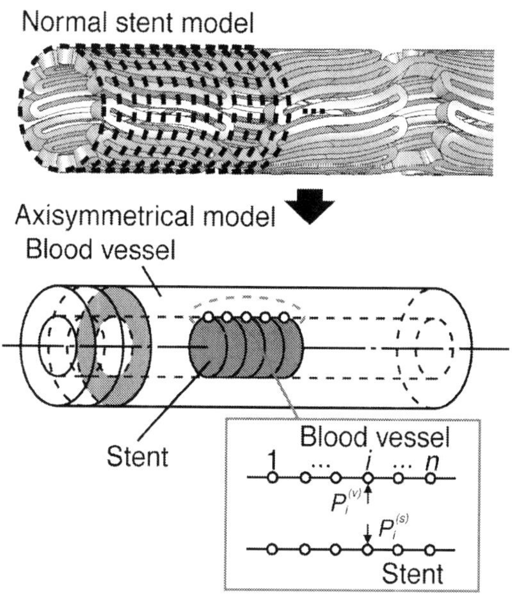

Fig. 11. Model used to compute the contact force between the stent and vascular wall. Thestent and blood vessel are simplified to axisymmetrical models.

Next, let us consider ring i (i = 1, 2, ..., n) in contact along the surfaces of stent and vascular wall. When a unit radial force is applied to ring i of model m (m = s, v, which correspond to the stent and the blood vessel, respectively), the radial displacement at ring j, denoted as $r_{ji}^{(m)}$, is calculated using the finite element method. The influence matrix $[C^{(m)}]$ is defined in terms of as $r_{ji}^{(m)}$ as follows:

$$\left[C^{(m)}\right]=\left[\left\{r_1^{(m)}\right\},\left\{r_2^{(m)}\right\},\ldots,\left\{r_n^{(m)}\right\}\right] \tag{6}$$

where

$$\left\{r_i^{(m)}\right\}=\left(r_{1i}^{(m)},r_{2i}^{(m)},\ldots,r_{ni}^{(m)}\right)^{\mathrm{T}} \tag{7}$$

The radial displacement $\{r^{(m)}\}=(r_1^{(m)}, r_2^{(m)}, \ldots, r_n^{(m)})^T$ due to unknown contact force $\{P^{(m)}\}=(P_1^{(m)}, P_2^{(m)}, \ldots, P_n^{(m)})^T$ is given as follows:

$$\left[C^{(m)}\right]\left\{P^{(m)}\right\}=\left\{r^{(m)}\right\} \tag{8}$$

Therefore, the following expressions, namely, the equilibrium equation and the condition ofcontact, are obtained:

$$\left.\begin{array}{l} P_i^{(v)}+P_i^{(s)}=0 \\ R_i^{(v)}+r_i^{(v)}=R_i^{(s)}+r_i^{(s)} \end{array}\right\} \text{(contact)} \tag{9}$$

$$\left.\begin{array}{l} P_i^{(v)}=P_i^{(s)}=0 \\ R_i^{(v)}+r_i^{(v)}>R_i^{(s)}+r_i^{(s)} \end{array}\right\} \text{(non}-\text{contact)} \tag{10}$$

where $R_i^{(s)}$ and $R_i^{(v)}$ denote the initial radii of the stent and the blood vessel, respectively, at i.

Then, by replacing $\{P^{(v)}\}$ and $-\{P^{(s)}\}$ with $\{P\}$, the equation of contact force is obtained as follows:

$$\left[C^{(v)}+C^{(s)}\right]\{P\}=\left\{R^{(s)}-R^{(v)}\right\} \tag{11}$$

Equation (11) is solved for $\{P\}$ to obtain the distribution of the contact force between the stent and the blood vessel wall. Since the rigidity of the strut section of the stent is considerably much lower than that of the wire section, $r_{ji}^{(s)}$ is negligible, except for the case in which i and j are located in the wire section of the stent. Therefore, in the present study, the influence matrix $[C^{(s)}]$ is composed by diagonally placing the partial matrix for the wire section, and the contact force on the strut section of the stent is not evaluated. The Gaussian elimination method is used to solve the simultaneous equations.

5.2 Calculation of distribution of straightening force on vascular wall

Straightening of the blood vessel occurs when a straight stent is inserted into a curved vessel. This insertion causes a straightening force to act on the vascular wall. The straightening effect should be evaluated while considering the interaction process of the expansion of the stent with the curved blood vessel. However, solving this problem is extremely complicated. Therefore, the straightening of the blood vessel is assumed to be independent from the expansion of the vessels by the stent, and it is simplified as shown in Fig. 12. The stent is approximated by a beam with the flexural rigidity obtained in the previous section, and the stent is modeled as a laminated beam by combining with a curved beam, which is used a model of the blood vessel.

The beam models are divided into n intervals. When a unit force is applied to point i of model m (m = s, v, which correspond to the stent and the vessel, respectively), the deflection at point j is denoted as $d_{ji}^{(m)}$. The n-order influence matrix $[D^{(m)}]$ is defined in terms of $d_{ji}^{(m)}$ as follows:

$$\left[D^{(m)}\right]=\left[\left\{d_1^{(m)}\right\},\left\{d_2^{(m)}\right\},\ldots,\left\{d_n^{(m)}\right\}\right] \tag{12}$$

where

$$\left\{d_i^{(m)}\right\}=\left(d_{1i}^{(m)},d_{2i}^{(m)},\ldots,d_{ni}^{(m)}\right)^{\mathrm{T}} \tag{13}$$

The force $\{F^{(m)}\}$ distributed along the beam is related to the deflection $\{d^{(m)}\}$ so as to satisfy the following equation:

$$\left[D^{(m)}\right]\left\{F^{(m)}\right\}=\left\{\delta^{(m)}\right\} \tag{14}$$

The equilibrium equation and contact condition are given by the following equations:

$$F_i^{(s)}+F_i^{(v)}=0 \tag{15}$$

$$\delta_i^{(s)}=\delta_i^{(i)}+\delta_i^{(v)} \tag{16}$$

where $d_i^{(i)}$ denotes the initial deflection of the blood vessel at point i, and $d_i^{(s)}$ and $d_i^{(v)}$ are the deflections of the stent and the blood vessel, respectively. Replacing -$\{F^{(v)}\}$ and $\{F^{(s)}\}$ with $\{F\}$, the equation of straightening force is obtained as follows:

$$\left[D^{(v)}+D^{(s)}\right]\{F\}=\left\{\delta^{(i)}\right\} \tag{17}$$

5.3 Limitations of these calculation methods

The methods will be useful for improving stent shape in order to reduce the peak force acting on the vascular wall. Although these methods are useful to calculate a contact force and a straightening force on a vascular wall, there are the following limitations:

1. The isotropic material property was assumed for the artery.
2. The cross-sectional distortions and thickness changes of the stent and blood vessel were ignored.
3. The friction on the contact surface between the stent and blood vessel was ignored.
4. It was consider that the straightening of the blood vessel and the expansion of the vessels by the stent were independent each other.
5. The stent and blood vessel were simplified by using axisymmetrical models in computation of the contact force and using the laminated beam in calculation of the straightening force, respectively.

6. Design method of stent suitable for diverse clinical manifestation

The stent must have the radial stiffness sufficient to expand the stenotic part in the blood vessel outward. Simultaneously, it must be sufficiently flexible to conform to the vascular wall. Neither the symptom nor the blood vessel shape is always in the same state. Therefore, it is more important to design stents suited to each unique symptom of every patient. (Colombo et al., 2002) made evaluations of stent 'deliverability,' 'scaffolding,' 'accurate positioning,' and so on for the average lesion of the coronary artery. They proposed a guideline for use in determining a suitable stent. This guideline was based on clinical trials. Therefore, it is governed by the doctor's sense. One more noteworthy point is that the best-suited stent cannot be actually available for every specific symptom because a stent must be chosen only from among commercially available stents.

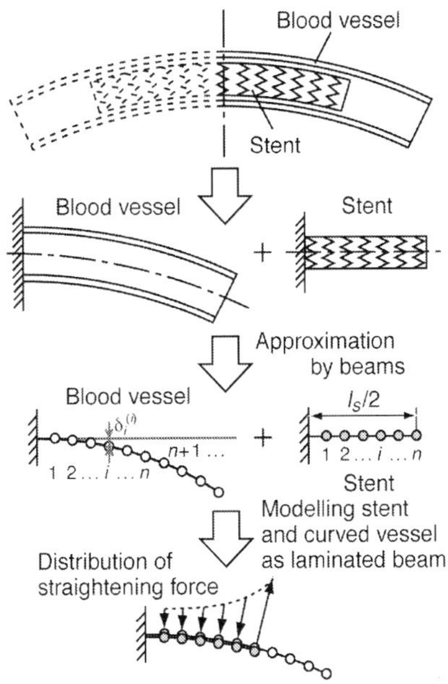

Fig. 12. Model for calculation of the straightening force on vascular wall. The stent andblood vessel are simply modeled as a laminated beam.

In this section, we describe a method for designing a stent that has good mechanical properties to suit diverse clinical manifestation. Figure 13 shows the flow of designing a stent suitable for diverse clinical manifestation. The first step is to determine the radial stiffness of the stent necessary to expand the stenotic part in the blood vessel based on symptom information. Next, based on the determined radial stiffness and the sensitivities of mechanical properties of the stent defined in Section 4, the design variables of a suitable stent are determined. In the second step, the force on the vascular wall by insertion of the designed stent is first evaluated by using the methods described in Section 5. This force is associated with the risk of in-stent restenosis. Next, based on the evaluation result, the designed stent is modified to be more suitable for the symptom. After modification, the force on the vascular wall is evaluated again. The effect of shape modification is confirmed by comparing the forces on the vascular wall between before and after modification. Finally, the modified stent shape is proposed as better suited stent shape. The detail of this proposed design method will be described in following sections.

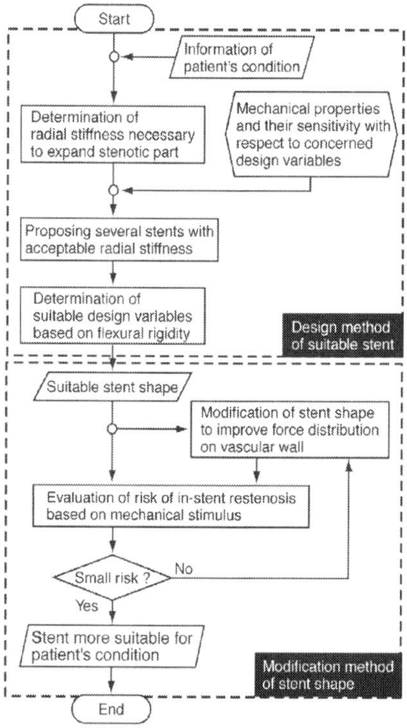

Fig. 13. Flow of designing stent suitable for diverse clinical manifestation. Designing a stent suitable for diverse clinical manifestation consists of two steps. In the first step, it is possible to design a stent necessary to expand the stenotic part of a blood vessel. In the second step, the designed stent is modified to suit the patient's symptom better.

6.1 Design method of stent having suitable mechanical properties to expand stenotic artery

6.1.1 Mechanical properties of artery model

When considering expanding the stenotic part in the blood vessel as presented in Fig. 14, it is important to know the vascular mechanical properties. Arteries in a living body are always pressured inside. Simultaneously, they are tensed with strain produced along its axial direction by several tens of percent. In this sense, it can be regarded as a thick cylinder being in a multiaxial stress state. It is necessary to assume such a multiaxial stress test to measure mechanical properties of the blood vessel. This means that it is extremely difficult to obtain mechanical properties of the blood vessel because the mechanical properties are associated with various levels and types of blood vessels.

In this study, the pressure strain elastic modulus propounded by (Peterson et al., 1960), which is easy to manipulate, is used. The pressure strain elastic modulus $E_{p,v}$ of a blood vessel can be expressed by the following equation

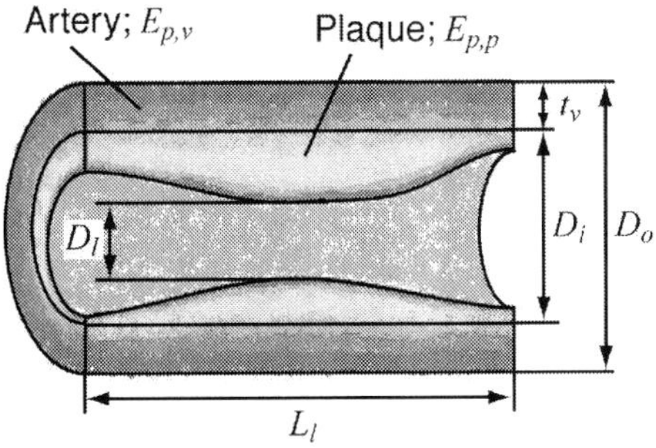

Fig. 14. Dimension of blood vessel with stenosis

$$E_{p,v} = \frac{\Delta p}{\Delta D_o / D_o} \tag{18}$$

where D_o is the blood vessel's outer diameter, and ΔD_o and Δp are the increase in the blood vessel diameter and the increase in the internal pressure respectively. The pressure strain elastic modulus has been used widely in clinical studies. Therefore, many reports on that subject are available. Table 3 presents value of the pressure strain elastic modulus for each type of human blood vessel, extracted from references (Hayashi et al., 1980; Hayashi, 2005; Stratouly et al., 1987; Gow & Hadfield, 1979), with modification.

Artery	$E_{p,v}$	Reporter
Arteria pulmonalis	0.016	Greenfield and Griggs (1963)
Ascending aorta	0.076	Patel et al. (1964)
Basilar artery	0.186 (β=14)	Hayashi et al. (1980)
Common carotid artery	0.049	Arndt et al. (1969)
Common iliac artery	0.120	Stratouly et al. (1987)
Coronary artery	0.602	Gow and Hadfield (1979)
Femoral artery	0.433	Patel et al. (1964)
Thoracic aorta	0.126	Luchsinger et al. (1962)

Table 3. Pressure strain elastic modulus for each type of human artery

6.1.2 Determination of radial stiffness necessary to expand stenotic part in artery

To determine the radial stiffness of the stent necessary to expand the stenotic part in the blood vessel, the requirements are, in addition to the pressure strain elastic modulus $E_{p,v}$ of the blood vessel, shown as follows. The outer and inner diameters, D_o and D_i, of the blood vessel in the normal state, the least diameter D_l produced by the stenotic part, and the length of the stenotic part L_l are required. The pressure strain elastic modulus of the plaque $E_{p,p}$ is required. Also required is the inner diameter after the treatment is made, D_t, which is an indicator to use as the target setting for the percentage by which to improve the blood flow level there. Given all the values listed above, the calculations for the necessary radial stiffness are made as follows.

By knowing that the increase in the blood vessel diameter is obtainable from circumferential strain of a cylinder, the blood vessel with stenosis can be modeled by simply using springs connected in parallel, as presented in Fig. 15. Thus, the internal pressure in blood vessel p^*, which is necessary to expand the stenotic part, can be obtained by

$$p^* = \left(E_{p,v} + E_{p,p}\right)\frac{D_t - D_l}{D_o} = E_{p,vl}\frac{D_t - D_l}{D_o} \tag{19}$$

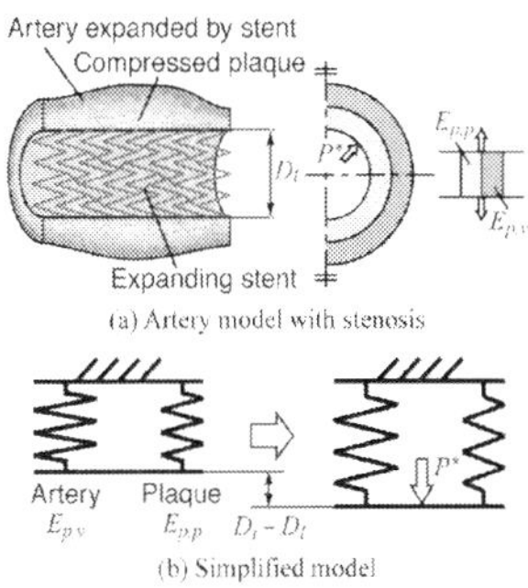

Fig. 15. Simplified modeling of expansion of stenosis in artery by insertion of stent

When measuring the pressure strain elastic modulus of the plaque $E_{p,p}$ is difficult, it is possible to replace the sum of the elastic moduli, $E_{p,v}$ + $E_{p,p}$, by the pressure strain elastic modulus of the diseased blood vessel $E_{p,vl}$, which can be easily measured.

As a matter of fact, the diameter of the stent, which is inserted into the stenotic part, is greater than the target vascular diameter after treatment. In conclusion, from the stent diameter d_s and the target vascular diameter D_t obtained after treatment, the radius reduction Δr in stent after inserting can be calculated by means of the following equation:

$$\Delta r = \frac{d_s - D_t}{2} \tag{20}$$

By substituting equations (19) and (20) into $K_p = p/\Delta r$, which was defined by (Yoshino et al., 2008) to obtain the radial stiffness, the radial stiffness K_p necessary to expand the stenotic part can be obtained from

$$K_p^* = 2\left(E_{p,v} + E_{p,p}\right)\frac{D_t - D_l}{D_o\left(d_s - D_t\right)} = 2E_{p,vl}\frac{D_t - D_l}{D_o\left(d_s - D_t\right)} \tag{21}$$

The obtained K_p^* is the least required stiffness to expand the stenotic part in the blood vessel. Therefore, when designing a stent, the chosen stiffness should be greater than this K_p^* value.

Now consider the case where a stent with a diameter of 6 mm is inserted into a coronary artery. The symptoms shown in Table 4 are examples based on references (Gow & Hadfield, 1979; Le Floc'h et al., 2009). Based on this data, the radial stiffness K_p^* necessary to expand the stenotic part in the blood vessel is calculated by equation (21) to be K_p^* = 366.7 MPa/m.

The required radial stiffness K_p^* is plotted as a broken line on the radial stiffness map shown in Fig. 16. From this line, it is possible to determine the design variables for a stent with sufficient radial stiffness to expand the stenotic part of the blood vessel. Based on these, the designer can proceed to work with some design propositions.

Coronary artery	Parameters
Outer diameter, D_o (mm)	4.90
Inner diameter, D_i (mm)	4.06
Least diameter of lesion, D_l (mm)	2.5
Length of lesion, L_l (mm)	10
Total flexion angle (deg.) (Flexion angle (deg.))	90 (45)
Rate of stenosis by ECST method (%)	38.4
Pressure strain elastic modulus of artery, $E_{p,v}$ (MPa)	0.602
Pressure strain elastic modulus of diseased artery, $E_{p,vl}$ (MPa)	0.628
Inner diameter after treatment, D_t (mm)	4.56

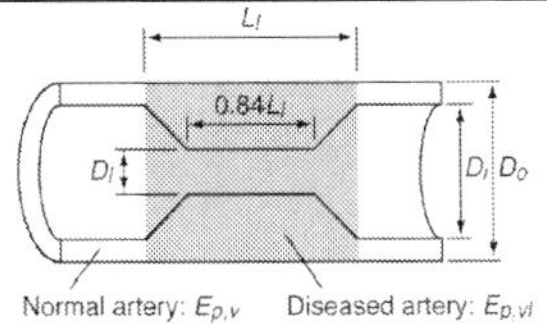

Schematic view of assumed symptom

Table 4. Assumed patient symptom information

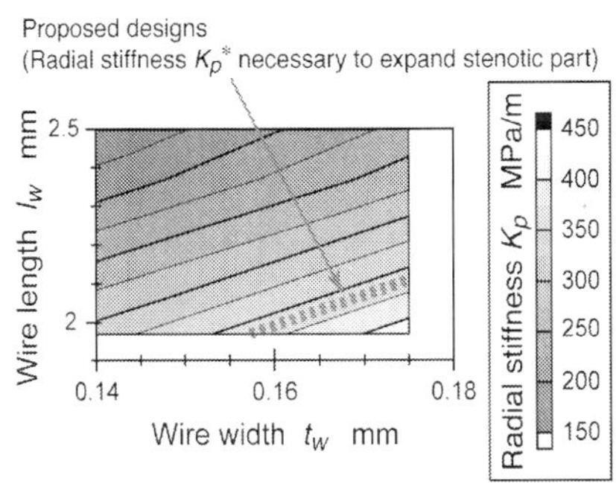

Fig. 16. Proposed designs having the radial stiffness necessary to expand the stenotic part of a blood vessel. By assuming that a stent is inserted into a coronary artery, a value of 0.602 MPa was used for the pressure strain elastic modulus $E_{p,v}$ of the normal part of the coronary artery, and a value of 0.628 MPa for the pressure strain elastic modulus $E_{p,vl}$ of the diseased artery. For the coronary artery model, D_o = 4.90 mm, D_i = 4.06 mm, and D_l = 2.5 mm were assumed. The corresponding part of the map is magnified and displayed.

6.1.3 Range of selectable flexural rigidity and the dilemma of selecting the design

After inserting an originally straight stent into a curved blood vessel and leaving it there, the stent generally conforms to the blood vessel shape. Nevertheless, because the flexural rigidity of the stent is greater than that of the blood vessel, the blood vessel tends to become straighter. This phenomenon of straightening of the blood vessel was previously described in detail, and a method to calculate the force resulting in the straightening of the blood vessel was described in previous section. It is apparent that this force depends on the flexural rigidity of the stent, and that greater flexural rigidity results in a larger force. A force too large can damage the vascular wall. Consequently, it is important to choose the most appropriate flexural rigidity of the stent.

By plotting the proposed designs obtained from the required radial stiffness K_p^* on a flexural rigidity map, the broken line shown in Fig. 17(a) is obtained. For one radial stiffness value, multiple flexural rigidity values can be selected as long as they are within the range given in the figure. None of the limiting values for the flexural rigidity, which might prevent damage on the vascular wall or neointimal thickening, are yet quantitatively available. Therefore, the flexural rigidity should be made smaller to decrease the force acting on the vascular wall. In other words, the designer should select the smallest flexural rigidity from the range of selectable values shown in Fig. 17(a).

Figure 17(b) shows the stent shape designed in consideration of the rules described above. The designed stent is referred to as SDCO and has a wire length l_w^* of 2.01 mm, and a wire width t_w^* of 0.16 mm. In addition, the length of the designed stent is determined so that it occupies the blood vessel from the stenotic part to the normal part at each end in consideration of actual clinical use. The length of the SDCO is 22.0 mm (the stent consists of 8 wire sections and 7 strut sections). The flexural rigidity of the SDCO is $K_b^* = 25.1\times10^{-6}$ Nm^2. Similarly, the shear rigidity of the SDCO, based on the map of the rigidity, is $K_s^* = 1.33$ N.

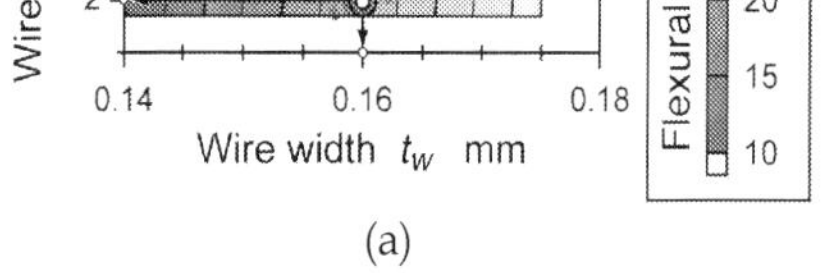

Fig. 17. Selection of the proposed design from the viewpoint of flexural rigidity. (a) The selectable range of flexural rigidities is indicated by the broken line. The corresponding part of the map is magnified and displayed as in Fig. 16. (b) A stent 6 mm in diameter is designed to suit the assumed symptom of the coronary artery.

6.2 Evaluation of the risk of in-stent restenosis based on the mechanical stimulus

Figure 18 shows the distributions of the contact force and the straightening force when the SDCO is inserted into the coronary artery. These distributions were calculated by the previously described method, which was proposed by (Yoshino et al., 2011).

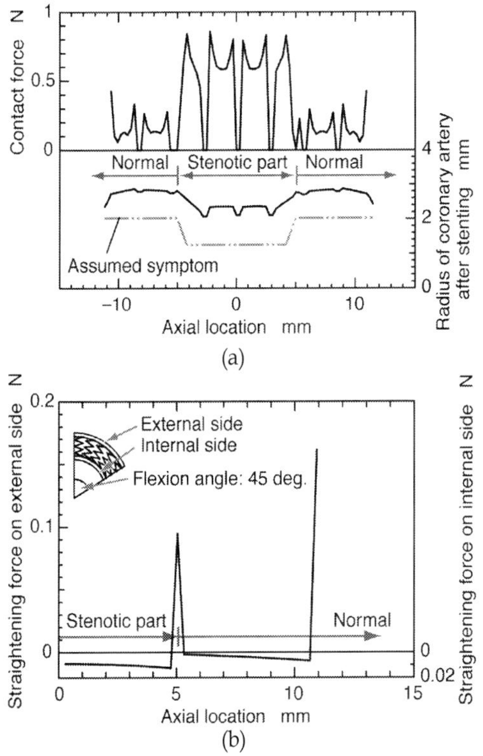

Fig. 18. Computational results of forces which are exerted on the coronary artery wall by insertion of the SDCO. (a) The distribution of the contact force between the SDCO and the coronary artery (upside), and the radius of the coronary artery after stenting (downside). The dot-dashed line indicates the initial shape of the artery wall with the assumed symptom. (b) The distribution of the straightening force on the coronary artery wall by insertion of the SDCO. The right side of the distribution is displayed from a consideration of the geometrical symmetry.

First, let us focus attention on the force on the vascular wall. Although the contact force is large at the stenotic part, the generation of this large force is unavoidable for expansion of the stenotic part. The contact force is also concentrated at both ends of the stent. The straightening force on the vascular wall is concentrated at the end of the stenotic part in addition to both ends of the stent.

It has been reported that the hyperplasia of smooth muscle cells and the neointimal proliferation occur at the stented part of an artery (Grewe et al., 2000; Clark et al., 2006). It is assumed that the hyperplasia and proliferation are caused by the mechanical stimulus acting on the vascular wall due to insertion of the stent. (Schweiger et al., 2006) reported that in-stent restenosis occurred primarily at both ends of the stent during the period one to three months after stenting. In the study of (Lal et al., 2007), the majority of patients in the target patient population showed in-stent restenosis at the stent ends. (Yazdani and Berry, 2009) cultured a stented native porcine carotid artery under physiologic pulsatile flow and pressure conditions for a week. They confirmed that the proliferation of smooth muscle cells occurred significantly at both ends of the stent. By summarizing these reports and evaluation results of the forces on the vascular wall, it is concluded that the force concentration provokes neointimal thickening due to the hyperplasia of smooth muscle cells, i.e., in-stent restenosis.

Next, the expansion of the vascular wall is examined. The normal part of the artery is expanded to become much larger than the target diameter. This excessive expansion causes expansion of the entire stented part. As a result, stagnation and vortices of blood flow are induced, further increasing the potential for stent thrombosis.

The designed stent has mechanical properties sufficient to expand the stenotic part of theartery, but is not suitable for the normal part of the artery. Thus, it requires modification.

6.3 Effective method to modify the stent shape in consideration of the risk of in-stent restenosis

6.3.1 Design objective for the shape modification

In the previous section, it was shown that although a stent suitable for the assumed symptoms could be designed using the proposed design method, further modifications were still required. Two kinds of design objectives are set up for shape modification.

Objective for the mechanical stimulus: as shown in Fig. 18, the contact force on the normal part of the artery is concentrated at both ends of the stent. This force concentration provokes neointimal thickening from the hyperplasia of smooth muscle cells. Therefore, the contact force must be reduced at both ends of the stent. However, this force should be larger than a certain limit so that stent functions on the lesion.

Objective for the blood flow: as stated above, the normal part of the artery is expanded to become much larger than the target diameter. This expansion state of the artery causes stagnation and vortices of blood flow. As a result, stent thrombosis may be induced. It is very important to reduce stagnation and vortex creation to decrease the risk of stent thrombosis. Therefore, the vascular wall should be expanded flatly by the insertion of the stent. The flat expansion of the artery can prevent the generation of stagnation and vortices in the stented artery.

6.3.2 Modification method of the stent shape to suit the clinical manifestation

The stent designed in the previous section has a uniform radial stiffness K_p^* along its axial direction. The radial stiffness K_p^* causes excessive expansion at the normal part of the artery because K_p^* was calculated for expansion of the stenotic part, which is generally stiffer than the normal part. The objective for the mechanical stimulus can be attained by decreasing the radial stiffness to a value corresponding to that of a normal artery. At the same time, the objective for blood flow must be attained. Therefore, the stent should have a radial stiffness K_p^{**} to expand the vascular wall at normal part only for the stent thickness t. This can achieve the flat expansion of the artery. By changing the stent radial stiffness at the normal part from K_p^* to K_p^{**}, the contact force can also decrease. In this study, it was decided to adopt the radial stiffness K_p^{**} as a compromise between the two design objectives.

The methods using the influence matrix described in Section 5 are used for the shape modification. Figure 19 shows the concept for modification of the stent shape. The distributed contact force P_i^* on the stenotic part can be calculated based on the radial stiffness K_p^*, as follows:

$$P_i^* = \frac{\pi D_t K_p^* \Delta r_t l_w^*}{n_{\mathrm{CP}}} \tag{22}$$

where D_t is the inner diameter of the blood vessel after treatment, Δr_t is the increase in the vascular radius of the target, l_w^* is the wire length of the stent designed based on the radial stiffness K_p^*. Also, n_{CP} represents the number of calculation points on the wire section. For simplicity, it was assumed that a uniform contact force was exerted on the vascular wall.

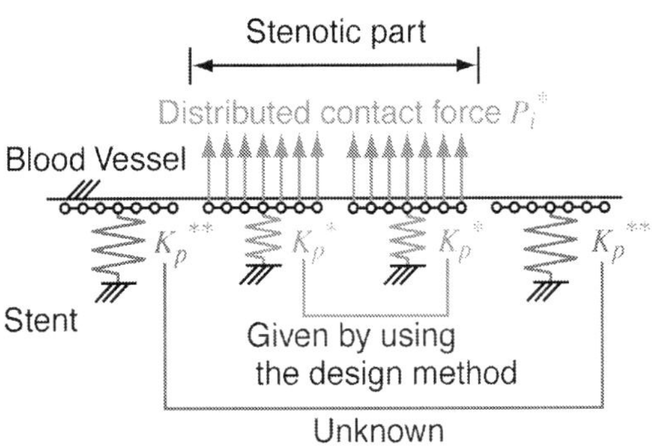

Fig. 19. Concept for modification of the stent shape. The distributed contact force is obtained from the radial stiffness of the stent on the stenotic part by using the design method. The force that should be applied on the normal artery is calculated by using the obtained distributed contact force on the stenotic part.

As described in Section 5, the radial displacement of the vascular wall $\{r^{(v)}\} = (r_1^{(v)}, r_2^{(v)}, \ldots, r_n^{(v)})^{\mathrm{T}}$ due to the unknown contact force $\{P\} = (P_1, P_2, \ldots, P_n)^{\mathrm{T}}$ is given as follows:

$$\left[C^{(v)}\right]\{P\} = \left\{r^{(v)}\right\} \tag{23}$$

where $[C^{(v)}]$ is the influence matrix of the blood vessel, defined by the radial displacement of the vascular wall due to the unit radial force. The calculated contact force P_i^* is substituted into equation (23), and the influence matrix $[C^{(v)}]$ is downsized to the matrix $[C_{normal}^{(v)}]$ of the normal blood vessel.

$$\left[C_{\text{normal}}^{(\text{v})}\right]\left\{P^{**}\right\}=\left\{r_{\text{normal}}^{(\text{v})}\right\}+\left[C_{\text{stenosis}}^{(\text{v})}\right]\left\{P^{*}\right\} \tag{24}$$

$[C_{stenosis}^{(v)}]$ is the influence matrix of the stenotic part, and $\{r_{normal}^{(v)}\}$ is the radial displacement of the normal part of the blood vessel wall. Equation (24) is solved for $\{P^{**}\}$ to obtain the distributed force P_i^{**} that can expand the normal part to the stent thickness t.

Here, it is assumed that the wire length after modification is l_w^{**} and the calculation points from k to l are included in the modified wire section. The required radial stiffness K_p^{**} is defined as follows:

$$K_p^{**}=\frac{\sum_{i=k}^{l} P_i^{**}}{\pi D_t l_w^{**} t} \tag{25}$$

The K_p^{**} value calculated by equation (25) is plotted on the radial stiffness map. As a result, the wire width t_w^{**} after modification is determined from the l_w^{**} value and the curve of K_p^{**} on the map. Therefore, the designed stent has to be modified to the shape of l_w^{**} and t_w^{**} at the normal part of the blood vessel. Considering that an increase in the stent length is undesirable, the designer should keep the wire length l_w^{**} equal to l_w^{**} after the modification. However, it is possible that the wire width t_w^{**} after the modification cannot be obtained because of the mismatch between the assumed l_w^{**} value and the curve of K_p^{**}. In this case, the designer can increase the wire length l_w^{**} , perform the same procedures, and determine the l_w^{**} and t_w^{**} values.

The required radial stiffness K_{p1}^{**} at both ends of the SDCO is 50.3 MPa/m. The radial stiffness K_{p2}^{**} at the normal part of the artery except for both stent ends is also calculated as 155.2 MPa/m. From the dot-dashed curves of K_{p1}^{**} and K_{p2}^{**} shown in Fig. 20(a), it is determined that l_{w1}^{**} is 2.52 mm, t_{w1}^{**} is 0.087 mm, l_{w2}^{**} is 2.01 mm, and t_{w2}^{**} is 0.094 mm. Figure 20(b) shows the modified shape of the SDCO.

The flexural rigidity K_{b1}^{**} at both ends of the SDCO is 5.75×10^{-6} Nm^2, and the shear rigidity K_{s1}^{**} is 0.20 N. At the normal part of the artery except for both stent ends, the flexural rigidity K_{b2}^{**} of the SDCO is 6.00×10^{-6} Nm^2, and the shear rigidity K_{s2}^{**} is 0.42 N.

6.3.3 Confirmation of the effect of the shape modification

Figure 21 shows a comparison of the force on the vascular wall by insertion of the stent before and after the modifications. After the modifications of the SDCO, an approximately 80 % reduction in the concentrated contact force was attained (Fig. 21(a)). Furthermore. the concentrated straightening force at the stent ends after modification was reduced to approximately 35 % of that before modification of the SDCO (Fig. 21(b)).

On the other hand, it is recognized that straightening force increases at the stenotic-healthy tissue interface (the axial location is 5 mm in Fig. 21(b)). The concentration of straightening force also occurs at the axial location of 8 mm. After modification of the stent shape, the flexural and shear rigidities of the stent vary with the axial location. The bending state of the stent changes at the changing point of the rigidities, which corresponds to the turn of the stent shape. Therefore, straightening force increases due to changing of the stent bending state based on rigidities changing. Although the risk of the vessel rupture slightly increases at the stenotic-healthy tissue interface, it is achieved that the straightening force is significantly reduced at both ends of the stent, where the hyperplasia of smooth muscle cells is frequently reported.

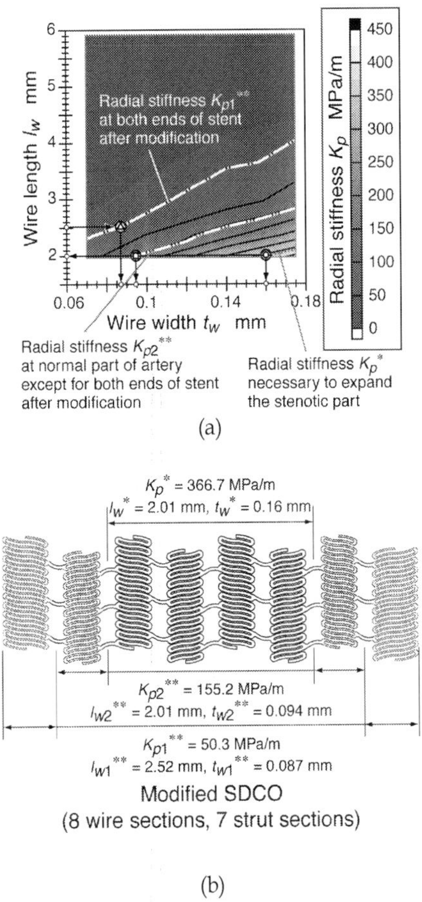

Fig. 20. The SDCO stent 6 mm in diameter is modified to suit the assumed symptom of the coronary artery. (a) Design variables after modification are determined by using the map of the radial stiffness. (b) The modified SDCO has a nonuniform shape along its axial direction.

The modified stent can expand the vascular wall in a more flat manner. Therefore, modification of the designed stent by using the proposed method can relax the force concentration at both ends of the stent by attaining flat expansion of the vascular wall.

7. Conclusion

This chapter described evaluating mechanical properties of a stent and designing/selecting a stent suitable for diverse clinical manifestation by efficiently using the evaluated results. By rounding up the results described above, it can be confirmed that two objectives, which are establishments of designing/selecting a stent suitable for diverse clinical manifestation, were achieved. The stent, which is designed by using the method described in this chapter, has nonuniform shapes along its axial direction. This shape is radically new compared to the conventional stent shape that is uniform along its axial direction. Thus, the stent shape designed by using the method is considered to be a new-generation type.

8. References

Babapulle, M. N.; Joseph, L., Bélisle, P., Brophy, J. M., & Eisenberg, M. J. (2004). A hierarchical Bayesian meta-analysis of randomised clinical trials of drug-eluting stents. *Lancet*, Vol. 364, No. 9434, (August 2004) 583-591, ISSN 0140-6736 (print version).

Carnelli, D.; Pennati, G., Villa, T., Baglioni, L., Reimers, B., & Migliavacca, F. (2010). Mechanical properties of open-cell, self-expandable shape memory alloy carotid stents. *Artificial Organs*, Vol. 35, No. 1, (January 2011) 74-80, ISSN 0160-564X (print version).

Clark, D. J.; Lessio, S., O'Donoghue, M., Tsalamandris, C., Schainfeld, R., & Rosenfield, K. (2006). Mechanisms and predictors of carotid artery stent restenosis--A serial intravascular ultrasound study. *Journal of the American College of Cardiology*, Vol. 47, No. 12, (June 2006) 2390-2396, ISSN 0735-1097.

Colombo, A.; Stanlovic, G., & Moses, J. W. (2002). Selection of coronary stents. *Journal of the American College of Cardiology*, Vol. 40, No. 6, (September 2002) 1021-1033, ISSN 0735-1097.

Duda, S. H.; Wiskirchen, J. Tepe, G., Bitzer, M., Kaulich, T. W., Stoeckel, D., & Claussen, C. D. (2000). Physical properties of endovascular stents: an experimental comparison. *Journal of Vascular and Interventional Radiology*, Vol. 11, No. 5, (May 2000) 645-654, ISSN 1051-0443.

Gow, B. S. & Hadfield C. D. (1979). The elasticity of canine and human coronary arteries with reference to postmortem changes. *Circulation Research*, Vol. 45, No. 5, (November 1979) 588-594, ISSN 0009-7330.

Grewe, P. H.; Deneke, T., Machraoui, A., Barmeyer, J., & Müller, K.-M. (2000). Acute and chronic tissue response to coronary stent implantation: pathologic findings in human specimen. *Journal of the American College of Cardiology*, Vol. 35, No. 1, (January 2000) 157-163, ISSN 0735-1097.

Hayashi, K. (2005). *Biomechanics*, Corona Publishing, Tokyo, Japan. (in Japanese)

Hayashi, K.; Handa, H., Nagasawa, S., Okumura, A., & Moritake, K. (1980). Stiffness and elastic behavior of human intracranial and extracranial arteries. *Journal of Biomechanics*, Vol. 13, No. 2, 175-184, ISSN 0021-9290.

Kastrati, A.; Mehilli, J., Pache, J., Kaiser, C., Valgimigli, M., Kelbaek, H., Menichelli, M., Sabaté, M., Suttorp, M. J., Baumgart, D., Seyfarth, M., Pfisterer, M. E., & Schömig, A. (2007). Analysis of 14 trials comparing sirolimus-eluting stents with bare-metal stents. *The New England Journal of Medicine*, Vol. 356, No.10, (March 2007) 1030-1039, ISSN 0028-4793 (print version).

Lagerqvist, B.; James, S. K., Stenestrand, U., Lindbäck, J., Nilsson, T., & Wallentin, L. (2007). Long-term outcomes with drug-eluting stents versus bare-metal stents in Sweden. *The New England Journal of Medicine*, Vol. 356, No. 10, (March 2007) 1009-1019, ISSN 0028-4793 (print version).

Lal, B. K.; Kaperonis, E. A., Cuadra, S., Kapadia, I., & Hobson, R. W. 2nd. (2007). Patterns of in-stent restenosis after carotid artery stenting: classification and implications for long-term outcome. *Journal of Vascular Surgery*, Vol. 46, No. 5, (November 2007) 833-840, ISSN 0741-5214.

Le Floc'h, S.; Ohayon, J., Tracqui, P., Finet, G., Gharib, A. M., Maurice, R. L., Cloutier, G., & Pettigrew, R. I. (2009). Vulnerable atherosclerotic plaque elasticity reconstruction based on a segmentation-driven optimization procedure using strain measurements: theoretical framework. IEEE transactions on medical imaging. Vol. 28, No. 7, (July 2009) 1126-1137, ISSN 0278-0062.

Mori, K. & Saito, T. (2005). Effects of stent structure on stent flexibility measurements. *Annals of Biomedical Engineering*, Vol. 33, No. 6, (June 2005) 733-742, ISSN 0090-6964 (print version).

Morice, M.-C.; Serruys, P. W., Sousa, J. E., Fajadet, J., Hayashi, E. B., Perin, M., Colombo, A., Schuler, G., Barragan, P., Guagliumi, G., Molnar, F., & Falotico, R. (2002) A randomized comparison of a sirolimus-eluting stent with a standard stent for coronary revascularization. *The New England Journal of Medicine*, Vol. 346, No. 23, (June 2002) 1773-1780, ISSN 0028-4793 (print version).

Nordmann, A. J.; Briel, M., & Bucher, H. C. (2006). Mortality in randomized controlled trials comparing drug eluting vs. bare metal stents in coronary artery disease: a meta-analysis. *European Heart Journal*, Vol. 27, No. 23, (December 2006) 2784-2814, ISSN 0195-668X (print version).

Peterson, L. H.; Jensen, R. E., & Parnell, J. (1960). Mechanical properties of arteries in vivo. *Circulation Research*, Vol. 8, No. 3, (May 1960) 622-639, ISSN 0009-7330.

Schweiger, M. J.; Ansari, E., Giugliano, G. R., Mathew, J., Islam, A., Morrison, J., & Cook, J. R. (2006). Morphology and location of restenosis following bare metal coronary stenting. *The Journal of Invasive Cardiology*, Vol. 18, No. 4, (April 2006) 165-168, ISSN 1042-3931.

Stratouly, L. I.; Cardullo, P. A., Anderson, F. A. Jr., Durgin, W. W., & Wheeler, H. B. (1987). The use of ultrasound imaging in the in-vivo determination of normal human arterial compliance. *Proceedings of the 13th Annual Northeast Bioengineering Conference*, pp.435-437, Philadelphia, PA, U.S., March 1987, Defense Technical Information Center, Fort Belvoir.

Tamakawa, N.; Sakai, H., & Nishimura, Y. (2008). Prediction of restenosis progression after carotid artery stenting using virtual histology IVUS. *Journal of Neuroendovascular Therapy*, Vol. 2, No. 3, (December 2008) 193-200, ISSN 1882-4072.

Yazdani, S. K. & Berry, J. L. (2009). Development of an in vitro system to assess stent-induced smooth muscle cell proliferation: a feasibility study. *Journal of Vascular and Interventional Radiology*, Vol. 20, No. 1, (January 2009) 101-106, ISSN 1535-7732.

Yoshino, D. & Inoue, K. (2010). Design method of self-expanding stents suitable for the patient's condition. *Proceedings of the Institution of Mechanical Engineers, Part H: Journal of Engineering in Medicine*, Vol. 224, No. 9, (September 2010) 1019-1038, ISSN 0954-4119 (print version).

Yoshino, D.; Inoue, K., & Narita, Y. (2008) Mechanical properties of self-expandable stents: a key to product design of suitable stents. *Proceedings of the 7th International Symposium on Tools and Methods of Competitive Engineering*, Vol. 1, pp. 659-672, ISBN 978-90-5155-044-3, Izmir, Turkey, April 2008.

Yoshino, D.; M. Sato, & K. Inoue. (2011). Estimation of force on vascular wall caused by insertion of self-expanding stents. *Proceedings of the Institution of Mechanical Engineers, Part H: Journal of Engineering in Medicine*, Vol. 224, No. 8, (August 2011) 831-842, ISSN 0954-4119 (print version).

Spin and Spin Recovery

Dragan Cvetković[1], Duško Radaković[2], Časlav Mitrović[3] and Aleksandar Bengin[3]
[1]*University Singidunum, Belgrade*
[2]*College of Professional Studies "Belgrade Politehnica", Belgrade*
[3]*Faculty of Mechanical Engineering, Belgrade University*
Serbia

1. Introduction

Spin is a very complex movement of an aircraft. It is, in fact, a curvilinear unsteady flight regime, where the rotation of the aircraft is followed by simultaneous rotation of linear movements in the direction of all three axes, i.e. it is a movement with six degrees of freedom. As a result, there are no fully developed and accurate analytical methods for this type of problem.

2. Types of spin

Unwanted complex movements of aircraft are shown in Fig.1. In the study of these regimes, one should pay attention to the conditions that lead to their occurrence. Attention should be made to the behavior of aircraft and to determination of the most optimal way of recovering the aircraft from these regimes. Depending on the position of the pilot during a spin, the

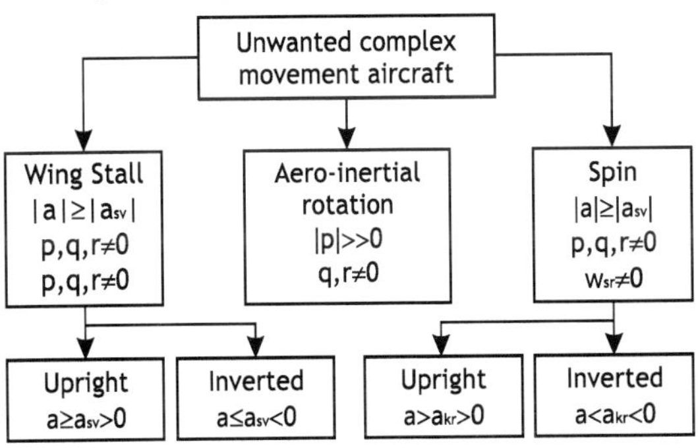

Fig. 1. Unwanted rotations of aircraft

spin can be divided into upright spin and inverted spin. During a upright spin, the pilot is in position head up, whilst in an inverted spin his position is head down.

The upright spin is carried out at positive supercritical attack angles, and the inverted spin at negative supercritical attack angles. According to the slope angle of the aircraft longitudinal axis against the horizon, spin can be steep, oblique and flat spin (Fig.2). During a steep spin, the absolute value of the aircraft slope angle is greater than 50 degrees, i.e. angle $|\upsilon| > 50°$, during an oblique spin $30° \leq |\upsilon| \leq 50°$, and during a flat spin $|\upsilon| < 30°$. According to

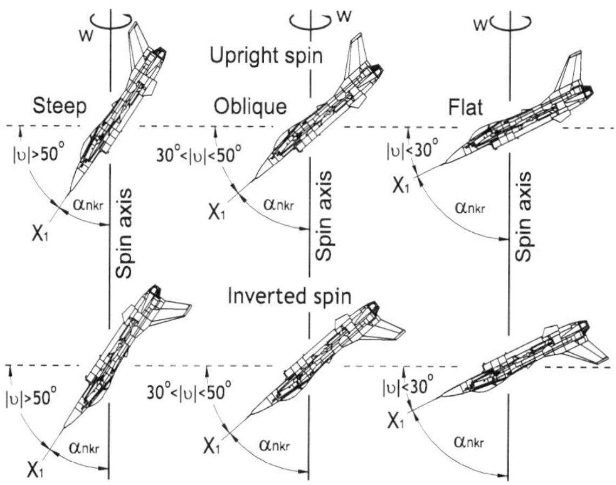

Fig. 2. The positions of aircraft in the vertical plane at entry and during the upright and inverted spin

direction of aircraft rotation, spin can be divided into a left and right spin. In a left spin (upright and inverted), the aircraft is rotating leftward, and during a right spin (upright and inverted) rightward. If the aircraft is observed from above, in a right upright and in a left inverted spin, the center of gravity of the aircraft will move in a clockwise direction, and vice versa while at a left upright and a right inverted spin.

The axis of spin is a an axis of a spiral by which the center of gravity is moving during a spin, and the spin radius is the radius of the horizontal projection of that spiral.

There exists inward and outward sideslip of the aircraft during a spin. The inward sideslip is when the air stream encounters the aircraft from the side of the inner wing, the wing in which direction the aircraft is rotating during a spin. The outward sideslip occurs when the air stream encounters the aircraft from the side of the outer wing.

Modern supersonic aircrafts (unlike older supersonic and even more the subsonic aircrafts) are characterized by a greater diversity of spin. This can be explained by the influence of constructive-aerodynamic properties of such aircrafts. Even for the same supersonic aircraft, the characteristics of upright and inverted spin can substantially differ depending on the initial angles of initiation (height, centering, etc.), regime length, position of rudder and ailerons during spin, etc. As a rule, these aircrafts have a distinct unevenness of motion and great fluctuations during spin.

The pilot has to study and reliably recognize the characteristic features of every spin type. Thus, he can quickly and accurately determine the type of spin and properly recover the aircraft from such a dangerous and complex motion. In modern aircrafts, spin can be divided into several types similar by characteristic features. The magnitude and character of changes in the angular velocities and load during a spin are taken into account as characteristic features. These determine the conditions for recovery from this regime, i.e. the value and sequence of rudder deflection. The given type includes all regimes of spin where in order to recover the aircraft the same control has to be applied. According to this principle, spin regime classifications are given on Figs. 3 and 4. In compliance with Figures 3 and 4, upright spin has four, and the inverted spin has three regimes for recovery. On these images, letters "N" and "L", alongside numerated ways of recovery, denote the upright and inverted spin, respectfully.

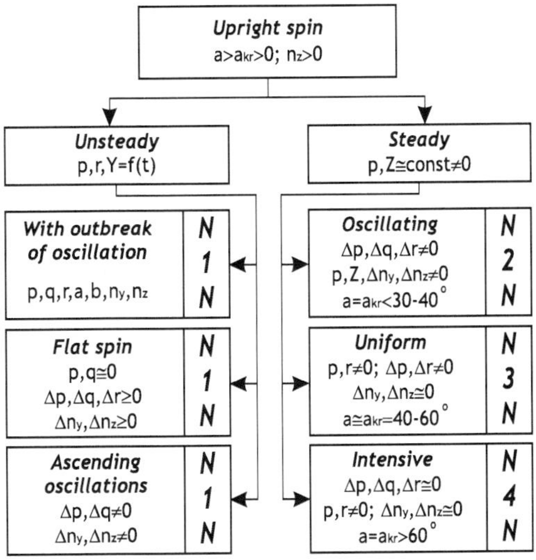

Fig. 3. Types of upright spin in modern aircrafts

Four basic methods can be applied for recovering from a upright spin (letter "N" denotes that the method refers to a upright spin):

- Method N1N - recovery from spin with simultaneous positioning of elevator and rudder in neutral position at neutral position of ailerons;
- Method N2N - recovery from spin by full rudder deflection opposite to spin with a delayed setting (2 - 4 sec) of the elevator in neutral position at neutral position of ailerons;
- Method N3N - recovery from spin by rudder deflection, and a delayed (3 to 6 sec.) full elevator deflection opposite to spin at neutral aileron position.
- Method N4N - recovery from spin as by method N3N but with simultaneous rudder and aileron deflection, possibly by full for recovery (for supersonic aircrafts aileron deflection is the appropriate deflection in the side of spin)

Recovery of modern aircrafts from inverted spin is carried out by three basic methods (letter "L" denotes that the method applies to recovery from inverted spin):

- Method N1L - recovery from spin by simultaneous elevator and rudder positioning in neutral with neutral aileron position;
- Method N2L - recovery from spin by rudder deflection totally opposite to spin with delayed (2 to 4 sec.) elevator deflection into neutral position at neutral aileron position;
- Method N3L - recovery from spin by rudder deflection and an elevator deflection (delayed 2 to 4 sec.) totally opposite to spin at neutral aileron position.

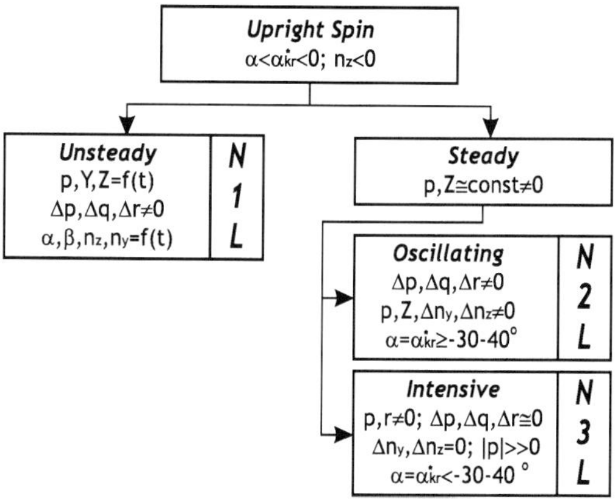

Fig. 4. Inverted spin types in modern aircrafts

3. Steady spin

A steady spin is a spin in which the aircraft does not change direction of rotation neither in roll or yaw (parameters p and r are unchanged sign). Rotation is very intense, and the average values of angular velocities do not change. Upright and inverted steady spins can be oscillating and uniform, and the upright can also be intensive.

The oscillating steady spin is characterized by very large changes in the amplitude of the rolling angular velocity and angular velocity slight changes yaw and pitch. It is usually a steep spin with average angles of attack of 30° or 40°. Changes of basic parameters during a left upright steady oscillating spin are shown on figure 5. Figure reflects an aircraft induced into spin at an altitude of 12.5 km at a velocity of 250 $\frac{km}{h}$. After a time interval $t = 5\,s$ a deflection of commands is carried out according to spin. Pilot has completely pulled the yoke to himself which has caused a negative rudder deflection ($\delta_{hk} = -20°$).

A positive deflection of the rudder is carried out $\delta_{vk} = +20°$, whilst ailerons are in neutral position. After a time interval $t = 6\,s$ parameters become stable so the average angular velocity of yawing is $r = -5\,\frac{rad}{s}$ and of rolling $p = -0.75\,\frac{rad}{s}$, the average normal load coefficient is $n_z = 1.25$. However, after a time interval $t = 8\,s$, the aircraft starts to oscillate and the following values are obtained: $n_z = 1.4$, $p = 3.6\,\frac{rad}{s}$ and $r = 0.6\,\frac{rad}{s}$. During that period the velocity on path V has oscillated between stabilized angular velocities of yawing and rolling, and an increase in velocity, which caused an increase in the normal load coefficient n_z, so that during the interval $t = (20 \div 25)\,s$ the normal load coefficient was $n_z = 2 \div 3$.

Unlike the oscillatory spin, the steady uniform spin is characterized by small amplitude oscillations of the aircraft, as well as intensive rotation of invariant direction. Figure 6 shows an example of a right-hand steady uniform spin. It can be seen that the aircraft was induced into spin at an altitude of $H = 10.500\,m$ and a velocity of $V = 320\,\frac{km}{h}$. After an interval $t = 6\,s$, commanding surfaces were deflected. First, a negative aileron deflection ($\delta_k = -10°$) can be observed resulting in a negative rolling moment, i.e. to lowering the right and lifting the left wing. Then the direction control was deflected in a negative direction ($\delta_{vk} = -20°$) which lead to a yaw to right. In addition, the yoke was drawn onto the pilot and thus a negative deflection ($\delta_{hk} = -20°$) was carried out on the elevator. At all time during spin, i.e. during

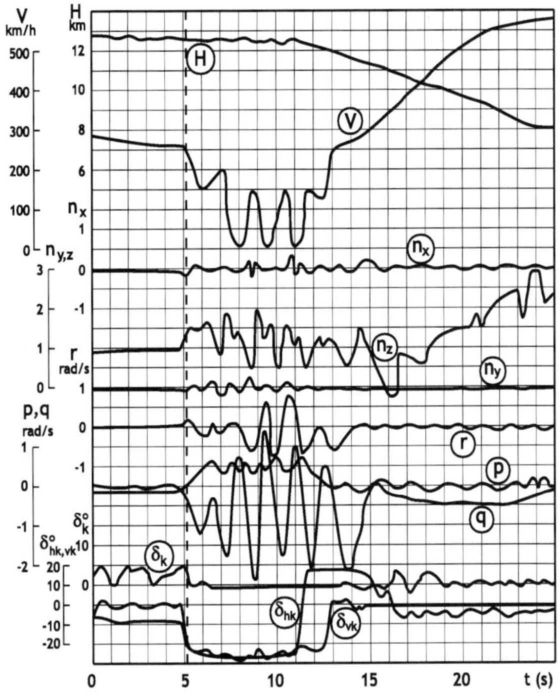

Fig. 5. Left upright steady oscillating spin

interval $t = (5 \div 40)\,s$, parameters (angular velocity of rolling p, angular velocity of yaw r, and velocity on path V) oscillated very little around the following values: $p_{sr} = -0.5\,\frac{rad}{s}$, $r_{sr} = -0.5\,\frac{rad}{s}$ i $V_{sr} = 250\,\frac{km}{h}$. In addition, there was a noteworthy smaller loss of altitude of $\Delta H = 2.500\,m$, than in the oscillating spin.

4. Unsteady spin

In some aircrafts, spin can periodically change the direction of rotation relative to normal and longitudinal axis. In some aircrafts, spin might even stop to oscillate, i.e. it will continue with certain oscillations. Spin at which the aircraft periodically changes direction of rotation regarding the normal and longitudinal axis, or even stops oscillating, is called unsteady.

During an unsteady spin, there is a non-uniform rotation with large amplitudes of aircraft parameter change. During the regime, usually there is a tendency towards an arbitrary transition of the aircraft from a spin of one direction into a spin of another direction, or from upright into inverted spin.

Unsteady spin in modern aircrafts has three forms: spin continuing with outbreak oscillations (diagram shown in Figure 7), spin continuing as a shape of a falling leaf on a spiral path (diagram shown in Figure 8), and spin during which oscillations of the aircraft are ascending (diagram shown in Figure 9).

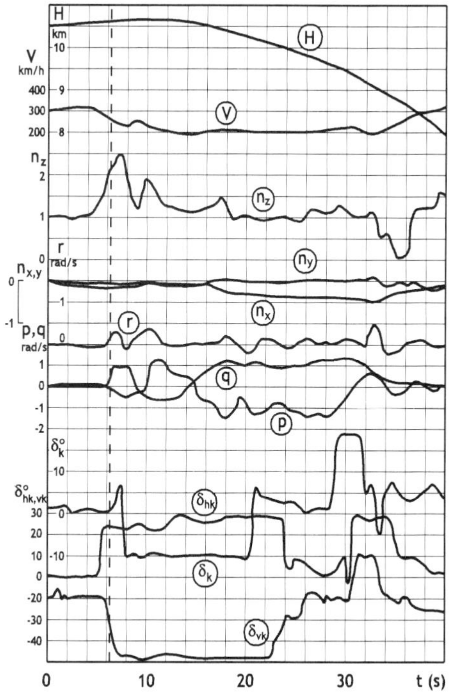

Fig. 6. Right-hand steady uniform spin

4.1 Spin continuing with outbreak oscillations

Diagram of parameter change for this shape of spin is shown in Figure 7. From this diagram, it is obvious that this is a left upright spin. In addition, it can be noticed that there is a periodic change of parameters, especially the angular velocity of roll p, angular velocity of yaw r, and the coefficient of normal load n_z. Every cycle of outbreak (change of angular velocities in roll and yaw) stands for about 15 s. Between these cycles the aircraft has stopped rotating around it's normal axis $r = 0$. Period of oscillations in every cycle was $t = 2.5\,s$. Velocity on path (V), during spin was approximately constant $V = 280\,\frac{km}{h}$, whilst altitude loss during an interval of $t = 90\,s$ was $\Delta H = 8.000\,m$.

4.2 Spin continuing in the shape of a falling leaf on a spiral path - flat spin

The parameter change in this shape of spin is shown in Fig. 8. During this spin there were periodic changes in magnitude and direction of angular velocities (p, r), and with harsh changes of the aircraft position in space. The aircraft was leaning from one wing to the other, with a nose deflection to the left and right, so that it resembles the movement of falling leaves. During this regime, the aircraft's center of gravity shifted according to a spiral path. The diagram shows that the normal load coefficient n_z has a significant oscillation. During a period of $t = 40\,s$ loss in altitude was $\Delta H = 4.000\,m$.

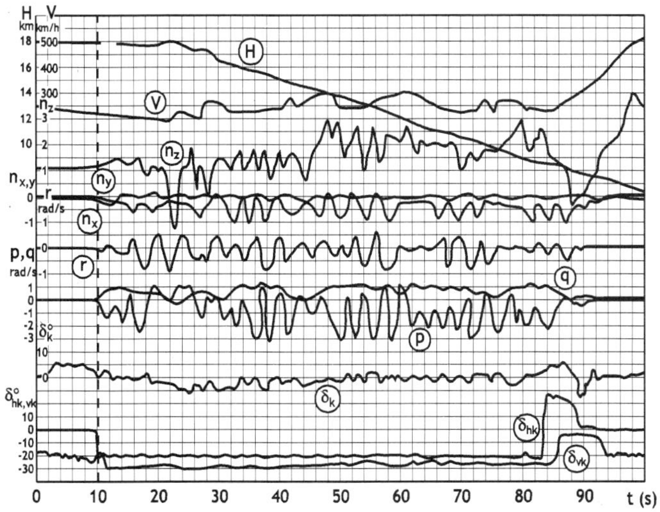

Fig. 7. Spin with outbreak oscillations

4.3 Spin continuing with ascending (intensifying) oscillations

In this form of spin there is a significant increase of oscillations in pitch that lead to an increase in oscillations of the normal load coefficient nz, that began in $t = 10\,s$. At time $t = 55\,s$ there is a rise in the normal load coefficient $n_z = 6$. Also, present are oscillations with a high frequency in angular velocity of rolling p, and in the oscillation of the vertical rudder in the interval $t = (40 \div 60)\,s$. Figure 9 shows the diagram of parameter change for this shape of spin. It is important to note that on these diagrams the velocity on the path V and altitude H are not trustworthy because of the inaccuracy of the measuring device (Pitot tube). This came as a result of the flight regime at high angles of attack where, because of the separated flow, the flow parameters were significantly altered.

5. Phases of spin

A typical spin can be divided into two phases shown in Fig.10, and they are: spin entry, incipient spin, steady spin and recovery from spin.

5.1 The spin entry phase

The spin entry phase begins with the aircraft being at attack angles higher than the critical angle of attack. This is the condition known as wing stall. Influenced by many factors (geometrical or aircraft aerodynamic asymmetry, rudder or aileron deflection, etc.) or disturbances (vertical wind stroke), the flow around the aircraft is asymmetrical. Due to this asymmetrical flow, there are aerodynamic rolling and/or yawing moments and angular rolling and/or yawing velocities. This means that the aircraft holds an uncontrolled rotation with respect to all three abiding axes, i.e., an autorotation arises. During this rotation, the angle of attack can periodically be less than the critical angle of attack, i.e. the nose of the aircraft can periodically ascend and/or descend. At this stage, it is not possible to determine what type of spin will develop.

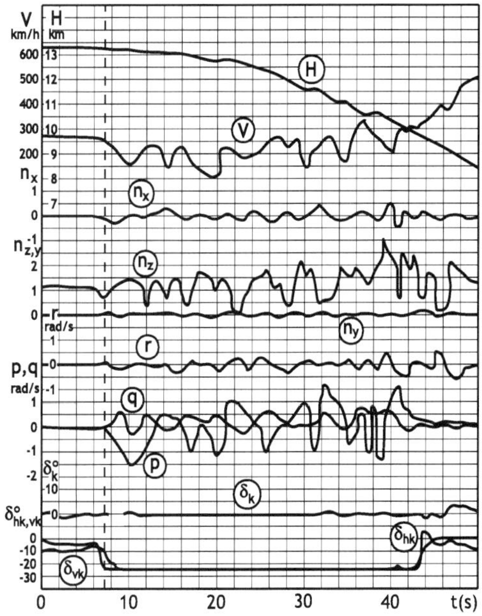

Fig. 8. Spin with outbreak oscillations

5.2 Incipient spin phase

The aircraft motion during the incipient spin phase is unsteady. Forces and moments acting on the aircraft are not in equilibrium. There are also, linear and angular accelerations. The aircraft rotates around the inclined axis, which is changing direction from horizontal to vertical. At this stage, the spin type can be determined.

5.3 Steady spin phase

During this phase, the aerodynamic and inertial forces and moments come into equilibrium. The aircraft is rotating downward around the vertical axis. The motion is steady, i.e. all motion parameters (attack angle, angular velocity, altitude loss per turn, time interval per turn, etc.) are constant.

5.4 Spin recovery phase

This stage begins by deflecting aircraft commanding surfaces into position for spin recovery. Autorotation stops and the aircraft enters dive as to increase flight speed. When the aircraft "accumulates" sufficient speed reserve, the pilot starts to pull out the aircraft from dive. This phase ends with the aircraft transitioning into horizontal flight.

6. Methods for upright spin recovery

As stated before, spin can only occur at overcritical attack angles. Due to this, for the aircraft to recover from spin it is necessary to decrease the attack angle, convert the aircraft to below

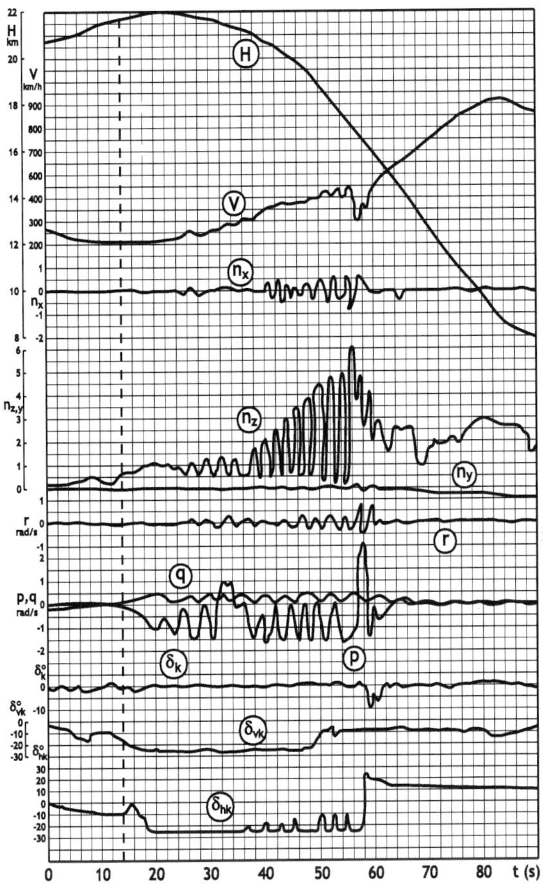

Fig. 9. Spin with ascending oscillations

critical attack angles, at which autorotation stops. This comprises the basic task for spin recovery.

For some time after the discovery of the physical image of spin, propositions were made to decrease the attack angle with an appropriate deflection of the elevator. This method proved to be efficient only for some cases, when aerodynamic pitching moments, produced by elevator deflection, were greater than the inertial pitching moments by absolute value.

However, when using such a control method, in most cases the aircraft did not recover from spin, even at full deflection of elevators. The spin became only steeper, but it did not stop, especially for aircrafts with rear centering.

In development of aircrafts their mass increased, which meant an increase in inertial moments, specially the inertial pitching moment. Therefore, in order to recover from spin it was necessary to increase the aerodynamic pitching moments, as well. All the same, it became

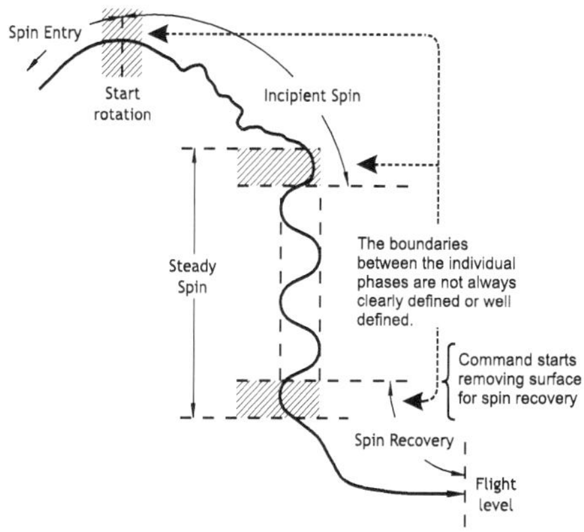

Fig. 10. Phases of a typical spin

clear that merely elevators could not achieve sufficient aerodynamic pitching moments for the purpose. Investigations have shown that in order to facilitate (sometimes, even to secure) spin recovery, previously the angular rate of rotation should be reduced (inertial moments decreased) for which an inward sideslip should be introduced, i.e. an aerodynamic yawing moment. This was achieved by deflecting the rudder opposite to spin.

As a result, the first scientifically based method for spin recovery was developed, by which a deflection was made opposite to spin, first by the rudder, then, after a delay (required because the inward sideslip, created by rudder deflection, could reduce the angular rate of autorotation), by the elevator. This was a, so-called standard method for spin recovery. However, it turned out that one standard method was not sufficient for modern aircrafts, characterized by vast diversity in spin regime.

Creating aerodynamic yawing moments, which in turn cause inward sideslip, is a powerful means for stopping, or at least greatly reducing the autorotation. Characteristics of aircraft recovery depend on the capability to achieve the best ratio between aerodynamic pitching and yawing moments, and between aerodynamic and appropriate inertial moments.

Four basic methods of spin recovery for modern aircrafts are as follows (letter "N" denotes method related to upright spin):

- Method N1N - spin recovery by simultaneous rudder and elevator positioning in neutral, with ailerons in neutral position;
- Method N2N - spin recovery by a full rudder deflection opposite to spin and a sequential elevator positioning in neutral(delayed by 2 - 4 sec), with ailerons in neutral position;
- Method N3N - spin recovery by rudder deflection, and sequential elevator deflection after 3 - 6 sec, completely opposite to spin, with ailerons in neutral position; and
- Method N4N - spin recovery by N3N method, but with simultaneous deflection of rudder and ailerons possibly at full for recovery (for supersonic aircrafts aileron deflection for recovery corresponds to a deflection to the side of spin.)

Deflection of controls and commanding surfaces during recovery from a left upright spin by the four basic methods is shown in Fig.11. Conditional labeling of yoke and pedal deflection that were adopted are shown in Fig.12.

The stated methods of spin recovery are structured by their efficiency or "force" increase, i.e. by the increase in aerodynamic moments created with rudder and elevator deflections for spin recovery. Therefore, the "weakest" method will be N1N and the "strongest" N4N.

Method N1N is recommended for aircraft recovery from upright unstable spin, method N2N for recovery from upright stable wavering spin, method N3N for recovery from upright stable uniform spin, and method N4N for recovery from upright stable intensive spin. These methods, as a rule, allow faster (with minimum time and altitude loss during recovery) and safer recovery of modern aircrafts from all possible upright spin regimes.

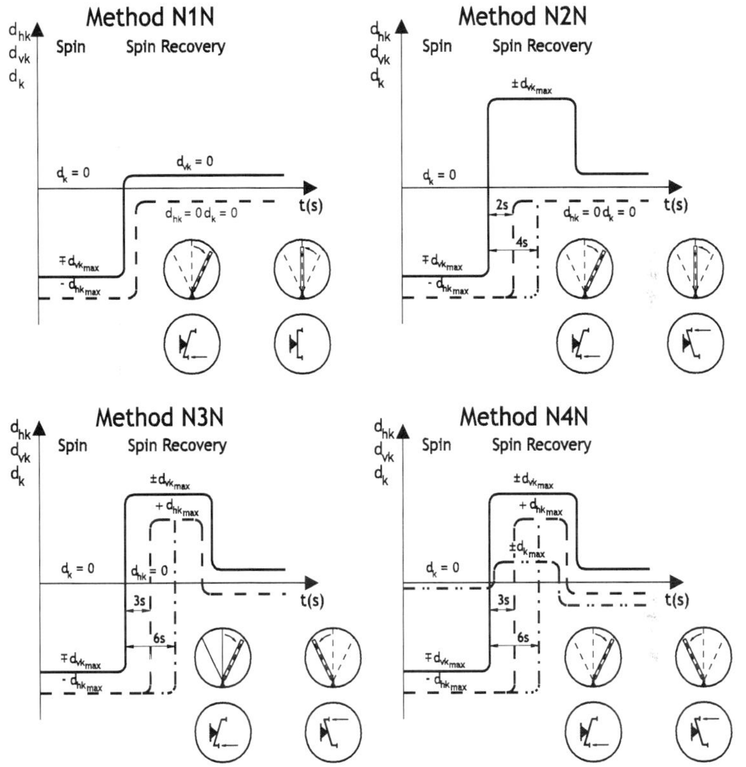

Fig. 11. Spin recovery methods for modern aircrafts from left upright spin (conditional notation of pedal and yoke position is shown in Fig.12)

When the aircraft falls into spin, usually it is necessary to position rudder completely on the side of spin so that their efficiency will be best while deflecting for recovery. In this case, first, a maximum rudder deflection (greatest stroke) is obtained, and second, a dynamic ("shock") effect is used during an abrupt deflection of the rudder from one end position to the other. This kind of method applies only when recovering aircraft from sufficiently stable spin regimes. However, since the pilot cannot know beforehand the type of spin that will be created (stable, unstable,...), as a rule, he has to completely position the rudders onto the side of spin.

The discussed methods for modern aircrafts' upright spin recovery (as well as method for inverted spin recovery), were developed by special investigations during flight. The possibility of applying four methods instead of a standard one (method N3N) substantially increases the assurance of spin recovery, which means a safe flight. On the other hand, this requires additional attention of the pilot, because now he has to choose among the four required methods to recover from spin and to remember the sequence of rudder activities for every method.

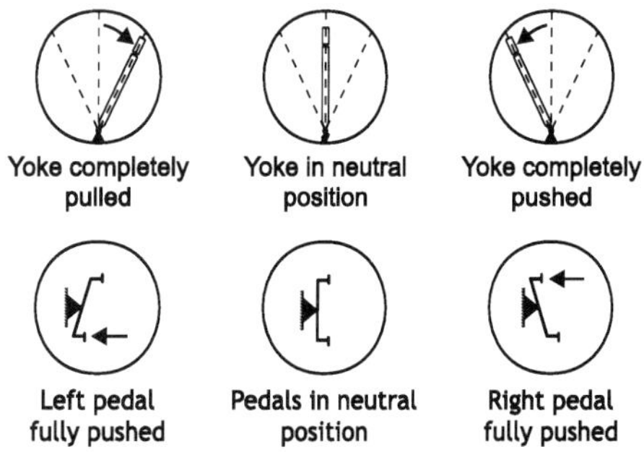

Fig. 12. Conditional annotation of deflections for yoke and pedals

However, in opinion of many pilots of high and medium proficiency, this problem does not cause extra caution, as it may seem at first glance. First, these spin recovery methods differ from one another only by "force", i.e. by the magnitude of required deflections and the interval of delay between them. Second, same rudders are used for spin recovery (only spin recovery by method N4N requires additional deflection of ailerons), and the direction of controls deflections is the same.

For subsonic aircrafts, just the use of "strong" method was generally appropriate, since there was danger of a "lack" of rudder for spin recovery. Hence, the required efficiency of subsonic aircraft rudders was conditioned, mostly by the need to ensure spin recovery, i.e. the ability to create sufficient aerodynamic moments to stop autorotation.

On the other hand, supersonic airplanes have the risk of excessive rudder "delivery" during recovery, i.e. creating excessively large aerodynamic moments. The possibility of exhibiting this danger has to do with that that the required efficiency of the supersonic airplane rudder is chosen starting from the requirement to provide maneuverability in flight at high Mach numbers, contrary to subsonic aircrafts. Hence, as a rule, the efficiency is more than sufficient for a spin recovery. Excessive rudder deflection can significantly impair characteristics of recovery (for example, unnecessarily increase the steepness of dive after spin, and consequently loose altitude for spin recovery, etc.), or lead to impossible spin recovery (the plane changes from upright to inverted spin, from left to right spin, etc.). This explains the necessity to apply the "weak" methods (N1N and N2N) for spin recovery.

However, only "strong" methods should be used for recovery of supersonic airplanes from stable intensive and uniform spins.

Therefore, for recovery of modern supersonic aircrafts from upright spin it is necessary to use "weak", as well as "strong" methods. It should be in mind at all times that, "strong" methods in any case are not above nor can they replace the "weak" methods. Each method has its area of application.

When choosing a recovery method the pilot must be guided only by the character of spin at the time he decided to recover from spin. Other data (for example, flight altitude) can be used only as supplementary for refining and hastening the determination of regime characteristics.

In all cases, it is necessary to deflect the rudders, if possible, more abrupt. Slow, sluggish deflecting will worsen the characteristics of spin recovery, and, sometimes, make it impossible to recover.

If recovery of aircraft is particularly difficult, the pilot must carefully choose the moment to start recovering, i.e. moment when to deflect the first rudder for recovery. The best moment for the recovery start is considered to be the moment when the plane stops yawing, the nose starts declining (in a inverted spin - nose starts inclining) and similar.

It is better for the pilot to count seconds instead of turns when deflecting rudder and elevator. Practice has shown that even small changes in the regime (accelerating or decelerating the aircraft) make it difficult and often impossible to properly count turns, especially at the notable positions of the aircraft in space (during periodic switching to inverted, etc.). It is not advantageous to count turns of the aircraft during spin, not only for recovery but during the regime, as well. There are several reasons for this. First, in an unstable spin regime (for example, in a flat spin where the aircraft's motion acts like a "falling leaf") the term "turn" is meaningless. Second, for the pilot, in order to evaluate the situation, it is of more importance that he knows the time of the acting regime and the altitude loss during that time. In addition, it is always easier to count seconds of duration (also much safer), and often, sometimes, it could be the only possible way to determine the mentioned interval and delay of aircraft recovery from spin. The interval and delay are usually determined count seconds aloud.

The control of altitude change during a spin is one of the most important conditions of a safe flight, especially entering spin at low altitudes.

Therefore, the pilot must follow the altimeter reading during a spin. The altimeter shows absolute values of altitude with great errors. This is due to a greater change in the flow over the Pitot-tube at higher attack angles and yaw angles, and at high rotation rates during spin. Even so, this instrument allows a proper determination of the altitude loss (altitude difference).

Flight practice shows that the most often applied methods for spin recovery for supersonic aircraft are N1N and N2N. If the first attempt to recover from a spin fails (for example, with the method N1N), i.e. the self-rotation did not stop, the pilot has to reset flight commands into position for spin, and, after 2 - 4 sec repeat recovery but with a "stronger" method (N2N). Obviously, in the first attempt the pilot applied a "weak" method due to an improper determination of the character of the spin.

The choice of the required method for spin recovery greatly depends on the weight composition of the aircraft - mass distribution and centering.

A known fact is that subsonic aircrafts, with rectangular wings, characteristically have a large weight distribution along the wingspan, which in turn adds up to generating significant inertial rolling and yawing moments during spin. On the other hand, modern supersonic aircrafts have a characteristically large mass distribution along the fuselage axis, which contributes to generation of significant inertial pitching and yawing moments.

The sequence of actions with the rudders for spin recovery depends on the nature of the interaction of inertial rolling, yawing, and pitching moments, with the inertial moments created by rudder deflection during autorotation.

7. Methods for recovery from inverted spin

To recover modern aircrafts from an inverted spin there are three basic methods (letter "L" denotes the method for recovery from an inverted spin):

- Method N1L - spin recovery by simultaneous positioning elevator and rudder into neutral, with ailerons in neutral position;
- Method N2L - spin recovery by deflecting the rudder fully opposite, followed by a delayed (2 - 4 sec) elevator positioning into neutral, with ailerons in neutral position;
- Method N3L - spin recovery by fully deflecting the both rudder and elevator (delayed for 2 - 4 sec) opposite to spin direction, with ailerons in neutral position.

The N1L method is recommended for recovery from an unstable inverted spin, the N2L method from a stable wavering spin, and method N3L from an inverted stable uniform spin.

Deflection of controls and commanding surfaces during inverted spin recovery according to the three basic methods is shown in Fig.13, with the adopted notation shown in Fig.12.

For supersonic aircrafts, most often in use is the method N2L, because these aircrafts are most characterized by an unstable wavering inverted spin. In general, supersonic airplanes rarely fall into inverted spin; more often, it is a upright spin.

As a rule, airplanes of usual geometry more easily recover from an inverted spin than from a upright spin. This is explained by the fact that during this regime the autorotation is weaker, the rudder is more efficient (practically it is out of the wing and stabilizer flow), the efficient arrow angle of the vertical surfaces is decreased, and the average absolute values of attack angles are reduced.

However, despite of everything already said, for the pilot the inverted spin is always more difficult than the upright spin. This is conditioned by the unusual position of the pilot: hanging on restraint harnesses, head down, and a negative load ($n_z < 0$) tends to detach him from the seat. In such conditions, the pilot could drop the yoke and release the pedals (he could lose control of the aircraft, especially if he is not firmly seated).

Sometimes, during aircraft wavering the pilot has great difficulty to visually determine type of spin - upright or inverted. This is expressed if the aircrafts longitudinal axis is close to the vertical axis - the plane will "swirl" at low, by absolute value, negative overcritical attack angles, which is typical for a stable inverted spin. In this case, in order to stop autorotation the rudders have to be put in neutral position.

In the absence of or inability to use visual landmarks (also, to have control over a classical spin), the pilot can easily determine the type of spin by sensation: if the seat is pressuring the pilot - it is a upright spin, if the pilot is detaching from his seat, i.e. hanging on restraint harnesses - it is an inverted spin). However, if the aircraft exits at high negative attack angles that randomly change during the inverted spin regime, determining spin type by sensation is inapplicable.

During an inverted spin, it is more difficult to determine the direction of rotation (whether the plane is rotating to the right or to the left). In a classical spin, when the position of the

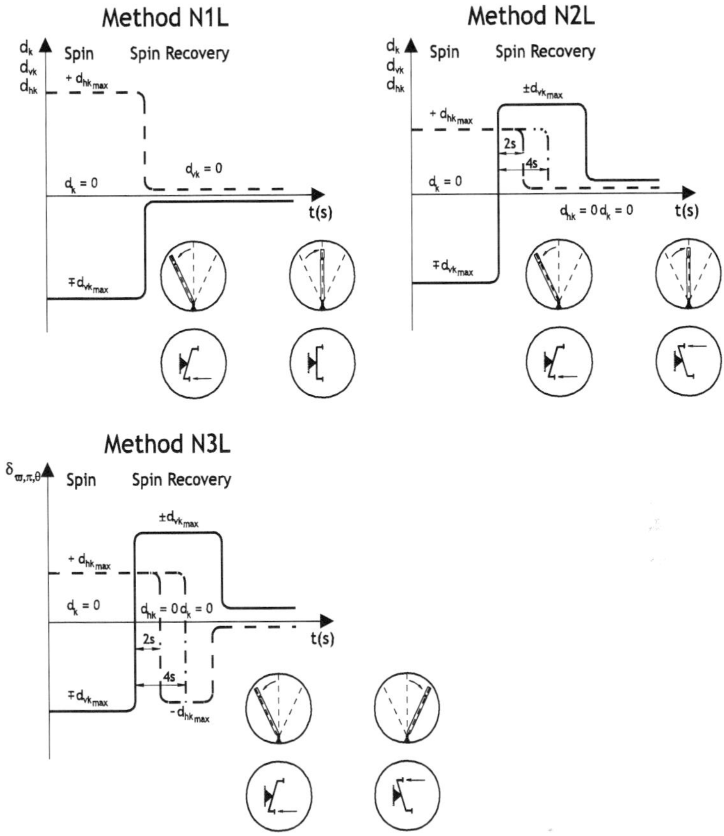

Fig. 13. Three basic methods of recovery from an inverted left spin for modern aircrafts (conditional annotation of yoke and pedal position is same as in Fig.12.)

nose of the aircraft to the horizon is practically unchanged, the direction of rotation can easily be determined according to the angular speed of rotation. However, when in inverted spin at high angular velocities of rolling, uneven motion of roll and pitch, it is impossible to determine the aircraft's direction of rotation with the mentioned method. The situation becomes more complex, because during inverted spin rolling motion is opposite to rotation. For a pilot seated in front cockpit look forward this means that, for example, in a right-hand spin the direction of rotation will into the left.

In a upright spin, the situation is opposite: directions of roll and yaw are the same. As so, for example, in a right upright spin the pilot can see the nose of the aircraft turning to the right and plane leaning to the same side. A more experienced pilot in this matter can distinguish a upright spin from an inverted.

Pilots lacking of sufficient flights for spin recovery training often determine spin direction according to rolling direction, but not yawing direction, because the rolling angular velocity is usually higher than the yawing angular velocity (except for a flat spin). Determination of spin direction in this manner is applied more often (rolling angular velocity increasing). Use of this manner of determining spin during an inverted spin will only disorient an insufficiently trained pilot.

Therefore, when recovering from this regime, it is necessary to have safe means of control to facilitate easy conservation of spatial orientation and assure possibilities for effective and proper actions with wings. Such means could be the yaw indicator (it hand always turns in yaw direction regardless on spin type) and attack angle indicator, and if it is not present - a normal load indicator. Attack angle indicator allows the pilot to reliably determine the spin type (upright or inverted), and the yaw indicator - its direction (left or right).

8. Spin modeling

Modeling is the most accurate graph-analytical method for determining aircraft characteristics prior to flight tests. Modeling of flight conditions, with initial data correction, quite properly reflects timely development of aircraft motion and enables more complete conclusions about flight test results.

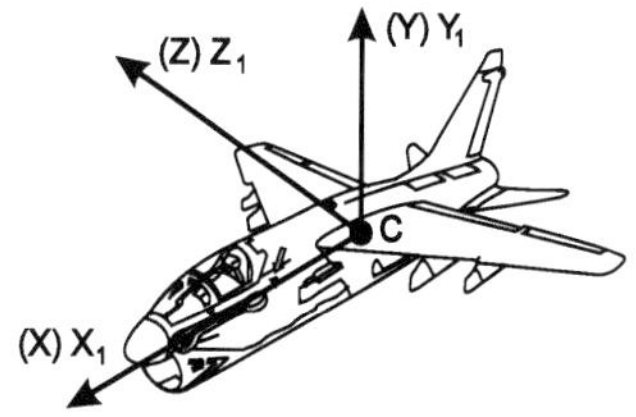

Fig. 14. Attached Coordinate System)

The differential equations of aircraft motion relative to its center of gravity is obtained from the Law of conversation of momentum. This is the moment equation. Projecting these equations to axes of the attached coordinate system (X_1, Y_1, Z_1), shown in Fig.14, whose axes we denote as (X, Y, Z) for simplicity, the following system of differential equations is derived:

$$\begin{aligned} m\left(\frac{dV_x}{dt} + q\,V_z - r\,V_y\right) &= R_x + G_x \\ m\left(\frac{dV_y}{dt} + r\,V_x - p\,V_z\right) &= R_y + G_y \\ m\left(\frac{dV_z}{dt} + p\,V_y - q\,V_x\right) &= R_z + G_z \end{aligned} \qquad (1)$$

$$\begin{aligned} I_x\frac{dp}{dt} + (I_z - I_y)\,q\,r + I_{xy}\left(p\,r - \frac{dp}{dt}\right) &= \mathcal{M}_x \\ I_y\frac{dq}{dt} + (I_x - I_z)\,p\,r + I_{xy}\left(q\,r - \frac{dp}{dt}\right) &= \mathcal{M}_y \\ I_z\frac{dr}{dt} + (I_y - I_x)\,p\,q + I_{xy}\left(p^2 - q^2\right) &= \mathcal{M}_z \end{aligned} \qquad (2)$$

where:

- V_x, V_y, V_z correspond to projections of the velocity of aircraft center of gravity with respect to the axes of the adopted coordinate system;
- p, q, and r represent projections of the aircraft angular velocities to the axes of the adopted coordinate system;
- I_x, I_y, I_z - aircraft axial inertia moments with respect to axes of the adopted coordinate system;
- I_{xy} - aircraft centrifugal inertia moment;
- R_x, R_y, R_z - projections of aerodynamic forces acting on the aircraft with respect to axes of the adopted coordinate system;
- G_x, G_y, G_z - projection of aircraft weight with respect to axes of the adopted coordinate system;
- M_x, M_y, M_z - projection of resulting moments from external forces acting on aircraft, with respect to axes of the adopted coordinate system.

8.1 Motion equations for modeling

In order to simplify the task during tests, following assumptions are made: the angle of sideslip is small so that $sin\beta \approx \beta$ and $cos\beta \approx 1$, effects of Mach and Reynolds number are ignored, and motion is investigated with engines turn off. This study uses known kinematic relations that are, with the adopted simplifications, equal to:

$$\begin{aligned}
\dot{\theta} &= r\,\sin\gamma + q\,\cos\gamma \\
\dot{\gamma} &= p + (q\,\sin\gamma - r\,\cos\gamma)\,\tan\theta \\
\dot{H} &= V\,\cos\alpha\,\sin\theta - V\,\sin\alpha\,\cos\gamma\,\cos\theta - V\,\beta\,\sin\gamma\,\cos\theta
\end{aligned} \tag{3}$$

where:

- θ - pitch angle, angle between X axis and horizontal plane,
- γ - angle of transverse inclination, angle between Y axis and vertical plane.

If it is assumed that velocity and altitude are unchanged, i.e. $V = const$ and $H = const$, and if the right-hand sides of Eq.(3) are approximated by a Taylor polynomial, the following simplified system of equations is obtained, a so-called **System I**:

$$\begin{aligned}
\dot{\alpha} &= a_{11}\,\frac{1}{\cos\alpha} + a_{12}\,\frac{\beta\,p}{\cos\alpha} + a_{13}\,q + a_{14}\,\frac{\cos\theta\,\cos\gamma}{\cos\alpha} + a_{15} \\
\dot{\beta} &= a_{21}\,\beta + a_{22}\,r\,\cos\alpha + a_{23}\,p\,\sin\alpha + a_{24}\,\cos\theta\,\sin\gamma + a_{25} \\
\dot{p} &= a_{31}\,r\,q + a_{32}\,\beta + a_{33}\,p + a_{34}\,r + a_{35} + a_{36} \\
\dot{r} &= a_{41}\,p\,q + a_{42}\,\beta + a_{43}\,p + a_{44}\,r + a_{45}\,q + a_{46} + a_{47} \\
\dot{q} &= a_{51}\,p\,r + a_{52}\,|\,\beta\,| + a_{53}\,r + a_{54}\,q + a_{55} + a_{56}\,\dot{\alpha} + a_{57} \\
\dot{\theta} &= a_{61}\,r\,\sin\gamma + a_{62}\,q\,\cos\gamma \\
\dot{\gamma} &= a_{71}\,p + a_{72}\,r\,\cos\gamma\,\tan\theta + a_{73}\,q\,\sin\gamma\,\tan\theta
\end{aligned} \tag{4}$$

The coefficients involved in the system of equations (4) are defined by expressions:

$$a_{11} = -\frac{S\rho V}{2m} C_z(\alpha) \qquad a_{12} = -1 \qquad a_{13} = 1 \qquad a_{14} = \frac{g}{V}$$

$$a_{15} = -\frac{S\rho V}{2m} C_{z_{\delta_{hk}}}(\alpha)\, \Delta\delta_{hk}(t) \qquad a_{21} = \frac{S\rho V}{2m} C_{y_\beta}(\alpha)$$

$$a_{22} = 1 \qquad a_{23} = 1 \qquad a_{24} = \frac{g}{V} \qquad a_{25} = \frac{S\rho V}{2m} C_{y_{\delta_{vk}}}(\alpha)\, \Delta\delta_{vk}(t)$$

$$a_{31} = \frac{I_y - I_z}{I_x} \qquad a_{32} = \frac{S\,l\,\rho V}{2\,I_x} C_{l_\beta}(\alpha) \qquad a_{33} = \frac{S\,l\,\rho V^2}{2\,I_x} C_{l_p}(\alpha)$$

$$a_{34} = \frac{S\,l\,\rho V^2}{2\,I_x} C_{l_r}(\alpha) \qquad a_{35} = \frac{S\,l\,\rho V^2}{2\,I_x} C_{l_{\delta_{vk}}}(\alpha)\, \Delta\delta_{vk}(t)$$

$$a_{36} = \frac{S\,l\,\rho V^2}{2\,I_x} C_{l_{\delta_k}}(\alpha)\, \Delta\delta_k(t) \qquad a_{41} = \frac{I_z - I_x}{I_y}$$

$$a_{42} = \frac{S\,l\,\rho V^2}{2\,I_y} C_{n_\beta}(\alpha) \qquad a_{43} = \frac{S\,l\,\rho V^2}{2\,I_y} C_{n_p}(\alpha)$$

$$a_{44} = \frac{S\,l\,\rho V^2}{2\,I_y} C_{n_r}(\alpha) \qquad a_{45} = -\frac{I_p\,\omega_p}{I_y}$$

$$a_{46} = \frac{S\,l\,\rho V^2}{2\,I_y} C_{n_{\delta_{vk}}}(\alpha)\, \Delta\delta_{vk}(t) \qquad a_{47} = \frac{S\,l\,\rho V^2}{2\,I_y} C_{n_{\delta_k}}(\alpha)\, \Delta\delta_k(t)$$

$$a_{51} = \frac{I_x - I_y}{I_z} \qquad a_{52} = \frac{S\,b\,\rho V^2}{2\,I_z} C_{m_\beta}(\alpha) \qquad a_{53} = \frac{I_p\,\omega_p}{I_z}$$

$$a_{54} = \frac{S\,b\,\rho V^2}{2\,I_z} C_{m_q}(\alpha) \qquad a_{55} = \frac{S\,b\,\rho V^2}{2\,I_z} C_m(\alpha)$$

$$a_{56} = \frac{S\,b\,\rho V^2}{2\,I_z} C_{m_{\dot\alpha}}(\alpha) \qquad a_{57} = \frac{S\,b\,\rho V^2}{2\,I_z} C_{m_{\delta_{hk}}}(\alpha)\, \Delta\delta_{hk}(t)$$

$$a_{61} = 1 \qquad a_{62} = 1 \qquad a_{71} = 1 \qquad a_{72} = -1 \qquad a_{73} = 1 \tag{5}$$

If assumed that velocity and altitude are changeable over time, i.e. $V = f(t)$ and $H = f(t)$, and if the right-hand side of Eq.(1) and (3) (projections of aerodynamic forces, weights and moments) is approximated by Taylor polynomial, the following simplified system of equations can be obtained, a so-called **System II**:

$$\dot\alpha = b_{11}\frac{\rho V}{\cos\alpha} + b_{12}\frac{p\,\beta}{\cos\alpha} + b_{13}\,q + b_{14}\frac{\cos\theta\,\cos\gamma}{V\,\cos\alpha} + b_{15}\frac{\dot V\,\tan\alpha}{V} + b_{16}\frac{\rho V}{\cos\alpha}$$

$$\dot\beta = b_{21}\,\rho V \beta + b_{22}\,r\,\cos\alpha + b_{23}\,p\,\sin\alpha + b_{24}\frac{\cos\theta\,\cos\gamma}{V} + b_{25}\frac{\dot V\,\beta}{V} + b_{26}\,\rho V$$

$$\dot p = b_{31}\,r\,q + b_{32}\,\rho V^2 \beta - b_{33}\,\rho V^2 p + b_{34}\,\rho V^2 r + b_{35}\,\rho V^2 + b_{36}\,\rho V^2$$

$$
\begin{aligned}
\dot{r} &= b_{41}\, p\, q + b_{42}\, \rho\, V^2\, \beta + b_{43}\, \rho\, V^2\, p + b_{44}\, \rho\, V^2\, r + b_{45}\, q + b_{46}\, \rho\, V^2 + b_{47}\, \rho\, V^2 \\
\dot{q} &= b_{51}\, p\, r + b_{52}\, \rho\, V^2\, \mid \beta \mid + b_{53}\, r + b_{54}\, \rho\, V^2\, q + b_{55}\, \rho\, V^2 + b_{56}\, \rho\, V^2\, \dot{\alpha} + b_{57}\, \rho\, V^2 \\
\dot{\theta} &= b_{61}\, r\, \sin\gamma + b_{62}\, q\, \cos\gamma \\
\dot{\gamma} &= b_{71}\, p + b_{72}\, r\, \cos\gamma\, \tan\theta + b_{73}\, q\, \sin\gamma\, \tan\theta \\
\dot{V} &= b_{81}\, \frac{\rho\, V^2}{\cos\alpha} + b_{82}\, V\, \dot{\alpha}\, \tan\alpha + b_{83}\, V\, q\, \tan\alpha + b_{84}\, \frac{V\, r\, \beta}{\cos\alpha} + b_{85}\, \frac{\sin\theta}{\cos\alpha} + b_{86}\, \frac{\rho\, V^2}{\cos\alpha} \\
\dot{H} &= b_{91}\, V\, \cos\alpha\, \sin\theta + b_{92}\, V\, \sin\alpha\, \cos\theta\, \cos\gamma + b_{93}\, V\, \beta\, \cos\theta\, \sin\gamma
\end{aligned} \tag{6}
$$

The coefficients involved in the system of equations (6) are defined by expressions:

$$
\begin{aligned}
& b_{11} = -\frac{S}{2\,m}\, C_z(\alpha) \qquad b_{12} = -1 \qquad b_{13} = 1 \qquad b_{14} = g \\
& b_{15} = -1 \qquad b_{16} = -\frac{S}{2\,m}\, C_{z_{\delta_{hk}}}(\alpha)\, \Delta\delta_{hk}(t) \qquad b_{21} = \frac{S}{2\,m}\, C_{y_\beta}(\alpha) \\
& b_{22} = 1 \qquad b_{23} = 1 \qquad b_{24} = g \qquad b_{25} = -1 \\
& b_{26} = \frac{S}{2\,m}\, C_{y_{\delta_{vk}}}(\alpha)\, \Delta\delta_{vk}(t) \qquad b_{31} = \frac{I_y - I_z}{I_x} \\
& b_{32} = \frac{S\, l}{2\, I_x}\, C_{l_\beta}(\alpha) \qquad b_{33} = \frac{S\, l}{2\, I_x}\, C_{l_p}(\alpha) \qquad b_{34} = \frac{S\, l}{2\, I_x}\, C_{l_r}(\alpha) \\
& b_{35} = \frac{S\, l}{2\, I_x}\, C_{l_{\delta_{vk}}}(\alpha)\, \Delta\delta_{vk}(t) \qquad b_{36} = \frac{S\, l}{2\, I_x}\, C_{l_{\delta_k}}(\alpha)\, \Delta\delta_k(t) \\
& b_{41} = \frac{I_z - I_x}{I_y} \qquad b_{42} = \frac{S\, l}{2\, I_y}\, C_{n_\beta}(\alpha) \qquad b_{43} = \frac{S\, l}{2\, I_y}\, C_{n_p}(\alpha) \\
& b_{44} = \frac{S\, l}{2\, I_y}\, C_{n_r}(\alpha) \qquad b_{45} = -\frac{I_p\, \omega_p}{I_y} \\
& b_{46} = \frac{S\, l}{2\, I_y}\, C_{n_{\delta_{vk}}}(\alpha)\, \Delta\delta_{vk}(t) \qquad b_{47} = \frac{S\, l}{2\, I_y}\, C_{n_{\delta_k}}(\alpha)\, \Delta\delta_k(t) \\
& b_{51} = \frac{I_x - I_y}{I_z} \qquad b_{52} = \frac{S\, b}{2\, I_z}\, C_{m_\beta}(\alpha) \qquad b_{53} = \frac{I_p\, \omega_p}{I_z} \\
& b_{54} = \frac{S\, b}{2\, I_z}\, C_{m_q}(\alpha) \qquad b_{55} = \frac{S\, b}{2\, I_z}\, C_m(\alpha) \qquad b_{56} = \frac{S\, b}{2\, I_z}\, C_{m_{\dot{\alpha}}}(\alpha) \\
& b_{57} = \frac{S\, b}{2\, I_z}\, C_{m_{\delta_{hk}}}(\alpha)\, \Delta\delta_{hk}(t) \qquad b_{61} = 1 \qquad b_{62} = 1 \\
& b_{71} = 1 \qquad b_{72} = -1 \qquad b_{73} = 1 \qquad b_{81} = -\frac{S}{2\,m}\, C_x(\alpha) \\
& b_{82} = 1 \qquad b_{83} = -1 \qquad b_{84} = -1 \qquad b_{85} = -g \\
& b_{86} = -\frac{S}{2\,m}\, C_{x_{\delta_{hk}}}(\alpha) \Delta\delta_{hk}(t) \qquad b_{91} = 1 \\
& b_{92} = -1 \qquad b_{93} = -1
\end{aligned} \tag{7}
$$

Notation in previous equations are:

- C_z - lift coefficient;
- C_x - drag coefficient;

- C_m - pitching moment coefficient;
- $C_{z_{\delta_{hk}}}$ - derivative of lift coefficient with respect to angle of deflection of the elevator;
- C_{y_β} - derivative of sideslip force with respect to sideslip angle;
- $C_{y_{\delta_{vk}}}$ - derivative of sideslip force with respect to angle of deflection of the rudder;
- C_{l_β} - derivative of rolling moment coefficient with respect to angle of sideslip;
- C_{l_p} - derivative of rolling moment coefficient with respect to rolling angular velocity;
- C_{l_r} - derivative of rolling moment coefficient with respect to yawing angular velocity;
- $C_{l_{\delta_{vk}}}$ - derivative of rolling moment coefficient with respect to rudder angle of deflection;
- $C_{l_{\delta_k}}$ - derivative rolling moment coefficient with respect to aileron angle of deflection;
- C_{n_β} - derivative of yawing moment coefficient with respect to angle of sideslip;
- C_{n_p} - derivative of yawing moment coefficient with respect to rolling angular velocity;
- C_{n_r} - derivative of yawing moment coefficient with respect to yawing angular velocity;
- $C_{n_{\delta_{vk}}}$ - derivative of yawing moment coefficient with respect to rudder angle of deflection;
- $C_{n_{\delta_k}}$ - derivative of yawing moment coefficient with respect to aileron angle of deflection;
- C_{m_β} - derivative of pitching moment coefficient with respect to angle of sideslip;
- C_{m_q} - derivative of pitching moment coefficient with respect to pitching angular velocity;
- $C_{m_{\dot{\alpha}}}$ - derivative of pitching moment coefficient with respect to derivative of angle of attack over time;
- $C_{m_{\delta_{hk}}}$ - derivative of pitching moment coefficient with respect to elevator angle of deflection;
- $C_{x_{\delta_{hk}}}$ - derivative of drag coefficient with respect to elevator angle of deflection;
- I_p - polar moment of inertia of engine rotor;
- ω_p - angular velocity of engine rotor;
- $\Delta\delta_{hk}$ - change in elevator angle of deflection;
- $\Delta\delta_{vk}$ - change in rudder angle of deflection;
- $\Delta\delta_k$ - change in aileron angle of deflection.

These coefficients are entirely determined with the aid of DATCOM[1] reference, and in some additional literature[2] their values are defined for the category of light aircrafts.

[1] D. E. Hoak: *USAF Stability and Control DATCOM*, (N76-73204), Flight Control Devision, Air Force Flight Dynamics Laboratory, Wright-Patterson Air Force Base, Ohio, 1975.

[2] D. Cvetković: *The adaptive approach to modeling and simulation of spin and spin recovery*, PhD Thesis, Faculty of Mechanical Engineering, Belgrade University, 1997.

8.2 Methods of modeling

It is obvious that most of the coefficients in Exp.(6) are not constant, but vary with time, i.e. angle of attack.

If assumptions that speed and altitude are constant are rejected, a complete system of equations is obtained. In order to perform modeling a computer is used with its memory loaded with the simplified or complete system of equations, which depends on the regime that has to be studied or the accuracy of results. The computer loaded with data necessary to calculate the coefficients and other values in Eq.(6), for a given time t_1, or the given angle of attack α_1. Data is obtained by testing models in wind tunnel or in flight. Then, the data is entered for moment t_2 or attack angle α_2, and so on. Data is entered until the end of the observed time interval, and it is entered point-by-point. The accuracy of obtained results depends on the magnitude of the time change (time difference between two points), i.e. the angle of attack. The more points are entered, the more accurate the results will be. The computer will integrate and associate values for angle of attack, angle of sideslip, pitching angle and angular velocities for the observed time interval, for which data is entered, and results are obtained as shown in Fig.15, i.e. following functions are defined: $\alpha = f(t), \theta = f(t)$, $\beta = f(t)$, $p = f(t)$, $q = f(t)$ i $r = f(t)$. For System II (Eq.7), following the same analogy, the

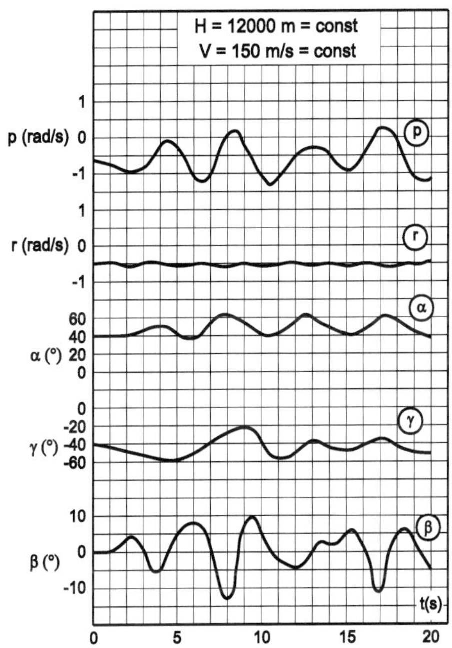

Fig. 15. Diagram for a timely development of a left-hand spin (System I)

computer will integrate and associate values from the observed time interval, for which data was entered, and results are obtained as shown in Fig.16.

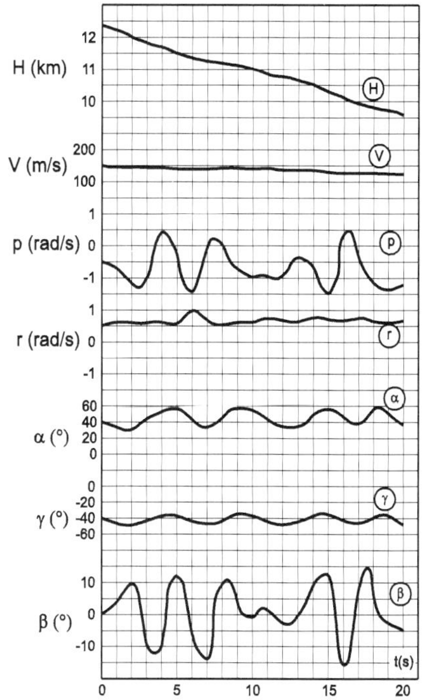

Fig. 16. Diagram for a timely development of a left-hand spin (System II)

8.3 Analysis of results from modeled flat spin (spin ongoing as a falling leaf On a spiral path)

An analysis of relations is made if effects from individual terms in equations of motion need to be determined (Eq.6). To do this, each member will be designated by the letter "A" with appropriate numerical indexes i, j $(A_{i,j})$, with "i" being the ordinal number of the equation $i = (1, n)$, and "j" being the ordinal number of the term in the equation $j = (1, m)$. For example, the first equation of Eq.6 will look:

$$\dot{\alpha} = A_{11} + A_{12} + A_{13} + A_{14} + A_{15},$$

where:

$$A_{11} = a_{11}\frac{1}{\cos\alpha}, \quad A_{12} = a_{12}\frac{\beta p}{\cos\alpha}, \quad A_{13} = a_{13}\,q,$$
$$A_{14} = a_{14}\frac{\cos\theta\,\cos\gamma}{\cos\alpha}, \quad A_{15} = a_{15}$$

Ratios of absolute values of terms in Eq.6 are assessed, as of these depend the magnitudes of effects of terms $\alpha, \beta, \theta, p, q$ and r,, and therefore are shown as fractions:

$$\bar{A}_{ij} = \frac{\mid A_{ij} \mid}{\sum_{j=1}^{m} \mid A_{ij} \mid} 100 \qquad [\%]$$

In case of modeling spin with the complete system of equations, terms A_{ij} and $\bar{A}_{ij}$ are introduced in the same manner.

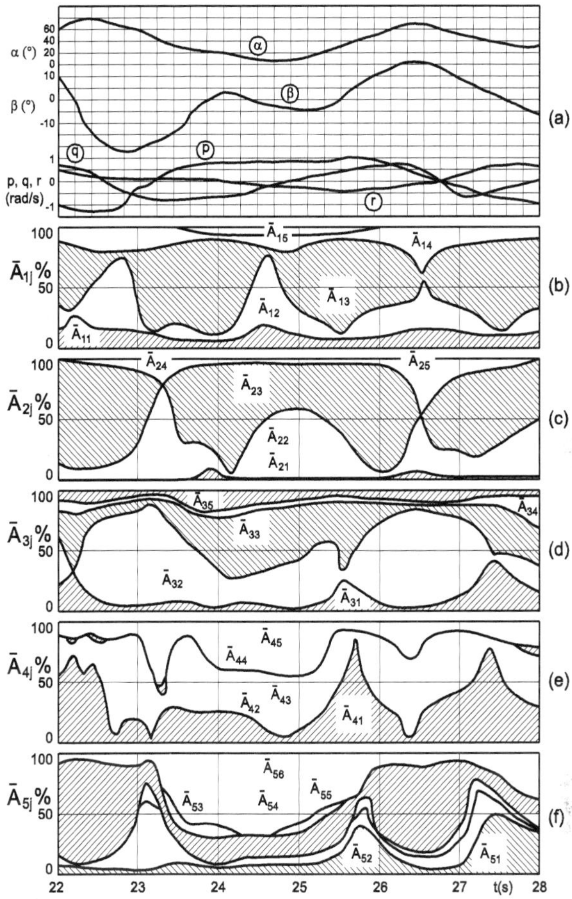

Fig. 17. Diagram of development for a flat spin (spin ongoing as a falling leaf)

Figure 17 shows results obtained for a numerically modeled spin. From diagrams (b), (c), (d), (e), and (f), it can be seen which terms have the most effect on results given in diagram (a). Diagrams (b), (c), (d), (e), and (f), were constituted by showing time on the abscissa, while values of $\bar{A}_{ij}$ $(i = const)$ are "stacked" on the ordinate, one over the other. For example, value $\bar{A}_{i1}$ is imposed from 0, value $\bar{A}_{i2}$ is imposed from point where $\bar{A}_{i1}$ ends, value $\bar{A}_{i3}$ is imposed from point where $\bar{A}_{i2}$, ..., ends, value $\bar{A}_{im}$ is imposed from point where $\bar{A}_{i,m-1}$ ends, and ends at 100%. By doing so, these diagrams show which term $\bar{A}_{ij}$ has the most impact on values of α, β, θ, p, q and r. For example, in diagram (b) it is shown that at moment $t = 23\,s$, value of $\bar{A}_{13}$ is higher than of $\bar{A}_{11}$, $\bar{A}_{12}$, $\bar{A}_{14}$ and $\bar{A}_{15}$. This means that $\bar{A}_{13}$ has the most effect on values of function $\alpha = f(t)$ at the given time. When observed what A_{13} is equal to, it can be noticed that the value of angle of attack, at this moment, mostly depends on q, since $a_{13} = 1$. In this way, every term A_{ij} on the right-hand side of Eq.(6) can be analyzed on its impact on the left-hand side of same equations. Analysis of every term A_{ij}, brings about a deeper understanding of the physical image of the studied regime. In addition, it enables the determination of terms that have the most effect on such a state, in cases when the aircraft does not fulfill necessary requirements for spin, and by appropriate modifications obtain an aircraft with necessary technical characteristics for spin.

9. References

[1] R.H. Barnard; D.R. Philpott; A.C. Kermode (2006). *Mechanics of Flight* (11th Edition), Prentice Hall, ISBN: 1405823593

[2] Warren F. Phillips (2009). *Mechanics of Flight* (2nd Edition), Wiley, ISBN: 0470539755

[3] B. Pamadi (2004). *Performance, Stability, Dynamics, and Control of Airplanes* (2nd Edition), AIAA Education, ISBN: 1563475839

[4] Barnes W. McCormick (1994). *Aerodynamics, Aeronautics, and Flight Mechanics*, (2nd Edition), Wiley, ISBN: 0471575062

[5] John D. Anderson (2001). *Fundamentals of Aerodynamics* (3rd Edition), McGraw-Hill Science/Engineering/Math, ISBN: 0072373350

[6] J. H. Blakelock (1991). *Automatic Control of Aircraft and Missiles*, John Wiley & Sons, Inc., New York, ISBN: 0471506516

[7] Bernard Etkin; Lloyd Duff Reid (1995). *Dynamics of Flight: Stability and Control* (3rd Edition), Wiley, ISBN: 0471034185

[8] D. E. Hoak (1975). *USAF Stability and Control DATCOM*, N76-73204, Flight Control Division, Air Force Flight Dynamics Laboratory, Wright-Patterson Air Force Base, Ohio

[9] D. Raymer (2006). *Aircraft Design: A Conceptual Approach* (4th Edition), AIAA Education Series, ISBN: 1563478293

[10] Roger D. Schaufele (2000). *The Elements of Aircraft Preliminary Design*, Aries Pubns, ISBN: 0970198604

[11] E. L. Houghton; P. W. Carpenter (2003). *Aerodynamics for Engineering Students* (5th Edition), Butterworth-Heinemann, ISBN: 0750651113

10

Surface Welding as a Way of Railway Maintenance

Olivera Popovic and Radica Prokic-Cvetkovic
Faculty of Mechanical Engineering, University of Belgrade
Serbia

1. Introduction

Since its early days the development of railway systems has been an important driving force for technological progress. From the 1840s onward a dense railroad network was spread all over the world. Within a few decades railway became the predominant traffic system carrying a steadily increasing volume of goods and number of passengers. This rapid development was accompanied by substantial developments in many areas such as steel production, engine construction, civil engineering, communication, etc (Zerbst et al., 2005). The railway industry worldwide is introducing heavier axle loads, higher vehicle speeds, and larger traffic volumes for economic transportation of goods and passengers. Increasing demands for high-speed services and higher axle loads at the turn of the 21st century account for quite new challenges with respect of material and technology as well as safety issues. The main factors controlling rail degradation are wear and fatigue, which cause rails to become unfit for service due to unacceptable rail profiles, cracking, spalling and rail breaks. Degradation of rail is microstructure and macrostructure sensitive and there is a complicated interaction between wear mechanisms, wear rates, fatigue crack initiation and growth rates, which affect rail life (Eden et al.,2005; Kapoor et al.,2002). Defects such as squats and wheelburns occur even in the most modern and well maintained railway networks and, as a broad general rule, every network develops one such defect each year, every two kilometers. At least one European railway network suffers almost 4000 rail fractures every year. Although such fractures are rarely dangerous when actively managed, they entail a high replacement cost and can be disruptive to the network (Bhadeshia,2002). The replacement of such defects with a short rail section is expensive and not always desirable as it introduces two new discontinuities in the track in the form of two aluminothermic weld that destroy the advantages obtained with long hot-rolled rail.

Given that an average cost per repair or short replacement rail can run into several thousands of euros and that the occurrence of wheel rail interface defects is likely to increase with the evident increase in levels of traffic on most railways, the importance of the surface welding is easy to understand. Growing need for reparation due to large financial demands, have imposed research in this field.

Based on up-to date theoretical grounds and referencial facts, the aim of this paper is to show the possibilities of surface welding of the pearlitic high-carbon steel and the properties of the obtained joint. Discussion of the aquired results and conclusions indicate superior properties of reparation welded layers in comparison to base steel. In repaired rail, maximal stresses are induced in newly deposited layer, i.e. new layer becomes area of future crack initiation, that in turn will delay its initiation and provide secure and reliable exploitation. This results open further possibilities for cheaper and reliable rail maintenance in future. Finally, this work shows clearly that repaired rails, due to improved microstructure and crack initiation resistance, have dominant mechanical properties in comparison to the original rails.

2. Rail degradation

There are many kinds of loadings which can adversely affect the life of rails; amongst these, wear and plastic deformation induced by contact stresses can combine to cause unacceptable changes in the rail head profile. Rails are subjected to complex stress state. There are many stresses that operate in a rail and can influence rail defects and rail failure. As bending and shear stresses arised principally from the gross vehicle load, the rail is also subjected to contact stresses, thermal stresses and residual stresses. Residual stresses in rails are introduced by different mechanisms. Primarily they stem from the manufacturing process, namely from heat treatment and roller straightening (Schleinzer & Fischer, 2000; Schleinzer & Fischer, 2001). A special case of residual stresses is welding residual stresses at rail joints. Since the loading conditions at the tread of a wheel and at the running surface of a rail have a number of features in common the appearance of cracks will also be similar. Cracks may be induced at or below the surface. Surface cracks are initiated due to high traction forces at high speed rails and they will propagate under the influence of a lubricant in an inclined angle in the direction of the motion of the applied load for rails operated in one direction. Transverse branching may then lead to the complete fracture of the rail. Sub-surface cracks are reported to initiate beneath the gauge corner 10-15 mm below the running surface and 6-10 mm from the gauge face (Clayton,1994). They seem to propagate towards the rail surface and to behave like original surface cracks after penetration.

Note, that cracks close to or at the surface are a rather new problem connected with high speed operating. In former times rails experienced enough wear to permanently remove the surface layer containing the new emerging cracks. In order to fulfil the increasing demands for higher axial and dynamic loads modern rail steels tend to exhibit much higher resistance to wearing with the disadvantage that the surface layer removed is not any more large enough to prevent small cracks from extending into the rail (with respect to the development of rail steels (Muders & Rotthauser,2000)).

A typical development of a rail crack is illustrated schematically in Fig. 1 (Ishida &Abe, 1996). Originating from a small surface or sub-surface crack, a dark spot is developing at the surface accompanied by crack growth in an inclined angle below the surface. At a certain point this crack branches into a horizontal and a transverse crack. The transverse crack will extend down into the rail and finally cause its fracture.

Today's rail failures can be divided into three broad groups as follows: those originating from rail manufacturing defects; those originating from defects or damage caused by inappropriate handling, installation and use and those caused by the exhaustion of the rail steel's inherent resistance to fatigue damage (Cannon et al.,2003).

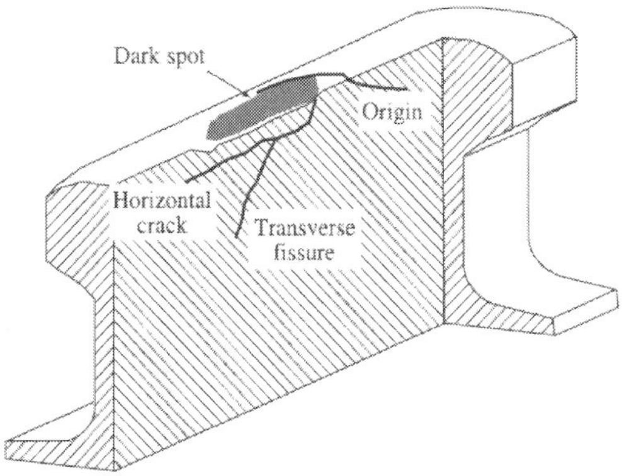

Fig. 1. Typical development of a rail crack (schematically) (Ishida &Abe, 1996).

Rolling contact fatigue (RCF) is likely to be a major future concern as business demands for higher speed, higher axle loads, higher traffic density and higher tractive forces increase. Head checks, gauge-corner cracks and squats are all names for surface-initiated RCF defects. They are caused by a combination of high normal and tangential stresses between the wheel and rail, which cause severe shearing of the surface layer of the rail and either fatigue or exhaustion of ductility of the material. The microscopic crack produced propagates through the heavily deformed (and orthotropic) surface layers of steel at a shallow angle to the rail running surface (about 10°) until it reaches a depth where the steel retains its original isotropic properties. At this stage the crack is a few millimetres deep into the rail head. At this point the crack may simply lead to spalling of material from the rail surface. However, for reasons still not clearly understood, isolated cracks can turn down into the rail, and, if not detected, cause the rail to break. These events appear to be rare, but are highly dangerous since RCF cracks tend to form almost continuously at a given site. Fracture at one crack increases stress in the nearby rail, increasing the risk of further breaks and disintegration of the rail (Cannon et al.,2003).

RCF initiation is not normally associated with any specific metallurgical, mechanical or thermal fault; it is simply a result of the steel's inability to sustain the imposed operating conditions. The problem is known to occur in most of the rail-steel types in common use today.

While wear has been reduced, rolling contact fatigue defects have become more prominent on busy routes where the rails are highly stressed. Although its wear reserve may not be used up, rail may have to be replaced because such defects quickly become critical for safety (Pointner & Frank, 1999). The relationship between RCF and mechanical wear is not well undersood, as for example zero (or minimal) mechanical wear leads to significant microcrack propagation and thus RCF failure. On the other hand, excessive mechanical wear eliminates RCF but leads to unrealistically short rail life (Kapoor et al., 2001).

The rate of rail degradation depends also on the location; rail head erosion is at a maximum in regions where the track curves. In Fig.2 is shown damage of the inner edge of rail head, caused by centrifugal force which tends to expel vehicle towards the outside of the track. Such damage can be repaired by surface welding, Fig 2b.

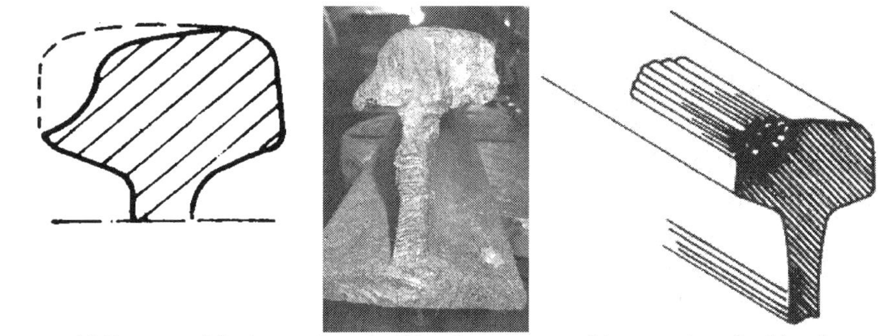

(a) Damage of the inner edge of rail head (b) reparation of rail head

Fig. 2. Rail head degradation (Popovic et al., 2006).

2.1 Fracture control concepts

A few different fracture control concepts are applied in railway systems, and one of them is damage tolerance concept (Zerbst et al., 2005). Within the frame of this concept, the possibility of fatigue crack growth is basically accepted. The aim is to prevent the crack to grow to its critical size during the lifetime of the component, i.e. to estimate number of cycles to critical crack size. In fatigue, crack extension is expressed as a function of stress intensity range ΔK and the crack extension rate, da/dN, whereby da denotes an infinitesimal crack extension due to an infinitesimal number of loading cycles dN. The basic idea is that the largest crack that could escape detection is presupposed as existent. After that, the initial crack can extend due to various mechanisms such as fatigue, stress corrosion cracking, high temperature creep, or combinations of these mechanisms. Such a failure process is visible, and catastrophic rail failure can be prevented by regular examination of the top surface of the railhead. Maintenance methods (lubrication and grinding) help combat the wear and rolling contact fatigue phenomena referred to in local parameters. By applying these methods appropriately, maintenance costs can be reduced (Vitez et al.,2005). Rail grinding prolongs rail service life by preventing the emergence of defects or by delaying their development, preventive grinding to improve the quality of the running surface of newly-laid rails and corrective grinding to remove rail defects that have already developed by reprofiling the rail to optimize wheel/rail contact.

3. Rail steels

Choice of material for rail steels is of fundamental importance. This is because the rail's behaviour in service depends critically on the properties of the metal. Much effort and a considerable amount of research has already been undertaken in the search for the ideal rail steel (Pointner & Frank,1999). In recent years rail steel production has improved as manufacturers have developed steels with increased hardness and better wear resistance.

There are many criteria which determine the suitability of a steel for rail track applications. The primary requirement is structural integrity, which can be compromised by a variety of fatigue mechanisms, by a lack of resistance to brittle failure, by localised plasticity and by excessive wear. All of these depend on interactions between engineering parameters, material properties and the environment. The track material must obviously be capable of being manufactured into rails with a high standard of straightness and flatness in order to avoid surface and internal defects which may cause failure. Track installation requires that the steel should be weldable and that procedures be developed to enable its maintenance and repair. Commercial success depends also on material and life time costs.

Since steel has one of the highest values of elastic modulus and shows superb strength, ductility and wear resistance, most modern rails have pearlitic microstructures and carbon-manganese chemistries similar to those produced in rails in 1900. Ordinary rail steels contain about 0.7 wt% of carbon and are pearlitic. Pearlite consists of a mixture of soft ferrite and a hard, relatively brittle iron carbide called cementite, Fig. 3a. Pearlite presumably achieves a high resistance to wear because of the hard cementite and its containment by the more plastic ferrite, but pearlitic steels are not therefore tough. In pealite, altering lamellae of iron and iron carbide are aranged, and lamella spacing has a large effect on hardness. Naturally cooled standard rails have coarse lamella spacing and relatively low values of about 300 Brinell hardness (HB). Control-cooled premium rails have finer lamella spacing and thus higher hardness of 340-390 HB (Lee & Polycarpou, 2005).

Raising carbon content and refining pearlite spacing increases the hardness of pearlitic steel, and this has been shown to lead to improved wear resistance. Hence rail manufacturers have worked to produce pearlitic steels with higher carbon contents (now achieving approximately 1 wt%) and finer structure (using head-hardening processes). Even though hardness generally has a positive effect on rail wear, there is a limit to the hardness that can be reached with pearlitic steels, and this hardness has been reached in modern rails (Lee & Polycarpou, 2005).

There has been considerable effort devoted to finding alternatives to the pearlitic rails, but with alterable results. In an attempt to develop rail steels with higher hardness and alternative microstructures, several types of bainitic steel were developed. While pearlitic steels obtain their strength from the fine grains of pearlite, bainitic steels (Fig. 3b) derive their strength from ultra-fine structures with a lot dislocations which are harmless but confer high strength (Aglan et al.,2004). Bainitic steel is easy to be cast, welded and inspected by ultrasonic methods. The new generation of bainitic steels achieved higher tensile and fatigue strengths and performed well in service.

(a)

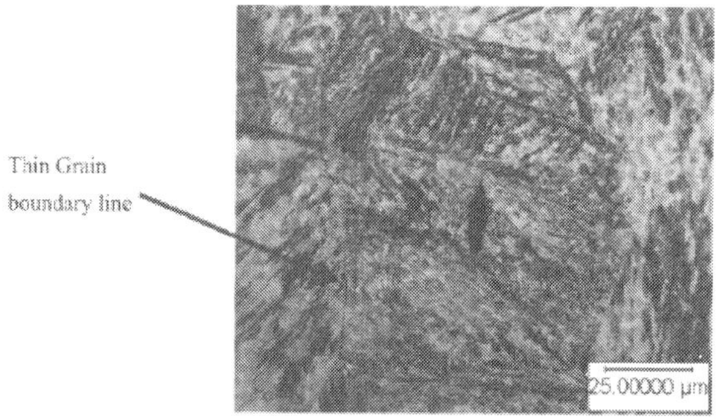

(b)

Fig. 3. Optical microstructures of rail steels: (a) pealite; (b) bainite (Aglan et al.,2004).

4. Weldability of rails and types of filler materials

Main problem in welding of pearlitic steels is their poor weldability, i.e. susceptibility to welding defects, due to its high carbon equivalent. Since the rail is produced from this type of steel and subjected to complex strain state, leading to its degradation, surface welding is presently the dominant maintenance way to prolong exploitation life. Damaged parts produced from pearlitic high-carbon steel can be surface welded, in spite of their poor weldability, and by properly choice of welding technology, it is possible to get improved structure with dominant properties comparing to the original part (Popovic et al.,2010). To achieve that, it is necessary that obtained morphology corresponds to the new steel generation, i.e. bainitic microstructure.

For surface welding are mostly in use semi-automatic arc welding processes, with flux-cored and self-shielded wires. Basic difference between them is the first requires an external shielding gas, and the second does not. In both cases, core material acts as a deoxidizer, helping to purify the weld metal, generate slag formers and by adding alloying elements to the core, it is possible to increase the strength and provide other desirable weld metal properties (Lee,2001; Sadler,1997). These processes have replaced slowly MMA process and they almost ideal for outdoors in heavy winds. The key strength of these processes lies in the replacement of those aspects of the conventional MMA process that often results in variability in the quality of the repair with automatic and more controlled operations. Although the MMA process is used many industries, it is heavily reliant on the competence of the welder, is time consuming, and is prone to internal defects such as porosity that subsequently grow through fatigue, and if not detected by ultrasonic inspection, result in rail breaks.

The result of flux-cored wire application is higher quality welds, faster welding and maximizing a certain area of welding performance (Popovic et al.,2010). The number of layers in surface welded joint depends of the damage degree, most frequently it's three, sometimes with buffer layer. The buffer layer is applied at the crack sensitive materials, what high carbon steel certainly is (high CE). The function of buffer layer is to slow down the growth of initiated crack with its own plasticity. Constructions, like railways, are exposed to cyclic load and wear in exploatation life, so crack must be initiated. Sometimes in these cases it is necessary to use buffer layer, what besides the good affects, has some drawbacks. Namely, the use of buffer layer significantly slows down surface welding process, due to replacement of wires and settings of other welding parameters. Since, as already noted, for surface welding are mainly in use semi-automatic and automatic processes, it significantly extends the working time. The new classes of flux-cored and self-shielded wires are recently developed, and it is possible to achieve the requested properties of welded joints without buffer layer (Popovic et al.,2011).

5. Experimental procedure

The material used in present work is pearlitic steel, received in the form of rails, type UIC 860 S49, what is the most common rail type on domestic railroads. It's chemical composition and mechanical properties are given in Table 1.

Chemical composition, %							Tensile strength R_m (N/mm^2)	Elongation A_c (%)
C	Si	Mn	P	S	Cu	Al		
0.52	0.39	1.06	0.042	0.038	0.011	0.006	680-830	≥14

Table 1. Chemical composition and mechanical properties of base metal.

The surface welding of the testing plates was perfomed by semi-automatic process. As the filler material, the self-shielded wire (FCAW-S) and flux-cored wires (FCAW) were used, whose chemical compositions and mechanical properties are given in Table 2. The plates were surface welded in three layers; sample 1 with FCAW-S without buffer layer; sample 2 with FCAW with buffer layer (according to Table 2).

Sample No.	Wire designation		Wire diam. mm	Chemical composition							Hard-ness, HRC
				C	Si	Mn	Cr	Mo	Ni	Al	
Sample 1	OK Tubrodur 15.43 (self-shielded wire)		1.6	0.15	<0.5	1.1	1.0	0.5	2.3	1.6	30-40
Sample 2	1.layer (buffer layer)	Filtub 12B (flux-cored wire)	1.2	0.05	0.35	1.4	-	-	-	-	-
	2. and 3. layer	Filtub dur 12 (flux-cored wire)	1.6	0.12	0.6	1.5	5.5	1.0	-	-	37-42

Table 2. Chemical composition of filler materials.

Heat input during welding was 10 kJ/cm and preheating temperature was 230^{0}C, since the CE equivalent was CE=0.64 (Popovic et al.,2010). Controlled interpass temperature was 250^{0}C. Sample 1 is surfaced with one type of filler material (self-shielded wire), while for surfacing of sample 2 were used two types of wires, but both flux-cored: one for buffer layer and the second one for last two layers. As shieleded gas for welding of sample 2, CO_2 was used. To evaluate the mechanical properties, specimens for further investigation were cut from surface welded rail head, according to Fig.4.

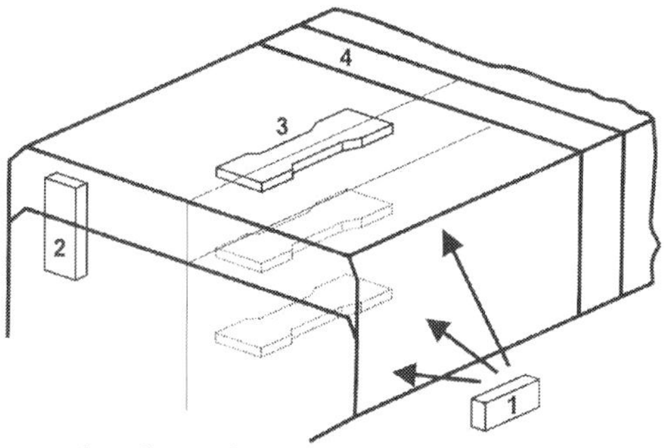

1- specimen for toughness and crack growth resistance estimation
2- specimen for microstructural analysis
3- tensile specimens
4- specimen for hardness measurements

Fig. 4. Specimens from surface welded rail head (Popovic et al.,2010).

5.1 Hardness

Hardness measurements were performed using a load of 100Pa. Hardness profiles of surface welded joints are shown in Fig. 5. The lowest hardness is in the base metal (250-300 HV), being the hardness of naturally cooled standard rails(Lee & Polycarpou, 2005; Singh et al., 2001). In HAZ hardness increase is noticable in both samples, due to complex heat treatment and grain refinement (Popovic et al.,2010). In sample 2 comes to a sharp decrease of hardness in first surfaced layer, i.e. in buffer layer. The function of buffer layer is to stop the growth of initiated crack with its own plasticity and reduced hardness. The hardness of II and III welded layers of both samples are the highest and similar, due to influence of alloying elements in filler materials, which shift transformation points to bainitic region[4]. Maximum hardness level of 350-390 HV is reached in surface welded layers and it provides improvement of mechanical properties and wear resistance.

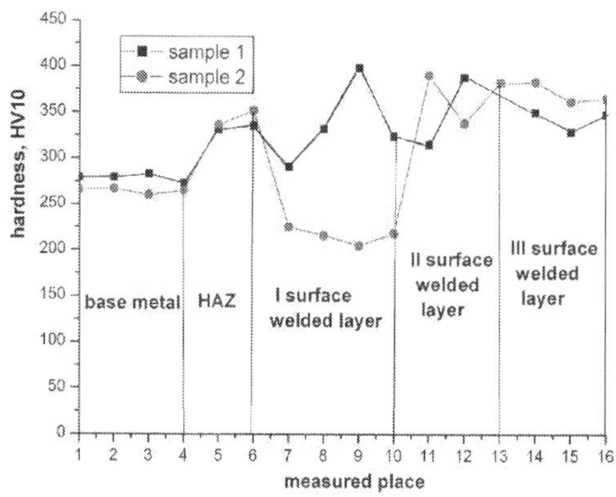

Fig. 5. Hardness profiles along the joint cross-section of samples (Popovic et al.,2011).

5.2 Microstructure

Microstructural analisys of all characteristical zones of welded layer has been done. Heat affected zone (HAZ) also has pearlitic microstructure, but with finer grain, than base metal (Figure 6), so its structure is improved and it is not a critical place in weldment. That is result of thermomechanical treatment of HAZ which is re-heated three times. Structural compatibility between deposite metal and base metal was achieved and martensitic layer wasn't formated.

The greatest differences appear in first layer microstructure, Fig.7. First layer microstructure of sample 1 consists of ferrite, pearlite and bainite, what is result of mixing of low-alloyed filler material with high-carbon base metal. For first layer deposition of sample 2 is used low-carbon wire alloyed with Mn, as a function of buffer layer, so characteristical structure consist of great fraction of ferrite with relatively large primary grains. Beside proeutectoid ferrite, microstructure contains Widmanstatten and acicular ferrite (Popovic et al.,2007).

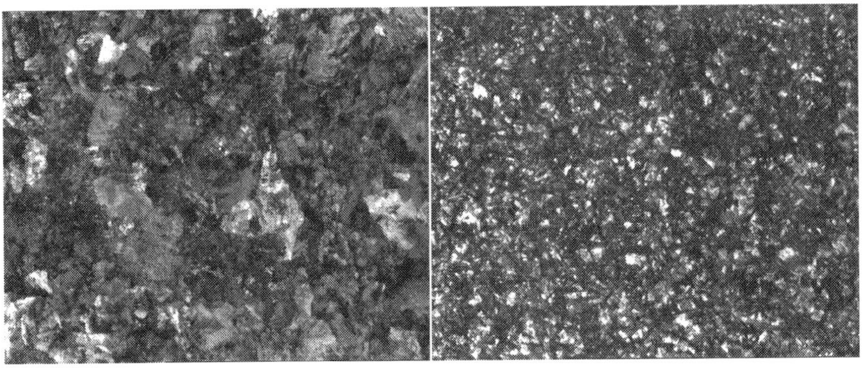

(a) 200x (b) 200x

Fig. 6. Microstructure of a) base metal and b) HAZ of both samples (Popovic et al.,2007).

The second layer microstructure is the most important in surface welded joint, because it has the greatest influence on mechanical and technological properties and exploatation behavior of repaired parts. For this structure is characteristic larger fraction of bainite, consequence to the less mixing with base metal. In second layer of sample 2 occurs fine grain ferritic structure with low content of bainite. This structure has finer grain compare to first layer, what is result of heat treatment and chemical composition (presence of Mo in filler material).

The third layer of sample 1 has some coarser grain structure, with higher content of bainite, compare to previous layer, what is consequence of re-heating absence. For third layer of sample 2 is characteristical bainitic microstructure with small amount of martensite and locally zones of proeutectoid ferrite.

Though used filler materials are different type, alloying concepts, sort of protection, buffer layer, as final result is obtained desirable bainitic microstructure with superior properties compare to base metal (Popovic et al.,2007). Except metallography examination, this is confirmed by other detail tests (Popovic,2006).

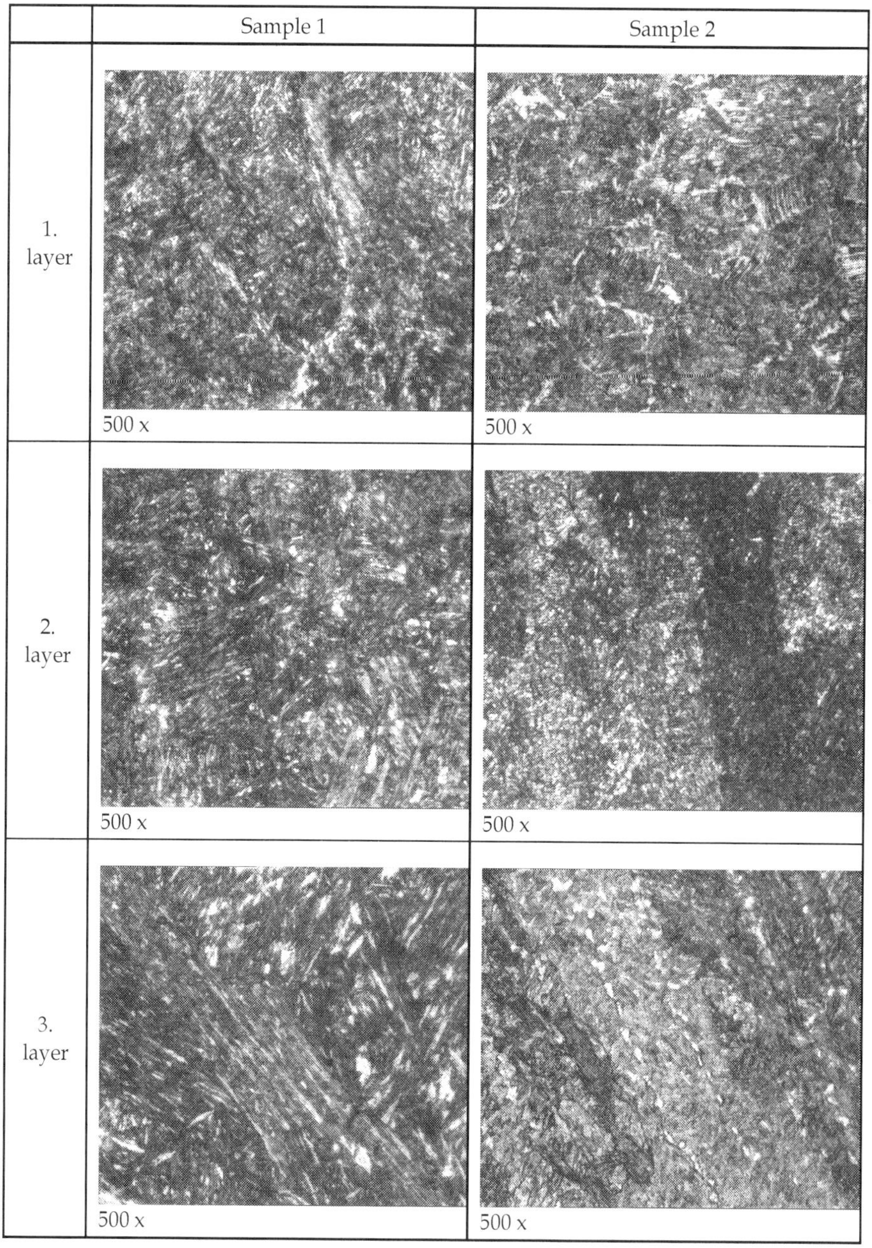

Fig. 7. Microstructure of all surface welded layers (Popovic et al.,2007).

5.3. Tensile tests

The tensile tests were conducted on a 2 mm thick specimens. The room temperature mechanical properties (ultimate tensile strength, UTS) of the surface welded joint are shown in Figure 8. The basic requirement in welded structures design is to assure the required strength. In most welded structures this is achieved with superior strength of WM compared to BM (overmatching effect), and in tested case this is achieved (Burzic & Adamovic,2008; Manjgo et al.,2010). The highest UTS is in weld metal of sample 2 (1210 MPa), due to solid state strengthening by alloying elements.

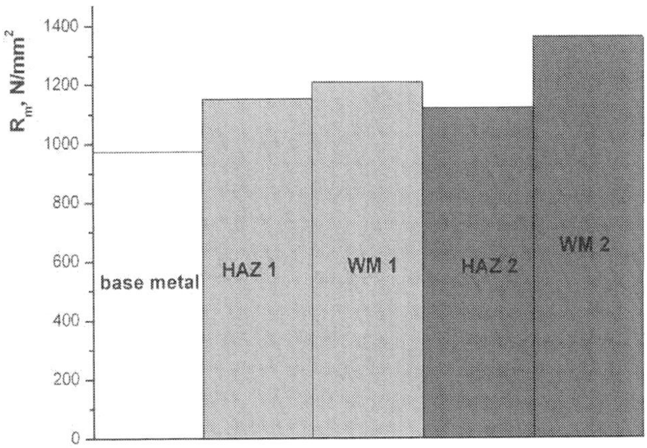

Fig. 8. Ultimate tensile strength of the surface welded joints (Popovic et al.,2011).

5.4 Impact testing

Impact testing is performed according to EN 10045-1, i.e ASTM E23-95, with Charpy V notched specimens, on the instrumented machine SCHENCK TREBEL 150 J. Impact testing results are given in Table 3 for base metal and HAZ at all testing temperatures. Total impact energy, as well as crack initiation and crack propagation energies, for weld metal of both samples at all testing temperatures (20°C, -20°C and -40°C) are presented in Table 4 and in Figure 9.

The total energy of base metal is very low (5 J), due to very hard and very brittle cementite lamellae in pearlite microstructure (Popovic et al.,2011), while the toughness of HAZ is higher (11-12 J) and is similar for both samples at all testing temperatures.

	Total impact energy, E_u, J		
	20°C	-20°C	-40°C
base metal	5	3	3
sample 1-HAZ	12	11	10
sample 2-HAZ	11	10	9

Table 3. Instrumented impact testing results of Charpy V specimens for base metal and HAZ at all testing temperatures.

	sample 1-WM			sample 2- WM		
	20^{0}C	-20^{0}C	-40^{0}C	20^{0}C	-20^{0}C	-40^{0}C
Total impact energy, E_u, J	29	23	17	34	14	11
Crack initiation energy, E_{in}, J	20	16	15	12	10	10
Crack propagation energy, E_{pr}, J	9	7	2	22	4	1

Table 4. Instrumented impact testing results of Charpy V surface weld metal specimens at all testing temperatures.

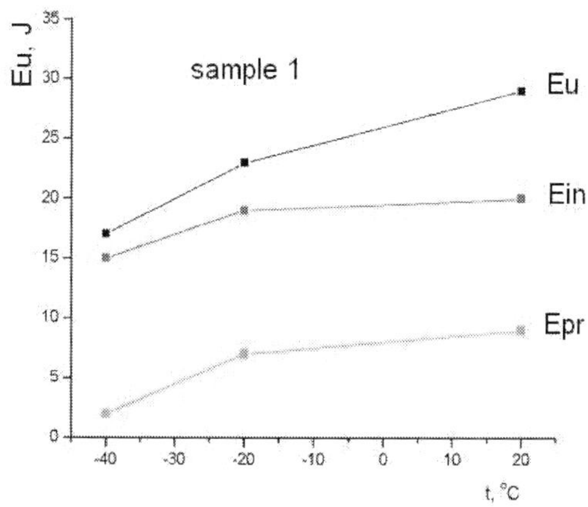

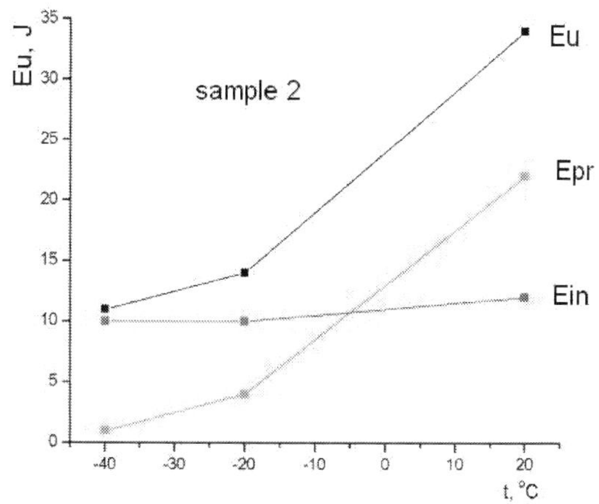

Fig. 9. Dependence total impact energy, crack initiation and crack propagation energy vs.temperature for (a) weld metal of sample 1 and (b)weld metal of sample 2 (Popovic et al.,2011).

Values of total impact energy of samples 1 and 2 at room temperature are significantly higher (29 J and 34 J) than in base metal (5 J), as a consequence of appropriate choice of alloying elements in the filler material. The presence of Ni, Mn and Mo promotes the formation of needled bainitic microstructure and grain refinements, and increases the strength and toughness also(Popovic, 2006). By analyzing the impact energy values of sample 1, a change of toughness in continuity is observed, with no marked drop of toughness, and for all tested temperatures, crack initiation energy is higher than crack propagation energy. This is the reason for the absence of significant decrease of toughness. The highest value of total impact energy was found for the sample 2 at room temperaure (34 J), which is the only case when the initiation energy is lower than propagation energy (12 J and 22 J, respectivelly). This shown practically the buffer layer function. Namely, the initiated crack during propagation comes to plastic buffer layer, which slows down crack further growth. For this reason, the crack propagation energy is the largest part of total impact energy. However, at -20^{0}C, significant drop of total impact energy is noticable (14 J) due to losing of buffer layer plastic properties at lower temperatures. The low-carbon wire (0.05%C i 1.4%Mn) has excellent toughness, but and marked rapid drop on S-curve (dependence toughness vs. temperature). Transition temperature of this material above -20^{0}C is confirmed by the obtained impact toughness results. The use of buffer layer is reasonable if the exploatation temperature is above -5^{0}C; on the contrary, at lower temperatures, buffer layer is losing its function and the toughess is decreased (Popovic et al.,2011).

Diagrams force-time, obtained by instrumented Charpy pendulum, are given in Figure 10. As can be seen, for the sample 1 the character of diagrams force-time changed little by lower temperature. Namely, this material at room temperature has diagram with marked rapid drop, as consequence of unstable crack growth. After the maximum load, a very fast crack growth is started, and it is confirmed by the low value of crack propagation energy(Grabulov et al.,2008). On the contrary, on the sample 2 diagram at room temperature, the presence of buffer layer is clearly shown. The initiated crack, during its growth, comes to buffer layer which temporary stops the further crack growth and changes crack growth rate. The obtained experimental diagram doesn't belong to any type, according to standard EN 10045-1. This leads to toughness increase, primarily crack propagation energy, and it is also here the only case when the crack initiation energy is lower than crack propagation energy.

t, ^{0}C	sample 1	sample 2 (BL)
20^{0}C	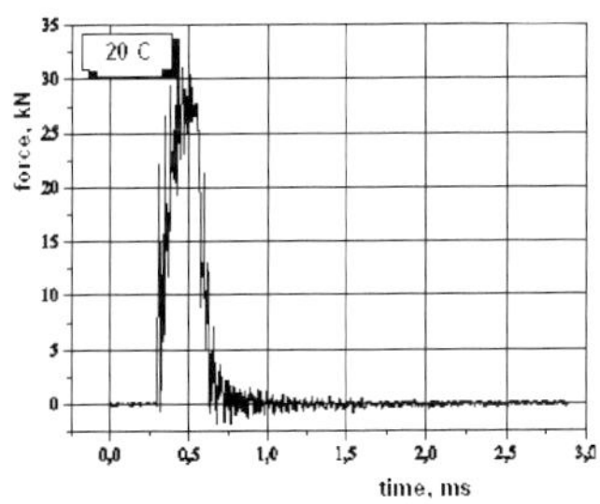	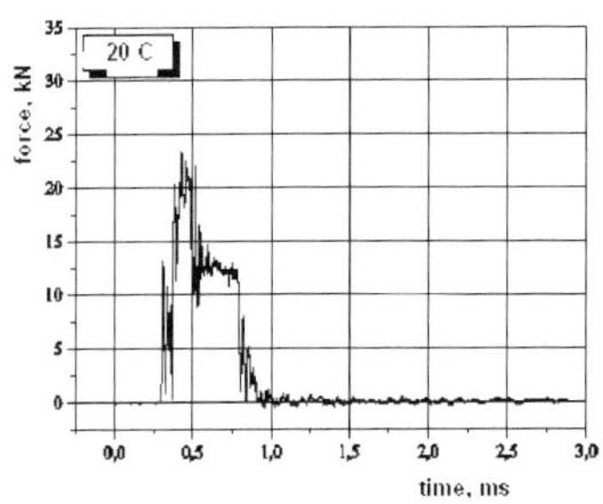

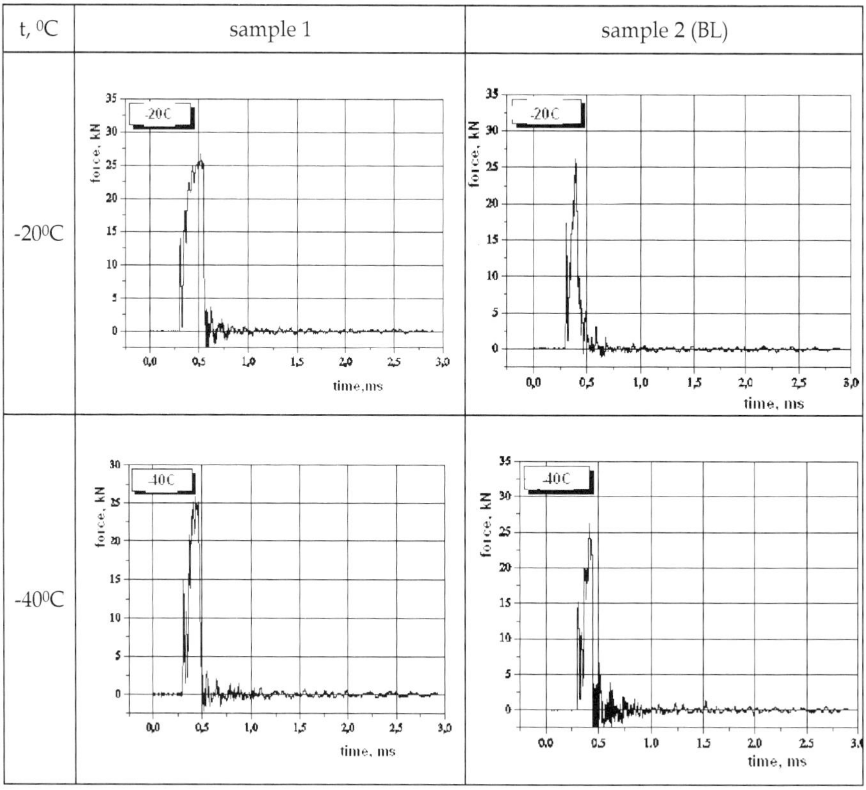

Fig. 10. Diagrams force-time, obtained by instrumented Charpy pendulum for sample 1 and sample 2 (Popovic et al.,2011).

5.5 Crack growth rate

A basic contribution of fracture mechanics in fatigue analysis is the division of fracture process to crack initiation period and the growth period to critical size for fast fracture (Burzic & Adamovic,2008). Fatigue crack growth tests had been performed on the CRACKTRONIC dynamic testing device in FRACTOMAT system, with standard Charpy size specimens, at room temperature, and the ratio R=0.1. A standard 2 mm V notch was located in base metal and in third layer of WM, for the estimation of parameters for BM, WM and HAZ, since initiated crack will propagate through those zones. Crack was initiated from surface (WM) and propagated into HAZ, enabling calculation of crack growth rate da/dN and fatigue treshold ΔK_{th}. The results of crack growth resistance parameters, i.e., obtained relationship da/dN vs. ΔK for base metal, sample 1 and for sample 2 are given in Figure 11 and 12. Parameters C and m in Paris law, fatigue threshold ΔK_{th} and crack growth rate values are given in Table 5 for all samples as obtained from relationships given in Figures 11 and 12, for corresponding ΔK values.

The behaviour of welded joint and its constituents should affect the change of curve slope in validity part of Paris law. Materials of lower fatigue-crack growth rate have lower slope in the diagram da/dN *vs.* ΔK. For comparison of the properties of surface welded joint constituents the crack growth rates are calculated for different values of stress-intensity factor range ΔK.

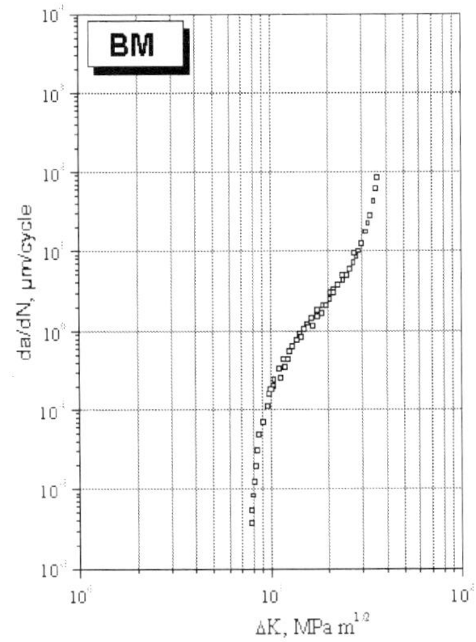

Fig. 11. Diagram da/dN vs. ΔK for base metal.

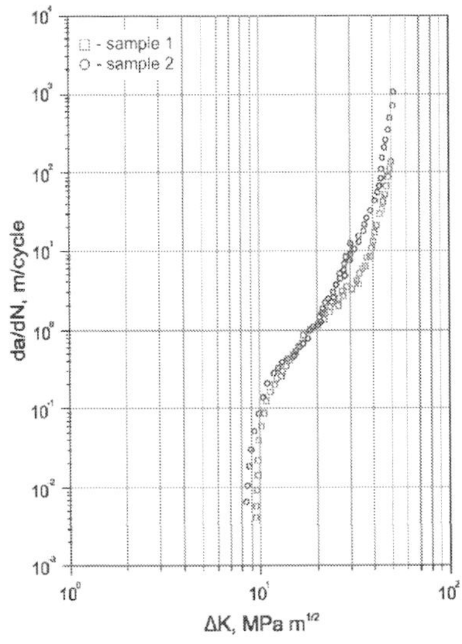

Fig. 12. Diagram da/dN vs. ΔK for sample 1 and sample 2.

	Zone of surface welded joint	Fatigue threshold ΔK_{th}, MPa $m^{1/2}$	Parameter C	Para-meter m	Crack growth rate da/dN, m/cycle			
					ΔK=10 $MPam^{1/2}$	ΔK=15 $MPam^{1/2}$	ΔK=20 $MPam^{1/2}$	ΔK=30 $MPam^{1/2}$
Base metal	BM	8.0	$3.31 \cdot 10^{-11}$	3.28	$6.31 \cdot 10^{-08}$			
sample 1	WM 1	9,5	$4.45 \cdot 10^{-13}$	3.74		$1.11 \cdot 10^{-08}$	-	-
	WM 2		$3.78 \cdot 10^{-13}$	3.61		-	$1.88 \cdot 10^{-08}$	-
	HAZ		$4.07 \cdot 10^{-13}$	3.79		-	-	$1,61 \cdot 10^{-07}$
sample 2	WM 1	8,9	$4.63 \cdot 10^{-13}$	3.87		$1.65 \cdot 10^{-08}$	-	-
	WM 2		$3.85 \cdot 10^{-13}$	3.88		-	$2.07 \cdot 10^{-07}$	-
	HAZ		$3.76 \cdot 10^{-13}$	3.93		-	-	$1.18 \cdot 10^{-06}$

Table 5. Parameters C, m, ΔK_{th} and crack growth rate values for all zones of surface welded joints.

Bearing in mind that weld metal consists of two layers (third layer is used for V notch), as referent values of ΔK were taken: ΔK =10 MPa $m^{1/2}$ for BM, ΔK =15 MPa $m^{1/2}$ for WM1, ΔK =20 MPa $m^{1/2}$ for WM2, and ΔK =30 MPa $m^{1/2}$ for HAZ. It's important that all selected values are within a middle part of the diagram, where Paris law is applied. The crack growth rate in base metal is 3-4 times higher than in both weld metal layers, i.e. the growth of initiated crack will be slower in weld metal layers. This means that for the same value of stress intensity factor rang ΔK, base metal specimen needs less number of cycles of variable amplitude than weld metal specimen, for the same crack increment.

In all three zones of surface weleded joint (WM2, WM1 and HAZ), sample 2 with buffer layer has higher crack growth rate than sample 1, i.e. the growth of initiated crack will be slower in sample 1. This means that for the same value of stress intensity factor rang ΔK, specimen of sample 2 needs less number of cycles of variable amplitude than specimen of sample 1, for the same crack increment. The maximum fatigue crack growth rate is achieved in HAZ for both samples, when stress intensity factor range approaches to plane strain fracture toughness.

If a structural component is continuously exposed to variable loads, fatigue crack may initiate and propagate from severe stress raisers if the stress intensity factor range at fatigue threshold, ΔK_{th}, is exceeded(Burzic & Adamovic,2008). Fatigue treshold value ΔK_{th} in base metal (ΔK_{th} =8 MPa $m^{1/2}$) is lower than fatigue treshold value ΔK_{th} in weld metal of both metal. Fatigue treshold value ΔK_{th} for sample 2 (ΔK_{th} =8.9 MPa $m^{1/2}$) is lower than that for sample 1 (ΔK_{th} =9.5 MPa $m^{1/2}$). This means that crack in sample 2 will be initiated earlier, i.e. after less number of cycles, than in sample 1.

Values of fatigue threshold and crack growth rates corespond to initiation and propagation energies in impact testing, and in this case, good corelation is achieved (Popovic, 2006). Sample 1 has higher crack initiation energy (20 J) and higher ΔK_{th} ((ΔK_{th} =9.5 MPa $m^{1/2}$ for sample 1 i ΔK_{th} =8.9 MPa $m^{1/2}$ for sample 2). With comparation of crack propagation energy and crack growth rate, it is hard to establish the precise analogy, as toughness was estimated for the surface weld metal, whereas crack growth rate for each surface welded layer. Generally, buffer layer didn't show slow the initiated crack growth, with aspect of crack growth rate, while this effect is obvious in the case of toughness, i.e. crack propagation energy (Popovic et al, 2011).

6. Conclusion

On the base of obtained experimental results and their analysis, the following is concluded:

1. The experimental investigation of surface welded joints with different weld procedures has shown, as expected, significant differences on their performance in terms of mechanical properties. But, in both cases, it was shown, that in spite of poor weldability of high carbon steel, they can be successfully welded. Structural compatibility between deposite metal and base metal was achieved and martensitic layer wasn't formated. Obtained HAZ has better structure compare to base metal.
2. The filler material is relevant parameter which affects on deposite layer quality. Work with self-shielded wires is more simple, specially for outdoor applications. Both used wires are on high technological level and can be recomended for reparation of high-carbon steel damaged parts. Final microstructure is the result of different influences: type of filler material, heat input, degree of mixture with previous layer and post heat treatment with subsequent surface layer. It is necessary to know all these factors and also to know the way of affect. Though applied wires are with different alloying concepts, result in both cases is that initial pearlitic moprhology is replaced by final desirable bainitic microstructure. It was shown that, by selecting corresponding parameters, it is possible to obtain the morphology of the best properties.
3. The maximal hardness level of 350-390 HV is reached in surface welded layers of both samples, with equal hardness of base metal (250-300 HV). The main difference appears in the first deposition layer, where as expected, in sample 2 the hardness is significantly lower (buffer layer). The obtained hardness values ensure simultaneously the improvement of mechanical and wear properties, and in the case of a rail, represents maximal hardness preventing the wheel wear (Popovic et al., 2010). Similar results are obtained by tensile testing. Sample 2 has slightly higher ultimate tensile strength (1360 MPa) than sample 1 (1210 MPa) due to solid solution strengthening by alloying elements.
4. The most improved results are obtained for impact properties. The toughness of base metal is 6-7 times lower than the toughness of weld metal, and more than twice lower than toughness of HAZ. For welding with buffer layer, at -20^{0}C, the drop of total impact energy is significant, due to lowering of buffer layer plastic properties at lower temperatures. The transition temperature of this material is above -20^{0}C, and it was confirmed by obtained impact toughness results. The use of buffer layer is beneficial for exploatation temperature above -5^{0}C. On the contrary, at lower temperatures, buffer layer loses its function and toughess decreases. On the contrary, for sample 1 the change of toughness is continous and without marked drop of toughness. At all tested temperatures, the crack initiation energy is higher than crack propagation energy. This

technology of magnesium matrix composites is a new method. The semi-solid material forming technology has advantages such as lower deformation resistance, good material mobility and so on [Flemings 1991, Yan et al., 2005]. It was composed of three processes such as: semi-solid billet fabrication [Yan et al., 2005], partial remelting [Yan et al., 2006] and thixoforming [Yan et al., 2008]. For this reason, the research on the basic theory of semi-solid stirring melting fabrication method and thixoforming process for the advanced MMCs is studied in this item. The works include the study of semi-solid stirring melting fabrication method [Yan & Fu et al., 2007; Yan & Lin et al., 2008] and reheating process [Yan & Zhang et al., 2008; Zhang et al., 2011] for the particle-reinforced MMCs. The material constitutive relation will be proposed [Yan & Wang et al., 2011]. Then the finite element model coupled with multi-physical fields will be built. The simulation will be gone based on the developed analytical program. The forming performances and deformed laws in the thixoforming for the particle-reinforced MMCs will be studied by the way of combining theoretical analysis with experimental method [Yan & Huang, 2011]. The results will play an important function to bulid the theoretical and technological fundament for the thixoforming process of the particle-reinforced MMCs applied the industry area.

2. Study on fabrication methods and various properties for magnesium matrix composites

2.1 Fabrication methods

The AZ61 alloy was used as the matrix material. The chemical composition of AZ61 was 5.8%~7.2%Al, 0.15% Mn, 0.40%~1.5%Zn, 0.10% Si, 0.05%Cu, 0.05%Ni, 0.005%Fe, and the rest is Mg. Its solidus temperature was 525°C, and the liquidus temperature was 625°C. The reinforcement was green α-SiC particle whose average diameter was $10_{\mu m}$. A self-manufactured electric resistance furnace was used for melting Mg alloy (shown in Fig.1). The liquid metal was stirred with the mechanical stirrer driven by the timing electrical machine in the melting process. In order to improve the accuracy of controlling temperature, the thermocouple was inserted directly into the liquid metal, and combined with the artificial aptitude modulator BT608 that adopted the industrial micro-processor, whose precision was only ±10°C. The MMCs specimens were sampled by the pipette connecting to a vacuum pump in this experiment.

The fabrication processes of SiCp/AZ61 composites were described as follow. AZ61 alloy matrix was heated to melt, gas was gotten rid of and slag was removed. Then SiC reinforcement was added into the molten. There were the three addition processes. In the fully-liquid stirring casting process (about 680°C), SiC particles were introduced into the fully-liquid molten, and then sampled after stirred to reach predetermined time. (2) In the stirring-melt casting process (590°C), SiC particles were introduced at the semi-solid state, then sampled after reached to a fully-liquid temperature of 680°C. (3) In semi-solid stirring casting process, SiC particles were introduced at the semi-solid state, then sampled after stirred to reach predetermined time. In above experiments, their volume fractions of SiC particles were 3%, 6%, and 9% respectively, whose preheated temperature was 500°C with holding time 2h. The stirring rate was 500r/min with holding time 10min. Then the specimen was made to the metallurgical phase sample and corrupted with 0.5% ammonium HF liquor, and its microstructural changes were observed under the optical microscope. Finally, the Vickers hardness was measured in a micro-sclerometer HXS-1000AK.

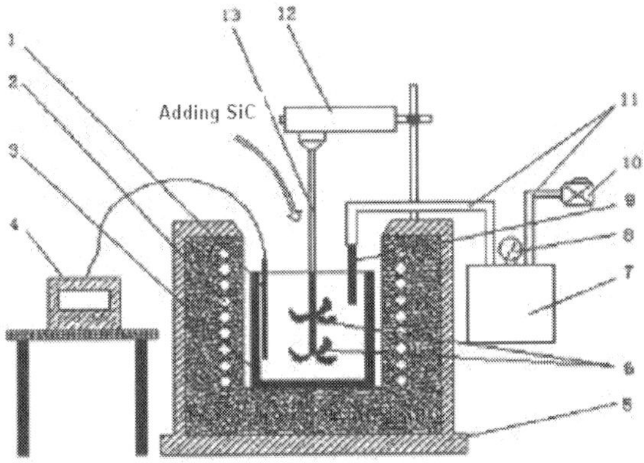

Fig. 1. Schematic diagram of SiCp/AZ61 composites fabricated in stirring casting process 1. thermocouple 2. resistance thread 3. crucible 4. BT608 (artificial aptitude modulator) 5. resistance furnace 6. strring lamina 7. vacuum jar 8. pressure meter 9. pipette 10. vacuum pump 11. guiding windpipe 12. timing electrical machine 13. stirring bar

The microstructures of SiCp/AZ61 composites in three casting processes were shown in Fig.2. The variations of influence of three casting processes on the microstructures of SiCp/AZ61 composites were shown in Fig.2- Fig.4. The distribution of SiC particles was a little uniform in the fully-liquid casting process where a lot of gas cavities and slacks were presented, and SiC particles were easy to sink and float. There were a few gas cavities in the semi-solid casting process where the distribution of SiC particles was inhomogeneous. SiCp/AZ61 composites fabricated by the stirring-melt casting method possessed not only few gas cavities but fairly uniform distribution.

The existence of gas cavities and slacks was attributed to the following factors: (1) Gas was involved in the molten during the mechanical stirring process. (2) Their non-uniform volume shrinkages presented in the composites solidification process due to the differences of their thermal expansion coefficient and heat conduction between matrix and reinforcement. (3) Hydrogen produced in the chemical reactions between Mg and H_2O was dissolved in the molten and formed gas cavities during solidification. (4) The formation of gas cavities was resulted from the particles clustering.

The main problem in the stirring casting process was the inhomogeneous distribution of reinforcement phase. The major reasons were followed as. (1) Due to having the different densities between matrix and reinforcement, SiC particles were settled down. (2) The higher surface tension and poor wettability between SiC and matrix presented, a few SiC particles were floated on the surface of the molten.

SiC particles were introduced at the semi-solid state during the semi-solid stirring casting process where the high viscosity semi-solid alloy can help withstand SiC particles from sinking and floating, but the uniform distribution can not be solved (shown in Fig.3). During the stirring-melt casting process the reinforcement were added at the semi-solid state, and the composites were poured immediately after reached 690°C (liquidus). The quite uniform SiCp/AZ61 composites can be obtained in this method (shown in Fig.4).

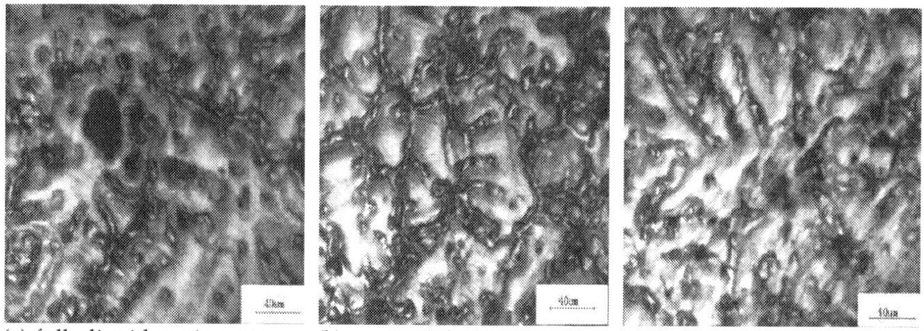

(a) fully-liquid casting process (b) semi-solid casting process (c) stirring-melt casting process

Fig. 2. Microstructures of SiCp/AZ61 composites in three casting processes

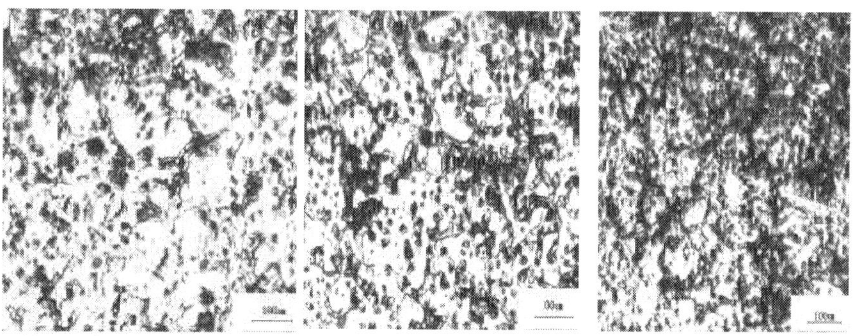

(a) 500r/min, 10min, 3Vol.% (b) 500r/min, 10min, 6Vol.% (c) 500r/min, 10min, 9Vol.%

Fig. 3. Microstructures of SiCp/AZ61 composites with various volume fractions of SiC particles in semi-solid stirring casting process

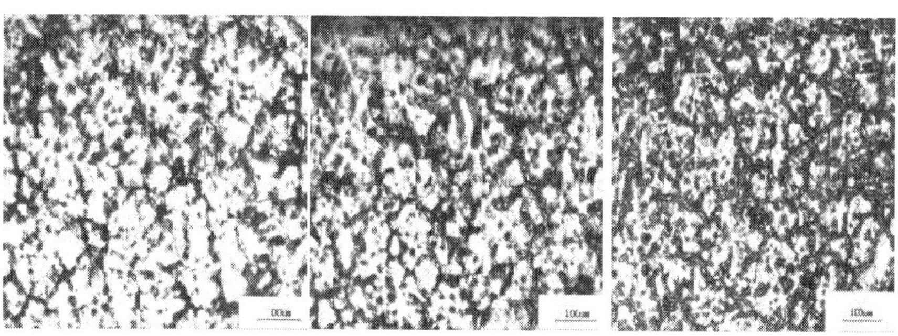

(a) 500r/min, 10min, 3Vol.% (b) 500r/min, 10min, 6Vol.% (c) 500r/min, 10min, 9Vol.%

Fig. 4. Microstructures of SiCp/AZ61 composites with various volume fractions of SiC particles in stirring-melt casting process

2.2 Optimization on stirring melt casting process

In this study, the composites were fabricated by a stirring melt casting method. The effects of volume fraction of SiC particles, stirring temperature and stirring time on the mechanical properties and microstructure of SiCp/AZ61 composites were investigated. The main technological parameters of preparing SiCp/AZ61 composites were optimized, which was helpful for obtaining its good properties.

The effects of volume fraction of SiC particles, stirring temperature and stirring time on the mechanical properties of SiCp/AZ61 composites were investigated by an orthogonal experimental method, in which average particle size and stirring speed were maintained the same. The orthogonal test table with three factors and three levers is shown in Table 1. According to design of the primary experiment, volume fractions of SiC particles were 3%, 6% and 9%, stirring temperatures were 580°C, 587°C and 595°C, and stirring times were 3min, 5min and 7min. Three factors were volume fraction of SiC particles, stirring temperature and stirring time. Two targets were tensile strength and elongation.

level	volume fraction of SiC A (*%)	stirring temperature B (°C)	stirring time C (min)
1	3	580	3
2	6	587	5
3	9	595	7

Table 1. Factors and levels of test

The effects of volume fraction of SiC particles, stirring temperature and stirring time on the tensile strength and elongation at room temperature of SiCp/AZ61 composites are shown in Table 2. (1) Tensile Strength Analysis. The level two (the volume fraction of SiC 6%) was

No.	volume fraction of SiC A	stirring temperature B	stirring time C	tensile strength /MPa	elongation /%
1	1	1	1	172	3.8
2	1	2	2	179	4.2
3	1	3	3	163	5.2
4	2	1	2	184	3.5
5	2	2	3	176	4.1
6	2	3	1	189	3.9
7	3	1	3	164	1.3
8	3	2	1	153	1.9
9	3	3	2	170	2.1
I	171.3	173.3	171.3		
II	183.0	169.3	177.7		
III	162.3	174.0	167.7		
R (tensile strength)	20.7	4.7	10.0		
I	4.4	2.9	3.2		
II	3.8	3.4	2.1		
III	1.7	3.7	3.5		
R (elongation)	2.7	0.8	1.4		

Table 2. The table of three factors and three levels in the orthogonal experiment

the best among the levels of factor A. The level three (595°C) was the best among the levels of factor B. The level 2 (5 min) was the best among the levels of factor C. Thus the optimum combination was $A_2B_3C_2$. (2) Elongation analysis. The level two (the volume fraction of SiC 3%) was the best among the levels of factor A. The level 3 (595°C) was the best among the levels of factor B. The level three (7 min) was the best among the levels of factor C. Thus the optimum combination was $A_1B_3C_3$. (3) Range Comprehensive Analysis. The greater the Range (R), the greater the effect of the lever change of the factor on the test target. This factor was more important. From Table 2, the sequence of tensile strength was $R_A>R_C>R_B$. So the sequence of primary and secondary in factors of A, B, C was the volume fraction of SiC particles, stirring time and stirring temperature. The sequence of elongation was $R_A>R_C>R_B$. So the sequence of primary and secondary in factor of A, B, C was also the volume fraction of SiC particles, stirring time and stirring temperature. Besides, from the ranges(R) in table 1, factor A has a more notable impact for tensile strength and elongation, and other factors do not have great impact. After comprehensive analysis, a better combination of factors was $A_2B_3C_2$, namely, the optimum processing plan of SiCp/AZ61 composites in the experimental condition was volume fraction of SiC particles 6%, stirring temperature 595°C and stirring time 5 min.

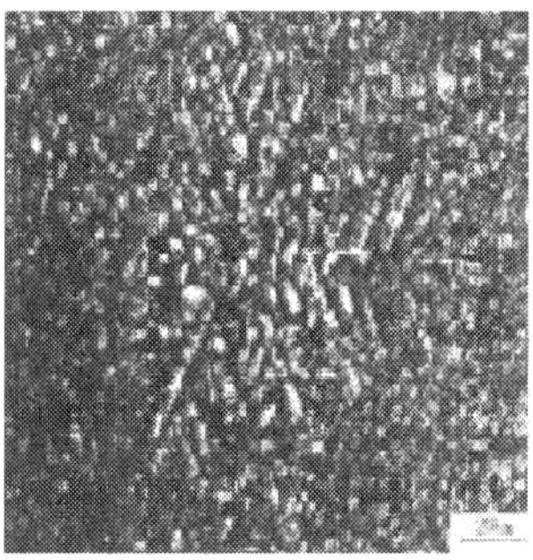

(a) φ (SiC)=0%

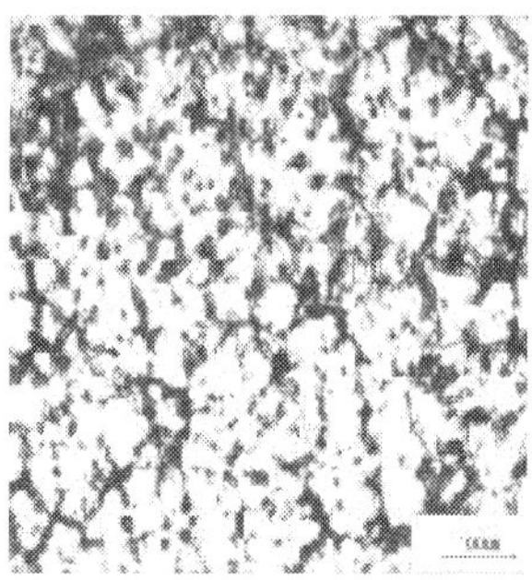

(b) φ (SiC)=3%

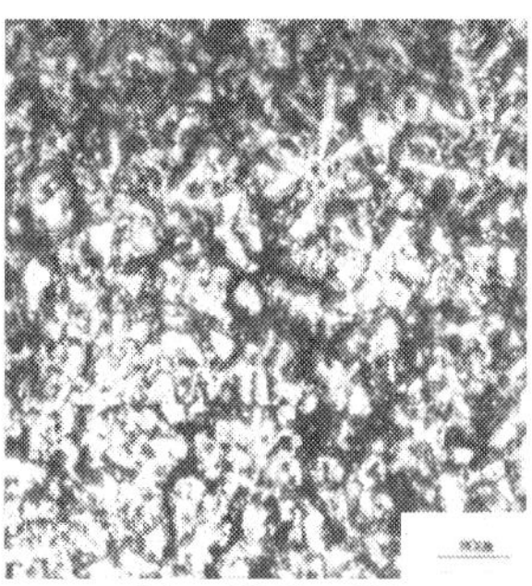

(c) φ (SiC)=6%

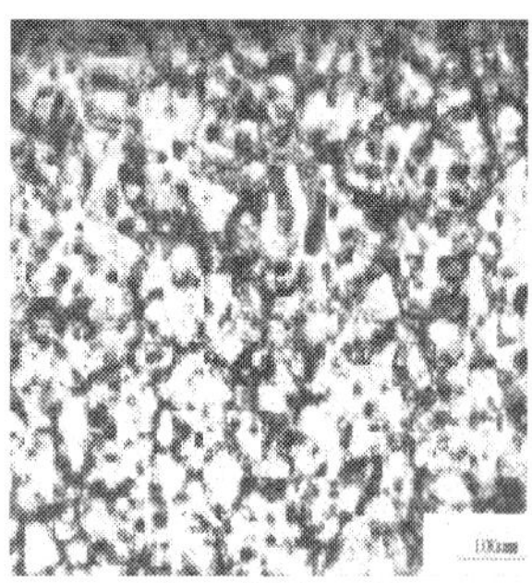

(d) φ (SiC)=9%

Fig. 5. Microstructures of SiCp/AZ61 composites with various volume fractions of SiC particles

The test results showed that the tensile strength of SiCp/AZ61 composites, that increased as the increasing of volume fraction of SiC particles increasing, and were higher than that of AZ61 (about 165MPa). The tensile strength was up to the maximum 189MPa when the volume fraction of SiC particles was 6%. However, the tensile strength decreased as the volume fraction of SiC particles increased continuously. Comparison with the volume fractions of SiC particles, the change trend of elongation decreased gradually with addition of SiC particles. The reason was a mass of rigid second phase existence in the matrix of SiCp/AZ61 composites, which could improve its rigidity and tensile strength. With the increasing volume fraction of SiC particles, problems of particle packing, agglomerating and clustering were presented in the matrix (Fig. 5), which caused tensile strength to decrease. Decreasing elongation was due to the non-uniform distribution of SiC particles and weak cracks in boundaries between the reinforcement and matrix.

The fracture morphology of AZ61 matrix at the ambient temperature is shown in Fig.6a. Ductility dimples existed, and cleavage cracks were present in a part. Fig.6b showed the fracture morphology of SiCp/AZ61 composites at the ambient temperature by a better processing plan ($A_2B_3C_2$). Compared with the fracture morphology of the matrix, the fracture morphology of SiCp/AZ61 composites at the ambient temperature were brittle where the fractured SiCp particles were found (Fig. 7a).

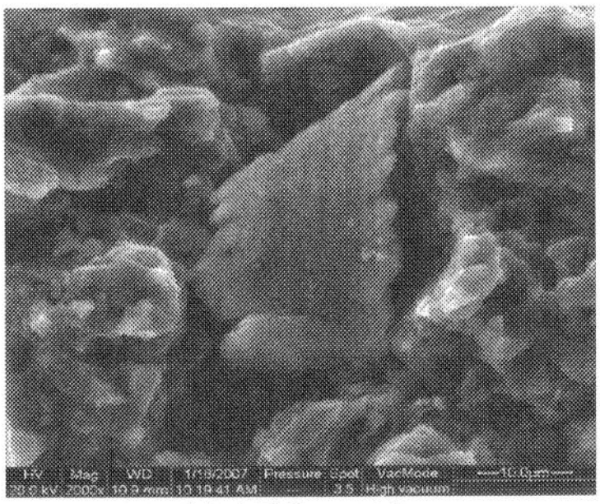

(a) AZ61 matrix

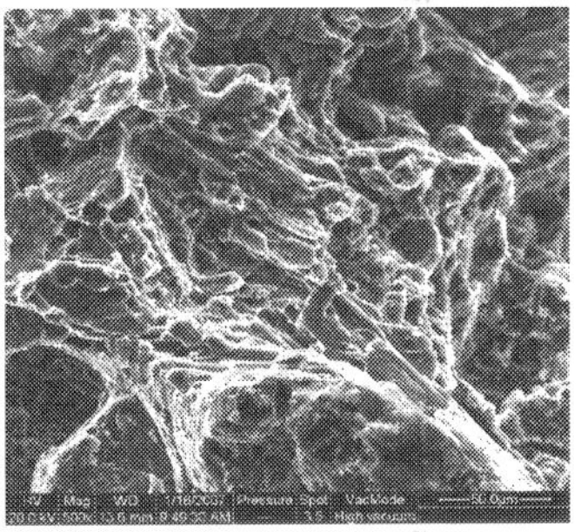

(b) SiCp/AZ61 composites

Fig. 6. SEM of tensile fracture surfaces

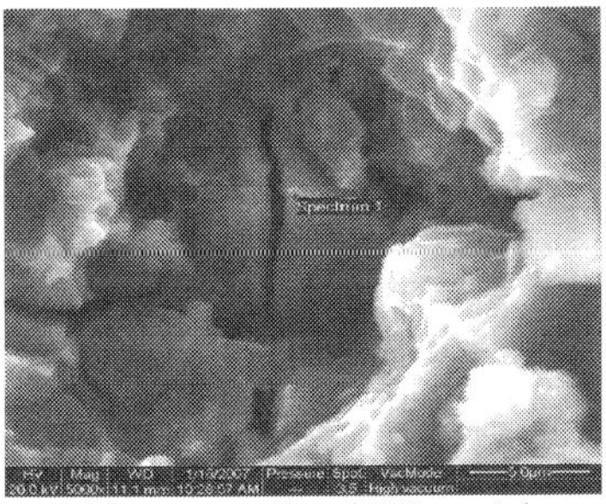

(a) Fractographs of SiC particle

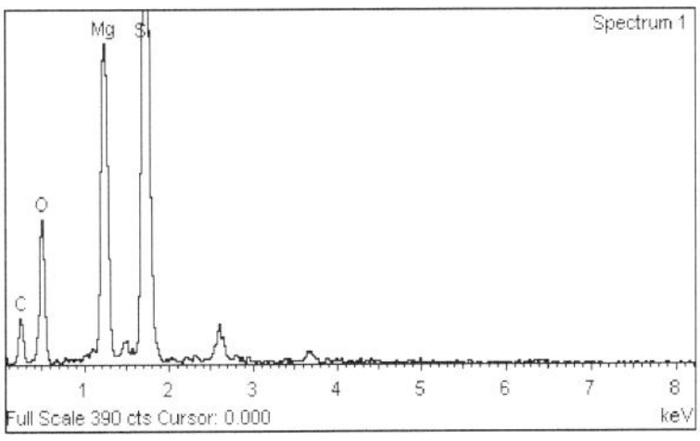

(b) EDS analysis

Fig. 7. Fractographs of SiC particle and EDS analysis

3. Semi-solid isothermal heat treatment technology for the partial remelting of composites

In this study, semi-solid isothermal heat treatment technology was used for the partial remelting of composites. The round semi-solid microstructure had been obtained by controlling the reheating processing parameters such as heating temperature and isothermal holding time. The law of microstructural evolution in the remelting process of SiCp/AZ61 composites was investigated, which was expected to offer some theoretical references for the design of thixoforming technology.

The reheating temperatures were taken as 590°C, 595°C, 600°C and 610°C respectively with isothermal temperature heat treatment times of 15min, 30min and 60min. When the scheduled time and temperature were reached, the specimen was taken out and water quenched. Then the specimens were made and etched with 4% nitic acid liquor, and its microstructure change was observed under the optical microscope. The Image-pro Plus software was used to measure the diameter of equal-area of microstructure. The average radius of grain microstructure was then calculated.

According to the Sheil equation (1) and equation (2), the liquid phase volume fraction of the partial remelting structure was calculated (shown in Fig.8).

$$f_L = \left(\frac{T_m - T}{T_m - T_L} \right)^{-1/1-K_0} \tag{1}$$

$$f_E = f_S(1 - f_P) - f_P \tag{2}$$

Where f_L, f_E and f_P represent the liquid phase volume fraction of matrix alloy, effective liquid phase volume fractions of composites and enforcing particles. T_m and T_L are melting point of the pure metal and the liquidus temperature of the alloy. K_0 is represented for the coefficient of distribution.

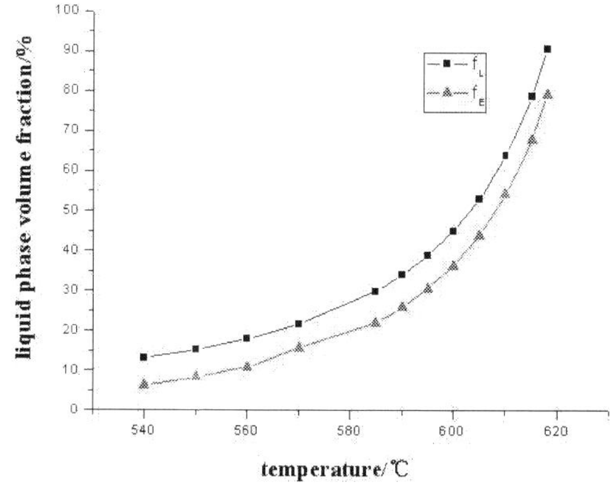

Fig. 8. Relationship between liquidphase volume fraction and temperature

Fig. 9. shows the microstructural evolution of SiCp/AZ61 composites during partial remelting. When the heating temperature reached 590°C with isothermal holding time of 15min, the grain boundaries had almostly been merged and could not be seen clearly. At the same time, SiC particles were inside the grains away from the grain boundaries (Fig. 9a). A separating tendency in the grains of coalescence emerged with the prolongation of isothermal holding time (Fig. 9b). While the holding time reached 60 min, a few grain boundaries became clear. A few globular grains appeared with SiC particles presented in the grain boundaries, but the liquid volume fraction was lower (Fig.9c). When the reheating temperature increased to about 595°C with holding time of 15 min, the grain microstructure evolved quickly, and a globular microstructure appeared, then the eutectic structure began to melt (Fig.9d). The grain boundaries appeared completely with holding time 30 min, and fine globular grains emerged. The effective liquid fraction of SiCp/AZ61 composites was about 31%, and the mean diameter of grains was approximately 60μm (Fig.9e). When the isothermal holding time was further increased to 60min, the grain microstructure was entirely spheroidized, which became more clear and round, and SiC particulate returned to the grain boundaries from interior of grains (Fig.9a,b). At the same time, the mean diameter of grains was about 85μm (Fig.9f). As the reheating temperature increased to 600°C with holding time of 15 min, the microstructural evolution of the sample during remelting was rapid. Some of grains began to spheroidize (Fig.9g). When the holding time reached 30 min, all grains had been spheroidized, whose sizes became relatively fine (Fig.9h). With the prolongation of holding time to 60 min, the grain microstructure tended to spheroidize and increase in size, and the effective liquid fraction was about 37% (Fig.9i). When the reheating temperature was above 610°C, the semi-solid microstructure began to dissolve and disappear. The specimens were susceptible to serious deformation, the liquid flow emerged from the sample, which would prevent semi-solid microstructure from partial remelting (Fig.9j). Therefore the optimal technological parameters of SiCp/AZ61 composites were the reheating temperature of 595°C~600°C and isothermal holding time of 30min~60min.This temperature interval was suitable for semi-solid thixoforming of SiCp/AZ61 magnesium matrix composites.

The microstructures of SiCp/AZ61 composites during partial remelting (Fig.9e, f) were compared with that of AZ61 alloy (Fig.10). It was observed that the microstructures of SiCp/AZ61 composites coalescenced basically before isothermal holding time at the predetermined temperature for 15min, and a separating tendency in the grains didn't appear obviously. After isothermal holding at the predetermined temperature for 25min, the grain microstructure began separating and spheroidizing. However the rate of separation and spheroidization for AZ61 alloy increased. When the reheating temperature reached 595°C with holding time of 0min,the grain microstructure was separated completely, and a few globular grains had appeared. With the prolongation of holding time from 20min to 40min, the mean diameter of the globules was 85μm and 110μm respectively. In addition, compared with AZ61 alloy, the microstructures of SiCp/AZ61 composites were finer during partial remelting due to addition of SiC particulates. Coalescence was restricted since the globules were isolated one with respect to the other by the presence of SiC particulates. At the same time the effective diffusion coefficient of the liquid phase was also reduced because of the presence of reinforced particulates, and during the subsequent isothermal holding process coalescence of α phase was hindered, and Ostwald ripening was also restricted.

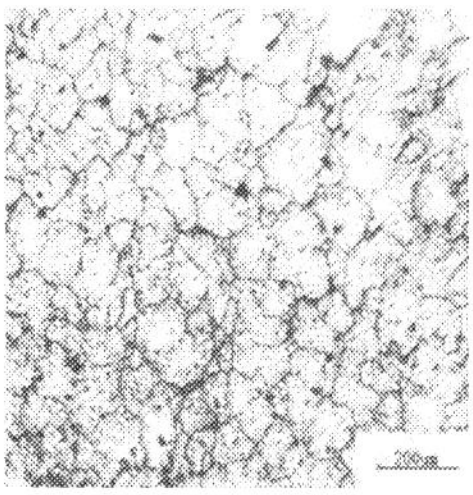

(a) 590 ^{0}C, 15min

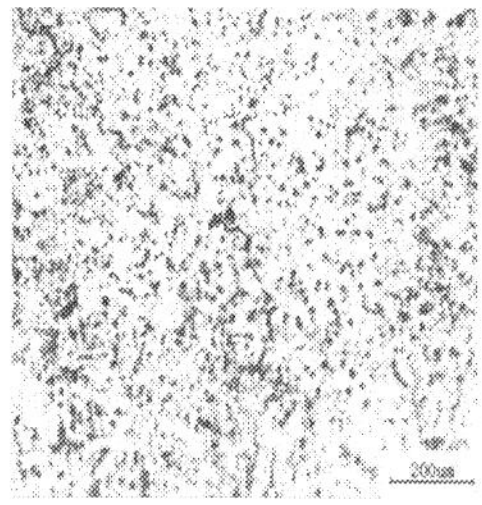

(b) 590 ^{0}C, 30min

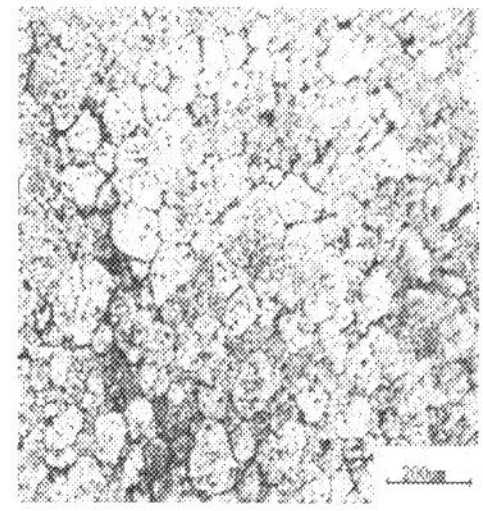

(c) 590 ^{0}C, 60min

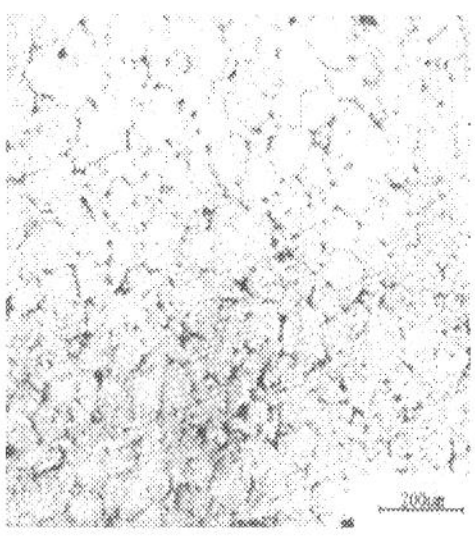

(d) 590 ^{0}C, 15min

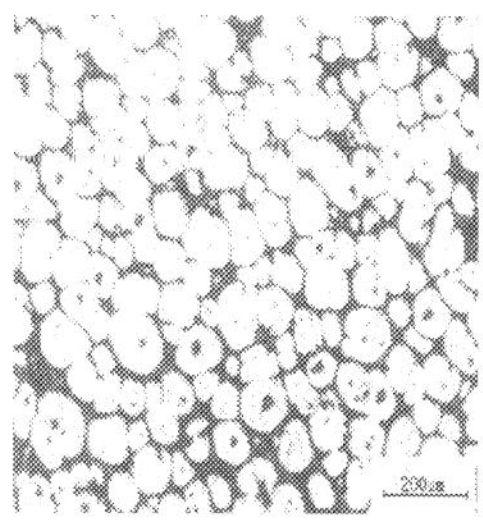

(e) 590 ^{0}C, 30min

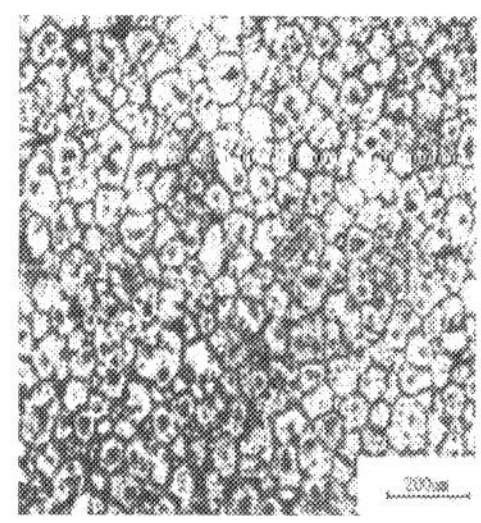

(f) 590 ^{0}C, 60min

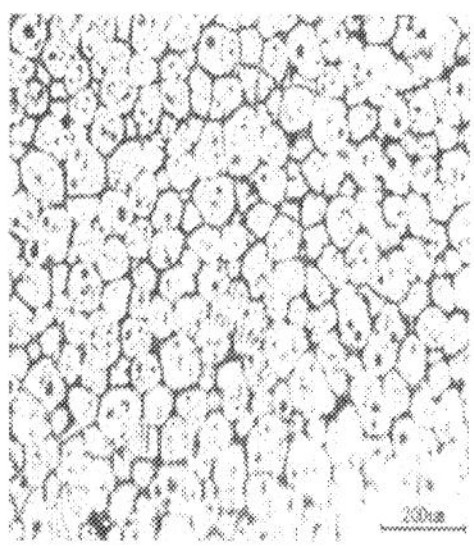

(g) 600 ^{0}C,15min

(h) 600 ^{0}C, 30min

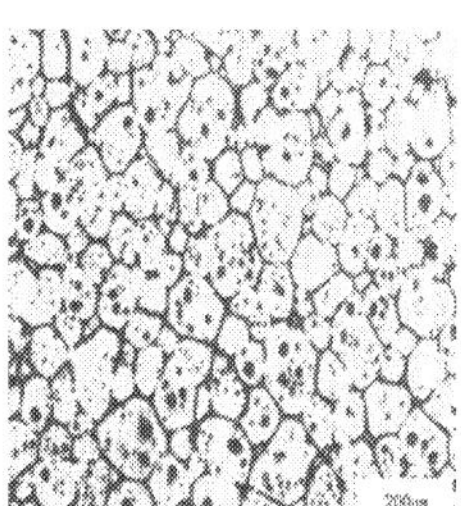

(i) 600 ^{0}C, 60min

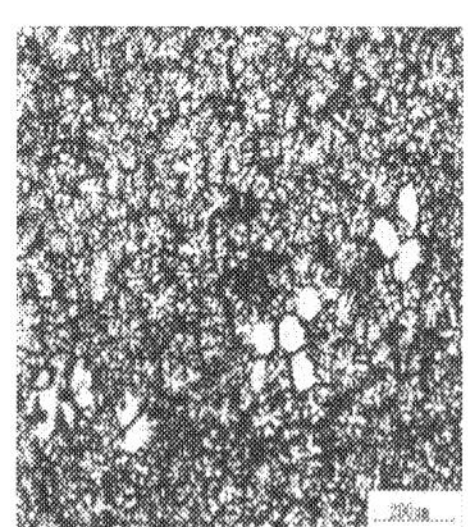

(j) 610 ^{0}C, 60min

Fig. 9. The microstructural evolution of SiCp/AZ61 composites during partial remelting

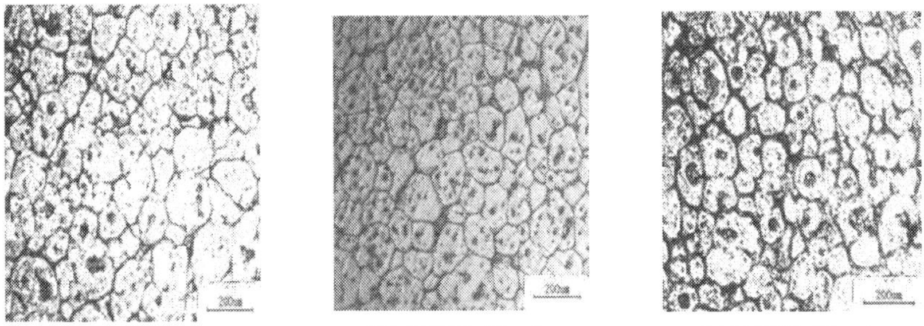

(a) 595 ^{0}C, 0min (b) 595 ^{0}C, 20 min (c) 595 ^{0}C, 40 min

Fig. 10. Microstructures of semi-solid AZ61 magnesium alloy billet during partial remelting

4. Thixotropic compression deformation behavior of composites

The characteristics of semi-solid composites deformed mechanism can be understood well only when the relationships between stress and strain are described. So the semi-solid compression tests for SiCp/AZ61 composites were conducted, whose mechanical properties and destruction model were investigated.

The experiments were conducted in a Thermecmastor-Z dynamic material testing machine, whose set-up was shown in Fig.11. The specimen was heated by electromagnetic wave, whose temperature was monitored by thermocouples. The graphite slices were placed between the specimen and the compression heads for reducing the influence of friction on experiment. In order to study and master the characteristic mechanics of semi-solid magnesium matrix composites at high solid volume fractions, the deformation temperatures were taken as 530°C, 545°C, 560°C and 570°C respectively. According to the heating procedure shown in Fig.12, the initial heating rate was 10°C/s; when the specimen temperature reached 500°C, the temperature rate was down to 1°C/s. Then the semi-solid compression experiments were done under the strain rates of $0.1s^{-1}$, $0.5\ s^{-1}$, $5.0\ s^{-1}$and $10\ s^{-1}$ respectively, in which the total strain was 0.6.

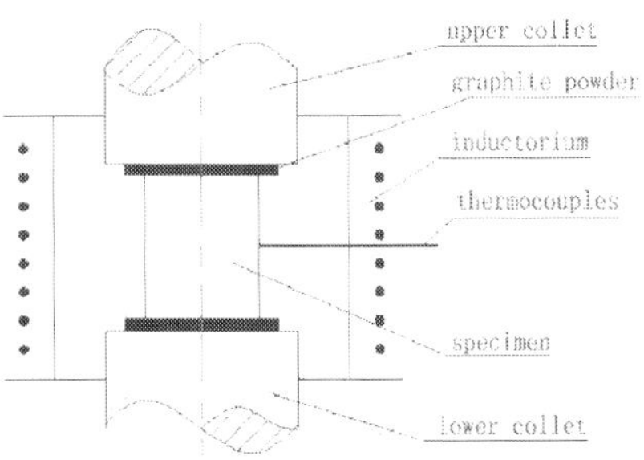

Fig. 11. Schematic diagram for the compressive tests

The stress-strain curves of semi-solid SiCp/AZ61 composites with various volume fractions of SiC particles are shown in Fig.13. The tendency of curves implies that the deformation temperature has a significant effect on the flow stress. It is observed that for a constant strain rate and constant volume fraction of SiC in the composites, the flow stresses and peak stresses decrease with the increasing of deformation temperature, which presents that the thixotropic plastic deformation of the composites is highly sensitive to temperature. The tendency is thought to be the result of variation of volume fractions of solid α phase. When the specimens has high solid volume fractions, the solid grains contact with each other and form a net, sliding and rotation of grains become hard. The plastic deformation of solid particles is the main mechanisms. With the increasing of temperature, the solid volume fractions decrease, and the solid grains are surrounded by liquid phase, which makes the solid grains to slid and rotate easily. Thus the sliding and rotation of grains plays a more significant role in the thixotropic plastic deformation.

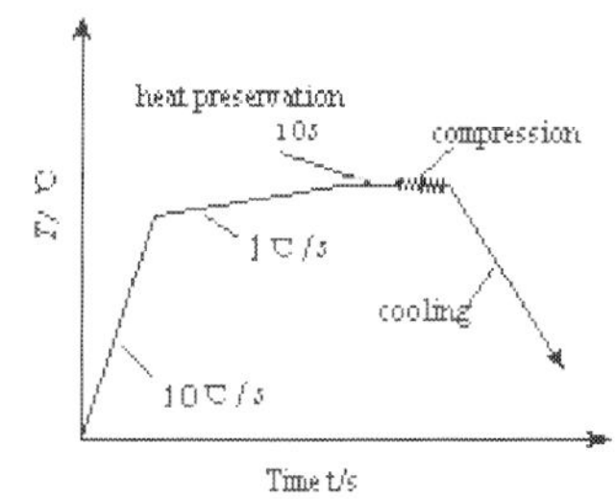

Fig. 12. Heating processing

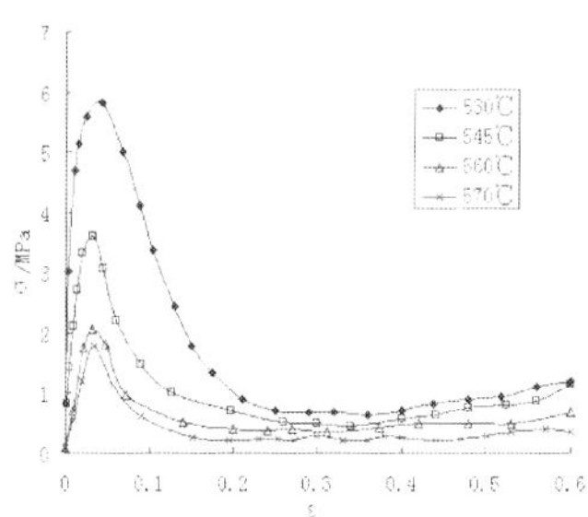

(a) f_p=3%, $\dot{\varepsilon}$ =0.1 s^{-1}

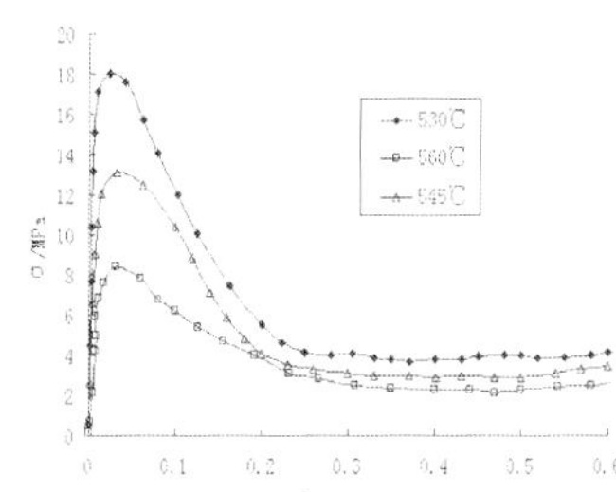

(b) f_p=6%, $\dot{\varepsilon}$ =5s^{-1}

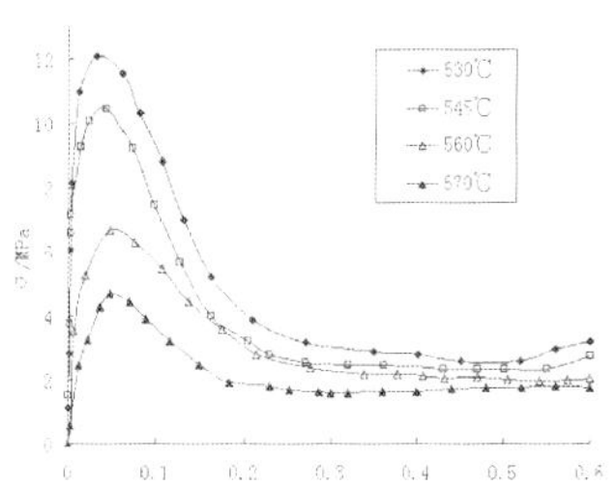

(c) f_p=9%, $\dot{\varepsilon}$ =10s^{-1}

Fig. 13. Curves of stress-strain relation at various temperatures for SiCp/AZ61 composites

Relationships between peak stress and temperature of composites with strain rate of $10s^{-1}$ are shown in Fig.14. It can also be seen that the variation of volume fractions of SiC has a significant effect on the peak stress of the composites. When the deformation temperature raises slightly higher than the solid phase line in semi-solid zone, the volume fractions of liquid phase is low in the matrix, and the peak stress decreases rapidly almost as linear form with the increasing of temperature initially, and then slows as the temperature increases further. The results are thought to be of the lower volume fractions of liquid phase. When the specimens has high volume fractions of solid phase , the solid grains contact with each other to form a net, the sliding and rotation of grains become hard. The plastic deformation of solid particles is the main mechanisms. Besides SiC particles increase the resistance to grain boundaries sliding and impose barriers to dislocation motion, which leads to higher resistance for plastic deformation. With the increasing of temperature, the volume fractions of liquid phase increase, and the solid grains are surrounded by liquid phase, which requires smaller force for the solid grains to slid and rotate. Thus the peak flow stress decrease rapidly, and then slowed as the temperature increases further.

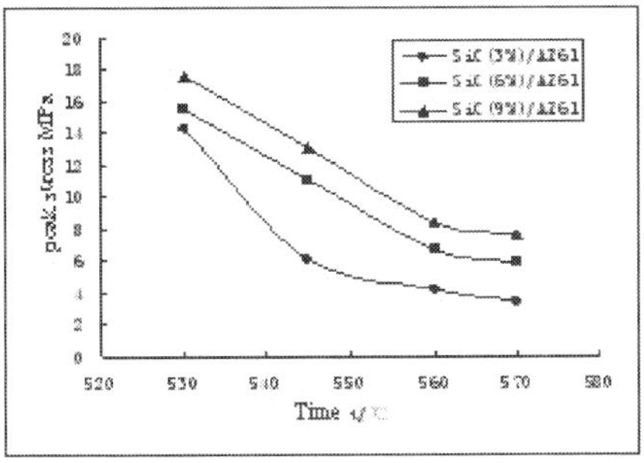

Fig. 14. Relationships between the peak stress and temperature of three SiCp/AZ61composites in semi-solid thixotropic compression

The relations of stress-strain rate at various strain rates are shown in Fig. 15. It can be seen that the peak stress increases as the strain rate increases at the constant temperature. When the composites are compressed at high strain rate (for example $10\ s^{-1}$), the liquid phase can not be squeezed out timely, in which the flow stress is very high during the initial deformation stage, and then decreases as the result of high shearing rate. With the further deformation, the liquid is squeezed out and pushed together, in which the solid particles are smashed. During this stage, the grain boundaries sliding and flow become easy and the flow stress decreases rapidly. At lower strain rate the solid grains are surrounded by liquid phase, which requires smaller force for sliding and rotation. So the flow stress is lower.

Fig. 16 presents stress-strain curves of the SiCp/AZ61 composite with different SiC fractions at 545°Cand 560°C and constant strain rate of $0.1s^{-1}$ and $10s^{-1}$. The fractions of SiC have a significant effect on the flow stress. The compression stress increases with the increasing of volume fractions of SiC particles. The reason is that SiC particles are mainly located in the inter-granular and boundary regions in the composites. The SiC particles impose barriers to dislocation motion and resistance to the solid grains sliding during the steady-state compressive deformation. So it increases the resistance for dislocation and grain boundaries sliding with the increasing of volume fractions of SiC particles.

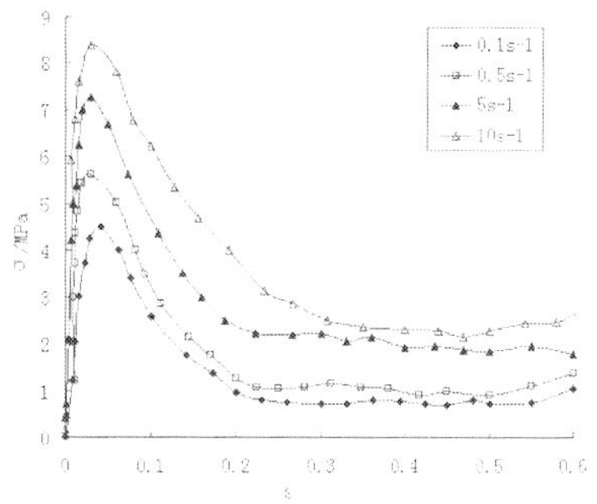

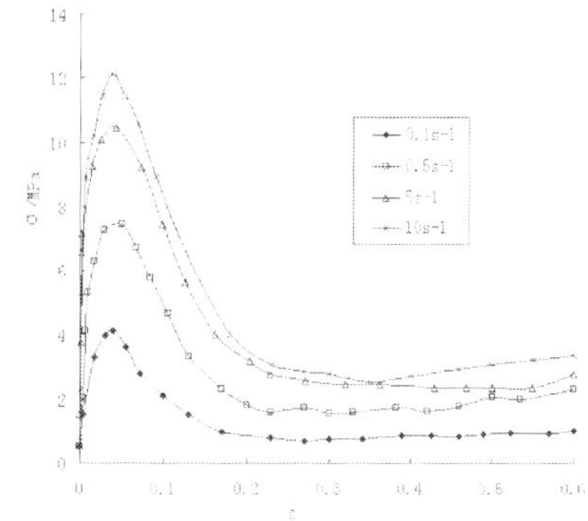

(a) f_p=3%,T=530°C (b) f_p=6%,T=545°C

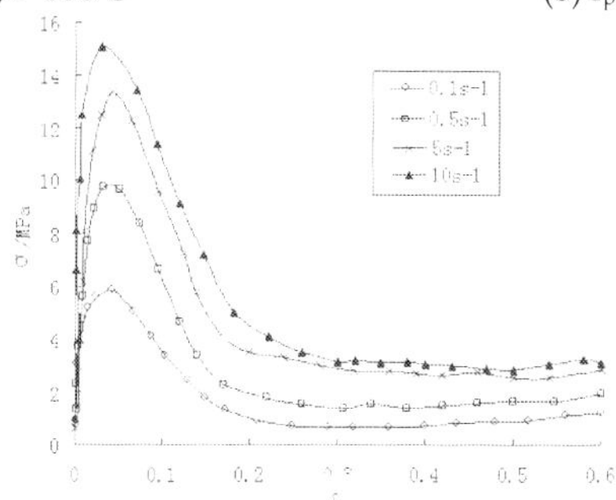

(c) f_p=9%,T=570°C

Fig. 15. Curves of stress-strain relations at various strain rates for SiCp/AZ61 composites

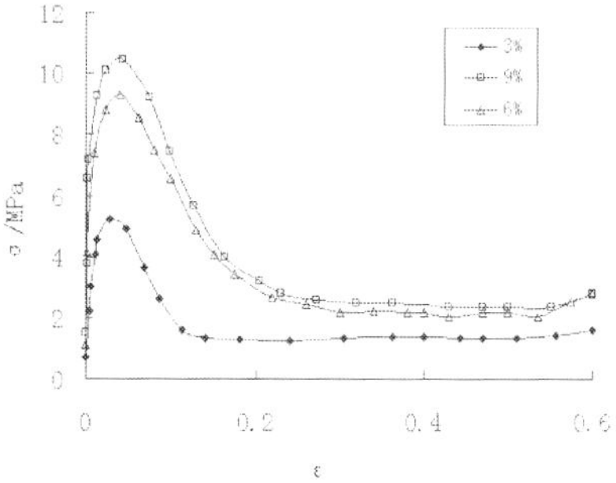

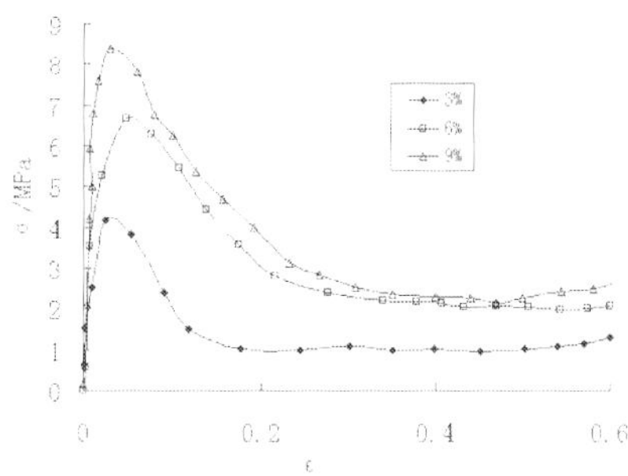

(a) T=545°C, $\dot{\varepsilon}$ =0.1s^{-1} (b) T=560°C, $\dot{\varepsilon}$ =10s^{-1}

Fig. 16. Curves of stress-strain relation at 545℃ and 560℃ for SiCp/AZ61 composites with different SiC fractions

The appearances of specimens compressed at semi-solid state are shown in Fig. 17. In can be seen that the surface longitudinal cracks of specimen happen at compression ratio of 20% and the liquid phase is squeezed out along the cracks. When the compression ratio reaches up 30%, the volume fractions of liquid phase squeezed out of the surface increases, which generates the mixed liquid-solid outer surfaces. The surface strength is very low and generates easily cracks. When all of liquid is squeezed out, the flow stress starts to ascend. At last the compressed specimen looks like popcorn.

(a) 0% (b) 20% (c) 30% (d) 60%

Fig. 17. Appearances of specimens compressed at semi-solid state

5. Constitutive model for thixotropic plastic forming of composites

On the basis of analysis of behavior of thixotropic plastic deformation of composites in compression process, its constitutive model is established. Then the model parameters are determined using the multiple nonlinear regression method.

Based on the experimental analysis of axial compression of composites in semi-solid state, there is a certain non-liner relationship among stress σ and strain rate $\dot{\varepsilon}_z$, strain ε_z, temperature T, liquid phase rate f_L, as well as the volume fraction of reinforcement f_p [Yan & Wang, 2011]. At the same time Hong Yan present the constitutive relationship of semi-solid magnesium alloy as follow [Yan & Zhou, 2006]:

$$\sigma \propto \exp(1/T)\dot{\varepsilon}^{a_1}\varepsilon^{a_2}(1-\beta f_L)^{a_3} \tag{3}$$

Where σ - stress, ε - strain, $\dot{\varepsilon}$ - strain rate, T - temperature, β - geometric parameters(β=1.5), f_L - liquid phase rate.

In the study of deformation behavior of composites under high strain rate [Bao & Lin, 1996] and [Li & Ramesh, 2000] found that the influence of volume fraction of reinforcement on the mechanical behavior of the material was present as following:

$$\sigma(f_p) = \sigma(\varepsilon, \dot{\varepsilon}) \cdot g(f_p) \cdot [1 + (\alpha\dot{\varepsilon})^m f_p] \tag{4}$$

Where σ- stress, ε- strain, $\dot{\varepsilon}$ - strain rate, $\sigma(\varepsilon,\dot{\varepsilon})$ - function of strain and strain rate, fp- volume fraction of reinforcement.

So the constitutive model of thixotropic plastic deformation of composites reinforced with particles is proposed.

$$\sigma = \exp(d/T)\cdot \varepsilon^{n} \cdot \dot{\varepsilon}^{m} \cdot [1-\beta f_L]^{a_1} \cdot g(f_p)\cdot [1+(\alpha\dot{\varepsilon})^{m} f_p]^{a_2} \tag{5}$$

Under assumption of $g(f_p) = e^{a+bf_p+cf_p^2}$ the constitutive model is established in the following form.

$$\sigma = \exp(a+bf_p+cf_p^2+d/T)\cdot \varepsilon^{n} \cdot \dot{\varepsilon}^{m} \cdot [1-\beta f_L]^{a_1} \cdot [1+(\alpha\dot{\varepsilon})^{m} f_p]^{a_2} \tag{6}$$

Where a, b, c, d, a_1, a_2 - constant, n—strain hardening index, m—strain rate sensitivity index,β—constant (β=1.5), α - correction coefficient , fp - volume fraction of reinforcement, f_L - liquid phase rate. $f_L = (\frac{T_M - T_L}{T_M - T})^{\frac{1}{1-K}}$, T_M - the melting point of pure metal, T_L - liquidus temperature of alloy, k - balance coefficient.

The parameters in proposed constitutive model were determined by the multiple nonlinear regression method. The nonlinear equation is transformed into linear one using legarithms for Esq.(3).

$$\ln\sigma = a + bf_p + cf_p^2 + d/T + n\ln\varepsilon + m\ln\dot{\varepsilon} + a_1\ln(1-\beta f_L) + a_2\ln[1+(\alpha\dot{\varepsilon})^{m} f_p] \tag{7}$$

Where

$$\begin{aligned} &y = \ln\sigma, X_1 = f_p, X_2 = f_p^2, X_3 = 1/T, X_4 = \ln\varepsilon, X_5 = \ln\dot{\varepsilon}, X_6 = \ln(1-\beta f_L), \\ &X_7 = \ln[1+(\alpha\dot{\varepsilon})^{m} f_p] \\ &A_0 = a, A_1 = b, A_2 = c, A_3 = d, A_4 = n, A_5 = m, A_6 = a_1, A_7 = a_2 \end{aligned} \tag{8}$$

Esq.(12) is changed as follow

$$y = A_0 + A_1X_1 + A_2X_2 + A_3X_3 + A_4X_4 + A_5X_5 + A_6X_6 + A_7X_7 \tag{9}$$

Table 3 shows the common statistic values. The correlation coefficient R = 0.974, determination coefficient $\overline{R}^2$ = 0.949, the adjustment determination coefficient $\overline{R}^2$ = 0.931, the Std. Error of the Estimate S =0.0670. As the equation has a number of explained variables, the determination should be based on the adjustment determination coefficient $\overline{R}^2$. As can be seen from the output that $\overline{R}^2$ is close to 1, the fit degree is high. So the representativeness of proposed constitutive model is strong.

Model	R	R Square	Adjusted R Square	Error of the Estimate
1	0.974(a)	0.949	0.931	0.0671449560

a Predictors:(Constant), x_7,x_6,x_4,x_2,x_3,x_5,x_1 b Dependent Variable:y

Table 3. Model Summary

The analysis is listed in Table 4. The significant test of regression equation is based on the table. Total in Sum of Squares is 835.005, Regression in Sum of Squares and Regression in

Mean Square are 413.512 and 59.073 respectively. Residual in Sum of Squares and Residual in Mean Square are 421.492 and 0.451 respectively. The test statistic observations F = 130.902. The concomitant probability p is approximately 0. The linear relationship between variables x and y is significant, which create a linear model.

Model		Sum of Squares	df	Mean Square	F	Sig.
1	Regression	413.512	7	59.073	130.902	0.000(a)
	Residual	421.492	934	0.451		
	Total	835.005	941			

a Predictors: (Constant),x_7,x_6,x_4,x_2,x_3,x_5,x_1 b Dependent Variable:y

Table 4. ANOVA (b)

Table 5 shows the regression coefficient analysis. As can be seen from the table and the estimated value of the test results, the corresponding variable regression coefficient A0=-8.27366, A1=50.158,A2=-296.555,A3 =14253.359,A4 =-0.053,A5=0.242,A6 =2.316,A7 =-0.505. The concomitant probability p is 0, whose regression is a significant. From comparison of regression coefficients, those indicate that the constitutive model has a significant meaning. The sensitivity coefficient m A5=0.242 of strain rate resulted from regression is good close to the replaced m value.

Model		Unstandardized Coefficients		Standardized Coefficients	t	Sig.
		B	Std. Error	Beta		
1	(Constant)	-8.27366	1.818		-9.728	0.000
	X_1	50.158	7.917	1.303	6.335	0.000
	X_2	-296.555	53.608	-0.934	-5.532	0.000
	X_3	14253.359	1402.791	0.342	10.161	0.000
	X_4	-0.053	0.012	-0.108	-4.601	0.000
	X_5	0.242	0.045	0.474	5.382	0.000
	X_6	2.316	0.553	0.143	4.186	0.000
	X_7	-0.505	0.463	-0.135	-1.091	0.000

a Dependent Variable：y

Table 5. Coefficients (a)

The analysis of the regression equation is a meaningful, and the following relation is got.

$$y = -8.27366 + 50.158X_1 - 296.555X_2 + 14253.359X_3 - 0.053X_4 + 0.242X_5 + 2.316X_6 - 0.505X_7 \quad (10)$$

From the inverse transform of equations (9) and (10), equation (6) becomes:

$$\begin{aligned} &\sigma = \exp(-8.27366+50.158f_p-296.555f_p^2 + 14253.359 / T)\cdot \varepsilon^{-0.053}\cdot \\ &\cdot\dot{\varepsilon}^{0.242}\cdot[1-\beta f_L]^{2.316}\cdot[1+(2.1\times 10^4\dot{\varepsilon})^{0.242} f_p]^{-0.505} \end{aligned} \quad (11)$$

Equation (11) is a constitutive relationship of thixotropic plastic forming of SiCp/AZ61 composites.

Fig.18 is the real stress test - a true strain curves and regression curve of the results of the comparison, Solid line is the experimental curve, dotted line is the calculation of one. The results calculated by multiple non-linear regression method are good agreement with experimental ones. So the proposed constitutive model has the higher forecast precision and practical significance.

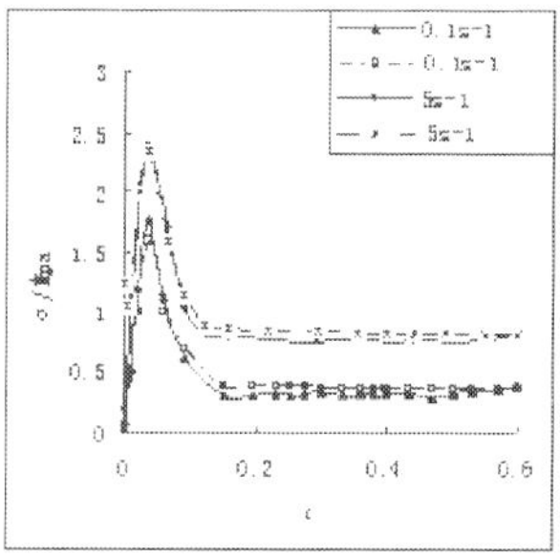

(a) 3vol.% SiCP/AZ61,570°C

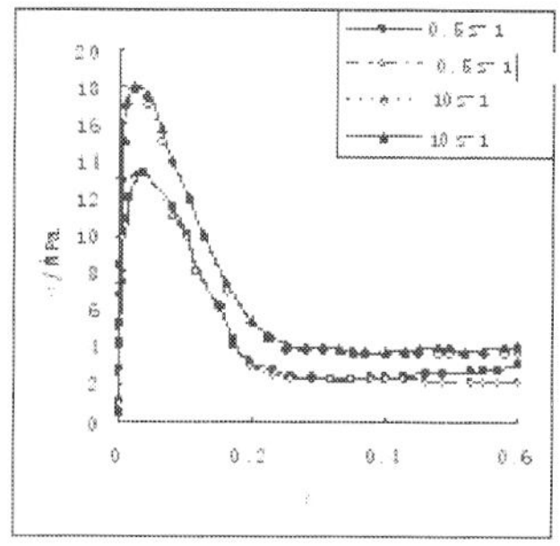

(b) 6vol.% SiCP/AZ61,560°C

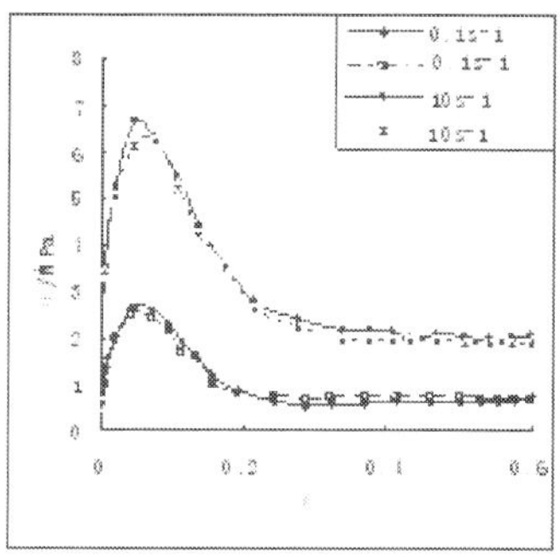

(c) 9vol.% SiCP/AZ61,530°C

Fig. 18. A comparison between true strain—stress curves of the test and regression curves

6. Numerical simulation for thixotropic plastic forming of composites

To investigate thixoforming process with numerical simulation method, which is a nonlinear system, some assumptions are taken as follow: (1) The semi-solid material is assumed as a continuous and incompressible one. (2) The solid grains in semi-solid metal are uniformly distributed in liquid phase, and because of the large deformation in forming, the semi-solid material is considered as an isotropy uniform medium. According to the above assumptions, the material deformation in thixoforming is supposed as a rigid viscoplastic one.

The material adopted in this paper was SiCp/AZ61 composite, and the simulations were performed in thixo-forging and forging. The flow stress model of SiCp/AZ61 composite in thixo-forging is expressed as follow [Yan &Wang, 2011].

$$\sigma = \exp(-8.27366 + 50.158 f_p \; - 296.55 f_p^{\,2} + 14253.359 / T)$$

$$\cdot\, \varepsilon \text{ -0.053} \cdot \dot{\varepsilon}\ 0.242 \cdot (1 - \beta f_L)\ 2.316 \cdot [1 + (2.1 \times 10^4\, \dot{\varepsilon})\ 0.242\, f_p] \text{-0.505} \qquad (12)$$

where σ is the stress; ε the strain; z $\dot{\varepsilon}$ the strain rate; T temperature; β constant(β=1.5); f_p is Volume fraction of SiC particle; f_L is liquid volume fraction

For establishing material modal of SiCp/AZ61 composite in forging, true stress-strain curves at various temperature and strain rates were performed by mean of isothermal compression experiments.

In this study, the workpiece is formed by the close-forge method. The experiment set-up was shown in Fig.19. Fig.20 shows the workpiece, whose structure and flow character are complicated. Comparisons between forging and thixo-forging of the workpiece will be done and predicted in advance using numerical simulation. This is an effective method to instruct application of semi-solid forming technology into its practice production.

The same simulated parameters are used to analyze the differences of mechanics properties and flow rule between forging and thixo-forging processes. The materials are normal and semi-solid SiCp/AZ61 composite respectively. Environment temperature is 20°C, warm-up temperature of the die is 320°C. The friction model is constant shearing stress model, whose coefficient is 0.25. Billet size is ø50×18.5mm, which is meshed to 50000 tetrahedron elements. Stroke of up-die is 14mm.

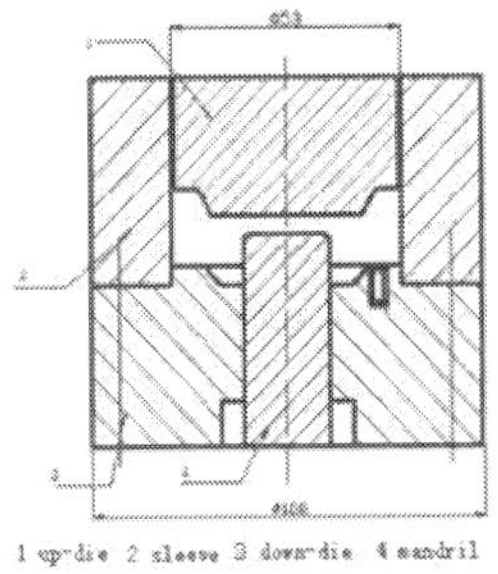

Fig. 19. Experiment set-up

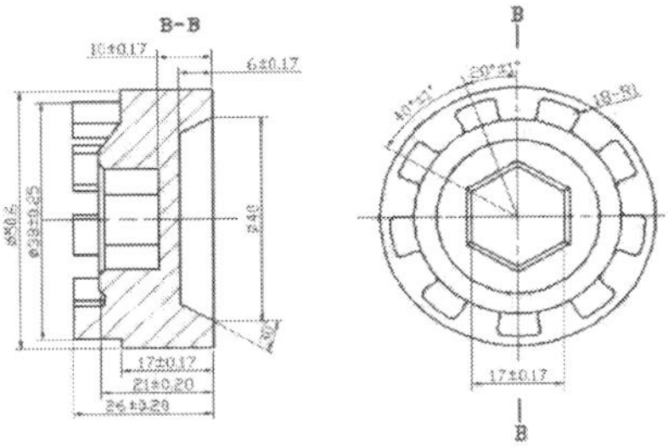

Fig. 20. SiCp/AZ61 composite workpiece

Fig.21 and Fig.22 give the filling stages simulated results in forging and thixo-forging processes respectively. Compared with the two kind of forming processes, it can be concluded that both had the basically identical deformation processes. In the initial stage, the hexagon hole in central section of workpiece was extruded and the rest moved in the rigid motion shown in Fig.21a, Fig.22a. As the stroke increased, metal deformation entered into the second stage, in which metal flowed from central to around in the extrusion pressure, and the cetral protruded and bottom platforms were formed (Seen Fig.21b, Fig.22b). In the last stage, the metal could be filled up claw easily in thixo-forging process, and could not be filled up claw in forging process (Seen Fig.21c, Fig.22c). Therefore, forging was more difficult in filling cavity than thixo-forging.

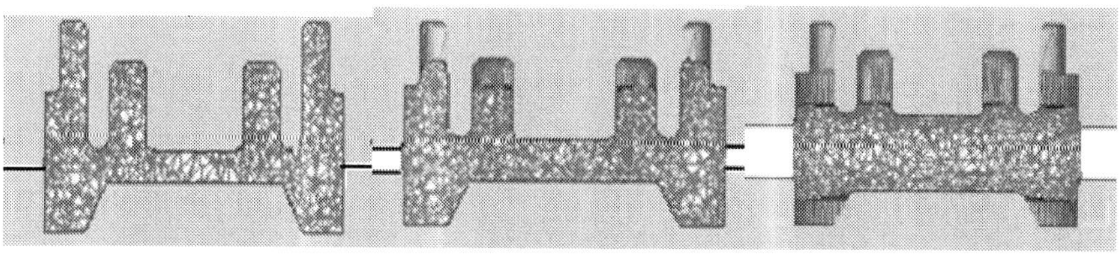

Fig. 21. Filling stages simulated results in thixo-forging process

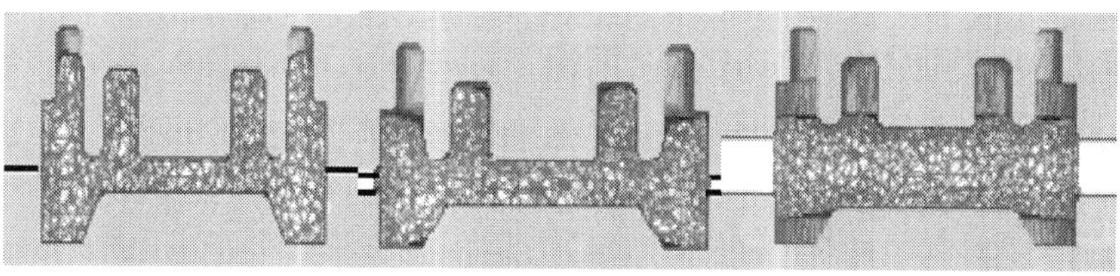

Fig. 22. Filling stages simulated results in forging process

Fig. 23. shows the effective stress distributions at different temperatures in thixo-forging process. The effective stress distribution was more uniform and its value was smaller with the increasing of forming temperature, which was contributed from the excellent fluidity of semi-solid composite.

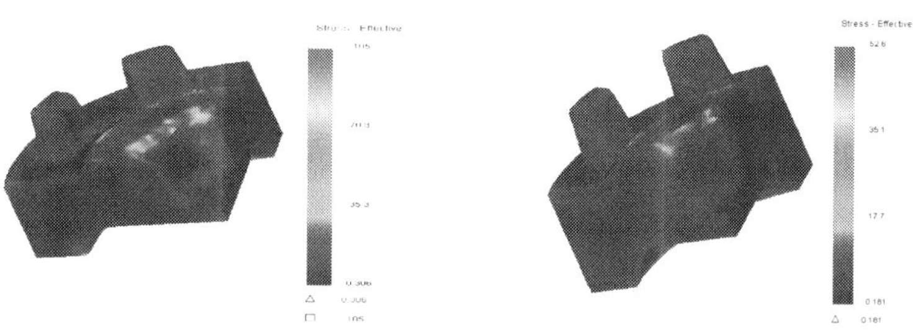

(a) 530°C (b) 560°C

Fig. 23. Effective stress distributions at different temperatures in thixo-forging process

Fig. 24. shows the effective stress distributions at different volume fraction of SiC particle in thixo-forging process. The effective stress was increased with the increasing of volume fraction of SiC particle.

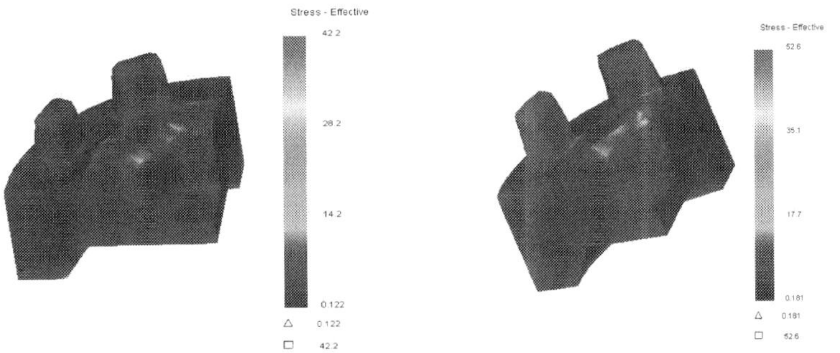

(a) Vol.3% SiC_P/AZ61, 560°C (b) Vol.6% SiC_P/AZ61, 560°C

Fig. 24. Effective stress distributions at different volume fraction of SiC particle in thixo-forging process

Fig. 25 shows temperature distributions at different volume fraction of SiC particle in thixo-forging process. When the volume fraction of SiC particle was 3%, the fluctuation period of temperature was 558～561℃, whose changed value was small. When the volume fraction of SiC particle was 6%, the fluctuation period of temperature was 558～572℃, whose changed value was more greater than that of the former. It could be gained that the temperature distribution in the latter was worse than that in the former.

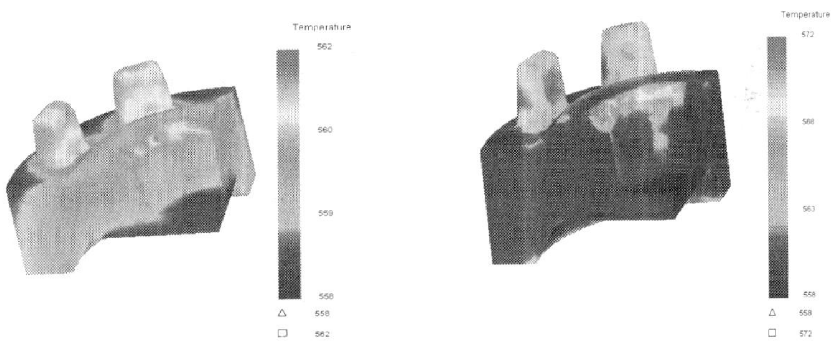

(a) Vol.3% SiC_P/AZ61, 560°C (b) Vol.6% SiC_P/AZ61, 560°C

Fig. 25. Temperature distributions at different volume fraction of SiC particle in thixo-forging process

Fig.26 shows the traditional forging and thixo-forging workpieces of SiCp/AZ61 composite. The thixo-forging has better fill effect and surface finish quality of workpiece than the traditional forging, which could achieve near-end deforming with high quality of workpiece in the former. Those coincide with the simulation results, which indicate that semi-solid

SiCp/AZ61 composite has good flow property, and can be used to form complicated workpiece.

(a) Traditional forging workpiece

(b) Thixo-forging workpiece

Fig. 26. Traditional forging and thixo-forging workpieces of composite

7. Conclusions

The microstructural structures of magnesium matrix composite were studied in three different casting processes. The results indicated that SiCp/AZ61 composites fabricated in stirring melt casting process, compared to those in fully liquid stirring casting process and in semi-solid stirring casting process, possessed fairly uniform distribution of SiC particulates and few porosity rate. It was an ideal metal matrix composites fabricated process.

Under the experimental conditions, the optimum processing plan of SiCp/AZ61 composites fabricated by a stirring melt casting method were the volume fraction of SiC particles 6%, stirring temperature 595°C and stirring time 5 min. In addition, the effects of volume fraction of SiC particles on the mechanical properties of SiCp/AZ61 composites was the most important among three factors (volume fraction of SiC particles, stirring temperature and stirring time), the second were stirring time and stirring temperature.

Semi-solid isothermal heat treatment technology was used for the partial remelting of SiCp/AZ61 composites. A fine semi-solid microstructure was obtained, whose equal-area diameter size was between 60μm and 85μm, and the effective liquid volume fraction was about 31%~38%. The optimal technological parameters of SiCp/AZ61 composites were the reheating temperature of 595℃~600℃ and an isothermal holding time of 30min~60min.

Compression tests on semi-solid SiCp/AZ61 magnesium matrix composites were carried out. Influences of strain-rate, strain, temperature and volume fraction of SiC particles on flow stress were analyzed. The results show that the flow stress of semi-solid SiCp/AZ61 composites is sensitive to temperature and strain rate. Meanwhile the flow stress increases with the increasing of the volume fraction of SiC particles.

The influence of deformation temperature, strain rate, strain, liquid volume fraction, volume fraction of reinforcement on flow stress in composites thixotropic plastic deformation process was considered. A new constitutive model of composites in thixotropic plastic deformation process was proposed. The constitutive equation of SiCp/AZ61 composites was obtained with the multiple nonlinear regression method based on data of thixotropic compression test. The calculated results were good agreement with the experimental ones. It is used to guide composites thixotropic plastic deformation process.

Numerical simulation can provide a help for the analysis of thixoforging process, and behavior of metal flow has been obtained. The effective stress distribution was more uniform and its value was more smaller with the increasing of forming temperature. The effective stress was increased with the increasing of volume fraction of SiC particle. The temperature distrubition was worse with the increasing of volume fraction of SiC particle. The differences between traditional forging and thixo-forging processes were analyzed. Results indicated that thixo-forging was better in filling cavity than forging. So the complicated workpiece can be done once in thixo-forging. Numerical simulation results are accorded with experimantal ones.

8. Acknowledgement

This research was supported jointly by grant # 50465003, # 50765005 and # 51165032 from the National Natural Science Foundation of China, Innovative Group of Science and Technology of College of Jiangxi Province and the Jiangxi Province Education Commission Foundation.

9. References

Hai, Z. Y., Xing, Y. L. (2004). Review of recent studies in magnesium matrix composites. *Journal of Materials Science*, Vol. 39: 6153-6171

Zhou, W., Xu, Z. M. (1997). Casting of SiC reinforced metal matrix composites. *Journal of Materials Processing Technology*, Vol. 63 , No.3: 358-363

Kang, C. G., Choi, J. S., Kim, K. H. (1999). The effect of strain rate on macroscopic behavior in the compression forming of semi-solid aluminum alloy. *Materials Processing Technology*, Vol.88, No.1-3: 159-168

Zhang, X. Q., Wang, H. W., Liao, L. H. (2004). In situ synthesis method and damping characterization of magnesium matrix composites. *Composites Science and Technology*, Vol. 67:720-727

Mordike, B. L. (2002). Development of highly creep resistant magnesium alloys. *Journal of Materials Processing Technology*, Vol. 117, No.3: 391-394

Flemings, M. C. (1991). Behavior of metal alloy in the semi-solid state. *Metall Trans A*, Vol.22, No.5: 957-981

Yan, H., Xia, J. C. (2005). Theoretical analysis of plastic forming process for semi-solid material. *Materials Science Forum*, Vol.488-489, 389-392

Yan, H., Zhang, F. Y. (2005). Structure evolution of AZ61 magnesium alloy in SIMA process. *Transactions of Nonferrous Metals Society of China*, Vol.15, No.3, 560-564

Yan, H., Zhang, F. Y. (2006). Microstructural evolution of semi-solid AZ61 magnesium alloy during reheating process. *Solid State Phenomena*, Vol.116-117, 275-278

Yan, H., Zhou, B. F. (2006). Thixotrpic deformation behavior of semi-solid AZ61 magnesium alloy during compression process, *Materials Science and Engineering B*, Vol.132, No.1-2, 179-182

Yan, H., Zhou, B. F. (2006). Constitutive model of thixotropic plastic forming for semi-solid AZ61 magnesium alloy. *Solid State Phenomena*, Vol.116-117, 577-682

Yan, H., Zhou, B. F. (2008). Study on thixo-forging of AZ61 wrought magnesium alloy. *Solid State Phenomena*, Vol.141-143, 577-682

Yan, H., Fu, M.F., Zhang, F. Y., Chen, G. X. (2007). Research on properties of SICp/AZ61 magnesium matrix composites in fabrication processes. *Materials Science Forum*, Vols. 561-565: 945-948

Yan, H., Lin, L. S. B., Pan, W. (2008). Study of SICp/AZ61 composites. *Solid State Phenomena*, Vols. 141-143 :551-555

Yan, H., Zhang, F. Y., Pan, W. (2008). Microstructural evolution of SiCp/AZ61 composites during partial remelting. *Solid State Phenomena*, Vols. 141-143 :545-549

Zhang, F. Y., Ye, J. X., Yan, H. (2011). Effects of SiC particle and holding time on semi-solid microstructure of SICp/AZ61 composites. *Advanced Materials Research*, Vols. 152-153: 628-633

Yan, H., Wang, J. J. (2011). Thixotropic compression deformation behavior of SICp/AZ61 magnesium matrix composites. *Transations of Nonferrous Metals Society of China*, Vols. 20: s811-s814

Yan, H., Wang, J. J.,Zhang, F. Y. (2011). A constitutive model for thixotropic plastic forming of semi-solid composites. *Advanced Materials Research*, Vols. 154-155:690-693

Yan, H., Huang, W. X. (2011). Numerical simulation on thixo-forging of magnesium matrix composite. *Advanced Materials Research*, Vols. 189-193:2535-2538

Bao, G., Lin, Z.(1996). High strain rate deformation in particle reinforced metal matrix composites. *Acta Materialia*, Vol.44, No. 3: 1011-1019.

Li, Y., Ramesh, K.T., Chin, E.S.C.(2000). The compressive viscoplastic response of an A359/SiCp metal—matrix composite and of the A359 aluminum alloy matrix. *International Journal of Solids and Structures*, Vol. 37, No. 51:7547-7562.

Yan, H., Huang, X. ,Hu, Q. (2010). Damping capacity of SICp/AZ61 composites. *Advanced Materials Research*, Vols. 123-125 : 35-38

Yan, H., Huang, Z. M. (2011). Study on creep properties of SiCp/AZ61 composites. *Advanced Materials Research*, Vols. 189-193: 4227-4230

12

Development of a Winding Mechanism for Amorphous Ribbon Used in Transformer Cores

Marcelo Ruben Pagnola[1] and Rodrigo Ezequiel Katabian[2]
[1]*Universidad de Buenos Aires, Facultad de Ingeniería,*
INTECIN (UBA-CONICET)
Laboratorio de Sólidos Amorfos (LSA)
[2]*Universidad de Buenos Aires, Facultad de Ingeniería,*
Departamento de Ingeniería Mecánica
Argentina

1. Introduction

In recent years, the application range of available soft magnetic materials has increased significantly due to the development of amorphous and nano-crystalized systems. Certain ferromagnetic alloys can be obtained as vitreous phases by rapid quenching techniques; some of them partially crystallize by certain heat treatments achieving structures composed by 10 to 40 nanometre long grains surrounded by a vitreous phase. One of these rapid quenching techniques is the melt-spinning, from which it is obtained amorphous metal strips that are, later, wound up into rolls.

The later-use of the wound rolls is the conformation of electric transformer cores showing meaningful improvement in its overall outputs, as well as an increment in the efficiency and fewer environmental impacts. In the past, these cores have been produced with grain-oriented and non-grain-oriented silicon steel sheets, ferrite sheets, Ni-Fe and Co-Fe alloys sheets produced by conventional casting processes, which require several mechanical and thermal processes, which some of them, have a high cost (Gelinas, 2000). The fabrication of nano-structured magnetic packages can be done, in this particular case, by the direct-employment of melt-spinning´s strips into different kinds of heat treatments, where it can also be adjusted the hysteresis cycle. Furthermore, its uses can be extended to complex geometries introducing a milling stage after the melt-spinning process, obtaining refined elemental powder particles (Nowacki, 2006; Byoung et al., 2007), which its dimensions can be modified by the control of the milling stage time (Dobrzanskia et al, 2004). The connotations of using soft magnetic alloys affect not only transformer cores but also AC motors (Pagnola et al., 2009; Pagnola, 2009). These new amorphous and nano-crystalized materials are currently sold up to 3 times the price of conventional materials (Condes, 2008).

Magnetic cores lose energy through two independent mechanisms: hysteresis (dissipated energy during the re-orientation cycle of magnetic domains) and Foucault current (eddy or parasitic current). These losses can rise up to 5% and 15% of the entire produced energy, which fluctuates over the manufacturing technique employed. Own research and other authors confirm that these losses can be reduced almost 80 % from those that appear in devices built with traditional steel (De Cristofaro, 1998; Douglas, 1988; Richardson, 1990). In table 1 and figure 1, it can be seen how much smaller these losses are, and what is more important the amount of energy saved. The LSA implemented Melt-Spinning technique through the project called "Advanced technology magnetic materials production" (PICT-2007-02018), and it aims reducing energy losses to the values given. Amorphous ribbons, similar to FINEMET®, were obtained by preliminary tests, these ribbons were 1mm wide and 20µm thick, and they were quenched straight up on the copper wheel in an air atmosphere, reaching a 10^6 K/sec cooling rate (Muraca et al., 2009).

Power [kVA]	Core Losses [W]		Saving percentage [%]	Manufacturer
	Fe - Si	Amorphous		
10	40	13,5	66	Osaka Transformer
10	40	11	72	Westinghouse
15	50	14	72	Allied and MIT
25	85	28	67	General Electric
25	85	16	81	Prototype Allied

Table 1. Core losses in regular Fe-Si cores and Amorphous alloys cores refer to Fe-Si (100%).

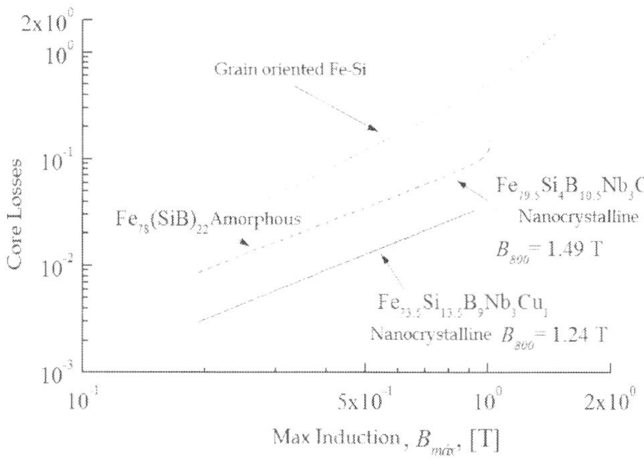

Fig. 1. Core losses with different alloys (Dobrzanskia et al., 2004).

2. Melt-spinning

One of the most common rapid quenching techniques to produce amorphous metals is the one called melt-spinning. Using this technique, the molten alloy is jetted on the surface of a high speed spinning copper wheel through a nozzle. The casting wheel acts as a heat sink reaching one million degrees per second cooling rate (Praisner et al., 1995) necessary to achieve the vitreous phase instead of a crystalline structure (see figure 2). In figure 3 it is shown a diagram of the melt-spinning apparatus, where it can be seen the small and weak linkage between the ribbon and the casting wheel.

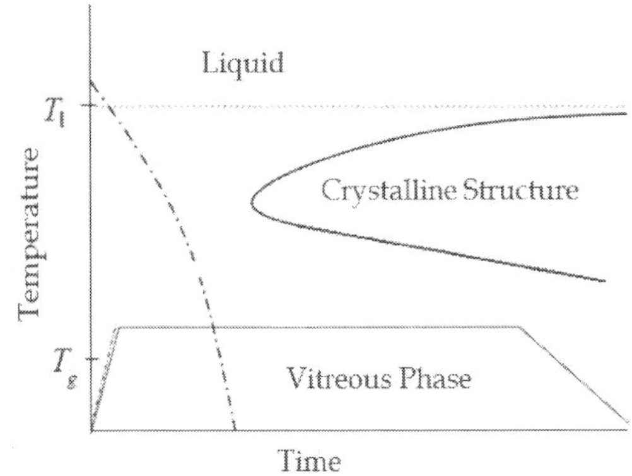

Fig. 2. Cooling procedure to avoid crystalline structure. (Moya, 2009).

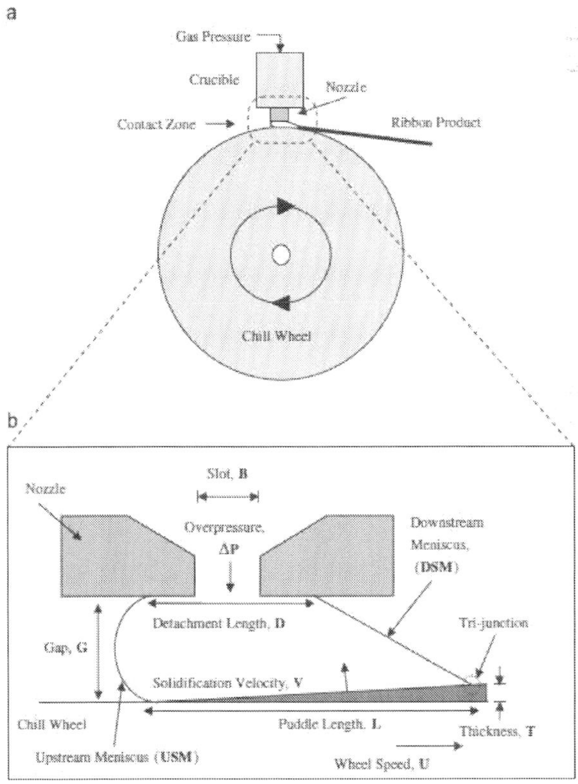

Fig. 3. (a) Schematic of a melt-spinning apparatus, (b) Blow up of the contact zone. (Theisen et al., 2010).

The amorphous alloy is obtained from a crystalline alloy, called mother alloy, which has the same chemical composition as the amorphous one. The way to get to the mother alloy is melting the proper quantity of the different components into an induction heater several times in order to insure a homogeneous alloy. Afterwards, the alloy is introduced into a quartz crucible with an induction coil which heats the alloy over the melting point; then, an argon over-pressure expulses the alloy through the nozzle on the high speed spinning wheel. As a result, a continuous amorphous ribbon is obtained; its thickness (≈ 20 - 100 μm) is a function of the injection pressure, the gap between the nozzle and the wheel and the cooling rate. Depending on the alloy and its corrosion susceptibility, the process should be in a controlled atmosphere, in a vacuum chamber or even in environmental conditions.

3. Winding system

The winding mechanism designed is capable of working at high winding speeds (an order of magnitude higher than those used in paper winding and steel-making, see table 2), it also insures the quality of the product as it´s has been solidified at the wheel without changing its surface roughness generated in the previous stage, since any aspect that has influence on its surface integrity during this stage has a direct impact on the magnetic package´s performance. This winding system is assembled next to the cylindrical sleeve by the casting wheel seen in Figure 4.

Two problems hold the design back at the first stage of the process:

1. Thread of the strip into the winding reel.
2. Tension control of the roll.

Both of these issues mainly appear because of the intrinsic characteristics of the melt-spinning technique. Due to the speed of the process and the fact that the strip has no fixed point at the casting wheel the solutions given are rarely similar to those found in regular winding machines. First of all, an automatic threading system was designed due to the impossibility to count on the proper time to thread the strip into the winding reel by a human (~ 5 to 10 seconds). This time implies an excessive collection of material (250 to 500m) by the casting wheel that can be wrecked by its own weight or successive folding.

With regard of the tension control, it's critical not only because of the typical problems in every wound roll, but also because an over-tension can separate the strip from the casting wheel where the material is still in a liquid state. Therefore, two zones were established in the machine, a free-tension zone and another one where it is controlled up to a set-point determined by the tension profile.

Figure 5 and 6 shows the proposed design, where it can be seen a set of guiding belts (1), that generate an air flow capable of dragging the strip from the casting wheel up to the pinch rollers (2). These rollers are powered by an asynchronous motor and variable frequency drive, where the strip purely rolls over them; the control of their speed and the winding velocity of the reel are the manipulated variables of the tension control system. Between the guiding belts and the pinch rollers, there are a set of idle rollers (8 & 9), which provide the system a stock of material in order to prevent an unwanted detachment of it at the solidification meniscus. Next to the pinch rollers there is a deflector (4) that simply guides the strip towards the winding reel (5). At last, the winding reel is surrounded by a wrapping belt that ensures the thread of the strip into the reel during the startup.

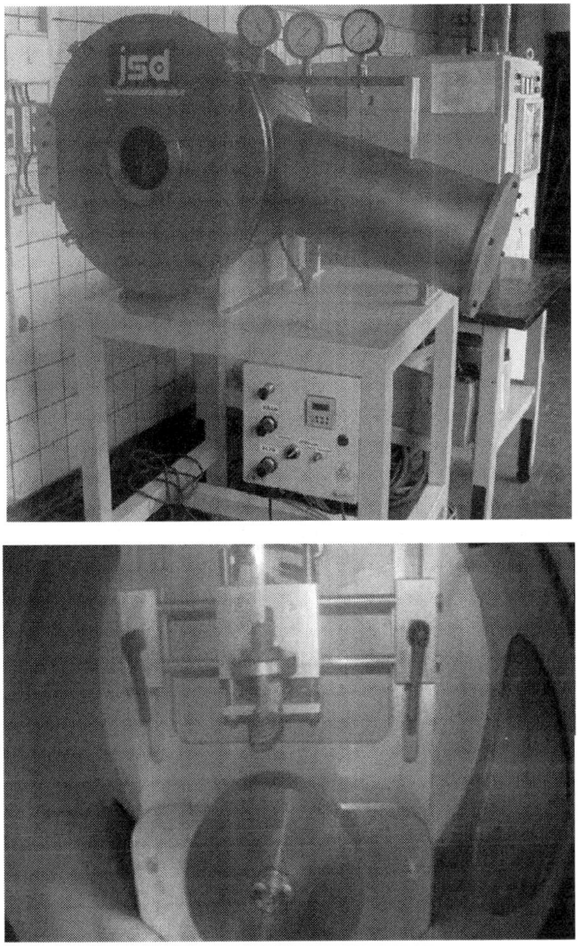

Fig. 4. Melt-spinning Equipment developed at LSA. Crucible and copper wheel.

Material	Winding Speed [m/s]	Winding Tension [Mpa]	Source
Paper	≈ 8	< 10	Liu, 2009
Steel	≈ 8	15 - 75	Liu, 2009
Plastic Films	≈ 15	< 4	Lee et al., 2002
Magnetic Tape	< 5	< 4	Liu, 2009
Amorphous strip	25 – 50	15 - 30	Own development

Table 2. State of the art - Winding parameters found in different industries.

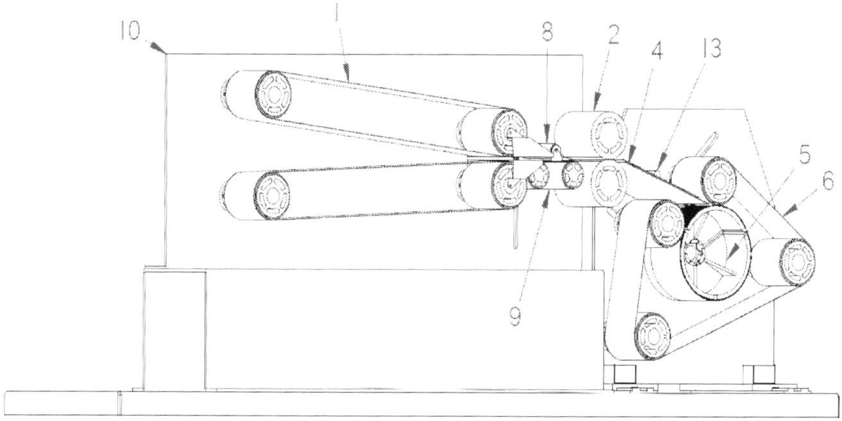

Fig. 5. Winding machine at startup: 1. Guiding belts; 2. Pinch rollers; 3. Initial deflector; 4. Main deflector; 5. Winding reel; 6. Wrapping belts; 7.Crosslide; 8. Stocking system - dancer roller; 9. Stocking system - Fixed roller; 10. Chassis; 11. Main motor; 12. Slide guides; 13.Tension sensor.

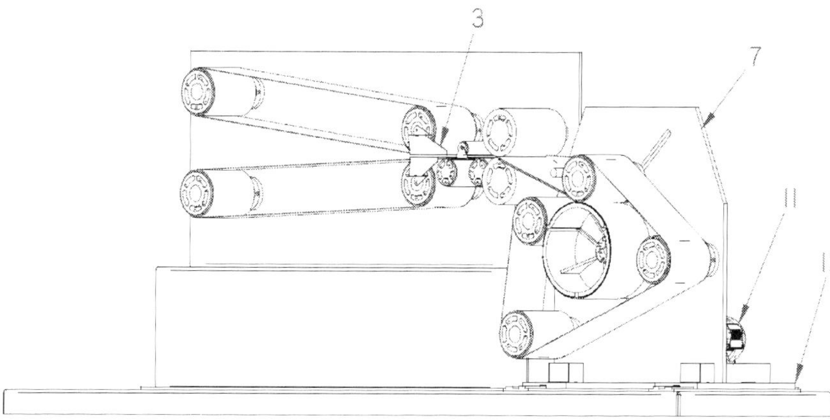

Fig. 6. Winding machine at startup.

3.1 Winding tension profile

Several winding stress models have been developed in order to find the proper winding tension profile (Li et al., 2009; Liu, 2009; Lee et al., 2002). Following the model proposed by Liu, every new wound lap is considered a collection of concentric laps of web material. In every one of them it is formulated the differential equations of internal equilibrium to find out stress, strain, displacement and pressures developed during the winding. Finally, the profiles shown in figure 7 were obtained.

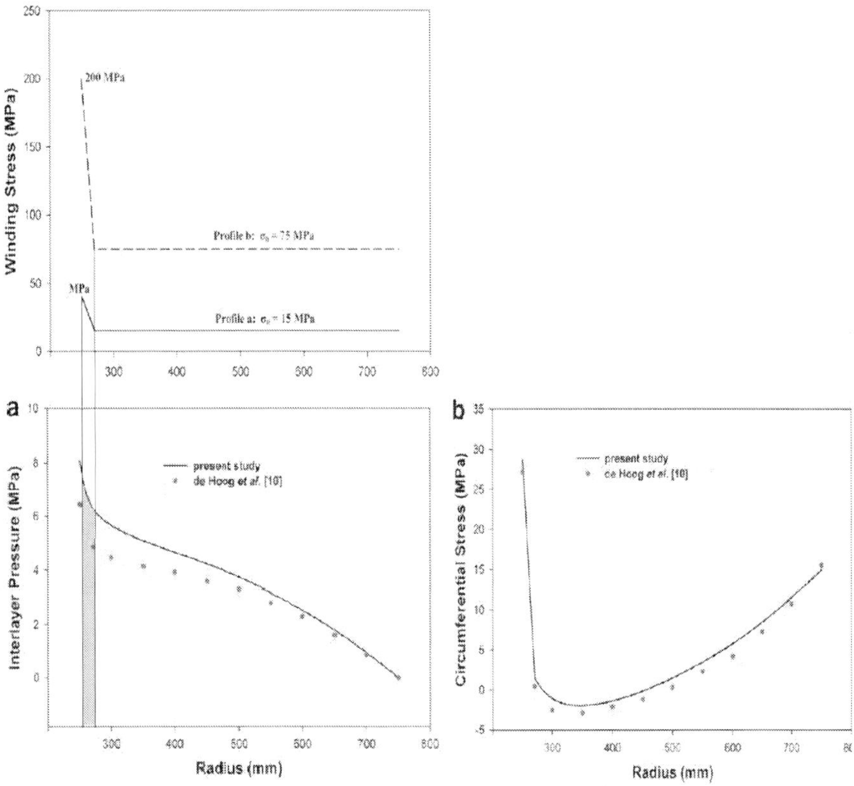

Fig. 7. Interlayer pressure (a) and circumferential stress (b) for the winding stress profile in the above figure (Liu, 2009).

As it can be seen, the winding stress profile starts at a higher value and then it starts decreasing through a ramp towards the regime value. During this ramp the roll is setting up its linkage with the winding reel, which will determine the end of the startup stage and the beginning of the working regime stage. To insure that the threading is complete the friction force generated by the internal pressure times the surface of every wound lap must be higher than the inertial force plus the winding tension. From now on, the roll is fixed to the winding reel and no slip between them will be found.

$$\mu_s \cdot \int_{r_o}^{r'} (p_i \cdot S) \cdot dr = F_{friction} > F_{inertia} + T_{winding} \tag{1}$$

Once established the tension profile, it can be calculated the power necessary for the winding reel motor and for the pinch rollers motor. On one hand, the winding reel motor (main motor) is going to take most of the torque necessary for the winding, on the other hand the pinch rollers motor will be working almost as a brake because upstream it is a free-tension zone and downstream the tension is provided by the main motor. This is why the power of the main motor is considered to take this torque times a service factor to make up for the startup situation. Taking into account these considerations and the dimensions of the strip and the reel, it is needed a 5HP AC motor for the winding reel.

3.2 Startup

When the casting wheel starts throwing the first cuts of amorphous strip, the guiding belts (1) drag it towards the pinch rollers (2) where a first thread is done. At this point, it´s the first contact with the winding machine and it is found pure rolling friction between the strip and the pinch rollers, where these jog the strip forward with its own tangential speed. As a consequence, it is needed a high precision in the mounting of this rollers in order to preserve the clearance between them; if it is bigger than the designed one the strip would slip between them. But if it is smaller, the material would be damage by an operation similar to a laminate. Due to the constant contact between the strip and these rollers, it is recommended to pay close attention to the hardness during the material selection of the rollers in order not to damage the surface quality of the strip; several options appear like copper, brass or even PTFE (TEFLON®) inserts. Next to the pinch rollers (2) the strip is guided towards the winding reel (5) by the main deflector (4). The winding reel is spinning to a higher tangential speed than the speed of the strip in order to ensure the threaded. Additionally, the reel is surrounded by a wrapping belt (6) which guarantees the strip to follow the profile of the reel until there is enough friction between the successive wounds of strip, so to create a bond within the reel and the strip, but this won´t happen until several wounds of ribbon had already been rolled over the reel.

3.3 Working regime

Once insured the threaded, the wrapping belts are completely removed, as can be seen in Figure 8, to look after the surface integrity of the strip. This action is performed by a pneumatic actuated scissor mechanism. Meanwhile, the spinning speed of the reel is reduced because during the threaded it was significantly higher; also, from this moment on, the dancer roller (8) can freely move in the vertical axis, and the tension sensor (13) is disposed as it is shown in Figure 8. From now on, the winder is at a working regime and we must proceed to the tension control of the roll.

The basic principle of the tension control is the small difference between the pinch roller´s speed and winding speed (represented by the tangential speed of the reel) which is slightly higher (He et al., 2010). The structure of the device for controlling the tension is shown in Figure 10. The tension measurement is used to tune up the spinning speed of the reel, by an asynchronous motor, variable frequency drive and encoder. The control set point establishes a tighter roll at the beginning and looser at the end, known as taper tension control (Good et al., 2008). With this system it is intended to obtain an optimum tension of the roll without inflicting any damage to the material.

A storage system is incorporated in order to prevent flaws on the tension control system, such as response time, slipping of the strip on the pinch rollers, lack of precision on electric and electronic components. Every difference between the pinch rollers and the casting wheel speed, overcomes into an over-tension of the strip or an excessive storage of material, which may cause a possible detachment of the solidification meniscus.

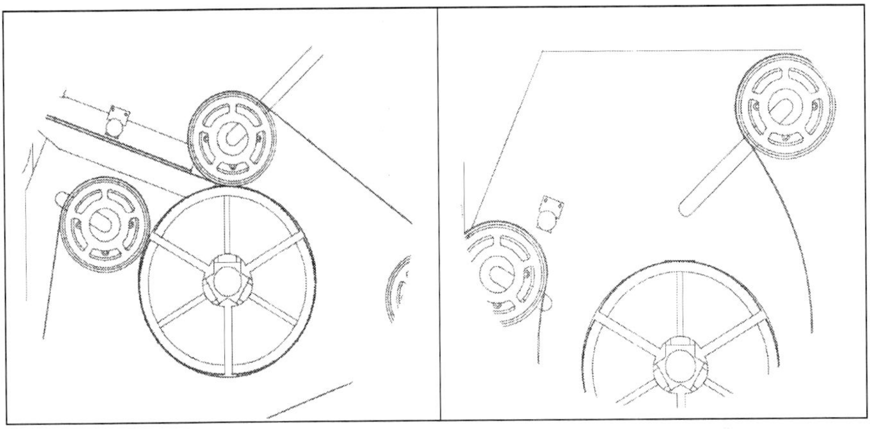

Fig. 8. Wrapping belts and tension sensor at startup condition (3.1) in working regime (3.2).

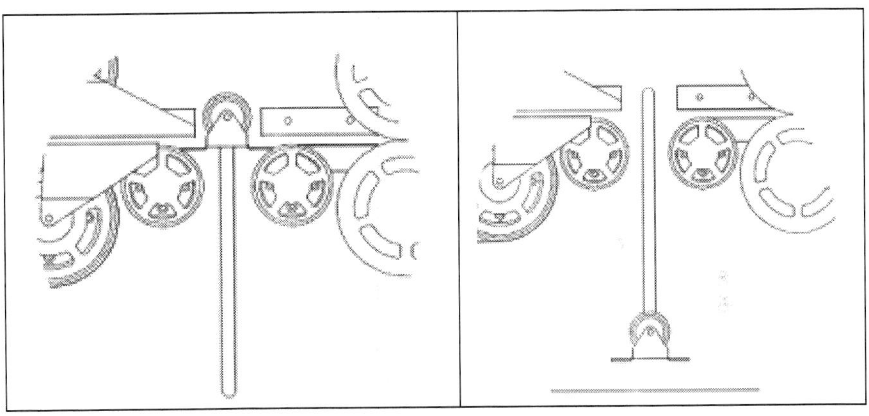

Fig. 9. Stocking system at startup and during the working regime.

The mechanism is composed by 3 idle rollers (Figure 9), two of them (9) are fixed, while the centered one (8) can move along the vertical axis forcing the strip to take a larger profile instead of the straight line from the startup. Sensing the position of this roller, it is tuned up the pinch rollers speed, only when the stock is excessive or insufficient. So, this variation will be considered as a transitory regime for the tension control system, in order to assume the pinch rollers´ speed as a constant. With this system a little stock of ribbon is created in order to absorb every produced over-tension.

The winding reel (5) is provided with a mandrel for a quick demounting of the finished roll. Moreover, the reel along with the wrapping belt is mounted on a cross slide (7) which can be moved over the sliding guides (12) in a cross direction, as can be seen in Figure 5 and 6. With this system several angular defects during the tuned-up before the startup of the equipment are corrected.

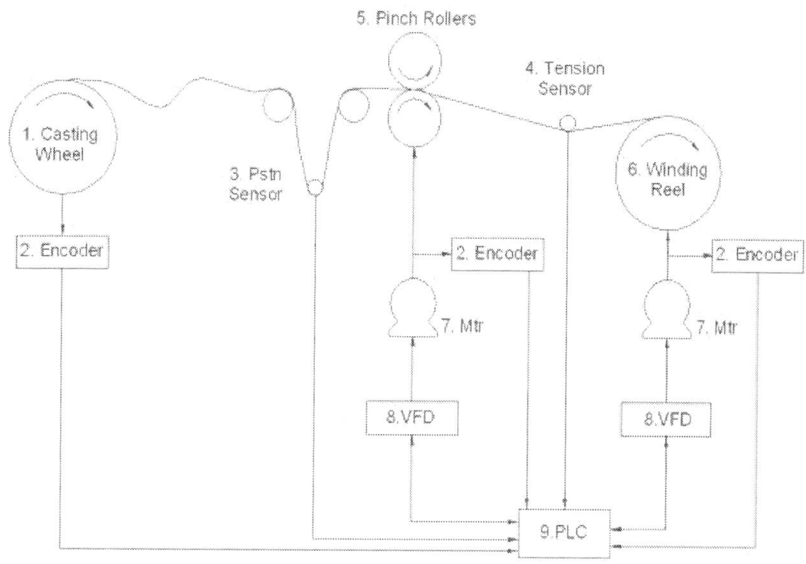

Fig. 10. Control System Structure: 1. Casting Wheel; 2. Encoder; 3. Position Sensor; 4. Tension Sensor; 5. Pinch Rollers; 6. Winding Reel; 7. Asynchronous Motor; 8. Variable Frequency Drive; 9. PLC.

4. Results and discussion

The initial investment and operating cost for the 25-30 lifetime-years of the transformer will be called Total Cost of Ownership. Within the Operating Cost of the device is included the cost of the dissipated electric energy at the windings (Cu) and at the core (Fe). Consequently, had the core losses been diminished (by using amorphous metal cores), the Total Cost of Ownership of a device produced by this technology would be reduced in comparison to those produced by traditional technology; and would profit a considerable economic gain to the owner of this machine. From this point of view, the implementation of an accurate winding system as the one proposed, presents an optimum solution to the formerly described process, not only to behoove the handling of the final product, but also to simplify the post-melt spinning heat treatments, such as isotermic annealing which is used to obtain nano-crystalized ribbons (Muraca . et al. 2009).

An accurate control of the material and design of the nozzle´s orifices (Saito, 2010; Kurokawa, et al. 1999), working pressures of the chamber and ejection temperature are crucial to prevent unwanted flaws (Saito, 2010; Marashi et al 2009) which include the absence of the formation of the strip on the casting wheel as can be seen in Figure 11. For this reason it is highly recommended to ensure the working conditions described along this article.

(a)

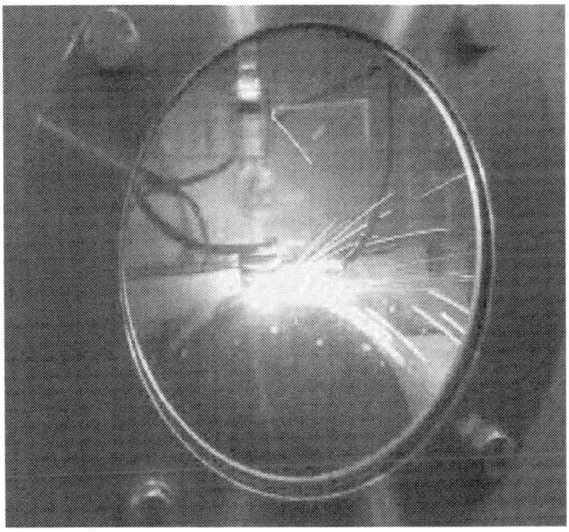

(b)

Fig. 11. (a) Ejection of molten material on the wheel in a non-operational regime, (b) Ejection of molten material on the wheel in a operational regime; both photos on its own equipment in LSA.

5. Conclusion

It is proposed in this paper the design of a winding mechanism for amorphous strips used in magnetic transformer´s cores, its general dimensions are specified in the drawing in Figure 12, and it´s assemble with the personally designed equipment is completely possible.

The components and parts designs are based on our own experience in building these equipments and on our investigations on the production of micro and nano-materials (Ozols et al., 1999; Pagnola, 2009; Muraca et al., 2009), as well as other author's technical considerations in the fabrication of different products for industrial magnetic packages as shown in Figure N. 13. (Croat, 1992; Kurokawa, et al. 1999) were considered.

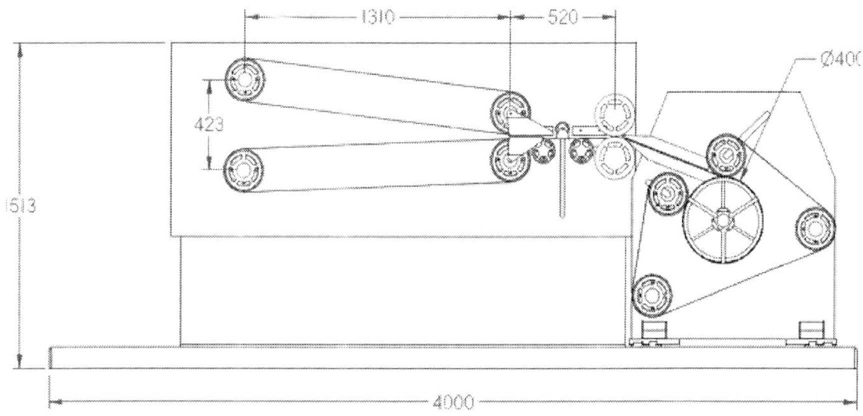

Fig. 12. General drawing of mechanical parts.

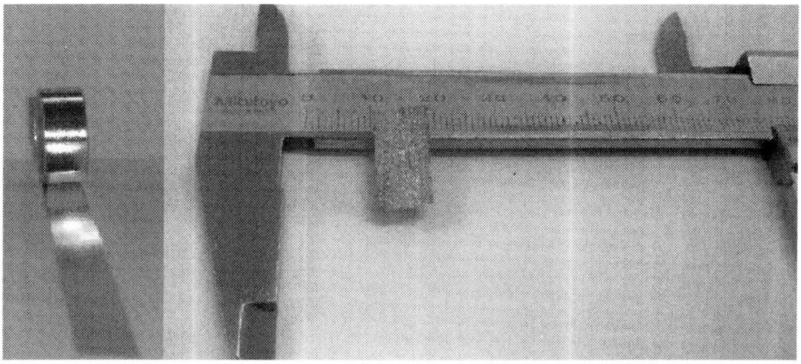

Fig. 13. $Fe_{78}Si_{13}B_9$ amorphous strips used in magnetic cores, and industrial magnetic package.

6. References

Bedell J.R., Method of and apparatus for casting metal strip employing a localized conditioning shoe, United States Patent: 4649984, March 17, 1987

Budzyn B., Carlson C., Two piece casting wheel, United States Patent: 4537239, August 27, 1985.

Byoung-Gi M., Yong Sohn K., Won-Wook P., Taek-Dong L., Effect of milling on the soft magnetic behavior of nanocrystalline alloy cores, Materials Science and Engineering A, pp. 449 451, 2007.

ConDes (Consultora Conexiones para el Desarrollo), Mercado de Materiales Magnético, en el marco proyecto FAN (www.con-des.com.ar), 2008.-

Croat J., Energy product rare earth-iron magnet alloys, United States Patent: 5172751, December 22, 1992.

De Cristofaro N., Amorphous Metals in Electric-Power Distribution Applications, Materials Research Society, MRS Bulletin, Vol. 23, No. 5, 1998, pp. 50-56.

Dobrzanskia L.A. , Nowosielskia R., Koniecznya J., Wysłockib J. , Przybyłb A., Properties and structure of the toroidal magnetically soft cores made from the amorphous strips, powder, and composite materials, Journal of Materials Processing Technology, 2004, pp.157-158.

Douglas J., Transformers with lower losses, Power Engineering Review, IEEE, Vol. 8, No. 3, 1988, pp. 12-13.

Frissora A.., Kojak A., Chilled casting wheel, United States Patent: 4502528, March 5, 1985.

Gelinas C., Soft magnetic Powders for AC Magnetic Applications. Euro PM2000,Soft Magnetic Materials Workshop, Berlin, 2000, pp. 1-8.

Good J., Roisum D., Winding: machines, mechanics and measurements, DEStech Publications Inc., ISBN: 978-1-932078-69-5, 2008.

He F., Zhao H. & Wang Q., Constant Linear Speed Control of Motor Based on Fuzzy PI in Strip Winding System, DBTA 2010, 978-1-4244-6977-2, pp 1-4.

Kurokawa K., Suhara S., Matsukawa T., Ishizuka H., Sato T., Method of manufacturing a wide metal thin strip, United States Patent: 5908068,June 1, 1999.

Kurokawa, Method of manufacturing a wide metal thin strip, United States Patent: 5908068, June 1, 1999.

Lee Y., Wickert J., Stress field width axisymetric wound rolls, Journal of applied mechanics, ASME, Vol 69, March 2002.

Li S., Cao J., A hybrid approach for quantifying the winding process and material effects on sheet coil deformation, Journal of Engineering Materials and technology, ASME,Vol. 126, July 2004.

Liu M., A nonlinear model of center-wound rolls incorporating refined boundary conditions, ELSEVIER, Computers and structures 87 (2009) 552-563.

Marashi S P H, Abedi A., Kaviani S., Aboutalebi S H., Rainforth M. and Davies H A, Effect of melt-spinning roll speed on the nanostructure and magnetic properties of stoichiometric and near stoichiometric Nd–Fe–B alloy ribbons, J. Phys. D: Appl. Phys. 42 11, 2009.

Muraca D., Silveyra J., Pagnola M., Cremaschi M., Nanocrystals magnetic contribution to FINEMET-type soft magnetic materials with Ge addition, Journal of Magnetism and Magnetic Materials, 321, 2009, pp. 3640–3645.

Moya J., "Vidrios Metálicos y Aleaciones Nanocristalinas: Nuevos Materiales de Estructura Avanzada", Cuadernos de Facultad n. 4, UBA 2009

Nowacki J., Polyphase sintering and properties of metal matrix composites, Journal of Materials Processing Technology 175, 2006, pp 316 323.

Ozols A., Sirkin H. and Vicente E.E., Segregation in Stellite powders produced by the plasma rotating electrode process Materials Science and Engineering: A Volume 262, Issues 1-2, 1999, pp. 64-69.

Pagnola M., Tesis Doctoral UBA: Desarrollo de Composites Ferromagnéticos, 2009.

Pagnola M., Saccone F., Ozols A., Sirkin H,, Improvement to approximation of second order function of hysteresis in magnetic materials, COMPEL: The International Journal for Computation and Mathematics in Electrical and Electronic Engineering, Volume 28 Issue 6, 2009, pp. 1579-1589.

Praisner T., Chen J., Tseng A., An experimental study of process behavior in planar flow melt spinning, Metallurgical and materials transactions B, Volume 26B, December 1995.

Richardson B., Amorphous metal cored transformers-justifying their use, IEE Colloquium on Magnetic Ribbons and Wires in Power, Electronic and Automotive Applications, 1990, pp.1-4.

Saito T., Electrical resistivity and magnetic properties of Nd–Fe–B alloys produced by Melt spinning technique, Journal of Alloys and Compounds Volume 505, Issue 1, 2010, pp. 23-28.

Theisen E., Davis M., Weinstein S., SteenP., Transient behavior of the planar-flow melt Spinning, ELSEVIER, Chemical engineering science 65 (2010) 3249-3259.

Wang C., Yan M., Surface quality, microstructure and magnetic properties of Nd2(Fe,Zr,Co)14B/Fe alloys prepared by different melt-spinning equipments, Materials Science and Engineering B 164, 2009, pp. 71-75.

13

Free Vibration Analysis of Centrifugally Stiffened Non Uniform Timoshenko Beams

Diana V. Bambill, Daniel H. Felix, Raúl E. Rossi and Alejandro R. Ratazzi
Universidad Nacional del Sur, UNS, Departamento de Ingeniería,
Instituto de Mecánica Aplicada, IMA,
Consejo Nacional de Investigaciones Científicas y Técnicas, CONICET,
Argentina

1. Introduction

Rotating beams - like structures are widely used in many engineering fields and are of great interest as they can be used to model blades of wind turbines, helicopter rotors, robotic manipulators, turbo-machinery and aircraft propellers. The governing differential equations of motion in free vibration of a non-uniform rotating Timoshenko beam, with general elastic restraints at the ends are solved using the differential quadrature method, (Bellman & Roth, 1986; Felix et al., 2008, 2009). The equations of motion are derived to include the effects of shear deformation, rotary inertia, hub radius, ends elastically restrained and non-uniform variation of the cross-sectional area of the beam. The presence of a centrifugal force due to the rotational motion is considered as Banerjee has developed, using Hamilton's principle to capture the centrifugal stiffening arising in fast rotating structures, (Banerjee, 2001). With the proposed model, a great number of different situations are admitted to be solved. Particular cases with classical restraints can be deduced for limiting values of the rigidities. Also step changes in cross-section are considered (Naguleswaran, 2004).

The natural vibration frequencies and mode shapes of rotating beams have been a topic of interest and have received considerable attention. A large number of researchers have studied the dynamic behavior of rotating uniform or tapered Euler-Bernoulli beams. (Yang el al., 2004; Özdemir & Kaya, 2006; Lin & Hsiao, 2001). Banerjee derived the dynamic stiffness matrix of a rotating Bernoulli-Euler beam using the Frobenius method of solution in power series and he includes the presence of an axial force at the outboard end of the beam in addition to the existence of the usual centrifugal (Banerjee, 2000).

Not so many studies have tackled the problem of rotating beams taking into account rotary inertia, shear deformation and their combined effects, hub radius and ends elastically restrained, (Bambill et al., 2010). In applications where the rotary inertia and the shear deformation effects are not significant, an analysis based on the Euler-Bernoulli beam theory can be used. However, Timoshenko theory allows describing the vibration of short beams, sandwich composite beams or high modes of a slender beam, (Rossi et al., 1991; Seon et al., 1999). (Banerjee et al., 2006) investigated the free bending vibration of rotating tapered Timoshenko beams by the dynamic stiffness method. (Ozgumus & Kaya, 2010) used the Differential Transform Method for free vibration analysis of a rotating, tapered Timoshenko beam.

$K_{\psi d}$	K_{Wd}	Method	Ω_1	Ω_2	Ω_3	Ω_4	Ω_5
0	0	DQM	9.94295	21.1789	37.1287	54.4936	77.1422
		FEM	9.97877	21.2145	37.1668	54.5377	77.1717
	0.1	DQM	12.3497	26.6244	45.1340	63.3202	86.7544
		FEM	12.3950	26.6574	45.1757	63.3506	86.7801
	1	DQM	12.6335	27.5522	46.9619	67.3571	93.8220
		FEM	12.6802	27.5867	47.0072	67.3872	93.8529
	10	DQM	12.6651	27.6547	47.1481	67.7705	94.4732
		FEM	12.7119	27.6894	47.1937	67.8007	94.5049
	$\to \infty$	DQM	12.6686	27.6662	47.1687	67.8159	94.5425
		FEM	12.7155	27.7009	47.2144	67.8461	94.5743
1	0	DQM	9.96265	21.5532	38.7860	57.0231	81.5044
		FEM	9.99874	21.5890	38.8248	57.0634	81.5315
	0.1	DQM	13.0297	27.9504	45.9575	63.5235	86.9102
		FEM	13.0763	27.9817	45.9976	63.5530	86.9362
	1	DQM	13.5409	29.8282	49.4887	70.9743	97.0122
		FEM	13.5897	29.8623	49.5335	71.0012	97.0437
	10	DQM	13.5994	30.0410	49.8608	71.8782	98.3059
		FEM	13.6484	30.0754	49.9060	71.9052	98.3387
	$\to \infty$	DQM	13.6060	30.0648	49.9016	71.9759	98.4391
		FEM	13.6550	30.0993	49.9469	72.0030	98.4720
10	0	DQM	9.96324	21.5641	38.8347	57.1056	81.6606
		FEM	9.99933	21.5999	38.8736	57.1459	81.6876
	0.1	DQM	13.0583	28.0142	46.0008	63.5342	86.9175
		FEM	13.1049	28.0454	46.0408	63.5635	86.9436
	1	DQM	13.5802	29.9428	49.6333	71.2194	97.2490
		FEM	13.6291	29.9769	49.6781	71.2462	97.2806
	10	DQM	13.6399	30.1611	50.0148	72.1525	98.5813
		FEM	13.6890	30.1956	50.0600	72.1794	98.6142
	$\to \infty$	DQM	13.6467	30.1856	50.0566	72.2531	98.7177
		FEM	13.6958	30.2201	50.1018	72.2800	98.7508
$\to \infty$	0	DQM	9.96331	21.5653	38.8403	57.1151	81.6785
		FEM	9.99940	21.6011	38.8792	57.1553	81.7056
	0.1	DQM	13.0616	28.0216	46.0058	63.5354	86.9184
		FEM	13.1082	28.0529	46.0459	63.5648	86.9444
	1	DQM	13.5848	29.9563	49.6504	71.2486	97.2773
		FEM	13.6337	29.9904	49.6952	71.2753	97.3090
	10	DQM	13.6447	30.1752	50.0329	72.1851	98.6141
		FEM	13.6938	30.2097	50.0781	72.2120	98.6471
	$\to \infty$	DQM	13.6514	30.1997	50.0748	72.2860	98.7509
		FEM	13.7006	30.2342	50.1201	72.3129	98.7840

Table 10. First natural frequencies $\Omega_i = \sqrt{\rho A_1(0) / EI_1(0)}\, L^2 \omega_i$ for a two-span elastically restrained rotating Timoshenko beam, with elliptical cross section and quadratic height variation along the axis. $l_1 / L = 1/2$ $l_2 / L = 1/2$, $h_{B1} / h_{A1} = 1/2$, $h'_{B1} = 0$, $h_{A2} / h_{B1} = 1/2$, $h_{B2} / h_{A2} = 1/2$, $h'_{A2} = 0$, $K_{W1} = 10$, $K_{\psi 1} = 0.1$, $\eta = 10$.

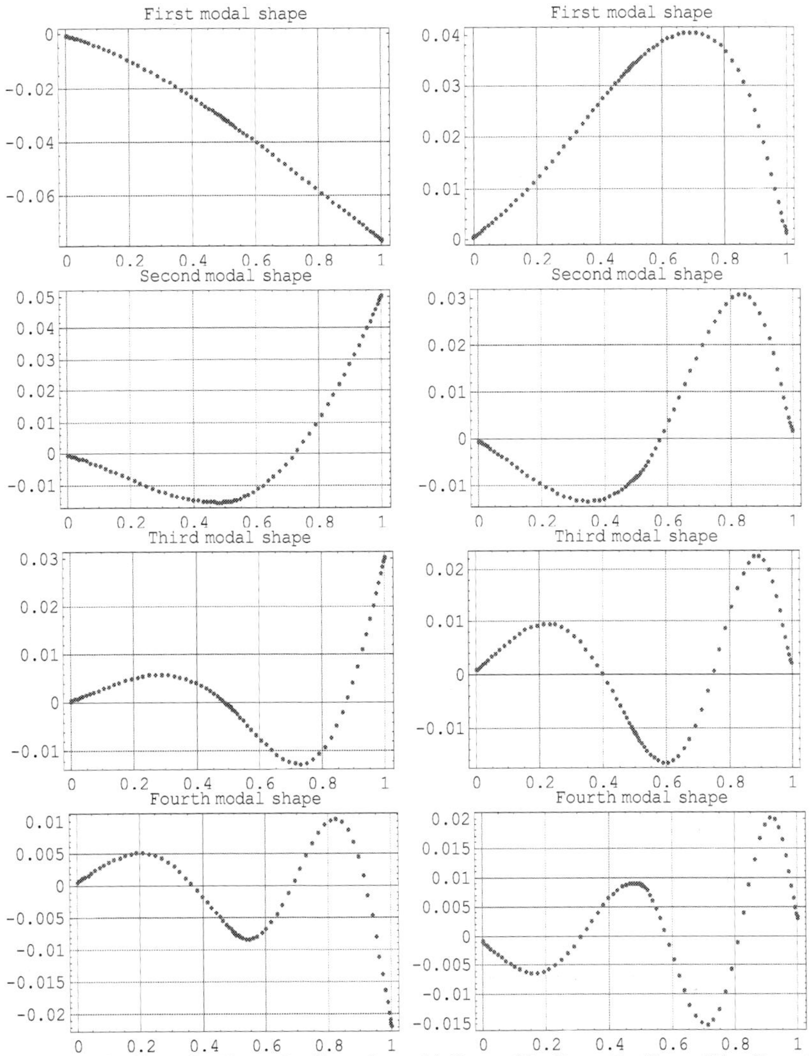

a) $K_{W1}=10$, $K_{\psi 1}=5$, $K_{Wd}=0$, $K_{\psi d}=0$ b) $K_{W1}=10$, $K_{\psi 1}=5$, $K_{Wd}=0.1$, $K_{\psi d}=1$

Fig. 4. Natural frequencies mode shapes for a two-span elastically restrained rotating Timoshenko beams, with elliptical cross section and quadratic height variation along the axis. $l_1/L=l_2/L=1/2$; $h_{B1}/h_{A1}=1/2$; $h'_{B1}=0$; $h_{A2}/h_{B1}=1/2$; $h_{B2}/h_{A2}=1/2$; $h'_{A2}=0$; $\eta=10$

Figure 4 shows the first four natural frequency mode shapes for beams, with two different kinds of boundary conditions: a) corresponds to $K_{W1}=10$, $K_{\psi 1}=5$, $K_{Wd}=0$, $K_{\psi d}=0$, while b) corresponds to $K_{W1}=10$, $K_{\psi 1}=5$, $K_{Wd}=0.1$, $K_{\psi d}=1$.

The next Figures, 5 and 6, present the variation of the fundament frequency parameter Ω_1 with the variation of the non-dimensional rotational speed η and the spring constant $K\psi_1$.

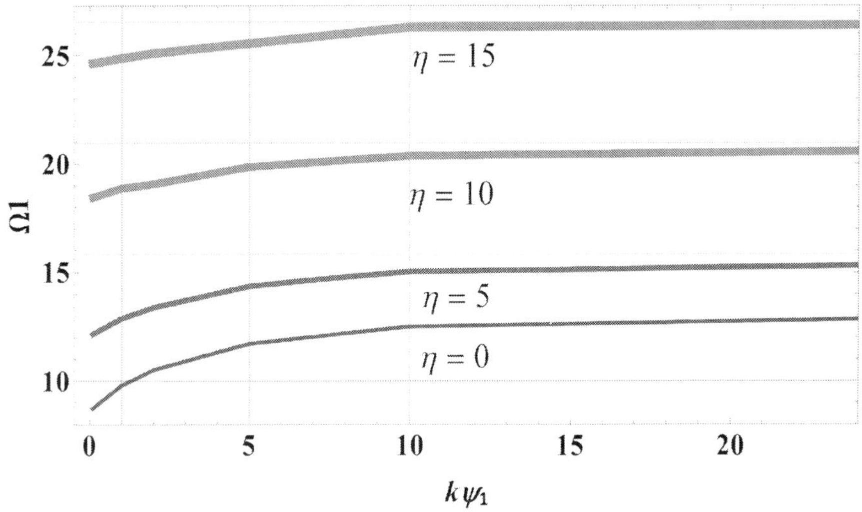

Fig. 5. The fundamental frequency coefficient Ω_1 of a one-span elastically restrained rotating Timoshenko beam versus the spring constant parameter of the rotational spring $K_{\psi 1}$, for different rotational speed parameters η. K_{w1}= 10; K_{wd}=1; $K_{\psi d}$=10

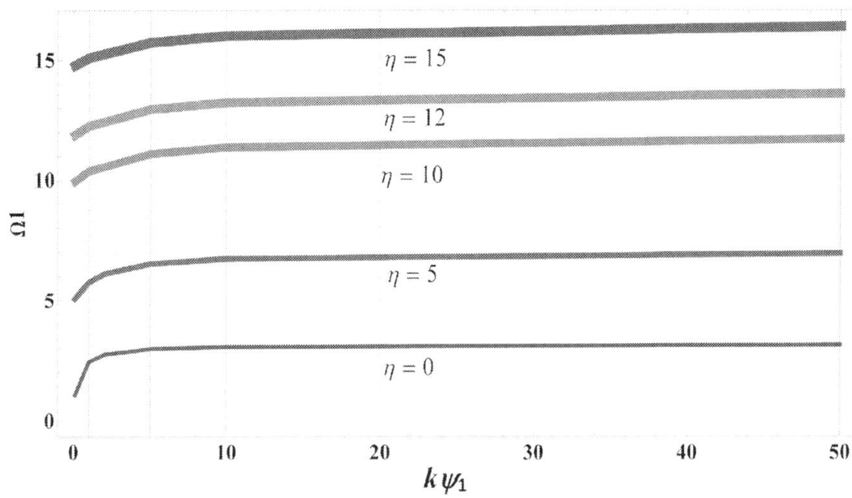

Fig. 6. The fundamental frequency coefficient Ω_1 of a two-span elastically restrained rotating Timoshenko beam versus the spring constant parameter of the rotational spring $K_{\psi 1}$, for different rotational speed parameters η. K_{w1}= 10; K_{wd}=0; $K_{\psi d}$=0

6. Conclusion

The differential quadrature method proves to be very efficient to obtain frequencies and mode shapes of natural vibration, for the rotating Timoshenko beam model.

The versatility of the proposed beam model (variable cross section, step change in cross section, elastic restraints at both ends) allows to solve a large number of individual cases.

Something interesting to point out is that because the method directly solves two ordinary differential equations, additional restrictions are not generated. This does not happen in other methodologies, such as the dynamic stiffness method (Banerjee, 2000, 2001).

As a matter of fact, the differential quadrature method has the same advantage as the finite element method and it needs less computer memory requirements than the FEM.

In particular the present results show that the frequency coefficients vary more significantly when the translational spring stiffness changes at the end of the beam farthest from the axis of rotation $K_{\psi d}$.

7. Appendix A

As Shu presents in his book (Shu, 2000), the differential quadrature method, DQM, is a numerical technique for solving differential equations.

In order to obtain the DQM analog equations to the governing equations of the rotating beam and its boundary conditions, the beam domain is discretized in a grid of points using the Chebyshev - Gauss - Lobato expression, (Shu & Chen, 1999):

$$x_i = \frac{1-\cos\left[(i-1)\pi/(n-1)\right]}{2} \quad ; \; i = 1,2,\ldots,n$$

where n is the number of discrete points or nodes and x_i is the coordinate of node i.

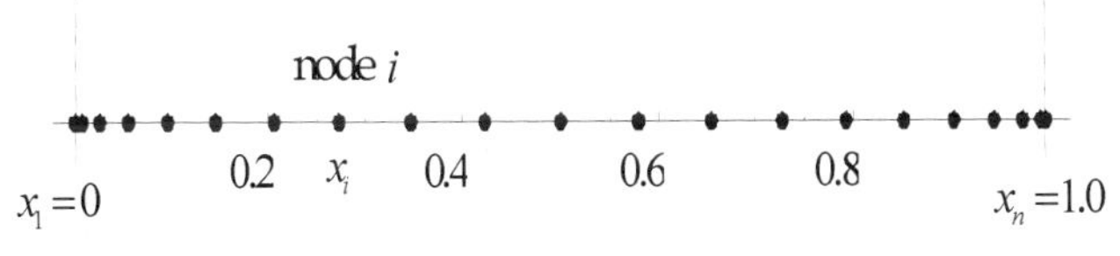

Fig. A1. Grid of n points

The weighting coefficients $A_{ij}^{(1)}$ and $A_{ij}^{(2)}$, which appeared in the linear algebraic equations of quadrature (28-35), were determined using the explicit expressions cited by (Bert & Malik, 1996).

The coefficients $A_{ij}^{(1)}$ correspond to first order derivatives and can be arranged in a square matrix of order n.

The matrix elements $A_{ij}^{(1)}$ with $i \neq j$, are determined by:

$$A_{ij}^{(1)} = \frac{\Pi(x_i)}{(x_i - x_j)\,\Pi(x_j)}$$

where

$$\Pi(x_i) = \prod_{\nu=1\ \nu\neq i}^{n} (x_i - x_\nu)\ ;\ \Pi(x_j) = \prod_{\nu=1\ \nu\neq i}^{n} (x_j - x_\nu)\ ;$$

The coefficients $A_{ij}^{(1)}$ with $i = j$, will tend to infinity and need to be calculated in another way.

The coefficients $A_{ij}^{(2)}$ correspond to second-order derivatives and are obtained from

$$A_{ij}^{(2)} = 2\left[A_{ii}^{(1)} \ast A_{ij}^{(1)} - \frac{A_{ij}^{(1)}}{x_i - x_j}\right]$$

with $i \neq j$ and $i,\ j = 1, 2, 3, \ldots, n$.
Because the sum of the weighting coefficients of a row of the matrix is zero, it is easy to calculate the diagonal terms of derivatives of any order q, using the following expression:

$$A_{ii}^{(q)} = -\sum_{j=1\ j\neq i}^{n} A_{ij}^{(q)}$$

And the equations for q equal to 1 and 2, corresponding to first and second order derivatives, are:

$$A_{ii}^{(1)} = -\sum_{j=1\ j\neq i}^{n} A_{ij}^{(1)}\ ;\ A_{ii}^{(2)} = -\sum_{j=1\ j\neq i}^{n} A_{ij}^{(2)}$$

8. Acknowledgment

The authors gratefully acknowledge the support of the Universidad Nacional del Sur (UNS) and the Consejo Nacional de Investigaciones Científicas y Técnicas (CONICET), Argentina.

9. References

Bambill, D.V.; Felix, D.H. & Rossi, R. E. (2010). Vibration analysis of rotating Timoshenko beams by means of the differential quadrature method. *Structural Engineering and Mechanics*, Vol. 34, No. 2, pp. 231-245, ISSN 12254568

Banerjee, J. (2000). Free vibration of centrifugally stiffened uniform and tapered beams using the dynamic stiffness method. *Journal of Sound and Vibration*, Vol.233, No.5, pp. 857-875, ISSN 0022-460X

Banerjee, J. (2001). Dynamic stiffness formulation and free vibration analysis of centrifugally stiffened Timoshenko beam. *Journal of Sound and Vibration*, Vol.247, pp. 97-115, ISSN 0022-460X

Banerjee, J.; Su, H, & Jackson, D. (2006). Free vibration of rotating tapered beams using the dynamic stiffness method. *Journal of Sound and Vibration*, Vol. 298, pp. 1034-1054, ISSN 0022-460X

Bellman, R. & Casti, J. (1971). Differential quadrature and long-term Integration. *J. Math. Anal.* Vol.34, pp. 235-238, ISSN 0022-247X

Bellman, R.E. & Roth, R.S. (1986). *Methods in approximation: techniques for mathematical modelling*, Editorial D. Reidel Publishing Company, ISBN 9-027-72188-2, Dordrecht, Holland

Bert, C. & Malik, M. (1996). Differential quadrature method in computational mechanics: A review. *Applied Mechanics Review* Vol.49, pp. 1-28, ISSN 0008-6900

Choi, S.; Wu J. & Chou Y. (2000). Dynamic analysis of a spinning Timoshenko beam by the differential quadrature method. *American Institute of Aeronautics and Astronautics* Vol.38, pp. 51-856, ISSN 0001-1452

Felix, D.H.; Rossi, R. E. & Bambill, D. V. (2008). Vibraciones transversales por el método de cuadratura diferencial de una viga Timoshenko rotante, escalonada y elásticamente vinculada, *Mecánica Computacional* Vol. XXVII, pp.1957-1973, ISBN 1666-6070

Felix, D. H.; Bambill, D. V. & Rossi, R. E. (2009). Análisis de vibración libre de una viga Timoshenko escalonada, centrífugamente rigidizada, mediante el método de cuadratura diferencial, *Revista Internacional de Métodos Numéricos para Cálculo y Diseño en Ingeniería.* Vol. 25, No. 2, pp. 111-132, ISSN 0213-1315

Ganesh, R and Ganguli, R. (2011). Physics based basis function for vibration analysis of high speed rotating beams. *Structural Engineering and Mechanics*, Vol.39, No.1, pp. 21-46, ISSN 1225-4568

Gunda, J. B. & Ganguli R. (2008). New rational interpolation functions for finite element analysis of rotating beams. *International Journal of Mechanical Sciences;* Vol. 50, pp. 578-588, ISSN 0020-7403

Gunda, J.B.; Singh, A.P.; Chhabra, P.S. & Ganguli, R. (2007). Free vibration analysis of rotating tapered blades using Fourier-p superelement, *Structural Engineering and Mechanics*, Vol.27, No.2, pp. 243-257, ISSN 1225-4568

Kumar A. & Ganguli R. (2009). Rotating Beams and Nonrotating Beams with Shared Eigenpair, *Journal of Applied Mechanics.* Vol.76. No.5, pp. 1-14, ISSN: 0021-8936

Hodges, D. H. & Rutkowski, M. J. (1981). Free vibration analysis of rotating beams by a variable order finite method, *American Institute of Aeronautics and Astronautics Journal.* Vol.19, No.11, pp. 1459-1466

Lin, S. C. & Hsiao, K. M. (2001). Vibration analysis of a rotating Timoshenko beam. *Journal of Sound and Vibration.* Vol. 240, pp. 303-322.

Liu, G. R. & Wu, T. Y. (2001). Vibration analysis of beams using the generalized differential quadrature rule and domain decomposition. *Journal of Sound and Vibration.* Vol.246, pp.461-481, ISSN 0022-460X

Naguleswaran, S. (2004). Transverse vibration and stability of an Euler-Bernoulli beam with step change in cross-section and in axial force. *Journal of Sound and Vibration.* Vol.270, pp.1045-1055, ISSN 0022-460X

Özdemir, Ö. & Kaya, M.O. (2006). Flapwise bending vibration analysis of a rotating tapered cantilever Bernoulli-Euler beam by differential transform method. *Journal of Sound and Vibration.* Vol.289, pp.413-420, ISSN 0022-460X

Ozgumus, O. & Kaya, M. O. (2010). Vibration analysis of rotating tapered Timoshenko beam using DTM. Meccanica. Vol. 45, pp. 33-42, ISSN 0025-6455

Rossi R.E. (2007). *Introducción al análisis de Vibraciones con el Método de Elementos Finitos.* EdiUNS, Universidad Nacional del Sur, IBSN 978-987-1171-71-2, Bahía Blanca, Argentina.

Rossi, R. E.; Gutiérrez R. H. & Laura P. A. A. (1991). Transverse vibrations of a Timoshenko beam of nonuniform cross section elastically restrained at one end and carrying a concentrated mass at the other. *J. Acoust. Soc. Am,* Vol.89, pp.2456-2458.

Seon, M. H.; Benaroya, H. & Wei, T. (1999). Dynamics of transversely vibrating beams using four engineering theories. *Journal of Sound and Vibration.* Vol.225, pp.35-988, ISSN 0022-460X

Singh, A.P.; Mani, V. & Ganguli, R. (2007). Genetic programming metamodel for rotating beams, *CMES - Computer modelling in Engineering and Sciences,* Vol.21. No.2, pp. 133-148.

Shu, C. (2000). *Differential Quadrature and Its Application in Engineering,* Springer-Verlag, ISBN 1852332093, London, England

Shu, C. & Chen, W. (1999). On optimal selection of interior points for applying discretized boundary conditions in DQ vibration analysis of beams and plates. *Journal of Sound and Vibration.* Vol.222, No.2, pp. 239-257, ISSN 0022-460X

Vinod, K. G., Gopalakrishnan, S. & Ganguli, R. (2007), Free vibration and wave propagation analysis of uniform and tapered rotating beams using spectrally formulated finite elements. *International Journal of Solids and Structures;* Vol.44, pp. 5875-5893, ISSN 0020-7683

Yang, J. B.; Jiang, L. J. & Chen, D. CH. (2004). Dynamic modelling and control of a rotating Euler-Bernoulli beam. *Journal of Sound and Vibration.* Vol.274, pp. 863-875, ISSN 0022-460X

14

Vibration-Based Diagnostics of Steam Turbines

Tomasz Gałka
Institute of Power Engineering,
Poland

1. Introduction

Of three general maintenance strategies - run-to-break, preventive maintenance and predictive maintenance - the latter, also referred to as condition-based maintenance, is becoming widely recognized as the most effective one (see e.g. Randall, 2011). To exploit its potential to the full, however, it has to be based on reliable condition assessment methods and procedures. This is particularly important for critical machines, characterized by high unit cost and serious consequences of a potential failure. Steam turbines provide here a good example.

In general, technical diagnostics may be defined as determining technical condition on the basis of objective methods and measures. The objectivity implies that technical condition assessment is based on measurable physical quantities. These quantities are sources of diagnostic symptoms. For any given class of objects, the development of technical diagnostics essentially involves four principal stages (Crocker, 2003), namely:

- measurement,
- qualitative diagnostics,
- quantitative diagnostics,
- prognosis (forecasting).

At the *measurement* stage we are able to measure physical quantities relevant to the object technical condition. On the basis of measurement data, at the *qualitative diagnostics* stage faults and malfunctions are identified and located with the aid of an appropriate diagnostic model. *Quantitative diagnostics* consists in estimating damage degree (advancement), for which a reference scale is necessary. Finally, *prognosis* is an estimation of the period remaining until an intervention is needed. Qualitative diagnostics may be viewed as being aimed at detecting hard (random) failures, while the aim of the quantitative diagnosis is to trace the soft (natural) fault evolution (Martin, 1994).

Complex objects, like steam turbines, are characterized by a number of residual processes (such as vibration, noise, heat radiation etc.) that accompany the basic process of energy transformation, and hence a number of condition symptom types. For all rotating machines, vibration-based symptoms are the most important ones for technical condition assessment, due to at least three reasons:

- high content of information,
- comparatively easy and non-intrusive measurement techniques,
- well-developed methods for data processing and diagnostic information extraction.

Of all vibration-based symptom types (see e.g. Morel, 1992; Orłowski, 2001), three are of particular importance for steam turbine diagnostics:

- absolute vibration spectra,
- relative vibration vectors,
- time evolution of spectral components.

These symptoms form the basis of diagnostic reasoning in both permanent (*on-line*) and intermittent (*off-line*) monitoring systems.

2. Vibration generation and vibrodiagnostic symptoms

Just like all rotating machines, steam turbines generate broadband vibration, so that power density spectra typically contain a number of distinct components. Due to different vibration generation mechanisms involved, it is convenient to divide the entire frequency range under consideration (typically from a few hertz up to some 10 to 20 kilohertz) into two sub-ranges, commonly referred to as the *harmonic* (or 'low') and *blade* (or 'high') frequency ranges, respectively. Sometimes the sub-harmonic range (below the fundamental frequency f_0 resulting from rotational speed) is also distinguished. This division is shown schematically in Fig.1.

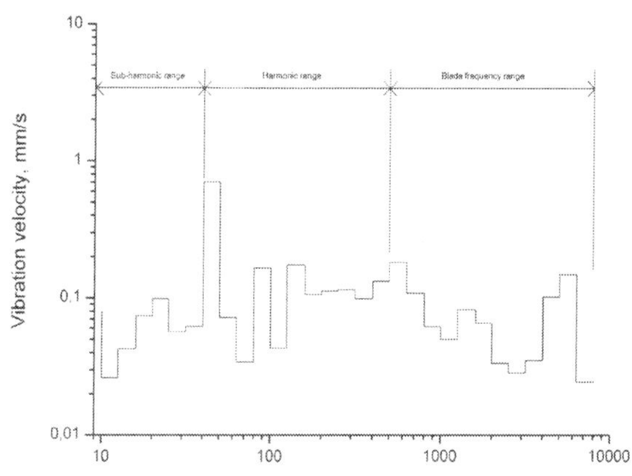

Fig. 1. Schematic representation of dividing the entire power density spectrum frequency range into sub-harmonic, harmonic and blade frequency ranges (after Gałka, 2009a).

Components from the harmonic range result directly from the rotary motion of turbine shaft and are related to malfunctions common to all rotating machines, such as:

- unbalance,
- shafts misalignment,
- bent or cracked rotors,
- magnetic phenomena in the generator.

Components that fall into the sub-harmonic range are typically determined mainly by the stability of the oil film in shaft bearings (Bently and Hatch, 2002; Kiciński, 2006). Those of very low frequencies (a few hertz) may be indicative of cracks in turbine casings and other non-rotating elements.

Individual components from the blade frequency range are produced as a result of interaction between steam flow and the fluid-flow system, and hence may be considered specific to steam turbines. There are three basic phenomena involved (Orłowski, 2001; Orłowski and Gałka, 1998), namely:

- flow disturbance caused by stationary and rotating blades edges,
- flow disturbance resulting from scatter of fluid-flow system elements dimensions.,
- flow disturbance by control valves opening.

First of these can be described in the following way: discharge edges of stationary and rotating blades introduce local interruptions of steam flow, thus reducing its thrust on a rotating blade and causing an instantaneous force of the opposite direction. Resulting force q_1 is thus periodic and can be expressed by

$$q_1 = \zeta_0 + \Sigma\zeta_k \cos k(n\omega t + \Psi_k) \qquad (1)$$

where ζ_0 is time-averaged thrust, ζ_k and Ψ_k are amplitude and phase of the k-th component, respectively, n is number of blades in a stage (stationary or rotating) under consideration and ω denotes angular frequency. This force can thus be expressed as a series of harmonic components with frequencies equal to $kn\omega = 2\pi knu$, where u is the rotational speed in s^{-1}. As for the second phenomenon, it results from the fact that manufacture of blades and their assembly into rotor stages or bladed diaphragms are not perfect, so for each blade the corresponding discharge cross-section is slightly different from the other ones. Resulting force has a form of a pulse generated once per rotation and thus may be expressed by

$$q_2 = \zeta_0 + \Sigma\zeta_k \cos k(\omega t + \Psi_k)\,. \qquad (2)$$

The third phenomenon is related to turbine control and shall be dealt with a little later. It should be mentioned, however, that – unlike the first two – the influence of control valves opening is usually limited to the vicinity of the control stage and diminishes as we move along the steam expansion path. Frequencies of basic spectral components resulting from interaction between steam flow and the fluid-flow system can be, on the basis of above considerations, expressed by

$$f_w = l \cdot u \qquad (3)$$

$$f_k = b \cdot u \qquad (4)$$

$$f_{(w+k)/2} = (l + b) \cdot u/2 \qquad (5)$$

$$f_{(w-k)/2} = (l - b) \cdot u/2 \qquad (6)$$

where l and b denote numbers of blades in rotor stages and bladed diaphragms, respectively. Components given by Eqs.(5) and (6) result from interactions between rotor stages and adjacent bladed diaphragms. Each turbine stage is thus in general characterized by as many as six individual vibration components.

Vibration signal that can be effectively measured in an accessible point of a turbine is influenced not only by relevant generation mechanisms, but also by its propagation to this point, as well as by operational parameters and interference (see e.g. Radkowski, 1995; Gałka, 2011b). In general terms it may be expressed as (Radkowski, 1995)

$$z(r,t) = h_p(r,t) * u_w(r,t) + \eta(r,t) \ , \tag{7}$$

where z denotes measured diagnostic signal, h_p is the response function for signal propagation from its origin to the measuring point and η denotes uncorrelated noise; all these quantities are functions of the spatial variable r and dynamic time t. $u_w(r,t)$ is given by

$$u_w(r,t) = \sum_{i=1}^{n} h_i(r,D_i,t) * x(t) + h(r,t) * x(t) \ , \tag{8}$$

where D_i describes development of the ith defect, h_i is the response function pertaining to this defect and h is the response function with no defect present; $x(t)$ is the input signal, generated by an elementary vibroacoustic signal source. This model is shown schematically in Fig.2.

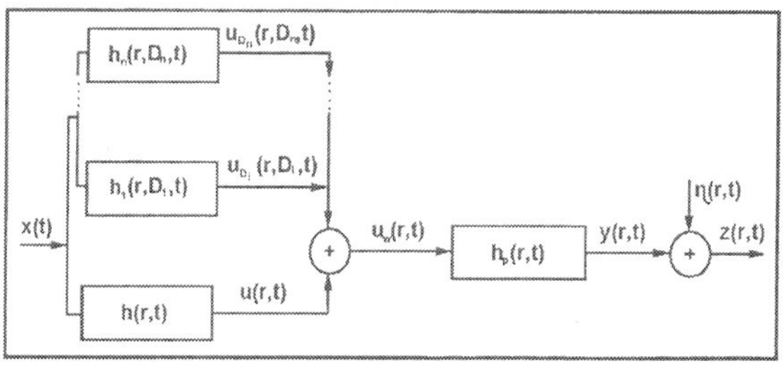

Fig. 2. Model of vibroacoustic signal generation and propagation (after Radkowski, 1995).

An alternative general relation, in a vector form, is provided by (Orłowski, 2001)

$$\mathbf{S}(\theta) = \mathbf{S}[\mathbf{X}(\theta), \mathbf{R}(\theta), \mathbf{Z}(\theta)] \ , \tag{9}$$

where **S**, **X**, **R** and **Z** denote vectors of symptoms, condition parameters, control parameters and interference, respectively, all varying with time θ.[1] Control parameters may be defined as resulting from object operator purposeful action, aimed at obtaining demanded performance (Gałka, 2011b). In steam turbines, usually (at least in power industry) the 'demanded performance' means demanded output power; active load P_u can thus be treated as a scalar measure of the vector **R**. As for the interference, two types can be distinguished: *external interference* (the source is outside the object) and *internal interference* (the source is within the object). With some reservations, the former can be identified with measurement errors, while the latter refers to all other contributions to the uncorrelated noise $\eta(t)$ in Eq.(7).

[1] The reason for using t and θ symbols for denoting time is that the former refers to the 'dynamic' time (e.g. that enters equations of motion), while the latter is for the 'operational' time – the argument of equations pertaining to technical condition evolution.

Let us assume that the influence of interference may be reduced to a point wherein it can be neglected. As control parameters are, at any given moment, known, there is obviously a possibility of symptom normalization with respect to them, either model-based or empirical. It has to be kept in mind, however, that normalization with respect to P_u, which seems most straightforward, in practice may be only approximate. P_u can be expressed as (Traupel, 2000)

$$P_u = (dm/dt)\Delta i \eta_t \quad , \tag{10}$$

where dm/dt denotes steam mass flow, Δi is the enthalpy drop and η_t is the turbine efficiency. Assuming that η_t remains constant (which is only an approximation), P_u may be controlled by changing Δi (qualitative control), dm/dt (quantitative control), or both. The latter method (known as group or nozzle control) is typically used in large steam turbines. Each control valve supplies steam to its own control stage section; the number of these valves in large steam turbines is usually from three to six and they are opened in a specific sequence. At the rated power the last valve is only partly open, or even almost closed, as it provides a reserve in a case of a sudden drop of steam parameters. Furthermore, Δi depends also on condenser vacuum, which for a given unit may change within certain limits depending on overall condenser condition, cooling water temperature, weather etc. Thus

$$P_u = f(r_1, r_2, \ldots, r_k, p_o) \quad , \tag{11}$$

where r_i denotes ith valve opening, k is the number of valves and p_o is the condenser pressure. In fact, r_i and p_o are the $\mathbf{R}(\theta)$ vector parameters, various combinations of which may yield the same value of P_u. In view of Eqs.(9) and (11), any $S_i(P_u)$ function ($S_i \in \mathbf{S}$) cannot thus be a single-valued one.

Some attention has been paid to developing experimental relations of the $S_i = f(P_u)$ type (see e.g. Gałka, 2001), bearing in mind that they are approximate and applicable to a given turbine type only. Such relations turn out to be strongly non-linear and differences between individual symptoms are considerable. In general, within the load range given by roughly $P_u = (0.85 \div 1.0)P_n$, where P_n is rated power, variations are quite small; thus, when dealing with large sets of data, the simplest approach is to reject those acquired at extremely low or high loads. It has to be added that the fact of vibration-based symptoms dependence on control parameters and interference may serve as a basis for developing certain diagnostic procedures; this issue shall be dealt with in Section 6.

3. Qualitative diagnosis

As already mentioned, qualitative diagnosis consists in determining what malfunctions or damages are present and localizing them. In this Section the influences of control and interference shall be neglected, i.e. it shall be assumed that symptoms under consideration are deterministic functions of condition parameters $X_i \in \mathbf{X}$.

For obvious reasons, the following review does not claim to be exhaustive and is concentrated on issues relevant to steam turbine applications. For comprehensive and detailed treatment the reader is referred e.g. to (Morel, 1992; Bently and Hatch, 2002; Randall, 2011).

3.1 Harmonic (low) frequency range

Basically this subsection deals with absolute vibration spectral components of frequencies determined by $f = nf_0$, where f_0 results from rotational speed, and to some extent also with relative vibration vectors or orbits. In practical applications, components corresponding to $n > 4$ are seldom accounted for; this means that we are dealing with first four harmonic and sub-harmonic ($n < 1$) components. As each of these is typically influenced by a number of condition parameters, it is convenient to speak in terms of possible malfunctions and faults rather than frequencies.

3.1.1 Unbalance

Unbalance is common to all rotating elements. Primary symptom of this malfunction is the $1 \times f_0$ component of absolute vibration in a direction perpendicular to the turbine shaft line. They are, however, many other possible malfunctions (some of them quite common) that produce similar vibration patterns; additional procedures are therefore usually needed for a correct diagnosis.

In general, a 'pure' unbalance, be it static, quasi-static or dynamic, produces a $1 \times f_0$ component that remains almost constant in amplitude and phase during steady-state operation and disappears at low rotational speed. As rotor systems are non-linear, this component is typically accompanied by higher harmonics ($n > 1$), with amplitudes decreasing as n increases. Shaft orbits usually are quite regular and nearly circular or slightly elliptical. If such vibration pattern is present, the probability of unbalance being the root cause is high. Proper rotor balancing will usually reduce the residual unbalance to an acceptable level.

Rotor systems will always respond to balancing. Step changes of the $1 \times f_0$ component not related to any maintenance activities (but occurring mainly after turbine shutdown and subsequent startup) may be indicative of a loose rotor disk. Similarly, sudden and dramatic change may result from a broken rotor blade; such step changes are often big enough to enforce turbine tripping. Much slower, but continuous increase is often indicative of a permanent rotor bow (see also sub-section 3.1.3). An example is shown in Fig.3; it is easily

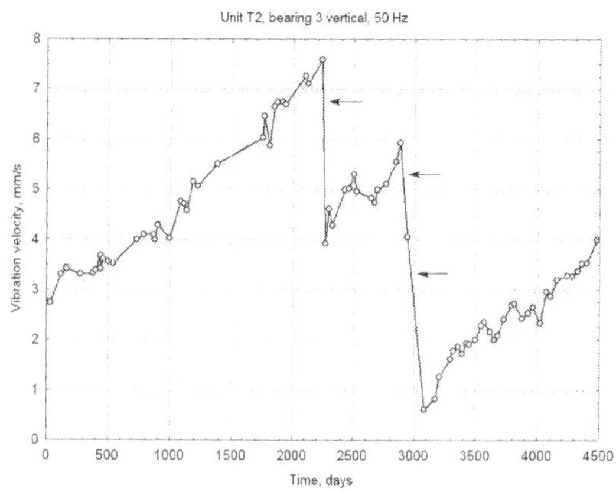

Fig. 3. Time history of the $1 \times f_0$ component with permanent rotor bow present: 230 MW unit, rear intermediate-pressure turbine bearing, vertical direction. Arrows indicate balancing sessions.

seen that balancing results in a considerable decrease of the 1 × f_0 component, but the improvement is only temporary. If this component is comparatively high at low rotational speed, coupling problem (offset rotor axles) is a possible root cause, especially in turbines with rigidly coupled rotors.

3.1.2 Misalignment

Ideally the entire turbine-generator unit shaft line (with overall length approaching 70 m in large units in nuclear power plants) should be a continuous and smooth curve; a departure from such condition is referred to as misalignment. The shape of this line is determined by shaft supports (journal bearings). As they displace during the transition from 'cold' to 'hot' condition, due to changing temperature field (this process may take even a few days to complete), at the assembly stage care has to be taken to ensure that the proper shape is maintained during normal operation. Relative vertical displacements may be even of the order of millimeters (Gałka, 2009a).

Misalignment modifies distribution of load between individual shaft bearings and therefore affects shaft orbits. With increasing misalignment magnitude they typically evolve from elongated elliptical shape through bent ('banana') and finally to highly flattened one (Bently and Hatch, 2002). High misalignment may lead to oil film instability, but in large steam turbines (especially modern ones, with only one bearing per coupling) this is a very rare occurrence. As for absolute vibration, 2 × f_0 component in directions perpendicular to the turbine axis is generally recognized as the basic misalignment symptom. Care, however, has to be taken when dealing with the turbine-generator coupling, as this component may be dominated by the influence of the generator (asymmetric position of rotor with respect to the stator electromagnetic field); in the latter case, dependence on the excitation current is usually conclusive. Marked misalignment is often accompanied with relatively high amplitudes of harmonic components in axial direction, but this symptom can by no means be considered specific.

3.1.3 Rotor bow

In general, three types of turbine rotor bow can be distinguished, namely:

- elastic bow, resulting from static load,
- temporary bow, caused by uneven temperature field and/or anisotropic rotor material properties, and
- permanent bow, wherein material yield strength has been exceeded (plastic deformation).

Permanent bow is obviously the most serious one. As it causes the center of gravity to move off from the shaft centerline, it basically produces an unbalance (cf. Fig.3). In general, rotor response vector may be expressed as (Bently and Hatch, 2002):

$$\mathbf{r} = r_e e^{j\delta} + \frac{M r_e \omega^2 e^{j\delta}}{[K - M\omega^2 + jD(1-\lambda)\omega]}, \tag{12}$$

where M denotes unbalance mass, shifted at the distance r_e in the direction determined by the angle δ. K and D are stiffness and damping coefficients, respectively; ω denotes rotor angular velocity and λ is the fluid circumferential velocity ratio ($\lambda = \nu/\omega$, where ν denotes average fluid angular velocity). First term describes the low-speed response (which, as mentioned earlier, is basically absent with 'plain' unbalance), while the second one refers to the dynamic synchronous response. It can be seen that for $\omega >> \omega_r$ (where ω_r is the resonance angular speed), when the first and the third term in the denominator can be neglected, rotor response is close to zero. This is a feature characteristic for this malfunction (colloquially speaking, the rotor 'balances itself out'), but in large steam turbines with heavy flexible rotors the $\omega >> \omega_r$ condition is seldom fulfilled.

It has been shown (Gałka, 2009b) that permanent rotor bow causes simultaneous increase of the $1 \times f_0$ component in vertical and axial directions, so that a developing bow should result in strong correlation between these components (see also Section 6). Available data seem to confirm this conclusion, in fact based on quite simple model considerations.

3.1.4 Rotor crack

As a very serious fault with potentially catastrophic consequences, rotor crack has received considerable attention (for perhaps the most comprehensive available review, see Bachschmid, Pennacchi and Tanzi, 2010). In general, crack reduces shaft stiffness and thus causes resonance to shift to a lower rotational speed. As a result, the $1 \times f_0$ component amplitude during steady-state operation will either increase or decrease. In large steam turbines, operated above the first critical speed, the latter may be the case. This effect may be combined with that of increasing rotor bow due to reduced bending stiffness. As a result of asymmetry introduced by a crack, the $2 \times f_0$ component may also increase substantially.

It is generally recognized that considerable continuous changes of first two harmonic components amplitudes (not necessarily both increasing!) and phases during steady-state operation indicate that a shaft crack is possibly present. Rates of these changes vary within broad limits, from the order of months to days or even hours - in the latter case, a catastrophic failure is most probably imminent. Such evolution of vibration patterns should serve as an alert. Presence of a crack may be confirmed by monitoring absolute and relative vibration during transients - typically after a turbine trip. Time histories of the $1 \times f_0$ and $2 \times f_0$ components, obtained in such manner, may be compared with reference data recorded after unit commissioning or a major overhaul. Significant reduction of critical speeds and increase of vibration amplitudes on passing through them are indicative of this malfunction, as well as is high overall relative vibration amplitude; the latter will sometimes render the startup impossible to complete, as the unit shall be tripped automatically below nominal rotational speed.

3.1.5 Bearing problems

A problem specific to shaft journal bearings is oil film instability that induces so-called self-excited vibrations. This issue has attracted considerable attention and detailed theoretical models have been developed (Bently and Hatch, 2002; Kiciński, 2006). It can be shown that threshold rotational speed for the onset of instability Ω_{th} is given by

$$\Omega_{th} = \frac{1}{\lambda}\sqrt{\frac{K}{M}}, \tag{13}$$

where K and M denote stiffness and mass, respectively, and λ is the oil circumferential velocity ratio. It is therefore obvious that a suitable stability margin should be provided by proper design and operation, which influence all three quantities that determine Ω_{th}.

Bearing instability is nicely demonstrated with laboratory-scale model rotor systems. For large steam turbines in power industry, operated at a fixed rotational speed, this is a rare occurrence. Most frequently it results from bearing vertical displacement, due to thermal deformation and/or foundation distortion. Downward displacement reduces bearing load and causes K to decrease, so that Ω_{th} may become lower than the nominal rotational speed. In such circumstances, instability is unavoidable. Most typical symptom of this malfunction is the increase of sub-harmonic spectral components, often of the 'hump' shape centered slightly below $0.5 \times f_0$. Shaft orbits typically exhibit loops. Strong instability results in high relative vibration that leads to bearing damage. Proper adjustment of bearing positions is the primary action to be taken; sometimes reduction of the bearing size (length), in order to increase specific load, is necessary for a permanent remedy (Orłowski and Gałka, 1995).

Due to strong non-linearity, journal bearings generate higher harmonic components which may be very sensitive to bearing condition, clearances and oil pressure. An example is shown in Fig.4, in the form of a time history of the $3 \times f_0$ absolute horizontal vibration component. Initially very low, it increased dramatically following a minor bearing damage and remained at a high level, exhibiting considerable variations that suggest a resonance nature of the phenomenon. Permanent improvement was achieved only after a major overhaul. It has to be noted that such behavior is to a large extent influenced by design features; therefore care has to be taken when generalizing the results over other turbine types. In any case, sensitivity of spectral components to oil pressure is decisive.

Typical malfunctions which have their representations in the low frequency range and their corresponding symptoms have been listed in Table 1, which summarizes this subsection.

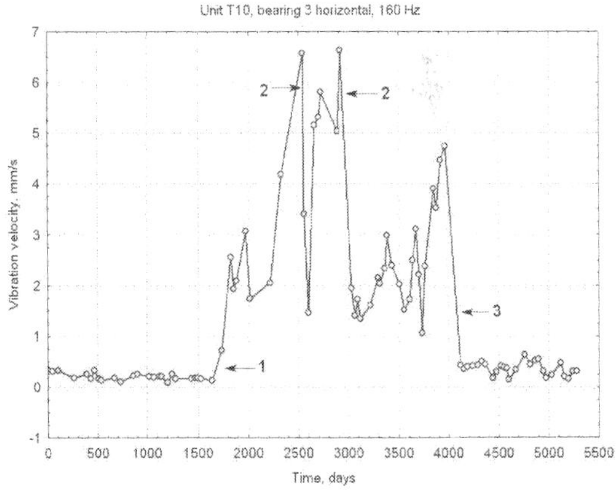

Fig. 4. Time history of the $3 \times f_0$ component: 200 MW unit, rear intermediate-pressure turbine bearing, horizontal direction. Arrows: 1, bearing damage; 2, bearing position and clearances adjustments; 3, major overhaul

Malfunction	Typical symptoms
Unbalance	$1 \times f_0$ component in vertical and horizontal directions, constant amplitude and phase, decreasing at low rotational speed
Misalignment	$2 \times f_0$ component in vertical and horizontal directions, 'banana-shaped' or flattened shaft orbits, high harmonic components in axial direction
Permanent rotor bow	$1 \times f_0$ component in vertical and horizontal direction (also at low rotational speed), strong correlation between $1 \times f_0$ components in vertical and axial directions,
Rotor crack	Continuous changes of $1 \times f_0$ and $2 \times f_0$ components amplitudes and phases during steady-state operation, reduction of critical speeds and increase of vibration amplitudes on passing through them
Bearing problems	Increase of sub-harmonic components (typically slightly below $0.5 \times f_0$), relative vibration increase, shaft orbits with loops, high and unstable amplitudes of higher harmonic components, sensitive to bearing oil pressure

Table 1. Typical steam turbine malfunctions and their representation in low-frequency vibration-based symptoms.

3.2 Blade (high) frequency range

So-called blade spectral components, with frequencies given by Eqs.(3) to (6), are usually low in amplitude. Typically they fall into the frequency range from a few hundred hertz to about 10÷20 kilohertz. In vibration displacement spectra they are undistinguishable, so velocity or acceleration spectra have to be employed. Constant-percentage bandwidth (CPB) analysis is the most convenient tool; 23% CPB yields satisfactory results.

Technical condition of the individual fluid-flow system components, i.e. rotor stages and bladed diaphragms, influences the ζ_k coefficients in Eqs.(1) and (2) and hence the vibration amplitudes in relevant frequency bands. Blade components are, however, highly sensitive to control and interference. Influence of control may be seen as a competition between two mechanisms. First, with nozzle control typical for large steam turbines, there is an asymmetry of steam pressure distribution over the turbine cross-section that depends on the control valve opening. This asymmetry affects forces resulting from the steam flow thrust, again via the ζ_k coefficients. As turbine load increases and consecutive valves are opening, pressure distribution becomes more uniform. Second, with increasing turbine load and steam mass flow, the ζ_0 coefficient also increases. As already mentioned, it may be expected that the former mechanism shall influence vibration patterns at points close to the control stage, as the asymmetry decreases as we move downstream the steam expansion path. The latter should be noticeable for last low-pressure turbine stages, with long blades and large cross-section area. In practice, influence of steam flow asymmetry on blade components is quite strong in points located at the high-pressure turbine; operation at extremely low loads[2] may cause them to increase even by a few times. Steam mass flow influence is usually much weaker.

[2] Load minimum is usually imposed by the steam generator (boiler or nuclear reactor) stable operation considerations.

Fig.5 shows relative standard deviation ($\sigma/\hat{S}$, where $\hat{S}$ denotes mean value) plotted against mid-frequency of 23% CPB spectrum bands, determined for a 120 MW steam turbine. It is immediately seen that for the harmonic range $\sigma/\hat{S}$ is below 0.1, while in the blade range it may be as high as about 0.6 to 0.8. Similar analysis for other turbine types has yielded quantitatively comparable results (Gałka, 2011b). In such circumstances, a time history of a blade spectral component has to be considered a monotonic curve with large fluctuations imposed; an example is shown in Fig.6. Therefore the very occurrence of a high amplitude cannot be unanimously considered as indicative of a fluid-flow system failure. From the point of view of measurement data processing, values heavily influenced by control and/or interference have to be treated as outliers.

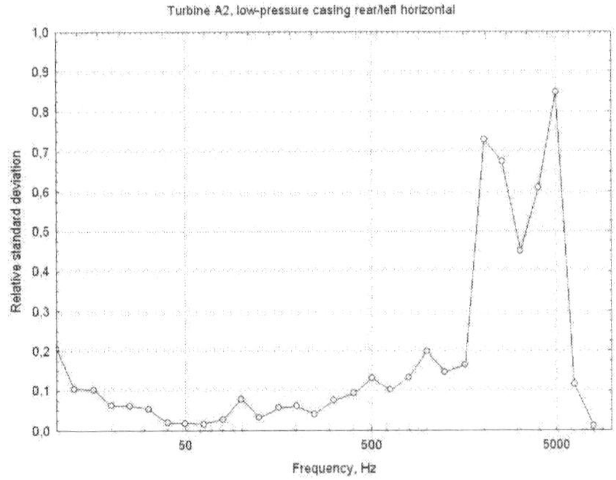

Fig. 5. Relative standard deviation vs. frequency: results for a 120 MW unit, low-pressure turbine casing rear/left side, horizontal direction; data obtained from 90 consecutive measurements (after Gałka, 2011b).

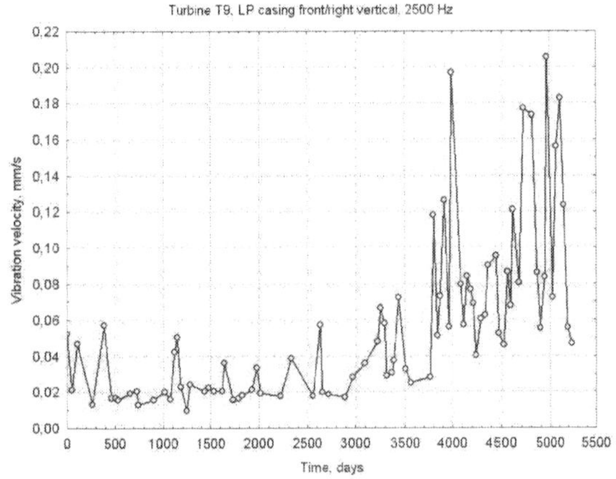

Fig. 6. Time history of the 2500 Hz component: 200 MW unit, low-pressure turbine casing front/right side, vertical direction

Salkind, N.J. (Ed.) (2007). *Encyclopedia of Measurements and Statistics,* SAGE Publications, ISBN 978-1-412-91611-0, Thousand Oaks, USA

Traupel, W. (2000). *Thermische Turbomaschinen. Zweiter Band: Geanderte Betriebsbedingungen, Regelung, mechanische Probleme, Temperaturprobleme,* ISBN 978-3-540-67376-7, Springer, Berlin

15

On the Mechanical Compliance of Technical Systems

Lena Zentner and Valter Böhm
Ilmenau University of Technology,
Germany

1. Introduction

In the safe physical human-machine interaction the compliance of technical systems is an elementary requirement (Zinn et al., 2004; Bicchi & Tonietti, 2004). The physical compliance of technical systems can be provided either by control functions implementation and/or intrinsic by structural configuration and material properties optimization (Beder & Suzumori, 1996; Wang et al., 1998). The latter is advantageous because of higher reliability as well as general simplicity of the design and production technologies (Beder & Suzumori, 1996; Ham et al., 2009). In the following we focus on mechanical systems with intrinsic mechanical compliance.

In general the deformability of structures is primarly characterised by their stiffness. Stiffness is the measure of the ability of a structure to resist deformation due to the action of external forces (IFToMM Terminology, 2010). Compliant mechanisms are mechanisms, whose functionality is based on its deformability. The mobility of these mechanisms results from their mostly elastic or plastic deformability (The definition is based on (Bögelsack, 1995; Howell, 2001; Christen & Pfefferkorn, 1998)). For the description of these mechanisms it is advisable to use the compliance instead of the stiffness. The compliance is the reciprocal of stiffness and is defined as the measure of the ability of a structure to exhibit a deformation due to the action of external forces (IFToMM Terminology, 2010). The goal of each engineer is by the design of mechanisms the setting of compliance depending upon the purpose of its application. It should be considered, that the compliance is dependent on a variety of parameters. The optimal design of these mechanisms can be realized only with precise knowledge of the influence parameters and possible types of compliance.

2. Influence factors of compliance

First, the factors will be considered that determine the compliance of mechanisms generally, without a specific application.

The compliance of a mechanism is determined with respect to a displacement of a specific selected reference point or area of the mechanism as a result of an external force. That approach is necessary because the deformation of a mechanism is usually associated with varying displacements for differing areas of a mechanism. Accordingly, the compliance of a mechanism depends on the location of the reference point. At the same mechanism for the same load and boundary conditions, the evaluation of compliance with respect to different points leads to different results. For the reference usually the force application point or area is chosen. Depending on the application, the amount of the displacement vector of the reference point or its components are used when specifying the compliance.

The compliance is not a pure structure-related property, defined only by the initial geometric configuration and initial material properties. Geometric boundary conditions (location, type), the loading situation (location, type, magnitude, direction and loading history) and environmental conditions (e.g. thermal, chemical) must be also considered in order to formulate the compliance of a mechanism. On the material side the compliance is influenced by actual and previous environmental conditions and by the loading history (elastic or plastic behaviour). The geometric configuration for a given load is dependent on the material properties and geometric boundary conditions. Therefore the compliance of the mechanism depends on its actual geometric configuration and actual material properties and is valid only for the considered reference point by the given actual load and boundary conditions (Figure 1).

In practical applications, the boundary conditions, load levels and the reference point are given. In this case, the adjustment of the compliance can be achieved by appropriate design and material selection. The effort for the design depends on the variety of possible future applications of the mechanism.

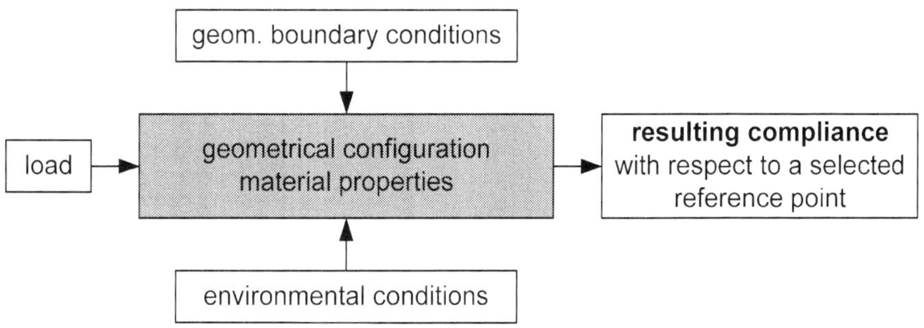

Fig. 1. Influence factors on the mechanical compliance of mechanisms

2.1 Variability of compliance

Generally, the compliance can be either constant or variable. The constant compliance is impossible in the nature. However, we can use the theoretical models with constant compliance, for example, in the linear theory of small bending of beams. In this case the force is linear proportional to the displacement. Table 1 shows deferent compliance for a compliant quarter-circle shaped beam with radius R=20 mm. For this problem we use the Castigliano's theorem for describing the displacement of the end of beam with the geometric linear theory:

$$u = \sqrt{u_1^{\,2} + u_2^{\,2}} = \frac{FR^3}{2EI_3}\sqrt{\frac{\pi^2}{4} + 1} \tag{1}$$

The compliance applied of the end of beam can be given by:

$$\frac{u}{F}=\frac{R^3\sqrt{\pi^2+4}}{4EI_3} \tag{2}$$

The compliance would be demonstrated by means of one point for constant bending stiffness in one dimension domain. Consequently, the constant compliance is the compliance of the zero degree.

The mathematical model for the large displacements is based on the theory of curved beams. The following equations are nonlinear equations of equilibrium and constitutive equations of a curved beam (Zentner, 2003).

$$\begin{aligned} &Q_1'-\kappa Q_2=0\\ &Q_2'+\kappa Q_1=0\\ &EI_3\kappa'+Q_2=0\\ &\theta'=\kappa-\frac{1}{R}\\ &u_1'=\cos\theta-1\\ &u_2'=\sin\theta \end{aligned} \tag{3}$$

Where κ - curvature of the loaded beam, Q_i - internal forces, EI_3 - bending stiffness, u_i - displacements on the directions of x_1 and x_2, θ - angle between the tangent and the axis x_1. All these parameters depend on the beam coordinate s ($0 \leq s \leq L$). This system of nonlinear equations was solved with the program MATHEMATICA with boundary conditions:

$$\begin{aligned} &Q_1(L)=0\\ &Q_2(L)=F\\ &\kappa(L)=0\\ &\theta(0)=0\\ &u_1(0)=0\\ &u_2(0)=0 \end{aligned} \tag{4}$$

As for the large displacements, we have different compliances depending on the particular position. To each position, a different force dF corresponds, which displaces a beam point on du. This situation is presented in Table 1 as a curve in 2D domain. The different compliance is characterized by $\partial u/\partial F$ dependent on u. By means of changing the prestressing of a compliant structure its compliance can be changed. Another example of this is a structure in the ring shape as a prestressed spring. While changing the clamping with the help of the parameter h, the compliance of the spring can be purposefully set onto the point P. Such compliance can be called the compliance of the first degree.

In addition to this, if we also take the temperature into consideration concerning a mechanism made of a temperature sensitive material, the compliance will depend on the two parameters, namely the displacement position and the temperature. In such a case we have compliance of the second degree. The surface F=F(u, T) would reflect the compliance depending on the two parameters, as a value for compliance ∂u/∂F for a definite temperature T and a particular displacement u.

The list of such parameters, which influence the compliance, can continue to be developed. If the compliance depends on the N parameter, we deal with the case of the compliance of the N degree.

Degree of compliance	Modelling	Figure	Compliance
N=0	Linear theory	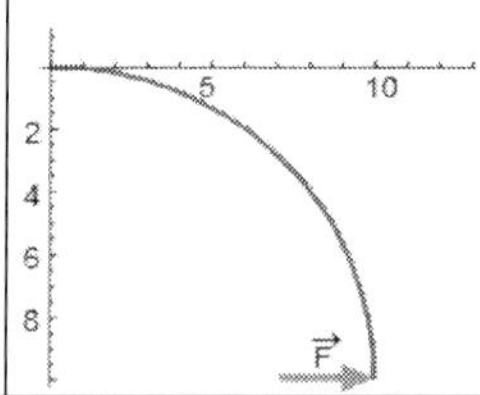	$\frac{u}{F}=\frac{R^3\sqrt{\pi^2+4}}{4EI_3}$
N=1	Non-linear theory (large deformations)	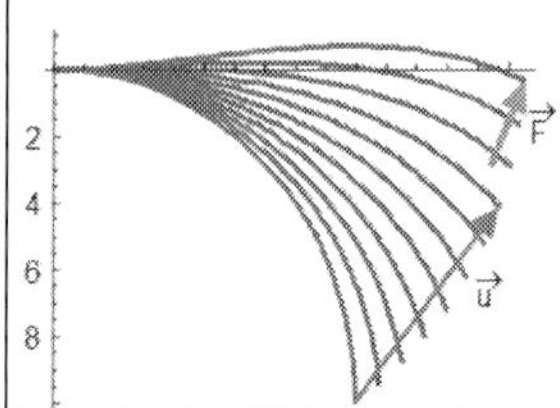	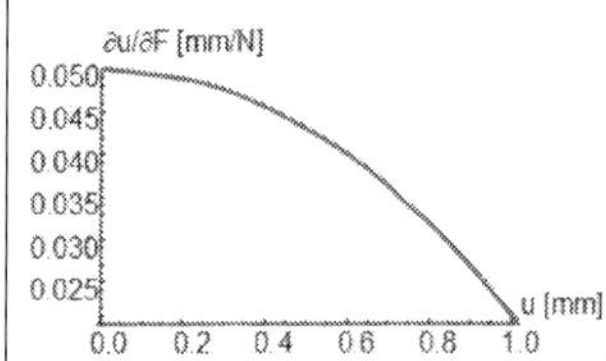
N=2	Non-linear theory and dependence of the temperature, e.g. E(T)	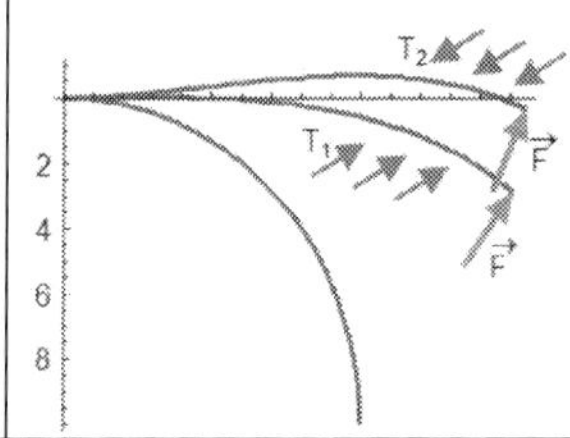	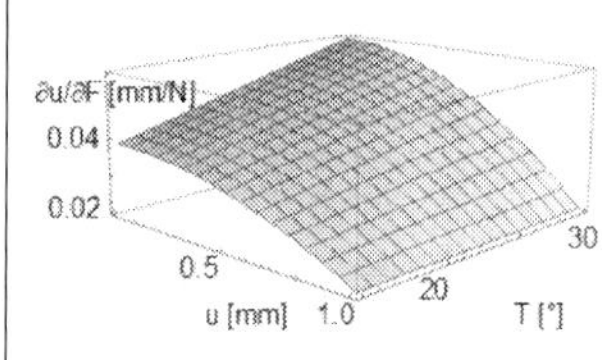

Table 1. Compliance of three degrees, from zero till two, for a compliant quarter-circle shaped beam (EI_3=100 Nmm^2, F=1N)

2.2 Distribution of compliance

The mostly deformable parts of a compliant mechanism are called as compliant joints. Compliant joints can be classified by their distribution of their compliance. The joints with concentrated, local compliance have a small deformable area with the reference to the dimension of a mechanism. In contrast, compliance joints with distributed compliance include a large area of the deformable part. The decision whether a deformable area is small or large, depends on the purpose of the modelling. For example, the installation of a substitute rigid body model for a compliant mechanism, a great role is played by extension of the joint.

In case of a joint with a local compliance, the rigid body joint is introduced in the most cases into the middle of the compliant part. With a compliant joint possessing the distributed compliance, it is important that the position of the rigid body joint is determined for a substitute model. It can be admitted that for the compliant joints the following conditions are available: if the extension of the joint is 10 or more times smaller than the biggest dimension of the whole mechanism, it is classified as a joint with a distributed compliance.

Mechanisms with concentrated compliance behave like classic rigid link mechanisms, where kinematic joints are replaced with flexible hinges, and in consequence methods conceived to design rigid body mechanisms can be modified and applied successfully in this case (Albanesi et al., 2010). Mechanisms with distributed compliance are treated as a continuum flexible mechanism, and Continuum Mechanics design methods are used instead of rigid body kinematics (Albanesi et al., 2010). An overview of calculation methods of compliant mechanisms is indicated in (Albanesi et al., 2010; Shuib et al., 2007).

3. Classification of compliant mechanisms concerning the deformation

In case the deformation-behaviour is chosen as a criterion for classifying compliant mechanisms, two subgroups - dynamic and static deformation - can be distinguished (Zentner & Böhm, 2009). Furthermore the static deformation behaviour of compliant mechanisms having a fixed compliance are considered, whereat influences caused by inertia are neglected. The static deformation behaviour is divided into stable and instable behaviour (Figure 2). Stable deformation behaviour is characterised by a surjective mapping of a particular load F on the deformation u. Thereby one can differentiate between a monotonic behaviour and the behaviour with a singular smooth reversion. In case of instable behaviour of compliant mechanisms snap-through (deformation-behaviour with jump-discontinuities) and bifurcation (local bifurcation of the behaviour) are possible.

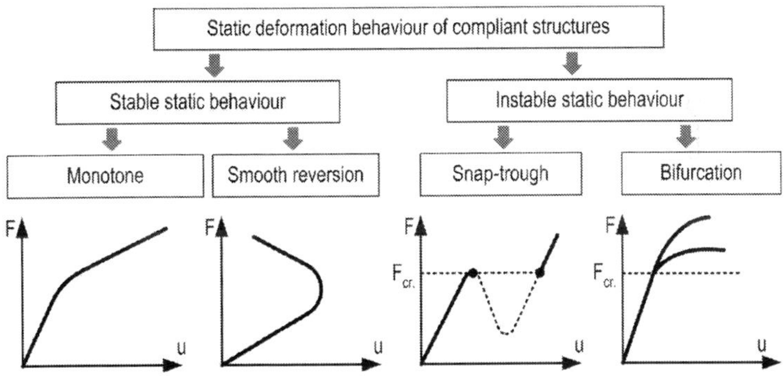

Fig. 2. Classification of the static deformation of compliant mechanisms.

3.1 Stable deformation-behaviour of compliant mechanisms: monotonic deformation

Figure 3 shows an example of the monotonic deformation behaviour of a pneumatically driven compliant mechanism. By increasing the load (here: internal pressure) the characteristic deformation parameters such as the angle between the longitudinal-axis of the rigid structural parts also increases. This mechanism is used as a finger of a gripper.

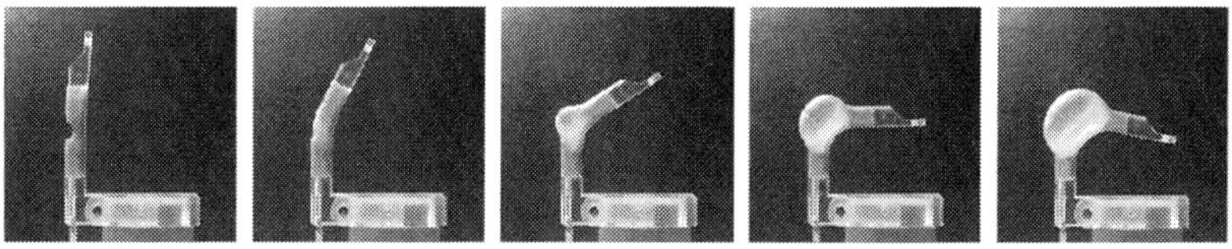

Fig. 3. Monotonic deformation behaviour of a pneumatically driven compliant mechanism made of silicone rubber

Another example concerns a compliant fluid driven structure, which is applied as a medical probe. The cross-section diameter of probe changes from at the fixed end (3 mm) to the free end (1 mm) linearly. In the probe model there is a hollow with constant diameter of 0.2 mm. An unstretchable thin fibre is embedded in the wall with constant distance h from the symmetry axes of probe-beam. Under inner pressure p in the hollow, with the cross-section of A, the probe structure will bend towards the embedded fibre. Linear material law is supposed. The equation for displacement of probe is calculated analytically in order to examine the possibilities to obtain the required bending.

$$\begin{aligned} &Q_1' - \kappa Q_2 = 0 \\ &Q_2' + \kappa(Q_1 - pA) = 0 \\ &EI_3\kappa' + Q_2 = 0 \\ &\theta' = \kappa \\ &u_1' = \cos\theta - 1 \\ &u_2' = \sin\theta \end{aligned} \quad (5)$$

Corresponding boundary conditions are:

$$\begin{aligned} &Q_1(L) = pA(L) \\ &Q_2(L) = 0 \\ &\kappa(L) = \frac{pA(L)h}{EI_3} \\ &\theta(0) = 0 \\ &u_1(0) = 0 \\ &u_2(0) = 0 \end{aligned} \quad (6)$$

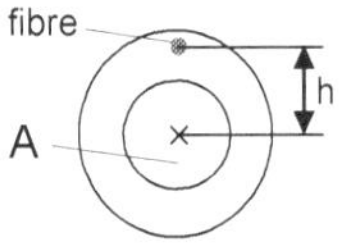

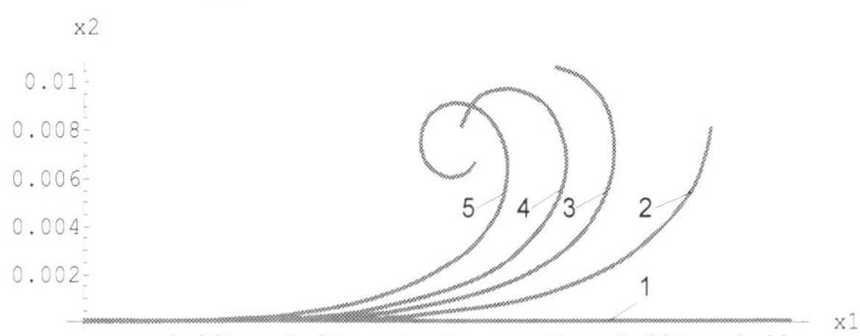

Fig. 4. Displacement behaviour of a compliant fluid driven structure by increasing internal pressure (1-5)

The characteristic parameter for such structure is the angle between the tangent of beam-end and the x_1-axe. This angle is strictly monotonic increasing, when the pressure rises.

3.2 Stable deformation-behaviour of compliant mechanisms: Behaviour with direction reversion

The deformation behaviour with direction reversion of a pneumatically driven compliant mechanism is shown in Figure 5. The geometry is optimised, to achieve a reversion by loading the internal pressure. The horizontal displacement of the working element was chosen as one characteristic parameter, whereat the vertical displacement and the distance of the working element to the clamping are also possible characteristic parameters which can be taken into account. In the present case the increasing load initially yields to an extension of the horizontal displacement of the working element and finally causes a smaller displacement.

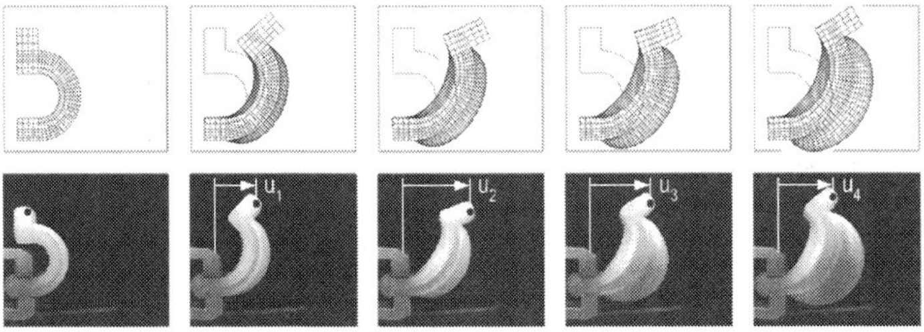

Fig. 5. Static deformation behaviour with direction reversion of a compliant mechanism: at the top - FEM-calculations; at the bottom - compliant mechanism made of silicone rubber with characteristic parameters u_i, i=1,..,3 (u_2>u_3>u_1)

Figure 6 a graphically presents the change of the characteristic parameters mentioned above. Therein the displacement of the working element u_3 equals the reversal point of the structure.

The considered deformation effect, referenced here as reversal effect, consists in reversion of movement, which the working element of the mechanism performs during unidirectional change of the internal pressure (Zentner et al., 2009). Thus movements in two opposite directions depend on the magnitude of the uniformly increasing pressure, and two opposite working directions follow one another. One of the important consequences of this effect is that the movement range of the mechanisms in the first working direction is not limited by the maximum pressure but by the material properties and geometric forming, whereby the sensors utilization can be minimized in grasping applications for example. The main characteristic criteria of the mechanisms considered are: (1) fulfilling movements in two opposite directions, where both movements are caused by unidirectional pressure-activating of the actuator; (2) the movement range in the first working direction being, independently from the size of the pressure load, bounded; (3) a given position of the working element of the mechanism is to be reached at two different pressure loads and therefore with different compliances. The reversal effect can be generally realized in two ways: (1) through cascading of several conventional structures; or (2) with a non-conventional actuator. In both cases such an effect can be achieved by both geometrical properties (geometrically asymmetrical actuators), and material properties (actuators with variation of the material properties). A combination of materials of different elasticity and/or anisotropic materials can fulfil this characteristic, too. The numerical calculations approved, that the mechanism behaviour is influenceable geometrically and materially. Hence it can be adjusted to specific tasks.

Compared to other mechanisms, more complex motion trajectories can be easily provided with unidirectional pressure change with the help of these mechanisms.

One application for a mechanism having the property of a direction reversion is using it as gripping fingers. Through model based optimisation a novel dependency of the load on the displacement could be achieved (Figure 6 a, broken line). Therein the characteristic displacement will not increase at a defined value of load irrespective of further increase of load. This property puts aside the sensory effort to monitor the gripping force. This force is already defined by the structure's mechanical properties. Such a structure is shown in Figure 6 b.

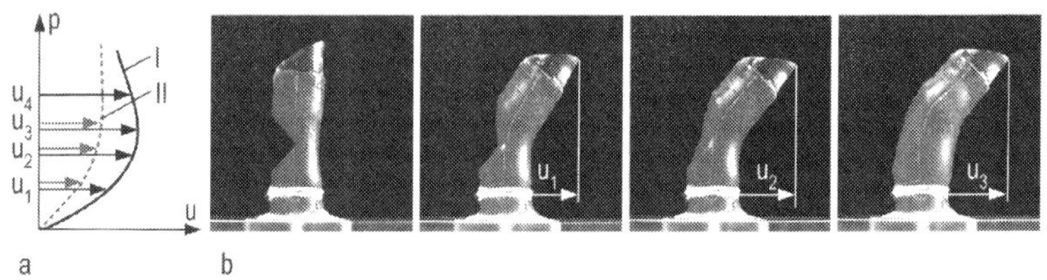

Fig. 6. a: I - Dependency of displacement u on internal pressure p of the mechanism introduced in Figure 4, II - p(u) for gripping-fingers with defined gripping-force; b: gripping-fingers with defined gripping-force made of silicone rubber

3.3 Instable deformation behaviour of compliant mechanisms: Snap-through

In contrast to the stable deformation behaviour of compliant mechanisms, the instable case has more than one equilibrium position for a particular load. The instable deformation behaviour shows snap-through or bifurcation.

In case of snap-though a sudden transition from one equilibrium position to another happens. Thereby a given load corresponds to several equilibrium positions. In Figure 7, a rotational structure is shown having a half-toric curve around the spherical curve in the origin state. One characteristic feature of such a mechanism is the potential bistable deformation behaviour, which can be enforced by the specified geometric parameters (shape, wall thickness, etc.).

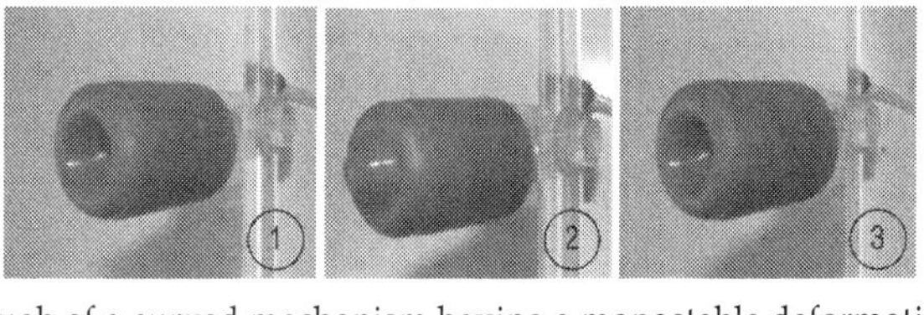

Fig. 7. Snap-through of a curved mechanism having a monostable deformation behaviour

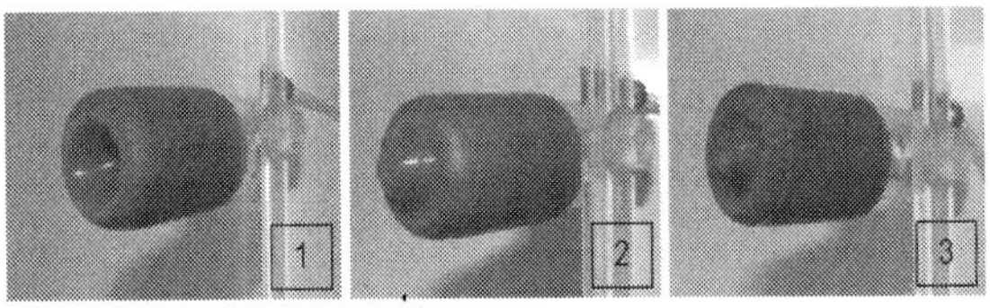

Fig. 8. Snap-through of a curved mechanism having a bistable deformation behaviour

Two different mechanisms with a big and small wall thickness are presented in Figure 7 and 8, respectively. By increasing the load, the angular point (centre point) of the median curvature of both mechanisms moves outwards up to the critical load (Figure 9 c). Herein the value of the critical load is different to the named structures. An arbitrary small rise of load causes a huge displacement, as soon as the critical load is reached. In this process the median curvature penetrates completely (state 2 in Figure 9). Removing the load causes the first mechanism to reverse to the original position (monostable deformation). This is demonstrated by state 3 in Figure 9 a. The second mechanism switches to another equilibrium position (bistable deformation), named in Figure 9 b with state 3. A sketch of both characteristics and the calculated positions by means of FEM is shown in Figure 9 c.

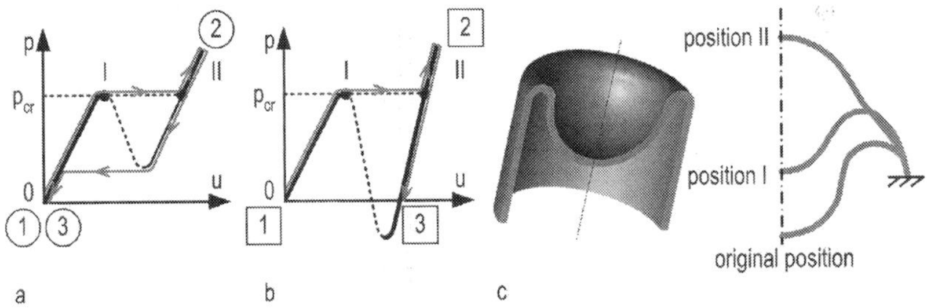

Fig. 9. Sketch of Snap-through behaviour of a curved mechanism: mechanism with monostable deformation (l.), mechanism with bistable deformation (r.)

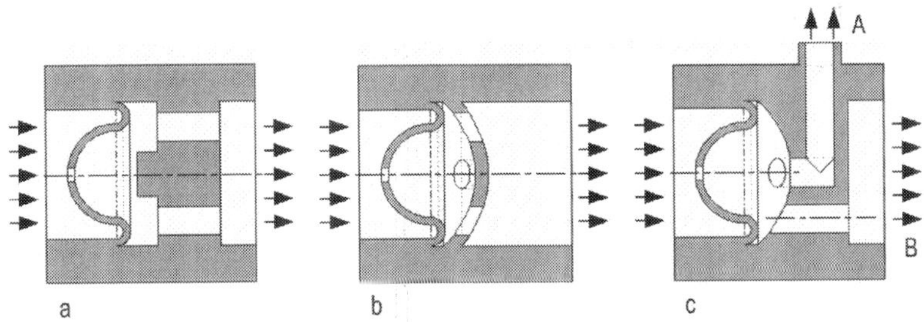

Fig. 10. Applications as mechanical valves demonstrated for double-curved rotational mechanisms: a, b - pipe with one output is disabled for $p=p_{cr}$, c - pipe with two outputs A and B, output B is closed if critical load is reached

Some applications of these mechanisms used as mechanical valves are shown in Figure 10. To generate bistable deformation (Figure 10 a) an opening is inserted in the centre point of the mechanism. Because of the critical dynamical pressure the mechanism is deformed and the flow is interrupted. In this case the current position guarantees the closure of the pipe. Low-pressure on the compliant part of the valve makes the flow possible again. The next Figure 10 b shows that the effect of the critical pressure yields to a deformation of the monostable mechanism, so the pipe is completely closed. If the pressure falls under the critical level, the original position is recaptured and the flow rate is reconstituted. The last example in Figure 10 c illustrates a valve installed in a pipe with two outputs. Output B is disabled, as soon as the critical pressure is reached. Decreasing the pressure enables this output.

3.4 Instable deformation behaviour of compliant mechanisms: Bifurcation

Situations with bifurcation of structures are avoided systematically in engineering. The following theoretical analyses reveal some opportunities to apply this behaviour profitably to a technical system.

The best known examples for the loss of stability under static loads are the Eulerian cases of stability. For loads under the critical level, the equilibrium is determinate, whereas at the critical level of loads, bifurcations in the solutions occur to state equations. The solutions are no longer bijective, one load situation may lead to more than one possible geometric configurations of the system. Such structures are shown in Figure 11. Herein, the load is generated by the attraction force of the filaments e.g. SMA-wires or by the low-pressure in cavities. If the wires are uniformly pulled or the cavities possess the same low-pressure, the classical Euler stability problem (bifurcation) is regarded as replacing the named rotationally drive configuration by an axial acting force.

The following statement explains how the bifurcation effect can be used profitably. The response of a systematically designed system with bifurcation behaviour (deformation or displacement in several directions) on external (e.g. temperature change) or on user-defined conditions leads to one preferred direction. The deformation direction is selected "autonomously" whereas the drive regime for each process always remains the same. The sensory and control effort are minimised enormously. The control of the system is partly adopted by "intelligent" mechanics.

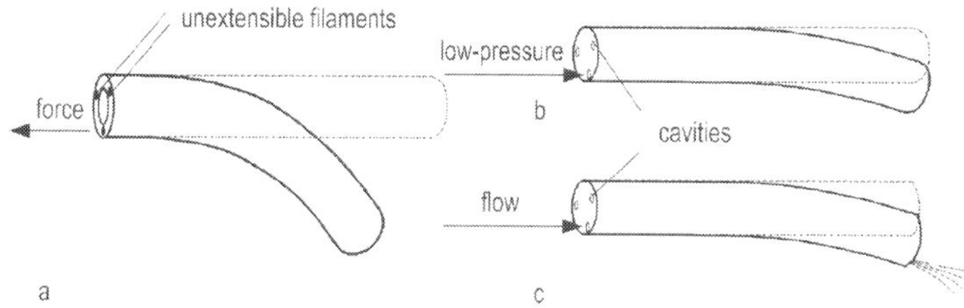

Fig. 11. Structures among the influence of an axial load which is generated by the attraction force of the filaments or Shape-Memory-Alloy-wires (a) and by the low-pressure in cavities (b, c)

Figure 12 exemplifies this phenomenon in the case of a half-cycle shaped bending beam subject to loading by a single force with constant direction (conservative force) but with a

2D-free floating location of the site of application under load. Two possible trajectories of this point and two realisations of the equilibrium are illustrated. Solutions have been determined numerically, a current application is the design of compliant grasping devices.

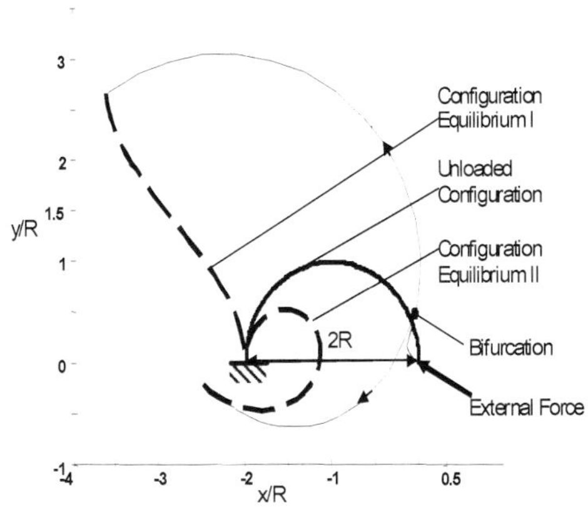

Fig. 12. Equilibrium situations of a half-cycle shaped beam under external load; A force constant in amount and directions traces the free end of the beam

4. Conclusion

The introduced classification which considers the deformation of compliant mechanisms is supposed to forward their development and to facilitate their implementation in rigid body systems or the functional expanded substitution of individual parts of the rigid body. The meaningful application of compliant mechanisms especially of such structures with instable static behaviour offers a great development potential. The role of the sensor system can be partly or completely adopted by "intelligent" mechanics. With the application of compliant mechanisms and structural elements, which show an instable static behaviour and therefore segue from one state to another depending on external conditions, elementary characteristics of the system can change (Risto et al., 2008; Linß et al., 2008; Risto et al., 2010; Griebel et al., 2010). Hence such systems will autonomously and directly adapt to the working conditions.

In relation with functional dominating compliant characteristics many application-oriented tasks, for example gripping-fingers with particular characteristics, medical structural elements and systems are conceivable.

5. References

Albanesi, A. E., Fachinotti, V. D., Pucheta, M. A. (2010). A review on design methods for compliant mechanisms. *Mecánica Computacional*, Vol.29, E. Dvorkin, M. Goldschmit, M. Storti (Eds.), Buenos Aires, (2010), pp. 59-72.

Beder, S., Suzumori, K. (1996) Elastic materials producing compliant robots. *Robotics and Autonomous Systems*, Vol.18, No.1, (July 1996), pp. 135-140, ISSN 0921-8890

Bicchi, A., Tonietti A. (2004) Fast and "soft-arm" tactics. *IEEE Robotics and Automation Magazine*, Vol. 11, No. 2, (June 2004), pp. 22-33, ISSN 1070-9932

Bögelsack, G. (1995). Nachgiebige Mechanismen in minaturisierten Bewegungssystemen, *Proceedings of the 9th World Congress on Theory of Mach. and Mech.*, Milano, August-September 1995

Christen, G., Pfefferkorn, H. (1998). Nachgiebige Mechanismen, *VDI Berichte Nr. 1423*, 1998

Griebel, S., Fiedler, P., Streng, A., Haueisen, J., Zentner L. (2010). Medical sensor placement with a screw motion. *Proceedings of Actuator 10 / International Conference on New Actuators*, ISBN 978-3-933339-12-6, Bremen, June 2010

Ham, R., Sugar, T., Vanderborght, B., Hollander, K., Lefeber, D. (2009) Compliant actuator designs. IEEE Robotics and Automation Magazine, Vol.16, No.3, (September 2009), pp. 81-94, ISSN 1070-9932

Howell, L. L. (2001). *Compliant Mechanisms*, John Wiley & Sons, ISBN 978-0471384786, New York

IFToMM Terminology, Version 2.3 - April. 2010, http://www.iftomm.3me.tudelft.nl/2057/frames.html

Linß, S., Zentner, L., Schilling, C., Voges, D., Griebel, S. (2008). Biological inspired development of suction cups. *Proceedings of 53. Internationales Wissenschaftliches Kolloquium, IWK. Technische Universität Ilmenau*, ISBN 978-3-938843-37-6, Ilmenau, September 2008

Risto, U., Zentner, L., Uhlig, R. (2008). Elastic structures with snap-through characteristic for closing devices. *Proceedings of 53. Internationales Wissenschaftliches Kolloquium, IWK. Technische Universität Ilmenau*, ISBN 978-3-938843-37-6, Ilmenau, September 2008

Risto, U., Uhlig, R., Zentner, L. (2010). Thermal controlled expansion actuator for valve applications. *Proceedings of Actuator 10 / International Conference on New Actuators*, ISBN 978-3-933339-12-6, Bremen, June 2010

Shuib, S., Ridzwan M. I. Z., Kadarman, H. (2007). Methodology of Compliant Mechanisms and its Current Developments in Applications: A Review. *American Journal of Applied Sciences*, Vol.4, No.3, (March 2007), pp. 160-167, ISSN 1546-9239

Wang, W., Loh, R. N. K., Gu, E. Y. (1998) Passive compliance versus active compliance in robot based automated assembly systems. *Industrial Robot*, Vol.25, No.1, (1998), pp. 48-57, ISSN: 0143-991X

Zentner, L. (2003). Untersuchung und Entwicklung nachgiebiger Strukturen basierend auf innendruckbelasteten Röhren mit stoffschlüssigen Gelenken, *Ilmenau ISLE Verlag*, 2003, ISBN 3-932633-77-6

Zentner, L., Böhm, V., Minchenya, V. (2009). On the new reversal effect in monolithic compliant bending mechanisms with fluid driven actuators. *Mechanism and Machine Theory*, Vol.44, No.5, (May 2009), pp. 1009–1018, ISSN 0094-114X

Zentner, L., Böhm, V. (2009). On the classification of compliant mechanisms. *Proceedings of EUCOMES 08 The Second European Conference on Mechanism Science*, ISBN 978-1-4020-8914-5, Cassino, September 2009

Zinn, M. Khatib, O., Roth, B., Salisbury, J. K. (2004). A new actuation approach for human friendly robot design. *The International Journal of Robotics Research*, Vol. 23, No.4-5, (2004), pp. 379 - 398, ISSN 0278-3649

Part 3

Thermo-Fluid Systems

16

Waste Heat Recycling for Fuel Reforming

Rong-Fang Horng[1] and Ming-Pin Lai[2]
[1]*Department of Mechanical Engineering, Kun Shan University, Tainan City,*
[2]*Department of Aeronautics and Astronautics, National Cheng Kung University, Tainan City, Taiwan*

1. Introduction

At the current rate of consumption, it is estimated that the crude oil reserves of the entire world will be depleted in less than 40 years. Due to an unstable international situation, the price of crude oil keeps rising, compelling the price of other fossil energies to soar. Because of its high dependency on imported energy, Taiwan is deeply affected by the crude oil price. Additionally, because of its low cost and high carbon number, fossil fuel has been the major fuel used in Taiwan. The enormous number of vehicles and motorcycles in Taiwan emit excess amounts of carbon dioxide, sulfur, and nitrogen; consequently, the level of greenhouse gas emissions in Taiwan far exceeds that of other countries. Because coal, natural gas, and petroleum will still be the major sources of energy in the short to medium term future; and since increasing energy efficiency is presently a globally acknowledged strategy; the use of new energy technologies for the transformation of traditional fuels into clean fuels - such as clean coal, fuel cells, and fuel reforming - is being promoted all over the world. Combining all the factors described, it has become necessary to quickly develop new energy and combustion techniques with existing equipment and resources to reduce air pollution produced by combustion.

Plasma-assisted reforming of hydrogen production is a promising energy utilization technology, and an increasing number of research units are conducting related experiments around the world. Plasma-assisted production of hydrogen is different from traditional hydrogen production approaches, and the equipment involved in the plasma-assisted production of hydrogen is small, easy to start, and economical; it can also increase overall thermal efficiency of internal combustion engine by combining with the hydrogen produced from the fuel reforming system.

The method of hydrogen production by plasma or catalytic converters with various hydrocarbon fuels has been studied by various researchers internationally. An arc generated plasma was used to facilitate the reforming reaction by heating. The heated mixture could be distributed to all of the passages in the reaction chamber by arc rotation driven by magnetic field. Water vapor could be directed into the system to lower the temperature of the electrodes via heat recycling to prevent the reaction region from over-heating, therefore extending the life of the system (Bromberg et al., 1997). The optimum settings and the cost of a plasma-catalyst natural gas reformer were analyzed. Preliminary results showed that the specific energy consumption of a small scale plasma reformer can be effectively reduced through efficient thermal management and heat recycling (Bromberg et al., 2000). The characteristics of hydrogen production from isooctane were studied and reported that hydrogen yield at catalyst temperatures as high as 800°C was significantly improved by using a combined plasma/catalytic system (Sobacchi et al., 2002). Some researchers investigated autothermal reforming of methane and propane by 2 wt%-catalysts on alumina support. They proposed the activity level of the catalysts in descending order as: Rh > Pd > Ni > Pt > Co. They demonstrated that high activity was achieved by loading suitable amount of Ni, while under the same loading, activity was higher for the Rh-based catalyst (Ayabe et al., 2003). A plasma catalytic reactor was employed for the combined oxygen-steam reforming of methane. It was found that by combining the dielectric barrier discharge (DBD) and a Ni catalyst the conversion of methane was not improved, but complete conversion of oxygen was achieved. Under a suitable temperature to maintain the Ni catalytically active, the product selectivities changed significantly (Pietruszka & Heintze, 2004). Preparation of intake mixture for reforming performance was investigated. It showed that feeding fuel via an intake swirl significantly facilitated the conversion efficiency of methane, and elevated the concentration of hydrogen. Both higher arc frequencies and longer retention time improving hydrogen concentration were also confirmed (Horng et al., 2007). The plasma converter on hydrogen production via methane reforming and the carbon deposit growth on the electrode surface were explored, and the microstructure of carbon was analyzed with scanning electron microscopy (SEM) and micro-Raman spectroscopy (Horng et al., 2006a). The reforming of aliphatic hydrocarbon such as methane, propane and neopentane, using non-thermal plasma was carried out. It demonstrated that reforming by combining carbon dioxide with methane and propane, yielded a higher H_2/CO ratio than using neopentane. When using steam reforming, methane yielded the highest H_2/CO ratio (Futamura et al., 2006). The literature on the field of hydrogen production assisted by non-thermal plasma reforming was reviewed by a group of researchers. They concluded that most of the existing reforming reactors were still being developed and advanced in research laboratories (Petitpas et al., 2007). The Tokyo Institute of Technology studied the plasma and the glid-arc reforming approaches and found that both of them required high temperature to produce hydrogen. In contrast, the dielectric barrier discharge (DBD) reforming approach could produce hydrogen at lower temperatures (between 400 and 600 °C) by adding water vapor in fuel (Nozaki & Okazaki, 2005).

Concerning the heat regeneration, a high energy level of thermal plasma to accelerate the reforming reaction was studied. It was suggested that the methods of heat insulation, heat regeneration and improved plasma catalysis could reduce the energy loss and improve methane conversion efficiency (Bromberg et al., 1999b). The utilization of phase change material (PCM) for heat-recovery was experimentally studied. Copper balls as the PCM were encapsulated by nickel film with/without an insertion of carbon or ruthenium as an inhibition layer. The results showed that the copper PCM with the thick film of nickel and an inactive layer between nickel and copper was available for producing hydrogen by high-temperature waste heat recovery (Maruoka et al., 2002). A high temperature plasma technique was applied on the reforming of methane or other hydrocarbon fuels to produce hydrogen-rich gas. To reduce energy loss and to increase methane conversion, heat could be insulated and exchanged at the plasma converter. Thermal efficiency of partial oxidation reforming using external heating from the perspective of thermodynamics was studied. It was revealed that higher O/C ratios resulted in low thermal efficiency as a significant amount of fuel was found to be oxidated in the reforming process (Lutz et al., 2004). The reforming of exhaust gas recirculation (EGR) for producing H_2 and CO was studied. By adding the contents of H_2 and CO, the in-cylinder premixed combustion of engine would be enhanced; and then a super lean burn process, low NOx operations could be achieved (Zheng et al., 2004). A numerical simulation on the technique of pure external heating and the method of reusing partial oxidation produced through external heating of hydrogen were used for investigating thermal efficiency. The former was found to be of higher efficiency, which was enhanced with increasing H_2O/Carbon ratio (Lutz et al., 2003). A small scale hydrogen producing system was designed with utilizing heat exchange to heat up the vapourising unit. The reforming thermal efficiency attained was as high as 78% while the methane conversion efficiency was 89%, capable of supplying a 1kW fuel cell (Seo et al., 2006). Dimethyl ether (DME) and methanol-reformed gas (MRG) were produced from methanol by onboard reformers utilising the exhaust heat from the engine. Because the reactions producing DME and MRG are endothermic, a part of exhaust heat energy can be recovered during the fuel reforming process. This research experimentally investigated characteristics of combustion, exhaust emissions, engine efficiency and overall thermal efficiency including the waste heat recovery through the fuel reforming in the HCCI combustion engine system (Shudo, 2006). Tri-reforming of natural gas using flue gas from power plants with the combination of heat transfer or heat exchange including reactor heat up and waste heat utilization was studied. The tri-reforming process could be applied to the production of synthetic gas by reforming of natural gas using gas mixtures (containing CO_2, H_2O, and O_2) as co-feeds. As a result, the mitigation of CO_2 could be achieved by waste heat recovery. However, various strategic considerations, technical approaches, and specific research directions have been presented. More research is necessary towards effective CO_2 conversion into useful substances using renewable sources of energy (Song, 2006). The hydrogen production from low carbon fuels by using the plasma-catalyst hybrid converters with energy conservation was investigated. The energy saving schemes, namely, heat recycling and heat insulation were applied to enhance the hydrogen production. The results showed that the energy saving systems enabled the O_2/C ratio to be decreased to reduce oxidation of hydrogen and carbon monoxide and thereby improving the concentration of hydrogen-rich gas. By heat recycling, the improvement in methane conversion efficiency in a lower fuel feeding rate was achieved; moreover, the hydrogen production increased significantly with the water-gas shifting reaction under the same operation parameters (Horng et al., 2008a, 2009). System analysis was conducted on a combined system for methane steam reforming comprising conventional hydrogen production and waste heat recovery from steelmaking. The results showed that the factors evaluated including the natural gas consumption, enthalpy flow, CO_2 emission, cost, and exergy loss were improved for the waste heat recovery system. This supports the feasibility of hydrogen production from recovered waste heat. Moreover, the proposed system is expected to produce less CO_2 emission due to less fuel consumption (Maruoka et al., 2010).

Combustion principles state that hydrogen has a high heating value and a high flame speed. If hydrogen can maintain the normal combustion efficiency in engine even at an extremely lean mixture, it would reduce fuel consumption and exhaust emissions. A small plasma fuel reformer was used to ionize the mixture of gasoline and air, and the reformate gas was directed into the internal combustion engine. The results clearly indicated that the low levels of NOx, CO and HC were obtained simultaneously (Cohn et al., 1996). The combustion characteristics of a hydrogen-gasoline mixture via mathematical models were explored. It was obtained that the best thermal efficiency was acquired at 8.0 % of hydrogen by mass, and the specific energy consumption was significantly improved at 10 % of hydrogen by mass. It also found that most of the pollutants emitted by engines were produced during the cold-start and warming-up processes. Using hydrogen-rich gas produced by fuel converters as an auxiliary fuel for the engine, or using it with catalysts in the exhaust pipes for rapidly heating the engine would greatly increase the energy using efficiency during the cold start and warming up; it would also reduce the emission of pollutants during the cold start of the engine, and reduce the emission of NOx, CO, and HC in the driving condition of the vehicle (Al-Janabi & Al-Baghdadi, 1999). The performance of hydrogen fueled carbureted and fuel-injected SI engine was compared. The results revealed that the fuel-injected engine gave better output power and had a lower risk of backfiring (Verhelst & Sierens, 2001). Recycled exhaust and added methane and air for reforming with a honeycomb catalyst was studied on an HCCI (Homogenous Charge Compression Ignition) engine. By reusing the heat from the exhaust and adding external heat for partial oxidation reforming, steam reforming and water-gas shifting, the effect of temperature on reformate gas concentration and fuel conversion efficiency was investigated (Peucheret et al., 2005).

A set of small plasma fuel converter was designed for hydrogen production and applied on a 4-stroke motorcycle with a heat storing catalyst in the exhaust pipe to reduce exhaust emission. The results demonstrated that the exhaust emissions produced by the engine were extremely low in the cold-start process (Horng et al., 2006b). The driving performance and exhaust emission characteristics of a 125cc motorcycle equipped with an onboard plasma reformer for producing hydrogen-rich gas was investigated. The produced hydrogen-rich gas was induced into the internal combustion engine as supplementary fuel. It showed that the NO_X emission was improved by 56.8% under a constant speed of 40 km/h. During transient driving condition, the improvement of 16%-41% in NO_X concentration was achieved. The emissions of the motorcycle were also analysed on a chassis dynamometer tracing an ECE-40 driving pattern. The NO_X emission was improved by 34%, as was the HC emission by 4.08%, although the CO emission was increased (Horng et al., 2008b). The performance of an engine by fuel injection with hydrogen was investigated, and the results showed that the knocking and backfiring were absent. The experiments included three operating parameters, namely, ignition timing, injection timing and equivalence ratio, all of which were optimised for the engine performance in terms of good thermal efficiency, good brake mean effective pressure, and low NOx emission (Mohammadi et al., 2007). The effect of hydrogen addition to a natural gas engine on engine performance and emission was studied. The results showed that adding hydrogen in lean burn condition could improve the thermal efficiency of engine with reducing the CO and HC emissions. However, an increase in NOx emission was obtained in the process due to the increase in the combustion temperature. They further showed that by modifying the ignition timing, the NOx emission could be reduced to that similar to the original fuelling system (Ma et al., 2007).

An on-board plasmatron converter was used for hydrocarbon fuel reforming. The study proved that both high carbon and low carbon fuels could efficiently generate hydrogen-rich gas; and once it was directed into the engine, the thermal efficiency and NOx and CO emissions of the engine were evidently improved (Rabinovich et al., 1994). The investigation on the exhaust gas reforming of gasoline was carried out. The reformed fuel was fed to an engine as an additive to gasoline. The study on the effects of reformed fuel addition on engine performance at extreme lean burn conditions was performed. Results showed that low levels of NOx and HC emissions were achieved with the improved thermal efficiency and extended lean burn operation (Jamal et al., 1996). The hydrogen producing systems for powering vehicles were explored. The characteristics of different fuels on reducing exhaust emissions were compared and reported that the shape design of a reaction chamber could improve thermal management, reduce heat lost and optimize incubation time (Bromberg et al., 1999a). Hydrogen-rich gas produced from a plasmatron fuel converter with the exhaust gas recirculation (EGR) approach was directed into an internal combustion engine. The results showed that the NOx emission was evidently reduced with low level of HC emission under cold-start of the engine (Bromberg et al., 2001).

In summary, a fuel reformer for hydrogen production is to be a promising way in the future; and it would be one of the directions for application in internal combustion engines and fuel cells. This study attempts to develop a set of plasma fuel converter with waste heat recycling. The reformation of methane for hydrogen production is investigated, and the results are anticipated to serve as a reference for future research on the reformation of other fuels, including natural gas, or biogas. The application of this system is expected to improve the thermal efficiency and the exhaust emissions of engines. Additionally, the applications of fuel converters on fuel cells would be becoming popular. The stationary fuel converters could be applied for producing hydrogen on solid oxide fuel cells (SOFC). As for the portable fuel cells, if the purity of reformate gas can be achieved, the portable hydrogen generating device could be carried onboard vehicles directly to power vehicles with proton exchange membrane fuel cells (PEMFC).

2. Experimental equipment and methodology

2.1 Experimental equipment

This study focuses on the energy conservative plasma fuel converters with waste heat recycling; a photo and a schematic diagram of the experimental equipment of the system are shown in Fig. 1 and Fig. 2, respectively. The reaction chamber was comprised of three sub-systems: the fuel and gas supply system, the arc generation system, and the reformate gas sampling and analyzing system. The fuel and gas supply system was comprised a fuel nozzle and a flow meter. For the arc generation system, a car ignition system was used with a signal generator and an induction coil to generate arc. Converting car ignition system into a small high voltage DC power supply is practical, as it enabled the system to be directly equipped on-board, and the use of the existing equipment reduced the required space and cost of power supply. The anode of the sparking electrode was constructed by removing the earth electrode of spark plug, while the cathode was the reaction chamber itself. The specifications of the arc generating system are shown in Table 1 (a).

(a) Plasma power supply unit

	Primary side	Secondary side
Voltage (V)	12	4000
Current (A)	3.0	0.004
Power (W)	36	16
Arc frequency (Hz)	200	Variable

(b) Catalytic converter

D × L (mm^2)	Φ46.2*50.0
Pt/Rh ratio	5/1
Catalyst loading (g/ft^3)	50
Cell density ($Cell/in^2$)	100

Table 1. The specifications for the plasma converter and the catalyst

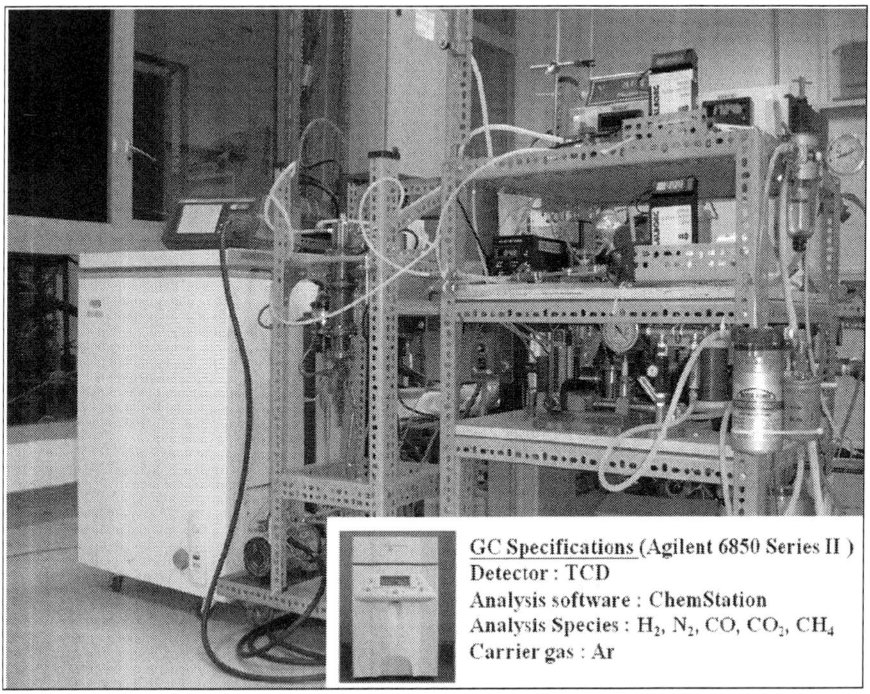

Fig. 1. Photo of the experimental equipment

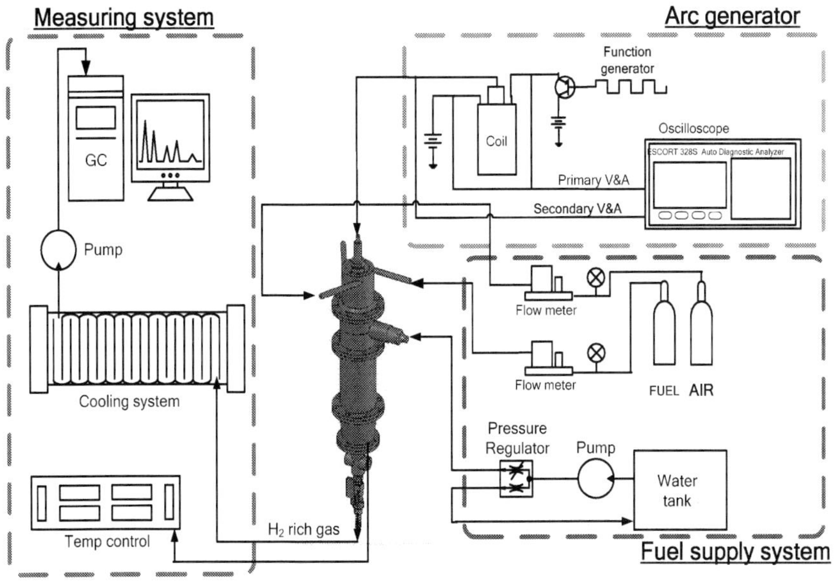

Fig. 2. A schematic diagram of the experimental set-up

The reformate gas sampling system of this experiment was designed to collect products from reforming reactions. A portion of the products was distributed to vehicle exhaust emission analyzer (Horiba MEXA-554JA) via pump to monitor the concentrations of the gas product. Another portion of the product was collected in the gas bags and directed into gas chromatography (Agilent 6850GC) for analysis. The reaction chamber was made of stainless steel, and could be further classified into the plasma reaction zone, the middle chamber, the catalyst section, the gas collecting chamber, and the fuel nozzle, as shown in Fig. 3. The internal surface of the reaction chamber was the ground electrode and the arc could be stretched freely with the intake flow. That is, the arcing system and the ground electrode were designed so as to enhance the uniform distribution of the discharge energy in the plasma reaction region. The main function of the catalyst section was to enhance the reforming reactions. The gas collection chamber was designed to thoroughly mix the reformate gas for sampling.

The commercialized catalysts were used for reforming, and the specifications of which are shown in Table 1 (b). This commercialized catalyst was chosen for its low cost, availability, small pressure drops, and great mechanical strength; in addition, the catalyst has a metal substrate, and therefore has an advantage of rapid cold-start.

The energy conservation measures of this study could be classified into waste heat recycling and heat insulation. For the heat recycling method, a stainless steel cover was added to form an air jacket on the catalyst outer wall, to direct the hot reformate gas back into the air jacket for heat recovery. For the heat insulation method, the outer surface of the reaction chamber was covered with glass fiber. The reaction temperature of the catalysts at the front, middle, and rear sections were measured via three K-type thermocouples. The average temperatures

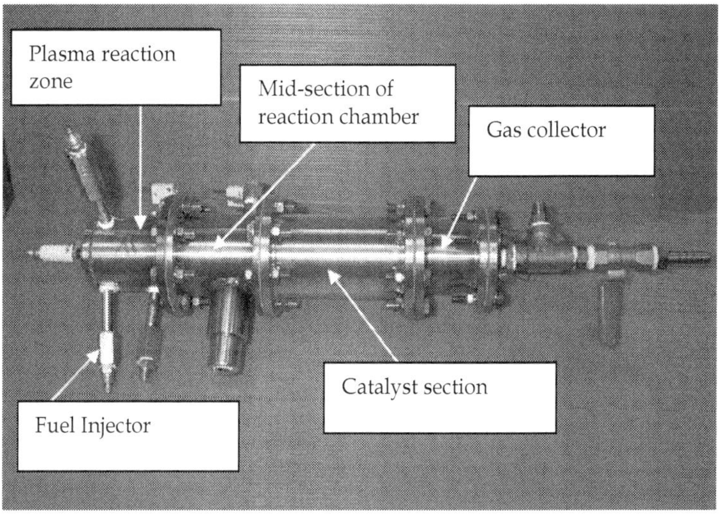

Fig. 3. The configuration of the reactor

at the three points were used to define the actual temperature of the reformate gas. This temperature was also used to correspond with the theoretical calculation to verify the trend of the experimental results.

2.2 Methodology

In this study, the fuel was reformed by coupling the arc energy from a spark discharge with the catalyst converter for partial oxidation reforming. The arc was the external energy source of the system to cause the formation of the plasma; and the arc was stretched by the kinetic energy induced by the tangential intake air flow which allowed for a more even discharge distribution. In addition, the production of hydrogen-rich gas was improved due to the use of catalysts. The product was collected in gas bags for analysis via gas chromatography.

Currently, the industries have several approaches for producing hydrogen; these approaches can be classified into three main categories, based on the energy sources: (1) Hydrogen production via hydrocarbon fuels; (2) hydrogen production via non-hydrocarbon fuels; and (3) hydrogen production via a combination of hydrocarbon and non-hydrocarbon fuels. Other classification for hydrogen production includes steam reforming (SR), partial oxidation reforming (POX), autothermal reforming (ATR), catalytic decomposition, and coal gasification. The reforming approaches and mechanisms for hydrogen production are shown in Fig. 4. This study generated hydrogen by the reformation of hydrocarbon fuels by POX and ATR.

Fig. 5 shows the variation of the reformate gas yields produced by methane under various O_2/CH_4 ratios at different reforming temperatures by theoretical calculation. It showed that the temperature of the reformate gas was crucial for the overall yield; higher reaction

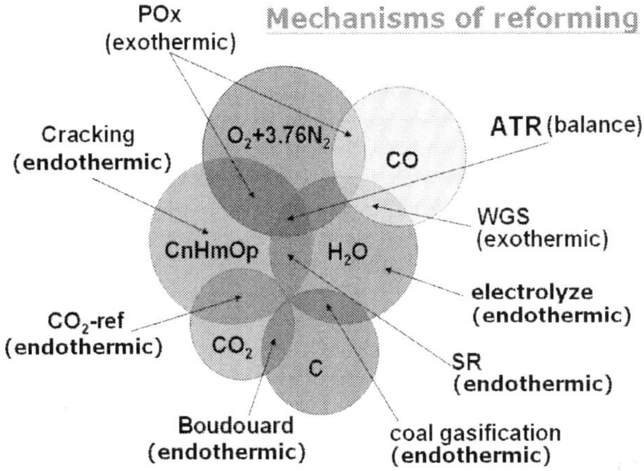

Fig. 4. A relationship chart of reforming mechanisms

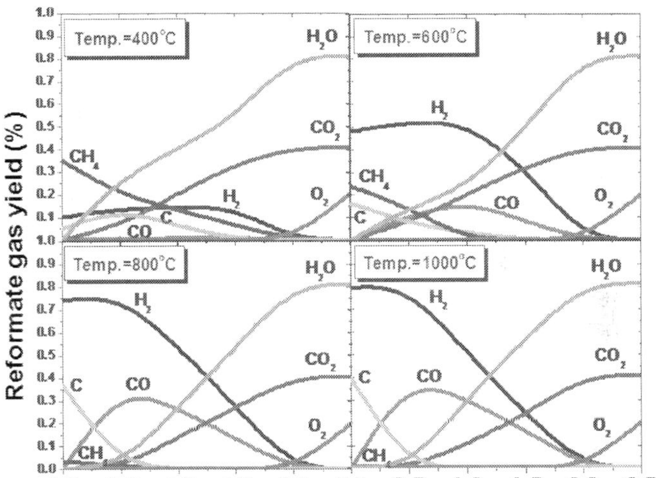

Fig. 5. The variation of the reformate gas yields produced by methane under various O_2/CH_4 ratios at different reforming temperatures

temperatures and lower O_2/CH_4 ratios gave better H_2 and CO yields. However, the appeal of this system was its portability, and thus external energy supply was not an option. Therefore, heat released during the oxidation reforming process was used to provide the high temperature required for the reactions. This approach reduced the need for external energy, and was also advantageous in minimizing the size of the converter.

Fig. 6 shows the enthalpy of reaction under different O_2/C ratios in methane reforming processes. The spots in the figure represent different reaction processes, and the processes are displayed sequentially as: the cracking process, the partial oxidation process, and the complete combustion. It could be observed that lower O_2/C ratios were involved in endothermic processes, as lower O_2/C ratios could not provide the required energy for the reformation processes. Therefore, methane of a higher O_2/C ratio was used in this study for reforming reactions. However, O_2/C could not be too high, or H_2 and CO would be oxidized into H_2O and CO_2. This study contained the O_2/C ratio between B and C, as shown in Fig. 6; methane reaction with O_2/C ratios within this range provided the required energy for reforming, and simultaneously prevented a reduction in catalyst activity due to the formation of carbon deposition in fuel rich when the O_2/C ratio was too low. Many possible reaction processes may happen for this study; the reactions listed in Table 2 are the possible routes, such as oxidation, shifting, and methanation, which could occur in this study.

Reaction	Equation	ΔH_{STP} (kJ/mol)
Partial oxidation	$CH_4 + 0.5O_2 \rightarrow CO + 2H_2$	-35.6
Water gas shifting	$CO + H_2O \rightarrow CO_2 + H_2$	-41.2
Complete oxidation	$CH_4 + 2O_2 \rightarrow CO_2 + 2H_2O$	-802.2
H_2 oxidation	$H_2 + 0.5O_2 \rightarrow H_2O$	-241.8
CO oxidation	$CO + 0.5O_2 \rightarrow CO_2$	-283.0
CO methanation	$CO + 3H_2 \rightarrow CH_4 + H_2O$	-206.2
CO_2 methanation	$CO_2 + 4H_2 \rightarrow CH_4 + 2H_2O$	-165.0

Table 2. The reaction in the reforming processes and the corresponding enthalpy of reaction

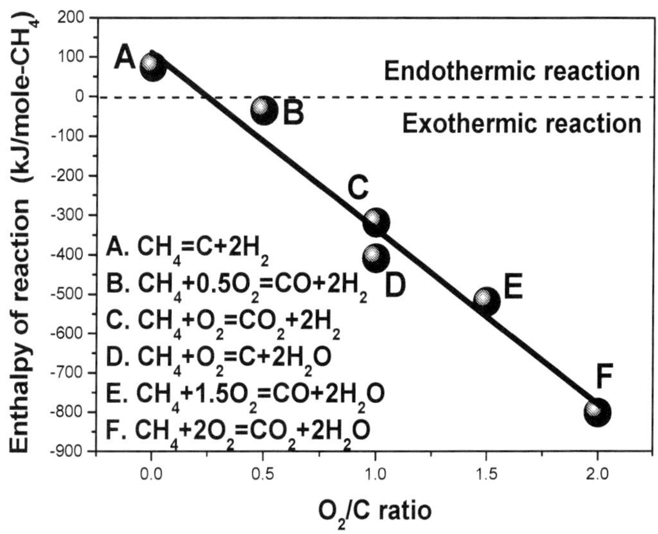

Fig. 6. A comparison of the enthalpy of reaction under different O_2/C ratios in methane reforming processes

2.3 Experimental parameters and the related calculations

In this study, the methane flow rate was in the range of 1 to 10 L/min, the O_2/C (Oxygen/Carbon) ratio was between 0.5 and 1.0, S/C (H_2O/Carbon) ratio was between 0 and 2, and the arc frequency was 200 Hz. Methane and air were supplied from tanks and were directed into the plasma reaction zone through the fuel nozzle via a tangential induction. The frequency of the arc was controlled by the signal generator, and was monitored on an oscilloscope. This study measured the voltage and current of the primary side (the low-voltage side), and monitored the arc frequency with an oscilloscope and a clamp tester. The voltage and current of the secondary side (high-voltage side) was measured with the oscilloscope and the clamp tester, in coordination with an automotive diagnostic analyzer.

Partial oxidation and autothermal reforming reactions were performed in this study. With an appropriate amount of oxygen, the heat released from the fuel oxidization dissociated carbon (C) and hydrogen (H) atoms from the fuel with water addition, and reformed the atoms into hydrogen gas (H_2) and carbon monoxide (CO). The equations (1) - (4) below represent the theoretical and actual reactions.

1. Theoretical reactions

Partial oxidation reforming

$$CH_4 + z_0(O_2 + 3.76N_2) \rightarrow 2z_0CO + 2H_2 + (1 - 2z_0)C + 3.76z_0N_2 \tag{1}$$

Autothermal reforming

$$CH_4 + z_0(O_2 + 3.76N_2) + (2 - 2z_0)H_2O \rightarrow CO_2 + (4 - 2z_0)H_2 + 3.76z_0N_2 \tag{2}$$

2. Actual reactions

Partial oxidation reforming

$$CH_4 + z(O_2 + 3.76N_2) \rightarrow x(aCO + bCO_2 + dH_2 + eO_2 + fCH_4 + gN_2) + hH_2O + jC \tag{3}$$

Autothermal reforming

$$CH_4 + z(O_2 + 3.76N_2) + yH_2O \rightarrow x(aCO + bCO_2 + dH_2 + eO_2 + fCH_4 + gN_2) + hH_2O + jC \tag{4}$$

z_0 is the equilibrium constant in the theoretical reaction; z, a, b, d, e, f, g, h, j and y are the equilibrium constants in the actual reaction, and x is the proportionality constant, respectively. Several expressions are used to describe the ratio of fuel and air in chemical equilibrium, and can be mainly classified as air-fuel mass ratio, and oxygen-carbon molar ratio. This study expressed the reforming reaction processes using the oxygen-carbon molar ratio. The equation for the oxygen-carbon (O_2/C) ratio related to volumetric flow rate is shown in equation (5).

$$\frac{O_2}{C} = \frac{\dot{n}_{O_2}}{\dot{n}_{CH_4}} = \frac{\dfrac{\dot{m}_{O_2}}{M_{O_2}}}{\dfrac{\dot{m}_{CH_4}}{M_{CH_4}}} = \frac{\rho_{O_2} \times Q_{O_2}}{\rho_{CH_4} \times Q_{CH_4}} \times \frac{M_{CH_4}}{M_{O_2}} = \frac{0.21 \times Q_{air}}{Q_{CH_4}} \tag{5}$$

process. In contrast, by comparing different fuel flow rates at the same O_2/C ratio, it is found that increasing the fuel flow rate would lower the enthalpy of reaction per molar fuel. This is because the fuel oxidation releases more heat and the reaction temperature is relatively high; therefore, in this condition, the system does not rely on higher O_2/C ratios to promote the conversion of fuel. As a result, the overall energy loss percentage decreases as the fuel flow rate increases.

3.2 The effect of fuel supply rate

Conversion efficiencies are crucial indicators of reforming processes and can be used to indicate the reformation efficiency. Higher conversion efficiency implies that more fuel participates in the reactions, therefore more reformate products are produced. As shown in Fig. 9, the methane flow rates of this study are 2, 4, 6, 8, and 10 L/min, and the O_2/C ratio is between 0.5 and 1.0. High fuel conversion efficiency is found at relatively high O_2/C ratios and high methane flow rates. This is because at the same O_2/C ratio, an increase in the flow rate of fuel would increase the heat release of oxidation; therefore the required energy for reaction is effectively obtained. It is also found that as the O_2/C ratio increases, the increase in conversion efficiency gradually slows down because it nearly reaches the maximum space velocity of the reforming system.

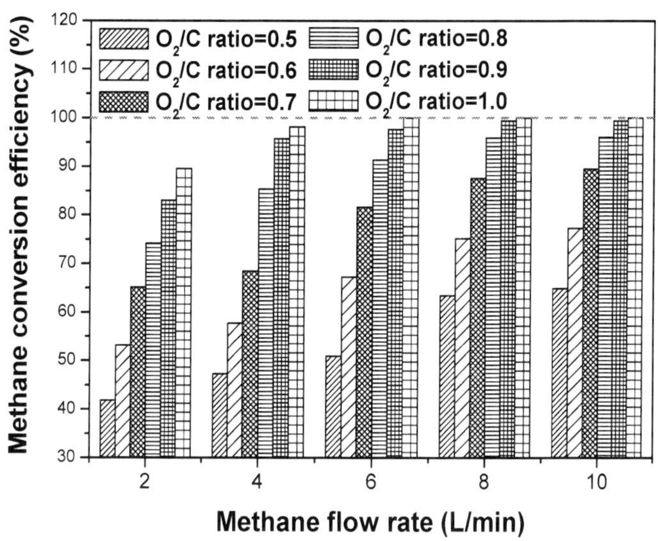

Fig. 9. The effect of methane flow rate on methane conversion efficiency at different O_2/C ratios

3.3 The effect of the reformate gas temperature

The following section discusses the effect of reformate gas temperature on the selectivity of H_2 and CO. The definition of selectivity (*S*) is shown in equations (7) and (8), and can be classified as carbon-based (C-base) and hydrogen (H-base) selectivity.

$$S_{C-base} = \frac{x}{n}(\frac{moles\ of\ CxHy\ formed}{moles\ of\ CnHm\ consumed}) \times 100\% \tag{7}$$

$$S_{H-base} = \frac{y}{m}(\frac{moles\ of\ CxHy\ formed}{moles\ of\ CnHm\ consumed}) \times 100\% \tag{8}$$

The selectivity is to demonstrate the form of the reformate gas existing after the reforming reactions. Briefly, hydrocarbon fuel reformation produces hydrogen-based and carbon-based products. The selectivity of hydrogen-based product may have the form of hydrogen (H_2), water (H_2O), or other hydrocarbons. Carbon-based products may exist as carbon monoxide (CO), carbon dioxide (CO_2), and other hydrocarbons. Fig. 10 shows that as the temperature increases, H_2 selectivity shows a quadratic increase, while CO selectivity has a logarithmic increase. Both H_2 and CO have the greatest selectivity when the temperature is between 600 °C and 700 °C. In Fig. 11, the yields of H_2 and CO are found to increase as the reformate gas temperature rises; the yield reaches the maximum value approximately between 750 and 800 °C, and is 77 % for H_2 and 73 % for CO. As performing the calculation by using the best parameters, the experimental results are found to be close to the theoretical values; except that once the reformate gas temperature exceeds 800 °C, the experimental values diverge from the theoretical ones. Around this temperature, the inconsistency between the experimental results and the theoretical calculations is thought to result from the combustion of a portion of the methane at such a high temperature, and H_2 and CO were oxidized to form the CO_2 and H_2O.

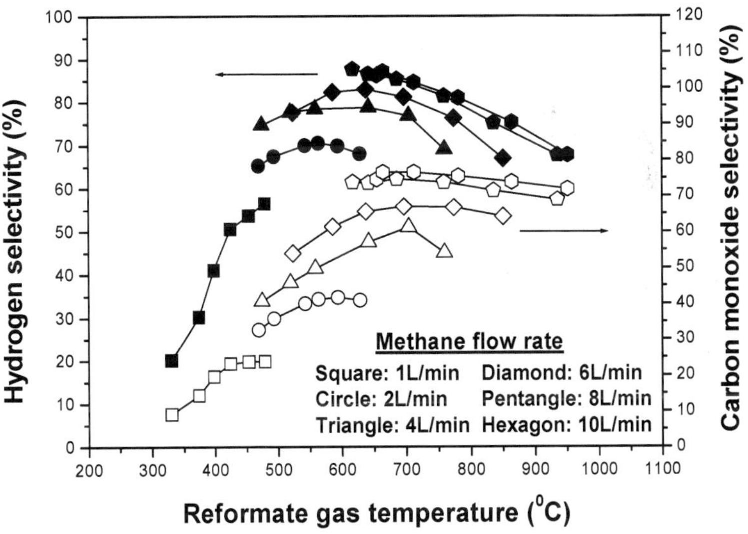

Fig. 10. The relationship between the reformate gas temperature and the selectivity of H_2 and CO at different methane flow rates

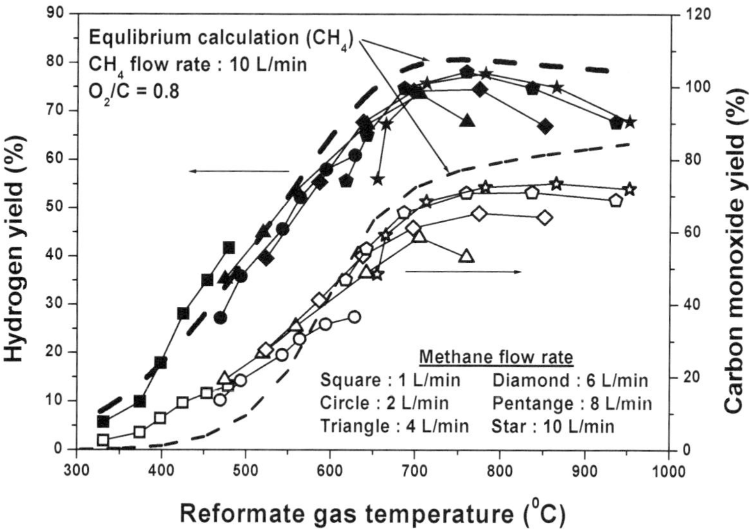

Fig. 11. The relationship between the reformate gas temperature and the yield of H_2 and CO at different methane flow rates

3.4 The effect of space velocity

Generally, space velocity is determined by the input mixture, the operating temperature, and the size of the catalyst. Space velocity mainly refers to the volumetric flow rate of input mixture per unit volume of catalyst bed, but it is also an indicator for the reforming capacity. Fig. 12 shows the effect of space velocity on thermal efficiency at different methane flow rates and O_2/C ratios. The definition for thermal efficiency (η_{th}) is as shown in equation (9).

$$\eta_{th} = \frac{\dot{m}_{H_2} LHV_{H_2} + \dot{m}_{CO} LHV_{CO}}{\dot{m}_{fuel} LHV_{fuel}} \times 100\% \tag{9}$$

Where $\dot{m}$ *and* LHV represent the mass flow rate and lower heating value of the species, respectively.

It is apparent from Fig. 12 that as space velocity increases, thermal efficiency initially increases abruptly, and the rate of increase slows down soon after; the trend indicates that the parameters used in the system did not exceed the amount that the reforming system could treat. At best operating parameters (methane flow rate of 10 L/min and an O_2/C ratio of 0.8), the greatest thermal efficiency of 72.25 % is obtained and the corresponding space velocity is approximately 35,000 h^{-1}. The best operating parameters were used as the input for calculation by using a commercialized HSC program, and the calculated values were compared with the experimented results obtained from the best parameter setting of the reforming system. It can be observed from Fig. 13 that the difference in the reformate gas concentration between the theoretical and the experimental results is within 1 %.

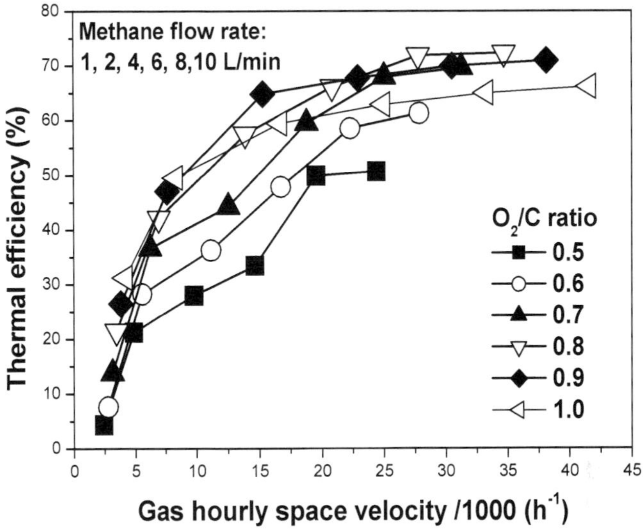

Fig. 12. The effect of space velocity on thermal efficiency at different methane flow rates and O_2/C ratios

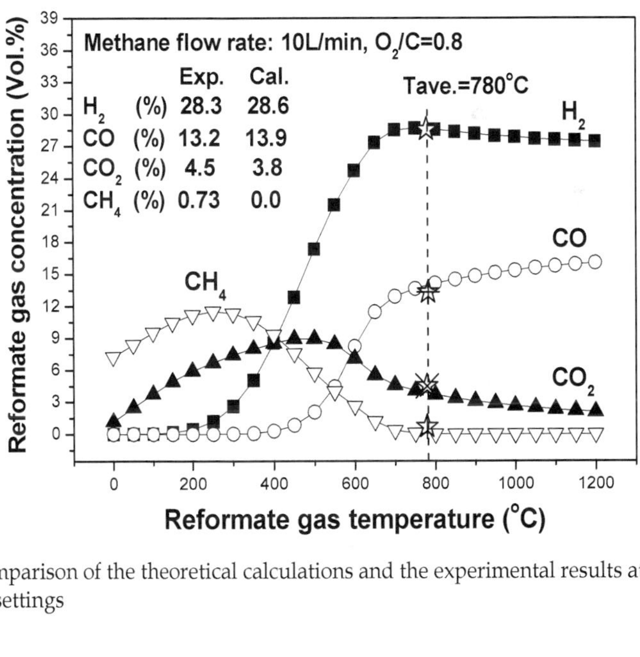

Fig. 13. Comparison of the theoretical calculations and the experimental results at the best parameter settings

The reforming performance for all the parameters at different equivalent power of input fuel is compared, as shown in Fig. 14. The input power of fuel directed into the reaction chamber is shown on the horizontal axis, and the output power derived from the output flow rate of H_2 and CO is shown on the vertical axis. The result of this study shows that when methane is used for reforming, the output power increases linearly with the input power. The linear regression result could also serve as a convenient reference for future parameter settings for reforming.

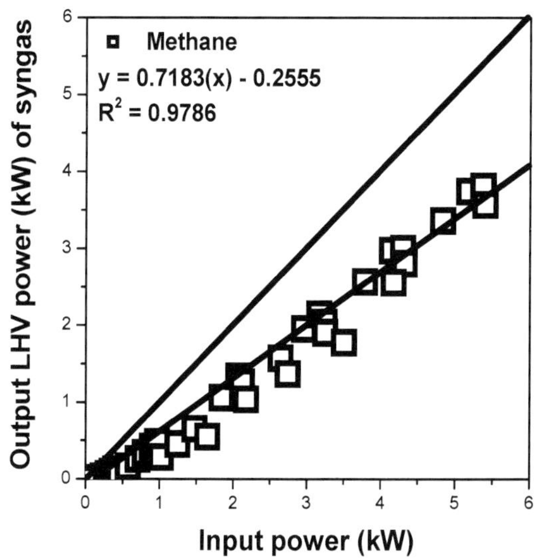

Fig. 14. Comparison of reforming performance at different input power

3.5 The effect of the operating conditions on the reformation efficiency

The following section discusses the variations in reformation efficiency under different operating conditions. The increase in hydrogen output and the reforming thermal efficiency were compared under the different water vapor addition, and under the use of different energy conservation approaches.

Table 3 shows the comparison of performance of the plasma fuel converter under different operating conditions, where symbol A represents the original system, i.e. the partial oxidation reforming, and B represents the autothermal reforming reaction with water vapor added. C and D use different energy conservation approaches; C represents the heat recycling approach, and D represents the heat insulation. The fuel flow rate is set at 10 L/min and the O_2/C ratio at 0.8. The summary of results shown in Table 3 indicates that the best operating condition of this study is the combination of A+D, which has a maximum thermal efficiency of 77.77 %; the combination can be applied to internal combustion engines to promote energy utilization efficiency in the future. However, from the perspective of hydrogen production, the combination of A+B+D with an appropriate amount of water vapor for reaction would increase the hydrogen yield by approximately 5 %; it would also

	S/C ratio	Fuel Conv. (%)	H_2 yield (%)	CO yield (%)	Energy loss (%)	η_{th} (%)
A	n/a	95.99	77.75	72.44	24.34	72.25
A + B	0.5	95.59	85.92	56.8	25.06	71.63
A + C	n/a	98.16	80.3	77.88	23.23	75.71
A + D	n/a	99.29	81.81	81.14	22.68	77.77
A+ B + D	0.5	98.71	86.26	70.28	23.27	76.6

A: Original system, B: ATR, C: Heat recycling, D: Heat insulation;
Fuel flow rate: 10 L/min, O_2/C ratio: 0.8

Table 3. Comparison of plasma converter performance under different operating conditions

increase the amount of power generated if applied to fuel cells. From this table, the variations in hydrogen and carbon monoxide yields are significantly with the combination of water gas shifting reaction, although the methane conversion efficiency is reduced. This is because the system did not adopt an external heating source to maintain its working temperature, and water addition would induce regional endothermic reactions due to the latent heat of water vaporization. The regional endothermic reaction consequently lowers the reformate gas temperature of the system and results in a reduction in the methane conversion efficiency. Another possibility for the reduction of conversion efficiency is that methanation leads to reverse reaction and increases the selectivity of methane.

This section will focus on the comparison of the theoretical calculation of the chemical equilibrium for methane reformation. The variations between the experimental results and the calculated values are compared. A comprehensive comparison of the experimental results with the theoretical calculation of the dry analysis on the output concentration of H2+CO is shown in Fig. 15; the values were measured under all of the parameters of the original system, the autothermal reforming reactions, and the energy conservative method. Linear regression analysis reveals that the R^2 value for the concentration of syngas (H_2+CO) is as high as 0.9179, indicating that the values obtained via experiments and the values obtained from theoretical calculation are very close.

The following section discusses the effect of plasma reforming parameters (PRP) on the overall reforming performance. The purpose is to compile all of the measured results of this study and to identify the correlations within the parameters. The reforming indicators, H_2+CO concentration and thermal efficiency will be discussed. Fig. 16 shows the effect of the first plasma reforming parameters (PRP1) on the concentration of H_2+CO. PRP1 is comprised of the methane molar flow rate, oxygen-carbon molar ratio, and the methane conversion efficiency; these reforming parameters are also closely associated with the concentration of H_2+CO. The R^2 value of the quadratic regression at different operating conditions can be as high as 0.9104. It is also shown in Fig. 16 that an increase of PRP1 signifies a greater methane flow rate, a more appropriate O_2/C ratio, and a better methane conversion efficiency; and the concentration of H_2+CO first increases as PRP1 increases, and then subsequently decreases. Thus, the data imply that there is an upper limit for the O_2/C ratio in this system, and exceeding this limit would lead to a decrease in hydrogen concentration due to the oxidization of H_2+CO into H_2O and CO_2 by excess oxygen.

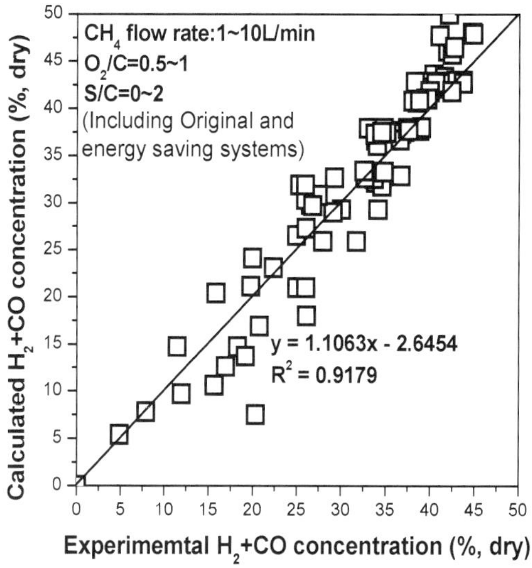

Fig. 15. Comparison of theoretical calculations and the experimental results of methane at different operational parameters

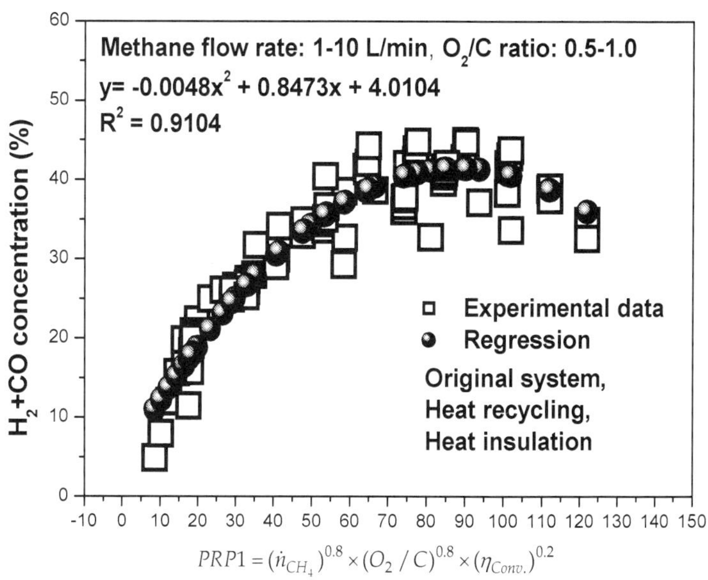

Fig. 16. The effect of the 1st plasma reforming parameters (PRP1) on H_2+CO concentration under different operation conditions

Fig. 17 shows the effect of the second plasma reforming parameters (PRP2) on thermal efficiency under the operational parameter settings. The plasma reforming parameters are comprised of the methane molar flow rate, the oxygen-carbon molar ratio, the reformate gas temperature, and the conversion efficiency of methane. The effects of these plasma reforming parameters on thermal efficiency are obvious, and the R^2 value of the regression calculation is as high as 0.9284. Of all the second plasma reforming parameters, PRP2 corresponds to higher thermal efficiency with higher methane flow rate, greater oxygen-carbon ratio, and a higher reformate gas temperature. Under these parameters, the catalysts are able to work at an appropriate temperature, and to improve the yield of hydrogen and carbon monoxide in the reforming. As the methane conversion efficiency increases, so does the overall energy utilization efficiency; the total thermal efficiency shows a logarithmic increase as PRP2 increases.

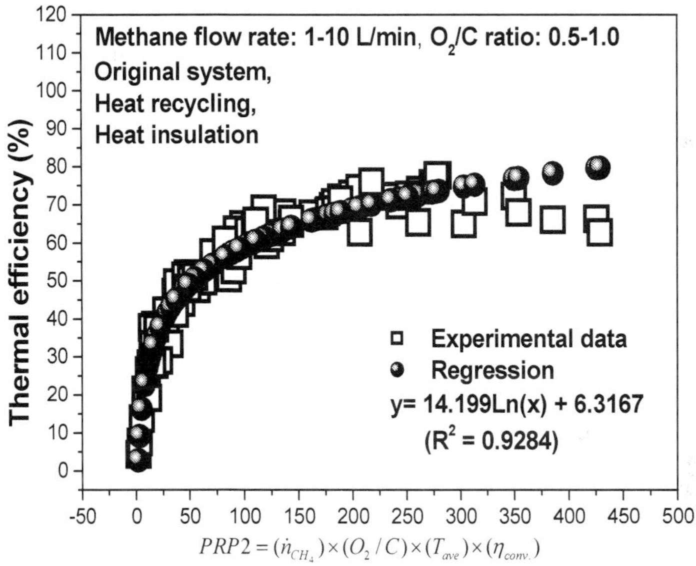

Fig. 17. The effect of the 2nd plasma reforming parameters (PRP2) on the total thermal efficiency under different operating conditions

3.6 A preliminary study on engine with hydrogen-rich gas as auxiliary fuel

The following section introduces the results of the preliminary study using hydrogen-rich gas as an auxiliary fuel for a 125 cc 4-stroke motorcycle engine. The specifications of the engine used in this experiment are shown in Table 4. Hydrogen-rich gas was supplied from different gas tanks at specific ratios; the gas mixture was directed into the engine as fuel. The performance and the pollutant emissions of the engine were measured to improve the parameters for the designing of the plasma fuel reformer in the future. Because the reformer of this study reforms gas via a partial oxidation reforming approach, the composition of the reformate gas is diverse. The typical gases in the reformate products are hydrogen, carbon

Engine type	Air cooled single cylinder 4-stroke engine		
Bore	52.4 mm		
Stroke	57.8 mm		
Displacement	124 cm^3		
Compression ratio	9.2:1		
Compression pressure	12/570 ($kg/cm^2/rpm$)		
Fuel supply system	CV type carburetor		
Intake valve open	0° BTDC	Intake valve close	25° ABDC
Exhaust valve open	33° BBDC	Exhaust valve close	0° ATDC
Fuel	Unleaded gasoline #92 with hydrogen-rich gas		

Table 4. The specifications of the motorcycle engine

monoxide, carbon dioxide, and nitrogen; thus, this experiment was carried out by supplying the mixture of these four gases as added fuel.

Both H_2 and CO have relatively high heating values and can be used as fuel for running engines. Additionally, hydrogen has a high flame speed, and is favorable for assisting combustion. In contrast, although CO_2 and N_2 do not provide heating value, they could be used for exhaust gas recirculation (EGR), and could effectively lower the combustion temperature to impede the formation of NOx. Fig. 18 shows the appropriate operating range of hydrogen-rich gas at various levels of throttle openings and engine speeds. The left panel

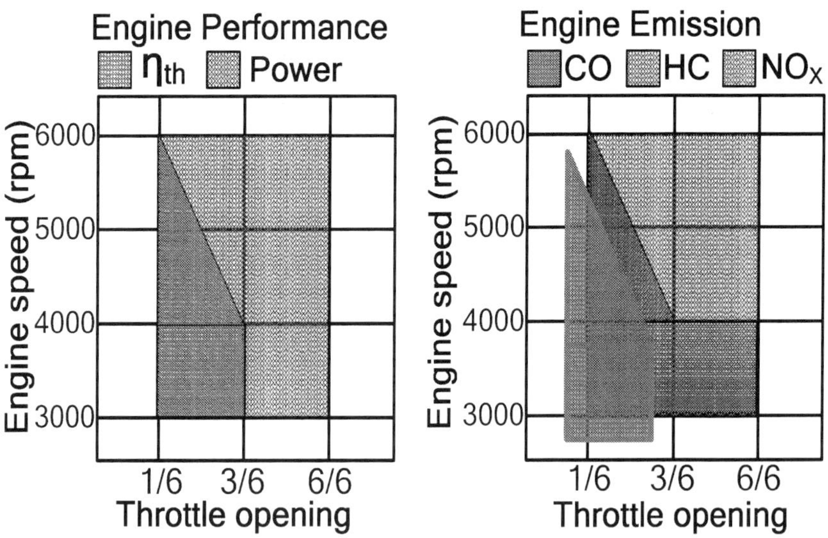

Fig. 18. Appropriate operating range of throttle openings and engine speeds by adding hydrogen-rich gas as an auxiliary fuel

shows the engine performance, and includes thermal efficiency (shown in red), and the output horsepower (shown in blue). The right panel shows the comparison of CO (green), HC (purple), and NOx (blue) emissions. The results shown in this figure indicate that engine performance and exhaust emissions could be improved simultaneously by adding an appropriate amount of hydrogen-rich gas; that is, hydrogen-rich gas has the characteristics of assisting and impeding combustion. The output horsepower can be improved at any level of throttle opening and engine speed; in contrast to the thermal efficiency of the engine, which has better result only at low throttle opening. It could be caused by the lower ratio of $[(H_2+CO)/(CO_2+N_2)]$ in hydrogen-rich gas, thus resulting in the less impact on the combustion-assisting effect in other throttle openings. For the exhaust emissions of engine, although the improvement of CO and HC are not as evident as expected, NOx was greatly improved, confirming the effect of impeding combustion. Therefore, in order to produce the appropriate composition of hydrogen-rich gas for simultaneously improving the thermal efficiency and exhaust emissions of engine, the operating parameter setting of plasma converter would be very important.

4. Conclusions

This study investigated the reformation of methane for hydrogen production by an energy conservative plasma fuel converter under different operation parameters and conditions. The experimental parameters include fuel flow rate, O_2/C ratio, and S/C ratio. Additionally, the operating conditions are categorized into autothermal reforming reaction and energy conserving approaches. By conducting a series of experiments, the conclusions were drawn as followed:

For methane reforming, the reforming temperature for the best hydrogen selectivity is between 600 and 700 °C. By taking energy conservation approaches, the hydrogen production and the thermal efficiency of reformer are improved. Under the best parameter setting, the total thermal efficiency can be as high as 77.77 % with the methane flow rate of 10 NL/min and the O_2/C ratio of 0.8. Furthermore, the hydrogen yield can be further increased to 86.26 % by adding an appropriate amount of water vapor (S/C ratio of 0.5), and the produced hydrogen flow rate is approximate 17.25 NL/min. And in the reforming process, the O_2/C ratio, the fuel flow rate, the space velocity, and the reformate gas temperature have a great influence on reforming performance. As the first plasma reforming parameter (PRP1) increases, the H_2+CO concentration shows a quadratic polynomial increase accordingly; and the thermal efficiency shows a logarithmic increase with the second plasma reforming parameters (PRP2). In this study, the HSC commercialized software was used for theoretical equilibrium calculation. The experimental results are very close to the value obtained through theoretical calculation. The regression coefficient (R^2) can be as high as 0.918 between the experimented and calculated H_2+CO concentrations.

5. Acknowledgments

This study was supported by the National Science Council Research Projects (project numbers: NSC 95-2622-E-168-008-CC3, NSC 95-2221-E-168-032-MY2 & NSC 97-2221-E-168-033-MY3) and Jellys Technology Inc.

6. References

Al-Janabi, H.A.S. & Al-Baghdadi, M.A.S. (1999). A prediction study of the effect of hydrogen blending on the performance and pollutions emission of a four stroke spark ignition engine. *International Journal of Hydrogen Energy*, Vol.24, No.4, (April 1999), pp. 363-375, ISSN 0360-3199

Ayabe, S., Omoto, H., Utaka, T., Kikuchi, R., Sasaki, K., Teraoka, Y. & Eguchi, K. (2003). Catalytic autothermal reforming of methane and propane over supported metal catalysts. *Applied Catalysis A: General*, Vol.241, No.1-2, (February 2003), pp. 261–269, ISSN 0926-860X

Bromberg, L., Cohn, D. R. & Rabinovich, A. (1997). Plasma reformer-fuel cell system for decentralized power applications. *International Journal of Hydrogen Energy*, Vol.22, No.1, (January 1997), pp. 83–94, ISSN 0360-3199

Bromberg, L., Cohn, D.R., Rabinovich, A., Surma, J.E. & Virden, J. (1999a). Compact plasmatron -boosted hydrogen generation technology for vehicular applications. *International Journal of Hydrogen Energy*, Vol.24, No.4, (April 1999), pp. 341–350, ISSN 0360-3199

Bromberg, L., Cohn, D. R., Rabinovich, A. & Alexeev, N. (1999b). Plasma catalytic reforming of methane. *International Journal of Hydrogen Energy*, Vol.24, No.12, (December 1999), pp.1131-1137, ISSN 0360-3199

Bromberg, L., Cohn, D.R., Rabinovich, A., Alexeev, N., Samokhin, A., Ramprasad, R. & Tamhankar, S. (2000). System optimization and cost analysis of plasma catalytic reforming of natural gas. *International Journal of Hydrogen Energy*, Vol.25, No.12, (December 2000), pp. 1157–1161, ISSN 0360-3199

Bromberg, L., Cohn, D. R., Rabinovich, A. & Heywood, J.B. (2001). Emissions reductions using hydrogen from plasmatron fuel converters. *International Journal of Hydrogen Energy*, Vol.26, No.10, (October 2001), pp. 1115–1121, ISSN 0360-3199

Cohn, D.R., Rabinovich, A. & Titus, C.H. (1996). Onboard plasmatron generation of hydrogen for extremely low emission vehicle with internal combustion engine. *International Journal of Vehicle Design*, Vol.17, No.5-6, (1996), pp. 550-561, ISSN 1477-5360

Futamura, S., Kabashima, H., Annadurai, G. (2006). Roles of CO_2 and H_2O as oxidants in the plasma reforming of aliphatic hydrocarbons. *Catalysis Today*, Vol. 115, No.1-4, (June 2006), pp. 211–226, ISSN 0920-5861

Horng, R. F., Chang, Y. P. & Chung, C. L. (2006a). Carbon deposit growth on the electrodes of a plasma converter in the generation of hydrogen from methane. *International Journal of Hydrogen Energy*, Vol.31, No.14, (November 2006a), pp. 2040-2051, ISSN 0360-3199

Horng, R.F., Chang, Y.P. & Wu, S.C. (2006b). Investigation on the production of hydrogen-rich gas in a plasma converter for motorcycle applications. *Energy Conversion and Management*, Vol.47, No.15-16, (September 2006b), pp. 2155-2166, ISSN 0196-8904

Horng, R. F., Chang, Y. P., Huang, H. H. & Lai, M. P. (2007). A study of the hydrogen production from a small plasma converter. *Fuel*, Vol.86, No.1-2, (January 2007), pp. 81-89, ISSN 0016-2361

Horng, R.F., Huang, H.H., Lai, M.P., Wen, C.S. & Chiu, W.C. (2008). Characteristics of hydrogen production by a plasma-catalyst hybrid converter with energy saving schemes under atmospheric pressure. *International Journal of Hydrogen Energy,* Vol. 33, No.14, (July 2008a), pp. 3719-3727, ISSN 0360-3199

Horng, R.F., Wen, C.S., Liauh, C.T., Chao, Y. & Huang, C.T. (2008). Driving characteristics of a motorcycle fuelled with hydrogen-rich gas produced by an onboard plasma reformer. *International Journal of Hydrogen Energy,* Vol. 33, No.24, (December 2008b), pp. 7619-7629, ISSN 0360-3199

Horng, R.F., Lai, M.P., Huang, H.H. & Chang, Y.P. (2009). Reforming performance of a plasma-catalyst hybrid converter using low carbon fuels. *Energy Conversion and Management,* Vol. 50, No.10, (October 2009), pp. 2632-2637, ISSN 0196-8904

Jamal, Y., Wagner T. & Wyszynski, M. L. (1996). Exhaust gas reforming of gasoline at moderate temperature. *International Journal of Hydrogen Energy,* Vol.21, No.6, (June 1996), pp. 507-519, ISSN 0360-3199

Lutz, A.E., Bradshaw, R.W., Keller, J.O. & Witmer, D.E. (2003). Thermodynamic analysis of hydrogen production by steam reforming. *International Journal of Hydrogen Energy,* Vol.28, No.2, (February 2003), pp. 159–167, ISSN 0360-3199

Lutz, A.E., Bradshaw, R.W., Bromberg, L. & Rabinovich, A. (2004). Thermodynamic analysis of hydrogen production by partial oxidation reforming. *International Journal of Hydrogen Energy,* Vol.29, No.8, (July 2004), pp. 809–816, ISSN 0360-3199

Ma, F., Wang, Y., Liu, H., Li, Y., Wang, J. & Zhao, S. (2007). Experimental study on thermal efficiency and emission characteristics of a lean burn hydrogen enriched natural gas engine. *International Journal of Hydrogen Energy,* Vol.32, No.18 (December 2007), pp. 5067–5075, ISSN 0360-3199

Maruoka, N., Sato, K., Yagi, J. & Akiyama, T. (2002). Development of PCM for recovering high temperature waste heat and utilization for producing hydrogen by reforming reaction of methane. *ISIJ international,* Vol.42, No.2, (May 2002), pp. 215-219, ISSN 0915-1559

Maruoka, N., Purwanto, H. & Akiyama, T. (2010). Exergy analysis of methane steam reformer utilizing steelmaking waste heat. *ISIJ international,* Vol.50, No. 9, (September 2010), pp. 1311-1318, ISSN 1347-5460

Mohammadi, A., Shioji, M., Nakai, Y., Ishikura, W. & Tabo, E. (2007). Performance and combustion characteristics of a direct injection SI hydrogen engine. *International Journal of Hydrogen Energy,* Vo.32, No.2, (February 2007), pp. 296-304, ISSN 0360-3199

Nozaki, T & Okazaki, K. (2005). Non-equilibrium reaction fields created by atmospheric pressure plasmas and application to advanced utilization of natural gas resource. *Journal of the Japan Institute of Energy,* Vol.84, No.6, (June 2005), pp. 462-467, ISSN 0916-8753

Outokumpu. (2002). Chemical reaction and equilibrium software with extensive thermochemical database. *HSC Chemistry® for Windows,* Version 5.1, ISBN 952-9507-08-9

Petitpas, G., Rollier, J.D., Darmon, A., Gonzalez-Aguilar, J., Metkemeijer, R. & Fulcheri, L. (2007). Review–a comparative study of non-thermal plasma assisted reforming technologies. *International Journal of Hydrogen Energy,* Vol. 32, No.14, (September 2007), pp. 2848–2867, ISSN 0360-3199

Peucheret, S., Wyszynski, M.L., Lehrle, R.S., Golunski, S. & Xu, H. (2005). Use of catalytic reforming to aid natural gas HCCI combustion in engines: experimental and modelling results of open-loop fuel reforming. *International Journal Hydrogen Energy*, Vol.30, No.15, (December 2005), pp. 1583–1594, ISSN 0360-3199

Pietruszka, B. & Heintze, M. (2004). Methane conversion at low temperature: the combined application of catalysis and non-equilibrium plasma. *Catalysis Today*, Vol.90, No.1-2, (June 2004), pp. 151–158, ISSN 0920-5861

Rabinovich, A., Cohn, D. R. & Bromberg, L. Plasmatron internal combustion engine system for vehicle pollution reduction. *International Journal of Vehicle Design*, Vol.15, No.3-5, (November 1994), pp. 234-242, ISSN 1477-5360

Seo, Y.T., Seo, D.J., Jeong, J.H. & Yoon, W.L. (2006). Design of an integrated fuel processor for residential PEMFCs applications. *Journal of Power Sources*, Vol.160, No.1, (September 2006), pp. 505–509, ISSN 0378-7753

Shudo, T. (2006). An HCCI combustion engine system using on-board reformed gases of methanol with waste heat recovery: ignition control by hydrogen. *International Journal of Vehicle Design*, Vol.41, No.1-4, (2006), pp. 206-226, ISSN 1477-5360

Sobacchi, M.G., Saveliev, A.V., Fridman, A.A., Kennedy, L.A., Ahmed, S. & Krause, T. (2002). Experimental assessment of a combined plasma/catalytic system for hydrogen production via partial oxidation of hydrocarbon fuels. *International Journal of Hydrogen Energy*, Vol.27, No.6, (June 2002), pp. 635–642, ISSN 0360-3199

Song, C. (2006). Global challenges and strategies for control, conversion and utilization of CO_2 for sustainable development involving energy, catalysis, adsorption and chemical processing. *Catalysis Today*, Vol.115, No.1-4, (June 2006), pp. 2–32, ISSN 0920-5861

Verhelst, S. & Sierens, R. (2001). Aspects concerning the optimisation of a hydrogen fueled engine. *International Journal of Hydrogen Energy*, Vol.26, No.9, (September 2001), pp. 981-985, ISSN 0360-3199

Zheng, M., Reader, G.T. & Hawley, J.G. (2004). Diesel engine exhaust gas recirculation-a review on advanced and novel concepts. *Energy Conversion and Management*, Vol. 45, No.6, (April 2004), pp. 883-900, ISSN 0196-8904

Steam Turbines Under Abnormal Frequency Conditions in Distributed Generation Systems

Fabrício A. M. Moura[1], José R. Camacho[2],
Geraldo C. Guimarães[2] and Marcelo L. R. Chaves[2]
[1]*Federal University of Triângulo Mineiro,*
[2]*Federal University of Uberlândia,*
Brazil

1. Introduction

Considerable efforts have been developed to encourage the installation of independent electricity producers in distribution systems. Obvious example of such efforts is demonstrated by the attempt to standardize their interconnection and protection requirements. However, this standardization can be difficult due to variations in the design of distribution circuits, the various types of generators coupled to the network and the particular requirements of each utility. However, a series of questions points to the development of further studies to ensure the quality of electric power, the system transient stabilty and lifetime of the blades of steam turbines (Moura et al.,2011).

Distributed generation shows frequent use of steam turbines as a primary machine to produce electricity. Such units have increased considerably due to a restructuring of the energy sector worldwide.

Most of the process of converting thermal energy into electrical energy occurs in the steam turbine. This is due to the numerous advantages of such turbines over other similar technologies. Among the main advantages stands the balanced construction, relatively high efficiency, few moving parts, ease of maintenance, and availability in large sizes. The industries that typically employ the technology of cogeneration are the sugar and ethanol, the pulp and paper and oil refining (Anderson & Fouad, 2003).

Recently, special attention has been devoted to turbo-alternators under abnormal conditions of operation of the electrical system concerning the frequency (over or under frequency). In particular, major research efforts have been spent with the main aim of assessing the possible damage they are subjected to the steam turbine when in operation under conditions of prolonged under frequency, during a severe overload condition imposed on the system when occurring a deficit of generation (Kundur, 1994).

Contingencies common to the distribution of electricity systems can cause serious damage to the generators installed in parallel to the network as well as steam turbines responsible for providing mechanical power to their axis. Thus, it is necessary to assess the possible impacts of steam turbines of the IP (independent producer) electrical machinery in system abnormal operating conditions.

For this it is essential to develop a computer model able to represent accurately the phenomena experienced.

2. Steam turbines control functions

The speed regulation system is the primary control loop of the Automatic Generation Control (*AGC*) of an electrical system. The control of active power in a power system is accomplished primarily by controlling the torque of the machines primary drive system (Anderson & Fouad, 2003).

Figure 1 illustrates a system for regulating the speed of a steam turbine. By controlling the position x_E of the control valve, a control can be exercised over the admission of steam (or water) at high pressure trough the turbine, and therefore the torque of the turbine. For a hydraulic amplifier, the input is the position of the pilot valve x_D and the output is position x_E of the main piston. A small downward movement of point E, the steam flow increases by an amount that, if measured in terms of power valve, represents an increment ΔP_v. This increased flow is rendered by the increase in turbine power, ΔP_T.

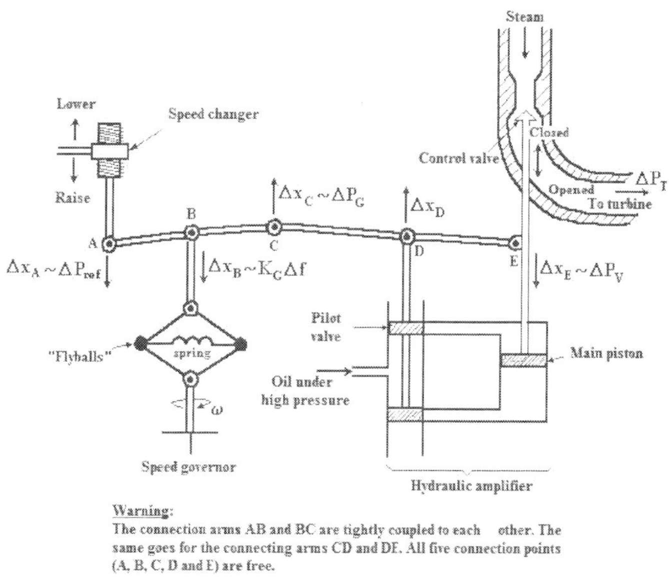

Fig. 1. Example illustrating the mechanical speed control system of a steam turbine

The pilot valve X_D, still in figure 1, can have its position changed, through the links or arms connection system, in three ways:

- Directly, by the speed changer. A small downward movement of the connecting point A (Δx_A) corresponds to an increase in the reference power, ΔP_{ref};
- Indirectly, by feedback, due to changes in the position of the connection point E of the main piston (Δx_E), resulting from variations in steam pressure;

- Indirectly, by feedback due to changes in the position of the connection point B (Δx_B), resulting from changes in speed.

Historically, *load-frequency primary controls* are part of the so-called *characteristics of speed governors*. However, the term "regulator" implies a mechanical speed sensor mechanically connected to control variations in input power, and since today most of the load-frequency controls are electro-hydraulic, speed control term does not describe generically this control function (Kundur, 1994).

2.1 Isochronous governor

Another simple mechanism of regulation is shown by figure 2. In this case, the deviation in speed, captured by the centrifugal body (*flyballs*), cause the displacement Δx in the pilot valve, which makes the oil to flow though the main servo motor. This servo motor, in turn, opens or closes the valve or turbine blades, depending on the direction of Δx.

It is important to observe that a displacement Δx in the pilot valve causes a rate of change of the servo piston, or a rate of change in the valve position.

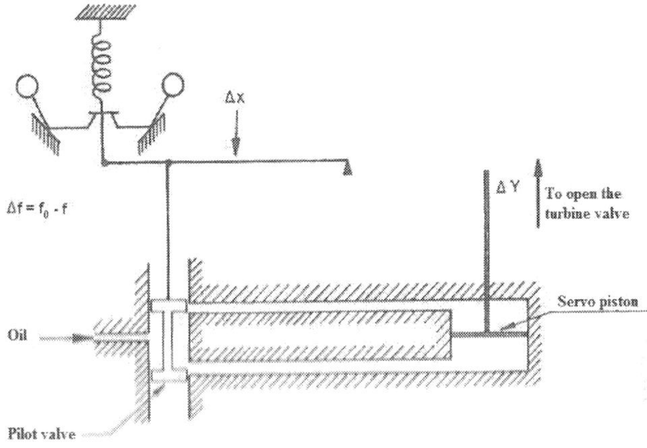

Fig. 2. "Isochronous" type speed governor

With reference to figure 2, can be written for the position of pilot valve (Δx):

$$\Delta x = k_1 \Delta f \tag{1}$$

where:
Δf = frequency deviation (f_0 - f);
k_1 = proportionality constant.

Also, for the turbine valve position (Δy), we have:

$$\frac{d}{dt}(\Delta y) = k_2 \Delta x \tag{2}$$

where:
k_2 = proportionality constant.

$$\frac{d}{dt}(\Delta y) = k_1 k_2 \Delta f$$

or

$$\Delta y = \int k_1 k_2 \Delta f \, dt \tag{3}$$

These equations are described in the block diagram in figure 3.

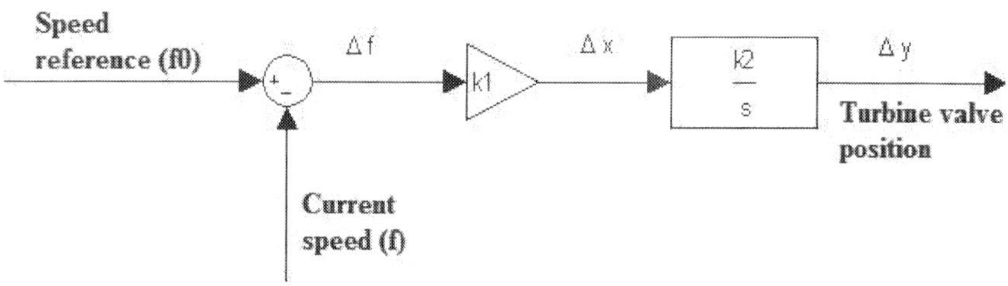

Fig. 3. Block diagram of the isochronous type speed governor

Due to the integrative action of the main servo motor, it is evident that Δy reach a new equilibrium state only when Δf reaches zero.

The governor with the configuration of figure 2 will therefore reduce the frequency error to zero in steady state and move the valve position, within physical limits, to any position that is necessary to make such reduction. It is known as an isochronous governor.

Note that a frequency shift may occur either by variation of the actual frequency, or change in the reference value f_0, which is accomplished through a change in the pivot point, usually via a screw mechanism, triggered by a "synchronization engine", or change of speed level motor.

Isochronous governors are not used in multiple machines systems, because of the need for appropriate division of load between the machines. The division itself is ensured by providing a regulation and speed decrease in the governor action or speed control (Kundur, 1994).

2.2 Governor with speed drop

Figure 4 shows the schematic of a governor with speed drop or adjustment of equilibrium. The characteristic drop is due to the connection between the pilot valve and the main servo motor

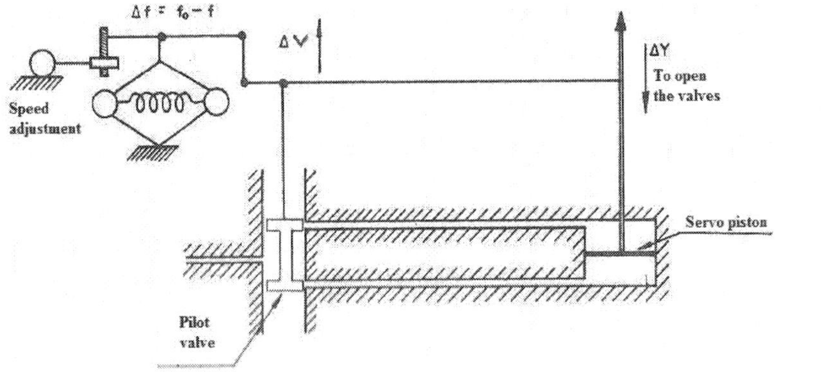

Fig. 4. Schematics of the *speed drop* governor

With reference to the block diagram of the isochronous governor in figure 2, the effect of the feedback connection, figure 4, is represented by inserting a gain K_3. This fact is illustrated in figure 5.

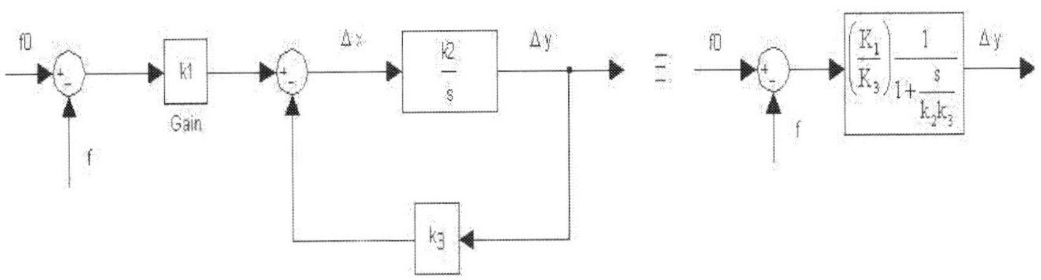

Fig. 5. Block diagram of the governor with speed drop

Figure 6 shows the time response of the valves controlled by the speed governor to a step change in the frequency deviation. TG is known as a time constant of speed governor, usually the order of a fraction of a second (0.5 s).

From the transfer function in figure 6, a function given by equation (4), one can notice that the value of the steady-state frequency shift Δf required to produce a variation of 1 p.u. at the valve of the turbine ΔV is R. This parameter is known as "speed control".

$$\frac{\Delta\text{valve (p.u.)}}{\Delta\text{frequency (p.u.)}} = \frac{1}{R(1+sT_G)} \tag{4}$$

The parameter R can also be expressed in percentage as

$$R\% = \frac{|N_0 - N|}{N_R} \times 100\%$$

where:
N_0 = speed in steady-state at no load (valves closed);

N = steady-state speed at full load (valves completely opened);
N_R = rated speed.

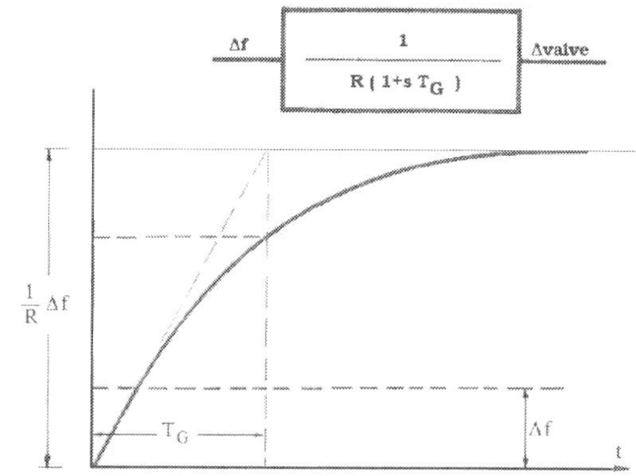

Fig. 6. Throttle response controlled by the speed governor in figure 4 to a step change in the frequency deviation.

Thus, a regulation of 5% means that a frequency deviation of 5% causes a 100% change in power or at the valve outlet. In the terminology of process control, the governor would be characterized as a proportional control of a proportional gain of 1/R, or a proportional band of R×100%.

In order to familiarize the reader with the methodology used to perform computer simulations of dynamic software analysis, it is presented through figure 7 the block diagram of a steam turbine with reheat, figure 7(a) and without reheating, figure 7(b). However, they are representations for a system that is being represented by a single equivalent electric machine.

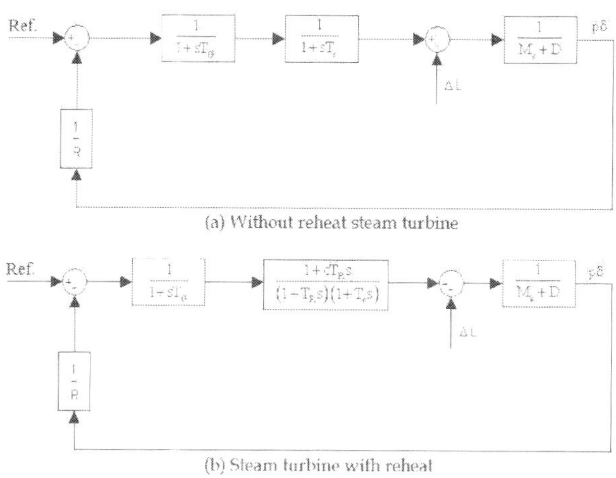

Fig. 7. Block diagrams for isolated systems represented by a single equivalent machine: (a) steam without reheat, (b) steam with reheat

In figure 7 was used the following notation:

M = Σ (MS + MR + MH) = combined inertia moment;
D = ∂PL / ∂ω = damping characteristics of the load;
Rs = Regulation of steam without reheating;
Rr = Regulation of steam with reheating;
T_R = Reheater time constant;
T_S = Charging time of the steam turbine;
T_G = Time constant of the governor.

However, figure 8 presents the block diagram of a steam turbine in a simplified form. It is one of the models adopted by the IEEE - Institute of Electrical and Electronic Engineers (Anderson & Fouad, 2003). It is noteworthy that for a steam unit T_4 = 0s.

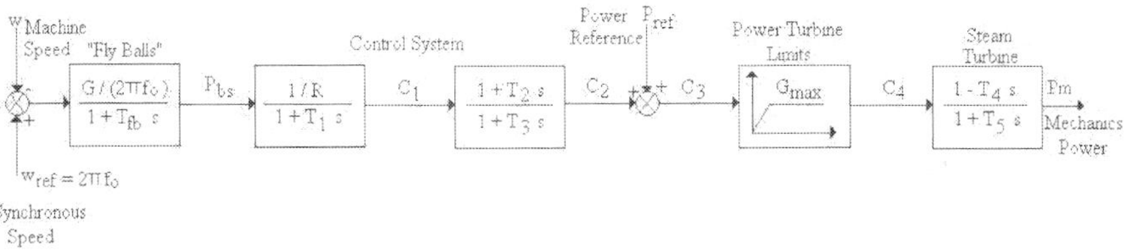

Fig. 8. Model for a thermal turbine speed governor

3. Steam turbine operation under abnormal conditions of frequency

Special attention has been given to the operation of turbo-generators under abnormal system operating conditions concerning the frequency (over or under frequency). In particular, more research efforts have been made to analyze the possible damage the steam turbines are the subject when operating in conditions of prolonged under frequency, during a severe overload imposed on the system with the occurrence of a generation deficit.

To avoid a total collapse of the system and minimize the damage to the equipment during such disturbances, considerable effort has been expended in the development and implementation of automatic load shedding. These load shedding programs are idealized to reject just enough loads to lighten the remaining generation system from overload to restore as soon as possible the system frequency close to nominal.

Both the turbines and generators have operational restrictions, to different degrees, under abnormal frequency conditions. On the other hand, steam turbines are more sensitive to the under frequency phenomena when compared with the attached generators. For this reason, the following discussion is related to the steam turbines (Anderson & Fouad, 2003).

3.1 Turbine limits

The steam turbine consists of a sequence of stages of increasing dimensions with each stage having more complicated geometry of stator and rotor blades. Each blade row has its own natural frequency. The turbines are carefully designed so that the resonance frequencies of the blades to the rated speed are sufficiently out of phase, to avoid vibration and excessive "stress" or fatigue.

Must be highlighted that here will be presented a discussion for systems operating in a nominal frequency of 60 Hz, like in Brazil.

Figure 9 illustrates the phenomenon that involves operations of the turbines under off-nominal frequency ("off-frequency operation"), showing the amplitude of vibrations fatigue for a number of stages of a turbine vane according to the nominal frequency. It is observed that when the turbine operates out of its nominal frequency, the amplitude of "stress" increases and some damage can be cumulative.

The Campbell diagram, illustrates how a change in turbine speed can provide steam excitation frequencies that coincide with the natural frequencies of a blade, how can be seen in the reference cited at final of this paragraph. At the figure 9 is demonstrated the relationship between blade failure due to fatigue and off-frequency operation. Below stress level A, the vibration stress amplitude is low enough that no damage results. Operation at stress level B would produce a failure at 10^4 cycles, and at level C failure would occur at 10^3 cycles (Kundur, 1994).

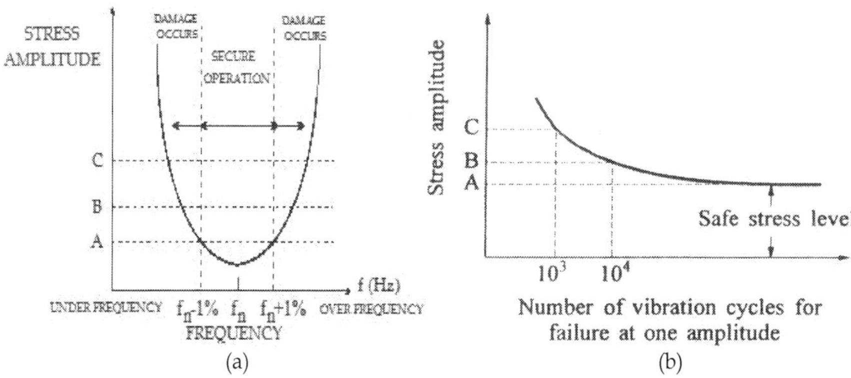

Fig. 9. Frequency x *stress* (fatigue) range for steam turbines (a) Increase in vibration amplitude with off-frequency operation; (b) Stress *x* number of cycles to failure

Research for a large amount of vibration frequency data leads to the time limits suggested for operations outside of the nominal frequency range and for different steam turbines that are represented in figure 10.

The diagram of figure 10 represents the estimated minimum time to the failure of some part of the structure of the blades, illustrating the limits of time for operation of steam turbines for both the under-frequency and for over-frequency. It is observed for a frequency deviation of 5% or more, the damage time becomes very small and it is not practical to operate a system more than a few seconds in this frequency range.

As can be seen, for the lower limit frequency, i.e., approximately f_n- 6% Hz (see figure 10), the permissible minimum time is one second for frequencies near the nominal time is undefined so that a variation frequency of $\pm 1\%$, will have no harmful effect on the blades.

It is important to note that the effect of operation outside the nominal frequency is cumulative, i.e., half a minute of operation under full load at f_n–4% Hz now leaves only approximately another half a minute of f_n–4% Hz for operation in the unit life.

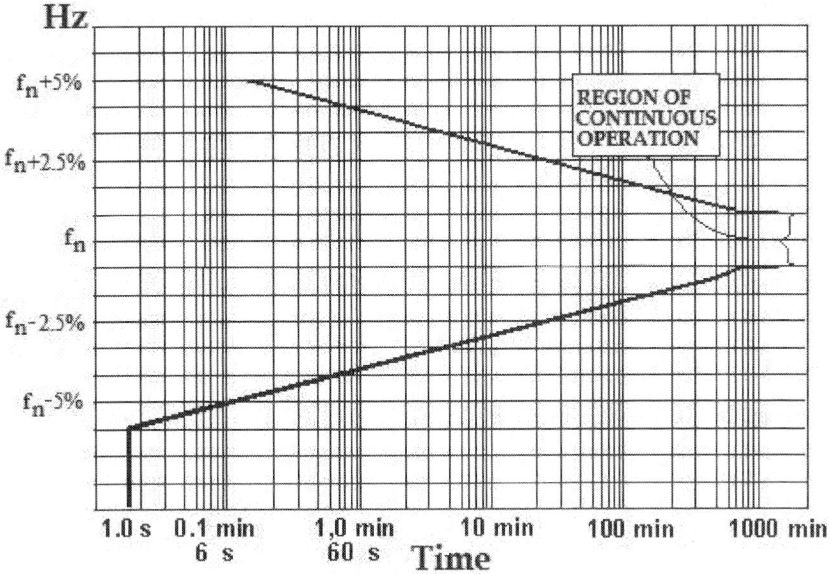

Fig. 10. Frequency limits of the operation of a steam turbine

From figure 10, we can build table 1 that shows the maximum time values of operating a steam turbine according to its frequency.

Frequency at full load (Hz)	Maximum operating time
$fn - 1\%$	Continuously
$f_n - 2\%$	100 minutes
$f_n - 3\%$	10 minutes
$f_n - 4\%$	1 minute
$f_n - 5\%$	0.1 minute
$f_n - 6\%$	1 second

Table 1. Maximum operating time of a steam turbine according to its frequency of operation at full load

The condition most often found corresponds to the turbine over-frequency, resulting in sudden shutdown of the generator by the action of the breaker. Under these conditions the characteristic of the speed governor will allow an over speed of around 5% and from there take immediate action to reduce the speed to close to the nominal (Anderson & Fouad, 2003).

3.2 Cascading shots

If a limit is reached for operating a turbine, it will be disconnected from the system by actions of its protection. This exacerbates the problem, since the power imbalance increases (greater deficit of generation), and consequently, the frequency drops quickly and the system enters a cascade process until it comes to the collapse.

Thus, we can define three characteristic zones of frequency for operation of turbines, as shown in table 2.

OPERATION ZONE	FREQUENCY - f (Hz)
Favorable	$f_n - 0.17\% < f_n < f_n + 0.17\%$
Tolerable	$f_n - 1\% < f_n < f_n + 1\%$
Extreme	$f_n - 6\% < f_n < f_n + 1.7\%$

Table 2. Characteristic zones of frequency for operation of turbines

The zone limits are practical, because they were obtained empirically. Table 2 indicates a favorable frequency zone of (60±0.1) Hz. As a reference, in the USA the limits are of the order of (60+0.05) Hz.

Figure 11 shows the characteristic ranges or zones of frequencies (based on Table 2) and the minimum values that can cause damage in the operation of steam turbines.

For this reason, emphasis is given to methods to protect the turbine under-frequency conditions, and to this end, a system must be equipped with a load shedding program, and this requires a prior knowledge of the system behavior.

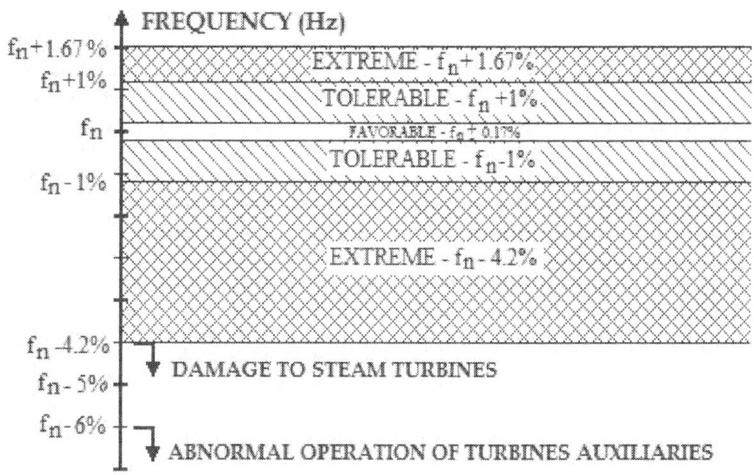

Fig. 11. Effect of frequency on the operation of the turbines

4. Case study

In order to assess the impacts on the steam turbine blades of a thermal unit, which steam is originated by the burning of bagasse and cane sugar straw, it is used the electrical system depicted in Figure 12 below.

Figure 12 illustrates the single line diagram of the electrical system to be used as reference for the study to be conducted. This single line diagram is very representative of the typical condition of independent generation interconnected with a larger electrical system. It is an

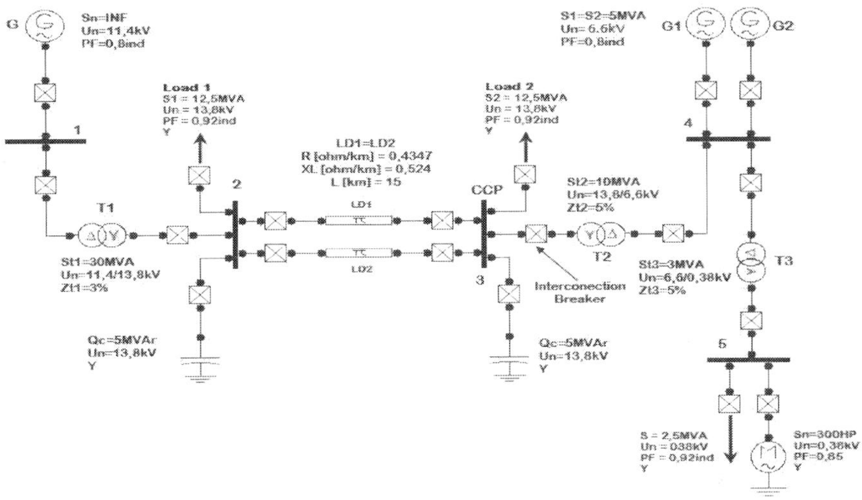

Fig. 12. Single line diagram for the electrical system in the case considered

electricity distribution network, connected to the busbar number 3 (Point of Common Coupling - PCC), through a transformer, the independent producer electrical system. The representation of the (IPP) Independent Power Producer consists of two synchronous generators (whose primary machines are steam turbines), a static electrical load as well as an electric motor, internal to this "independent" system. In addition to meeting its domestic demand for energy, these generators also provides power to the distribution network.

The three-phase distribution network has an ideal voltage source (infinite bus), connected to transformer T1, through busbar 1. This transformer T1 is connected to two power distribution lines of 13.8 kV through busbar 2. This busbar has a capacitor bank and load. The two referred distribution lines end at busbar 3, where it is connected to transformer T2. At busbar 3, there is a bank of capacitors and load. Transformer T2 is used to make the connection between the power utility and the independent power producer. The transformer T2 is also connected to busbar 4, where are connected the generators of the independent power producer (IEEE Std. 1547, 2003). These generators have the nominal values of 5 MVA, 6.6 kV. At busbar 4 there is also another transformer T3, which is connected to busbar 5, where we have the loads of the independent power producer.

The voltage source, which is the mains of the distribution system, was implemented as na ideal three-phase source. Therefore, the short-circuit level at busbar 1 is considered infinite.

The data needed for modeling the independent power producers generators (among others, the sub-transient, transient, synchronous reactance and time constants) are listed in Table 3.

The rated parameters obtained for the machine voltage and speed regulators, as well as data referred to the independent power producer synchronous generator, were obtained directly from manufacturers.

Data needed for G1 (G1 = G2)	
$S_n = 5MVA$	x_0=0.046pu
U_n= 6.6kV	T'_{d0}=1.754s
R_A= 0.004pu	T'_{q0}=0s
x_L= 0.1pu	T''_{d0}=0.019s
x_d= 1.8pu	T''_{q0}=0.164s
x_q=1.793pu	H_1=1s ;H_2=2s
x'_d=0.166pu	P=4 poles
x'_q=0.98pu	f=60 Hz
x''_d=0.119pu	ω_s = 188.5 rad/s
x''_q=0.17pu	---

Table 3. Synchronous machine parameters for the independent generator

It is highlighted that this independent generator is driven by a steam turbine with regulators and governors modeled in "ATP-Alternative Transients Program", through the use of TACS subroutines. Thus we adopted here the block diagram shown in Figure 8 in order to assess the dynamic response of the system of Figure 12, towards the contingency to be applied. Similar works using ATP were developed like (Saldaña et al., 2006) and an important work about distributed generation can be found in through (Freitas et al. 2006).

However, it is necessary the modeling of the voltage regulator to obtain a better and more reliable response from the model. In this way, it is presented through Figure 13 the block diagram used to represent the excitation regulator. This is the IEEE type I model (Guimarães, 1990).

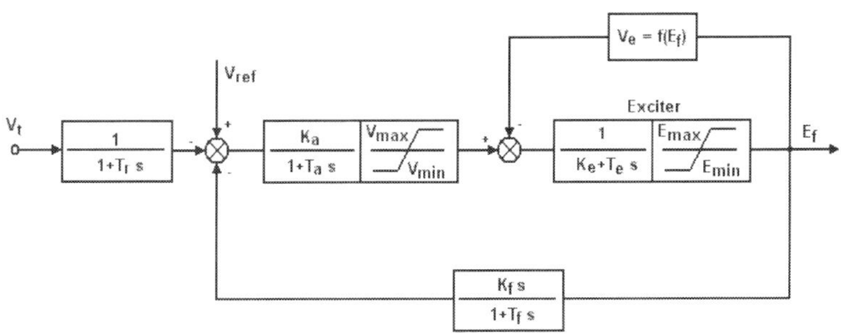

Fig. 13. Voltage regulator model

4.1 Sudden opening of the interconnection breaker

With the purpose of evaluating the influence of penetration of the independent power producer in the largest network and its close correlation with the inertia moment of the generator, with regard to the maintenance of transient stability was established the power flow illustrated in Table 4.

The machines of the IPP generate a total of 4.2 MW and 1.5 MVAr respectively of active and reactive three-phase power. Thus, the total of active power generated 1.7 MW is designed to the grid of the local electricity authority.

Source	P_G [MW]	Q_G [MVAr]
G1	0.7	0.25
G2	0.7	0.25

Table 4. Active and reactive power generated by the independent power sources

It is worth clarifying to the reader the pre-established conditions to the speed governors of the IPP machines, these where adjusted to operate with a 5% slope, in order to better meet the interconnection. Therefore, no machine installed in the IPP generation, has the characteristic to ensure that frequency oscillations will be inside the favorable frequency range, i.e., there is no need for the presence of an isochronous generator.

Faced by the sudden opening of the interconnection circuit breaker there is a considerable fluctuation in the speed of the electric machine G2, Figure 14(a). Such transient oscillation immediately affects the frequency in the electrical system belonging to the IPP, burdening strongly the quality of electricity, especially with regard to the efficiency of micro-computerized devices (Dugan et al., 2002).

However, slight changes in the operating speed of the machines originate variations in power demand, since the active and reactive powers, required by electrical loads vary dynamically with system frequency. Nevertheless, there is clearly and strongly the tendency of the synchronous machine speed G2 to stabilize, as shown in Figure 14(a) in the period after contingency, thus indicating the influence of the IPP generation penetration in the transient stability of its own machines.

As for the speed regulator, Figure 14(b) convincingly portrays the attempt of that regulator to meet the new situation imposed on the system.

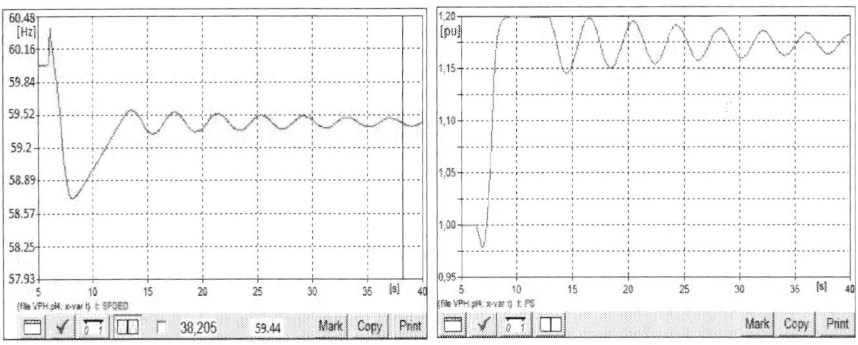

(a) (b)

Fig. 14. Response of the mechanical variables of the generator G3 (a): Operating frequency; (b) Action of the speed governor

In the moment immediately after the opening of the interconnection breaker, the machine is subjected to an immediate rejection of load. Thus, the operating speed rises gently to 189.63 rad/s (f = 60.4 Hz). However, at this moment the lower inertia machine G1 is taken out of the system, in order to restore the load-generation balance. It is worth mentioning the performance of the machine G2 speed regulator in Figure 14(b) to reduce the steam consumption to ensure the G2 synchronous speed.

With the withdrawal of G1, a new scenario is presented with respect to the power flow in the system. The machine G2 becomes responsible for all the electrical loads at the IPP. Thus, according to equation (5), the balancing equation, the electrical power becomes greater than the mechanical power, which leads to deceleration of the generator, taking the angular speed to ω = 184.45 rad/s (f = 58.71 Hz). At this moment, the speed governor acts to increase the steam consumption to restore the load-generation balance. However, the frequency of operation of the machine oscillates in a damped transient between f_1 = 59.6 Hz (ω = 187.15 rad/s) and f_2 = 59.33 Hz (ω = 186.42 rad/s), and tends to stabilize at a frequency f = 59.4 Hz, i.e., ω = 186.74 rad/s.

$$\frac{2.H.d^2\delta}{\omega_R.dt^2} = P_m - P_e = P_a \tag{5}$$

As for the speed governor, it follows the oscillations observed in the operating speed of the machine G2, tending to stabilize around 1.17 pu, in other words, it is necessary the production by the boilers of 17% more steam, to be supplied to the generators turbine to meet the new request of active power, as can be observed in Figure 14(b).

The moment immediately after the withdrawal of the machine G1, the remaining generator G2 reaches the frequency f = 58.71 Hz. That is classified as extreme in Table 2, burdening sharply the impeller blades of this generating set. However, during the damped transient period, G2 oscillates within the tolerable region of Table 2. The observation of such limits ensures the life time of the equipment, thus reducing the mechanical wear of turbine blades and reduces the number of stoppages for maintenance of this generating set.

In order to get a better response in operation for the IPP electrical system after the opening of the interconnection circuit breaker, it is possible the insertion of a flywheel in the axis of G2, in order to rise the generating set equivalent inertia.

However, it is necessary to reject the generating unit G1 inside the 100 ms after the application of contingency.

4.2 Application of a flywheel

The flywheel is a mechanical device with significant moment of inertia used for storing rotational energy. Has high resistance to variations in speed, which helps to ensure a small variation in rotation speed of the shaft (Kundur, 1994).

The energy is stored in the rotor as kinetic energy, or more specifically, rotational energy:

$$E_k = \frac{1}{2}.I.\omega^2 \tag{6}$$

where:

ω is the angular speed, and

I is the mass moment of inertia about the center of rotation.

An increase in angular speed greatly increases the energy stored in the flywheel, and the energy density of the KESS - Kinetic Energy Storage System. However, the no-load losses, viscous friction with air and by contact and friction in mechanical bearings, will be higher, hurting the efficiency of the system. In addition, the rotating parts of the system must support high mechanical stresses due to high speed at which they are submitted. These limitations found in KESS can now be properly solved because:

- the development of new materials such as fiberglass and carbon;
- the use of vacuum systems, which reduce the losses by viscous friction with the air;
- the use of magnetic bearings that eliminate friction by contact;
- the advancement of power electronics, which improved signal conditioning;
- the evolution of microelectronics, that allowed more sophisticated and less expensive control systems.

These factors enable the development and implementation of a new generation of KESSs, with superior performance, whose speed can reach tens of thousands of rpm (approximately 60,000 rpm). Thus, increases the number of possibilities for the use of flywheels, since this generation presents high energy density and power.

Among the applications for this equipment, we can mention:

- uninterruptible power supply (UPS systems);
- provision of extra energy in case of increased demand;
- compensation of voltage dips.

The moment of inertia is a measure of resistance to the applied torques in a rotating object, i.e., the higher the moment of inertia, the slower the wheel will spin after applying a certain force on its axis.

It is presented here the main ways to obtain the moment of inertia for a solid cylinder:

$$I = \frac{1}{2}.m.r^2 \tag{7}$$

However, for an empty cylinder with thin walls, one must employ the equation (8) detailed below:

$$I = m.r^2 \tag{8}$$

However, in the case of an empty cylinder with thick walls, one should use the equation 9 below:

$$I = \frac{1}{2}.m.(r_1^2 + r_2^2) \tag{9}$$

Where m is the mass and r is the radius of gyration.
It is worth to present equation (10) used to obtain the constant of inertia of a generator.

$$H = \frac{\frac{1}{2}.\omega_s^2.I}{S_n} \tag{10}$$

In a simplified perspective, the procedure is the analysis of the flywheel shown in Figure 15.

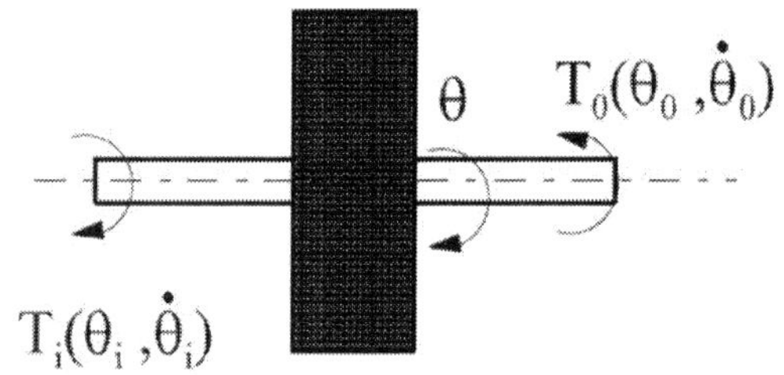

Fig. 15. Simplified representation of a flywheel.

Where:

I	is the inertia moment of the wheel;
θ	is the wheel angular position;
T_i	corresponding torque to an angular position θ_i ;
T_0	corresponding torque to an angular position θ_0 ;
$\dot{\theta}_i$	angular speed at the angular position θ_i ;
$\dot{\theta}_0$	angular speed at the angular position θ_0 .

Taking, arbitrarily, the torques T_i and T_0 as a positive and negative, respectively, we get the equation for the motion of the wheel:

$$M = T_i(\theta_i, \dot{\theta}_i) - T_0(\theta_0, \dot{\theta}_0) - I.\ddot{\theta} = 0 \tag{11}$$

It is presented here different situations for three specific inertia moments, and their corresponding implications on the speed response of the IPP generator G2 in Figure 16.

With the increase in the synchronous machine G2 equivalent moment of inertia, due to the insertion of a wheel, in Figure 16(a) for I = 600 kgm², there is the stabilization of the operating speed of the generator for the considered time period in the simulation at t = 40s. However, at the moment immediately after the untimely opening of the interconnection breaker, the operating speed of machine G2 reaches ω = 189.5 rad/s (f = 60.3 Hz). This is due to the instantaneous rejection of the applied load. Nevertheless, due to the withdrawal of machine G1, G2 provides a considerable reduction in its speed of operation due to the increase in electric power demand. The frequency reduces to 184.93 rad/s (f = 58.9 Hz), the extreme region of operation according Table 2. However, the machine varies in speed within the range between $186.53 < \omega < 187.07$ rad/s, or $59.4 < f < 59.5$ Hz. Note that, observing Table 2, the oscillation frequency remains within the favorable zone of operation, ensuring the life time of steam turbine blades.

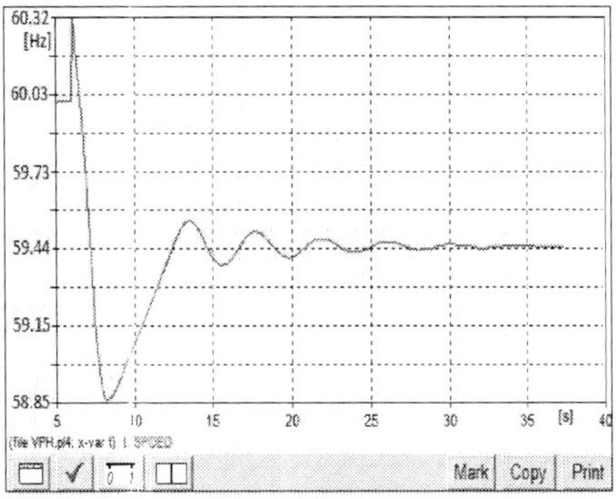

(a)

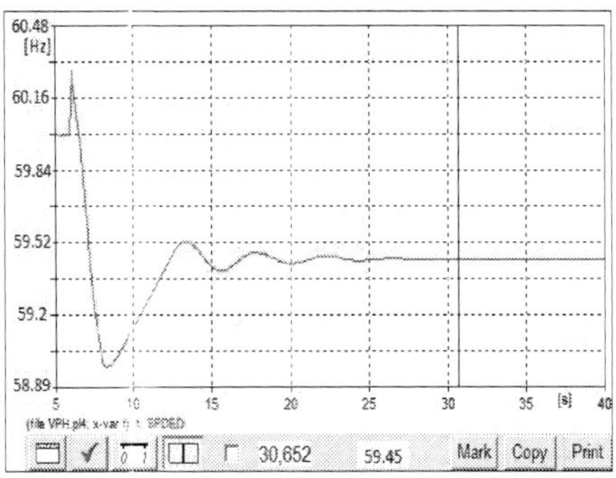

(b)

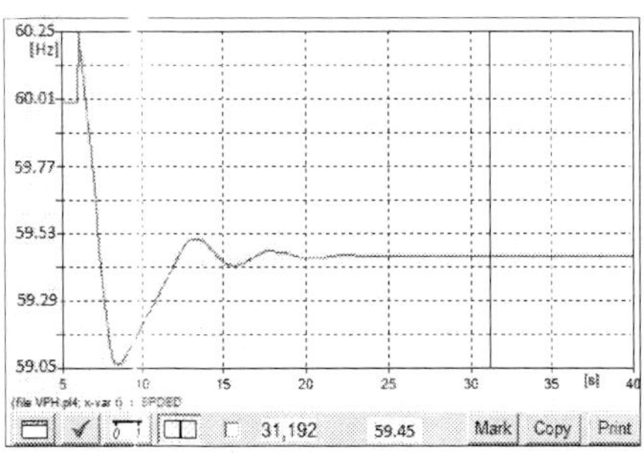

(c)

Fig. 16. Frequency response of machine G2 for different inertia moments (a): I = 600 kgm^2; (b): I = 700 kgm^2; (c): I = 800 kgm^2

Figure 16(b) shows the response to an equivalent moment of inertia I = 700 kgm^2. The system has more damping, reducing the time required to reach a new steady state speed at ω = 186.77 rad/s (f = 59.5 Hz). Therefore, according Table 2, the industrial frequency falls in the tolerable operating zone, contributing not only to maintain the transient stability of the system, but also ensures the integrity of the steam turbine blades, as well as presenting a significant contribution in terms of the quality of electric power. As for the oscillations of frequency in the transient period, it is verified a maximum speed of ω = 189.4 rad/s (f = 60.3 Hz), at the moment immediately after the contingency. Nevertheless, with the removal from operation of G1, machine G2 reaches a minimum speed of ω = 185.28 rad/s (f = 59.0 Hz) at the extreme operation zone.

However, the machine G2 varies in speed within the range between $186.61 < \omega < 187.01$ rad/s or $59.4 < f < 59.5$ Hz. It can be noticed that the frequency oscillation, according Table 2, remains in the favorable zone of operation, ensuring the expected life time for blades of the steam turbine.

With the main objective to reduce the oscillations, evident in the speed of operation, of the remaining synchronous generator (G2) it was applied an equivalent moment of inertia I = 800 kgm^2, the result can be seen in Figure 16(c). There, the damping obtained allows reaching the new steady state more efficiently. This helps to mitigate the harmful effects to the quality of electric power to ensure better response of electrical parameters of the system.

With regard to changes in speed, the maximum reaches the magnitude ω =189.3 rad/s (f = 60.3 Hz) at the tolerable zone of operation, but the minimum is settled at ω = 185.55 rad/s (f = 59.06 Hz) at the extreme operation zone. Thus, one can conclude here and for the other equivalent inertia moments simulated that, at the instant immediately after the removal of generator G1, the synchronous machine G2 presents a significant reduction in its operating speed. In this way, it is generated a sub-frequency costly to the physical and mechanical integrity of the turbine blades, since the obtained frequency, at this instant, fits the extreme region of operation at Table 2. However, oscillations in the transient period do not affect the impeller blades, as they pass inside the tolerable zone. Nevertheless, after the damped transient period, the machine reaches the new steady-state and stabilizes at a speed ω = 186.77 rad/s (f = 59.5 Hz), similar to that obtained for the case illustrated in Figure 16(b).

5. Conclusion

Through the discussions presented along this chapter, there is the need to make the correct and effective study of the synchronous machine speed governors, in order to obtain better dynamic response of the system. However, this goal is achieved from te transfer function of these governors, so it is essential to be aware of the topology of the turbine used, i.e., whether or not this is reheating . It shows that before the model transfer function, yet it is necessary to choose an appropriate software to implement the model. In this way, it can be seen that ATP, through TACS subroutines allows the correct and efficient computational modeling of both the speed governor and voltage regulator.

The answers regarding the performance of the speed governor, for the transfer function shown here, respond faithfully and reliably to dynamical requests imposed to the electrical system under analysis.

Due to untimely performance by interconnection break, the IPP machines face an immediate rejection of load, which burdens the system frequency, thereby compromising the integrity of turbine blades. However, it appears that even without the application of mitigation measures the system tends to stabilize. Nevertheless, the responses refer to a system such that the penetration of IPP does not take place significantly. The literature shows that higher levels of active power exported by the IPP mean larger damage to the turbines due to higher frequency oscillations. Thus, the use of the flywheel is shown as an effective alternative, since it damps transient oscillations, thys contributing to ensuring the life time of the impeller blades.

In short, it is extremely important the correct modeling and use of speed governors in studies relevant to the dynamic stability of power system analysis in terms of commitment to the life time of the blades of steam turbines.

6. References

Moura, F. A. M. ; Camacho, J. R. ; Chaves, M. L. R. & Guimarães, G. C. *"The Dynamic Interaction of Independent Power Producer Synchronous Machines Connected to a Distribution Network in ATP-EMTP"*, ICEM2010 - XIX International Conference on Electrical Machines, Rome, Italy, September 2010.

Moura, F. A. M. ; Camacho, J. R. ; Chaves, M. L. R. & Guimarães, G. C. *"Independent Power Producer Parallel Operation Modeling in Transient Network Simulations for Interconnected Distributed Generation Studies"*, Electric Power Systems Research, Elsevier, Article in Press, (2009),doi:10.1016/j.epsr.2009.08.016.

Anderson , P. M. & Fouad , A. A. *Power System Control and Stability*, vol. I. Iowa, 1977.

Kundur, P. *Power Systems Stability and Control*, McGraw-Hill, EPRI Power Systems Engineering Series, New York, 1994.

IEEE Standard for Interconnecting Distributed Resources with Electric Power Systems, IEEE Std. 1547, New York, USA, 2003.

LIPA-Long Island Power Authority, *Control and Protection Requirements for Independent Power Producers, Transmission Interconnections*, found at the internet in 22/11/2007, at http://tinyurl.com/33clq4.

Saldaña, C. ; Calzolari, G. & Cerecetto, G. *"ATP modeling and field tests of the ac voltage regulator in the Palmar hydroelectric power plant"*, Electric Power Systems Research, Elsevier, 76 (2006), pp.681-687.

Freitas, W. ; França, A. M.; Vieira Jr., J. C. M. & Silva, L. C. P. *"Comparative Analysis Between Synchronous and Squirrel Cage Induction Generators for Distributed Generation Applications"*. IEEE Trans. Power Systems, vol. 21,NO.1, FEBRUARY 2006.

Guimarães, G. C. *Computer Methods for Transient Stability Analysis of Isolated Power Generation Systems with Special Reference to Prime Mover and Induction Motor Modelling*, PhD Thesis, University of Aberdeen, 1990.

Dugan, R. C. ; McGranaghan, M. F. ; Santese, S. & Beaty, H. W. *Electrical Power Systems Quality*, McGraw-Hill, New York, 2002.

18

Aeronautical Engineering

Časlav Mitrović, Aleksandar Bengin,
Nebojša Petrović and Jovan Janković
*University of Belgrade, Faculty of Mechanical Engineering,
Serbia*

1. Introduction

Aerospace engineering is the primary branch of engineering concerned with the design, construction and science of flight vehicle. Consequently, they are usually the products of various technological and engineering disciplines including aerodynamics, propulsion, avionics, materials science, structural analysis and manufacturing. These technologies are collectively known as aerospace engineering. It is divided into two major and overlapping branches: aeronautical engineering and astronautical engineering.

It is typically a large combination of many disciplines that makes up aeronautical engineering. The development and manufacturing of a modern flight vehicle is an extremely complex process and demands careful balance and compromise between abilities, design, available technology and costs. Aeronautical engineers design, test, and supervise the manufacture of aircraft. They also develop new technologies for use in aviation.

Aeronautical Engineering is a chapter that encompasses challenging areas such as aircraft design, light-weight structures, stability and control of aeronautical vehicles, propulsion systems, and low and high speed aerodynamics. The field also covers their aerodynamic characteristics and behaviors, airfoil, control surfaces, lift, drag, and other properties.

The chapter will include all our research and published papers.

2. Aerodynamics

This chapter presents a selection of published scientific papers. The work was the subject of research in the following fields of aerodynamics: Subsonic, Transonic and Supersonic, High Angle of Attack, High Lift, Computational Fluid Dynamics, Wind Tunnel and Flight Testing and Helicopter Rotor Aerodynamics.

2.1 Unsteady motion of two dimensional airfoil in incompressible inviscid flow

In this subchapter we shall be looking at many ways in which to solve the problem of unsteady incompressible flow over an aerofoil. The flow being incompressible is a great simplifier to the problem, this allows to take many of the results of steady flow as read. It is still however, not a trivial problem.

2.1.1 Introduction

Numerical method for solution of unsteady flow over airfoil oscillating in incompressible, inviscid flow field is based on well known panel method in form developed for steady flow by Hess and Smith. Due to unsteadiness in addition to boundary conditions, additional conditions are necessary. Kelvin theorem and Kutta condition is to be applied in unsteady form. This results with a nonlinear problem.

2.1.2 Formulation of problem and boundary conditions

Inviscid incompressible fluid flow is described by one partial differential equation which satisfy law of fluid continuity (Laplace equation):

$$\frac{\partial^2 \Phi}{\partial x^2} + \frac{\partial^2 \Phi}{\partial y^2} = 0$$

Equation is the same for steady and unsteady flow. Hence, methods for steady flow can be applied in a solution procedure for unsteady flow. When is the panel method applied, the problem is reduced on determination of the velocity field which must satisfy boundary condition of zero normal flow at the surface of airfoil.

When the airfoil is moving relative to a steady undisturbed flow, this boundary condition must be expressed in form:

$$\left[\vec{V}_F(t) - \vec{V}_S(t)\right] \cdot \vec{n}(t) = 0$$

where $\vec{V}_F$ is the fluid velocity, $\vec{V}_S$ is the surface velocity and $\vec{n}$ is normal at the surface.

When the airfoil is moving relative to a steady incompressible inviscid flow, circulation $\Gamma(t)$ around airfoil varying with time t. Total circulation around airfoil and vortex wake must be zero (Kelvin theorem), which is possible to achieve only if the circulation with intensity $(\partial\Gamma/\partial t)\delta t$ is separated from trailing edge of airfoil in interval dt. This separate vorticity is carried out in downstream direction. There is direct relationship between this and Kutta condition for trailing edge. Kutta condition can be expressed as equality of airfoil upper and lower side pressures on the trailing edge, and on the basis of unsteady Bernoulli equation can be expressed as:

$$\frac{\partial\Gamma}{\partial t} = -\frac{1}{2}\left(V_U^2 - V_L^2\right) = -\frac{\left(V_U + V_L\right)}{2}\left(V_U - V_L\right)$$

where U and L denote upper and lower side of trailing edge, respectively ($x/l \to 1$).

Therefore, above mentioned consistence of Kutta condition and vortex shedding model can be clearly seen. Namely, during time above mentioned circulation $\Gamma(t)$ around airfoil is balanced by vortex shedding with intensity $\left(V_U - V_L\right)$, with mean velocity $\left(V_U + V_L\right)/2$.

Therefore, unsteadiness of the problem is taken into consideration through unsteady form of kinematics boundary condition, Kelvin theorem, unsteady Bernoulli equation, and unsteady form of Kutta condition for trailing edge.

2.1.3 Discretization and numerical solution procedure

Solution of flow over airfoil (moving arbitrary), depending on time and starting from time t=0 is calculated in successive time intervals t_k (t_o= 0, k=1,2,3...). Fig. 1 shows model for the time t_k.

Airfoil contour in time t_k is replaced by N linear elements. $(\sigma_i)_k$ and γ_k are uniform source and circulation distributions on i-th element (i=1,2,...,N), where $(\sigma_i)_k$ varies from one element to another, γ_k is the same for each element on airfoil and k denotes time t_k. Total circulation Γ_k is given as $\gamma_k\times$ (airfoil perimeter).

An elementary vortex wake with length of Δ_k and pitch angle of θ_k in regard to the x-axis (to the free stream direction) is attached to the trailing edge. Length Δ_k and angle θ_k are arbitrary in first iteration. Their values will be determined as the part of the solution. Circulation in trailing edge vortex wake element is $(\gamma_w)_k$, where:

$$\Delta_k(\gamma_w)_k = \Gamma_k - \Gamma_{k-1} \tag{1}$$

Hence, circulation on the element is equal to the difference between circulations around airfoil in times t_{k-1} and t_k, assuming that Γ_{k-1} has been already determined. Vortex wake consist of concentrated vortices formed by vorticity shed at earlier times, which is assumed to be transformed into discrete vortices. Concentrated vortices is moving with resulting velocity calculated in the center of each vortex at each successive time interval. Therefore, strength and positions of discrete vortices are regarded as known at time t_k, and there are N+3 unknowns $(\sigma_i)_k$ (i=1,2,...,N), γ_k, Δ_k and θ_k at the time t_k.

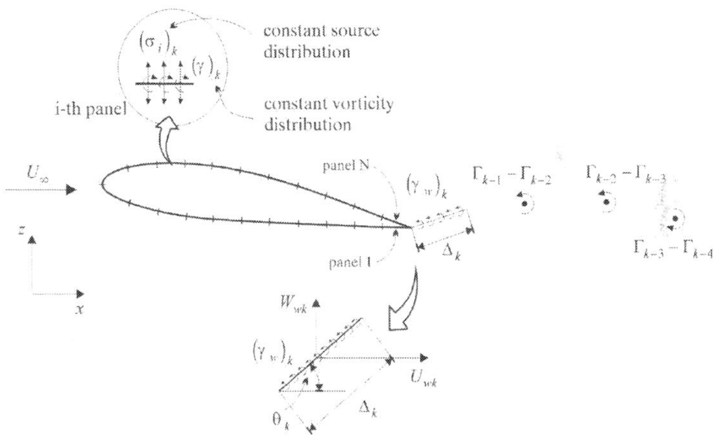

Fig. 1. Solution at time t_k

They are determined by satisfying following conditions:

- N conditions of zero normal velocity component in external middle point at each segment of airfoil

$$(V_{nj})_k = 0 \tag{2}$$

where $(V_{nj})_k$ is total normal velocity component at the external middle point of j-th. element in time t_k.

- condition of equal pressures in middle points of two elements on an airfoil on both sides of trailing edge (unsteady Kutta condition):

$$(V_{t1})_k^2 = (V_{tN})_k^2 + 2(\Gamma_k - \Gamma_{k-1}) / (t_k - t_{k-1}) \tag{3}$$

where $(V_{t1})_k$ is total tangent velocity component at middle point of first element and $(V_{tN})_k$ is total tangent velocity component at middle point of N-th element in time t_k.

- trailing edge vortex wake element length and direction (Δ_k and θ_k) are determined from condition that the element is tangent to the local resultant velocity and that its velocity is proportional to the local resulting velocity. If $(U_w)_k$ and $(W_w)_k$ are components of total velocity at middle point of trailing edge vortex wake element, excluding influence of the element on itself, then:

$$\tan\theta_k = (W_w)_k / (U_w)_k \quad , \quad \Delta_k = \left[(U_w)_k^2 + (W_w)_k^2 \right]^{\frac{1}{2}} \left[t_k - t_{k-1} \right] \tag{4}$$

By taking into the consideration that the problem considers incompressible flow, formulas for velocities induced from source and vorticity distributions are the same as these for the steady case.

Basic set of equations is nonlinear therefore following iterative procedure for its solution is accepted. Since Δ_k and θ_k are assumed it leaves N linear equations from (2) and square equation from (3). N linear equations are solved to give $(\sigma_i)_k$ (i=1,2,...,N) relatively to the γ_k. Then γ_k is determined from square equation (3). Once known $(\sigma_i)_k$ (i=1,2,...,N) and γ_k, $(U_w)_k$ and $(W_w)_k$ can be determined and replaced into (4) to obtain new values of Δ_k and θ_k. Procedure is repeated until Δ_k and θ_k are of requested accuracy.

When intensities are determined, source and vortex distribution is known. Pressure coefficient follows from unsteady form of Bernoulli equation:

$$c_p = 1 - \frac{V^2}{U_\infty^2} - \frac{2}{U_\infty^2}\frac{\partial\varphi}{\partial t} \tag{5}$$

where V is total velocity on external side of airfoil, and Φ is velocity potential. Forces and moment coefficients are obtained by direct integration of pressure distribution.

Since solution in time t_k is determined, model is applied for the time t_{k+1} in the same way as this for time t_k, with vortex vale computed in time t_k. Distributed vorticity in trailing edge vortex wake element in time t_k is now considered as concentrated in vortex with strength $(\gamma_w)_k\Delta_k$ at the position in time t_{k+1} determined by coordinates:

$$x = (x_{TE})_k + \frac{1}{2}\Delta_k \cos\theta_k + (U_w)_k (t_{k+1} - t_k) \quad , \quad z = (z_{TE})_k + \frac{1}{2}\Delta_k \sin\theta_k + (W_w)_k (t_{k+1} - t_k)$$

Resultant velocity in the center of each next concentrated vortex in vortex wake is calculated by solution at time t_k, and their position in time t_{k+1} directly follows.

2.1.4 Discussion of results

Computer program was developed from formulated numerical method. Flow over NACA 0012 airfoil was modeled by using 20 linear elements. Airfoil oscillating at high frequency trough certain angle of attack is considered flow characteristics are computed in time interval from 0 to 30 sec. Fig. 2 shows static and dynamic lift curve resulting from the application of this numerical method. Dynamic lift curve is obtained by airfoil oscillating through angle of attack of 6°, with amplitude 5° and reduced frequency k=0.5.

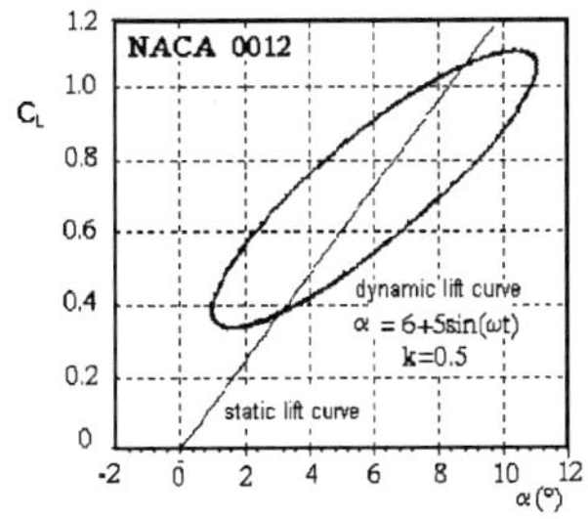

Fig. 2. Static and dynamic lift curve

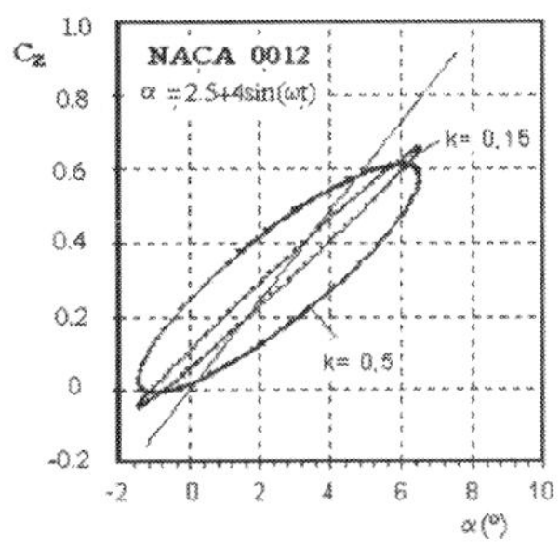

Fig. 3. Influence of reduced frequency

Dynamics lift curves for two reduced frequencies k=0.5 and k=0.15 based on obtained results are also shown (Fig. 3). In this case, airfoil oscillating through angle of attach of 2.5°, with amplitude of 4°. Dynamic lift curve shape depending on reduced frequency can be seen from the figure.

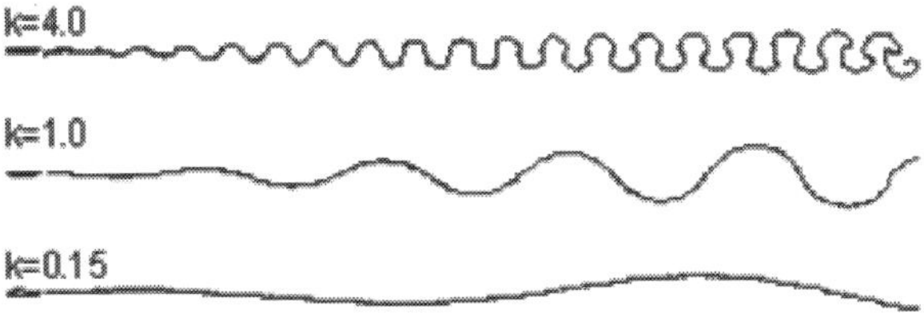

Fig. 4. Computed vortex traces as a function of the reduced frequency k, $\alpha = 2.5 + 4\sin(\omega t)$

Vortex wake shape of airfoil oscillating through same angle of attack, with same amplitude, but with different reduced frequency obtained by this method is shown at Fig. 4. It is similar with one obtained by methods of visualization in tunnels.

Numerical method presented in this paper leads to inviscid flow field calculation over airfoil moving arbitrary in time if it is supposed that flow stays attached and that it is separated art trailing edge. Method is completely general although only results of oscillating at high frequencies are presented.

2.2 An optimal main helicopter rotor projection model obtained by viscous effects and unsteady lift simulation

This subchapter presents a model of airfoil which allows obtaining a method for optimal main helicopter rotor projection by viscous effect and unsteady lift simulation through algorithm and set of program entireties, applicable to ideological and main project of helicopter rotor. Numerical analysis considerations in this paper can be applied, on basis of a real rotors theoretical consideration, with sufficient accuracy in analysis and constructive realizations of helicopter rotor in real conditions. The method for unsteady viscous flow simulations by inviscid techniques is developed. This allows the to define such optimal conception model of aerodynamic rotor projection which is corresponding to rotor behavior in real condition and with sufficiently quality from the aspect of engineer use.

2.2.1 Introduction

The basic aim is to determine the lift of helicopter rotor blades by using singularity method, that is to develop an optimal model of aerodynamic rotor planning by simulating unsteady flow and viscous effects. The model is considered corresponding to rotor behavior in real conditions and with high quality from the aspect of engineer use. The idea of unsteady lift modeling and viscous effects simulation by using singularity method is based on the need for avoidance of very expensive experiments in the first stage of rotor planning by using contemporary aerodynamic analysis applied to available computer technique.

It is necessary to modulate vortex wake, unsteady 2D flow field characteristics and blade dynamic characteristics for determination of aerodynamic forces acting upon helicopter rotor blade.

In the first part airfoil is approximated by vortex and source panels, and boundary layer by layer of vortices changing their position during time. Vortex trail and trail of separation are modulated by free vortices. The separation is modulated by free vortices. The dependence of coefficient of lift on angle of attack is determined for known motion of an airfoil. The behavior of one separated vortex in an article time model is spread into a series of positions at a certain number at different vortices. The separated flow velocity profile is approximated by superposition of displacement thickness of these vortices coupled with the potential solution model. After every time step the position of free vortices is changed, for what it is required generation of new vortices that would all together satisfy the airfoil contour boundary conditions.

In addition to a complex problem of analytical modeling these phenomena there is a problem of modeling interactions between effects of these phenomena. This approach to helicopter rotor planning allows experimental investigations in tunnels and in flight to be final examinations. In that way, a very useful interaction between numerical calculation and experimental results is achieved.

The finale result is obtaining of unsteady lift dependence of angle of attack. Model established in such a way is characteristic for the helicopter rotor blade airflow and it is based on the influence of the previous lifting surface's wake influence on the next coming blade.

2.2.2 Foundations of the irrotational 2-D flow

The planar potential flow of incompressible fluid can be treated in Cartesian coordinates x and y. Two dimensional potential incompressible flow is completely defined by the speed potential and stream function and is presented by Cauchy-Reimann equations:

$$\frac{\partial \varphi}{\partial x}+\frac{\partial \psi}{\partial y}=0 \quad \text{i} \quad \frac{\partial \psi}{\partial x}-\frac{\partial \varphi}{\partial y}=0$$

where u and v are velocities in x and y directions respectively.

The Laplace partial differential equations can also be introduced $\nabla^2\varphi=0$ and $\nabla^2\psi=0$. The fulfillment of Cauchy-Reimann conditions enables combining of the velocity potential and stream function $w(z)=\varphi(x,y)+i\psi(x,y)$. This complex function entirely defines the planar potential flow of incompressible fluid as a function of a complex coordinate. The complex analytical function w (z), called the complex flow potential, always has a unique value for the first derivative. This derivative of the complex potential is equal to the complex velocity at that point, i.e.:

$$\frac{dw}{dz}=u-iv=\overline{V}$$

where u is its real part (velocity component in x-direction) and v is the imaginary part (velocity component in y-direction).

Circulation and flow are equal to zero for any closed curve in complex plane. Complex potential has no singularities except at the stagnation point.

2.2.3 Simulation of the moving vortex

If we assume that the moving vortex of intensity Γ_0 is at the distance z_0, than the perturbance potential of such a flow is:

$$w(z)=V_\infty ze^{-i\alpha}+V_\infty\frac{a^2}{z}e^{+i\alpha}+i2a\cdot V_\infty\cdot sin(\alpha-\beta)\ln z+\frac{i\Gamma_0}{2\pi}\ln(z-z_0)-\frac{i\Gamma_0}{2\pi}\ln\left(\frac{a^2}{z}-\overline{z}_0\right)$$

and it's complex velocity:

$$\overline{V}=V_\infty e^{-i\alpha}-V_\infty\frac{a^2}{z^2}e^{+i\alpha}+\frac{i}{z}2a\cdot V_\infty\cdot sin(\alpha-\beta)+\frac{i\Gamma_0}{2\pi}\frac{1}{(z-z_0)}+\frac{i\Gamma_0(a^2/z^2)}{2\pi\left(\frac{a^2}{z}-\overline{z}_0\right)}$$

In this case displacement of the stagnation point appears, which is now at the distance $z = a \cdot e^{i\beta'}$. According to the request that complex velocity at the stagnation point must be zero:

$$\frac{dw}{dz} = \overline{V} = 0$$

we can determine position z and according to that β i.e. $\overline{V}(ae^{i\beta}) = 0$. The angle β defines new position of stagnation point, while the difference of angles is $\Delta\beta = \beta - \beta$. In order to keep the stagnation point at the steady position, a vortex of intensity Γ_1 must be released.Intensity of the vortices that are released can be determined according to the Kelvin's theorem $d\Gamma/dt = 0$.

So every change in circulation must be compensated by vortex inside of cylinder/airfoil of the opposite sign, whose intensity is a function of the variation of circulation around the cylinder $\Gamma_i = -(d\Gamma/dt)\Delta t$ and it is equal do the difference of the intensities before and after the free moving vortex is introduced $\Gamma_1 = \Delta\Gamma_1 = 4\pi a \cdot V_\infty\left[sin(\alpha - \beta_1') - sin(\alpha - \beta)\right]$.

Now the total complex potential has the value:

$$w(z) = w_0(z) + w_{\Gamma_0}(z) + w_{\Gamma_1}(z)$$

where:

$$w_0(z) = V_\infty z e^{-i\alpha} + V_\infty \frac{a^2}{z} e^{+i\alpha} + i2aV_\infty sin(\alpha - \beta)\ln z$$

$$w_{\Gamma_0}(z) = \frac{i\Gamma_0}{2\pi}\ln(z - z_0) - \frac{i\Gamma_0}{2\pi}\ln\left(\frac{a^2}{z} - \overline{z}_0\right) \quad , \quad w_{\Gamma_1}(z) = \frac{i\Gamma_1}{2\pi}\ln(z - z_1) - \frac{i\Gamma_1}{2\pi}\ln\left(\frac{a^2}{z} - \overline{z}_1\right)$$

Complex velocity is:

$$\begin{aligned} \overline{V} = {} & V_\infty e^{-i\alpha} - V_\infty \frac{a^2}{z^2} e^{+i\alpha} + \frac{i}{z} 2a \cdot V_\infty \cdot sin(\alpha - \beta) \\ & + \frac{i\Gamma_0}{2\pi}\frac{1}{(z - z_0)} + \frac{i\Gamma_0(a^2/z^2)}{2\pi\left(\frac{a^2}{z} - \overline{z}_0\right)} + \frac{i\Gamma_1}{2\pi}\frac{1}{(z - z_1)} + \frac{i\Gamma_1(a^2/z^2)}{2\pi\left(\frac{a^2}{z} - \overline{z}_1\right)} \end{aligned} \tag{6}$$

After a period of time Δt induced velocity at the trailing edge is a consequence of the disposition of both moving and released vortex.

Setting again the condition that a point at the circle is stagnation point and complex velocity at that point equal to zero, we determine a new position of the stagnation point by new angle β'. This new angle β' defines a new difference $\Delta\beta = \beta - \beta'$. So the new position of the stagnation point is defined by distance $z = a \cdot e^{i\beta'}$. This again causes generation of a new vortex Γ_2 which is equal do the difference of vortex intensity before and after displacing of the free moving vortex from time t_1 to time t_2.

$$\Gamma_2 = \Delta\Gamma = 4\pi a V_\infty sin(\alpha - \beta') - 4\pi a V_\infty sin(\alpha - \beta)$$

Now the total complex potential has the value:

$$w(z) = V_\infty z e^{-i\alpha} + V_\infty \frac{a^2}{z} e^{+i\alpha} + i2aV_\infty sin(\alpha-\beta)\ln z + \frac{i\Gamma_0}{2\pi}\ln(z-z_0) - \frac{i\Gamma_0}{2\pi}\ln\left(\frac{a^2}{z} - \overline{z}_0\right)$$
$$+\frac{i\Gamma_1}{2\pi}\ln(z-z_1) - \frac{i\Gamma_1}{2\pi}\ln\left(\frac{a^2}{z} - \overline{z}_1\right) + \frac{i\Gamma_2}{2\pi}\ln(z-z_2) - \frac{i\Gamma_2}{2\pi}\ln\left(\frac{a^2}{z} - \overline{z}_2\right) \quad (7)$$

Complex velocity is:

$$\overline{V} = V_\infty e^{-i\alpha} - V_\infty \frac{a^2}{z^2} e^{+i\alpha} + \frac{i}{z} 2a \cdot V_\infty sin(\alpha-\beta) + \frac{i\Gamma_0}{2\pi}\frac{1}{(z-z_0)} + \frac{i\Gamma_0 (a^2/z^2)}{2\pi\left(\frac{a^2}{z} - \overline{z}_0\right)}$$
$$+\frac{i\Gamma_1}{2\pi}\frac{1}{(z-z_1)} + \frac{i\Gamma_1 (a^2/z^2)}{2\pi\left(\frac{a^2}{z} - \overline{z}_1\right)} + \frac{i\Gamma_2}{2\pi}\frac{1}{(z-z_2)} + \frac{i\Gamma_2 (a^2/z^2)}{2\pi\left(\frac{a^2}{z} - \overline{z}_2\right)} \quad (8)$$

Vortices travel down the flowfield by its velocity so that their position is determined by solving the system of equations:

$$\frac{dx_i}{dt} = \left.\frac{\partial\varphi}{\partial x}\right|_i , \quad \frac{dy_i}{dt} = \left.\frac{\partial\varphi}{\partial y}\right|_i .$$

Position of the vortex at a new moment can be determined by $d\overline{z}_m = \overline{V}_m dt$ and its elementary displacement $\Delta\overline{z}_m = \overline{z}_m^n - \overline{z}_m^s = \Delta t \cdot \overline{V}$.

Now a new distance of each vortex as well as the trajectory of the vortex or any fluid particle is given by $\overline{z}_m^n = \overline{z}_m^s + \Delta t \cdot \overline{V}$, or:

$$\overline{z}_m^n = \overline{z}_m^s + \Delta t \cdot \left\{ V_\infty e^{-i\alpha} - V_\infty \frac{a^2}{z^2} e^{+i\alpha} + \frac{i(\Gamma - \Gamma_m)}{2\pi z} + \frac{i\Gamma_0}{2\pi}\frac{1}{(z-z_0)} + \frac{i\Gamma_0}{2\pi\left(z - \frac{z^2}{a^2}\overline{z}_0\right)} + \right.$$
$$\left. \sum_1^n \left[\frac{i\Gamma_m}{2\pi}\frac{1}{(z-z_m)} + \frac{i\Gamma_m}{2\pi\left(z - \frac{z^2}{a^2}\overline{z}_m\right)} \right] \right\} \quad (9)$$

2.2.4 Calculation

According to the whole mentioned analysis a numerical model was established. This model is based on the following:

- vortices move along the flowfield by the flowfield velocity
- trajectory of every vortex is defined
- for every point in the flowfield it is necessary to determine value of complex potential and complex velocity
- at a certain moment in a defined initial point a free moving vortex is simulated
- after every time interval Δt moving vortex changes the stagnation point position which must be compensated by introduction of new vortex which brings stagnation point back to its proper place
- new positions of all vortices are calculated by multiplying local velocities and adding these values to previous
- boundary condition of impermeability of the airfoil must be fulfilled
- vortex diffusion is simulated by variation of the vortex size and arbitrary step
- computer program is made so that flow parameters can be calculated for different angles of attack.

2.2.5 Analysis of the calculation

Program was run first for characteristic cases of flow around the cylinder and than flow around airfoil.

Due to computer limitations, a rather small number of streamlines is shown. Model is calculated for different angles of attack, different flowfield velocities, different intensities of the moving vortex and its starting position.

The result of complete calculation for 2/3D simulation (include compare with photos from water tunel) is shown at figs. 5-12.

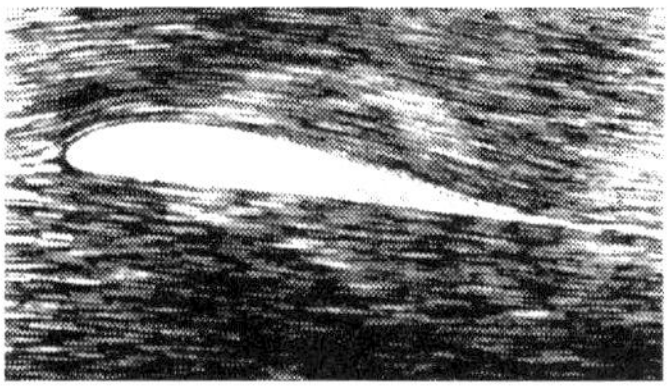

Fig. 5.

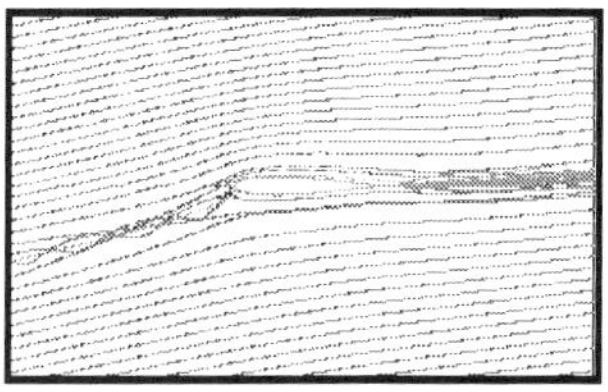

Fig. 6.

Fig. 7.

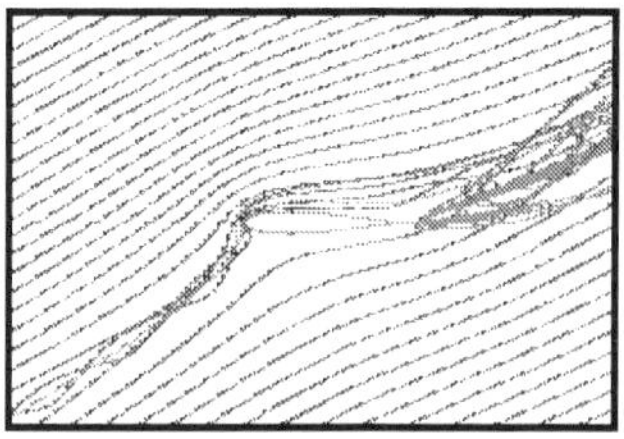

Fig. 8.

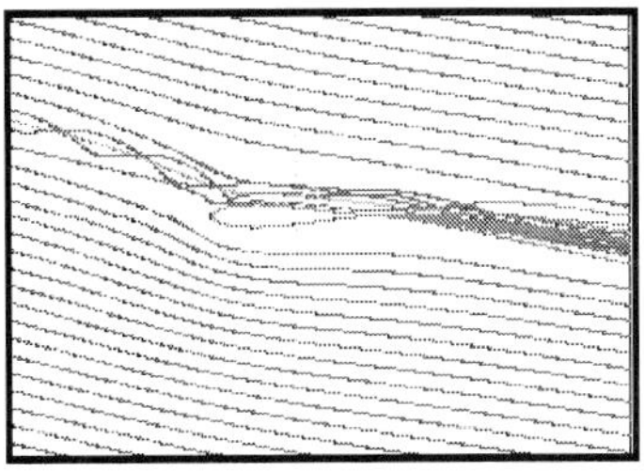
Fig. 9.

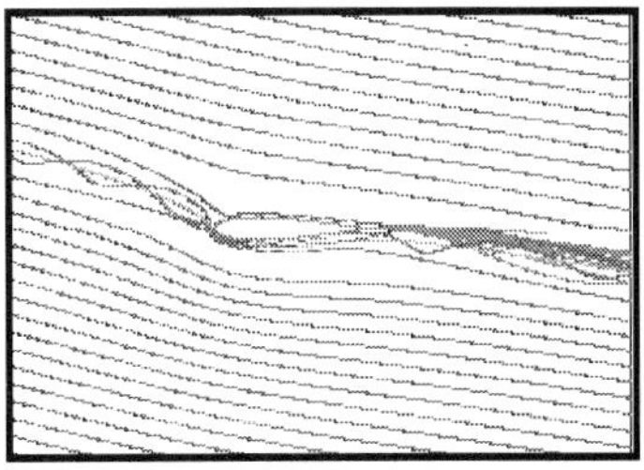
Fig. 10.

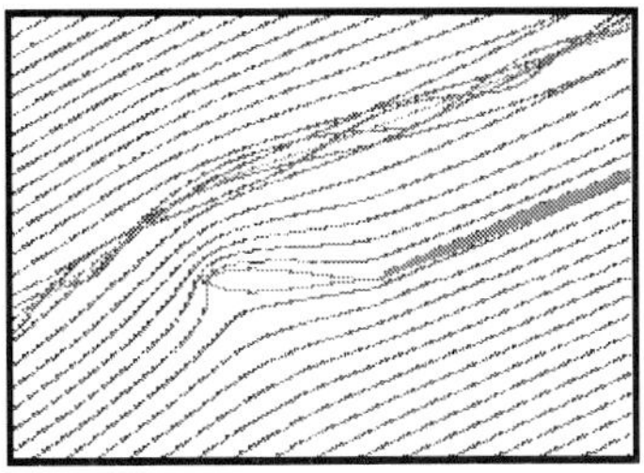
Fig. 11.

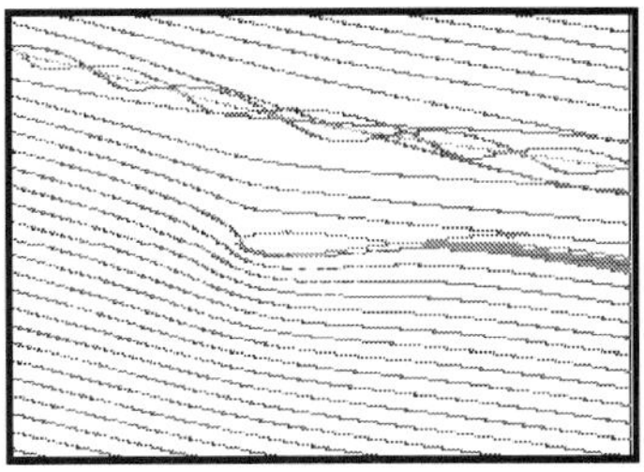
Fig. 12.

2.2.6 Conclusion remarks

Analysis of results achieved by unsteady lift modeling and viscous effects simulation method shows that they can be used with sufficient accuracy in rotor analysis and construction.

The aim of this particular simulation is to use advantages of vortex methods. For example, vortex methods use the description of flow field of the smallest range; aerodynamic forces can be obtained with small number of vortices. On the other hand, singular vortex distribution can be accurately determined by using data obtained in small time range. Vortex methods, also, permit boundary layer simulation at large Reynolds's number by local concentration of computational points.

Achieved results imply the direction for the further development of this program

- the model should be expanded for transonic flow in the aim of blade tip analysis
- a computational dispersion and gradual disappearing of vortex wake are possible by using viscous vortex shell, as more elegant solution than violent elimination of vortex wake used in this program
- an inclusion of curved vortex elements in aim of achieving better results from vortex wake self-induction aspect

On the basis of analysis of presented model and program package for viscous effects and unsteady lift simulation it can be concluded that this subchapter presents an original scientific contribution, applicable in aerodynamic analyses of practical problems in helicopter rotor projecting.

2.3 Improved solution approach for aerodynamics loads of helicopter rotor blade in forward flight

This subchapter presents the numerical model developed for rotor blade aerodynamics loads calculation. The model is unsteady and fully three-dimensional. Helicopter blade is assumed to be rigid, and its motion during rotation is modeled in the manner that rotor presents a model of rotor of helicopter Aerospatiale SA 341 "Gazelle".

2.3.1 Introduction

Helicopter rotor aerodynamic flow field is very complex, and it is characterized by remarkably unsteady behavior. The most significant unsteadiness appears during the forward flight. In that case, the progressive motion of helicopter coupled with rotary motion of rotor blades causes drastic variations of local velocity vectors over the blades, where the advancing or retreating blade position is of great significance. In first case, the local tip transonic flow generates, while in second, speed reversal appears. In addition, in forward flight, blades encounter wakes generated by forerunning blades and so encounter non-uniform inflow. The wake passing by the blade induces high velocities close to it causes changes in lifting force. Besides that, in horizontal flight blades constantly change pitch, i.e. angle of attack at different azimuths. Such angle of attack variations are very rapid, so that dynamic stall occurs, especially in case of retreating blades.

2.3.2 Dynamics

In this paper the rotor of helicopter SA 341 "Gazelle" is modeled, at which the blades are attached to the hub by flap, pitch and pseudo lead-lag hinges. Blade motion in lead-lag plane is limited by dynamic damper, which permits very small maximum blade deflection.

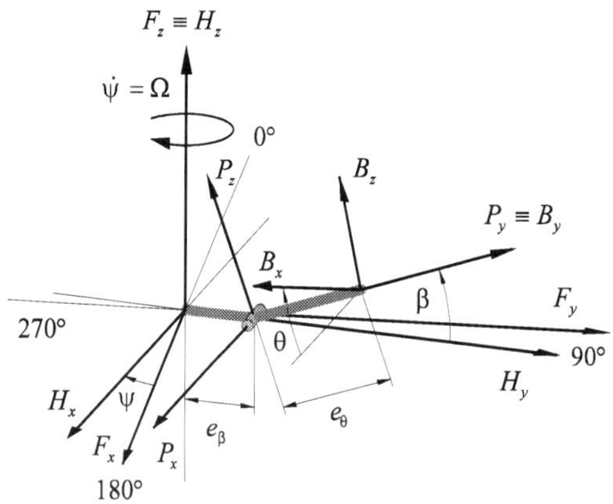

Fig. 13. Coordinate systems

Due to such small angular freedom of motion, we can assume that there is no led-lag motion at all. Pitch hinge is placed between flap hinge and pseudo lead-lag hinge.

According to that, the following frames have been selected for use:

- fixed frame F, with x-axis in the direction of flight, and z-axis oriented upwards;
- frame H, connected with the rotor and rotates together with it; it is obtained by rotating the F frame for a certain azimuth angle ψ, keeping the common z-axis;
- frame P, connected to the flapping hinge, so that the y-axis is oriented along the blade; its origin is displaced from the rotating axis for the value e_β, while it is rotated for the angle β with respect to the frame H;
- frame B, connected to the blade, displaced for the value e_θ from the origin of P, and tilted for the value θ (pitch angle) with respect to the P.

In derivation of the equations of motion, the following was assumed:

- the rotor does not vibrate, and its rotation velocity Ω is constant,
- the blade is considered absolutely rigid.

With assumptions above mentioned, equation of blade flapping motion is:

$$B\ddot{\beta} + \Omega^2\left(B\cos\beta + m_b e_\beta x_g R\right)\sin\beta = M_A$$

where B is moment of inertia about P_x, m_b mass of the blade, x_g position of blade center of gravity in P frame, and M_A is aerodynamic moment.

2.3.3 Aerodynamics

The flow field is assumed to be potential (inviscid and irrotational) and incompressible. In that case, velocity potential satisfies the Laplace equation $\Delta\Phi = 0$.

The equation is the same, both for steady and unsteady flows. Owing to that, methods for steady cases can be applied for the solution of unsteady flow problems, as well. Unsteadiness is introduced by:

- unsteady boundary condition: $\left(\vec{V} - \vec{V}_T\right)\cdot\vec{n} = 0$
- Kelvin theorem: $\dfrac{D\tilde{\Gamma}}{Dt} = 0$
- unsteady form of the Bernoulli equation: $\dfrac{p - p_\infty}{0.5\rho_\infty} = V_\infty^2 - V^2 - 2\dfrac{\partial\Phi}{\partial t}$

where: Φ - is velocity potential, $\vec{V}$ - is absolute fluid velocity, $\vec{V}_T$ - is lifting surface velocity, $\vec{n}$ - is the normal of the lifting surface at a certain point, and $\tilde{\Gamma}$ - is the bound circulation.

In case of inviscid problems, it is necessary to satisfy Kutta condition at the trailing edge. Based on unsteady Bernoulli equation, the difference between upper and lower surface pressure coefficients, in case of the thin lifting surface, is:

$$\Delta C_P = C_{P_U} - C_{P_L} = -\frac{V_U^2 - V_L^2}{V_\infty^2} - \frac{2}{V_\infty^2}\frac{\partial}{\partial t}\left(\int_{LE}^{M} \gamma \, dl\right) \qquad (10)$$

where integral should be calculated from leading edge to a certain point M at the surface, and γ is the local bound vortex distribution. If we assume that spanwise velocities are small, the difference of velocity squares can be calculated as:

$$V_U^2 - V_L^2 \approx 2V_\infty \gamma \tag{11}$$

By substituting (11) in (10), the following equation can be obtained:

$$\Delta C_P = -\frac{2}{V_\infty^2}\left(V_\infty \gamma + \frac{\partial}{\partial t}\int_{LE}^{M} \gamma dl\right)$$

The Kutta condition can be expressed as the uniqueness of pressure coefficients at the trailing edge:

$$\Delta C_p = \frac{2}{V_\infty^2}\left(V_\infty \gamma_{TE} + \frac{\partial}{\partial t}\int_{LE}^{TE} \gamma dl\right) = \frac{2}{V_\infty^2}\left(V_\infty \gamma_{TE} + \frac{\partial \tilde{\Gamma}}{\partial t}\right) = 0$$

where $\tilde{\Gamma}$ is the contour circulation which covers the lifting surface. Since it is impossible to be $V_\infty = \infty$, the relation within the parentheses must be equal to zero and the expression for unsteady Kutta condition comes out directly as:

$$\frac{\partial \tilde{\Gamma}}{\partial t} = -V_\infty \gamma_{TE}$$

If the right hand-side part is substituted with (11) written for the trailing edge, we obtain:

$$\frac{\partial \tilde{\Gamma}}{\partial t} = -\frac{V_{U_{TE}}^2 - V_{L_{TE}}^2}{2} = -\left(V_{U_{TE}} - V_{L_{TE}}\right)\frac{V_{U_{TE}} + V_{L_{TE}}}{2}$$

From this equation, it can be clearly seen that the variation of the lifting surface circulation in time can be compensated by releasing vortices of magnitude $\left(V_{U_{TE}} - V_{L_{TE}}\right)$ at the velocity $\left(V_{U_{TE}} + V_{L_{TE}}\right)/2$.

2.3.4 Discretization and numerical solution procedure

The method for the solution of this problem is based on the coupling of the dynamic equations of blade motion with the equations of aerodynamics.

Discretization in time is done by observing the flow around the blade in a series of positions that it takes at certain times t_k, which are spaced by finite time intervals Δt at different azimuths. Also, discretization of the thin lifting surface is done by using the panel approach. By this method, the lifting surface is divided in a finite number of quadrilateral surfaces - panels. Vorticity distribution is discretized in a finite number of concentrated, closed quadrilateral linear vortices, in such a way that one side of the linear vortex is placed at the first quarter chord of the panel, and represents the bound vortex of the corresponding panel.

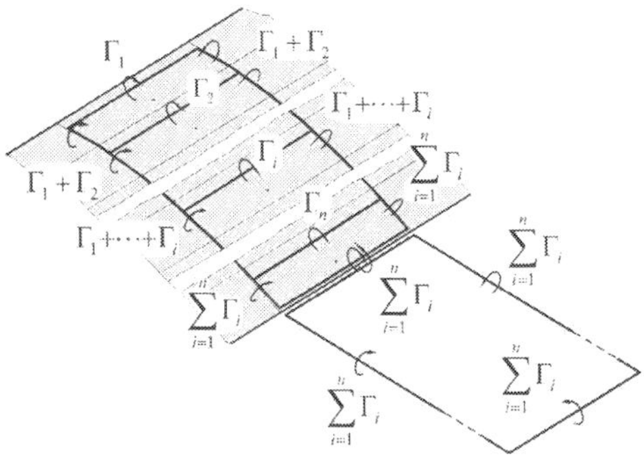

Fig. 14. The steady panel scheme

The opposite side of the vortex is always placed at the trailing edge, while the other two sides are parallel to the flow. The wake is represented by quadrilateral vortex in the airflow behind the lifting surface. One side of it is connected to the trailing edge, while the opposite one is at the infinity. The other two sides (trailing vortices), which actually represent the wake, are placed parallel to the airflow. The vorticity of the quadrilateral vortex is equal to the sum of the vorticities of all bound vortices of the panels that correspond a certain lifting surface chord, but opposite in direction. Then the trailing edge vorticity is equal to zero.

Model established in such a manner corresponds to the steady flow case. On the other hand, it can be very easily spread in order to include the unsteady effects.

2.3.4.1 Vortex releasing model

The variation of the lifting surface position in time induces variation of circulation around the lifting surface as well. According to the Kelvin theorem, this variation in circulation must also induce the variation around the wake. According to the unsteady Kutta condition, this can be achieved by successive releasing of the vortices in the airflow.

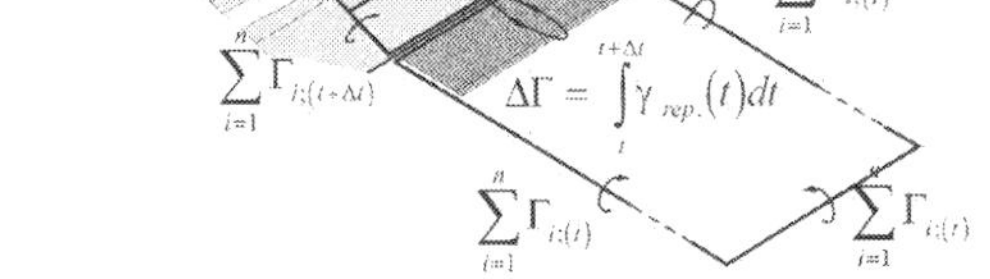

Fig. 15. Vortex releasing

Suppose that the lifting surface has been at rest until the moment t, when it started with the relative motion with respect to the undisturbed airflow. The vortex releasing, as a way of circulation balancing, is done continually, and in such a way a vortex surface of intensity

$\gamma(t)$ is formed. At the next moment $t+\Delta t$, the flow model will look like in Figure 15. The circulation of the vortex element joined to the trailing edges is equal to the difference in circulations at moments $t+\Delta t$ and t.

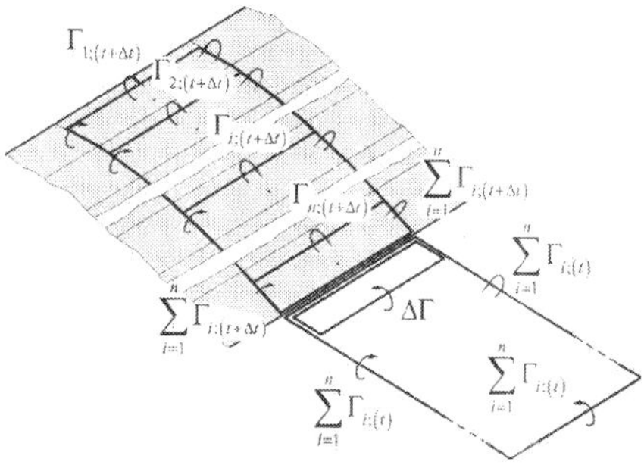

Fig. 16. Unsteady panel scheme

We will discretize the vortex "tail" by replacing it with the quadrilateral vortex loop, whose one side is at the trailing edge, and the opposite side is at the finite distance from the trailing edge (shed vortex). By this, we can obtain the final model for unsteady case.

2.3.4.2 Discretization of wake

The established vortex-releasing model is appropriate for the wake modeling using the "free wake" approach. During the time, by continuous releasing of the quadrilateral vortex loops, the vortex lattice formed of linear trailed and shed vortices is created. The wake distortion is achieved by altering the positions of node points in time, by application of kinematics relation $\vec{r}_i(t+\Delta t)=\vec{r}_i(t)+\vec{V}_i(t)\Delta t$.

The velocities of the collocation points are obtained as sums of the undisturbed flow velocity and velocities induced by other vortex elements of the flow field. Induced velocities are calculated using the Biot-Savart law. In order to avoid the problems of velocity singularities, line vortex elements are modeled with core. The core radius varies with the gradient of the bound circulation at the position where vortex is released.

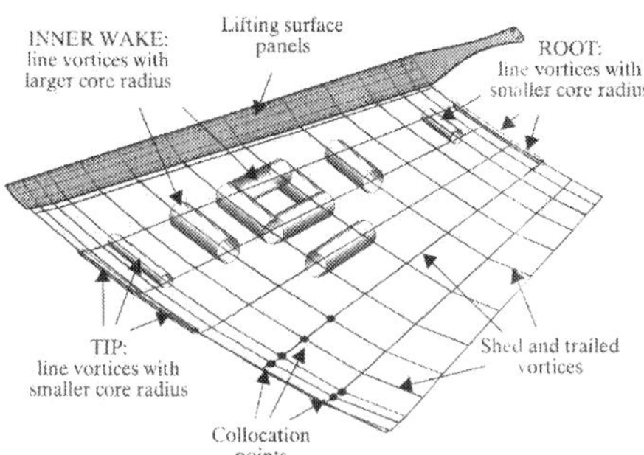

Fig. 17.Discretized wake

The wake influence at large distances from the blade is negligible, so it is possible to neglect the wake distortion at a sufficient distance from the rotor. The number of revolutions for which it is necessary to calculate the wake distortion can be determined as $m = 0.4/\mu$, where μ is the advance ratio. After that, the wake shape is "frozen" in achieved state, and it moves by flow field velocity, keeping it for the rest of the time. The "frozen" part of the wake still influences the adjacent area in which it is still being distorted. After some distance, even the influence of the "frozen" part becomes negligible, and then it is eliminated from the model.

2.3.4.3 Definition of equation set

The boundary condition of impermeability of the lifting surface should be satisfied at any moment of time, $t_k\,(k = 0,1,2\ldots)$ in a finite number of points of lifting surface

$$\left(\vec{V}_i - \vec{V}_{Ti}\right)\cdot \vec{n}_i = 0; \quad i = 1,2,\ldots,n$$

Points at which this condition must be satisfied are called the control points. One of them is placed on each panel, at the three-quarter chord panel positions. By this, at every moment of time, the number of lifting surface impermeability conditions is equal to the number of unknown values of circulations of bound vortices. The equations of motion of the lifting surface are known, as well as the velocities $\vec{V}_{Ti}$ of all characteristic points, and their normals $\vec{n}_i$ as well. At each flow field point, velocity can be divided to the free stream velocity and perturbation velocity: $\vec{V}_i = \vec{V}_\infty + \vec{w}_i$. The perturbation velocity is induced by lifting surface and wake vortex elements. It is calculated by Biot-Savart law. At every moment, the wake shape and circulations of its vortex lines are known, and so the wake-induced velocity at every flow field point is known as well. On the other hand, the circulations of the bound vortices are unknowns (their positions are defined by the lifting surface shape). The boundary condition for the i-th control point can be written as:

$$\sum_{i=1}^{n} a_{i,j}\Gamma_{ji}(t) = b_i$$

where $a_{i,j}$ are the coefficients depending of the blade geometry, and b_i are the coefficients containing the influence of the wake and free-stream flows. This way, by writing equations for all control points, the equation set of the unknown bound circulations is obtained.

Along with this equation set, the Kutta condition must be satisfied. In case of numerical solutions, it is customary to satisfy Kutta condition in vicinity of the trailing edge. According to that, we obtain discretized form of unsteady Kutta condition:

$$V_\infty \frac{\Gamma_n}{l_n} + \frac{\tilde{\Gamma}(t+\Delta t) - \tilde{\Gamma}(t)}{\Delta t} = 0$$

where Γ_n/l_n is intensity of the distributed vorticity at the trailing edge panel γ_n, and l_n is the trailing edge panel cord length. By adding the Kutta conditions to the equation set, an over-determined equation set is obtained. It can be reduced to the determined system by the method of least squares. After that, it can be solved by some of the usual approaches, by which the unknown values of circulations Γ_i at the time t are obtained.

2.3.4.4 Determination of the aerodynamic force

After unknown circulations Γ_i are obtained, velocity at every point of the flow field is known, and we can use them for the determination of aerodynamic forces that act on the blade. The calculation aerodynamic force is necessary for the defining of the blade position at the next moment of time. The total aerodynamic force is calculated as the sum of forces acting on all panels. Aerodynamic force acting on a single panel can be defined by:

$$\vec{F}_i = \rho \vec{V}_\infty \times \Gamma_{i(ef.)} \overrightarrow{BC}$$

where $\overrightarrow{BC}$ is bound circulation vector and effective circulation can be defined by using:

$$\Gamma_{i(ef.)} = \Gamma_i + \frac{1}{V_\infty}\frac{\partial}{\partial t}\int_{LE}^{M_i} \gamma dl = \Gamma_i + \frac{1}{V_\infty}\frac{\partial}{\partial t}\left(\sum_{k=1}^{i-1}\Gamma_k + \frac{\Gamma_i}{4}\right)$$

where M_i denote point at quarter-chord position on the i-th panel.

After determination of aerodynamic forces, moment of aerodynamic forces M_A, necessary for blade flapping equation, can be calculated as sum of moments acting flapping hinge.

2.3.5 Results

Presented method for blade air-loads calculation has been implemented in a computer code. Forward flight application is presented for SA 341 "Gazelle" helicopter at moderate forward speed condition (advance ratio μ is 0.35).

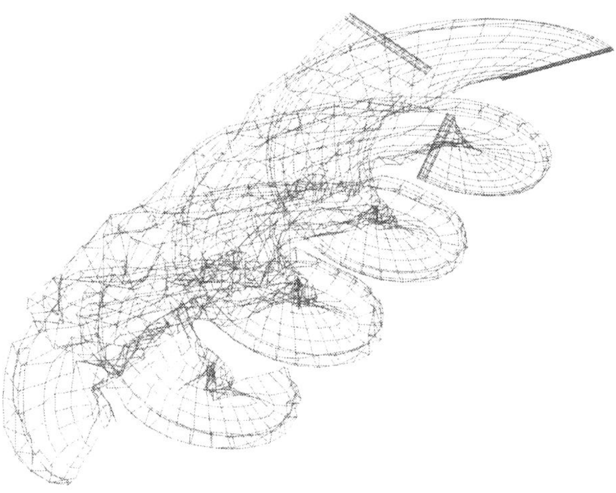

Fig. 18. Calculated free wake geometry; μ = 0.35 (3D view)

By analyzing the drawing of the blade wakes, it can be concluded that model applied in this paper gives reasonable simulation of actual wake behavior, specially in the domain of wake boundaries, where wake roll-up occurs (although it is slightly underestimated compared with existing experimental data). In addition, larger wake distortion in the domains of the forerunning blades or their wakes is noticeable.

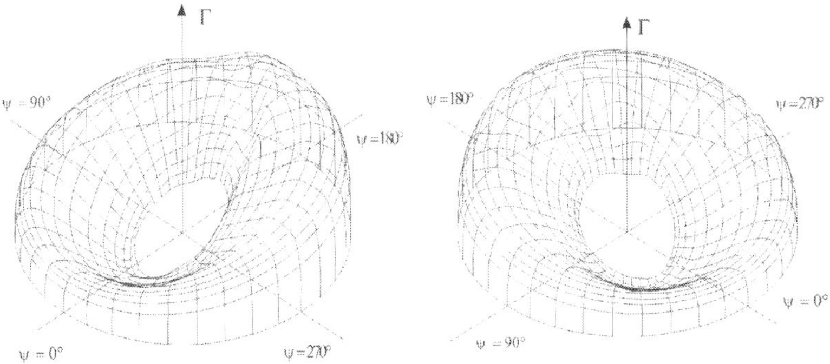

Fig. 19. Circulation distribution on the rotor disk; μ=0.35 (different points of view)

The program results (Figure. 19), show the difference in circulation distributions at different azimuths, as well as the disturbances caused when blades are passing the wakes of other blades, and the characteristic reversal flow domains.

2.3.6 Conclusion

Obtained results can define suggestions for the future solution improvement. Firstly, by incorporating the transonic flow calculations, the advancing blade tip simulation would be more appropriate. Secondly, viscous interaction should be included as well, which would improve the wake roll-up simulation. The viscous vortex core simulation in time would improve the results concerning the wake vanishing effects far enough from the blade. Finally, introduction of the curvilinear vortex elements would give better results from the aspect of wake self-induction. With these enhancements, this solution should prove to be a useful and efficient tool to the rotorcraft performance evaluation.

3. Flight dynamics and control

The flight dynamics mathematical model of an aircraft, which would be strictly determined, would comprise of a system of non-linear, non-stationary, partial differential equations. To simplify these equations we introduce a number of assumptions. Ignored are the elastic characteristics of the aircraft so the aircraft can be thought as a rigid body and, in that way, the dispersal of parameters is eliminated. Also, fuel consumption is disregarded and so the non-stationarity due to temporal change in helicopter mass is eliminated.

3.1 Mathematical models of helicopter flight dynamics

The helicopter is specific in regards to other traffic-transportation means, not just by its structure but also by its motion possibilities. The helicopter can move vertically, float in the air,

Once the axis is chosen, the remaining axis of the coordinate system will be normal to it and pointed lateral, that is, to the tail of the helicopter. Customarily, the force component along the chosen axis is said to be the tow force, the force component pointed towards the tail is said to be H force, and the force component pointed lateral is said to be Y force. If the force components are designated without subscripts, it is assumed they are determined relative to control axis, whereas subscripts "D" and "S" are used when relating to rotor axis, that is, shaft axis. Since flutter and mount angles are usually small (amounts greater than 10° are considered extreme), the relation between these components can be obtained:

$$T \approx T_D \approx T_S \quad , \quad H \approx H_D + T_D\, a_1 \cong H_S + T_S\, B_1$$

3.1.2.1 Longitudinal equilibrium of forces

Angle B_1 is the longitudinal amplitude of a cyclic change in blade step; angle a_{1s} is the angle between shaft and axis of rotor disc. After extensive calculus the expression for longitudinal amplitude of cyclic change in blade step is obtained:

$$B_1 = \frac{M_f - G \cdot fR + H \cdot hR + M_S \cdot a_1}{T \cdot hR + M_S}$$

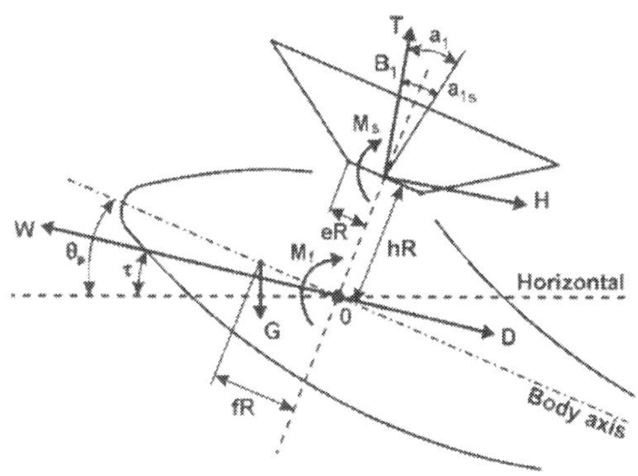

Fig. 23. Drawing for determining longitudinal equilibrium of forces

For e=0, it can be adopted that M_s=0 and M_f=0, and since T=G, follows:

$$B_1 = -\frac{f}{h} + \frac{H}{G} \quad , \quad \theta_p = -\frac{D}{G}\cos\tau - \frac{f}{h} + \frac{M_f}{G \cdot hR}$$

Above equation has a simple physical interpretation: the amplitude of the longitudinal cyclic control must have such a value in order to position the direction of the resultant rotor force through the center of mass.

3.1.2.2 Lateral equilibrium of forces

Angle A_1 presents the amplitude of lateral cyclic change in blade step of the supporting helicopter rotor:

$$A_1 = -\frac{G \cdot f_1 R + M_S \cdot b_1 + T_t \cdot h_t R}{G \cdot hR + M_S}$$

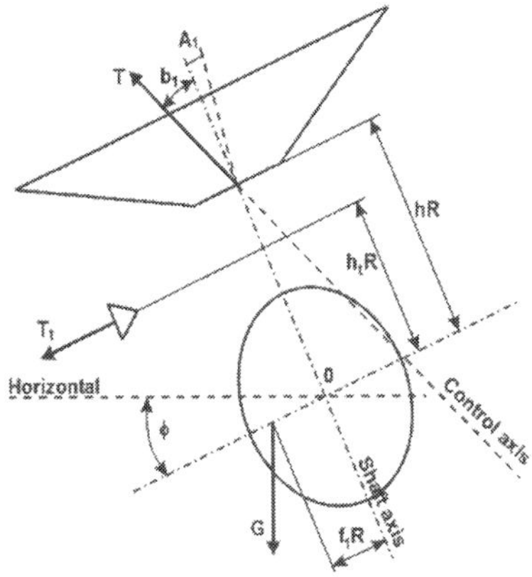

Fig. 24. Drawing for determining lateral equilibrium of forces

By replacing value A_1 into corresponding equations, we obtain the value of angle ϕ, which determines the position of the fuselage.

$$\varphi = -\frac{T_t}{G} + \frac{G \cdot f_1 R + M_S \cdot b_1 + T_t \cdot h_t R}{G \cdot hR + M_S}$$

If M_S=0 and h_t=h, which can often be assumed, follows $\varphi \approx \frac{f_1}{h}$, which means the rotor hub is positioned vertically above the center of mass. All values of these determined angles are so called *trimmed values*.

3.1.3 Non-linear mathematical model of flight dynamics

Mathematical modeling of helicopter motion is a very complex task and, therefore, it is necessary to introduce series of assumptions and approximations. Knowledge of motion of individual helicopter blades is not necessary for investigating dynamic characteristics of the helicopter, except in a special case, but rather for defining forces and moments in a disturbed flight it is sufficient enough to view the rotor as whole. Because of a great number of different helicopters, in this paper a single rotor helicopter was studied, that has its blades connected to the hub by hinges. As mentioned before, the helicopter can perform different motions and it would be very difficult to make a mathematical model that would combine all those motions. It is assumed the helicopter is airborne and in straightforward flight. It is required that the helicopter, at straight forward flight, has following velocity components: W_x, W_y, and W_z, at nominal values, and angle of turn ψ, angle of roll ϕ, angle of climb θ, as long as the intensity of disturbance is in permitted limits. Figure 25 presents a schematic of the helicopter with a floating coordinate system tied to its center of mass, and Figure 26 presents a helicopter block diagram.

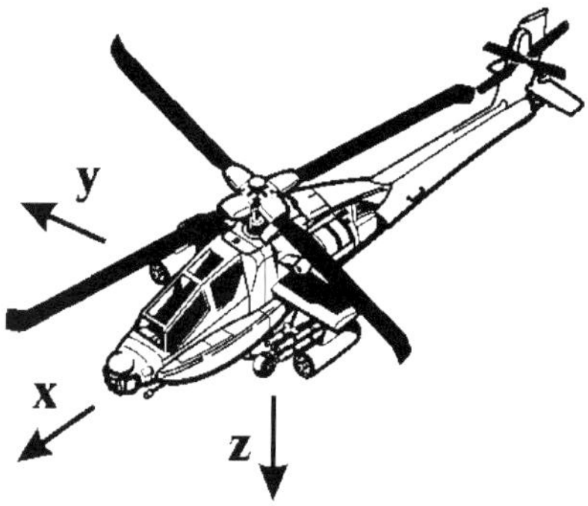

Fig. 25. Schematic of helicopter

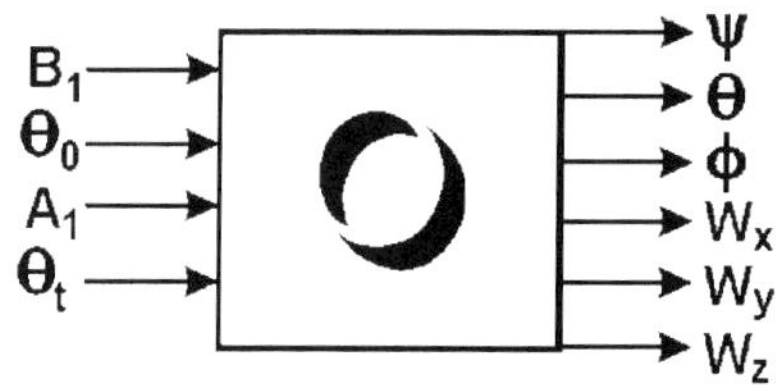

Fig. 26. Helicopter dynamics block diagram

After introducing a series of assumptions, such as:

- helicopter mass is a constant value,
- helicopter is a rigid body,
- 0xz is a plane of symmetry,
- angle increments ΔΨ, Δθ, Δϕ are too small, and so on;

and we come to a non-linear mathematical model by deviations in the form:

$$\frac{d(\Delta W_x)}{dt} = \frac{1}{m}\left[f_1(\Delta W_x, \Delta W_z, \Delta\dot{\theta}, u_1, u_2) - (mg\cos\tau)\Delta\theta\right]$$

$$\frac{d(\Delta W_z)}{dt} = \frac{1}{m}\left[f_2(\Delta W_x, \Delta W_z, \Delta\dot{\theta}, u_1, u_2) + W_{zN} m\Delta\dot{\theta} - (mg\sin\tau)\Delta\theta\right]$$

$$\frac{\mathrm{d}(\Delta\theta)}{\mathrm{dt}} = \Delta\dot{\theta} \quad , \quad \frac{\mathrm{d}(\Delta\dot{\theta})}{dt} = \frac{1}{J_y} f_3(\Delta W_x, \Delta W_z, \Delta\dot{W}_z, \Delta\dot{\theta}, u_1, u_2)$$

$$\frac{d(\Delta W_y)}{dt} = \frac{1}{m}\left[f_4(\Delta W_y, \Delta\dot{\varphi}, \Delta\dot{\psi}, u_3, u_4) + W_{zN} m\Delta\dot{\psi} + mg\cos\tau\Delta\theta + mg\sin\tau\Delta\psi\right]$$

$$\frac{\mathrm{d}(\Delta\varphi)}{\mathrm{dt}} = \Delta\dot{\varphi} \quad , \quad \frac{\mathrm{d}(\Delta\dot{\varphi})}{dt} = \frac{1}{J_x}\left[f_5(\Delta W_y, \Delta\dot{\varphi}, \Delta\dot{\psi}, u_3, u_4) + J_{xz}\Delta\dot{\psi}\right]$$

$$\frac{\mathrm{d}(\Delta\psi)}{\mathrm{dt}} = \Delta\dot{\psi} \quad , \quad \frac{\mathrm{d}(\Delta\dot{\psi})}{dt} = \frac{1}{J_z}\left[f_6(\Delta W_y, \Delta\dot{\varphi}, \Delta\dot{\psi}, u_3, u_4) + J_{xz}\Delta\dot{\varphi}\right]$$

Where:

- $U_1=\Delta B_1$ - amplitude of cyclic change in step in the longitudinal direction (regarding to longitudinal motion),
- $U_2=\Delta\theta_0$ - change of collective step of the helicopter rotor blade (regarding to longitudinal motion),
- $U_3=\Delta A_1$ - amplitude of cyclic change in step in the lateral direction (regarding to lateral motion), and
- $U_4=\Delta\theta_t$ - change of collective step of the tail rotor (regarding to lateral motion).

3.1.4 Linearized mathematical model of flight dynamics

In technical applications it has been shown that, with an acceptable accuracy, linearized mathematical models may be used under the condition that deviations of physical quantities from their nominal values are small.

The outcome of adopted presumptions is that the output values, input values, and the vector of state for both longitudinal and lateral motion will be:

$$\underline{X}=(X_1 \ldots X_9)^T \ , \ \underline{X_i}=(X_{i1} \ldots X_{i6})^T , \ \underline{u}=(u_1 \ldots u_4)^T \quad (13)$$

The vector equation of state for the linearized mathematical model with non-dimensional quantities, deviations, that is, quantities of state is shown in equation 13. Equation 14 present the equation at exit.

$$\underline{\dot{X}}=\begin{bmatrix} a_{11} & a_{12} & a_{13} & a_{14} & 0 & 0 & 0 & 0 & 0 \\ a_{21} & a_{22} & a_{23} & a_{24} & 0 & 0 & 0 & 0 & 0 \\ 0 & 0 & 0 & 1 & 0 & 0 & 0 & 0 & 0 \\ a_{41} & a_{42} & a_{43} & a_{44} & 0 & 0 & 0 & 0 & 0 \\ 0 & 0 & 0 & 0 & a_{55} & a_{56} & 0 & a_{58} & a_{59} \\ 0 & 0 & 0 & 0 & 0 & 0 & 1 & 0 & 0 \\ 0 & 0 & 0 & 0 & a_{75} & 0 & a_{77} & 0 & a_{79} \\ 0 & 0 & 0 & 0 & 0 & 0 & 0 & 0 & 1 \\ 0 & 0 & 0 & 0 & a_{95} & 0 & a_{97} & 0 & a_{99} \end{bmatrix} \underline{X}+\begin{bmatrix} b_{11} & b_{12} & 0 & 0 \\ b_{21} & b_{22} & 0 & 0 \\ 0 & 0 & 0 & 0 \\ b_{41} & b_{42} & 0 & 0 \\ 0 & 0 & b_{53} & b_{54} \\ 0 & 0 & 0 & 0 \\ 0 & 0 & b_{73} & b_{74} \\ 0 & 0 & 0 & 0 \\ 0 & 0 & b_{93} & b_{94} \end{bmatrix} \underline{u} \quad (14)$$

$$\underline{X_i}=\begin{bmatrix} 1 & 0 & 0 & 0 & 0 & 0 & 0 & 0 & 0 \\ 0 & 1 & 0 & 0 & 0 & 0 & 0 & 0 & 0 \\ 0 & 0 & 1 & 0 & 0 & 0 & 0 & 0 & 0 \\ 0 & 0 & 0 & 0 & 1 & 0 & 0 & 0 & 0 \\ 0 & 0 & 0 & 0 & 0 & 1 & 0 & 0 & 0 \\ 0 & 0 & 0 & 0 & 0 & 0 & 1 & 0 & 0 \end{bmatrix} \underline{X}$$

Besides the way this is presented, in a form of common matrix, also should be noted that longitudinal and lateral motions are separated, because this was the condition for deriving this mathematical model.

movements of the airplane while spinning. It is form of "auto-rotation", which means that there is a natural tendency for the airplane to rotate of its own accord.

In order to reach height, a sailplane pilot has to circle inside a thermal column in constant turns of a very small radius. If for any reason (severe turbulence, pilot's error, etc.) the speed of sailplane drops below the stalling speed, at such high angles of bank the sailplane will most probably fall into a spin. Spin is a very dangerous and unpleasant maneuver. Good spin recovery characteristics are the imperative for any modern sailplane.

Fig. 27. VUK-T sailplane

Vuk-T sailplane (Fig. 27) is a modern single seat, all composite sailplane for advanced pilot training and competitions, designed at the Belgrade Faculty of Mechanical Engineering. Beside the other complex analyses, spin characteristics have been analyzed thoroughly.

Major assumptions that were starting points in this subject are the adopted aerodynamic concepts of the sailplane and the full system of equations for the aircraft motion in case of spin. Evaluation of aerodynamic quotients and their derivatives, and thus, the equation system as a whole, is based on defined aircraft geometric characteristics and atmospheric conditions. The method developed for simulation of the real flight situation is based on the first and second non-linear equation systems.

3.2.2 Non-linear mathematical models of flight dynamics

Mathematical modelling of spin motion is a very complex task and, therefore, it is necessary to introduce series of assumptions and approximations. Mathematical model is based on the first and second non-linear equations systems, shown bellow. The first system (V=const and H=const) is:

$$\dot{\alpha} = a_{11}\frac{1}{\cos\alpha} + \frac{\beta p}{\cos\alpha} + a_{13}q + a_{14}\frac{\cos\theta\cos\gamma}{\cos\alpha} + a_{15}$$

$$\dot{\beta} = a_{21}\beta + a_{22}r\cos\alpha + a_{23}p\sin\alpha + a_{24}\cos\theta\sin\gamma + a_{25}$$

$$\dot{p} = a_{31}rq + a_{32}\beta + a_{33}p + a_{34}r + a_{35} + a_{36}$$

$$\dot{r} = a_{41}pq + a_{42}\beta + a_{43}p + a_{44}r + a_{45}q + a_{46} + a_{47}$$

$$\dot{q} = a_{51}pr + a_{52}|\ \beta\ | + a_{53}r + a_{54}q + a_{55} + a_{56}\dot{\alpha} + a_{57}$$

$$\dot{\theta} = a_{61} r \sin\gamma + a_{62} q \cos\gamma$$

$$\dot{\gamma} = a_{71} p + a_{72} r \cos\gamma \tan\theta + a_{73} q \sin\gamma \tan\theta$$

The coefficients are given by:

$$a_{11} = -\frac{S\rho V}{2m} C_z(\alpha) \ a_{12} = -1 \ a_{13} = 1 \ a_{14} = \frac{q}{V} \ a_{15} = -\frac{S\rho V}{2m} C_{z\delta_{nk}}(\alpha)\Delta\delta_{nk}(t)$$

$$a_{21} = \frac{S\rho V}{2m} C_{y\beta}(\alpha) \ a_{22} = 1 \ a_{23} = 1 \ a_{24} = \frac{g}{V} \ a_{25} = \frac{S\rho V}{2m} C_{y\delta_{yk}}(\alpha)\Delta\delta_{vk}(t)$$

$$a_{31} = \frac{I_y - I_z}{I_x} \ a_{32} = \frac{Sl\rho V^2}{2I_x} C_{l\beta}(\alpha)$$

$$a_{33} = \frac{Sl\rho V^2}{2I_x} C_{lp}(\alpha) \ a_{34} = \frac{Sl\rho V^2}{2I_x} C_{lr}(\alpha) \ a_{35} = \frac{S\rho V^2}{2I_x} C_{l\delta_{vk}}(\alpha)\Delta\delta_{vk}(t) \ a_{36} = \frac{Sl\rho V^2}{2I_x} C_{l\delta_k}(\alpha)\Delta\delta_k(t)$$

$$a_{41} = \frac{I_z - I_x}{I_y} \ a_{42} = \frac{Sl\rho V^2}{2I_y} C_{n\beta}(\alpha) \ a_{43} = \frac{Sl\rho V^2}{2I_y} C_{n_p}(\alpha) \ a_{44} = \frac{Sl\rho V^2}{2I_y} C_{n_r}(\alpha) \ a_{45} = -\frac{I_p \omega_p}{I_y}$$

$$a_{46} = \frac{Sl\rho V^2}{2I_y} C_{n\delta_{vk}}(\alpha)\Delta\delta_{vk}(t) \ a_{47} = \frac{Sl\rho V^2}{2I_y} C_{n\delta_k}(\alpha)\Delta\delta_k(t)$$

$$a_{51} = \frac{I_x - I_y}{I_z} \ a_{52} = \frac{Sb\rho V^2}{2I_z} C_{m_\beta}(\alpha) \ a_{53} = \frac{I_p \omega_p}{I_z} \ a_{54} = \frac{Sb\rho V^2}{2I_z} C_{m_q}(\alpha) \ a_{55} = \frac{Sb\rho V^2}{2I_z} C_m(\alpha)$$

$$a_{56} = \frac{Sb\rho V^2}{2I_z} C_{m_\alpha}(\alpha) \ a_{57} = \frac{Sb\rho V^2}{2I_z} C_{m\delta_{nk}}(\alpha)\Delta\delta_{nk}(t) \ a_{61} = 1 \ a_{62} = 1 \ a_{71} = 1 \ a_{72} = -1 \ a_{73} = 1$$

The second system (V and H are not constant) is:

$$\dot{\alpha} = b_{11}\frac{\rho V}{\cos\alpha} + b_{12}\frac{p\beta}{\cos\alpha} + b_{13} q + b_{14}\frac{\cos\theta\cos\gamma}{V\cos\alpha} + b_{15}\frac{V\tan\alpha}{V} + b_{16}\frac{\rho\dot{V}}{\cos\alpha}$$

$$\dot{\beta} = b_{21}\rho V\beta + b_{22} r\cos\alpha + b_{23} p\sin\alpha + b_{24}\frac{\cos\theta\cos\gamma}{V} + b_{25}\frac{\dot{V}\beta}{V} + b_{26}\rho V$$

$$\dot{p} = b_{31} rq + b_{32}\rho V^2\beta - b_{33}\rho V^2 p + b_{34}\rho V^2 r + b_{35}\rho V^2 + b_{36}\rho V^2$$

$$\dot{r} = b_{41} pq + b_{42}\rho V^2\beta + b_{43}\rho V^2 p + b_{44}\rho V^2 r + b_{45} q + b_{46}\rho V^2 + b_{47}\rho V^2$$

$$\dot{q} = b_{51} pr + b_{52}\rho V^2 |\ \beta\ | + b_{53} r + b_{54}\rho V^2 q + b_{55} q + b_{56}\rho V^2\alpha + b_{57}\rho V^2$$

$$\dot{\theta} = b_{61} r\sin\gamma + b_{62} q\cos\gamma$$

$$\dot{\gamma} = b_{71}p + b_{72}r\cos\gamma\tan\theta + b_{73}q\sin\gamma\tan\theta$$

$$\dot{V} = b_{81}\frac{\rho V^2}{\cos\alpha} + b_{82}V\dot{\alpha}\tan\alpha + b_{83}Vq\tan\alpha + b_{84}\frac{Vr\beta}{\cos\alpha} + b_{85}\frac{\sin\theta}{\cos\alpha} + b_{86}\frac{\rho V^2}{\cos\alpha}$$

$$\dot{H} = b_{91}V\cos\alpha\sin\theta + b_{92}V\sin\alpha\cos\theta\cos\gamma + b_{93}V\beta\cos\theta\sin\gamma$$

and the applied coefficients are:

$$b_{11} = -\frac{S}{2m}C_z(\alpha)\ b_{12} = -1\ b_{13} = 1\ \ b_{14} = g\ \ b_{15} = -1\ b_{16} = -\frac{S}{2m}C_{z\delta_{vk}}(\alpha)\Delta\delta_{hk}(t)$$

$$b_{21} = \frac{S}{2m}C_{y\beta}(\alpha)\ b_{22} = 1\ b_{23} = 1\ b_{24} = g\ \ b_{25} = -1\ b_{26} = \frac{S}{2m}C_{y\delta_{vk}}(\alpha)\Delta\delta_{vk}(t)$$

$$b_{31} = \frac{I_y - I_z}{I_x}\ b_{32} = \frac{Sl}{2I_x}C_{l_\beta}(\alpha)\ \ b_{33} = \frac{Sl}{2I_x}C_{l_p}(\alpha)\ b_{34} = \frac{Sl}{2I_x}C_{l_r}(\alpha)$$

$$b_{35} = \frac{Sl}{2I_x}C_{l\delta_{vk}}(\alpha)\Delta\delta_{vk}(t)\ b_{36} = \frac{Sl}{2I_x}C_{l\delta_k}(\alpha)\Delta\delta_{vk}(t)$$

$$b_{41} = \frac{I_z - I_x}{I_y}\ b_{42} = \frac{Sl}{2I_y}C_{n_\beta}(\alpha)\ b_{43} = \frac{Sl}{2I_y}C_{n_p}(\alpha)$$

$$b_{44} = \frac{Sl}{2I_y}C_{n_r}(\alpha)\ b_{45} = -\frac{I_p\omega_p}{I_z}\ \ b_{46} = \frac{Sl}{2I_y}C_{n\delta_{vk}}(\alpha)\Delta\delta_{vk}(t)\ b_{47} = \frac{Sl}{2I_y}C_{n\delta_k}(\alpha)\Delta\delta_k(t)$$

$$b_{51} = \frac{I_x - I_y}{I_z}\ b_{52} = \frac{Sb}{2I_z}C_{m\beta}(\alpha)\ b_{53} = \frac{I_p\omega_p}{I_z}\ \ b_{54} = \frac{Sb}{2I_z}C_{mn}(\alpha)\ b_{55} = \frac{Sb}{2I_z}C_m(\alpha)\ b_{56} = \frac{Sb}{2I_z}C_{m\alpha}(\alpha)$$

$$b_{57} = \frac{Sb}{2I_z}C_{m\delta_{nk}}(\alpha)\Delta\delta_{nk}(t)\ \ b_{61} = 1\ b_{62} = 1\ \ b_{71} = 1\ b_{72} = -1\ b_{73} = 1$$

$$b_{81} = -\frac{S}{2m}C_x(\alpha)\ b_{82} = 1\ b_{83} = -1\ \ b_{84} = -1\ b_{85} = g\ b_{86} = \frac{S}{2m}C_{x\delta_{nk}}(\alpha)\Delta\delta_{hk}(t)$$

$$b_{91} = 1\ b_{92} = -1\ b_{93} = -1$$

3.2.3 The general model

Modeling was carried out using the software package MATLAB. Simulation was performed by using the SIMULINK module, having the special feature to simulate a dynamic system within a graphic mode, where the linear, non-linear, time-continuous or discrete multi-variable systems having concentrated parameters can be analyzed. Simulation is achieved by creating the SIMULINK model and implementing appropriate functions for the numeric solving differential equations.

3.2.4 Program results

The program has been tested on the VUK-T sailplane project developed at the Institute of Aeronautics, at Faculty of Mechanical Engineering in Belgrade. The obtained numerical results fully agree, and in some cases are complementing with the flight test results obtained for this sailplane.

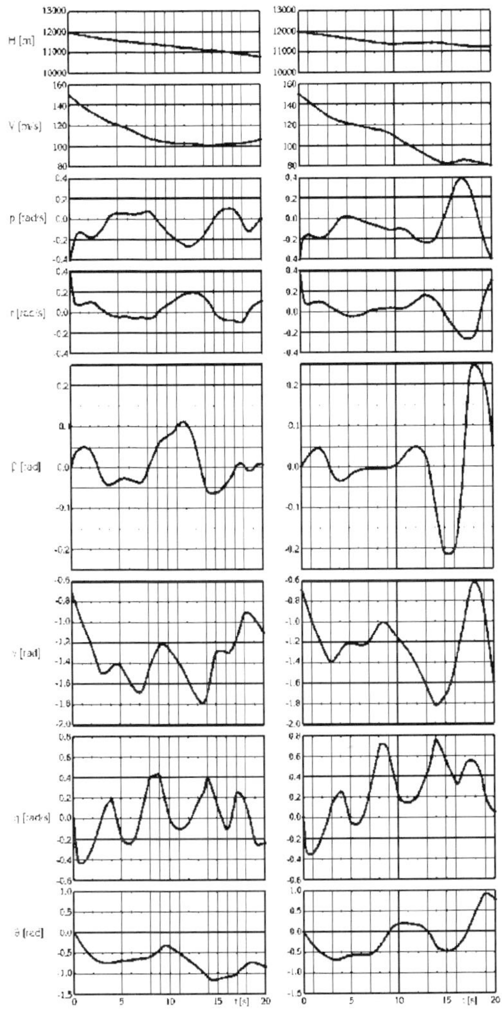

Fig. 28. Basic diagram (left), S_E greater 40% (right)

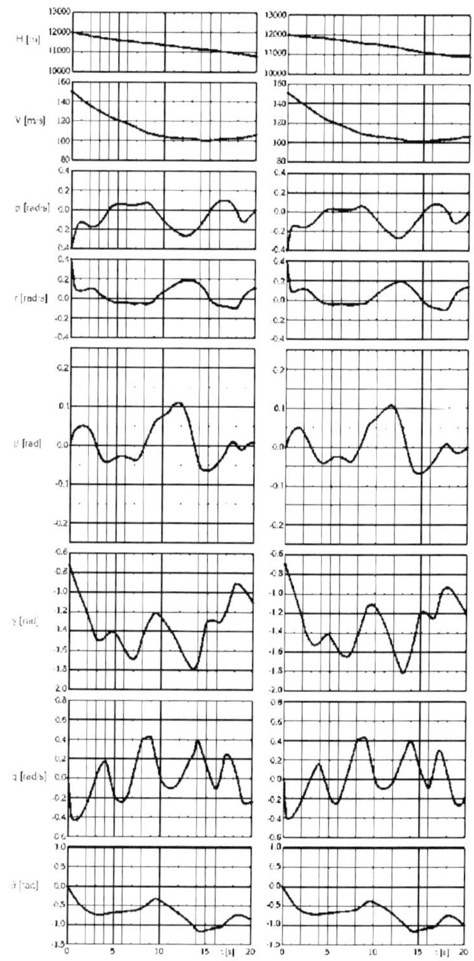

Fig. 29. Basic diagram (left), S_R smaller 40% (right)

In the process of sailplane analysis, influence of variation of elevator and ruder areas has been tested very effectively through the previously described model. The results obtained for two different elevator (S_E) and rudder (S_R) areas are presented in Fig's. 28 & 29 respectively.

The results present variations of height, speed, and the characteristic angular velocities and angles of the sailplane during the first 20 seconds of spin. Practical application of this software has shown that computational design in flight mechanics gives exceptional results, allowing the designer to follow the development of calculation with a large number parameters in detail, and continuously be able to fulfill the criteria of technical demands, performing all necessary optimizations.

4. Systems, subsystems and equipments

4.1 Modeling of transient state and pressure recovering in aircraft proportional hydraulic servo-actuators

In the paper is presented computer analysis of hydraulic servo-actuator transient state between its engage and full pressure recovering. In different mathematical forms hydraulic actuator dynamic model is assumed without any geometrical and physical discontinuities and ambiguity of initial pressure conditions. Hydraulic actuator can be assumed with two serial connected compressible fluid flows controlled by supply and return variable fluid flow restrictors enclosed in control servo-valve and separated by actuator piston, expressed by equivalent mass, viscous damping and arbitrary external force. Presented mathematical model includes transient state of hydraulic actuator, which can be described and determined by ambiguity of the initial pressure conditions in actuator chambers as result of the final state of its previous operations and existing fluid leakage which can produce arbitrary value of initial pressure. Initial condition of pressure is primarily caused by external force, which is arbitrary value during actuator operations. Pressure surge in the moment of change direction of piston motion as result of geometric and flow asymmetry of actuator and its control servo-valve is included in the model also. Hydraulic actuator is usually assumed with compressible fluid flow including the effects of its viscosity. Fluid compressibility is assumed as quasi-static change of its density depending of static pressure. Each of the mentioned effects produces local pressure drop and surge and corresponding actuator operational time delay, which is the limiting factor of its cyclic velocity. Dominant influence on actuators time delay (less than 3% of unit step discrete control piston stroke time) is caused by fluid volumetric compressibility. In the paper following problems are treated: pressure discontinues changes in reverse of piston motion direction; nonlinear effects of actuator behavior; actuator dynamic model linearization.

Any direction change of actuator motion produces pressure discontinuity in its source and return pipelines. This is caused by inversion of fluid flow which produce connection change between supply pipeline and actuator chambers. In the moment of fluid flow direction change each of actuator chambers inter-change connections with system pump and return pipeline and produce corresponding discrete change of pressure in actuator chambers. Possible pressure drop or surge is also caused by geometric asymmetry of servo valve.

4.1.1 Effects of system discontinuities

Any change of direction of actuator motion produces pressure discontinuity in its both pipelines. This discontinuity is caused by inversion of fluid flow which produce the change of connection between supply pipeline and both actuator chambers. In the moment of change of fluid flow direction each of actuator chambers change connections with system pump and return pipeline, producing corresponding discrete changes of pressure in actuator chambers. Possible pressure drop or surge is also caused by geometric asymmetry of servo valve. These effects are explained on the following figures.

On figure 30 is shown actuator motion asymmetry between direct and reverse modes. This asymmetry is result of pressure distribution along supply and return streamlines, shown on diagrams a) and c) on figure 32. Diagram a) corresponds to direct mode of actuator function represented by symmetrical pressure drops at supply and return branches of its servo-valve. The third step of pressure drop correspond to the actuator piston position and corresponds to the applied external force. On diagram c) is shown pressure distribution for reverse actuator mode. Main difference between these modes is in opposite directions of external force related to the streamline of fluid flow. For reverse mode external force support system pump like additional serial connected system sources. This fact appears on figure 30 as different curve gradient for direct and reverse modes. However, absolute value of gradient is greater for reverse mode. More data about gradient value will be shown corresponding to the figure 35. On figure 31 is presented actuator output on servo-valve control input assumed as transient step unit function. This approximation is very close to the real situation for digitally controlled actuators.

On diagrams b) and d) of figure 32 are shown, respectively, equivalent pressure drops for direct and reverse actuator modes corresponding to the conventional mathematical modeling of hydraulic actuator with total pressure drop on servo-valve by neglecting effects at its supply and return parts as two separated fluid flows. This usual approximation cannot be accepted if in system are included effects of hydraulic pressure drop and surge caused by fluid compressibility.

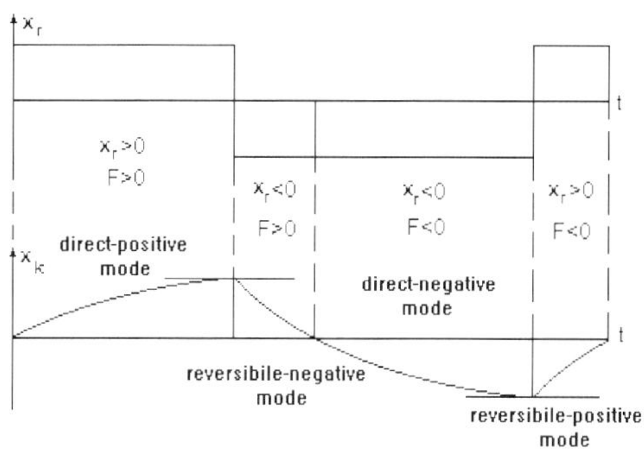

Fig. 30.

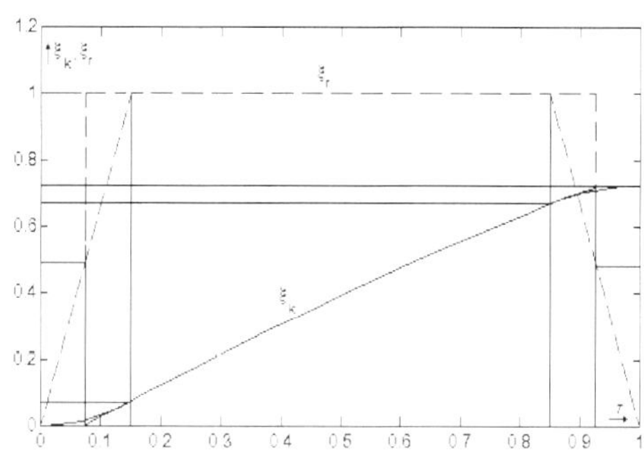

Fig. 31.

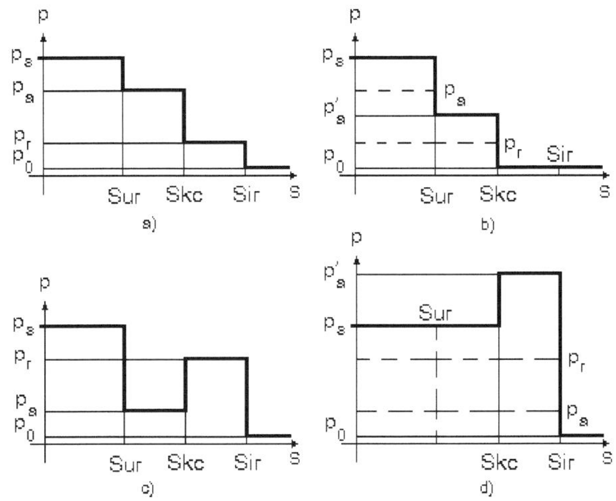

Fig. 32.

4.1.2 Conventional system modeling

Real state is described by pressure drops at supply and return parts of control servo valve and corresponding pressure difference caused by external force:

$$\Delta p_{ul} = p_s - p_a \quad \Delta p_{iz} = p_r - p_0 \quad p_a - p_r = \frac{|F|}{A_k}\mathrm{sgn}(Fx_r) \tag{15}$$

Function sgn denotes both directions of external load action, represented as direct and reverse modes of actuator function. Equations (15) defines basic formulation of system dynamic model. Closed formulation of mentioned pressure drops cannot be determined without additional approximations. If we assume that fluid flow through servo-valve is a turbulent, approximate expressions of corresponding equivalent pressure differences in accordance to the figures b) and d) are defined for incompressible fluid flow by following relations:

$$Q_s = Q_0 = A_k \dot{x}_k = \mu^+ b_r^+ x_r \sqrt{\frac{2}{\rho}(p_s - p_a^{'})} \quad Q_s = Q_0 = -A_k \dot{x}_k = \mu^- b_r^- x_r \sqrt{\frac{2}{\rho}(p_r^{'} - p_0)} \tag{16}$$

where $p_r^{'} = \frac{k_a}{A_k}|x_k| + p_s$ and $p_a^{'} = \frac{k_a}{A_k}|x_k| + p_0$

4.1.3 Separate flow modeling

Previous relations (15) can be expanded for approximately symmetric supply and return flow characteristics of servo-valve in the following form (with assumed value of p0=0) for direct and reverse modes:

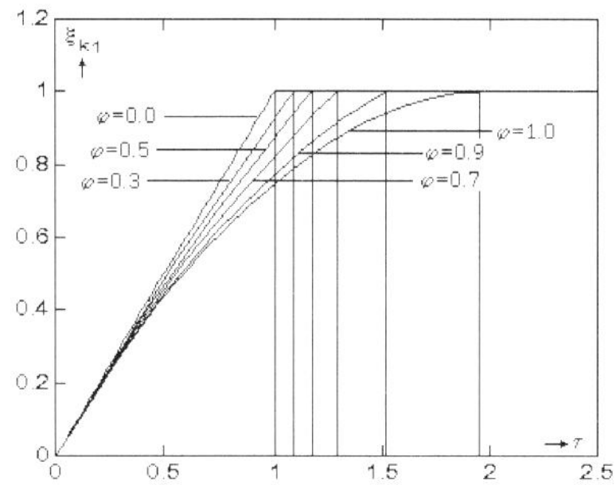

Fig. 34.

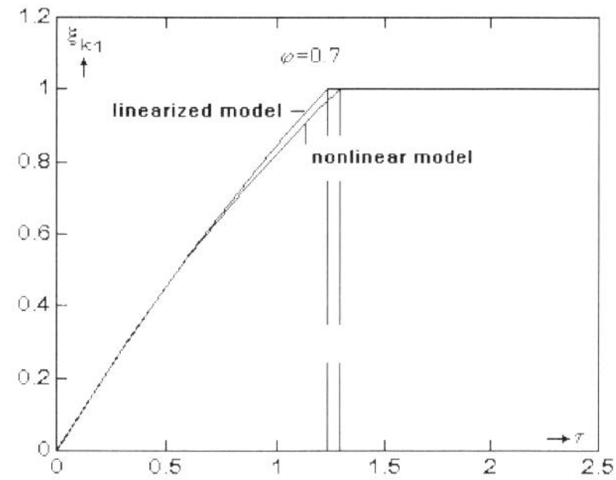

Fig. 35.

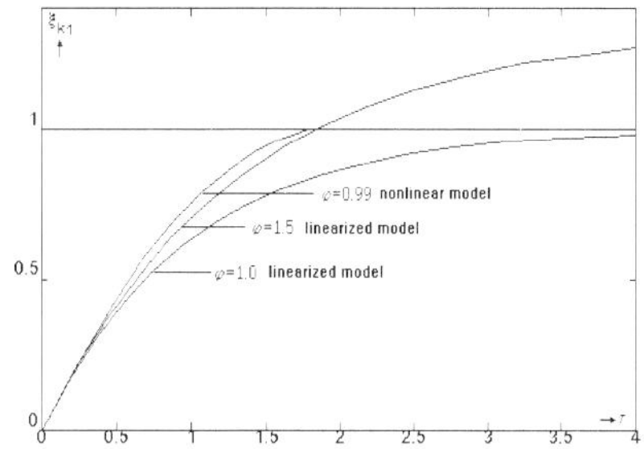

Fig. 36.

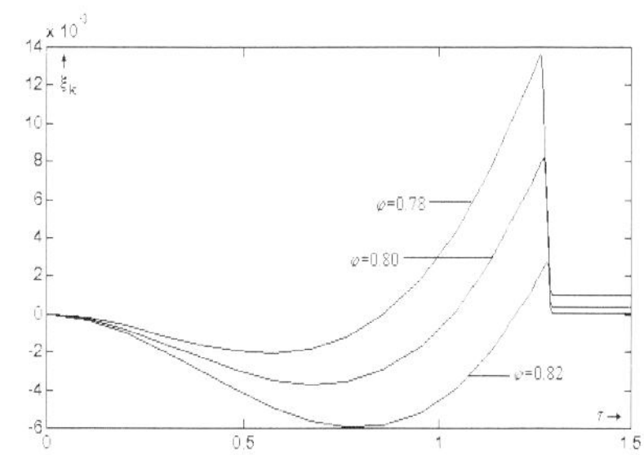

Fig. 37.

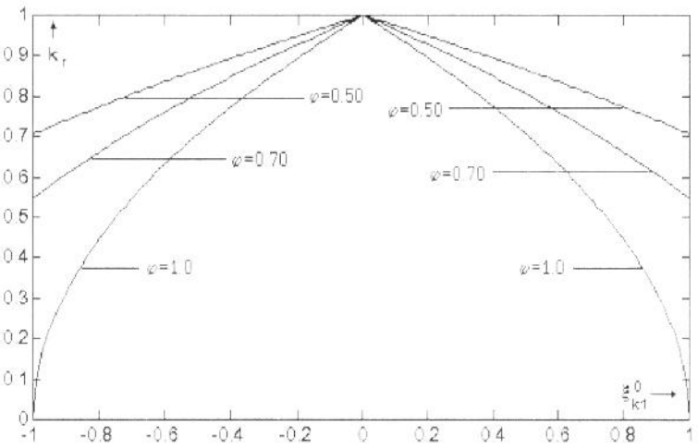

Fig. 38.

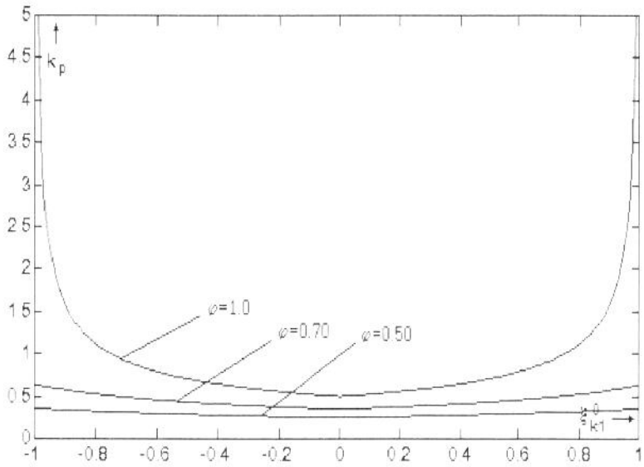

Fig. 39.

On figures 38 and 39 are shown diagrams of non dimensional coefficients of model linearisation corresponding to its nominal regimes.

4.1.6 Actuator transfer characteristics

In this paper actuator simulation is presented for step and frequent harmonic input of actuator control servo-valve. Step input is used for actuator quasi stationary characteristics and parameters identification. Harmonic impute can be used for high cyclic actuator identification.

On diagram of figure 40 is presented computer simulation of nonlinear periodic actuator output piston stroke for harmonic control input, both in non dimensional form. If corresponding relative amplitude of servo-valve throttle is equal 0.50 depending of relative time, equal 1 for total piston stroke for its maximal possible velocity. It is easy to see that approximate harmonic output is recovered approximately after 5 cycles. System nonlinearity is shown on diagram of figure 41, which represents inverse simulation of system relative control input for harmonic relative output.

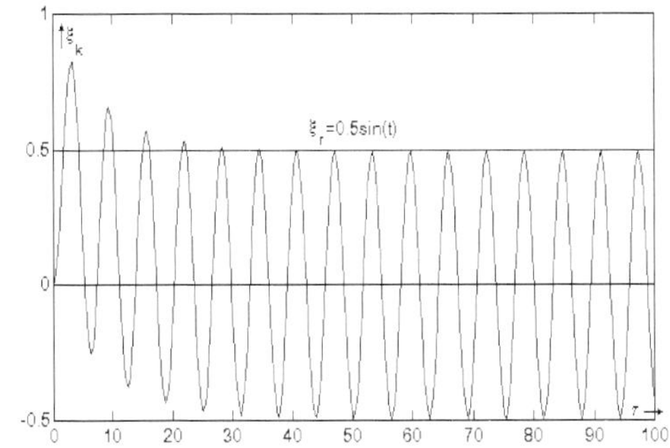

Fig. 40.

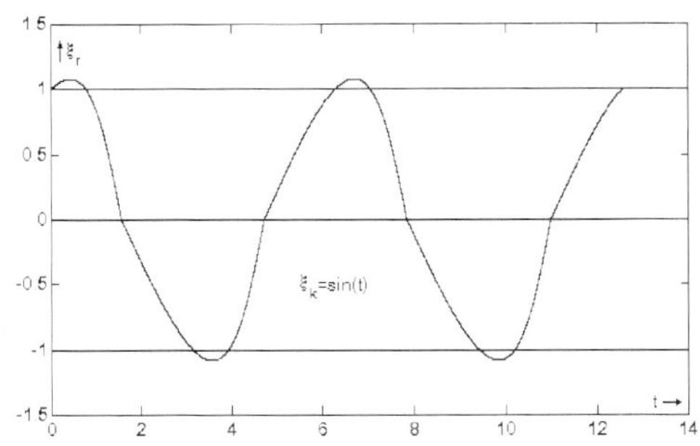

Fig. 41.

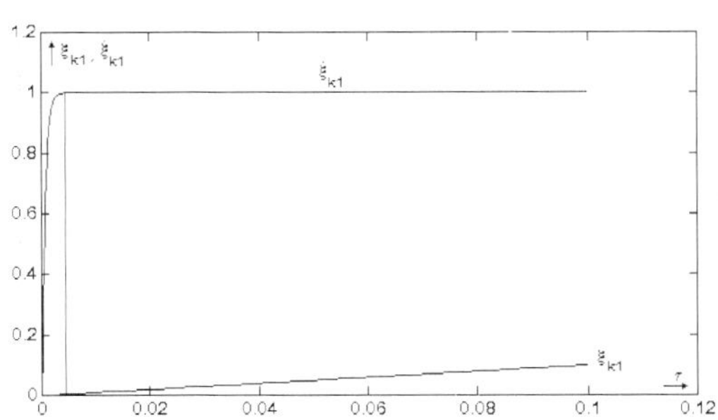

Fig. 42.

On diagram of figure 42 are presented simulation curves of relative actuator piston stroke position and its corresponding relative velocity as actuator system outputs for unit step control relative input if exists only inertial actuator load including equivalent total mass of aerodynamic control surfaces. Corresponding actuating relative time delay is less than 0,004.

4.1.7 Actuator modeling with assumed quasi-static fluid compressibility

Pressure drop in hydraulic systems can be caused by small external load or local increasing of fluid flow to be greater than maximal possible pump source flow. To prevent this it is suggested to separate corresponding branch of actuator supply by corresponding reduction valve. In these cases possible pressure surge are not of high influence and is determined by equivalent actuator stiffness together with potential external load. Pressure increases proportionally with increasing of piston displacement corresponding to its velocity. This pressure increasing is too slower than for the cases of pressure surge caused by fluid compressibility. If actuator is assumed with quasi static compressible fluid flow, system model can be presented for symmetric supply and return branch of fluid flow in the following form:

$$\mu^0 b^{r0}(\pm x_r)\sqrt{\frac{2}{\rho}(p_s - p_a)} = \pm A_k \dot{x}_k + \beta A_k (H_{cil}\theta + x_k)\dot{p}_a + c(p_a - p_r)$$
$$\mu^0 b^{r0}(\pm x_r)\sqrt{\frac{2}{\rho}(p_r - p_0)} = \pm A_k \dot{x}_k - \beta A_k (H_{cil}\theta - x_k)\dot{p}_a + c(p_a - p_r) \quad (23)$$
$$F = A_k(p_a - p_r)$$

where are: μ flow coefficient, b^r equivalent geometric wide, x_r position of control valve throttle, p_s supply pressure of hydraulic system pump, p_a static pressure in supply chamber of actuator cylinder, p_r static pressure in return chamber of actuator cylinder, p_0 static pressure in return pipeline, A_k area of actuator piston, β coefficient of fluid compressibility, c coefficient of fluid leakage, H_{cil} piston stroke, θ coefficient of parasite volume of connected pipeline to actuator cylinder, x_k position of piston and F applied external force (including inertial forces) to the actuator piston. Both signs in equations corresponds to the direct and reverse modes of actuator function.

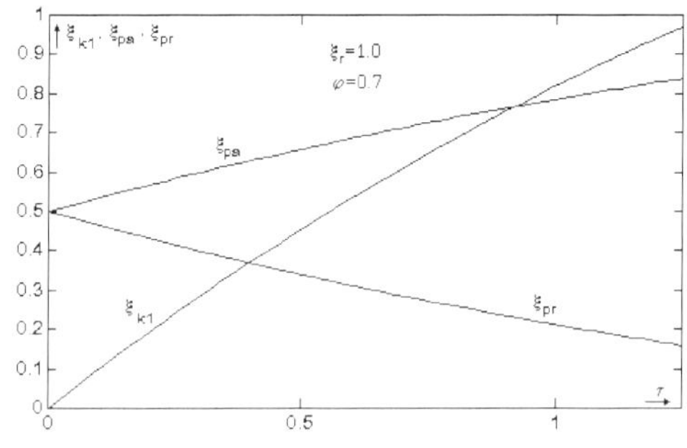

Fig. 43.

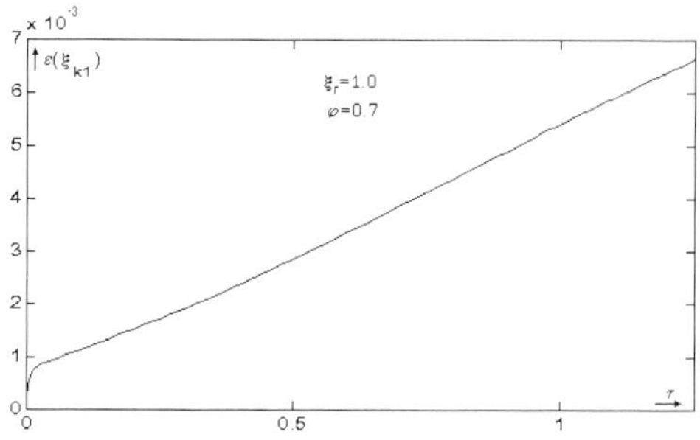

Fig. 44.

On figure 43 are presented simulation of actuator piston relative stroke and relative static pressure in supply and return actuator chambers for usual mathematical form of actuator dynamic model. As it is explained in abstract, initial relative pressure values (equal 0.5) can exists in ideal model only. In real cases, initial values of relative pressure in actuator chambers is the result of actuator history and fluid leakage, which produces it's ambiguity. Extreme case corresponds to the zero values of initial relative pressures. Corresponding simulation of supply and return relative pressures are presented on figures 45 and 46.

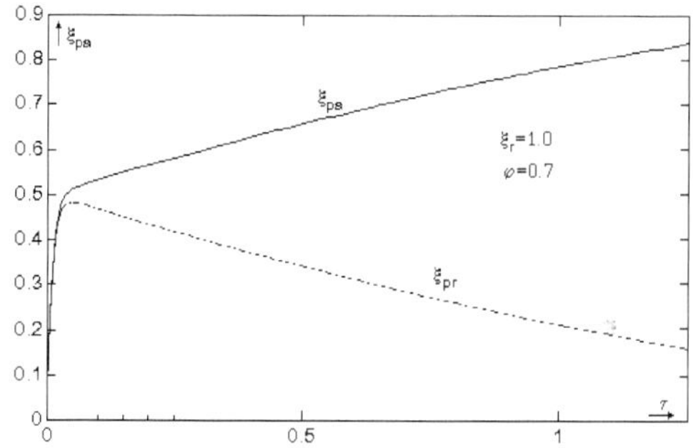

Fig. 45.

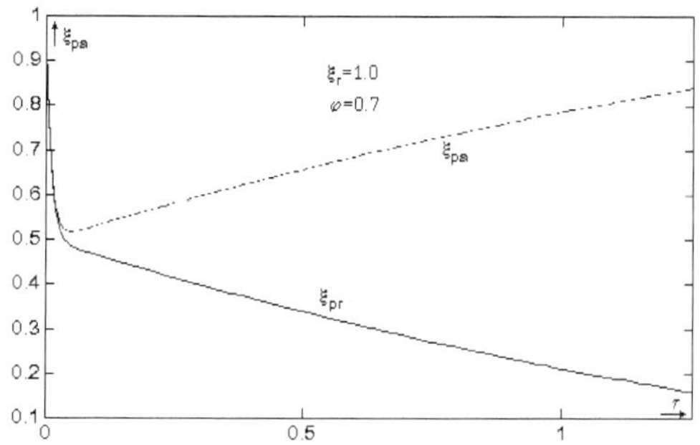

Fig. 46.

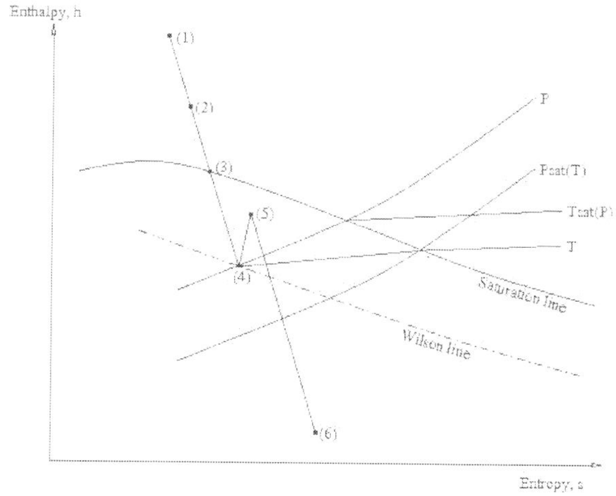

Fig. 2. State line for expanding steam with spontaneous condensation.

its maximum at point (4). This area is termed nucleating zone and is terminated by the Wilson point, which is the point of maximum supercooling and is defined as:

$$\Delta T = T_S(P) - T_G \tag{2}$$

Downstream of this point, nucleation ceases effectively and the number of droplets in the flow remains constant. The nuclei grow rapidly between points (4) and (5) and restore the system to thermodynamic equilibrium. Further expansion of the flow between points (5) and (6) takes place close to equilibrium conditions.

3. Governing equations for two-phase wet steam flow

Assuming no inter-phase slip, the volume occupied by the liquid to be very small and that the vapour phase behaves as a perfect gas, the fundamental equations of motion for steady 1-D flow of a condensing vapour over an incremental distance dx along duct axis shown in Fig. 3 can be written as:

Continuity:

$$m = m_G + m_L = constant \tag{3}$$

And

$$m_G = \rho_G \, A \, u \tag{4}$$

where m is the mass flow rate, ρ is the density, A is the area and u is the velocity. The suffixes G and L denote the vapour and liquid phase respectively. Using the definition of Eq. (4), Eq. (3) can be written in differential form as:

$$\frac{d\rho_G}{\rho_G} + \frac{dA}{A} + \frac{du}{u} + \frac{dm_L}{m - m_L} = 0 \tag{5}$$

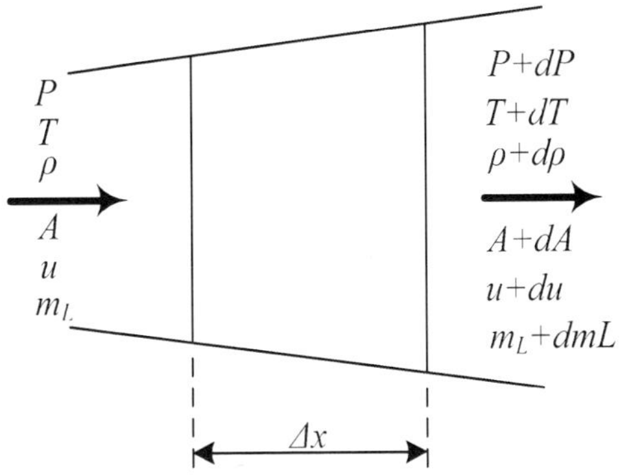

Fig. 3. Control volume for 1D flow analysis.

Momentum:

$$\frac{dP}{P} + \frac{\rho_G\, u^2}{\rho}\frac{f\, dx}{2\, d_e} + \frac{\rho_G\, u^2}{(1-\omega)P}\frac{du}{u} = 0 \tag{6}$$

Where P is the pressure, f is the friction factor, ω is the wetness fraction and d_e is the hydraulic mean diameter for the duct section.

Energy:

$$\frac{u^2}{c_P\, T_G}\frac{du}{u} + \frac{dT_G}{T_G} + \frac{d(\ell\, m_L)}{c_P\, T_G\, m} = 0 \tag{7}$$

Where c_P is the specific heat at constant pressure, T is the temperature and ℓ denotes the energy given up by unit mass of vapour on condensing.

Equation of state:

$$\frac{dP}{P} - \frac{d\rho_G}{\rho_G} - \frac{dT_G}{T_G} = 0 \tag{8}$$

Condensing flow of steam may be regarded as a special case of the flow of a compressible fluid in a duct with friction and heat addition in which heat added to the flow is the result of the phase change, which also affects the mass flow rate. In the analysis of one-dimensional condensing flow, steam is assumed to consist of liquid droplets of specified sizes and uniformly distributed in parent vapour which fills the space between them. The treatment combines the droplet formation and growth equations with one-dimensional gas dynamics equation. The relationship between the vapour pressures over a curved surface to its radius of curvature at low pressure is given by the Kelvin-Helmholtz equation. The equation is modified to include higher virial coefficients and is given by:

$$r^* = \frac{2\sigma}{\rho_L\, R\, T_G \ln\left\{{}^{P}/_{p_s\,(T_G)} + 2B\,[\rho - \rho_s]\right\}} \tag{9}$$

where r^* is the critical droplet radius, σ is the surface tension of liquid droplet, R is the gas constant, $p_s(T_G)$ is the saturation pressure corresponding to local vapour temperature T_G and B is the virial coefficient. For given vapour conditions, droplets with radius r^* will be in unstable equilibrium with the vapour condition. For equilibrium, larger droplets need a lower supersaturation and will grow. On the other hand, smaller droplets will find the surrounding supercooling state insufficient and therefore tend to evaporate. In order to condense, the molecules must form droplet of radius r^* which is against their natural tendency. The only route to the formation of super-critical droplets is through collision within the body of the vapour. From eq. (7), it can be seen that the critical radius is inversely proportional to the supersaturation (p/p_s) and therefore, the lower the supersaturation, the larger the size of critical droplets and the smaller the chance that new droplet embryo will be formed.

Original investigation of the rate of formation of critical clusters within the supercooled vapour was carried out by (Volmer & Weber, 1996, Farkas, 1927, Becker & Doring, 1935, Frenkel, 1946, Zeldovich, 1942). This was later refined, modified and reviewed by numerous investigators. The expression for the nucleation rate as the number of droplets formed per unit volume and time as given by the classical nucleation theory is:

$$J \;= q \left(\frac{2\sigma}{\pi\, m^3}\right)^{1/2} \frac{\rho_G{}^2}{\rho_L} exp\left[-\ \frac{4\pi\, r^*\, \sigma}{3\, k\, T_G}\right] \tag{10}$$

where J is the nucleation rate, k is the Boltzmann's constant and q is the condensation coefficient, which is defined as the fraction of molecular collisions which results in condensation. Substituting for critical radius, r^* in terms of the supersaturation ratio, it will be seen that very small changes in the supersaturation of the fluid can influence the nucleation rate drastically.

Condensation occurs by the nucleation of new droplets and the growth of any existing droplets within specified incremental step. The incremental mass of liquid formed, dm_L can be determined by using the nucleation and droplet growth equations. The equation for nucleation rate is derived from classical nucleation theory. Once the mass of liquid over an incremental step is known and regarding the terms $\frac{fdx}{2d_e}, \frac{dA}{A}$ and dm_L as independent variables, Eqns. 5-8 can be solved for four unknowns, $\frac{du}{u}, \frac{dT_G}{T_G}, \frac{dP}{P}$ and $\frac{d\rho_G}{\rho_G}$. The resulting expression can be integrated using fourth order Runge-kutta technique to yield the changes in the flow properties over the step. But to increase the accuracy of the coupling between the main flow equation and the equations describing droplet behavior, the calculation is carried out in the following steps:

1. The droplet growth equations are integrated at constant inlet pressure, P_{in} and temperature, $T_{G,in}$ to the element Δx to obtain the first approximation to the values of m_L and dm_L.
2. The values of m_L *and* dm_L found in Eq. 3 are inserted into the flow equations which are then integrated to give a first approximation for the exit consition, P_{out} and $T_{G,out}$.
3. The droplet growth equation are integrated for a second time over Δx, assuming constant vapour condition, P_{out} and $T_{G,out}$ and new values of m_L and dm_L are calculated.

4. The values of m_L and dm_L found from Eq. 3 and Eq. 5 are averaged and assumed to represent the variables over incremental step, Δx.
5. Using the average values of m_L and dm_L found in Eq. 6, the flow equations are integrated to obtain the final values of P_{out} and $T_{G,out}$.
6. To ensure that all properties are compatible, the droplet temperature at exit is adjusted so that it correspond to the mean droplet radius.

Then starting at inlet to the nozzle, the flow equations are integrated step by step until the end of nozzle is reached.

In this investigation, upon completion of the nucleation phase, the properties of fluids are calculated based on the two methods. The first method is carried out by combining all droplet groups into a single population and their properties are averaged when there is no new droplet embyo forms in the subsequent step length. This is called the "Average" method. The other method of calculation is when every single droplet groups and their properties are retained even after the completion of nucleation phase. This is called the "Non-Averaged" method. The droplet radius calculation using this method is based on r.m.s. values for each droplet group. In theory, this method should results in more accurate solution but it requires an extremely huge processing power to complete the whole calculation.

4. Results, analysis and discussion

4.1 Calculation on dry, steady flow

In order to test for the accuracy of the numerical code, the model was applied to a few test cases at dry condition in a convergent-divergent nozzle. Three cases were calculated namely the subsonic-supersonic flow, purely subsonic flow and the flow involving shock. In this test, the inlet temperature and total pressure for all cases are set to 421.3K and 72,700 Pa respectively while the back pressure is varied according to the desired cases. For purely subsonic case, the back pressure is set to 68,000 Pa (giving a pressure ratio of 1.069) while for the supersonic flow, linear interpolation is employed for the determination of the back pressure. On the other hand, the back pressure for the shock case is set to 60,400 Pa.

Fig. 4 illustrates the general Mach number distribution along x direction for all cases. The numerical results are compared with the exact solutions which are given by solid lines close to each case respectively. It can be clearly seen that all numerical results show a slight deviation from the exact solutions except for the supersonic case where both numerical and exact solutions agrees very well with each other. For the pure subsonic flow case, the Mach number increases with distance until it reaches a peak value of 0.57 at the throat. Downstream of the throat, the Mach number decreases in the divergent section due to the decrease in the flow velocity. Comparison with the exact solution shows a slightly higher value and this could be attributed to the numerical error presented in the model. For the subsonic-supersonic case, the Mach number increases along the nozzle. The numerical solution for this case shows good agreement with the exact solution. The Mach number increases starting from the nozzle inlet to the nozzle exit. In order to emulate the shock case, the exit pressure is reduced slightly below the pressure imposed on the subsonic case. A normal shock-wave was found slightly downstream of the throat in the divergent section of the nozzle. The flow is supersonic upstream the normal shock-wave, whereas it becomes subsonic downstream the normal shock-wave.

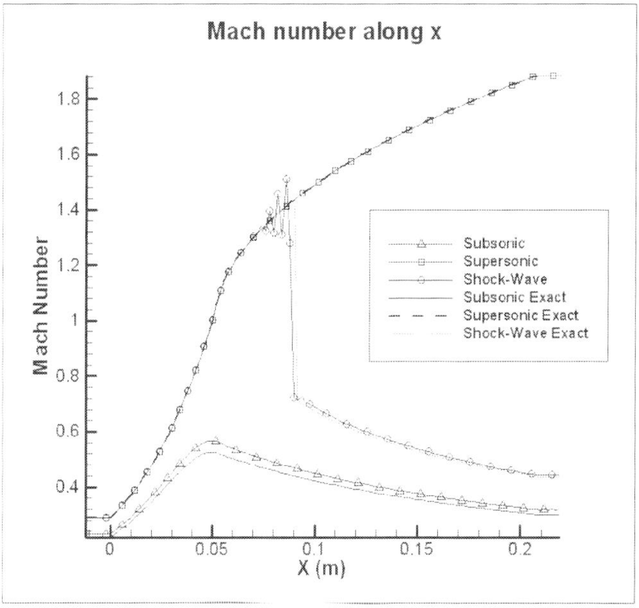

Fig. 4. Comparisons of Mach number using numerical solution and exact solution (for dry case).

In addition to the Mach number, the pressure variation along the distance is also investigated and this is illustrated in Fig. 5. For the subsonic case, the pressure decreases along the nozzle length due to the reduction in flow area (where the velocity increases). Minimum pressure is calculated at the throat and the pressure increases again downstream of the throat due to the increase in flow area. For the supersonic case however, the pressure decreases starting from the nozzle inlet towards the nozzle exit. Downstream of the throat where choking occurs, the velocity of the fluid further increases thus reducing the pressure further. On the other hand, for the shock case, the pressure distribution is similar to the supersonic case except for a location downstream of the throat where sudden jump in pressure is observed indicating shock. It is also worth mentioned that prior to the shock, obvious oscillations can be seen due to the adoption of the central discretization scheme in the calculation which is highly unstable. Although artificial viscosity has been added to the equations, it is believed that the value is either too small or too large to improve the stability of the solutions. Thus further investigation is needed to eliminate or at least minimize the oscillation. Downstream the shock position, the pressure further increases towards the nozzle exit.

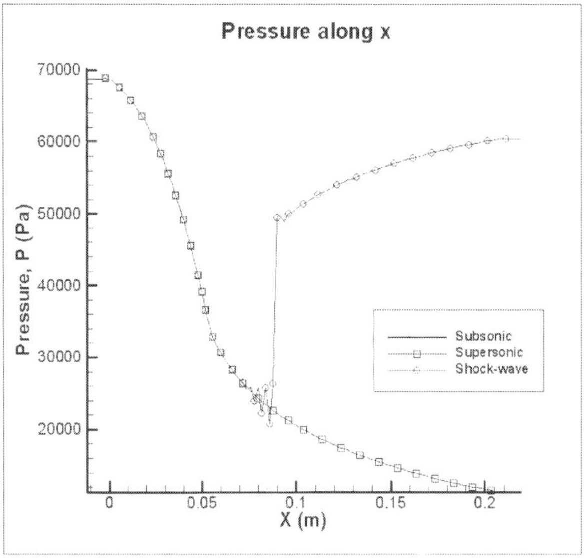

Fig. 5. Pressure variations using numerical solution and exact solution (dry case).

4.2 Steady flow wet steam simulation

Based on the prediction made on dry condition on a few test cases, it has been demonstrated that the numerical scheme is able to calculate the compressible flow properties with good accuracy. The next step involves the calculation of two-phase flow properties using similar scheme but with the addition of the wetness treatment. The numerical scheme was applied to three cases involving nucleating steam flows in one-dimensional nozzles. These nozzles are those of (Binnie and Wood, 1938), (Krol, 1971) and (Skillings, 1987). The exact boundary conditions imposed for these cases are summarized in Tab. 1. For all tests, the flow with condensation takes place.

Nozzle		P_0 (Pa)	T_0 (K)	$T_{sat}(P)$ (K)
Binnie and Wood (1938)		143,980	391.87	383.28
Krol (1971)	Case 1	221,000	423.00	396.60
	Case2	294,000	453.00	405.96
Skillings (1987)	Case 1	32,510	357.20	344.60
	Case 2	35,440	349.00	346.13

Table 1. Boundary conditions for different test cases.

In the one dimensional calculation, the number of node used in all cases is 125. In addition to the comparison made with the experimental data (where available), the 2D results of the same method are also plotted for comparisons. The 2D calculation was carried out in a few nozzle configurations namely (Binnie & Wood 1938), (Kroll, 1971) and (Skilling, 1987). For each of these nozzles, structured mesh with consistent size of 125 × 15 is adopted and is illustrated in Fig. 6.

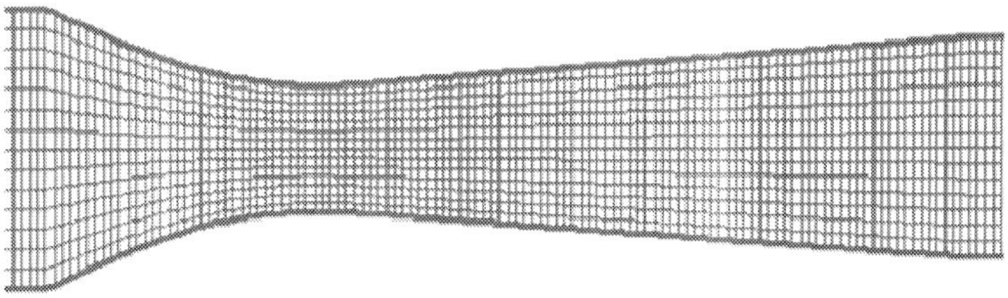

(a)

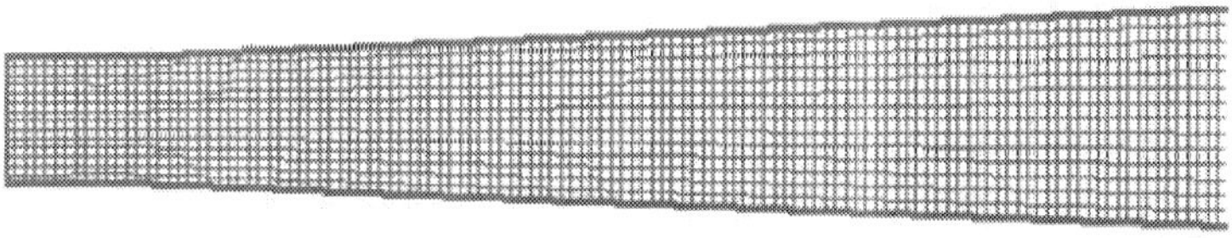

(b)

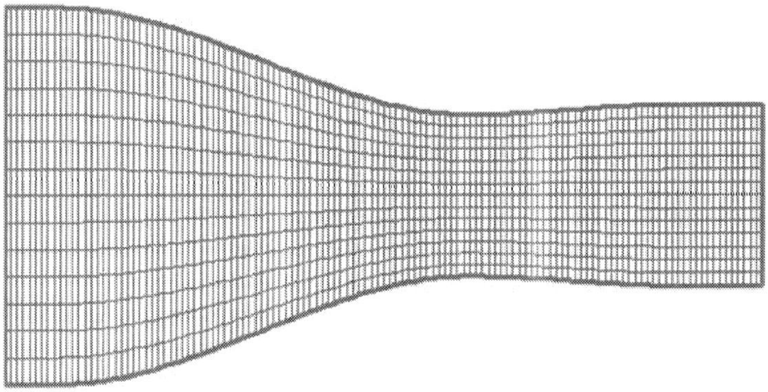

(c)

Fig. 6. 2D mesh for different nozzles (a) Binnie & Wood; (b) Kroll (c) Skilling.

Fig. 7 shows the pressure ratio distribution along nozzle length based on (Binnie &Wood, 1938). The general distribution on the pressure shows a decreasing trend due to the increase in velocity. The onset of condensation can also be observed by the sudden jump in pressure when condensation occurs and is typical for wet supersonic expansions. This is due to the release of latent heat at supersonic conditions which tends to retard the supersonic flow. The trend continues to decrease downstream of the throat towards the nozzle exit. The distributions of pressure along the nozzle predicted are in good agreement with experimental data.

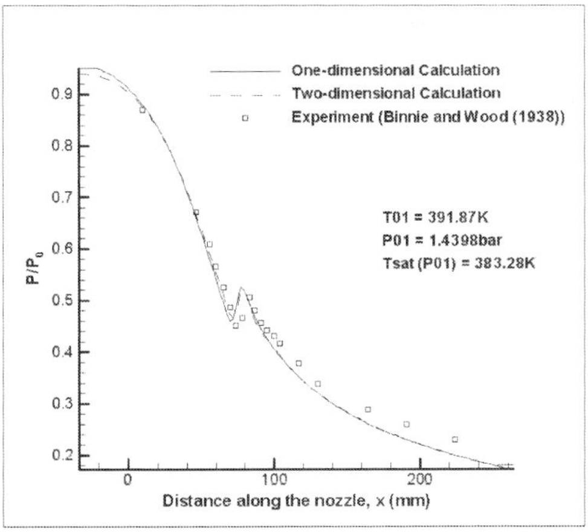

Fig. 7. Comparison of pressure ratio along Binnie and Wood nozzle.

Fig. 8(a) and 8(b) shows the pressure distribution along Krol nozzle for Case 1 and Case 2 respectively. The calculated pressure distribution for 1D and 2D cases show an excellent agreement for Case 1. However, experimental validation could not be made for this case due to the absent of experimental data. In order to increase confidence in the current scheme however, the model was validated with Case 2 where detail measurement was present. The comparison between the 1D and 2D calculation against experimental data is illustrated in Fig. 8(b). Despite the absent of data upstream of the nozzle throat, reasonable agreement was achieved downstream of the throat.

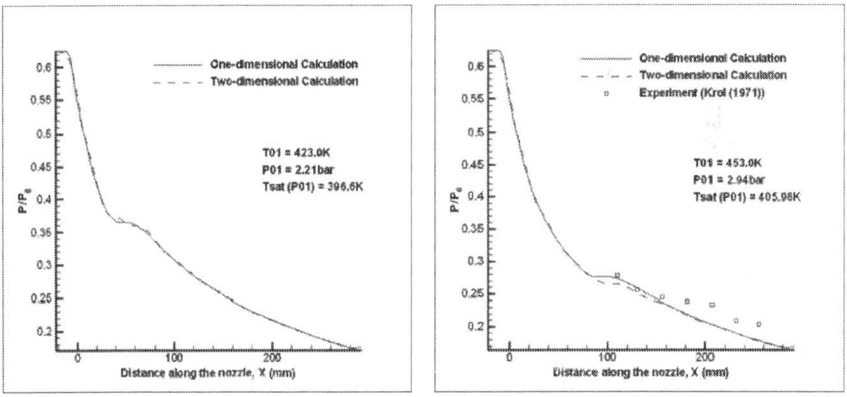

(a) (b)

Fig. 8. Comparison of pressure distribution along Krol nozzle for (a) Case 1 (b) Case 2.

In addition to the Binnie & Wood and Krol nozzles, the mathematical model is also tested on Skilling nozzle (Skilling, 1987) at two different boundary conditions (Case 1 and 2). This is shown in Figs. 9(a) and 9(b) respectively. The general pressure distribution shows a

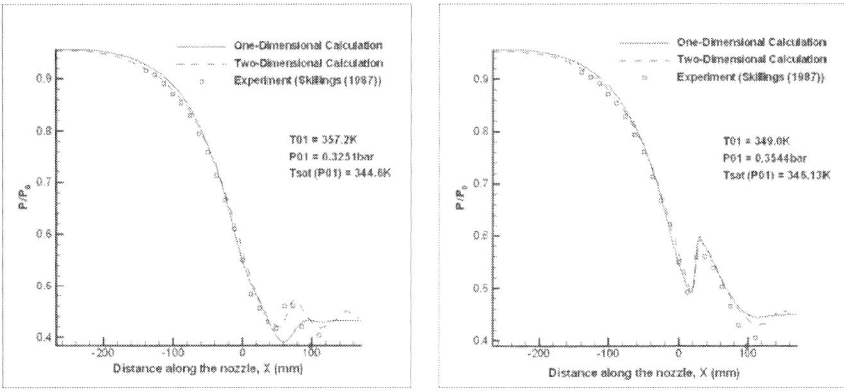

(a) (b)

Fig. 9. Comparison of pressure distribution along Skilling nozzle for (a) Case 1 (b) Case 2.

decreasing trend when the flow was expected to increase in reduced cross sectional area. Comparison with the experimental result reveals good agreement. In addition, the onset of condensation was predicted correctly for both cases. From this calculation, it is evident that the location where condensation begins is highly influenced by the inlet temperature imposed on the inlet boundary.

To quantitatively validate the calculation model, comparison is also made based on the droplet size and its distribution along the duct. Fig. 10 shows the distributions of droplet radius for 1D and 2D calculations in Binnie & Wood nozzle. It can be seen that the droplet starts to form slightly downstream of the throat, grows rapidly if the surrounding parent vapour permits, and continue to grow towards the exit. It is these droplets that contribute to the wetness inside turbine channel. The droplet size information for the 2D case was taken at stream centre with the assumption of constant fluid properties along the y-direction. It is believed that the actual properties vary at least slightly in this direction, thus contributing to the slight deviation as compared to the 1D result. The 2D calculation shows slightly higher droplet radius after rapid growth downstream of the throat and this can be attributed to the cumulative contribution of different sets of droplet groups from the region in y-direction which was not accounted for the calculation.

Fig. 11(a) and 11(b) show the comparison made on the calculated droplet radius with experimental data on Krol nozzle (Krol, 1971) for Case 1 and Case 2 respectively. The general trend on the size distribution is almost identical for all nozzles, i.e. droplet grows rapidly once it is formed and continues to grow at a slower rate towards the nozzle exit. For Case 1, reasonable agreement was achieved for the 2D calculation. However, the 1D model was unable to predict the droplet size accurately despite showing similar growth profile. This discrepancy could be attributed to the two-dimensional effect in nozzle flow where the property variation in y-direction is important and needs to be taken into account in the calculation. In this case, the 1D calculation has over-predict the droplet radius due to the faster expansion rate for Kroll nozzle as compared to the Binnie & Wood nozzle. For Case 2, similar trend was shown at which better prediction was made by the 2D model as compared

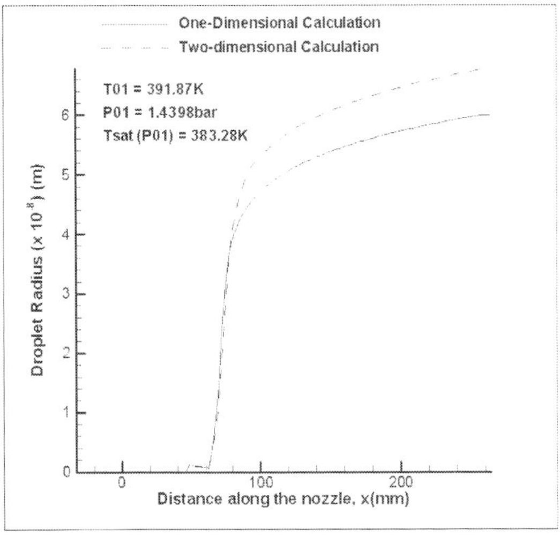

Fig. 10. Comparison of droplet size along Binnie & Wood nozzle (Binnie and Wood, 1938).

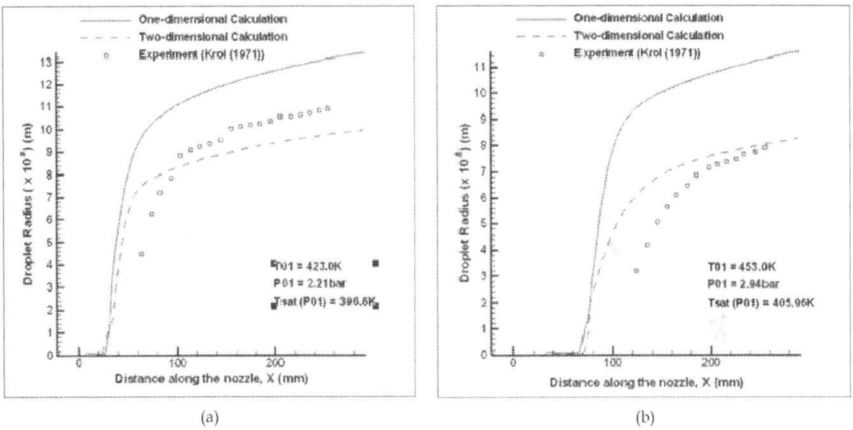

Fig. 11. Comparison of droplet size distribution in Krol nozzle for (a) Case 1 (b) Case 2.

to the one dimensional model. Based on these figures, it is also interesting to note that the temperature imposed at the inlet boundary has significant effect in delaying the onset of condensation. This has been demonstrated by Case 1 which inlet total temperature is lower than Case 2, thus results in the faster droplet formation and growth.

In addition to Binnie & Wood and Krol nozzles, further test is also carried out to investigate the ability of the model to calculate the droplet size distribution inside the channel. The third test case was done by adopting the Skilling nozzle where droplet size measurement was available. Fig. 12(a) and 12(b) illustrate the comparison on the droplet size along the nozzle for Case 1 and Case 2 respectively. In Case 1, the inlet total temperature was set

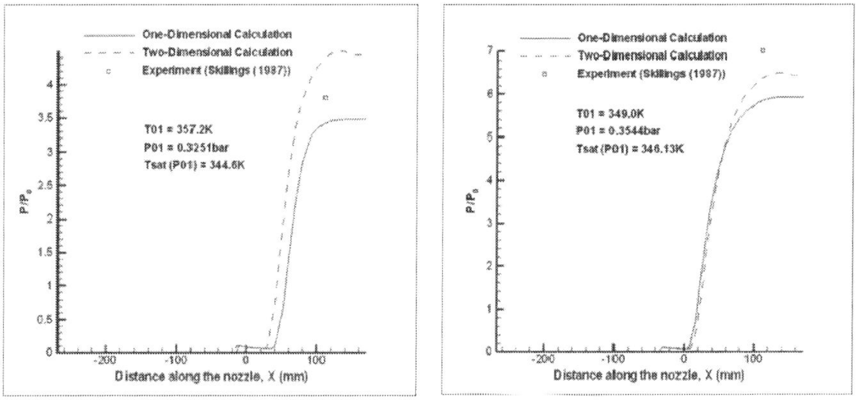

(a) (b)

Fig. 12. Comparison of droplet size distribution along Skilling nozzle for (a) Case 1 (b) Case 2.

higher than Case 2 with the value of 357.2 K, thus there is a slight delay in the formation of droplet embryo. In this case, the droplet starts to form slightly downstream of the throat region, undergoes rapid growth and terminates at nozzle exit. Based on Fig. 12(a), the 1D model has under-predict the droplet size, while the 2D model has over-predict the size. For Case 2, the inlet total temperature is set lower (349 K), thus the formation of droplet embryo occurs earlier somewhere in the throat region and grows rapidly once the saturation condition is met. Reasonable accuracy was achieved both for 1D and 2D cases. In this case, both 1D and 2D calculations have under-predict the droplet size. Similar to Case 1, only one experimental reading was available thus qualitative comparison on droplet size distribution along the nozzle was not possible. It is difficult to quantitatively compare the distribution of droplet size along the nozzle since only one reading was made available (Skilling, 1987) for each case.

4.3 Unsteady supercritical heat addition

In wet steam flows, condensation in a nozzle usually occurs in the divergent section just downstream of the throat with the consequence of the considerable release of latent heat, Q, to the flow. In addition, this phenomenon has the effect of moving the flow Mach number to unity. This is due to the addition of heat to the supersonic flow, which causes it to decelerate. The effect produced when Mach number equals to unity was described as "thermal choking" (Pouring, 1965) and when this happens, the flow can no longer sustain the additional heat and becomes unstable. The excess quantity of this additional heat is termed supercritical heat addition and has caused the flow to be unstable. This phenomenon commonly occurs in high speed condensing flow has been studied in detail by (Barschdoff, 1970) and (Skillings, 1987). Attempt was also made by (Guha, 1990) to model the supercritical heat addition effect by using Denton's method (Denton, 1983). The method adopts the unsteady time-marching treatment for condensing flow of steam. However, the frequency was over-predicted and ranges from 540 Hz to 650 Hz. In recent decades, the study of supercritical heat addition in high speed condensing flow was also made by (White & Young, 1993) and (Yusoff et al., 2003). The predicted frequency was calculated to be 420 Hz and 393 Hz respectively.

In this work, the modified 1D model has been applied to calculate the effect of unsteady supercritical heat addition in Skilling nozzle (Skillings, 1987). The calculation is carried out in an unsteady mode until the solution converged into a steady oscillation. Fig. 13 illustrates the effect of supercritical heat addition to the static pressure in the nozzle. To investigate the unsteady effect accurately, the variations of static pressure at a point located just downstream of the nozzle's throat is plotted. In this case, higher oscillation was observed at the beginning of the iterations and the pressure variation settles at approximately 400 iterations. From this point, regular oscillation with a frequency of approximately 302 Hz is calculated. The inside figure illustrates the static pressure after stable and consistent oscillation is achieved. The pressure distribution is plotted against time for the calculation of the frequency. Tab. 2 shows the summary of frequencies calculated based on different authors using different numerical treatment. The prediction made by the current work shows reasonably good agreement as compared to the experimental data based on the Skillings nozzle.

Author	Case	Frequency, Hz	Pressure Difference, mbar
Skillings (1987)	Experiment	380	20
Skillings (1987)	1D program	640	-
Guha (1990)	1D program	540	-
White and Young (1993)	2D program	420	-
Yusoff et al. (2003)	2D program	393	18
Hasini et al. (2011)	1D program	302	17

Table 2. Frequency and pressure difference comparisons by different authors.

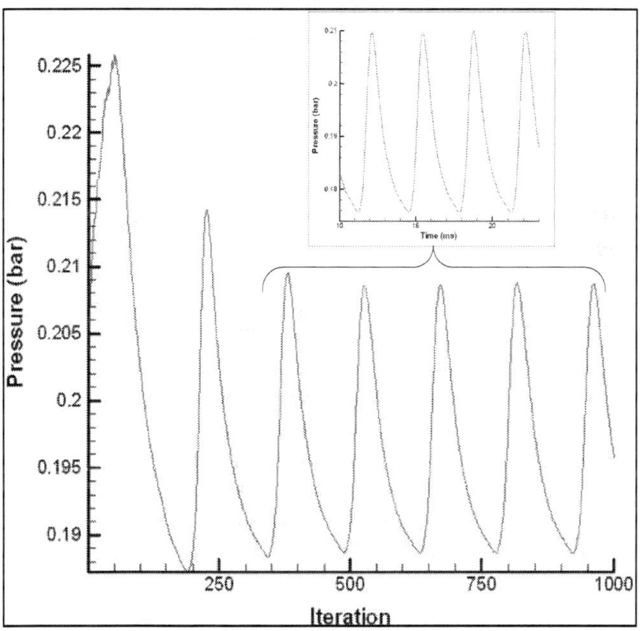

Fig. 13. Unstable supercritical heat addition effect to static pressure.

4.4 Polydispersed droplet calculations

Thus far, the calculation of droplet properties was averaged when a new droplet group formed with respect to time and space. This is done to simplify the calculation steps thus, reducing the computation time. In reality however, these droplet groups should be treated individually as their contribution to the total wetness inside the channel can be different. However, the calculation of the individual droplet spectrum is very complex and requires large computation power epecially when three-dimensional condensing flow is involved. Despite the simplications made in the calculation algorithm for the droplet properties, very little understanding was made on the effect of the calculation method in predicting the droplet properties. Thus this section aims to investigate the properties of wet steam given by individual droplet group tracking (called Non-Averaged method) and the average droplet property (called the Averaged method). The polydispersed droplet calculation is carried out where different droplet group formation and growth is tracked individually. In this study, the effect of droplet property averaging is carefully compared and investigated.

Fig. 14 shows the progressive growth of individual droplet group as superheated steam enters a constant area duct at 1 bar and initially at 44.8K supercooled. In order to observe their individual behavior, the droplets formed during each incremental step of the calculation have been regarded as a distinct group and treated separately. It can be seen that the variations in the critical droplet size, r^* is unaffected at the beginning but increases as the fluid become less supercooled. To investigate the behavior of the droplets, the variation of the radii of each individual droplet group is plotted separately. Taking droplet group (1) as

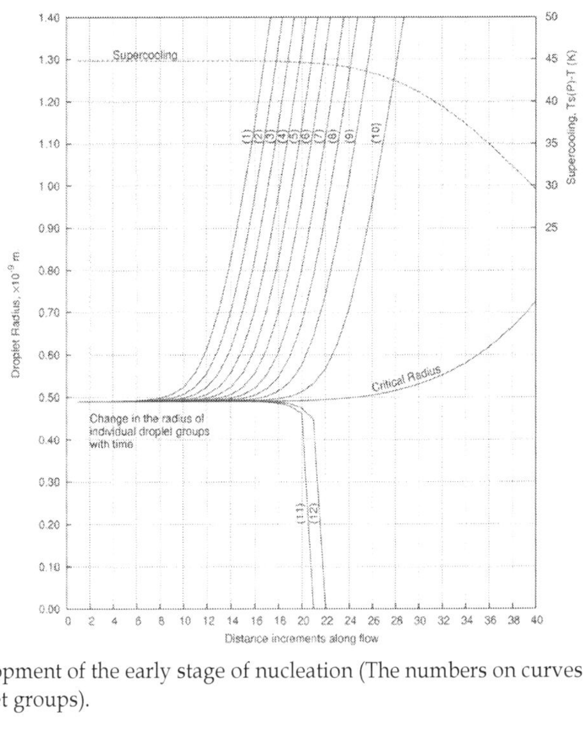

Fig. 14. Development of the early stage of nucleation (The numbers on curves refer to specific droplet groups).

an example, it will be seen that the rate of growth of this droplet group is initially small but rises rapidly as soon as they grow beyond critical radius. The pattern is similar for the next 9 droplet groups but a different behavior is observed in the case of groups (11) and (12). With the depletion of the supercooling, the critical radius increases. The newly formed droplet embryos in group (11) has a radius which is less than the critical radius. Consequently, this droplet group is unable to grow and survive (due to its initial radius which is less than the critical radius) and therefore, it will evaporate. This is represented by the sudden drop in radius as in Fig. 15. Thus, with the step size adopted in the calculation, the nucleation process can be assumed to be effectively over after the formation of droplet group (10). By selecting a shorter step size, the number of droplet groups will increase but the number of droplets in each group will fall, increasing both the accuracy and the volume of algebra to be calculated. In this work, the step length was adjusted to keep the number of droplet groups to approximately 10.

After the nucleation phase, the droplet groups are combined into a single population and their properties are averaged. For the example given in this work, the changes in relevant fluid properties after the nucleation phase are given in Fig. 15. It can be seen that, as expected, after the nucleation stage, the droplets grow rapidly at the beginning but with the depletion of supercooling, the rate of growth reduced and becomes extremely slow as the fluid approaches thermodynamic equilibrium. Similar profile can be seen for the wetness fraction where rapid increase in wetness is predicted at the beginning but as supercooling reduced, the rate at which water forms also reduced.

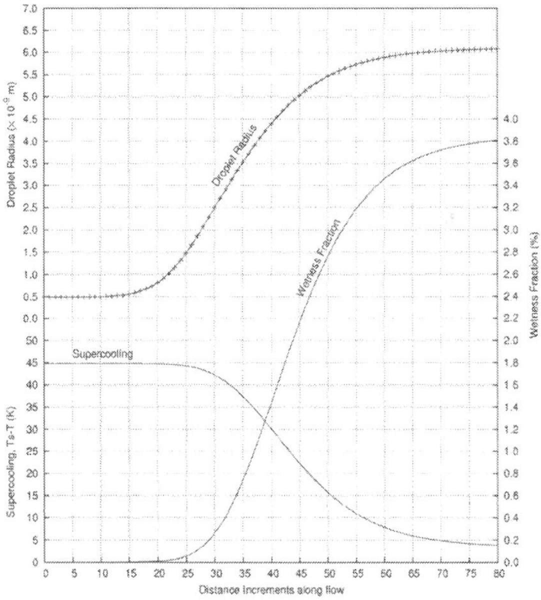

Fig. 15. Change in fluid properties during the return to equilibrium.

In order to check for the accuracy between the calculation procedures in estimating fluid properties after the completion of nucleation phase, comparison is made between the

"Averaged" and "Non-Average" methods. The comparison of the changes in the relevant fluid properties between the "Average" and "Non-Average" methods is shown in Fig. 16. In general, the supercooling and wetness fraction show similar trend for both methods. However, the average method seems to under-predict the droplet radius after the nucleation phase. This is probably due to the simplifications made to the governing equations and calculation procedure when numerical sensor created in the scheme automatically ignores the subsequent droplet embryos which grows for a very short period of time and later failed to survive. The cumulative error resulting from this simplification seems to increase at a very slow pace and later becomes constant. The maximum magnitude of this error is estimated to be approximately 5%. Taking into account the wetness fraction and supercooling properties predicted by both methods, and considering the amount of resources required by the Non-average method, it is anticipated that the Average method is capable of estimating the fluid properties in condensing flow. With some minor modification towards the numerical scheme, the scheme can be adopted to the real steam turbine passage for the prediction of various fluid properties in wet steam flow.

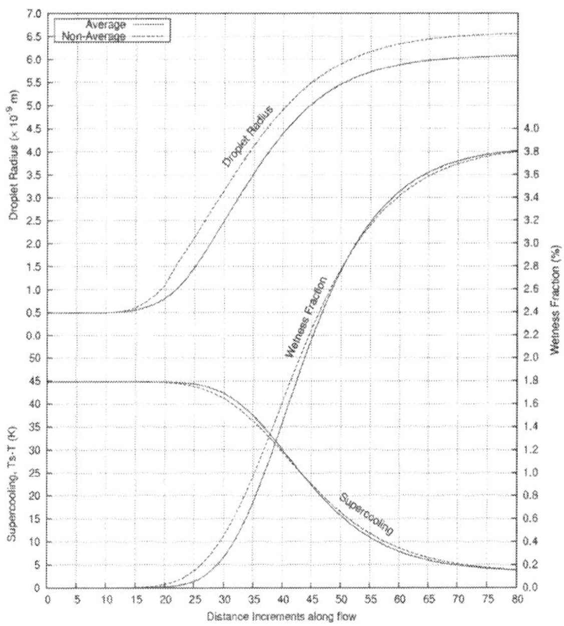

Fig. 16. Comparison on the changes of fluid properties between the "Average" and "Non-Average" methods.

5. Conclusion

The mathematical model for the calculation of high speed condensing flow has been developed. This model was applied to different test cases involving dry and wet flow expansion in converging-diverging nozzle. The mathematical prediction shows excellent agreement with experimental data for pressure along the nozzle. However, the calculated droplet size in a few test cases only shows reasonable agreement with the measure size. The case for unsteady supercritical heat addition also reveals promising results indicating the capability of the model to calculate this phenomenon which might cause instability in turbine channel. The polydispersed droplet calculation has also been carried out using the model. Two methods of calculation were tested and compared. The first method tracks the individual droplet groups at the beginning of nucleation process. Upon the completion of nucleation, the droplet groups are combined into a single population and its properties are averaged. The "averaged" properties are repeatedly used for the subsequent steps until equilibrium. The second method however, retains the calculation of individual droplet groups and their properties and is called the "Non-Average" method. Comparison between the two methods reveals similar trend and even though the average methods slightly under-predicts the droplet radius after the completion of nucleation, the method is thought to be more suitable and economical due to considerable saving in computation time and processing power as compared to the other method. In addition, it is also anticipated that the discrepancies between the two methods are relatively small and therefore can be regarded as negligible.

6. References

Aitken, J. (1881). On Dusts, Fogs and Clouds. *Trans. Royal Society (Edinburgh)*, Vol. 30, pp.337-368

Barschdorff, D. (1970). Droplet Formation, Influence of Shock Waves and Instationary Flow Patterns on Condensation Phenomena at Supersonic Speed, *Proc. 3rd Intl. Conf. on Rain Erision and Allied Phenomena*, Farnborough

Becker, R. & Doring, W. (1935). Kinetische Behandlung der Keimbildung in Ubersettigten Dampfen. *Ann. D. Phys.* (Leipzig), Vol. 24, pp. 719-752

Binnie, A.M. & Wood, M.W. (1938). The Pressure Distribution in a Convergent-Divergent Steam Nozzle. Proc. Instn. Mech. Engrs, Vol. 138, pp. 229-226

Binnie, A.M. & Greem, J.R. (1943). An Electrical Detertor of Condensation in High-Velocity Steam. *Proc. Royal Society (London), A*, Vol. 181, pp. 134-154

Callender, H.L. (1915). On the Steady Flow of Steam Through a Nozzle or Throttle, *Proc. Instn. Mech. Engrs.*, pp. 53-77

Campbell, B.A. & Bakhtar, F. (1970). Condensation Phenomena in High Speed Flow of Steam. *Proc. Instn. Mech. Engrs.*, Vol. 185, pp. 395-404

Deich, M.E.; Tsiklauri, G.V.; Shanin, V.K. & Danilin, V.S. (1972). Investigation of Flows of Wet Steam inNozzles. *Teplofizika Vysokikh Temperatur*, Vol. 1, Part. 1, pp. 122-129

Denton, J.D. (1983). An Improved Time-Marching Method for Turbomachinery Flow Calculation. *ASME Paper 82-GT-239*

Farkas, L. (1927). Keimbildungsgeschwindigketit in Ubersettingten Dampfen, *Z. Phys. Chem.*, Vol. 125, pp. 236-242

Filippov, G.A. & Povarov, O.A. (1980). Moisture Separation in the Turbines of Nuclear Power Station. *Energy*, Moscow, pp. 329

Frenkel, J. (1946). *Kinetics Theory of Liquid*. Oxford University Press, London

Gerber, A.G. & Mousavi, A. (2007). Application of Quadrature Method of Moments to the Polydispersed Droplet Spectrum in Transonic Steam Flows with Primary and Secondary Nucleation. *Applied Mathematical Modelling*, Vol. 31, pp. 1518-1533

Guha, A. (1990). The Fluid Mechanics of Two-Phase Vapour Droplet Flow with Application to Steam Turbine, *Ph.D Thesis*, University of Cambridge, UK

Gyarmathy, G. (1962). Grundlagen einer Theorie der Nassdampfturbine. *Dissertation ETH, Zurich, Juris-Verlag* (English Translation: CEGB (London) Rept T-781 (1963))

Gyarmathy, G. & Meyer, H. (1965). Spontane Kondensation. *VDI Forschungsheft 508*, VDI-Verlag, Dusseldorf

Henderson, J.B. (1913). Theory and Experiment in the Flow of Steam Through Nozzle, *Proc. Instn. Mech. Engrs.*, Part 1, Vol. 2, pp. 253-314

Hill, P.G. (1966). Condensation of Water Vapour During Supersonic Expansion in Nozzles. *J. Fluid Mech.*, Vol. 25, Part 3, pp. 593-620

Krol, T. (1971). Results of Optical Measurement of Diameters of Drops Formed due to Condensation of Steam in a Laval Nozzle. *Trans. Instn. Fluid Flow Machinery (Poland)*, Vol. 57, pp. 19-30

Martin, H.M. (1918). A New Theory of the Steam Turbine, *Engineering*, Vol. 106, pp. 1-3, 53-55, 107-108, 161-162, 189-191, 245-246

Nikkhahi, B.; Shams, M. & Ziabasharhagh, M. (2009). A Numerical Investigation of Two-Phase Steam Flow Around a 2D Turbine's Rotor Tip. *Int. Comm. Heat and Mass Transfer*, Vol. 36, pp.632-639

Oswatitcsh, K.L. (1942). Condensation Phenomena in Supersonic Nozzle. *Z. Angew. Math. Mech.*, Vol. 22, Part 1, pp. 1-14 (RTP Translation no. 1905)

Pouring, A.A. (1965). Thermal Choking and Condensation in Nozzle, *Physics of Fluids*, Vol. 8, Part 10, pp. 1802-1810

Puzyrewski, R. (1969). Condensation of Water Vapour in a Laval Nozzle. *Instn. Fluid Flow Machinery*, Gdansk (Poland)

Rettaliata, J.T. (1938). Undercooling in Steam Nozzle. *Transaction of ASME*, Vol. 58, pp. 599-605

Skilling, S.A. (1987). An Analysis of the Condensation Phenomena Occuring in Wet Steam turbine, *Ph.D Thesis*, University of Birmingham, UK

Stodola, A. (1915). Undercooling of Steam Through Nozzle. *Engineering*, pp.643-646

Volmer, M. & Weber, A. (1996). Keimbildung in Ubersettingten Gebilden, *Z. Phys. Chem.*, Vol. 119, pp. 227-301

Von Helmholtz (1886). Untersuchungen über Dämpfe und Nebal, Besonders über Solche von Losungen. *Ann. D. Phys.*, Vol. 27. Pp. 508

Wegener, P.P. (1969). Gas Dynamics of Expansion Flows with Condensation and Homogeneous Nucleation of Water Vapour. *Non-Equilibrium Flows* (P.P. Wegener ed.), Marcel-Dekker, New York, Vol. 1, pp. 163-243

Wilson, W.T.R. (1897). Condensation of Water Vapour in the Presence of Dust-free Air and Other Gases. *Phil. Trans. Royal Society (London) A*, Vol. 189, pp. 265-307

White, A.J. & Young, J.B. (1993). Time Marching Method for the Prediction of Two-Dimensional Unsteady Flow of Condensing Steam, J. Propulsion and Power, Vol. 9, No. 4

Wroblewski, W.; Dykas, S. & Gepert, A. (2009). Steam Condensing Flow Modeling in Turbine Channels. *Int. J. Multiphase Flow*, Vol. 35, pp. 498-506

Yellot, J.L. (1934). Supersaturated Steam. *Transaction of ASME*, Vol. 56, pp. 411-430

Yellot, J.L & Holland, C.K. (1937). The Condensation of Flowing Steam, Part 1 - Condensation in Diverging Nozzles. *Transaction of ASME*, Vol. 59, pp. 171-183

Yusoff, M.Z.; Hussein, I.; Boosroh, M.H.; El-Awad, M.M. & Ahmad, Z. Numerical Simulation of Unsteady Flow Due to Supercritical Heat Addition during Condensation, *Advanced Technology Congress*, 2003.

Zeldovich, J. (1942). Theory of Nucleation and Condensation. Soviet *Physics - JETP (English Translation)*, Vol. 12, pp. 525

20

Experimental Study on Generation of Single Cavitation Bubble Collapse Behavior by a High Speed Camera Record

Sheng-Hsueh Yang[1], Shenq-Yuh Jaw[2] and Keh-Chia Yeh[1]
[1]*National Chiao Tung University Hsinchu, Taiwan,*
[2]*National Taiwan Ocean University Keelung, Taiwan,*
R.O.C.

1. Introduction

It has been known that the collapse of the cavitation bubbles could cause serious destruction of pressure pipes, hydraulic machineries and turbine structures. After the cavitation bubble is generated, the variation of its surrounding velocity and pressure field could result in its collapse. If the process of the collapse of a cavitation bubble appears near the solid boundary, its impact to the boundary could generate an immense water-hammer pressure effect (Plesset and Chapman, 1971). The shock wave generated in this process of bubble collapse could possibly impact or even destroy the solid boundary of structure. The bubble collapse studies include the understanding of the shock wave, the characteristics of the resultant luminescence, and the jet related fields. If the cavitation bubble is located near the solid boundary at certain suitable distance, it is more possible for the production of counter-jet in the process of bubble collapse. There has not been a firm conclusion for the exact characteristics which causes the destruction of the interface on the solid boundary.

Rayleigh (1917) studied the corrosion of high speed blade subjected to the effect of cavitation bubble. He mentioned that the bubble collapse is able to produce a high speed flow jet which damages the solid surface. Plesset (1949) further considered the influence of the physical characteristics of fluid viscosity and surface tension and derived the Rayleigh-Plesset equation. Kornfeld and Suvorov (1944) brought up the theory of bubble collapse near a solid boundary. They proposed that the bubble would be deformed to a non-spherical shape with the bubble surface tension penetrated subsequently to generate the phenomenon of flow jet. This phenomenon was proved in the experiment carried out by Naude and Ellis (1961). The numerical model in Plesset and Chapman's research (1971)also revealed this phenomenon. If the solid boundary is located on the right side of the bubble, the jet flow would be formed on the left side of the bubble and penetrates it before arriving at the right side interface of the bubble. The damage of the solid boundary might be caused by the impact of this jet flow during the bubble collapse. Benjamin and Ellis(1966)and Philipp and Lauterborn (1998) also detected the bubble collapse phenomenon and its consequent behavior of damage at the solid boundary. Recent research results reveled that the destructive power of the jet flow was not the main factor for the damage of the solid boundary. However, the jet flow influence which causes the collapse of the bubble is still an important element for the research of the hydrodynamics of the flow field.

Rayleigh (1917) first analyzed the theoretical pressure variation of the flow field of the bubble collapse. The bubble collapse results in a very high pressure, forming a shock wave which is sent towards the outside of the bubble. Harrison (1952) in his experimental results proved the existence of a noise generated by the collapse of bubble at its surrounding rigid boundary. Vogel and Lauterborn (1988) found a close relationship between the strength of the wave pulse and the distance between the position of the bubble and the rigid boundary. This wave pulse could then generate a series of shock waves. This phenomenon was studied and revealed in the experiments carried out by Tomita and Shima (1986); Ward and Emmony (1991) ; Ohl et al.; Shaw et al. (1995); Lindau and Lauterborn (2003).

Light could be emitted in the process of the bubble collapse when the volume of the bubble is compressed to its minimum radius during which the gas inside is heated in a heat-insulated process. For bubbles under low viscosity and high pressure, it is easier for the emission of light. This is because at high viscosity, the time for bubble collapse is increased and the gas inside is not heated to the sufficient temperature to emit light. Ohl et al. (1998) also found the emission of light near the solid boundary under specific conditions in the process of bubble collapse. This phenomenon is called the "Single Cavitation Bubble Luminescence (SCBL)". Buzukov and Teslenko (1971) and Akmanov et al. (1974) also had similar research reports.

Counter jet could be generated when the bubble is located near the solid boundary. The initial formation and increment of the size of the counter jet is very rapid and it could exist for a while. Experiments related to the counter jet are found in Harrison (1952) and Kling and Hammitt's (1972) researches but it is until Lauterborn (1974) who first described the counter jet phenomenon. There has not been a final conclusion for the cause of the generation of the counter jet. Counter jet did not appear in the numerical simulations carried out by Best (1993); Zhang et al. (1993); Blake et al. (1997). However, it appeared in the experiments carried out by Tomita and Shima (1986); Vogel et al. (1989); Ward and Emmony (1991); Philipp and Lauterborn (1998); Kodama and Tomita (2000). The discrepancy between the numerical simulations and the experimental results leads to the assumption that the counter jet flow field is not part of the bubble collapse process. Its formation might be generated by a complicated mechanism in the fluid during the bubble collapse. For example, if the bubble is in contact with the solid boundary, the counter jet would not be generated. The shock wave generated appears at the final stage of the process of bubble collapse. Since the counter jets also appear at the final stage of the bubble collapse, there are speculations for their possible formation due to the shock wave structure.

According to Rayleigh's equation, when the effect of the surrounding solid boundary is excluded, the relationship between the time of bubble collapse and its radius is:

$$R_{max} = 1.09\sqrt{\frac{p - p_v}{\rho}}\, t_c \tag{1}$$

where R_{max} is the maximum radius, p and ρ are the pressure of the flow field and the fluid density at ambient temperature respectively, p_v is the vapor pressure, t_c is the bubble collapse time. If the solid boundary condition is put into consideration, a greater bubble collapse time is required. Generally the size of the cavitation bubble produced in the laboratory is about 1.5 mm in radius. Under ambient temperature, the bubble collapse time ranges from $100\mu s \sim 200\mu s$. It is not easy to generate cavitation bubbles for their small volumes, short collapse time, and complicated flow fields; all of which contribute to a great difficulty of the measurement. In order to record and analyze the characteristics of the flow field of the bubble collapse, common experimental setup includes a high speed camera with framing rates ranging between several thousand to 100 million frames per second. Some researchers used the method of particle image velocimetry (PIV) to measure the velocity flow field of the process of bubble collapse (Vogel et al., 1989). However since the volume of the bubble was small and its collapse time was too short, only a rough sketch of the flow field was obtained. Lawson et al. (1999) applied the PIV method to measure the flow filed of the collapse of a 80 mm diameter rubber balloon and compared it with the numerical simulation. Although these results obtained agreement, there is great discrepancy between the flow field features of the collapse of a balloon and a bubble. Jaw et al. (2007) obtained sound experimental results using soap bubbles filled with smog particle and applied the PIV method to measure flow fields at different phases during the process of bubble collapse.

In laboratory, a single cavitation bubble could be generated in a test tube using a high energy laser beam to focus on a single point (Lauterborn, 1972). In the following years, many related studies utilized this method to generate a single cavitation bubble. Since these bubbles were generated by the high energy laser beam which causes fluid aeration, it was restricted by the strength of the energy provided by the laser. Usually the bubble created using this method has small volume with 1.5 mm in radius. In addition, the inside pressure of the bubble was not equivalent to the vapor pressure at ambient temperature. Moreover, since the bubbles were formed by fluid aeration which parted the fluid molecules, there is no re-congealable vapor inside the bubbles to repeat the experiment. Some other researchers used the method of electrolysis to generate a bubble on a platinum electrode at the bottom of a box. However, this method has a defect of disturbing the flow field during the bubble collapse. Another method for forming the bubble is through the use of a needle to inject air into the test tube before using a lithotripter to generate a shock wave up to 94 MPa to break the bubble (Philipp et al.1993). Sankin et al.(2005) also used a lithotripter to generate a 39 MPa shock wave to break the laser induced bubble in order to measure the flow field of the interaction between the bubble collapse and the shock wave.

From the paper reviews presented above, it is perceived that the cavitation bubble collapse flow is very difficult to measure due to the facts that the bubble size is small, the collapse time is very short, and the flow induced is very complicate. In addition, as mentioned before, the bubble generated by the optical breakdown is different from a true cavitation bubble. A cavitation bubble containing re-condensable vapor, when collapsed, will produce greater energy than the ones without re-condensable vapor (Zhu and Zhong, 1999). To resolve these problems, a simpler method for the generation of a true cavitation bubble is proposed in this study. By rotating a L tube filled with tap water, a single cavitation bubble is generated and stayed at the center of the rotational axis due to the effect of centrifugal force. The cinematographic analysis of bubble collapse flows induced by pressure waves of different strengths can thus be performed easily. By lowering the strength of the pressure wave, the bubble collapsed in a longer period of time, the characteristics of the true cavitation bubble collapse flow are clearly manifested. Improvement in the further used the PIV method that can be clear revealed velocity flow field feature during the bubble collapse. The present study focuses on the investigation of the formation of the liquid jet and the counter jet, at different stand-off distances to the boundary, and their consequent influences on the bubble collapse flow.

2. Experiment Instrument and cavitation bubble generator

The experimental setup for the flow field measurement of cavitation bubble collapse is shown in Figure 1. This device is consisted of an insulated optical platform, a motor, a rotatable U-shape platform, a transparent cylindrical tube, a set of light sources, a shock wave pressure generator, a high speed camera and a pressure sensor. The DC brushless motor is capable to produce a maximum controlled rotational velocity up to 2,000 RPM, to supply a maximum power up to two horsepower, and to stop the rotational motion in a relatively short period of time.

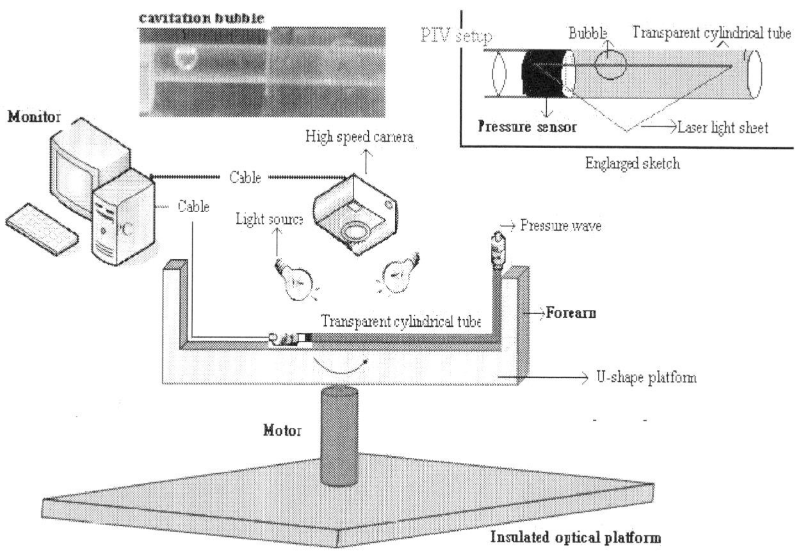

Fig. 1. Schematic diagram of the experiment setup.

The U-shape platform was made up of an acrylic platform of 20 mm in thickness. Centered at the rotational axis of the motor, its rotatable arm has a radius of 250 mm, which resulted in a total horizontal length of 500 mm. Two vertical forearms each of 150 mm in height are fixed to the edge of the platform. On the platform of the horizontal rotatable level arm sites the transparent cylindrical tube of 200 mm in length, with its internal and external diameter of 5mm and 8mm respectively. A soft PVC tube with an internal diameter of 5mm is fixed to the vertical forearm in order to conveniently exchange the experimental equipment. At one end, this tube is connected to the shock wave pressure generator with a piston while it is extended to connect the transparent cylindrical tube at the other end. At the extremity of the transparent cylindrical tube, a rigid boundary with a 1 mm drilled hole is set up to connect the highly sensitive pressure sensor that measures the shock wave pressure at different strengths during the process of the single bubble collapse (shown at the upper part of Figure 1). On the other hand, the cavitation bubble generation takes place at the site on the platform of the rotational axis where the pressure is at the lowest. Therefore, the transparent cylindrical tube must be located across the center of the rotational axis for easier cavitation bubble generation.

During the experiment of generating a single cavitation bubble, the transparent cylindrical tube on the U-shape platform is filled with tap water shown in Figure 2. The surface of the fluid at the part of the vertical forearm tube is in touch with air with one atmosphere pressure. Therefore, the center location of the L tube at initial condition has a hydrostatic pressure of p_0

$$p_0=p_{atm}+\rho g\Delta h, \quad (2)$$

where p_{atm} is the atmosphere pressure, g is the acceleration of gravity, and Δh is the water depth difference.

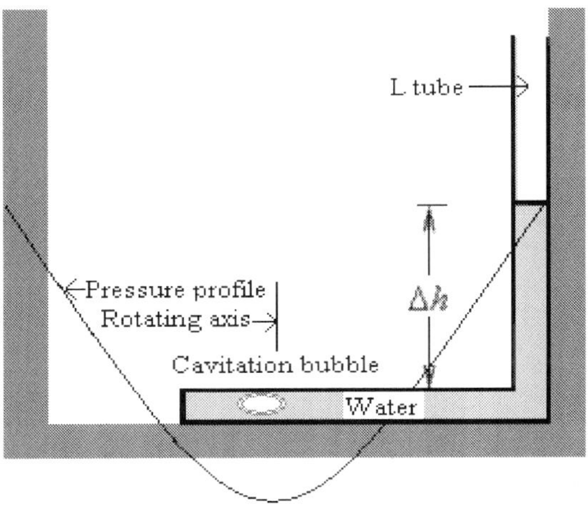

Fig. 2. The pressure distribution for a rotating U-shape platform.

When the U-shape platform is rotated by the motor, the fluid is subjected to a centrifugal force resulting in a parabolic fluid pressure distribution shown as the solid line in Figure 2 at different radius. At the vertical forearm, although the Δh is slightly increased, the hydrostatic water pressure is still kept at one atmospheric pressure because the surface interface is still in touch with the air. Therefore, the pressure difference between the free surface atmospheric pressure and the pressure at the center of rotation is p_c

$$P_c=\rho g\Delta h -1/2\rho r^2\omega^2, \quad (3)$$

where r is the rotational radius and ω is the rotational velocity. When ω is gradually increased, the pressure at the center of the rotation in the transparent cylindrical tube is gradually decreased to a saturated vapor pressure at local present water temperature. At this condition, a single cavitation bubble at the rotational center can be generated. The rotational speed needed for generating a cavitation bubble is related to the Δh . Greater Δh means a greater rotational velocity required for the production of cavitation bubble. If Δh is kept constant, an increasing rotational velocity would result in a greater size of cavitation bubble. Therefore by controlling the rotational velocity of the U-shape platform, a desirable size of a single cavitaiton bubble could be generated.

After the cavitation bubble is generated, the U-shape platform is stopped to restore the pressure back to the hydrostatic pressure instantly. This pressure difference alone is not enough to break the cavitation bubble. Therefore, in order to observe the flow field of the collapse of the cavitation bubble, this study uses a pulse setup to hit the piston of the PVC soft tube in contact with the free water surface and instantly generates a shock wave pressure sending an impact to cause the collapse of the cavitation bubble. The signal to propel the pulse setup impacting the piston device is triggered while the image data and the pressure profile are recorded and stored by the computer through the high speed camera and the pressure sensor respectively. This experimental setup allows the real-time recording of the time-series relationships between the flow field image data and the pressure change profile with their subsequent analysis.

A Fastec high speed camera is used to extract and record the experimental images. The speed of image extraction is determined by the size of the image. For example, an image extraction speed of 4,000 frame /second is used for an image size of 1280×128 pixels. A Kulite XTL-190 pressure sensor incorporating with the NI-6221 Analog I/O device are used for the measurement of the pressure profile. The NI-6221 Analog I/O device can send a 10 V signal to drive the pressure sensor and receive a 0 - 0.5 V pressure signal to record data which enables itself for the analysis of the pressure change profile in the transparent cylindrical tube.

On the other hand, PIV method is used Argun laser pass a transparent cylindrical glass to form a light sheet and in the liquid arranged TSI glass bead-hollow particle (8-12μm) to assist the camera catch the particle image during the cavitation bubble collapse process, as shown in upper right schematic diagram in Figure 1. The light sheet thickness is 1.5 mm pass the bubble location and the camera catch the bubble collapse image process then record a cinematograph file. After this file is transfer to several sequence particle image data. Using the particle images and the PIV analysis method can obtain the velocity flow field feature during the bubble collapse process. Therefore, a single cavitation bubble and the subsequent bubble collapse flows induced by pressure waves are easily generated by the experimental setup proposed in this study. Cinematographic analysis of the cavitation bubble collapse flows at different stand-off distances are performed and discussed in the following.

3. Flow field measurement of the collapse of cavitation bubbles

It is found that the presence of the solid boundary has distinct influence on the flow field of a pressurized cavitation bubble and its final collapse. A distance parameter $\gamma = d/R_{MAX}$ (where R_{MAX} is the maximum radius and d is the distance between the bubble center and the solid boundary) is assigned to represent the distance from the center of the bubble to the solid boundary. When γ is in the range of $1 < \gamma \approx 3$, counter jet could be observed. However, no counter jet is generated under the condition of $\gamma > 3$. The results are described below:

3.1 Flow field measurement of bubble collapse at $\gamma \approx 7$

Under this condition, the distance between the center of the cavitation bubble and the solid boundary is nearly seven time of its radius. The flow field of the process of cavitation bubble collapse is not affected by the solid boundary. Therefore the solid boundary is assumed to be insignificant to the process of bubble collapse. This process of the cavitation bubble being pressurized followed by its final collapse is shown in Figure 3. The pressure wave is sent from the left side of the bubble surface, impacting the bubble with peak strength up to 155kPa. The pressure wave caused a concaved deformation of the bubble shown in images from the first to the third rows of Figure 3.

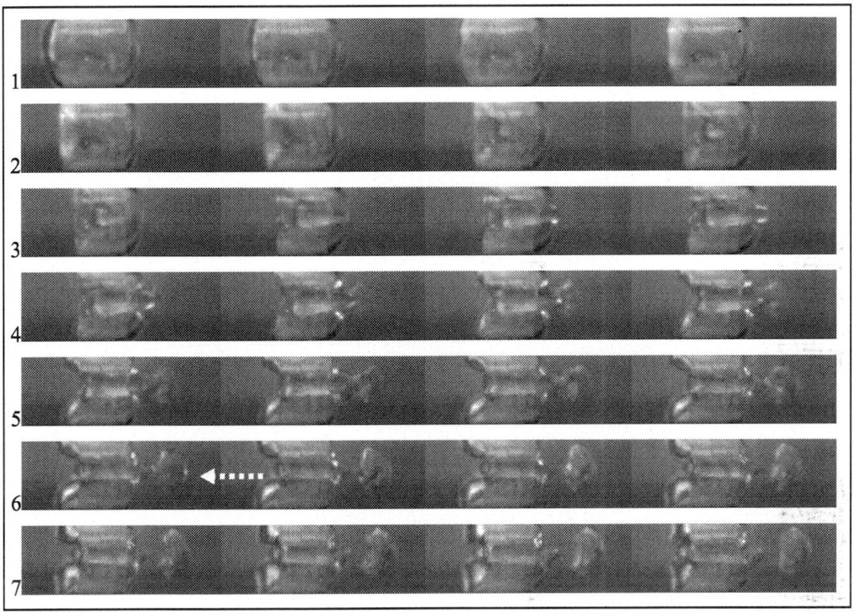

Fig. 3. Top view of images of the process of bubble collapse at $\gamma \approx 7$. From 1st row to 3rd row: image of the inward dent process; 4th and 6th rows: images of the Kelvin-Helmholtz vortex process (the Kelvin-Helmholtz vortex is indicated by a dotted line with an arrow). The peak strength of the pressure wave is 155 kPa. Image interval time is 1/4000 second. The size of each individual frame is 10.8 mm × 3.1 mm. The bubble R_{max} is 2.5 mm.

When sufficient energy is accumulated by the liquid jet during its continuous motion to the right side of the bubble, the overlaid surface is squeezed and subsequently spouted into a jet flow. When the jet flow extended to the static fluid at the right side of the bubble, rapid variation in the flow velocity is created which leaded to a Kelvin-Helmholtz vortex shown in images listed in images from the fourth to the sixth rows of Figure 3. Jaw et al. (2007) clearly described the Kelvin-Helmholtz vortex, indicating that the interaction between the pressure and the velocity variation is the main cause of this phenomenon.

The bubble collapse process is a complicated and three dimension flow structures. Using the 2D PIV analysis method was lacked a vertical direction motion measurement. In other word, during the bubble surface was pressured to touch the solid boundary, the pressure is uniformly distributed across the tube area, the bubble deformation was approximate a symmetrical development condition. Under this condition, using a high speed camera and 2D PIV method could be obtained flow field. Figure 4 show the velocity flow field of the Kelvin-Helmholtz vortex formation process that used the PIV method to obtain flow field variation during the liquid jet to form vortex formation. The jet flow instantaneously spouted into the static fluid that cause between the jet flow and static fluid shear force difference increased, then the Kelvin-Helmholtz vortex formation is generated, as shown in Figure 4. From these series of images, the features of the cavitation bubble collapse without solid boundary effect are clearly manifested.

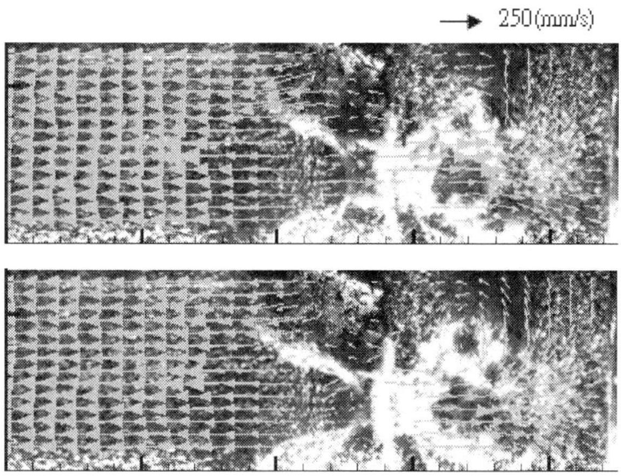

Fig. 4. Exhibit the PIV measurement results at $\gamma \approx 7$. (The velocity flow field of the Kelvin-Helmholtz vortex formation process) The peak strength of the pressure wave is 155 kPa. Image interval time is 1/4000 second. The size of each individual frame is 11.0 mm × 3.1 mm. The bubble R_{max} is 2. 3 mm.

3.2 Flow field measurement of bubble collapse at$\gamma \approx 2$

As described in the introduction, the counter jet would be generated when the distance between the center of the bubble and the solid boundary is within one to three times the bubble's radius ($1 < \gamma \approx 3$). The experiments conducted with $\gamma \approx 2$ falls within this range.

The distance from the right side of the bubble surface to the solid boundary is only one radius long. The Kelvin-Helmholtz vortex was generated after the bubble surface is broken and the jet flow is formed. This vortex would touch the solid boundary and subsequently form the stagnation ring on the solid boundary shown in the left front view diagram in Figure 5. After the stagnation ring touched the solid boundary, it was divided into two fluid flows. One of them was outside the stagnation ring splashing outwardly along the radial direction. The other fluid flow inside the stagnation ring was squeezed inwardly along the central direction to form a counter jet shown in the lower right side of the diagram in Figure 5. The preexistence of fluid between the bubble surface and the solid boundary allowed the fluid inside the stagnation ring to be squeezed towards the center

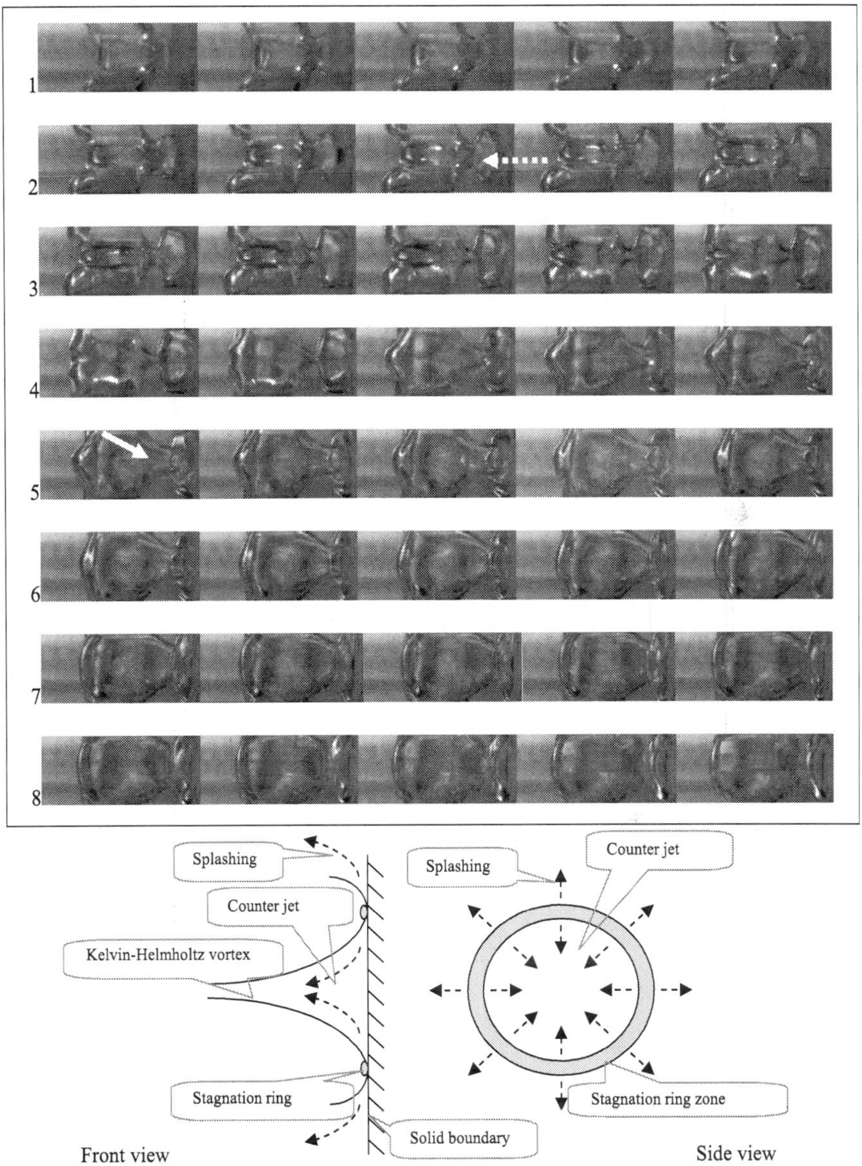

Fig. 5. Upper Part: The process of bubble collapse at $\gamma \approx 2$ (the Kelvin-Helmholtz vortex is indicated by a dotted line with an arrow, the counter jet indicated by a solid line with an arrow). The peak strength of the pressure wave is 250kPa. The image time interval is 1/4000 second. The size of each individual frame is 9.4mm × 3.1 mm. R_{max} is 2.5mm. (Lower Part: sketch of Kelvin-Helmholtz vortex forming the counter jet.

resulting in a counter jet. Therefore in order to generate a counter jet, the bubble should be located at γ> 1 so that there would be enough space between the bubble surface and the solid boundary.

On the other hand, after the bubble surface was penetrated to form the Kelvin-Helmholtz vortex, a zone with high velocity and low pressure was formed at the root of the central axis of the vortex where the bubble was stretched and deformed towards its right side shown in images from the first to the third rows of Figure 5. In the first image at the fifth row of Figure 5, a counter jet located at the central axis of the bubble could be clearly seen.

Many researchers who studied the counter jet have mentioned the existence of the stagnation ring. However, in these studies, the time for the collapse of the bubble was too short for the appearance of the Kelvin-Helmholtz vortex. The relationship between the stagnation ring and the Kelvin-Helmholtz vortex was still not clear. In this study, the process for the formation of the Kelvin-Helmholtz vortex and the counter jet was clearly revealed for a shock wave of lower pressure was utilized to impact the cavitaiton bubble. If the strength of the pressure wave is increased, the resultant counter jet could penetrate the cavitation bubble and subsequently separated the bubble into a number of small bubbles as shown in Figure 6. In the second image at the fourth row of Figure 6 a counter jet located at the central axis of the bubble could be clearly seen.

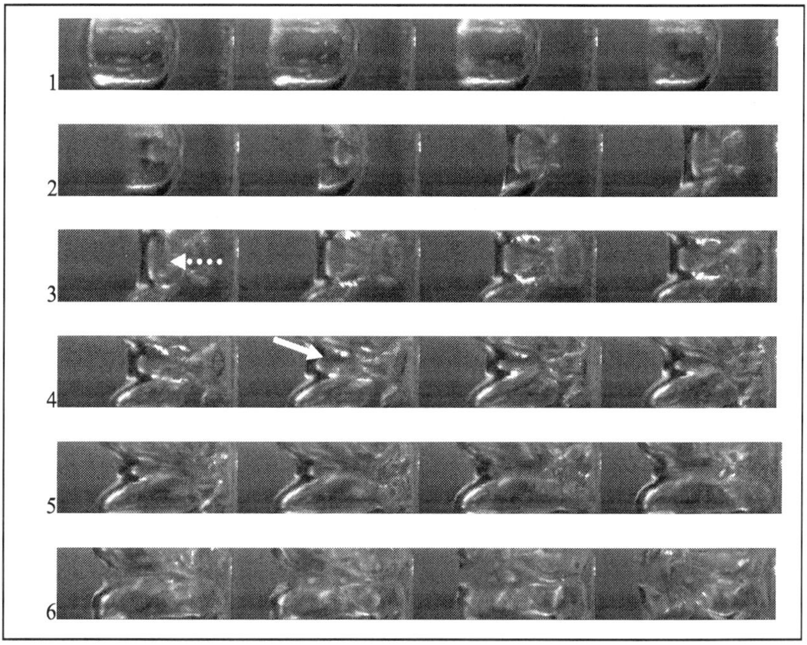

Fig. 6. Images of the process of bubble collapse at γ≈ 2 with image time interval of 1/4000 second. The peak strength of the pressure wave is 405 kPa. The size of each individual frame is 9.8 mm × 3.1 mm (the Kelvin-Helmholtz vortex is indicated by a dotted line with an arrow; the counter jet indicated by a solid line with an arrow). R_{max} is 2.5 mm.

Figure 7 show image and the velocity flow field from the Kelvin-Helmholtz vortex touch to the solid boundary and transfer to form the counter jet formation which used the PIV method to obtain velocity flow field variation. After the Kelvin-Helmholtz vortex touch solid boundary, the vortex is formed a planiform shape bubble along the solid boundary, and meantime the vortex formed a radial direction outward splashed out motion, as shown in upper left image and calculation result of Figure 7. Following by the bubble on the forward stretched effect and between the bubble and solid boundary space restriction conditions, the Kelvin-Helmholtz vortex at center part splash out are restricted and caused the velocity gradually decreased to stagnation at near the solid boundary center, as shown in upper right image and calculation result of Figure 7. Finally, the stagnation ring and counter jet are formed as shown in lower image and calculation results of Figure 7. They can be clear revealed that the stagnation ring location and counter jet motion that shown in Figure 8. The stagnation ring formation is located at a turning point of the velocity vectors

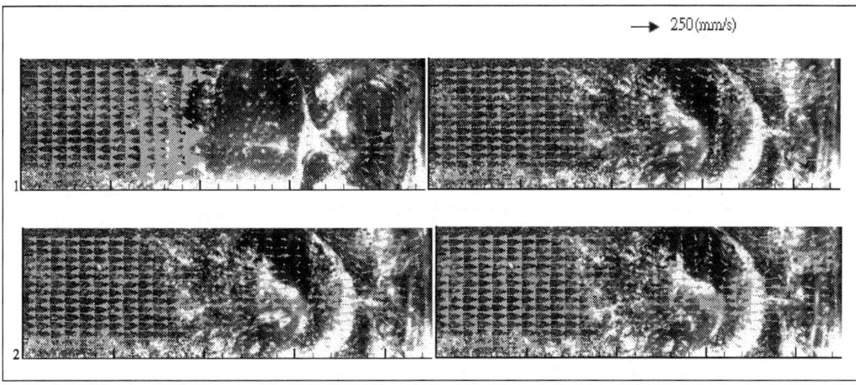

Fig. 7. Exhibit the PIV measurement results at $\gamma \approx 2$. (The velocity flow field of the Kelvin-Helmholtz vortex formation process) The peak strength of the pressure wave is 260 kPa. Image interval time is 1/4000 second. The size of each individual frame is 11.0 mm × 3.1 mm. The bubble R_{max} is 2. 3 mm.

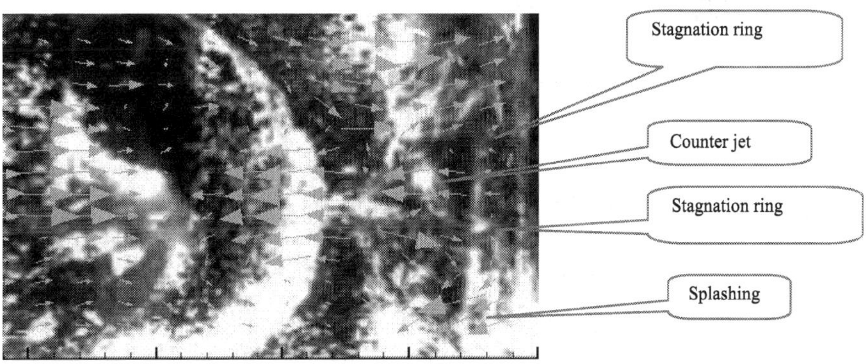

Fig. 8. The velocity flow field is the Figure 7 lower left diagram near solid boundary enlarged result that can reveal the stagnation ring location, splashing and the counter jet formation.

near the solid boundary. The counter jet formation is located at between two stagnation rings. The above-mentioned of the PIV calculation results reveal that the stagnation ring and the counter jet formation identically with Figure 5 lower part schematic diagram.

3.3 Flow field measurement of bubble collapse at $\gamma \approx 3$

The generation of the counter jet needs to satisfy the condition of $1 < \gamma \approx 3$. A critical value of $\gamma \approx 3$ is found to be a decisive value for the generation of a counter jet. In this study, three different strengths of pressure waves were used to trigger the breakdown the cavitation bubble. Flow field observation for this process of the collapse of cavitation bubble is carried out under this critical condition.

The images located at the first and second row of Figure 9 revealed the flow field of bubble collapse under a pressure wave of 200 kPa in strength. A liquid jet was formed followed by penetrating the bubble surface to produce the jet flow and the Kelvin-Helmholtz vortex. The bubble was divided into two small bubbles because the Kelvin-Helmholtz vortex did not touch the solid boundary. This process of collapse was similar to the case at $\gamma \approx 7$ where the counter jet was not generated.

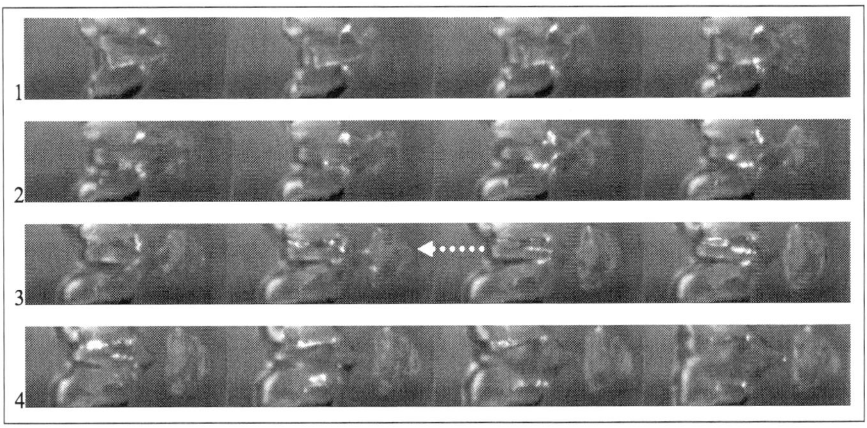

Fig. 9. Images of the process of bubble collapse at $\gamma \approx 3$. The peak strength of the pressure wave is 200 kPa. The size of each individual frame for is 10.5 mm × 3.1 mm, R_{max} : 2.55 mm. The image time interval is 1/4000 second. (The Kelvin-Helmholtz vortex is indicated by a dotted line with an arrow).

The process of the collapse of the bubble, with the strength of pressure wave increased to 300 kPa, is shown from the third to the sixth rows in Figure 10. Unlike the semi-hemispheric form of the Kelvin-Helmoholtz vortex shown in the third row of Figure 9, the vortex shown here was clearly influenced by the solid boundary when the liquid jet penetrated the bubble surface. The right side of the head of the vortex touched the solid boundary and turned into a planiform shape before splashing and spreading outwardly along its surrounding interface. On the other hand, before the head of the vortex touched the solid boundary, the outer ring of the vortex had already touched the tube wall and started spreading outwardly shown in the images at the first row in Figure 10. This spreading vortex kept

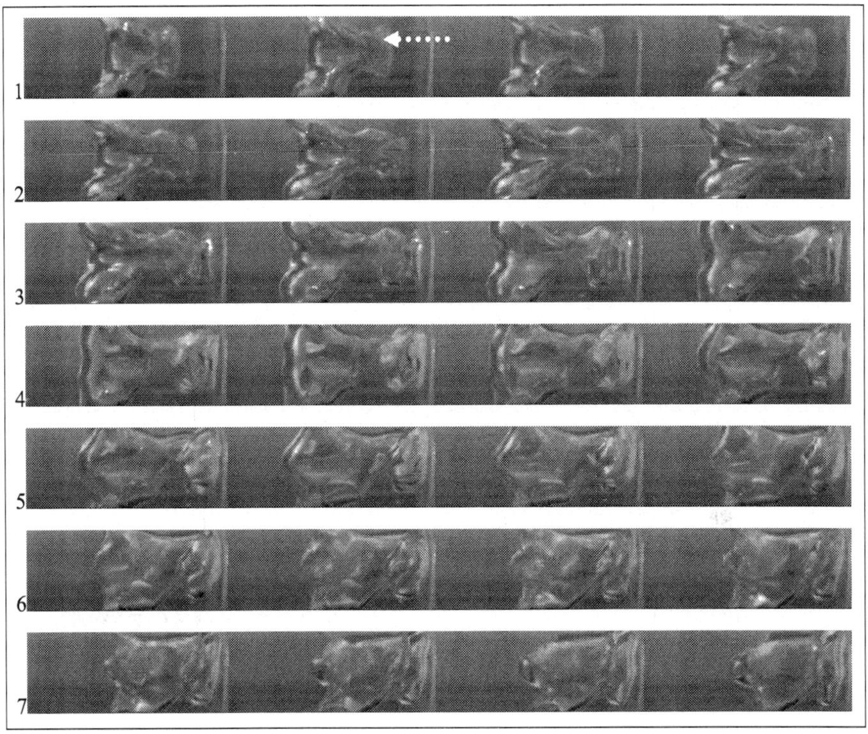

Fig. 10. Images of the process of bubble collapse at $\gamma \approx 3$. The peak strength of the pressure wave is 300kPa. The size of each individual frame is 9.6 mm × 3.1 mm. R_{max} : 2.35 mm. The image time interval is 1/4000 second. (The Kelvin -Helmholtz vortex is indicated by a dotted line with an arrow).

moving towards the right side until it touched the solid boundary and generated a subsequent shock wave which rebounded to produce the phenomenon of Richtmyer-Meshkov instability shown near the solid boundary in every image from the second to the third rows of Figure 10. Although the Kelvin-Helmholtz vortex could be generated under this strength of pressure wave, the vortex had already splashed and touched the surrounding solid boundary, disabling the vortex from forming the stagnation ring and the counter jet. At the end of this process, the bubble was divided by the liquid jet and the root of the vortex into two smaller bubbles shown in the images from the fifth to seventh rows in Figure 10.

If the strength of the pressure wave is increased to a peak value of 365 kPa, the Kelvin-Helmholtz vortex would touch the solid boundary before the formation of the stagnation ring and the counter jet. This process is shown in the image listed at the 5th and 6th rows of Figure 11.

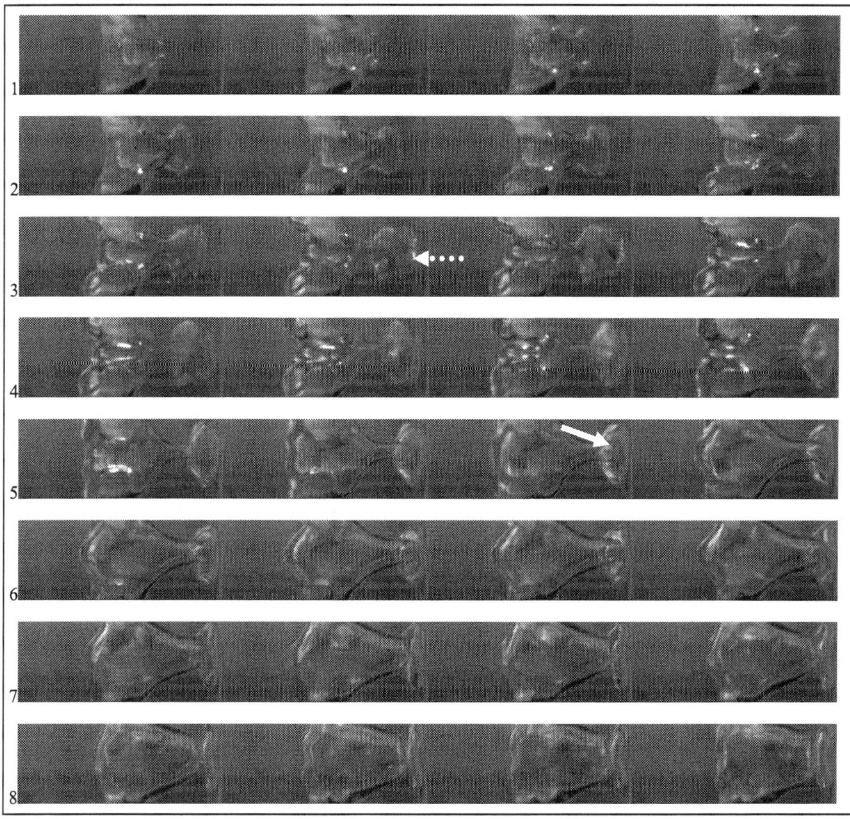

Fig. 11. Images of the process of bubble collapse at $\gamma \approx 3$. The peak strength of the pressure wave is 365 kPa. The size of each individual frame is 2.25 mm. The image time interval is 1/4000 second. (The Kelvin-Helmholtz vortex is indicated by a dotted line with an arrow; the counter jet is indicated by a solid line with an arrow).

3.4 Flow field measurement of bubble collapse at $\gamma \approx 1$ and $\gamma = 1$

The other critical value for the formation of the counter jet occurs at $\gamma \approx 1$ where the bubble is tightly close to the solid boundary. In this study, in order to understand the characteristics of the flow fields under this critical condition, measurements of flow fields at both locations where γ is slightly greater than and equal to one were carried out.

1. When the bubble is located at γ slightly greater than 1, there would be a small distance between the bubble surface and the rigid boundary. When the bubble was pressurized and concaved inward, the bubble became more planiform in shape for this deformation was caused by the solid boundary. The area of inward concaved bubble is larger than

the three cases mentioned before shown in Figure 12. After the liquid jet penetrated the bubble surface, there is not enough space to form a complete Kelvin-Helmholtz vortex. However, the space between the bubble surface and the solid boundary would still exist a gap allow the formation of stagnation ring after the liquid jet touches the solid boundary. This is followed by an outward splash along the radial direction while the inward stagnation ring was squeezed along the central direction to form the counter jet. Finally the bubble was divided into two smaller bubbles by the counter jet shown in the image of Figure 12 and diagram in Figure 13. In the further, using PIV calculation results shown in Figure 14. This result are clear revealed that the liquid jet direct touch the solid boundary and then form the stagnation ring and the counter jet formation.

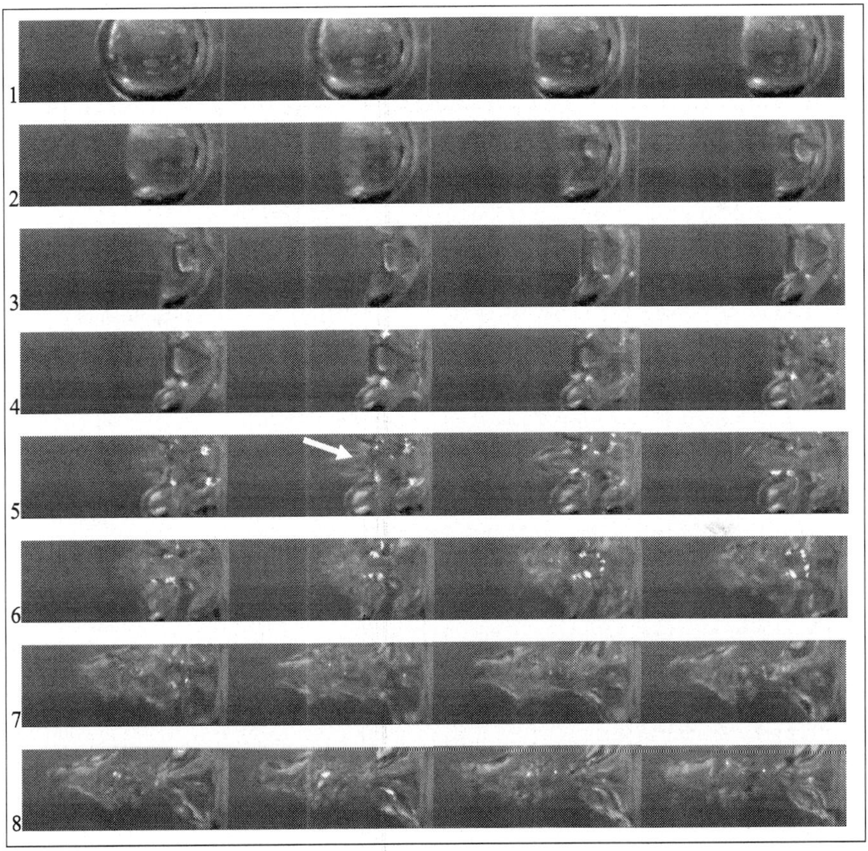

Fig. 12. Images of the process of bubble collapse at $\gamma = 1$; the peak strength of the pressure wave is 325 kPa; the image time interval is 1/4000 second. The size of each individual frame is 8.3mm × 3.1 mm. R_{max} is 2.4 mm. (The counter jet is indicated by a solid line with an arrow).

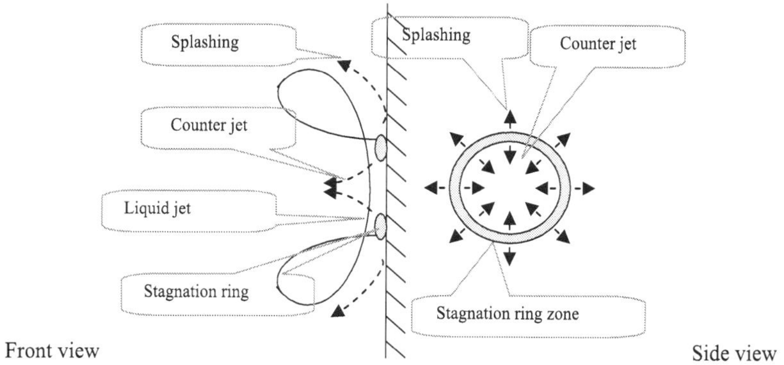

Fig. 13. sketch of the liquid jet position. (Note: Left diagram: the solid line is the bubble surface and the dotted line with an arrow is the splashing.).

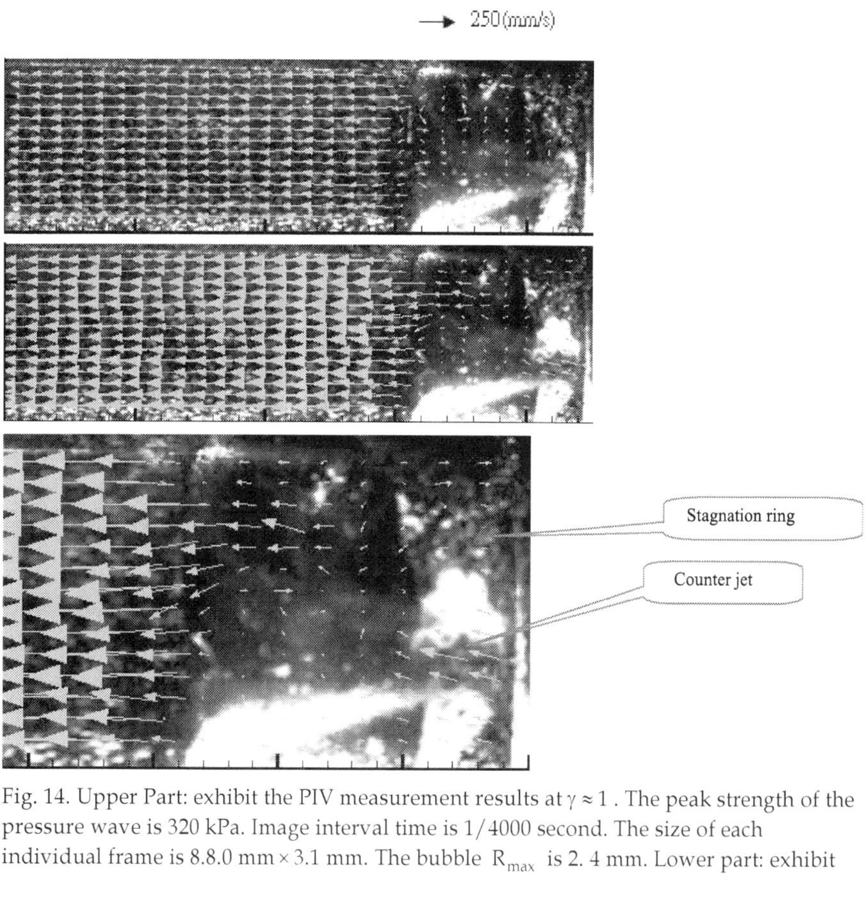

Fig. 14. Upper Part: exhibit the PIV measurement results at $\gamma \approx 1$. The peak strength of the pressure wave is 320 kPa. Image interval time is 1/4000 second. The size of each individual frame is 8.8.0 mm × 3.1 mm. The bubble R_{max} is 2. 4 mm. Lower part: exhibit

the upper right diagram near solid boundary enlarged the result that can reveal the stagnation ring location and the counter jet formation. (Note: the velocity vectors near the solid boundary are the particle motion results, not from the solid boundary extra velocity boundary condition).

2. Under the condition of $\gamma = 1$, the bubble interface was pressurized to form an inward concaved bubble. It was followed by the overlaying of the bubble interfaces on the solid boundary without any spaces left for the fluid. After the liquid jet impacted the solid boundary, it just moved outwardly as a splash along the radial direction. The bubble collapses subsequently on the radial trajectory without forming of the stagnation ring and the inwardly squeezed counter jet. This process of bubble collapse is shown in the images and diagrams in Figure 15.

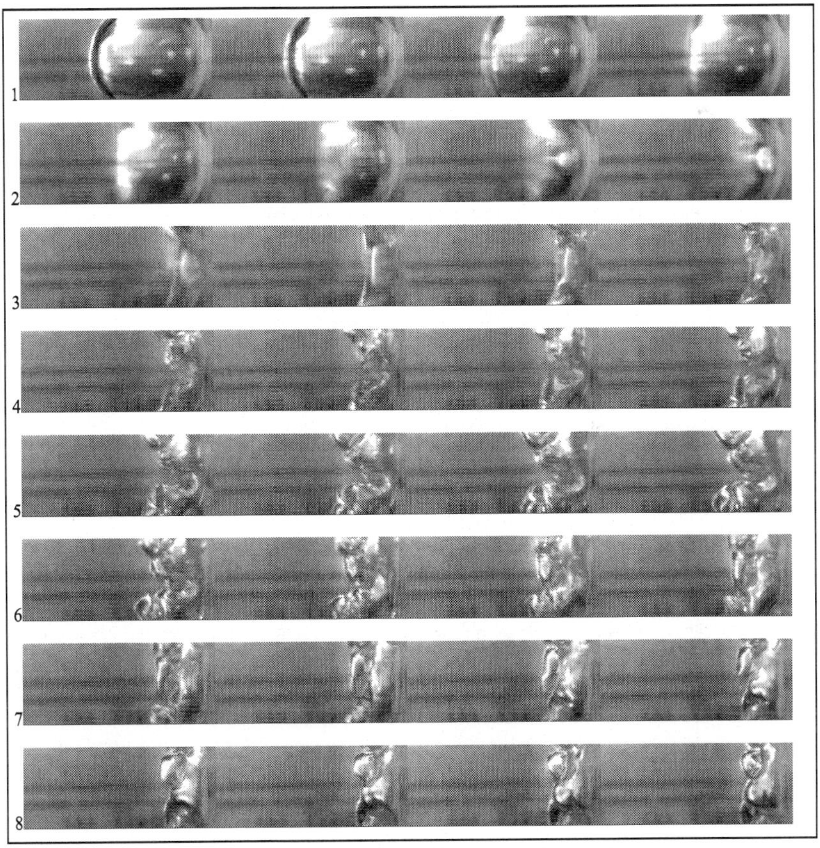

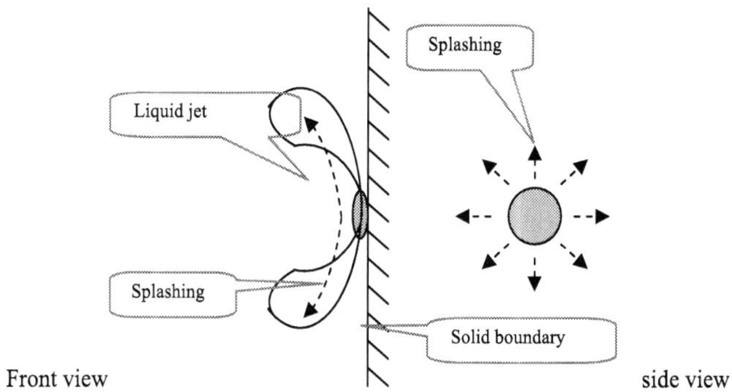

Fig. 15. Images of the process of bubble collapse at $\gamma = 1$; the peak strength of the pressure wave is 520 kPa; the image time interval is 1/4000 second. The size of each individual frame is 6.2 mm × 3.1 mm. R_{max} is 2.25 mm. Lower Part: sketch of the liquid jet position.

4. Conclusions

This study utilized a U-shape platform device to generate a single cavitation bubble and sequential images of the bubble collapse flow are recorded by a high speed camera for the detail analysis of the flow field characteristics and the cause of the counter jet during the process of bubble collapse induced by pressure wave. A series of bubble collapse flows induced by pressure waves of different strengths are investigated by positioning the cavitation bubble at different stand-off distances to the solid boundary. It is found that the Kelvin-Helmholtz vortices are formed when the liquid jet induced by the pressure wave penetrates the bubble surface. If the bubble center to the solid boundary is within one to three times the bubble's radius, a stagnation ring will form on the boundary when impacted by the penetrated jet. The liquid inside the stagnation ring is squeezed toward the center of the ring to form a counter jet after the bubble collapses. At the critical position, where the bubble center from the solid boundary is about three times the bubble's radius, the bubble collapse flows will vary. Depending on the strengths of the pressure waves applied, either just the Kelvin-Helmholtz vortices form around the penetrated jet or the penetrated jet impacts the boundary directly to generate the stagnation ring and the counter jet flow. This phenomenon used the particle image velocimetry method can be clearly revealed the flow field variation of the counter jet. If the bubble surface is in contact with the solid boundary, the liquid jet can only splash radially without producing the stagnation ring and the counter jet.

For all the experiments performed in this study, the strength of the pressure wave adopted to induce the bubble collapse flow was kept as low as possible so that the bubble collapsed in a longer period of time. The characteristics of the bubble collapse flows at different stand-off distances can thus be clearly manifested. However, different strengths of the pressure waves are needed to induce the bubble collapse flow at different γ locations. A lower strength of the pressure wave is needed for an increasing γ value and vice versa.

5. References

Akmanov, A. G., Ben'kovskii, V. G., Golubnichii, P. I., Maslennikov, S. I., and Shemanin, V. G. (1974). Laser sonoluminescence in a liquid, *Soviet Physics Acoustics*, 19, pp.417-418.

Benjamin, T. B. and Ellis, A. T. (1966). The collapse of cavitation bubbles and the pressures thereby produced against solid boundaries, *Philosophical Transactions of the Royal Society of London Series A, Mathematical and Physical Sciences*, 260, pp221-240.

Best, J. P. (1993). The formation of toroidal bubbles upon the collapse of transient cavities, *Journal of Fluid Mechanics*, 251, pp.79-107.

Blake, J. R., Hooton, M. C., Robinson, P. B. and Tong, R. P. (1997). Collapsing cavities, toroidal bubbles and jet impact, *Philosophical Transactions Mathematical, Physical and Engineering Sciences*, 355, pp.537-550.

Buzukov, A. A. and Teslenko, V. S. (1971). Sonoluminescence following focusing of laser radiation into liquid, *Journal of Experimental and Theoretical Physics Letters*, 14, pp. 189-191.

Harrison, M. (1952). An experimental study of single bubble cavitation noise, *The Journal of the Acoustical Society of America*, 24, pp.776-782.

Jaw, S. Y., Chen, C. J. and Hwang, R. R. (2007), Flow visualization of bubble collapse flow, *Journal of Visualization*, 10, pp.21-24.

Kling, C. L. and Hammitt, F. G. (1972). A photographic study of spark-induced cavitation bubble collapse, *Transactions of the ASME D: Journal of Basic Engineering*, 94, pp.825-833.

Kodama, T. and Tomita, Y., (2000). Cavitation bubble behavior and bubble-shock wave interaction near a gelatin surface as a study of in vivo bubble dynamics, *Applied Physical B*, 70, pp.139–149.

Kornfeld, M. and Suvorov, L. (1944). On the destructive action of cavitation, *Journal of Applied Physics*, 15, pp.495-506.

Lauterborn, W. (1974). Kavitation durch laserlicht, *Acustica*, 31, pp.52-78 .

Lauterborn, W. (1972). High-speed photography of laser-induced breakdown in liquids, *Applied Physical Letters*, 21, pp. 27-29.

Lawson, N. J., Rudman, A. Guerra, J. and Liow, J. L. (1999). Experimental and numerical comparisons of the break-up of a large bubble, *Experiments in fluids*, 26, pp.524-534.

Lindau, O. and Lauterborn, W. (2003). Cinematographic observation of the collapse and rebound of a laser-produced cavitation bubble near a wall, *Journal of Fluid Mechanics*, 479, pp.327–348.

Plesset, M. S. (1949). The dynamics of cavitation bubbles, *Trans. ASME: Journal of Applied Mechanics*, 16, pp.277-282.

Naude, C. F. and Elli, A. T. (1961). On the mechanism of cavitation damage by nonhemispherical cavities collapse in contact with a solid boundary, *Transactions of the ASME D: Journal of Basic Engineering*, 83, pp.648-656.

Ohl, C. D., Philipp, A., and Lauterborn, W. (1995). Cavitation bubble collapse studied at 20 million frames per second, *Annalen der Physik*, 4, pp.26-34.

Ohl, C. D., Lindau, O. and Lauterborn, W. (1998). Luminescence from spherically and aspherically collapsing laser induced bubbles, *Physical Review Letters*, 80, pp.393-397.

Plesset, M. S. and Chapman, R. B. (1971). Collapse of an initially spherical vapour cavity in the neighbourhood of a solid boundary, *Journal of Fluid Mechanics*, 47, pp.283-290.

Philipp, A., Delius, M., Scheffczyk, C., Vogel, A. and Lauterborn, W. (1993). Interaction of lithotripter- generated shock waves with air bubbles, *The Journal of the Acoustical Society of America*, 93, pp.2496-2509.

Philipp, A. and Lauterborn, W. (1998). Cavitation erosion by single laser-produced bubbles, *Journal of Fluid Mechanics*, 361, pp.75-116.

Raleigh, L., (1917). On the pressure developed in a liquid during the collapse of a spherical cavity, *Philosophical Magazine*, 34, pp.94-98.

Sankin, G. N., Simmons, W. N., Zhu, S. L. and Zhong, P. (2005). Shock wave interaction with laser- generated single bubble, *Physical Review Latter*, 034501, pp. 1-4.

Shaw, S. J., Jin, Y. H., Schiffers, W. P. and Emmony, D. C. (1996). The interaction of a single laser-generated cavity in water with a solid surface, *The Journal of the Acoustical Society of America*, 99, pp.2811-2824.

Tomita, Y. and Shima, A. (1986). Mechanisms of impulsive pressure generation and damage pit formation by bubble collapse, *Journal of Fluid Mechanics*, 169, pp.535-564.

Vogel, A. and Lauterborn, W. (1988). Time-resolved particle image velocimetry used in the investigation of cavitation bubble dynamics, *Applied Optics*, 29, pp.1869-1876.

Vogel, A., Lauterborn, W. and Timm, R. (1989). Optical and acoustic investigations of the dynamics of laser-produced cavitation bubbles near a solid boundary, *Journal of Fluid Mechanics*, 206, pp.299-338.

Ward, B. and Emmony, D. C. (1991). Direct observation of the pressure developed in a liquid during cavitation-bubble collapse, *Applied Physics Letters*, 59, pp.2228-2231.

Zhang, S. Duncan, J. H. and Chahine, G.L. (1993). The final stage of the collapse of a cavitation bubble near a rigid wall, *Journal of Fluid Mechanics*, 257, pp.147-181.

Zhu, S. and Zhong, P. (1999). Shock wave-inertial microbubble interaction: A theoretical study based on the Gilmore formulation for bubble dynamics, *The Journal of the Acoustical Society of America*, 106, pp.3024-2033.

21

Noise Reduction in Butterfly Valve Cavitation by Semicircular Fins and Visualization of Cavitation Flow

Kazuhiko Ogawa
Osaka Sangyo University,
Japan

1. Introduction

Butterfly valves have the advantage of being very compact and simple to install compared with other types of valves, and so they are widely used in industry. However, depending on the conditions, cavitation may occur around a butterfly valve. When severe noise and vibration occur because of cavitation around a butterfly valve, the valve body and pipe wall are subjected to erosion.

Butterfly valves are sometimes used inside the piping of air-conditioning facilities and the noise and vibration caused by cavitation can, in addition to making users uncomfortable, be mistaken for mechanical trouble. The need to prevent such noise and vibration is increasing from an environmental standpoint, and the prevention or suppression of cavitation itself is very important. Accordingly, many products have been proposed to prevent or control cavitation around many types of valves (Baumann,1991; Tullis,1989). As for research on the prevention of cavitation, the characteristics of a control valve with tortuous paths and an orifice with a multi-perforated cone to prevent cavitation from occurring around the orifice were reported.(Rahmaeyer et al.,1995; Kugou,1996)

These methods have already been applied to actual products, and those products have proved very successful in reducing noise. However, tortuous path valves, for example, are applicable only to cases wherein the fluids are clean and the shapes of the piping arrangements around the valves are complicated. Moreover, the air injection method that is very effective in reducing cavitation is limited to cases wherein the effect of air can be ignored. Hence, the authors proposed the sudden enlargement of a pipe downstream of a butterfly valve (Ogawa & Uchida,2005). This method was much simpler than the conventional methods. However, the sudden enlargement of the pipe is not adequate for flows containing particles because the particles accumulate in the enlarged section of the pipe.

The author has already proposed the attachment of fins to the valve body in order to further reduce cavitation noise around the butterfly valve (Ogawa & Uchida,2005, Ogawa,2008). This method can be used for flows containing particles because of the simple shape of the valve body. Cavitation occurs intensely around the butterfly valve because of the interference of the flow from the nozzle side with the flow from the orifice side (Itoh et al.,1988). To avoid this interference, semicircular fins were attached to the valve body. In this paper, it was confirmed based on the experimental results that the attachment of the fins was very effective in reducing cavitation noise.

In this study, cavitation bubbles were photographed by Hi-speed camera and the size and number were measured from those photographs and the effect of the fins and the upstream velocity distribution were investigated. In past studies, photographs focusing on the aspect of the butterfly valve cavitation have been reported in great numbers. For example, it was pointed out that the most erosive cavitation around a butterfly valve is the vortex cavitation on the orifice side by means of the pressure-sensitive films and high speed photography (Tani et al,1991). An observation of butterfly valve cavitation and the measurement of cavitation noise were performed to diagnose the cavitation condition (Kimura and Ogawa, 1986). However, since the measurements of number and size of the cavitation bubbles were not carried out, the details of the cavitation growth were not clear. In this study, close-up photographs of cavitation bubbles were taken and their number and size were analyzed. The difference between the normal valve and the valve with the fins is reported in the following.

In actual piping arrangements, straight lengths in front of butterfly valves are not sufficient to obtain normal velocity distribution in many cases. In extreme cases, more than two valves are installed in series or bends are installed just ahead of or behind those valves. Under such conditions, the upstream velocity distribution is different from the usual turbulent velocity distribution. Therefore, under the condition that the velocity distribution is biased, confirmation of whether or not the attachment of fins is useful for the noise reduction is necessary.

2. Reduction of cavitation noise by using fins

Figure 1 shows a typical case of intense cavitation. A butterfly valve is not usually used under such conditions, but this cavitation appears when a valve repeatedly opens and closes. The separation of the flow and the vortex region around the butterfly valve during light or moderate cavitation is easily presumed in this case.

In Fig. 1, the upstream side appears on the left. The leading edge of the butterfly valve is on the upper left and the trailing edge (nozzle side) is on the lower right. The flow from the lower left region (nozzle side) of this photograph passes along the wall surface and interferes intensely with the flow from the upper left (orifice side) at the top of the pipe. It has been pointed out (Itoh et al.,1988) that this interference from both flows causes intense cavitation and brings about the erosion of the wall surface. Therefore, the fins were attached to the valve body in this study to avoid the interference caused by these flows. The interference of the flows and the separation of the flow behind the valve were expected to be reduced when the fins were attached to the downstream surface of the valve.

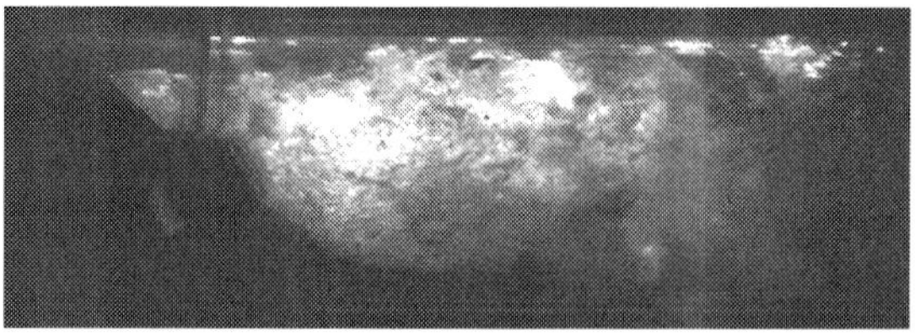

Fig. 1. Cavitation around a butterfly valve (Valve opening:45deg, Flashing Condition).

Figure 2 shows the test valves. The valve shown in Fig.2(a) is a normal valve without a fin

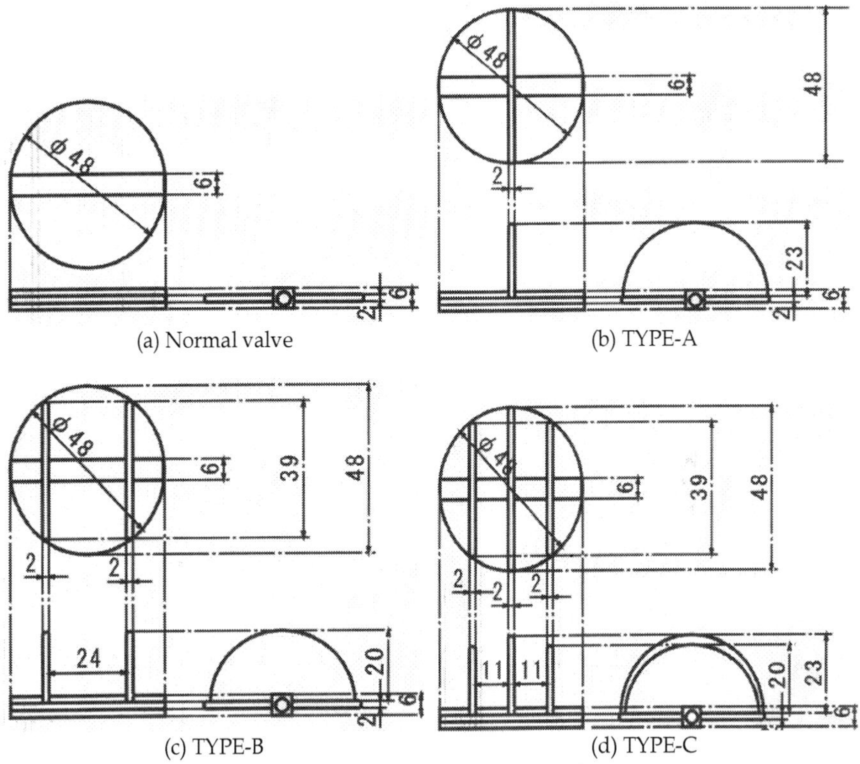

Fig. 2. The test valves with fins.

The author clarified that the fin must be installed in the downstream of the valve in order to reduce cavitation (Ogawa & Uchida,1995). In this study, a test valve is called "TYPE-A" when a semicircular fin is attached to the downstream surface of the valve. When two semicircular fins are attached to the downstream surface of a valve, the test valve is called "TYPE-B". On the "TYPE-C" test valve, three semicircular fins are attached to the downstream surfaces of the valve. For each valve, each fin was fixed perpendicular to the valve stem.

Cavitation noise measurements were performed in a closed-type cavitation tunnel, using water as the fluid. There was a pump on the downstream side of the test section. The upstream pressure was kept at atmospheric pressure and the valve opening was fixed during the experiment. The flow velocity was increased by controlling the frequency of the pump using an inverter. Cavitation noise was measured at each flow velocity and a visualization was created by a high-speed camera. Cavitation noise was measured using a noise meter placed close to the outside surface of the test section duct. The frequency range of this noise meter was 20-8000Hz.

3. Cavitation number and pressure loss coefficient

The cavitation number σ in this study is defined as follows:

$$\sigma = \frac{p - p_v}{\rho U^2 / 2} \tag{1}$$

where is upstream pressure, is saturated vapor pressure, is density of water, and U is average upstream flow velocity. The pressure loss coefficient ζ is defined by the following formula:

$$\zeta = \frac{\Delta P}{\frac{1}{2}\rho U^2} \tag{2}$$

where ΔP is the differential pressure across the test valve.

4. Effect of the fins on cavitation noise reduction (Comparison with constant valve opening)

Kimura and Ogawa(1986) clarified using measurements of cavitation erosion and noise that the cavitation around the butterfly valve occurs most intensely when the valve is halfway open (40°-45°). Therefore, noise was measured around halfway-open valves in this study, and the degree of the cavitation reduction was investigated based on these measurement results. In the experiments, cavitation noise was measured at each flow velocity, while flow velocity was increased gradually by controlling the frequency of the pump.

Figure 3 shows the results of noise measurement when the valve was kept open constantly. The effect of the fins on noise was clear in the case of TYPE-B. In the case of the normal valve without a fin, cavitation occurred at, but in the case of TYPE-B, cavitation occurred at. These results proved that cavitation occurrence was suppressed by the fins. Moreover, the maximum noise just before flashing was 5dB lower in the case of TYPE-B than in the case of the normal valve.

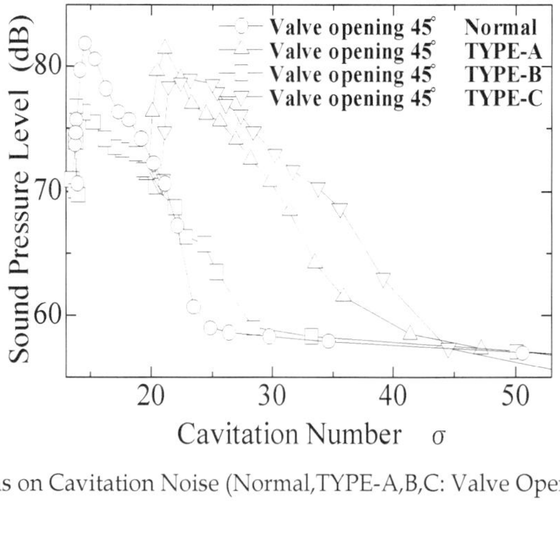

Fig. 3. Effect of Fins on Cavitation Noise (Normal,TYPE-A,B,C: Valve Opening=45deg).

In the cases of TYPE-A and TYPE-C, the cavitation number at cavitation inception was larger than that in the case of the normal valve. The fin promoted the occurrence of cavitation because the middle fin was fixed on the orifice side, where cavitation occurrence was intense.

5. Effect of the fins on cavitation noise reduction (Comparison with constant pressure loss coefficient)

The results shown in Fig. 3 were obtained by measuring cavitation noise at each flow velocity, while flow velocity was increased using a pump controlled by an inverter. The upstream pressure was kept at atmospheric pressure in each experiment, and the effect of the saturated steam pressure was minimal. Therefore, the flow velocity was also almost the same when the cavitation number was same. However, in many cases, the flow rate was not controlled by changes in the frequency of the pump, and the flow rate was controlled by adjusting the opening of the valve. Additionally, the flow rate in the actual plant was determined by the head curve of the pump and the pressure loss of the piping system as s whole. Accordingly, the effect of cavitation control should also be investigated under the constant pressure loss coefficient condition.

When a fin is attached to a valve body, the flow rate may be changed by the variation in the pressure loss of the valve. Figure 4 shows the pressure loss coefficients for each type of valve. For an example, the pressure loss of TYPE-C was larger than that of the normal valve at the same Reynolds number and at the same valve opening. When the butterfly valve is used to control the flow rate, the valve opening of TYPE-C must be larger than that of the normal valve in order to obtain the same flow rate. Therefore, the experimental results must be compared with a constant pressure loss coefficient. The following discussion is based on a comparison of cavitation noise under the condition that the pressure loss coefficients of the test valves are nearly equal.

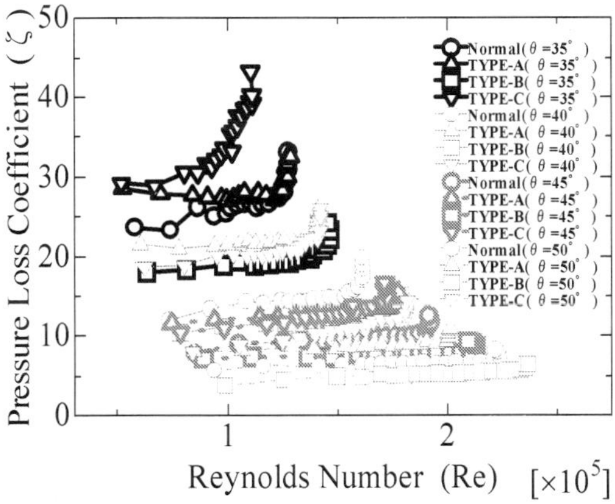

Fig. 4. Pressure Loss Coefficients for each Valve Type.

Figure 5 shows the pressure loss coefficients for each valve when σ was about 7. The pressure loss coefficient was almost constant for each valve within the range of the Reynolds number of the experiment. However, when the Reynolds number was $Re = 2.25 \times 10^5$ in the case of TYPE-C, the pressure loss coefficient began to increase and the noise began to decrease corresponding to flashing. According to the results shown in Fig.5, the noise levels around each valve were compared with a pressure loss coefficient of about 7.

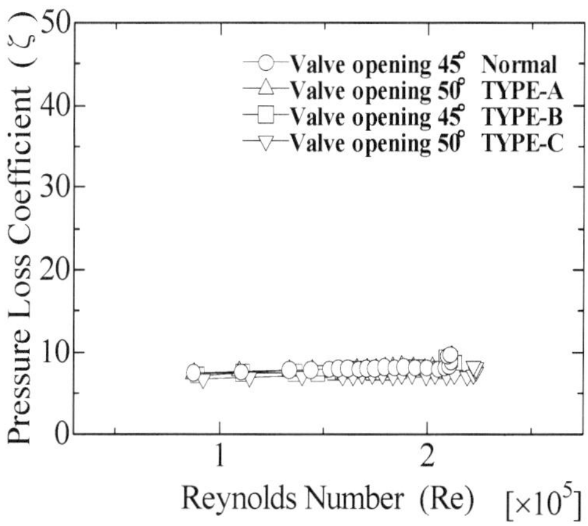

Fig. 5. Pressure Loss Coefficients ($\zeta \fallingdotseq 7$).

Figure 6 shows the effect of the fins on cavitation noise. The pressure loss coefficient for Type-B was about 7 at the valve opening of 45° and almost the same as that for the normal valve at the valve opening of 45°. Cavitation began to occur at about $\sigma = 30$ in the case of TYPE-B. On the contrary, in the case of the normal valve, light cavitation noise occurred

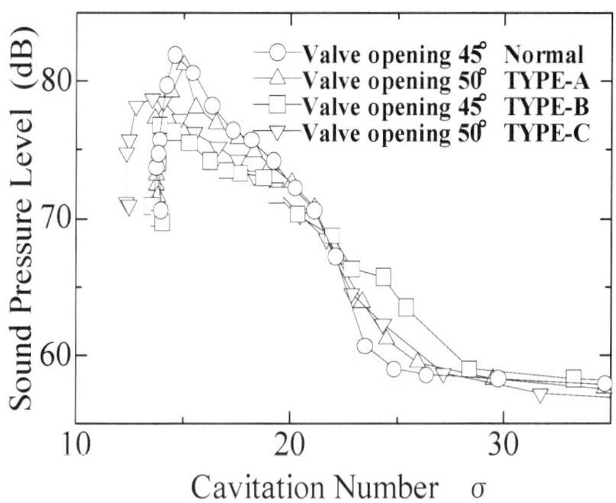

Fig. 6. Effect of Fins on Cavitation Noise under ($\zeta \fallingdotseq 7$).

around almost all the edges of the valve body at about $\sigma = 25$, where the sound pressure level of the cavitation began to rise abruptly. Accordingly, the inception of cavitation for TYPE-B was earlier than that for the normal valve. However, the results shown in Fig.6 indicate that the maximum cavitation noise value was suppressed in TYPE-B by the fins. Cavitation noise was reduced by about 5 dB when . The noise reduction effect in TYPE-B was remarkable just before flashing condition. Though cavitation occurred a little earlier than with the normal valve, the increase in the noise was milder in TYPE-B and the maximum sound pressure was suppressed.

The accuracy of the data mentioned above was ±1-1.5 dB near the inception of cavitation and ±0.5dB during cavitation growth and flashing. Though the differences among the valves were slight as shown in Fig.6, the noise reduction effect was clear for TYPE-B in flashing even taking into account the accuracy of the data.

6. The visualization of the effect of the fins

6.1 Visualization experiments

Figs. 7 and 8 show the photographs of the cavitation conditions of each valve taken with a high-speed camera. Figure 7 was taken through the upper surface of the transparent pipe (top view). Figure 8 was taken from the side surface (side view). These photographs were not taken simultaneously because only one camera was available to be used. As shown in Fig. 7(a), vortex cavitation clouds were clearly visible on the orifice side of the normal valve and 1 Dia and further downstream from the stem axis. These cavitation clouds were identified as vortices of an intensity which brings about the cavitation damage (Tani et al.,1991). Similar vortex cavitation also occurred in the TYPE-A valve with one fin. It was impossible to suppress flow interference using only a single fin. Moreover, it is obvious from Fig.7 (b) that the cavitation was further intensified since the fin existed in the part of the orifice side where the contraction flow was severe. However, such a vortex cavitation cloud is not clear around the TYPE-B valve as shown in Fig. 7(c). The interference of the flow from the orifice side with the flow from the nozzle side seemed to be suppressed by the fins. Accordingly, it can be said that the cavitation around the valve body in TYPE-B was

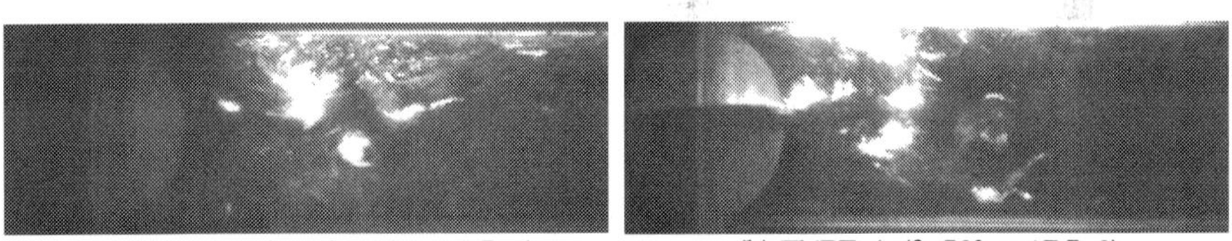

(a) Normal Valve (θ=45°,σ=17,ζ=8) (b) TYPE-A (θ=50°,σ=17,ζ=8)

(a) Normal Valve (θ=45°,σ=17,ζ=8) (b) TYPE-A (θ=50°,σ=17,ζ=8)

Fig. 7. Top Views of Cavitation Conditions.

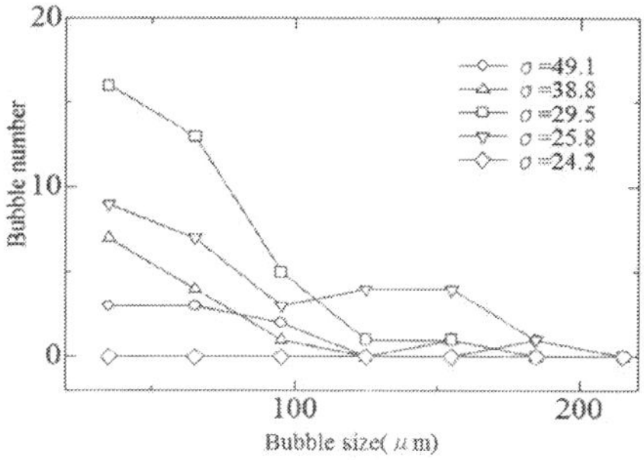

(a) Normal velocity distribution

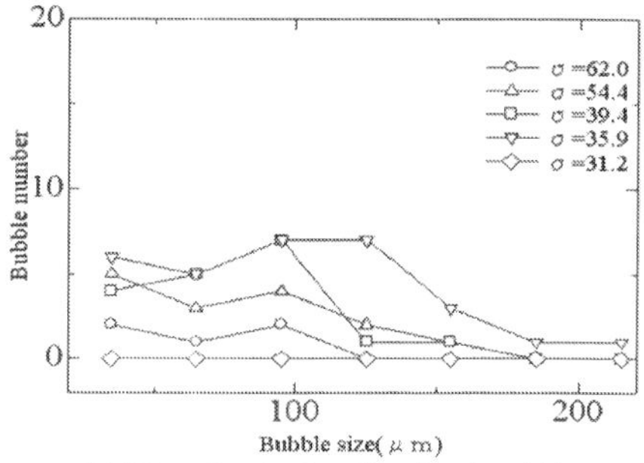

(b) Biased velocity distribution (A)

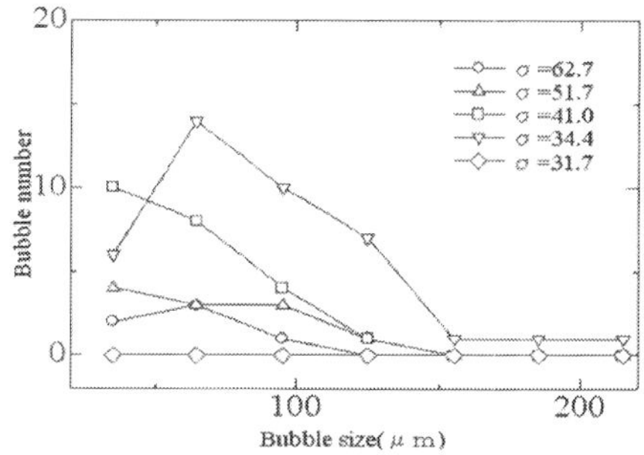

(c) Biased velocity distribution (B)

Fig. 15. Diameters and numbers of cavitation bubbles of the normal valve ($\theta = 45°$).

Fig.15(c) shows the diameters and numbers of cavitation bubbles under the biased velocity distribution (B). At $\sigma = 34.4$, the numbers of the cavitation bubbles with diameters between 65 μm and 125 μm increased remarkably comparing with the results of Fig.15(a). From the results of Fig.15(a),(b) and (c), when the bubbles with diameters from 20m to 30m increased, the cavitation noise tends to increase.

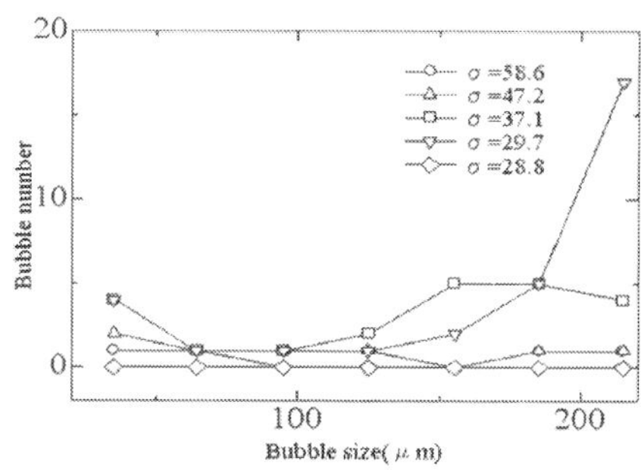

Fig. 16. Diameters and numbers of cavitation bubbles of the Type-B valve under the normal velocity distribution ($\theta = 45°$).

In the case of Type-B as shown in Fig.16, the cavitation numbers were less than the case of the normal valve. Though the cavitation bubbles with diameters larger than 200 μm increased greatly at $\sigma = 29.7$, this condition was just before the flashing and the number of bubbles was suppressed compared with the case of the normal valve in the range of diameters from 35μm to 125μm.

7.3 The aspect of cavitation bubbles

Figure 17 and Table 1 shows the positions of photographing. Position A was very near the leading edge of the valve body. Position B was in the wake region of the valve. Position C was on the centerline of the pipe and near the end of the cavitation clouds.

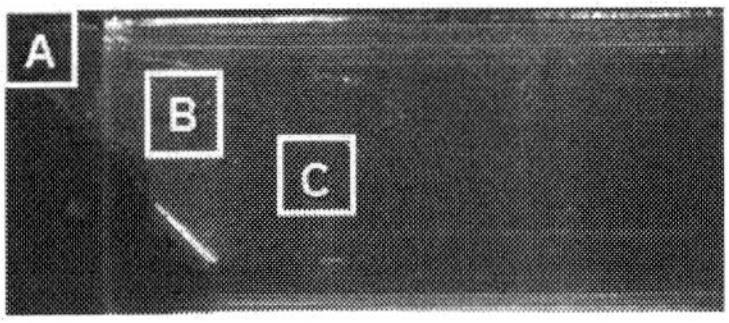

Fig. 17. The details of Photographing positions.

Position	x	y	Photographed Area a×b(mm)	Frame (per second)
A	-0.2 dia.	0.36 dia.	0.6×5.0	20000
B	0.3 dia.	0.3 dia.	1.7×6.8	10000
C	1.3 dia.	0 dia.	1.7×6.8	10000

Table 1. Photographing Positions.

Fig.18 shows the occurrence of the cavitation bubbles at position A. The five photographs in the image were of a series and photographed at 20000 frames/second. Position A was near the leading edge and the crescent in the central lower of the photograph is the one part of the leading edge of the valve body. It is very clear that the cavitation bubble occurred at the position on the right in the leading edge when photograph (c) is compared with photograph (d). The bubble seems to be oval and the size is about 200 μm . The cavitation bubble grows larger in the left side of the photograph.

Fig.19 shows the cavitation bubbles at position B. Oosition B was 0.3dia. downstream of the valve. This position is in the large separation region behind the valve. Though the mainstream is the right direction from the left, the flow circulates in the separation region, and the bubbles of the photograph move from the right to the left. The diameters of the bubbles range from 20μm to 200μm .

Fig.20 shows the cavitation bubbles at position C. Position C was 1.3dia. downstream of the valve. In this figure, the bubbles move from left to right and the number of the bubbles were relatively larger than that at position B. This vicinity is the position where the cavitation cloud ends, and the number of bubbles increases because the bubbles which have been flowing from the orifice and nozzle side flow together.

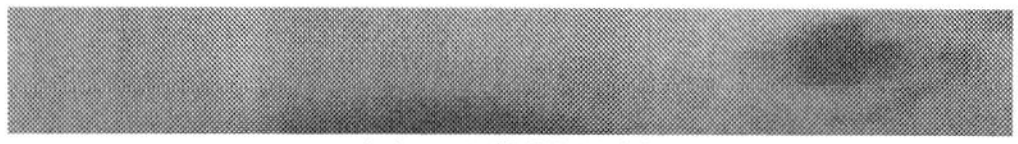

(a) t=1/20000(s)

(b) t=2/20000(s)

(c) t=3/20000(s)

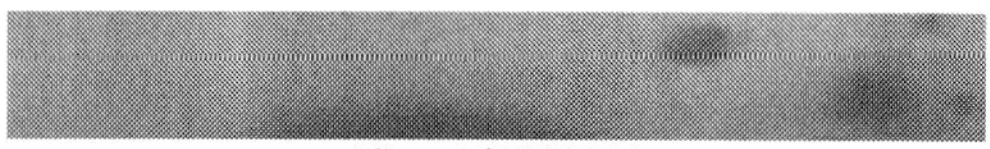

(d) t=4/20000(s)

(e) t=5/20000(s)

Fig. 18. Cavitation Bubbles at the Position A (valve opening= 45° , $\sigma = 48.6$).

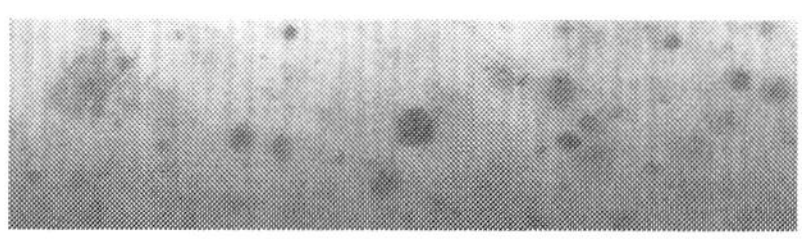

(a) t=1/10000(s)

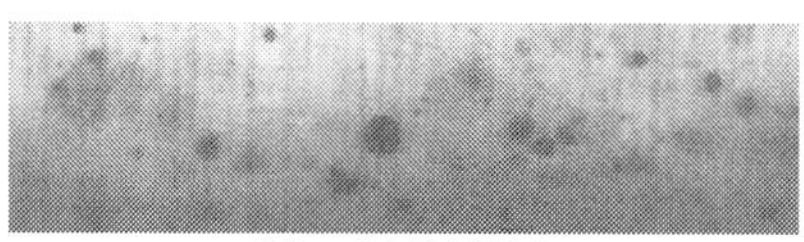

(b) t=2/10000(s)

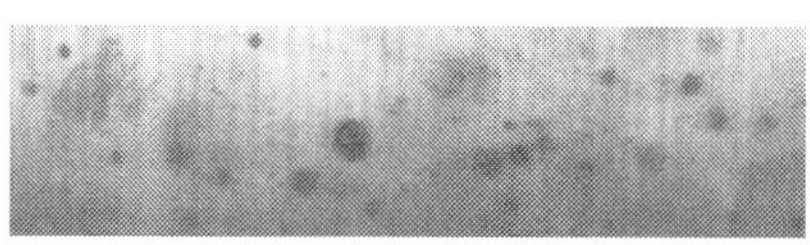

(c) t=3/10000(s)

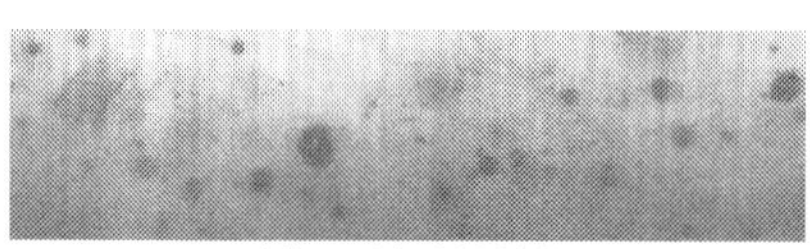

(d) t=4/10000(s)

Fig. 19. Cavitation Bubbles at the Position B (valve opening= 45° , $\sigma = 51.7$).

(a) t=1/10000(s)

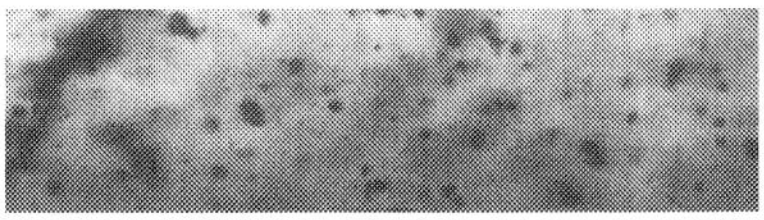

(b) t=2/10000(s)

(c) t=3/10000(s)

(d) t=4/10000(s)

Fig. 20. Cavitation Bubbles at the Position C (valve opening= $45°$, $\sigma = 44.6$).

8. Conclusions

The ability of fins to reduce the cavitation noise around a butterfly valve was investigated in this study. When two semicircular fins were attached to the downstream side of a valve body (TYPE-B), the experimental results indicated that the inception of cavitation was earlier than that found with a normal valve. This was discovered via comparison of noise given a constant pressure loss coefficient. However, cavitation noise increased gradually after inception and the maximum value of cavitation noise just before flashing was shown to be suppressed for the TYPE-B valve by the fins. Cavitation noise was reduced by about 5 dB when a valve opening of 45degrees was used. The effect of noise reduction around the TYPE-B valve was most remarkable just before flashing. In this study, it was found that fins are an adequate method for the reduction of cavitation noise. The optimum size and position of the fins should be investigated in future studies.

Visualizations created by a high-speed camera showed that intense vortex cavitation clouds were not clear in the TYPE-B valve with two fins. The interference of the flow from the orifice side with the flow from the nozzle side seemed to be suppressed by the fins. This was also confirmed by numerical analysis.

As for the visualization of cavaitation bubbles by using high-speed camera, it was found in this study that cavitation bubbles occur at the position before the leading edge and that the cavitation bubbles grow larger just behind the leading edge. It was also found that the maximum diameter occurs near the leading edge and that the bubbles become smaller through the pressure recovery of the flow. In this observation, the maximum diameter of the bubbles was $500\mu m$ near the leading edge. The diameter of the bubbles ranged from $20\mu m$

to 200μm at position B and position C. It is considered that the cavitation occurrence is dominant at the leading edge by the contraction flow and that the bubbles which occur at the leading edge become smaller in size through pressure recovery.

As for the effect of velocity distribution, it was clarified in this study that the cavitation noise around a butterfly valve becomes larger when the upstream velocity distribution was different from the normal velocity distribution. It was clear that the interference of the flow from the orifice side and the flow from the nozzle side was suppressed by the fins under not only the normal velocity distribution but also under the biased velocity distribution. The cavitation noise of Type-B was smaller than that of the normal valve in each upstream velocity distribution. Moreover, from the visualization results, it was found that the cavitation bubble diameters ranged from about 20 μm to about 200 μm and that the numbers of cavitation bubbles in Type-B was less than that of the normal valve.

9. References

Baumann, H.D.,(1991), *Control Valve Primer*, Instrument Society of America, pp.100-107.

Itoh, Y., Yamada,M., Oba, R. et al., (1988). *A peculiar Behavior of Cavitating Flow around a Butterfly Valve*, Transactions of the Japan Society of Mechanical Engineers, Series B, Vol.54, No.508, pp.3317-3224.

Kimura,T. and Ogawa,K., (1986), *Cavitation Vibration and Noise around a Butterfly Valve*, Vol.25, No.1, ISA Transactions, p.53-61.

Kimura,T. and Ogawa,K.,(1997), *Measurement of Cvaitation Erosion Using Fragile Material*, Transactions of the Japan Society of Mechanical Engineers, Series B, Vol.63, No.606, pp.360-365.

Kugou, N., Matsuda, H. et al.,(1996), *Cavitation Characteristics of Restriction Orifices (Experiment on Characteristics of Piping Vibration and Noise*, Fluids Engineering Conference, Volume 1, ASME FED-Vol.236, pp.457-462.

Ogawa,K. and Uchida,T., (2005), *Noise reduction of a butterfly valve cavitation by fins and improvement of shape of a piping arrangement*, ISA 2005 Technical Conference, (2005), ISA05-P179.

Ogawa,K., (2008), *Noise Reduction in Butterfly Valve Cavitation by Semicircular Fins and Visualization of Cavitation Flow*, 54th International Instrumentation Symposium,IIS-P047.

Rahmeyer,W.J.,Miller, H.L. and Sherikar, S.V.,(1995), *Cavitation Testing Results for a Tortuous Path Control Valve*, Cavitation and Multiphase Flow, ASME, FED-Vol.210, (1995), pp.63-67.

Tani, K., Ito,.Y, and Oba, R., (1991), *Spatial Distributions of Cavitaion Induced Pressure-Pulses around a Butterfly Valve*, Cavitation and Multiphase Flow Forum, ASME, FED-Vol.109,pp.143-147.

Tullis, J.P., (1989). *Hydraulics of Pipelines -Pumps, Valves, Cavitation, Transients-*, John Willey & Sons, pp.125-126.

Part 4

Simulations and Computer Applications

22

Computer Simulation of Involute Tooth Generation

Cuneyt Fetvaci
Istanbul University, Mechanical Engineering Department, Turkey

1. Introduction

Gearing is an essential component of many machines. From aerospace to high-speed automation, from missiles to submarines, few machines can operate without gears. Involute gears are the most popular power transmission devices for parallel axes owing to their simple geometry, easy manufacturing, and constant gear ratio even when the centre distance has been changed. Spur gears are the most popular form and the most efficient type of gearing for the cost when transmitting power and uniform rotary motion from one parallel shaft to another.

In mass production of gears, generating-type cutters are used. According to the type of relative motion between cutter and gear blank, generating cutters are classified as: rack cutters, hob cutters and gear shapers. Generation cutting is based on the fact that two involute gears of the same module and pitch mesh together - the gear blank and the cutter. This method makes to use one cutter for machining gears of the same module with a varying number of teeth. Rack-type cutters (rack or hob) can only generate external gears. Both external and internal gears can be generated by a pinion-type cutter. Figure 1 displays generating-type cutters.

For cylindrical gears in applications with uniform load-rotation conditions, an optimized and separate design of traction and thrust flank is desirable. This can be achieved by using different pressure angles for traction and thrust flank, which results in asymmetric tooth geometries. The load-carrying capacity of the gear mechanism can be improved without disturbing the material quality by using asymmetric profiles (Muni et al., 2007).

The computer simulation of gear cutting enables us to investigate the influence of design parameters on the generated profile before manufacturing. Undercutting and zero topland can be detected in design phase. Also the physical behaviour of the gear under operating conditions can be simulated and investigated. Therefore possible faults due to the inaccurate design can be detected for preventing time and material lost. An accurate geometrical representation of gear tooth surfaces is the fundamental starting point for developing a reliable computerized gear design which includes tooth contact analysis and stress analysis. Therefore, a good knowledge of the gear geometry is required.

Based on analytical mechanics of gears, parametric equations describing involute profile and root fillet profile of the gear teeth have been derived for hobbed and shaped gears (Buckingham, 1949; Colbourne, 1987; Litvin, 1994; Salamoun & Suchy, 1973). Litvin applied the vector analysis, differential geometry, matrix transformation and meshing equation to develop mathematical models for describing tooth profiles and their geometric properties (Litvin, 1994). Tsay proposed the mathematical model of the involute helical gears generated by rack-type cutters (Tsay, 1988). Chang and Tsay proposed a complete mathematical model of noncircular gears, including the fillets, bottom lands and working surfaces of tooth profiles manufactured by involute-shaped shaper cutters (Chang & Tsay, 1998). Figliolini and Angeles studied the generation of noncircular gears manufactured by the sharp-edge involute-shaped shaper cutters (Figliolini & Angeles, 2003). Chen and Tsay developed the mathematical models of helical gear sets with small numbers of teeth manufactured by modified rack- and pinion-type cutters (Chen & Tsay, 2005). Yang proposed the mathematical models of asymmetric helical external gears generated by rack-type cutters and internal gears generated by shaper cutters (Yang, 2005, 2007). Fetvaci has adopted Chang and Tsay's model to pinion-type shapers with asymmetric involute teeth for generating internal and external gears (Fetvaci, 2010a, 2011).

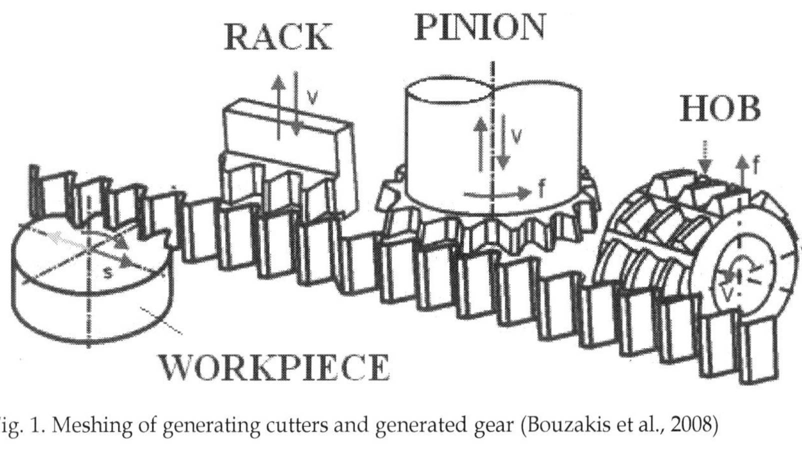

Fig. 1. Meshing of generating cutters and generated gear (Bouzakis et al., 2008)

The shape of the fillet has a direct effect on the motion/force transmission and eventual maximum bending stresses developed at the root of the gear tooth. The fillet curve of the gear belongs to the trochoid family and it is conjugate to the tip of the generating cutter. The equations for the root fillet of the spur and helical gears have been proposed by many authors in literature (Buckingham, 1949; Colbourne, 1987; Litvin, 1994; Salamoun & Suchy, 1973). Also the machining parameters of generating cutter have been given for rack-type and pinion-type cutters with two round edges or single round edge (Lin et al., 2007). During the generating process of spur gear tooth, the center of the rounded corner at the tip traces out a trochoid. Then the rounded tip envelopes another curve called as secondary trochoid, defining the root fillet. Su and Houser studied the application of trochoids to find exact fillet shapes generated by rack-type cutters (Su & Houser, 2000). Fetvaci and Imrak have adapted the equations of trochoids given by Su and Houser to Yang's mathematical model for spur gears with asymmetric involute teeth (Fetvaci & Imrak, 2008). Besides, simulated motion path of the rack cutter has been illustrated. Fetvaci has studied trochoidal paths of the pinion-type cutter during the generation of internal and external spur gears with asymmetric involute teeth (Fetvaci, 2010a, 2011).

The relative positions of the cutter during gear teeth generating process can be used for determining chip geometry for further analysis on tool wear and tool life. Tang et al. presented a computer simulation method of spur gear generating process with the sharp edge rack- and pinion-type cutters using Visual Lisp as programming language and AutoCAD as graphical display tools (Tang et al., 2008). Fetvaci presented computer simulation methods for generating asymmetric involute spur gears with rounded edge generating cutters (Fetvaci, 2010a, 2010b, 2011).

In this study, accurate mathematical models of the generating-type cutters for spur gear production are given, and trochoidal paths that determine the shape of generated tooth root fillet are investigated. Indirect gear design depending on the pre-selected set of cutting tool parameters is considered. Based on the mathematical models, computer codes have also been developed to compute the coordinates of the gear tooth profile generated by different types of cutters like cutter with sharp tip, partially round and full round tip for symmetric and asymmetric involute spur gears. Computer graphs are obtained to visualise the effect of tool parameters on generated gears before manufacturing.

The content of this chapter is organized as follows: In Section 2, the matematical models of generating cutter surfaces are studied according to Litvin's vector approach. The mathematical models: the locus of the rack- and pinion-type cutter surfaces, the equation of meshing and the generated gear tooth surfaces are given in Section 3. Trochoidal paths of the cutter tips are investigated in Section 4. As a result, Section 5 deals with computer simulation of the generating process for the verification and the validation of the matematical models. Simulated motion path of the cutter during generation process is also illustrated. The varieties of the cutter tip geometry are investigated. Finally, a conclusive summary of this study is given in Section 6.

2. Generating tooth surfaces

2.1 Rack cutter surfaces

For simplicity, the generation of spur gears with shaper cutters can be simplified into a two-dimensional problem. Due to the asymmetry of the rack cutter, left and right sides of the cutter are considered seperately. Figure 2. presents the design of the normal section of a rack cutter Σ_n, where regions $\overline{ac}$ and $\overline{bd}$ are the left- and right-side top lands, regions $\overline{ce}$ and $\overline{df}$ are the left- and right-side fillets and, regions $\overline{eg}$ and $\overline{fh}$ are the left- and right-side working regions.

The regions $\overline{ac}$ and $\overline{bd}$ are used to generate the bottomland of asymmetric spur gear and l_a and l_b represent design parameters of normal section of the rack cutter. In order to generate complete profile of the rack cutter surface a tooth of rack cutter will be repeated for $c_y = 0,1,2,..$. Equations of regions $\overline{ac}$ and $\overline{bd}$ of the rack cutter normal section can be represented in the coordinate system $S_n(X_n, Y_n, Z_n)$ by the following equations (Yang, 2005).

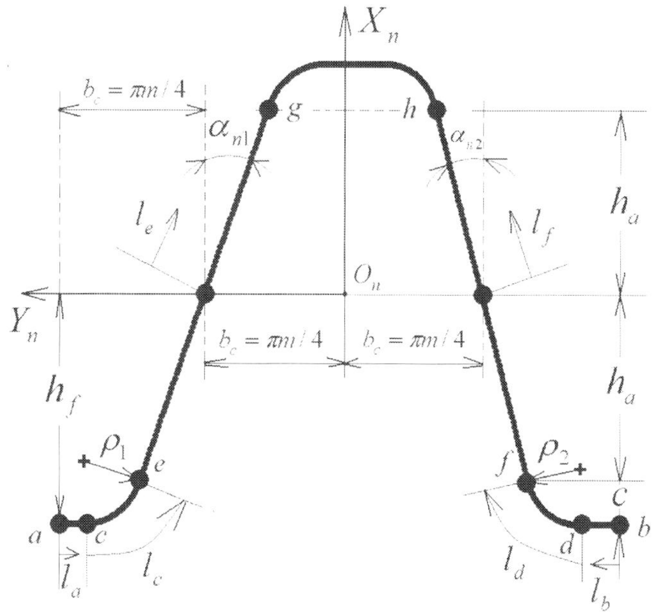

Fig. 2. Normal section of the rack cutter with asymmetric teeth (Fetvaci, 2011; Yang, 2005)

$$\mathbf{R}_n^{ac} = \begin{Bmatrix} x_n^{ac} \\ y_n^{ac} \end{Bmatrix} = \begin{Bmatrix} -h_a + \rho_1 \sin\alpha_{n1} - \rho_1 \\ (\frac{\pi m_n}{2} - l_a + c_y \pi m_n) \end{Bmatrix} \tag{1}$$

and

$$\mathbf{R}_n^{bd} = \begin{Bmatrix} x_n^{bd} \\ y_n^{bd} \end{Bmatrix} = \begin{Bmatrix} -h_a + \rho_2 \sin\alpha_{n2} - \rho_2 \\ (-\frac{\pi m_n}{2} + l_b + c_y \pi m_n) \end{Bmatrix} \tag{2}$$

where design parameters l_a and l_b are limited by $0 \le l_a \le b_c - h_a \tan\alpha_{n1} - \rho_1 \cos\alpha_{n1}$ and $0 \le l_b \le b_c - h_a \tan\alpha_{n2} - \rho_2 \cos\alpha_{n2}$ on the left- and right-side of the cutter respectively.

As depicted in Fig.1., regions $\overline{ce}$ and $\overline{df}$ on the normal section of the rack cutter generate different sides of the fillet surface of the gears. l_c and l_d are the design parameters of the rack cutter surface which determine the location of points on the fillets. The position vectors of regions $\overline{ce}$ and $\overline{df}$ are represented in the coordinate system S_n as follows (Yang, 2005) :

$$\mathbf{R}_n^{ce} = \begin{Bmatrix} x_n^{ce} \\ y_n^{ce} \end{Bmatrix} = \begin{Bmatrix} -h_a + \rho_1 \sin\alpha_{n1} - \rho_1 \cos l_c \\ (b_c + h_a \tan\alpha_{n1} + \rho_1 \cos\alpha_{n1} - \rho_1 \sin l_c + c_y \pi m_n) \end{Bmatrix} \tag{3}$$

and

$$\mathbf{R}_n^{df} = \begin{Bmatrix} x_n^{df} \\ y_n^{df} \end{Bmatrix} = \begin{Bmatrix} -h_a + \rho_2 \sin\alpha_{n2} - \rho_2 \cos l_d \\ (-b_c - h_a \tan\alpha_{n2} - \rho_2 \cos\alpha_{n2} + \rho_2 \sin l_d + c_y \pi m_n) \end{Bmatrix} \tag{4}$$

where design parameters l_c and l_d are limited by $0 \le l_c \le 90° - \alpha_{n1}$ and $0 \le l_d \le 90° - \alpha_{n2}$ respectively.

As shown in Fig. 1., two straight edges $\overline{eg}$ and $\overline{fh}$ of the rack cutter are used to generate the left- and right-side tooth surface of the asymmetric helical gear, respectively. The symbol m_n represents the normal module. The position vector of regions $\overline{eg}$ and $\overline{fh}$ are represented in the coordinate system S_n as follows (Yang, 2005):

$$\mathbf{R}_n^{eg} = \begin{Bmatrix} x_n^{eg} \\ y_n^{eg} \end{Bmatrix} = \begin{Bmatrix} l_e \cos\alpha_{n1} \\ (b_c - l_e \sin\alpha_{n1} + c_y \pi m_n) \end{Bmatrix} \tag{5}$$

and

$$\mathbf{R}_n^{fh} = \begin{Bmatrix} x_n^{fh} \\ y_n^{fh} \end{Bmatrix} = \begin{Bmatrix} l_f \cos\alpha_{n2} \\ (-b_c + l_f \sin\alpha_{n2} + c_y \pi m_n) \end{Bmatrix} \tag{6}$$

where l_e and l_f are the design parameters of the rack cutter surface which determine the location of points on the working surface. l_e and l_f are limited by $-h_a / \cos\alpha_{n1} \le l_e \le h_a / \cos\alpha_{n1}$ and $-h_a / \cos\alpha_{n2} \le l_f \le h_a / \cos\alpha_{n2}$ for the left- and right-side of the rack cutter respectively. The surface unit normals of the regions $\overline{ac}$ to $\overline{fh}$ of the rack cutter surfaces are represented by (Litvin, 1994),

$$\mathbf{n}_n^i = \frac{\dfrac{\partial \mathbf{R}_n^i}{\partial l_j} \times \mathbf{k}_n}{\left| \dfrac{\partial \mathbf{R}_n^i}{\partial l_j} \times \mathbf{k}_n \right|} \qquad \begin{matrix} (i = ac \sim fh) \\ (j = a \sim f) \end{matrix} \tag{7}$$

where $\mathbf{k}_n$ is the unit vector of the Z_n-axis.

2.2 Pinion-type shaper cutter surfaces

Pinion-type shaper cutters are designed consists of six generating regions as depicted in Fig.3. Regions 1 and 6 of the involute-shaped curves generate the working regions of involute spur gears, regions 2 and 5 of the circular arcs with centers at *E* and *G* generate the fillet surfaces, and regions 3 and 4 of the shaper cutter surfaces generate the bottom lands (Chang & Tsay, 1998). Based on (Figliolini & Angeles, 2003), nongenerating surfaces of the cutter are also shown for visual purposes only.

In Fig. 3, coordinates systems $S_s(X_s, Y_s)$ and $S_c(X_c, Y_c)$ represent the reference and the shaper cutter coordinate systems, respectively. According to the relationship between coordinate systems S_s and S_c, the position vector of region i can be transformed from

$$n_c = \frac{\dfrac{dR_c^i}{dl_j} \times k_c}{\left| \dfrac{dR_c^i}{dl_j} \times k_c \right|} \tag{15}$$

where k_c is the unit vector of the Z_c-axis. Parameter l_j represents ξ, θ and η, respectively.

By substituting Eq. (10) in Eq. (15), the unit normal vector of region 1 can be obtained as follows (Chang & Tsay, 1998) :

$$n_c^1 = \begin{Bmatrix} n_{xc}^1 \\ n_{yc}^1 \end{Bmatrix} = \begin{Bmatrix} -\sin(\xi - \psi) \\ \cos(\xi - \psi) \end{Bmatrix} \tag{16}$$

By substituting Eq. (12) in Eq. (15), the unit normal vector of region 2 can be obtained as follows (Chang & Tsay, 1998):

$$n_c^2 = \begin{Bmatrix} n_{xc}^2 \\ n_{yc}^2 \end{Bmatrix} = \begin{Bmatrix} -\sin(\theta + \xi_m - \psi) \\ -\cos(\theta + \xi_m - \psi) \end{Bmatrix} \tag{17}$$

By substituting Eq. (14) in Eq. (15), the unit normal vector of region 3 can be obtained as follows (Chang & Tsay, 1998):

$$n_c^3 = \begin{Bmatrix} n_{xc}^3 \\ n_{yc}^3 \end{Bmatrix} = \begin{Bmatrix} -\sin(\eta - \psi) \\ -\cos(\eta - \psi) \end{Bmatrix} \tag{18}$$

The equations for the right side of the cutter are similar to those of left's, provided that parameters are calculated according to corresponding pressure angle, and all equations corresponding to X_c coordinate are assigned an appropriate sign.

3. Generated gear tooth surfaces

3.1 Generating with rack-cuttter

To derive the mathematical model for the complete tooth profile of involute spur gears with asymmetric teeth, coordinate systems $S_n(X_n, Y_n, Z_n)$, $S_1(X_1, Y_1, Z_1)$ and $S_h(X_h, Y_h, Z_h)$ should be set up. The coordinate systems S_n, S_1 and S_h are attached to the rack cutter, involute gear, and gear housing, respectively as shown in Fig. 4. Z_1, Z_n and Z_h are determined by the right-hand co-ordinate system. During the generation process, the rack cutter translates a distance $S = r_{p1}\,\phi_1$ while the gear blank rotates rotates by an angle ϕ_1.

The mathematical model of the generated gear tooth surface is a combination of the meshing equation and the locus of the rack cutter surfaces according to gearing theory (Litvin, 1994). Applying the following homogeneous coordinate transformation matrix equation makes it possible to obtain the locus of the cutter represented in coordinate system S_1 as follows:

$$\mathbf{R}_1^i = [M_{1n}]\mathbf{R}_c^i \tag{19}$$

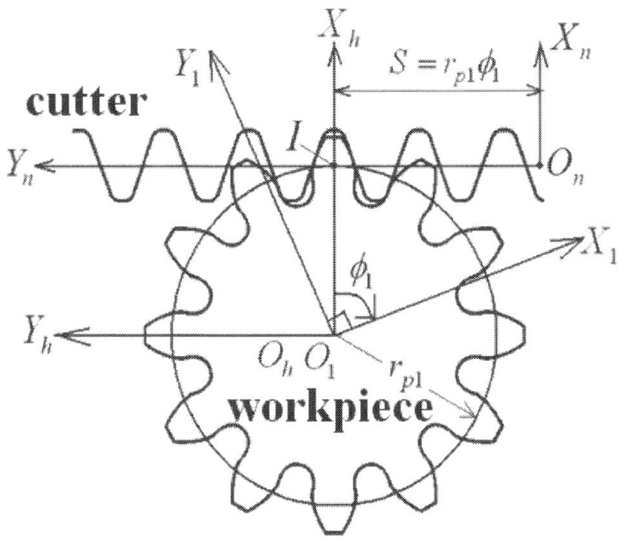

Fig. 4. Coordinate relationship between the rack cutter and the generated gear

where

$$[M_{1n}] = \begin{bmatrix} \cos\phi_1 & -\sin\phi & r_{p1}(\cos\phi_1 + \phi_1\sin\phi_1) \\ \sin\phi_1 & \cos\phi_1 & r_{p1}(\sin\phi_1 - \phi_1\cos\phi_1) \\ 0 & 0 & 1 \end{bmatrix}$$

According to the theory of gearing (Litvin, 1994), the common normal to the transverse section of the rack cutter and gear tooth surface must pass through the instantaneous center of rotation I. Thus, equation of meshing may be represented in coordinate system S_n as follows,

$$\frac{X_n^i - x_n^i}{n_{nx}^i} = \frac{Y_n^i - y_n^i}{n_{ny}^i} \tag{20}$$

Symbols X_c^i and Y_c^i represent the coordinates of a point on the instantaneous axis of gear rotation I-I in coordinate system S_c; x_c^i and y_c^i and are the coordinates of the instantaneous contact point on the rack cutter surface; n_{nx}^i and n_{ny}^i, are the direction cosines of the rack cutter surface unit normal $\mathbf{n}_n^i$. Angle ϕ_1 is the rolling parameter and the symbol r_{p1} denotes the radius of the gear pitch circle.

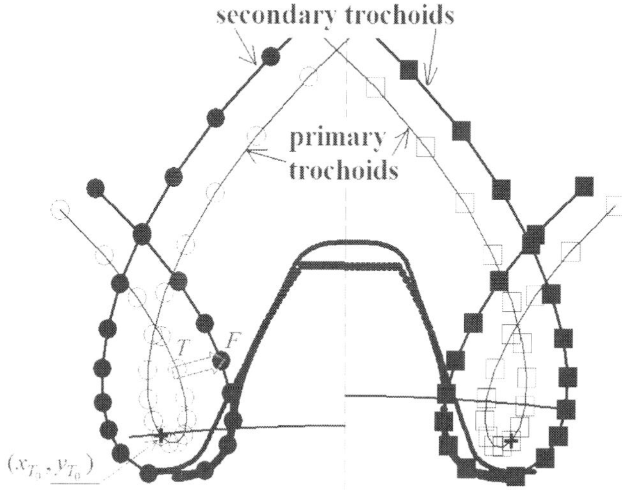

Fig. 6. Trochoidal paths of the rack-type cutter tip

$$\begin{Bmatrix} x_T \\ y_T \end{Bmatrix} = \begin{Bmatrix} x_{T_0} \cos\phi_1 - y_{T_0} \sin\phi_1 + r_{p1}(\phi_1 \sin\phi_1 + \cos\phi_1) \\ x_{T_0} \sin\phi_1 + y_{T_0} \cos\phi_1 + r_{p1}(-\phi_1 \cos\phi_1 + \sin\phi_1) \end{Bmatrix} \tag{25}$$

where angle ϕ_1 is the rolling parameter as stated before, r_{p1} is the radius of pitch circle (x_{T_0}, y_{T_0}) is the coordinate of point T_0 in the fixed coordinate system.

The actual form of gear tooth fillet is the envelope of the path of a series of circles equal in size to the rounding of the corner, and with their centers on the primary trochoidal path (Buckingham, 1988). This new path is called as secondary trochoid. The coordinate of the corresponding point F on the secondary trochoid can be expressed as:

$$\begin{Bmatrix} x_F \\ y_F \end{Bmatrix} = \begin{Bmatrix} x_T \\ y_T \end{Bmatrix} + \begin{Bmatrix} \rho \sin(\gamma - \phi_1) \\ \rho \cos(\gamma - \phi_1) \end{Bmatrix} \tag{26}$$

where $\gamma = \arctan(x_{T_0} / (y_{T_0} - r_{p_1}\phi_1))$ and ρ denotes the tip radius of the cutter. It can be cleary seen that the primary trochoid and the secondary trochoid are two equidistant curves.

4.2 Pinion cutter

The tooth fillet resulting from gear generation is in fact a trochoid which is created by the tool tip in its rolling movement. An epitrochoid curve determines the shape of the fillet of generated external gear tooth as a result of generation process by pinion-type shaper cutters. An epitrochoid is a curve traced by a point attached to a circle of radius r rolling around the outside of a fixed circle of radius R, where the point is a distance d from the center of the exterior circle. According to the analytical mechanics of gears, the rolling circle is the pitch circle of the generating shaper cutter, the fixed circle is the pitch circle of the machined gear and the distance d is measured from the origin of the cutter to the center of its rounded corner at the tip (point E). During the generating process of spur gear tooth presented in this paper, the center of the rounded corner at the tip traces out a trochoid. An equidistant curve with a distance of ρ defines the gear tooth root fillet.

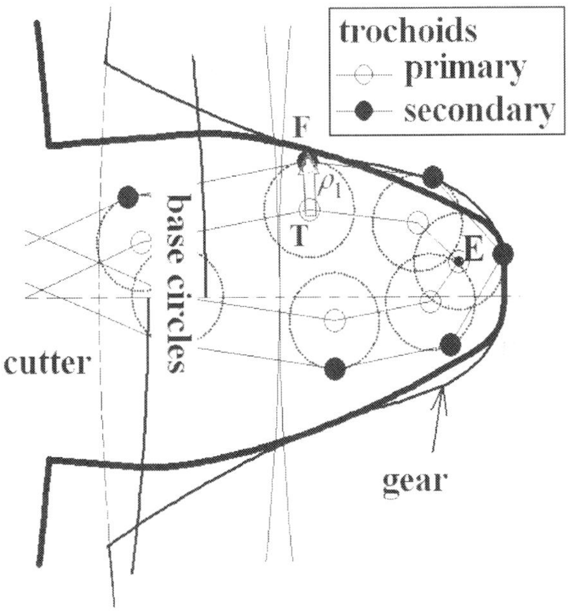

Fig. 7. Trochoidal paths of the pinion-type cutter tip

As depicted in Fig. (3) and Fig.(7), the rounded edge of the cutter is a circular arc and its center is located at point E. To ensure the tangents of the involute curve and circular arc at point A are the same and continuous, point E should be on the line $\overline{PA}$. It is first necessary to find the coordinates of points A and E (Colbourne, 1987).

The maximum involute extension angle at point A, denoted as ξ_m, can be evaluated from the following equation when the radius of tip circle r_B is given.

$$r_b \tan \xi_m = \sqrt{(r_B - \rho_1)^2 - r_b^2} + \rho_1 \tag{27}$$

According to involute geometry, the polar coordinates of point $A(r_A, \theta_A)$ are given by ,

$$r_A = r_b / \cos \xi_m \tag{28}$$

$$\theta_A = \pi / 2N_s + inv\alpha - inv\xi_m \tag{29}$$

The rectangular coordinates of point E can then be expressed in terms of x_E and y_E,

$$x_E = r_A \cos\theta_A - \rho_1 \sin(\xi_m - \theta_A) \tag{30}$$

$$y_E = r_A \sin\theta_A - \rho_1 \cos(\xi_m - \theta_A) \tag{31}$$

$$\theta_E = \tan^{-1}(y_E / x_E) \tag{32}$$

A general point on the primary trochoid which is the envelope of the center of round tip is depicted in Fig. 7. Applying the homogeneous coordinate transformation matrix given in Eq. (22), the equation of the primary trochoid (epitrochoid curve) can be written as follows:

$$\begin{Bmatrix} x_T \\ y_T \end{Bmatrix} = \begin{Bmatrix} x_E \cos(\phi_c + \phi_g) + y_E \sin(\phi_c + \phi_g) - (r_c + r_g)\cos\phi_g \\ -x_E \sin(\phi_c + \phi_g) + y_E \cos(\phi_c + \phi_g) + (r_c + r_g)\sin\phi_g \end{Bmatrix} \tag{33}$$

where (x_E, y_E) is the coordinate of point E, ϕ_c and ϕ_g are the rolling parameters, r_c and r_g are the pitch circle radius of the shaper and the machined gear, respectively.

The actual form of spur gear tooth fillet is the envelope of the path of a series of circles with their geometric centers on the primary trochoidal path. This new path is called as secondary trochoid which is the paralel curve of the primary trochoid. As a result, the coordinate of the corresponding point F on the secondary trochoid can be expressed as

$$x_F = x_T + \frac{\rho_1 y'_T}{\sqrt{x'^2_T + y'^2_T}} \tag{34}$$

$$y_F = y_T - \frac{\rho_1 x'_T}{\sqrt{x'^2_T + y'^2_T}} \tag{35}$$

where ρ_1 denotes the tip rounding radius of the shaper cutter, $x'_T = dx_T / d\phi_c$ and $y'_T = dy_T / d\phi_c$.

5. Computer graphs of tooth surfaces

Computer graphs of generating and generated surfaces can be obtained by using a programming language and graphic processor. In this study codes are developed by using GW-BASIC language to obtain the coordinates of the surfaces. GRAPHER 2-D Graphing System is used for displaying computer graphs of the cutters and gears. Also the ANSYS Preprocessor module is used for displaying gear generating process. Illustrative examples are given for both rack- and pinion-type cutters for different types of tool tip geometries.

For rack-type generation, types of tip fillet geometry are selected from the study proposed by Alipiev (Alipiev, 2009, 2011) and the related geometries displayed in the table are adopted to the present mathematical model. Table 1 displays the variation of tip geometry of the rack cutters.

		a	b
		$c_2 \rangle c_1$	$c_1 = c_2$
1	$mn \rangle \rho_1 \cos\alpha_1 + \rho_2 \cos\alpha_2$	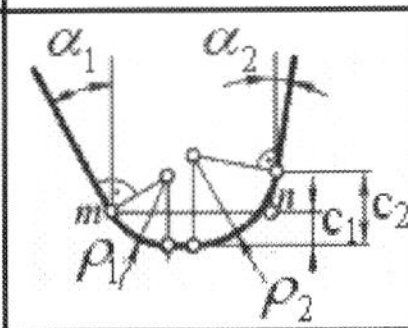	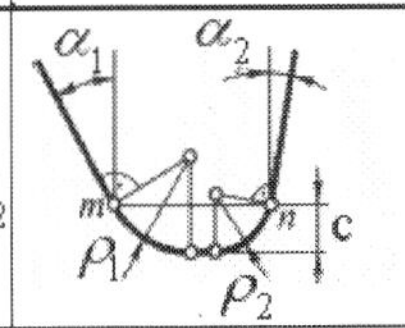
2	$mn = \rho_1 \cos\alpha_1 + \rho_2 \cos\alpha_2$	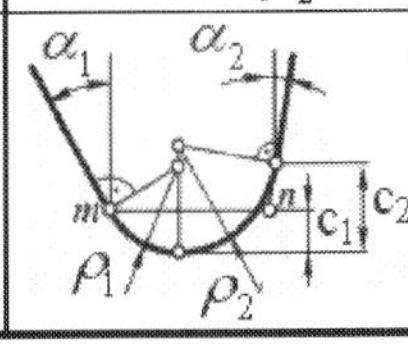	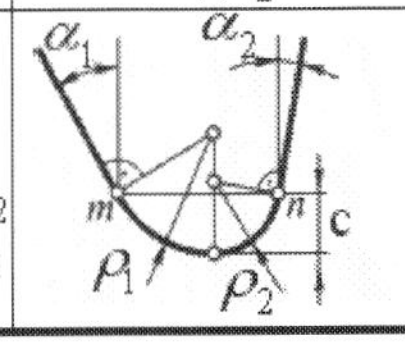

Table 1. Geometric varieties of rack tool tip (Alipiev, 2011)

As illustrated in Table 1, the rack cutter of type-1a has different clearances at its different sides. The side with a higher pressure angle has a lower radius of rounding and a lower clearance. The tooth semi-thicknesses at pitch line of the cutter are different from each other. Design parameters are selected as module $m = 2.5mm$, number of teeth $z = 24$, left side pressure angle $\alpha_1 = 20°$, right side pressure angle $\alpha_2 = 15°$, left side radius of rounding $\rho_1 = 0.2\times m$ and right side radius of rounding $\rho_2 = 0.3\times m$. Figure 8 displays the generating cutter of type-1a, generated surface and trochoidal paths of the tip.

As illustrated in Fig. 2. and classifed type-1b in Table 1, the cutter has a constant clearance for its all sides. The side with a higher pressure angle has a higher radius of rounding. The tooth semi-thicknesses at pitch line of the cutter are same. This type of cutter is adopted from the standard generating rack to asymmetric gearing. The relation ship between left and right side roundings is $\rho_1(1-\sin\alpha_1) = \rho_2(1-\sin\alpha_2)$. Design parameters are selected as module $m = 2.5mm$, number of teeth $z = 24$, left side pressure angle $\alpha_1 = 20°$, right side pressure angle $\alpha_2 = 15°$, left side radius of rounding $\rho_1 = 0.38\times m$ and right side radius of rounding $\rho_2 = 0.33\times m$. Generating and generated surfaces and trochoidal paths are illustrated in Fig 9.

Rack cutters with asymmetric teeth can also be designed with full rounded tips. The rack cutter of type-2a has a single rounded edge. The side with a higher pressure angle has a lower radius of rounding and a lower clearance. As depicted in Table 1 the centers of the rounded tip are at the center line of the cutter tooth. The tooth semi-thicknesses at pitch line of the cutter are same. Design parameters are selected as module $m = 2.5mm$, number of teeth $z = 24$, left side pressure angle $\alpha_1 = 22.5°$, right side pressure angle $\alpha_2 = 15°$, left side radius of rounding $\rho_1 = 0.4\times m$ and right side radius of rounding $\rho_2 = 0.587\times m$. Figure 10 displays the generating cutter of type-1a, generated surface and trochoidal paths of the tip. For visual clearity, only the corresponding halves (of secondary trochoids) that contribute to final formation of the generated tooth shape are shown.

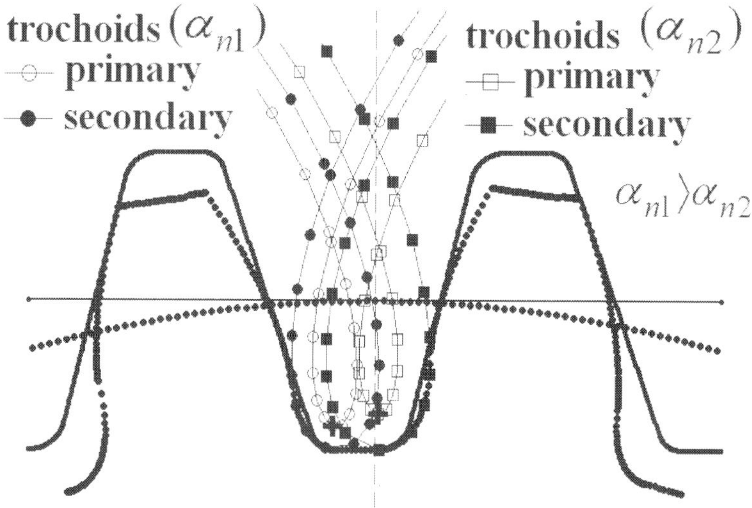

Fig. 8. Trochoidal paths of rack cutter of type-1a

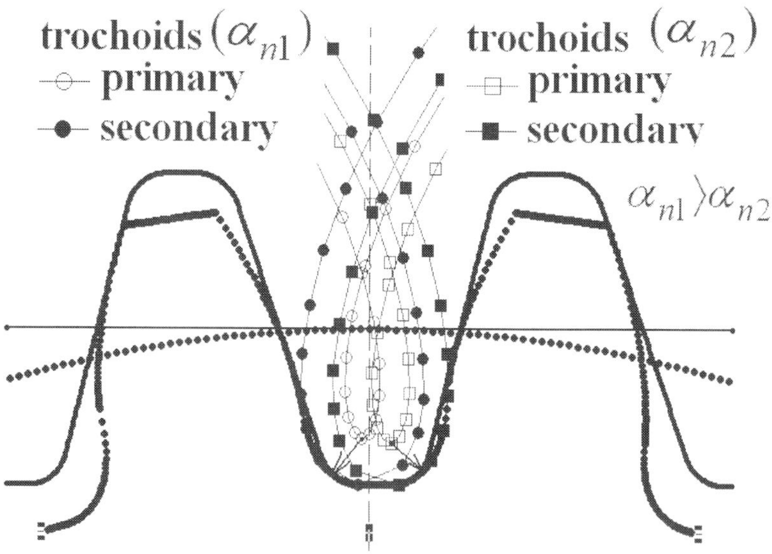

Fig. 9. Trochoidal paths of rack cutter with a rounded-tip for constant clearance

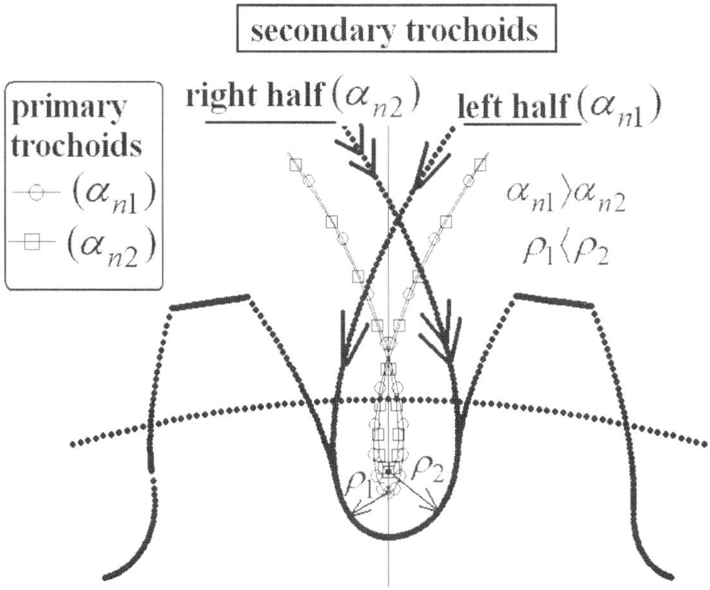

Fig. 10. Trochoidal paths of rack cutter with a fully rounded-tip

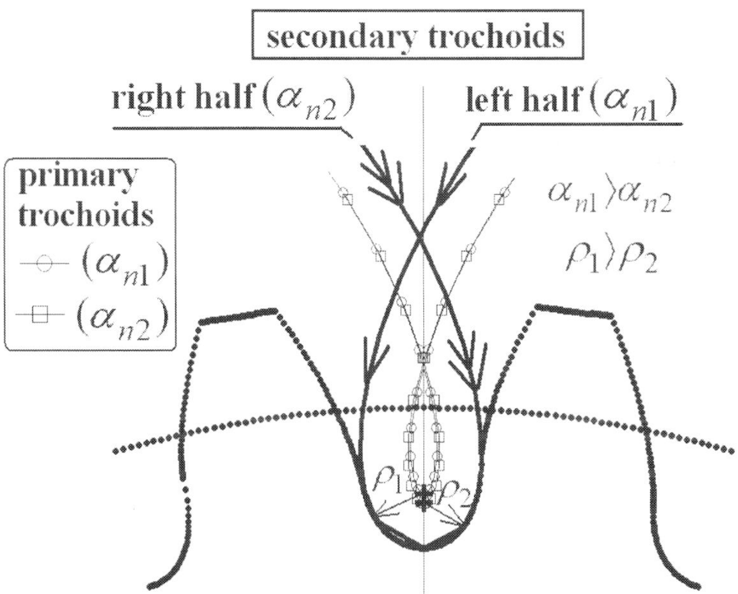

Fig. 11. Trochoidal paths of rack cutter with a fully rounded-tip for constant clearance

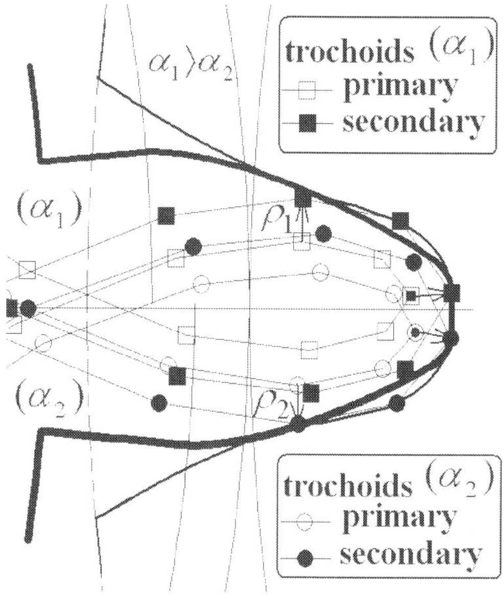

Fig. 13. Cutter with a larger rounding radius for higher pressure angle

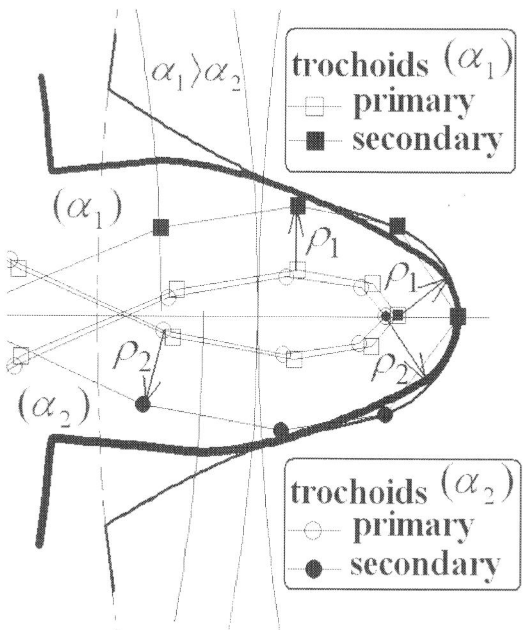

Fig. 14. Cutter with a full-rounded tip

The relative positions of the cutter during generating process can be visualized by using the mathematical of generating surfaces and transformation matrices. The present author used the locus equations of the cutters and obtained illustrations displaying simulated motion path of the cutter during generation by manipulating rolling parameter as $-\pi/4 \leq \phi_1 \leq \pi/4$ in the developed code. Each gear gap is produced through successive penetrations of the tool teeth into the workpiece, in the individual generating positions. This simulation can be used to determine the chip geometry (Bouzakis et al., 2008). Figure 15 displays the work gear and simulated motion path of the generating rack cutter with asymmetric teeth. Similiarly, Fig. 16. displays the work gear and simulated motion path of the generating pinion cutter.

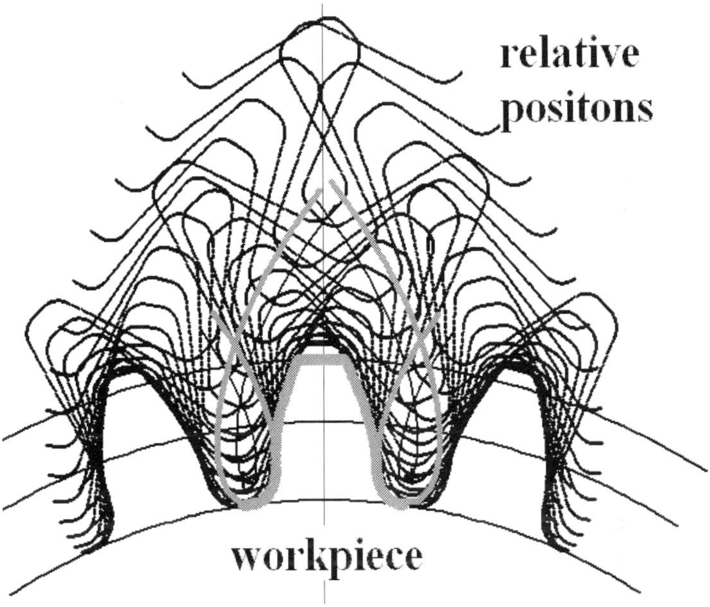

Fig. 15. Generated gear and generating positions of the rack-cutter with a rounded-tip

Figure 17 displays relative positions of the pinion cutter with symmetric involute teeth and a fully-rounded tip. The trochoidal curves exhibits symmetry according to center line of gear tooth space. Generating with a sharp-edge pinion cutter is depicted in Fig.18. In this case, primary trochoids determine the shape of the generated tooth fillet. The secondary trochoids do not exist.

Video files displaying generating positions of the cutter can be obtained with a proper software. In this study, ANSYS Parametric Design Language (APDL) is also used for obtaining graphic outputs and animation files displaying the simulated motion path of the generating cutters (ANSYS, 2009). Video files can be seen in the author's web page: http://www.istanbul.edu.tr/eng2/makina/cfetvaci/gearpage.htm

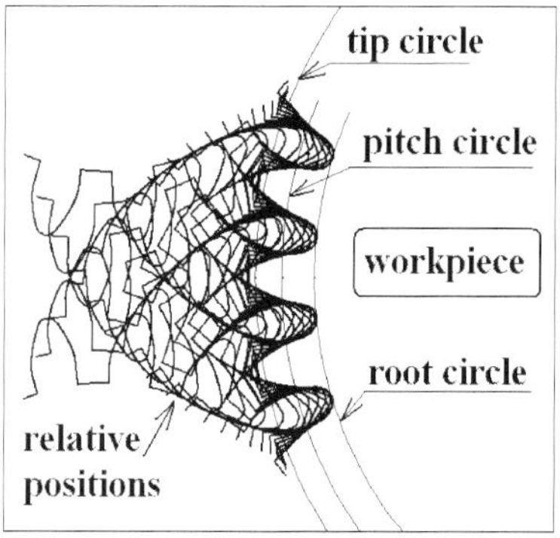

Fig. 16. Generated gear and generating positions of the pinion-cutter with a rounded-tip

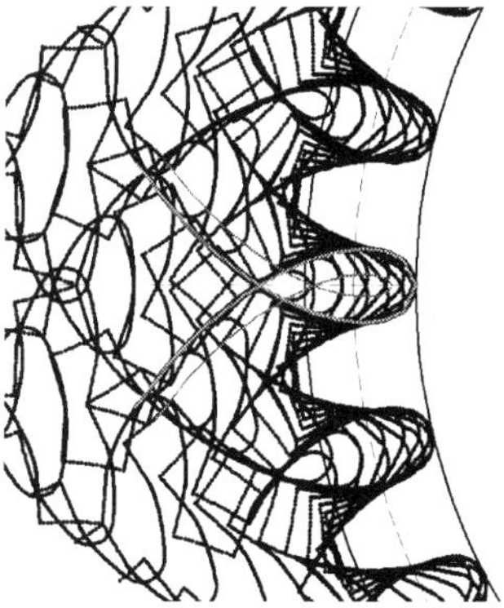

Fig. 17. Generated gear and generating positions of the pinion-cutter with symmetric teeth and a fully rounded-tip

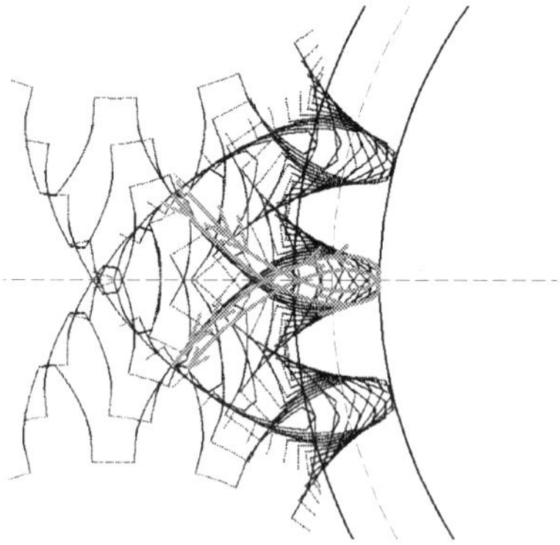

Fig. 18. Generated gear and generating positions of the pinion-cutter with symmetric teeth and a sharp-tip

6. Conclusion

In this study, computerized tooth profile generation of involute gears manufactured by rack- and pinion-type cutters are studied based on Litvin's vector method. Based on Yang's application mathematical model of rack cutter with asymmetric involute teeth is given. Trochoidal paths of the rack tool tip are investigated. For pinion-type generation Asymmetric involute teeth is adopted to Chang and Tsay's application. The developed computer program provides the investigation of the effect of tool parameters on the generated tool profile before manufactured. Trochoidal paths traced by the generating tool tip are investigated. It has been seen that geometric varieties of the rounded corner of pinion-type cutter determines the position of trochoidal paths relative to the center line of tooth space of the generated gear. Because of the position of the center of the tip rounding, there is a limitation on the geometric varieties of pinion-type cutter tip. Based on the given mathematical models, the simulated motion path of the generating cutters are also investigated. The relative position of the cutter to the workpiece has been illustrated. The simulation of shaper cutting action can be used to determine the chip geometry for further analysis about tool wear and tool life. The mathematical models can be extended to generalized mathematical model of the involute gears including spur and helical beveloid (involute conical) gears.

7. References

Alipiev, O. (2009). Geometric Synthesis of Symmetric and Asymmetric Involute Meshing using the Method of Realized Potential. *General Machine Design Conference*, p. 43-50, Ruse - Bulgaria, October 15-16, 2009

Alipiev, O. (2011). Geometric Design of Involute Spur Gear Drives with Symmetric and Asymmetric Teeth using the Realized Potential Method. *Mechanism and Machine Theory*, Vol. 46, No. 1, (January 2011), pp. 10-32, ISSN 0094-114X

ANSYS. (2009). *ANSYS Parametric Design Language Guide*. Available from, http://www1.ansys.com/customer/content/documentation/120/ans_apdl.pdf

Bouzakis, K.-D., Lili, E., Michailidis, N. & Friderik, O. 2008. Manufacturing of Cylindrical Gears by Generating Cutting Processes: A Critical Synthesis of Analysis Methods. *CIRP Annals - Manufacturing Technology*, Vol. 57, No.2, 2008, pp. 676-696.

Buckingham, E. (1949). *Analytical Mechanics of Gears*, McGraw-Hill, New York, USA

Chang, S.-L. & Tsay, C.-B. (1998). Computerized Tooth Profile Generation and Undercut Analysis of Noncircular Gears Manufactured with Shaper Cutters. *Journal of Mechanical Design*, Vol. 120, No. 1, (March 1998), pp. 92-99.

Chen, C.-F & Tsay, C.-B. (2005). Tooth Profile Design for the Manufacture of Helical Gear Sets with Small Numbers of Teeth, *International Journal of Machine Tools and Manufacture*, Vol. 45, No. 12-13, (October 2005), pp. 1531-1541

Colbourne, J.R. (1987). *The Geometry of Involute Gears*, Springer-Verlag, New Jersey, USA

Fetvaci, C. & İmrak, E. (2008). Mathematical Model of a Spur Gear with Asymmetric Involute Teeth and its Cutting Simulation. *Mechanics Based Design of Structures and Machines* , Vol. 36, No. 1, pp. 34- 46, ISSN 1539-7734

Fetvacı, C. (2010a). Definition of Involute Spur Gear Profiles Generated By Gear-Type Shaper Cutters. *Mechanics Based Design of Structures and Machines*. Vol.38, No. 4, pp. 481-492

Fetvaci, C. (2010b). Computer Simulation of Helical Gears with Asymmetric Involute Teeth. *Journal of The Faculty of Engineering and Architecture of Gazi University*, Vol. 25, No. 3, (September 2010), pp. 441-447, ISSN 1300-1884

Fetvaci, C. (2011). Computer Simulation of Asymmetric Involute Spur Gears Manufactured by Generating-Type Cutters . *Engineers and Machinery*, Vol. 52, No. 616, (May 2011), pp. 60-69

Figliolini, G. & Angeles, J. (2003). The Synthesis of Elliptical Gears Generated by Shaper-Cutters. *Journal of Mechanical Design*, Vol.125, No. 4, (December 2003), pp. 793-801, 2003.

Lin, T., Ou, H. & Li, R. (2007). A Finite Element Method for 3-D Static and Dynamic Contact/Impact Analysis of Gear Drives. *Computer Methods in Applied Mechanics and Engineering*. Vol. 196, No. 9-12, (February 2007), pp. 1716-1728.

Litvin, F.L. (1994). *Gear Geometry and Applied Theory*, Prentice Hall, New Jersey, USA

Muni, D.V., Kumar, S. & Muthuveerappan, G. (2007). Optimization of Asymmetric Spur Gear Drives for Maximum Bending Strength Using Direct Gear Design Method. *Mechanics Based Design of Structures and Machines*, Vol.35, No.2, pp. 127-145

Salamoun, C. & Suchy, M. (1973). Computation of Helical or Spur Gear Fillets. *Mechanism and Machine Theory*, Vol. 8, No. 3, (Autumn 1973), pp. 305-322, ISSN 0094-114X

Su, X. & Houser, D.R. (2000). Characteristics of Trochoids and their Application to Determining Gear Teeth Fillet Shapes. *Mechanism and Machine Theory*, Vol. 35, No. 2, (February 2000), pp. 291–304, ISSN 0094-114X

Tang, X., Ren, F., Jiang, Y. & Gao, S. (2008). Geometric Modeling and Dynamic Simulation of Involute Gear by Generating Method, *Proceedings 13th International Conference on Geometry and Graphics*, pp. 233-234, Dresden, Germany, August 4-8, 2008

Tsay, C.-B. (1988). Helical Gears with Involute Shaped Teeth: Geometry, Computer Simulation, Tooth Contact Analysis and Stress Analysis. *Journal of Mechanical Design*, Vol. 110, No. 4 , pp. 482-491

Yang, S.-C. (2005). Yang, Mathematical Model of a Helical Gear with Asymmetric Involute Teeth and its Analysis. *International Journal of Advanced Manufacturing Technology*, Vol. 26, No. 5-6, (September 2005), pp. 448-456, ISSN 0268-3768

Yang, S.-C. (2007). Study on an Internal Gear with Asymmetric Involute Teeth. *Mechanism and Machine Theory*, Vol. 42, No. 8, (August 2007), pp. 977-994

23

Applications of Computer Vision in Micro/Nano Observation

Yangjie Wei[1], Chengdong Wu[2] and Zaili Dong[3]
[1]*Graduate School of Chinese Academy of Sciences & The State Key Laboratory of Robotics,Institute of Automation, Chinese Academy of Sciences*
[2]*School of Information Science & Engineering, Northeast University*
[3]*State Key Laboratory of Robotics, Shenyang Institute of Automation, Chinese Academy of Sciences,*
China

1. Introduction

Nowadays, micro/nano science and technology has been one of the most attractive research fields. However, real time and accurate observation in micro/nano manipulation is a top important enabling technique. Most recently, with the great development of microscopes and computer vision techniques, real time visualization, including 2D motion measurement and 3D reconstruction, on micro/nano scale is becoming possible.

As for 2D motion measurement, visual motion measurement on micro/nano scale is still an open problem. Many researchers have designed different algorithms, and most of them are based on a block matching algorithm, which locates matching blocks in a researched digital image for the purposes of distance or similarity estimation. Usually, block matching based methods can achieve better performances when the texture is not relevant, or the aliasing problem in the derivative estimation, which is caused by the large inter-frame displacements (Giachetti & Torre, 1996). Images, however, are typically processed assuming a uniform grid of pixels. While straightforward, the uniform grid representation does not scale well in a multi-scale setting, because it requires an excessive amount of refinement to capture small details in a image, including sub-pixel resolution. The motion to be estimated is, on most situations in micro/nano manipulation, small and not integer. Therefore, it is necessary to improve the existing algorithms and obtain higher precision not limited by the pixel dimension, i.e., sub-pixel motion estimation. In 1989, Anandan reaches a sub-pixel precision by locally approximating the difference function with a quadratic surface and published (Horn, 1986; Horn & Schunck,1981; Singh, 1990), however, the sub-pixel estimation resolution usually introduces more computational burden.

As far as 3D reconstruction is concerned, depth measurement, i.e., methods to attain 3D information from 2D images, is an important research field in computer vision, and now it has been one of the key techniques in many fields, such as medicine, robotics, remote sensing and micro/nano manipulation. In recent years, there are various 3D reconstruction methods, including volumetric methods, depth from stereo (DFS), depth from focus (DFF) and depth from defocus (DFD)(Yin,1999), researched and used in real applications.

Volumetric methods usually reconstruct 3D models of external anatomical structures from 2D images. They represent the final volume using a finite set of 3D geometric primitives. Then, from an image sequence acquired around the object to reconstruct, the images are calibrated and the 3D models of the referred object are built using different approaches of volumetric methods. These methods work in the object volumetric space and do not require a matching process between the images used. Thus, typically, the 3D models are built from a sequence of images, acquired using a turntable device and an off-the shelf camera (Teresa et al,in press, 2008). However, in some real applications, we do not need to reconstruct the 3D model of objects, because depth is enough to understand the 3D relationship of scenes.

DFS estimates depth from two images of the same scene captured by cameras at different positions and with different postures (Wu, 1999). Because it needs to extract and match feature points in these images, the computational task is so huge. As for DFF, it uses a mapping relation between focus and depth to estimate depth. It obtains a sequence of images with different depth, measures the focus degree using a measurement operator (Bove 1993; Nayar,1992), and attains the desired depth when the measurement value is maximal or minimal. Compared to DFS, DFF is simple in principle, but its estimation accuracy is highly related to the number of images.

DFD is first introduced by Pentland in 1987(Pentland,1987). It has been proved to be an effective depth reconstruction method by using the concept of blurring degree of region images with limit depth of field (Girod & Scherock, 1989; Pentland et al, 1994; Navar et al, 1996). Usually, DFD algorithm captures two images obtained with different camera parameters, measures blurring degree of every point, and estimates depth using the point spread function. During the past years, DFD has become attractive because 1) it requires only two images; 2) it avoids matching and masking problems; 3) it is effective both in the frequency domain and in the spatial domain (Gokstorp,1994; Subbarao & Surya,1994). However, since all above DFD methods need to capture two defocused images with changed camera parameters, they can not be used in applications with high level magnification microscopes, such as micro/nano manipulation, because on these situations, it is destructive to change camera parameters. This is the main reason why DFD has not been used in micro/nano manipulation until now.

2. 2D motion measurement

Computer vision is one of the most important techniques used in motion measurements, especially 2D motion measurement, because the instruments used in computer vision are comparatively cheaper, the measurement process is simple and the result is direct. In recent years, with the development of revolution and sensitivity on visual sensors, the measurement scale of computer vision has reached micro/nano scale.

Block matching method (BMA) is one of the most widely applied methods to compute the visual 2D motion from images, i.e. to estimate the 2D motion projected on the image plane by the objects moving in the 3D scene, as it is less susceptible to a random error source than edge based or image moment methods.

2.1 BMA method

The foundational principle of BAM is to find a matching block from an image X in some other image Y, which may appear before or after X, and through comparing them to measure difference, such as distance or similarity, between two images. Therefore to select a criteria to determine whether a given block in image Y matches the search block in image X is top important, i.e., the object function.

BMA based techniques usually can be divided into two classes according to the measurement criterion, including the minimal difference and the maximal similarity. The widely used object functions based on difference measurements include Sum-of-Squared-Differences (SSD), Sum-of-Absolute-Differences (SAD) which can transferred into Local-SAD (LSAD) when its intensity is locally scaled and Zero-SAD (ZSAD) with setting the average gray level difference equal to zero. If the difference minimum is replaced by the maximum of a correlation measurement, some object functions can be got, such as Normalized-Cross-Correlation (NCC)(Qi & Michale,1987), Approximate-Maximum-Direct-Correlation (AMDC)(Kim & Meng,2007), or some other variations those are all approximate maximum likelihood estimators(Robinson & Milanfar,2004).

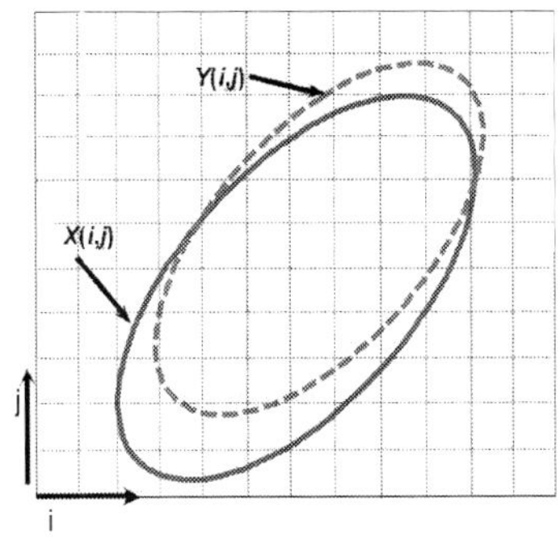

Fig. 1. Motion of the continuous image *F(i,j)* with respect to the pixel grid

Here, SSD, LSAD, ZSAD and NCC are all adopted to estimate the motion between two neighbor images in a same image sequence. The theory is shown in Fig. 1. Generally, *X(i,j)* and *Y(i,j)* are referred to as the model image and the target image respectively. F(*i,j*) is the continuous image function, $\varepsilon_{x,y}$ represents additive noise, $s=(s_x, s_y)$ is the shift between the model image and the target image.

$$X_{i,j} = F(i,j) + \varepsilon_x \tag{1}$$

$$Y_{i,j} = F(i - s_x, j - s_y) + \varepsilon_y \tag{2}$$

The object functions for SSD, LSAD, ZSAD and NCC are respectively defined as follow,

$$R(u,v)_{SSD} = \frac{1}{m \times n} \sum_{i=0}^{m-1} \sum_{j=0}^{n-1} [x_{i,j} - y_{i+u,j+v}]^2 \tag{3}$$

$$R(u,v)_{LSAD} = \frac{1}{m \times n} \sum_{i=0}^{m-1} \sum_{j=0}^{n-1} \left| x_{i,j} - \frac{\overline{X_{i,j}}}{\overline{Y_{i+u,j+v}}} y_{i+u,j+v} \right| \tag{4}$$

$$R(u,v)_{ZSAD} = \frac{1}{m \times n} \sum_{i=0}^{m-1} \sum_{j=0}^{n-1} [x_{i,j} - \overline{X_{i,j}} - y_{i+u,j+v} + \overline{Y_{i+u,j+v}}] \tag{5}$$

$$R(u,v)_{NCC} = \frac{\frac{1}{n \times m} \sum_{i=0}^{m-1} \sum_{j=0}^{n-1} X_{i,j} Y_{i+u,j+v}}{\sqrt{\frac{1}{n \times m} \sum_{i=0}^{m-1} \sum_{j=0}^{n-1} X_{i,j}^2} \sqrt{\frac{1}{n \times m} \sum_{i=0}^{m-1} \sum_{j=0}^{n-1} Y_{i+u,j+v}^2}} \tag{6}$$

$$X_{i,j} = x_{i,j} - \frac{1}{n \times m} \sum_{i=0}^{m-1} \sum_{j=0}^{n-1} x_{i,j} = x_{i,j} - \overline{X_{i,j}} \tag{7}$$

$$\begin{aligned} Y_{i+u,j+v} &= y_{i+u,j+v} - \frac{1}{n \times m} \sum_{i=0}^{m-1} \sum_{j=0}^{n-1} y_{i+u,j+v} \\ &= y_{i+u,j+v} - \overline{Y_{i+u.j+v}} \end{aligned} \tag{8}$$

where $x_{i,j}$, $y_{i,j}$ are the original gray intensity of each point in the model and target image respectively, $\overline{X_{i,j}}, \overline{Y_{i,j}}$ are the mean of each image inside their respective "block", u, v are coordinates of the model image block, and $R(u,v)$ is the object function between the model block and the target block. The shift $s = (s_x, s_y)$ is estimated by finding the peak of the objective function.

The shift between the model image and the target image can be denoted as,

$$s = s_n + s_\Delta \tag{9}$$

where $s_n = (n_x, n_y)$ is the integer shift and $s_\Delta = (\varepsilon_x, \varepsilon_y)$ is the sub-pixel shift. If the evaluation step is one pixel, s_n can be obtained from,

$$R(n_x, n_y) = \max\{R(u,v)\} \tag{10}$$

However, the peak position above can only be solved with pixel-level accuracy, and it is not enough in many applications, especially in micro/nano manipulation. In order to attain the sub-pixel resolution, a quadratic curve fitting around the peak s_n is usually used to estimate the sub-pixel shift s_Δ as follows,

$$f(x) = a_x x^2 + b_x x + c_x \tag{11}$$

$$f(y) = a_y y^2 + b_y y + c_y \tag{12}$$

where a_x, b_x, c_x, a_y, b_y, c_y are coefficients of the quadratic curves along x and y axis, which are the parameters to be estimated. The shift $s_\Delta = (\varepsilon_x, \varepsilon_y)$ is estimated by finding the peak of the following $f(x)$ and $f(y)$.

$$\widetilde{\varepsilon_x} = \max(f(x)) = 2a_x x + b_x \tag{13}$$

$$\widetilde{\varepsilon_y} = \max(f(y)) = 2a_y y + b_y \tag{14}$$

If three points are used, the estimation results of the shifts can be denoted,

$$\widetilde{\varepsilon_x} = \frac{R'(-1,0) - R'(1,0)}{2[R'(-1,0) + R'(1,0) - 2R'(0,0)]} \tag{15}$$

$$\widetilde{\varepsilon_x} = \frac{R'(0,-1) - R'(0,1)}{2[R'(0,-1) + R'(0,1) - 2R'(0,0)]} \tag{16}$$

where $R'(u - n_x, v - n_y)=R(u, v)$.

However, the precision of the method mentioned above is usually low as it has no additional points to be used. Thus, one simple and effective way to improve the estimation precision is to properly increase the interpolated points and the order of the fitting polynomials.

2.2 Improved block matching algorithm

2.2.1 The searching region

As is known, the searching region is the main factor which influences the computational cost and affects the performance of the BMA. Thus, in this section, with respect to an image sequence, an improved method is proposed to reduce the searching region effectively.

Since our aim here is to estimate the shift in the image sequence, and generally the motion is small between two neighbor images, it is not necessary to calculate $R(u, v)$ with blocks throughout the whole image. Furthermore, the computation procedure not only increases the computational burden but also adds some opportunities of wrong matching when the texture or gray level of the target image is very similar.

Assuming that the largest shift is known, our proposed new BMA can be denoted as following steps,

- First, define the initial position. Since the shift between every two images is very small, it is reasonable to take the position of the block in the model image as the initial position in the target image.
- Second, define the maximal distance that the block can move in target image as $step_{max\_x}$ and $step_{max\_y}$, where $step_{max\_x}$ and $step_{max\_y}$ can be determined through experience.
- Third, move step by step along x and y axis and calculate $R(u, v)$ at each step. During this process, the total distance along x axis should be always equal or lesser than $step_{max\_x}$, the total distance along y axis should be always equal or lesser than $step_{max\_y}$.
- Finally, find out the minimal or maximal value of the object function and the corresponding moving steps.

It is clear that using the improved algorithm the computational burden can be greatly reduced because of the reduced searching region. Besides, if the whole image is similar in texture and gray level, or the block is very small, the improved method can also effectively decrease casual matching errors.

2.2.2 The block size

In BMA, the size of a block is another important factor to influence the matching precision and the complexity. If the block is too small, the matching information is limited and it would influence the matching precision greatly. While if the block is too large, the deformation during movement cannot be omitted. Besides, the larger block can result in a large computational burden. Therefore, to research the relationship between the block size and the object function is meaningful.

A standard grid used as the target object is shown in Fig.2, the magnification of the microscope's objective lens is $60\times$, which means that the underlying image is very smooth. The black squares designate the "blocks". First, we tested the matching error of different object functions on different block sizes. The results are shown in Fig.3, where the vertical axis denotes the estimation errors, with unit of pixel; the horizontal axis denotes the block size, with unit of pixels.

1. The estimation error of NCC is sensitive to the block size and has no obvious rule. Therefore if one wants to estimate the sub-pixel shift between two images by using NCC, the block size should not be the main regulating parameter.
2. The estimation error of LSAD is sensitive to the block size too, the larger the block size is selected, the smaller the estimation errors. Thus, with respect to the LSAD sub-pixel estimation, it is often beneficial to increase the block size.
3. As far as other functions are concerned, including SSD and ZSAD, they have the similar estimation results which can achieve much smaller errors comparatively, and they are not sensitive to block size as NCC and LSAD.
4. As far as the standard grid block, the most appropriate size is 50×50 pixels, which includes an integral corner of a white grid and a little black background. It proves this kind of block both can express the internal difference of the block approximately and can eliminate the coupling between x and y direction.

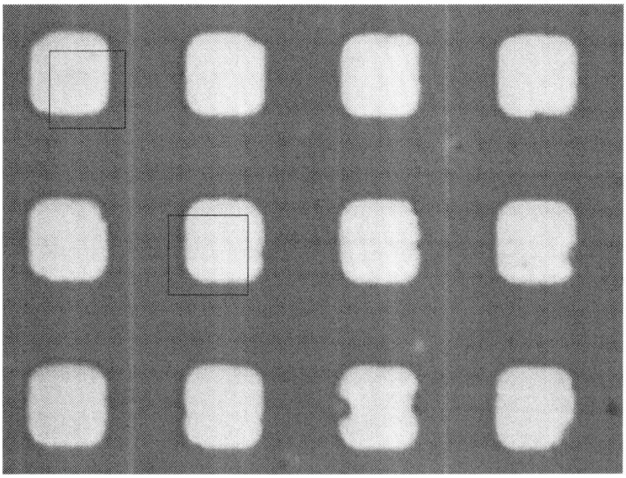

Fig. 2. The standard grid micrograph

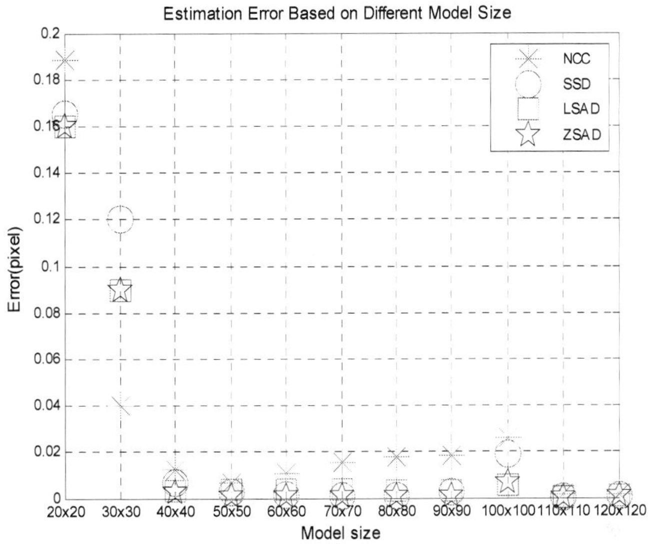

Fig. 3. Estimation errors with different block size

2.2.3 The sub-pixel fitting precision

Third, it is well known that estimation error of the curve fitting may become smaller when the order of a fitting polynomial increases. In this experiment, the comparison of the sub-pixel shift estimation error among different fitting polynomials including quadratic curve, cubic curve, quartic curve and spline curve, was conducted. Here, we used the same five fitting points, where the middle point is the peak and the other four points are on its both sides evenly. The main results are shown as Fig.5 to Fig.8, from which the following results can be obtained,

1. No matter what objective functions are used, the estimation error of the quadratic curve is always larger than the original method because of redundancy. While due to high smoothness, the spline curve can achieve an optimal result for all objective functions.
2. For NCC and SSD, the sub-pixel estimation is better enough when the fitting function is the cubic curve. That means, if the higher order fitting function is selected, the improvement in the estimation precision is unclear compared with the introduced computational burden. As far as the LSAD and ZSAD are concerned, the estimation precision improvement of the quartic fitting function is obvious. Thus, it can be concluded that the objective function is the main factor to decide the order of fitting equation.
3. Since the computational burden has been greatly reduced by using the new BMA proposed in section 2.2.1, much higher precision can be achieved by using higher order fitting functions and larger number of fitting points. That can be properly selected based on the preceding conclusions in real applications.

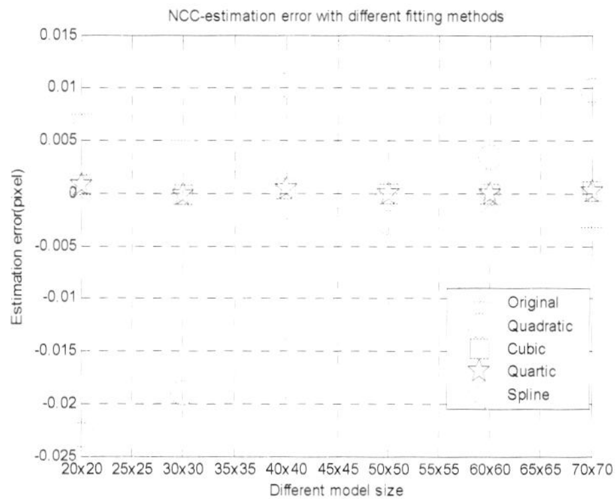

Fig. 4. Estimation errors with NCC

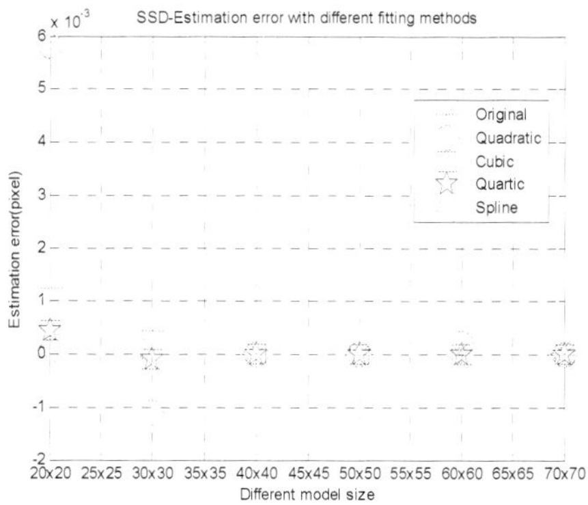

Fig. 5. Estimation error with SSD

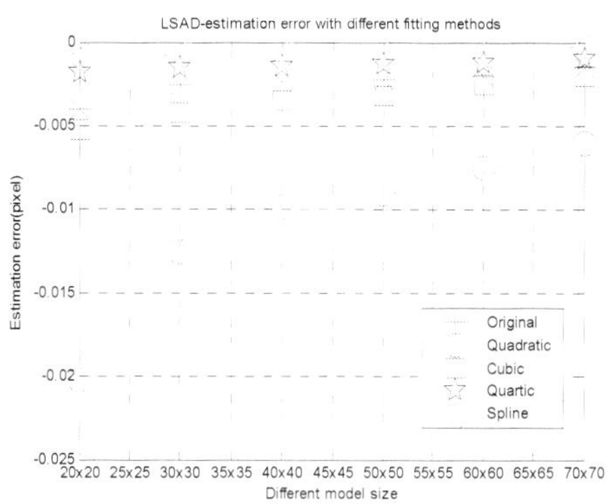

Fig. 6. Estimation error with LSAD

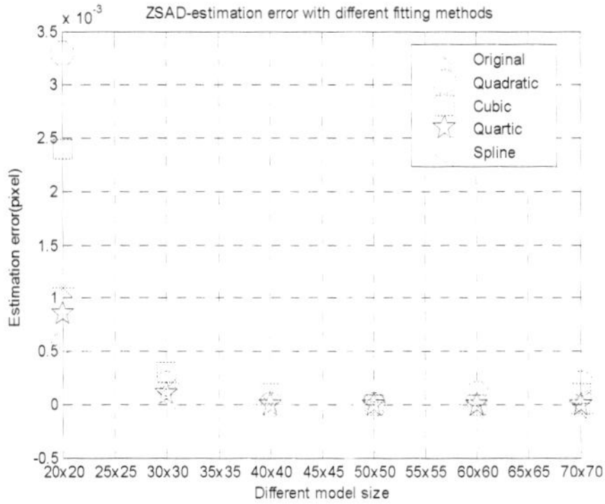

Fig. 7. Estimation error with ZSAD

Based on the theory and experiment results in 2.2, if we want to attain a matching result with high precision and low computational task using the grid block, the following parameters should be chosen: the reduced searching region, ZSAD object function, 50×50 pixels block size and the quartic sub-pixel fitting function.

2.3 The driving characteristic of a piezoelectric actuator

In order to validate the precision of our improved BAM practically, we used the PI nano platform to control the motion of the standard grid and calculated it with our method. Because the platform can output the nano motion, the shift of the grid is known. The KH-7700 microscope was used to capture the images of the grid, and the shift of each step was 50nm. Since the horizontal pixel is 57.47nm, the practical calculation result should be 0.89 pixels. Fig.9 is the result of the integral shift and Fig.10 is the shift of sub-pixel where the vertical axis denotes the motion, with unit of pixel, and the horizontal axis denotes the shift steps; the line with "*" is the true movement and the line with "o" is the calculation movement.

From Fig.9-10, we can see that the precision of our improved sub-pixel motion measurement method is very high, the integral pixel measurement is exactly equal to the true value and the sub-pixel measurement result is close to the true value. Therefore, the method can be used to measure the practical shifts in micro/nano manipulation.

Then, the sub-pixel block matching method of displacement measurement based on computer vision was used to measure the driving characteristic curve of a piezoelectric actuator practically. The result was shown in Fig.11, where the vertical axis denotes the motion, with unit of nm, and the horizontal axis denotes the driving voltage, with unit of V. Fig.11(a) is the driving curve when the voltage increases to 200V and then decreases to 0V smoothly; Fig.11(b) is the driving curve when the voltage increases to 150V and then decreases to 0V; Fig.11(c) is the driving curve when the changing routine of the voltage is 0V-200V-0V, 0V-150V-0V, 0V-100V-0V, 0V-50V-0V; Fig.11(d) is the driving curve when the changing routine of the voltage is 0V-200V-0V, 0V-160V-0V, 0V-120V-0V, 0V-80V-0V, 0V-40V-0V.

The measurement results of the piezoelectric actuator driving characteristic are consistent with the physics analysis. Furthermore, the proposed method, which is simple in manipulation and credible in measurement results, satisfies the requirement of the micro/nano measurement with high precision.

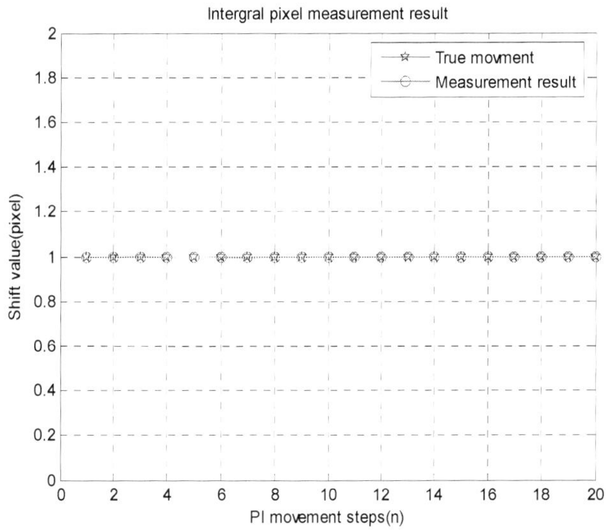

Fig. 8. The integral pixel measurement result

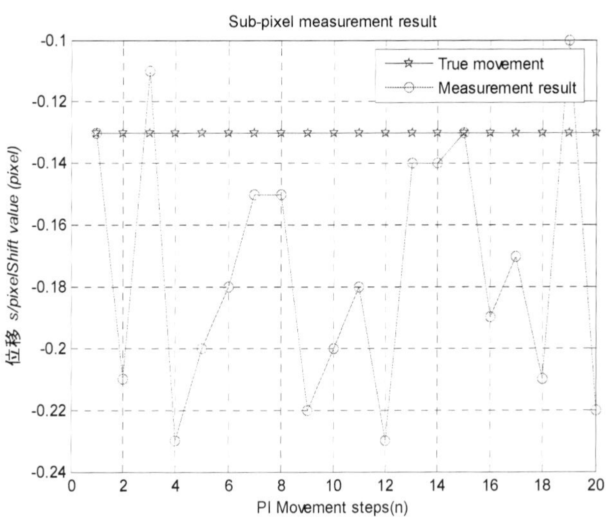

Fig. 9. The sub- pixel measurement result

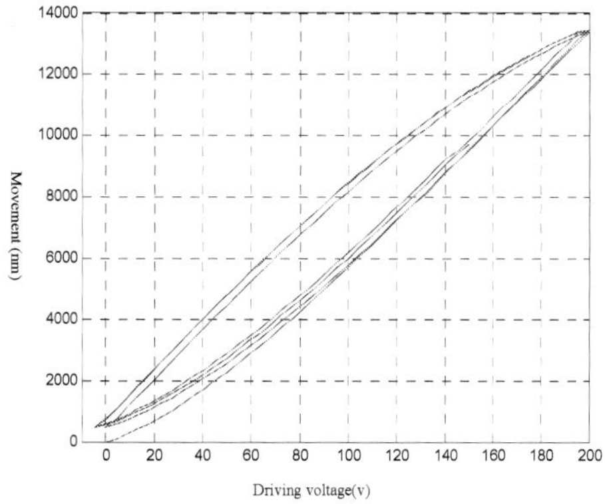

(a) 0V-200V-0V

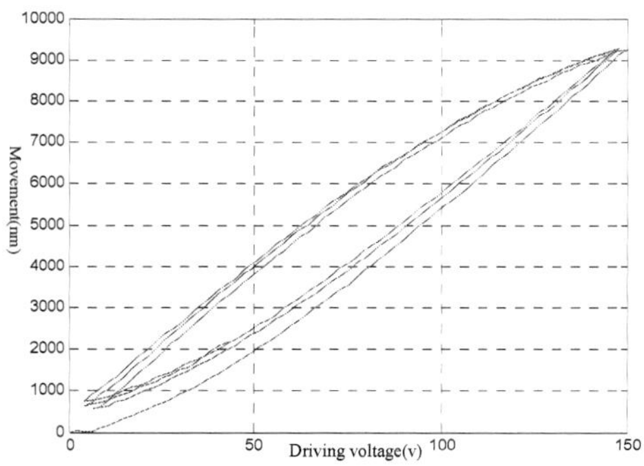

(b) 0V-150V-0V

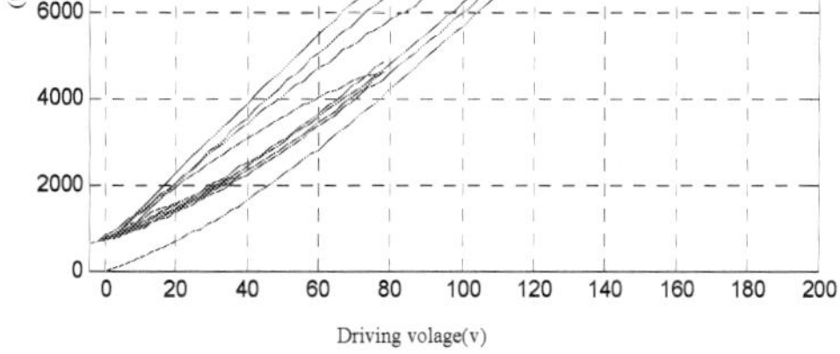

(c) 0V-200V-0V, 0V-150V-0V, 0V-100V-0V, 0V-50V-0V

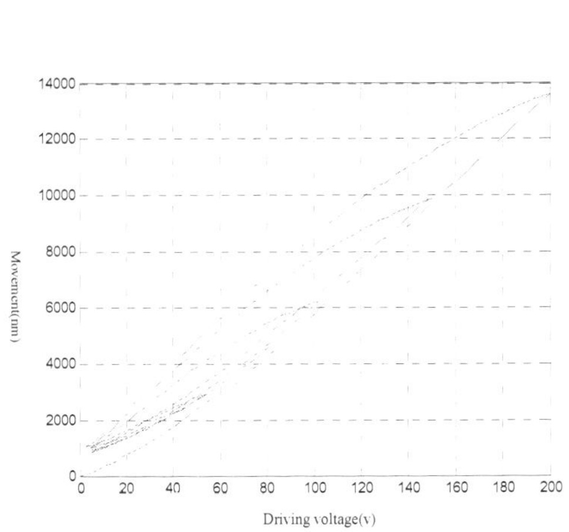

(d) 0V-200V-0V, 0V-160V-0V, 0V-120V-0V, 0V-80V-0V, 0V-40V-0V

Fig. 10. The driving characteristic of a piezoelectric actuator

3. 3D reconstruction

Shape, or depth profile, reconstruction, is based on measurement depth information from 2D images, and now it has been widely used in many fields, such as medicine, robotics, and remote sensing.

All existing DFD algorithms can be divided into two kinds, local DFD algorithm and global DFD algorithm. In local DFD, a window around every pixel point is predefined, and the point's blurring is defined as that of the window (Pentland,1987; Vinay & Subhasis,2007). However, the difficulty of selecting proper size of window is a well known disadvantage of DFD algorithm, because there is a trade-off between having a window that is as large as possible to average out noise, but as small as possible to guarantee that within it (Ens &Lawrence ,1993; Nair & Stewart,1992). As far as global DFD is concerned, its main idea is completely different with local DFD algorithm since it works on the entire image without information of its radiance, or the appearance of the surfaces, and depth. Therefore, it is necessary to construct the depth model and the radiance model simultaneously (Favaro et al 2008, 2003 2002). This, however, will bring the problem of huge computation cost. A general method to solve this problem is to simplify the imaging model, for example, assuming the scene contains "sharp edges", that is, there are discontinuities in the scene (Asada et al 1998). Another way is to use a cubic function or structure light to approach the radiance (Nayar et al 1996; Lagnado & Osher,1997). Unfortunately, both local DFD and global DFD are on the basis of attaining two defocused images with different camera parameters which may destroy the camera drastically if the camera's amplification level is high.

In this section a novel DFD method with single fixed optical microscope is proposed to reconstruct the shape of samples on micro/nano scale. In the method, the blurring image model is constructed with the relative blurring and the diffusion equation, and the relation between depth and blurring is discussed from four aspects. The method proposed needs only one microscope with unchanged camera parameters, so the reconstruction process is very simple. The experiments and error analysis results show that it can reconstruct shape on micr/nano scale.

3.1 The imaging model for defocus

In the defocus imaging model, a defocused image can be theoretically considered as the summation of some defocused points, and this process can be denoted by the following convolution function normally:

$$E(x,y) = I(x,y) * h(x,y) \tag{17}$$

where $E(x, y)$ and $I(x, y)$ are the defocused image and the focused image, respectively, $h(x, y)$ is the point spread function.

When the point spread function is approximated by a shift-invariant Gauss function, the imaging model in Eq.(17) can be formulated in terms of the isotropic heat equation:

$$\begin{cases} \dot{u}(x,y,t) = a\Delta u(x,y,t) \; a \in [0,\infty) \; t \in (0,\infty) \\ u(x,y,0) = E(x,y) \end{cases} \tag{18}$$

where a is the diffusion coefficient, $\dot{u} \doteq \frac{\partial u}{\partial t}$, " Δ "denotes the Laplacian operator, $\Delta u = \frac{\partial^2 u(x,y,t)}{\partial x^2} + \frac{\partial^2 u(x,y,t)}{\partial y^2}$.

If the depth map is an equifocal plane, a is constant. Otherwise, a is shift-variant, and the diffusion equation becomes:

$$\begin{cases} \dot{u}(x,y,t) = \nabla \cdot (a(x,y)\nabla u(x,y,t)) \; t \in (0,\infty) \\ u(x,y,0) = r(x,y) \end{cases} \tag{19}$$

where " ∇ "denotes the gradient operator and " $\nabla\cdot$ "is the divergence operator, $\nabla = \left[\frac{\partial}{\partial x} \; \frac{\partial}{\partial y}\right]^T, \nabla\cdot = \frac{\partial}{\partial x} + \frac{\partial}{\partial y}$.

It is also easy to verify that the variance σ is related to the diffusion coefficient a via:

$$\sigma^2 = 2ta \tag{20}$$

Suppose there are two images $E_1(x, y)$ and $E_2(x, y)$ for two different focus setting, also, $\sigma_1 < \sigma_2$ (that is, $E_1(x, y)$ is more defocused than $E_2(x, y)$), then $E_2(x, y)$ can be written as:

$$\begin{aligned} E_2(x,y) &= \iint \frac{1}{2\pi\sigma_2^2} \exp\left(-\frac{(x-u)^2+(y-v)^2}{2\sigma_2^2}\right) r(u,v)\,du\,dv \\ &= \iint \frac{1}{2\pi\Delta\sigma^2} \exp\left(-\frac{(x-u)^2+(y-v)^2}{2\Delta\sigma^2}\right) E_1(u,v)\,du\,dv \end{aligned} \tag{21}$$

where $\Delta\sigma^2 \triangleq \sigma_2^2 - \sigma_1^2$ is called the relative blurring(Favaro et al, 2008). So Eq.(18) can be written as:

$$\begin{cases} \dot{u}(x,y,t) = a\Delta u(x,y,t) \; a \in [0,\infty) \; t \in (0,\infty) \\ u(x,y,0) = E_1(x,y) \end{cases} \tag{22}$$

Eq.(19) becomes:

$$\begin{cases} \dot{u}(x,y,t) = \nabla \cdot (a(x,y)\nabla u(x,y,t)) \; t \in (0,\infty) \\ u(x,y,0) = E_1(x,y) \end{cases} \tag{23}$$

When the time-shifted is Δt ,the solution of the diffusion equation is $u(x,y,\Delta t) = E_2(x,y)$,and Δt can be defined as,

$$\Delta\sigma^2 = 2(t_2 - t_1)a \doteq 2\Delta t a \tag{24}$$

Thus, the relation between the relative blurring and the depth map can be denoted as:

$$\Delta\sigma^2 = \gamma^2(b_2^2 - b_1^2) \tag{25}$$

where γ is a constant between the blurring radius and the blurring degree, b_i (i=1,2) is the radius of the burring round :

$$b = \frac{Dv}{2}\left|\frac{1}{f} - \frac{1}{v} - \frac{1}{s}\right| \tag{26}$$

where s denotes depth of the blurring point and D denotes the radius of the lens.

3.2 The new shape reconstruction method

Suppose $E_1(x, y)$, whose depth map is $s_1(x, y)$, is the defocused image attained before depth variation, and $E_2(x, y)$ is another defocused image attained after depth variation, in this section, we will propose a new shape from defocus method in which the depth map $s_2(x, y)$ is attained through a depth change Δs, and the main theory is shown in Fig.12.

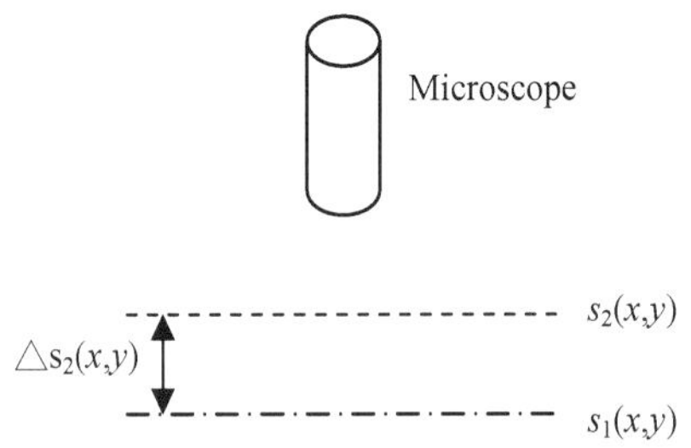

Fig. 11. The main theory of our method

Suppose s_0 is the focus depth, and $s_1(x,y)-s_2(x,y)=\Delta s(x,y)$. Based on the diffusion equations in section 2, the following functions can be given:

$$\begin{cases} \dot{u}(x,y,t)=\nabla\cdot(a(x,y)\nabla u(x,y,t))\ t\in(0,\infty) \\ u(x,y,0)=E_1(x,y) \\ u(x,y,\Delta t)=E_2(x,y) \end{cases} \quad (27)$$

where the relative blurring can be denoted as:

$$\begin{aligned} \Delta\sigma^2(x,y) &= \gamma^2(b_2^{\,2}(x,y)-b_1^2(x,y)) \\ &= \frac{\gamma^2 D^2 v^2}{4}\left[\left(\frac{1}{f}-\frac{1}{v}-\frac{1}{s_2(x,y)}\right)^2-\left(\frac{1}{f}-\frac{1}{v}-\frac{1}{s_1(x,y)}\right)^2\right] \\ &= \frac{\gamma^2 D^2 v^2}{4}\left[\left(\frac{1}{s_0}-\frac{1}{s_2(x,y)}\right)^2-\left(\frac{1}{s_0}-\frac{1}{s_1(x,y)}\right)^2\right] \end{aligned} \quad (28)$$

Define: $k=\frac{4\Delta\sigma^2}{\gamma^2 D^2 v^2}+\left(\frac{1}{s_0}-\frac{1}{s_1(x,y)}\right)^2$, thus the desired depth map is:

$$s_2(x,y)=1/(\frac{1}{s_0}\pm\sqrt{k}) \quad (29)$$

In real applications, it is reasonable to discuss the following four cases when the distance between the sample and the microscope is becoming shorter and shorter.

a. $s_1>s_2>s_0$

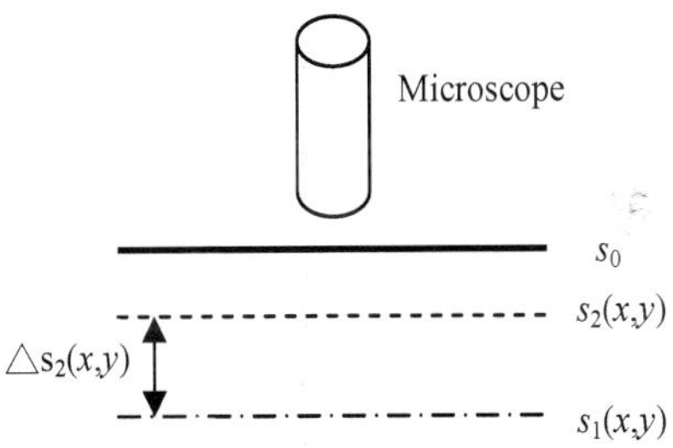

Fig. 12. The theory of case A

In this case, $E_1(x, y)$ and $E_2(x, y)$ are on the large side of s_0, and $E_1(x, y)$ is more defocused than $E_2(x, y)$, so it is backward diffusion process from $E_1(x, y)$ to $E_2(x, y)$, that is, the diffusion efficient a is negative. The theory is shown as Fig. 13 and the finial depth can be denoted as:

$$s_2(x,y)=1/(\frac{1}{s_0}-\sqrt{k}) \quad (30)$$

b. $s_0 > s_1 > s_2$

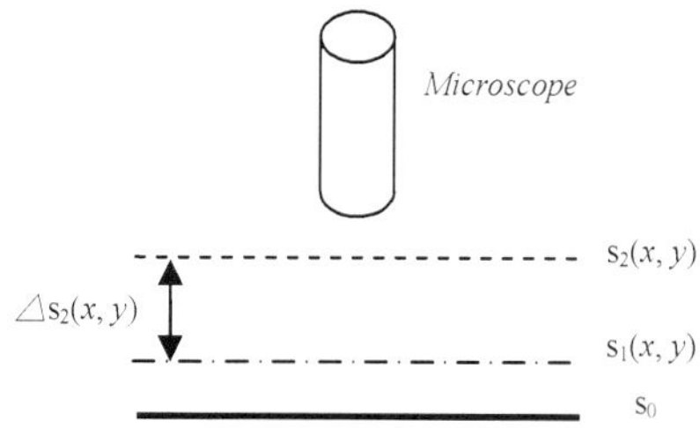

Fig. 13. The theory of case B

As is shown in Fig.14, here, $E_1(x, y)$ and $E_2(x, y)$ are on the small side of s_0, $E_1(x, y)$ is less defocused than $E_2(x, y)$, so it is afterward diffusion from $E_1(x, y)$ to $E_2(x, y)$, and the diffusion efficient a is positive. The finial depth can be denoted as:

$$s_2(x, y) = 1 / (\frac{1}{s_0} + \sqrt{k}) \tag{31}$$

c. $s_1 > s_0$, $s_2 < s_0$, $(s_0 - s_2) < (s_1 - s_0)$

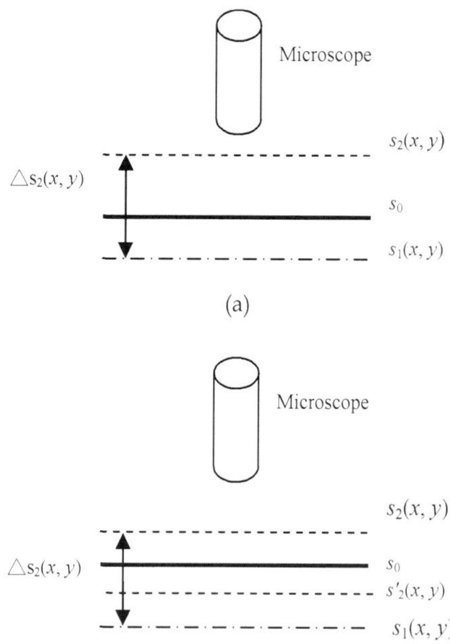

(a)

(b)

Fig. 14. The theory of case C

This case is a little more complicated than the first two scenarios. $E_1(x, y)$ is more defocused than $E_2(x, y)$, but they are not on the same side of s_0. Suppose $s'_2(x, y)$ is the symmetrical depth of $s_2(x, y)$ about s_0, the process can be transferred from Fig.15(a) to Fig.15(b), and the finial depth can be denoted as:

$$s'_2(x,y) = 1/(\frac{1}{s_0} - \sqrt{k}) \tag{32}$$

$$s_2(x,y) = s_0\text{-}(s'_2(x,y) - s_0) = 2s_0\text{-}s'_2(x,y) \tag{33}$$

d. $s_1 > s_0$, $s_2 < s_0$, $(s_0 - s_2) > (s_1 - s_0)$

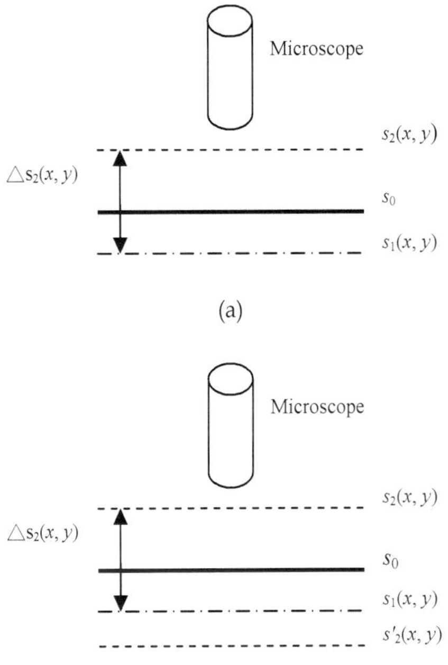

Fig. 15. The theory of case *D*

Here, $E_1(x, y)$ is less defocused than $E_2(x, y)$, and they are not on the same side of s_0. Suppose $s'_2(x, y)$ is the symmetrical depth of $s_2(x, y)$ about s_0, the process can be transferred to Fig.16(b), and the finial depth can be denoted as:

$$s'_2(x,y) = 1/(\frac{1}{s_0} + \sqrt{k}) \tag{34}$$

$$s_2(x,y) = s_0\text{-}(s'_2(x,y) - s_0) = 2s_0\text{-}s'_2(x,y) \tag{35}$$

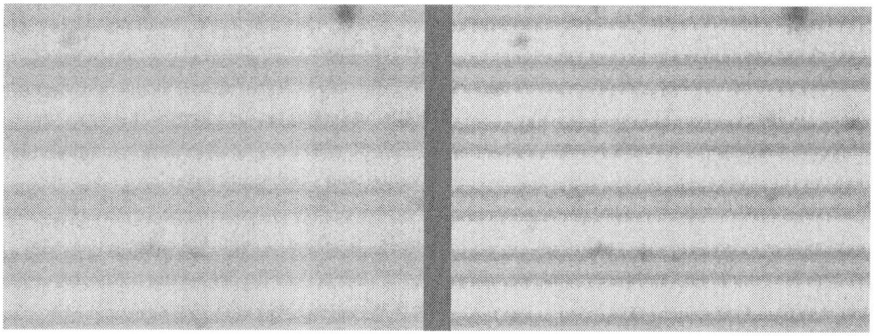

Fig. 18. The defocused images

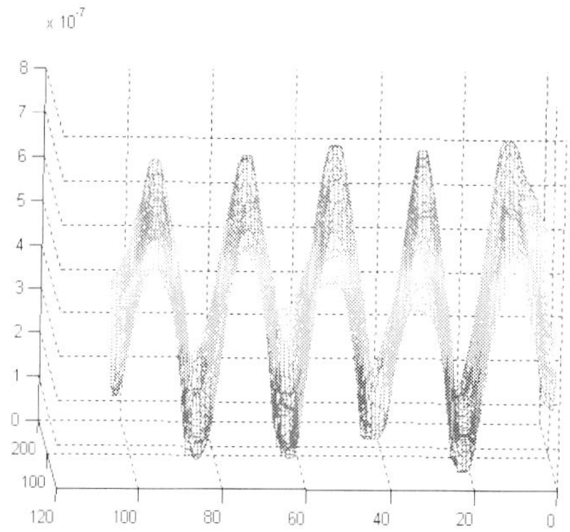

Fig. 19. The constructed 3D shape

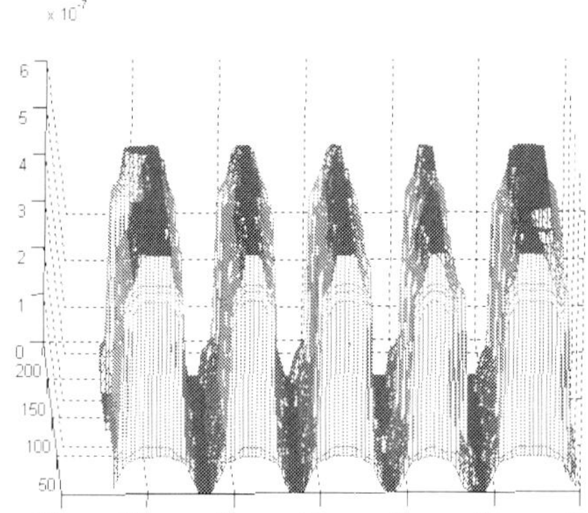

Fig. 20. The true 3D shape

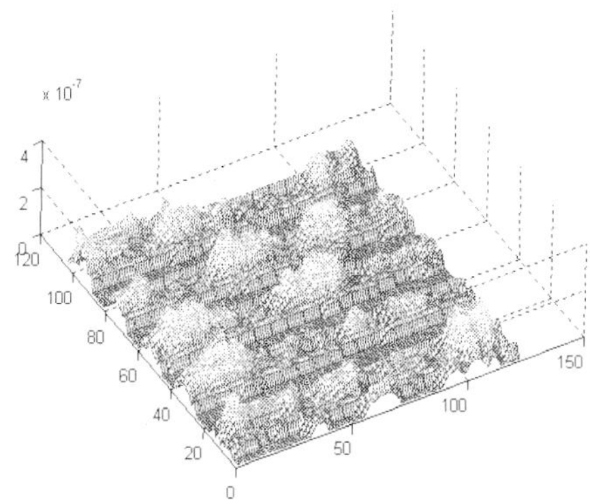

Fig. 21. The error map

From Fig. 10-21, we can see that the new algorithm can attain good results in constructing nano grid shape, and the precision of the proposed method is very high. The mean square error of the whole image is equal to -0.048, and the average error is -9.26 nm.

Secondly, we tested our algorithm on the grid region of 120×50 pixels, the results are shown as Fig.22- Fig.25. The mean square error is equal to -0.038 and the average error is 10.39 nm. From the figures, we can see that the error of our construction algorithm is slightly larger at the edge of the image and smaller at other region, this results from the optimization method. However, the average error is only about 2.3% and it can certainly satisfy the demand of micro/nano magnification.

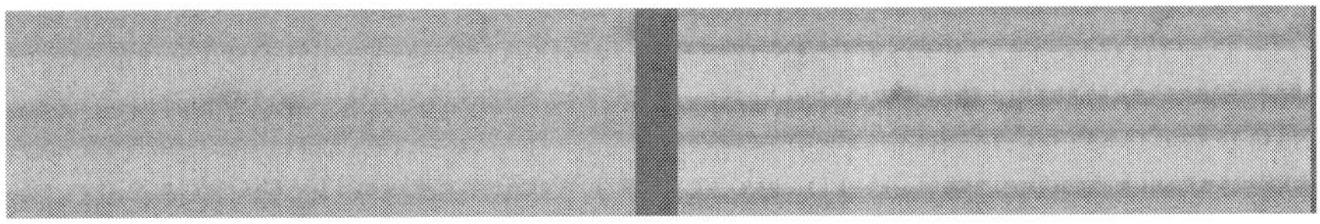

Fig. 22. The defocused images

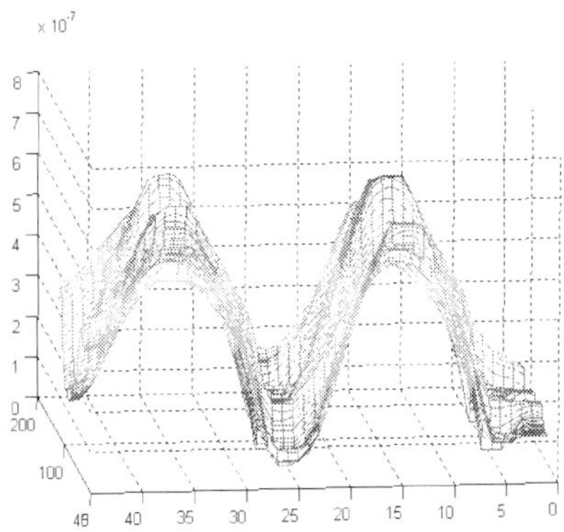

Fig. 23. The constructed 3D shape

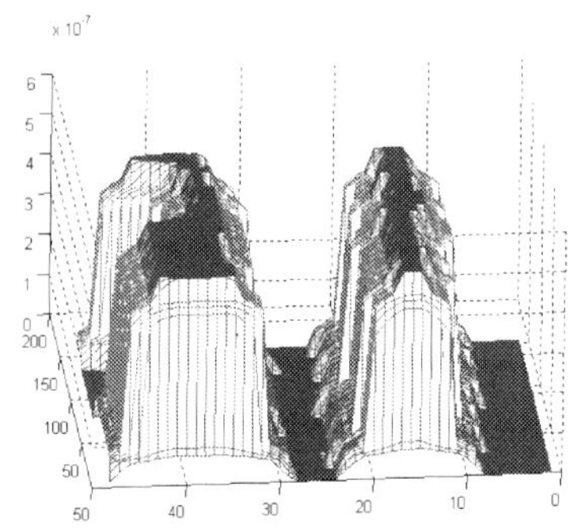

Fig. 24. The true 3D shape

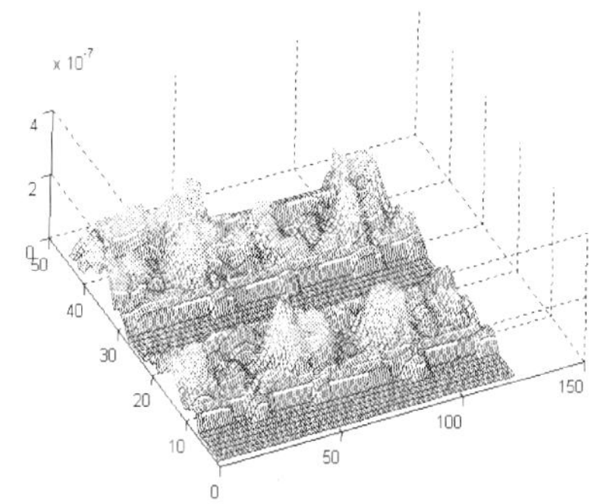

Fig. 25. The error map

3.3.2 Shape reconstruction of the AFM cantilever

The raise height of the cantilevers was controlled by the Iphysik Instrumente(PI) nano platform. Furthermore, we provided the performance of the algorithm on three kinds of raise height: 500nm, 300nm and 100nm. The theory of the experiment is shown as Fig.26.

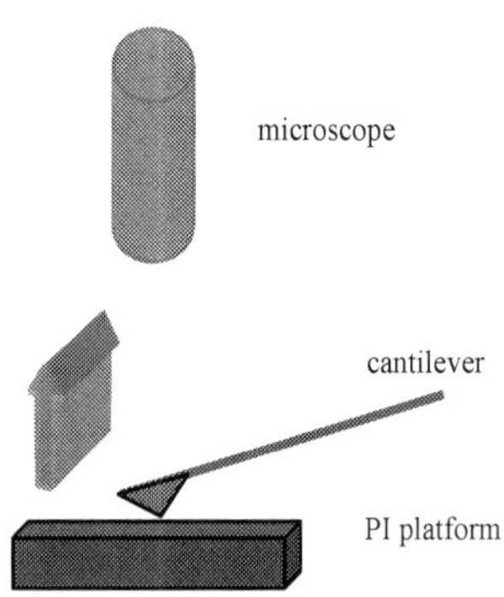

Fig. 26. The experiment theory of reconstruction AFM cantilever

Firstly, the experiment using the conductive AFM cantilever was conducted. Fig.27 is two defocused images, in which the left is the image before variation and the right is that after variation; Fig. 28(a)- Fig.28(c) are the constructed 3D shapes of the bended cantilever when the PI platform rises 500nm, 300nm and 100nm.

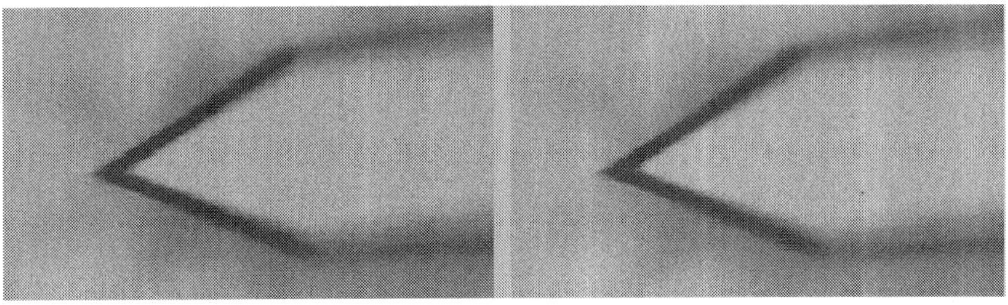

Fig. 27. The defocus images of the conductive cantilever

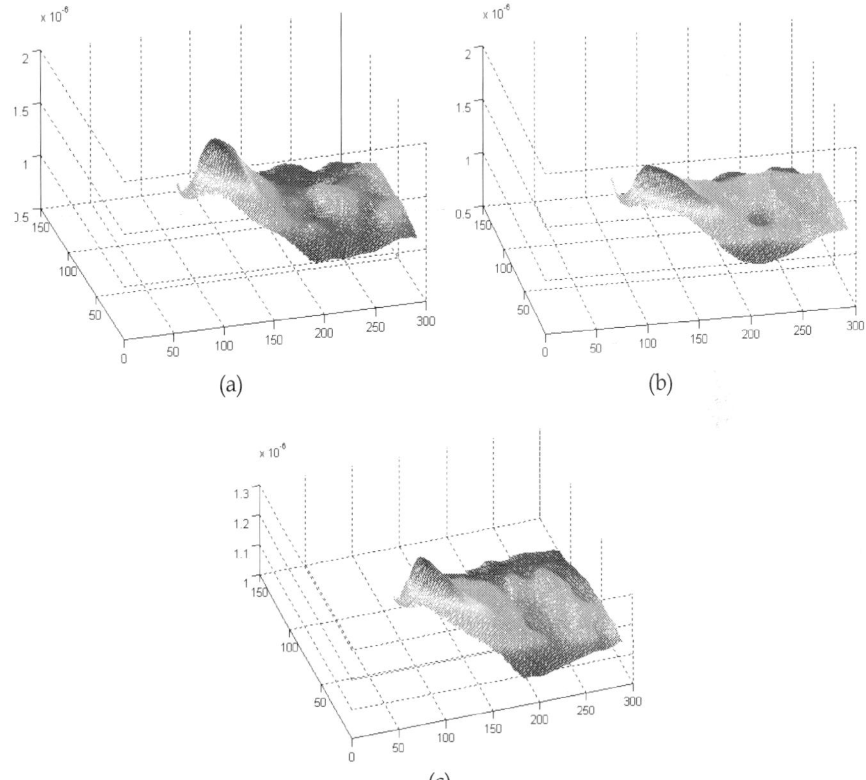

Fig. 28. The constructed 3D shape for 500nm, 300nm and 100nm

From Fig. 28., we can see that when the PI platform rises, the top end of the conductive cantilever bends obviously, and the deflection decreases gradually from the top end to the trailing end until close to a steady value; the bended degree is a monotonic function with the raise height. In order to contrast the bended precision, we choose the section image of them on the same position, and show them in Fig.29. From it, we can see that the deflection height is proportionately increases when the platform rises, and the height difference between the top end and the steady value is exactly equal to the raise height of the PI platform.

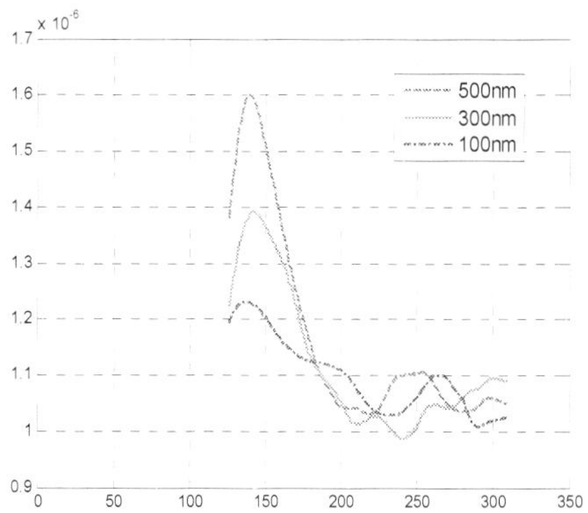

Fig. 29. The contrast of the conductive cantilever

Secondly, the experiment using the triangle cantilever was conducted. Fig.30 is two defocus images, in which the left is the image before variation and the right is that after variation; Fig.31(a)- Fig.31(c) are the constructed 3D shapes of bended cantilever when different raise height is 500nm, 300nm and 100nm; Fig.31 is the contrast image of these three sections. From them we can get the same conclusions as the last experiment, but the sensitivity of the triangle cantilever is lower because the reconstructed shapes are a little rough.

From these experiments, we can see that, regardless of the cantilever shape, our algorithm all can reconstruct the global bended shape exactly with only two defocused images. The following conclusion can be given:

1. The most obvious bend of the cantilever concentrates on the region near to the tip, it is reasonable because when the PI platform works up, the stress all concentrates on the tip due to our experiment theory.
2. The cantilever's original shape, material and illumination can influence the reconstruction result to some extent. For example, the conductive cantilever is thinner than the triangle cantilever, and the shape reconstruction of it is smoother due to its higher sensitivity; the black edge of the cantilever results in a little error in the result.
3. The raise height is larger, the calculation result is exacter, and the reconstruction image is smoother
4. No matter how much the raise height is, the reconstruction height tends to be steady finally. Furthermore, the height difference between the maximum value and the original value is equal to the raise height of the PI platform.

Fig. 30. The defocus images of the triangle cantilever

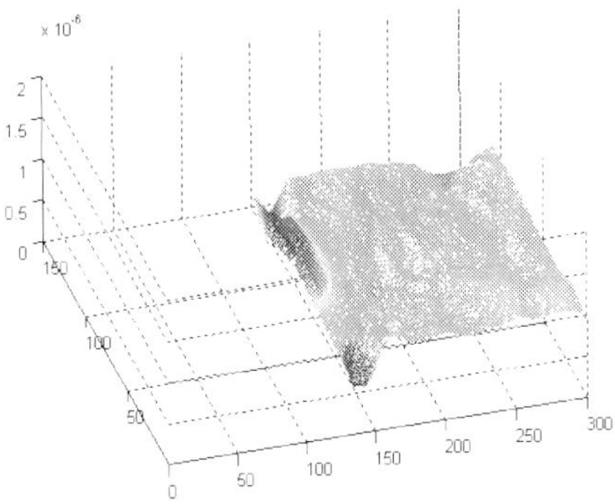

(a)

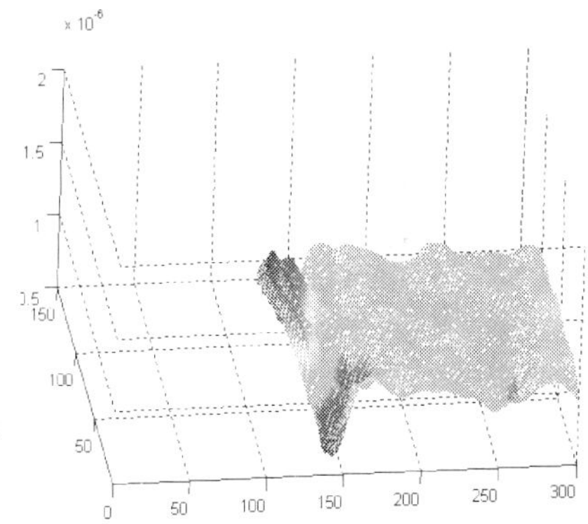

(b)

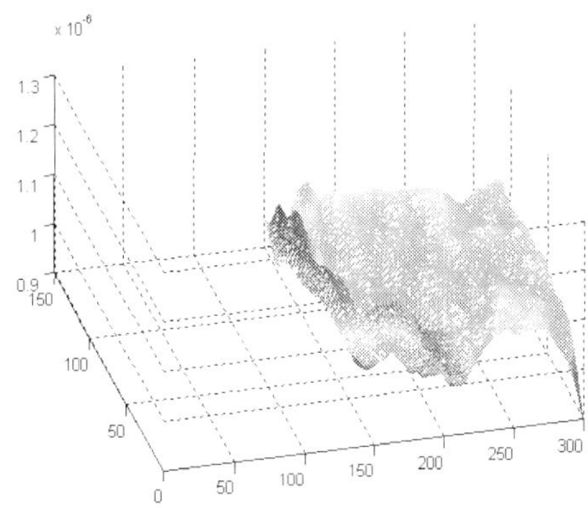

(c)

Fig. 31. The constructed 3D shape for 500nm, 300nm and 100nm

Singh, A.(1990). *An estimation-theoretic framework for image flow computation*, pp. 168-177, Proceedings of 3rd International Conference on Computer Vision, Osaka, Japan., December 4-7,1990.

Subbarao M.; Surya G.(1994). *Depth from defocus: A spatial domain approach*, International Journal of Computer Vision, Vol.13, No.3, pp. 271-294.

Teresa C. S. A; João M. R. S. T & Mário A. P. V. *Three-dimensional reconstruction and characterization of human external shapes from two-dimensional images using volumetric methods*, Computer Methods in Biomechanics and Biomedical Engineering, DOI: 10.1080/10255840903251288 (in press).

Teresa C. S. A; João Manuel R. S. T & Mário A. P. V.(2008). *3D Object Reconstruction from Uncalibrated Images using an Off-the-Shelf Camera*, Advances in Computational Vision and Medical Image Processing: Methods and Applications, pp. 117-136.

Vinay, P. N.; Subhasis C.(2007). *On defocus, diffusion and depth estimation*, Pattern Recognition Letters, Vol. 28, No.3, pp. 311-319.

Wu L. D. (1993).*Computer Vision*. Fudan University Press, 1993.

Yin C. Y.(1999). *Determining residual nonlinearity of a high-precision heterodyne interferometer*, Optical Engineering, Vol.38, No. 8, pp. 1361-1365.

24

Advanced Free Form Manufacturing by Computer Aided Systems – Cax

Adriano Fagali De Souza and Sabrina Bodziak
PROMOLDE Research Group,
Tupy Superior Institute (IST/SOCIESC), Santa Catarina, S.C.,
Brazil

1. Introduction

As part of an integrated system technology CIM (Computer Integrated Manufacturing), longed for since the 50's, the computational systems known the CAx (Computer Aided specific manufacturing phase) developed to help phases of the manufacturing of a product and its components. In order to complete integration of the chain of manufactures by computer, CAx can be integrated with other computational systems of management and planning of trials and output, such as MRP (Material Resource Planning), ERP (Enterprise Resource Planning), EDM (Electronic Document Management)/PDM (Product Dates Management).

In this context, this chapter discusses the computational CAx systems to aid the activities related to the manufacturing, in special, the manufacturing of free form geometries. Such geometries are frequently found in the industry automobile, industry of moulds and dies, aeronautic and need CAx systems for its making. Machines controlled by Computer Numerical Command (CNC) for making parts and prototypes as well as for aiding geometrical inspection, also take part on the family of CAx systems and are discussed in this chapter.

This chapter includes: i) a brief definition about free form geometries; ii) main concerns about the manufacturing process of free form geometries; iii) CAx for free form manufacturing; iv) advanced features of CAx for free form design and manufacturing.

2. A definition about free form geometries

First, a definition of free form geometry (also known as sculptured surface) has to be set: it is a non-linear and curved shape, with neither constant nor pattern curvature radius along the geometry

Initially these geometric forms were employed exclusively in situations where they were to be used such as the aeronautical and naval industry. Currently complex geometric forms are becoming popular and can be found in various components in the automobile industry; in the consumption industry; toys, packaging, electronic products , esthetical products and ergonomics. This definition is important once that any manufacturing planning depends upon its shape. For instance, rotational parts have to be turned and prismatic parts have to be milled. Free form geometry has its own particularity to be manufactured.

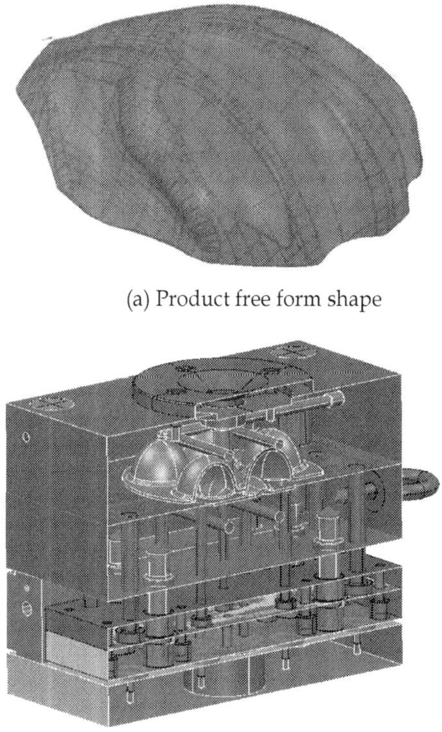

(a) Product free form shape

(b) Mould for plastic product

Fig. 1. Free form geometries

To represent computationally these sorts of geometry higher order polynomials are required. These equations were developed by Pierre Bézier, an French engineer who based on a Hermite proposal implemented the software Unisurf, in 1972. Such polynomials are known as Spline. The equation 1 is the Spline defined by Bézier. The curve is guided by a polygon (SOUZA et al, 2010). Using a mathematical software the Figure 2 was plotted.

$$p(u) = p_0\left(1 - 3u + 3u^2 - u^3\right) + p_1\left(3u - 6u^2 + 3u^3\right) + p_2\left(3u^2 - 3u^3\right) + p_3\left(u^3\right) \quad (1)$$

where:

p: is the control point of the polygon

u: parameter 0-1

This is a brief approach to understanding how to model a free form curve. There are many other equations to create a Spline curve by computer, like NURBS (non-uniform rational B-spline) and others. A 3D surface is created analogous to a curve, but using one more axes.

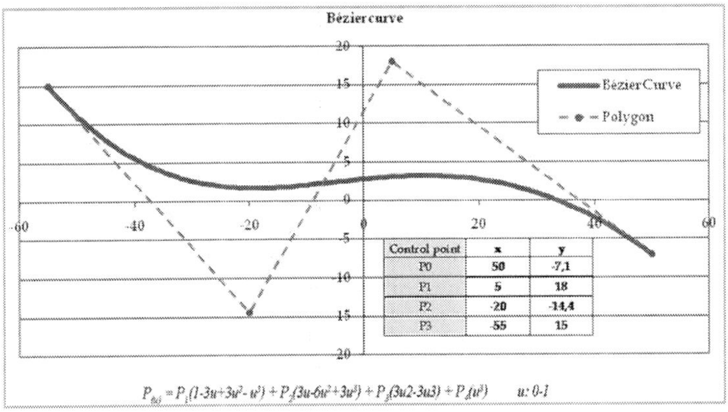

Fig. 2. A Bézier curve plotted in graphic software

It is also important to distinguish that free form geometries do not have to be associated with complex products. Many complex products have no complex shapes, for instance, a machine centre. It has many components, high precision assembling, and relatively high cost. However, in general, its components have prismatic or rotation shapes. Instead a simple bottle of pop soda has no assembly, it does not require such precision, and it is not an expensive product. However its shape are quite complex, and most of the time to manufacture complex geometries are much more expensive than simple forms, once it required high level of technology, like CAx.

3. Computer aided free form manufacturing

Due to its complexity, free form shapes cannot be machined by a conventional machines driven by human. It requires a CNC machine and the free form tool paths have to be calculated by software. After machined the geometry, geometric errors have to be accessed. Once again, due to its complexity it is not possible to make any inspection by ordinary metrology devices. Therefore, such cases require a measure machine coordinate (or any order technique to accessed geometric errors, like as laser or photometry).

So, the manufacturing steps of complex forms requires different CAx systems. Within these systems the following stand out:

- CAD (Computer Aided Design/Drafting). Software to aid the Project , design, mould industrially design the products.
- CAM (Computer Aided Manufacturing). Software to aid manufacturing activities.
- CAI (Computer Aided Inspection). Software to aid the inspection of geometric forms.
- CAE (Computer Aided Engineering). Software to aid the simulation of mechanical, strength, temperature, pressure etc.

The Figure 3 presents the integration of some CAx systems required to manufacture for free form geometries.

Besides, CAE (Computer Aided Engineering) software can be used to simulate the product mechanical requirements or the manufacturing process. However, in most of the times it is not essential for free form manufacturing such that mentioned above. You can find a detailed description about the CAx for free form modelling and manufacturing ahead in this chapter.

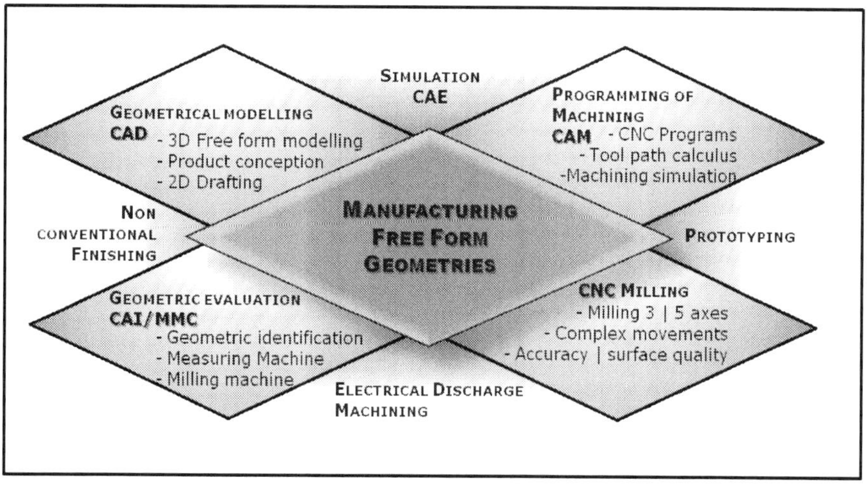

Fig. 3. Some CAx application on free form manufacturing

4. Main concerns about the manufacturing process of free form geometries

In the contemporary world the manufacture of free form geometries is carried out first by milling and electrical discharge machining (EDM) when necessary. Newer manufacturing process has been developed such as fabrication by deposition, like as the rapid prototyping, but yet the first both process are still the mainly ones and they are taken into account in this chapter.

Usually, milling is the first option due to the inconveniences of the EDM processes, because EDM it requires an extra effort to modelling the electrodes, CNC programming and milling the electrodes, time consumption for EDM machining itself, besides the EDM can injure the machined surface integrity. The EDM process is used where the milling tool cannot reach due to geometry limitation, or it may be required in specific cases. The 5 axis milling is a drive force for manufacturing free form shapes, once that the cost are reducing and the technology is getting raising.

For milling complex forms the generation of CNC programs should be fulfilled by a CAM system. The user should initially have a geometric model of CAD 3D, which should be used by CAM as a base to calculate the tool path. After the calculation the route of the tool should be simulated and subsequently coded in commands, which should be interpreted and transformed by the CNC machine in movements and actions. The Figure 4 illustrates the procedure to manufacture a free for geometry by CAM software.

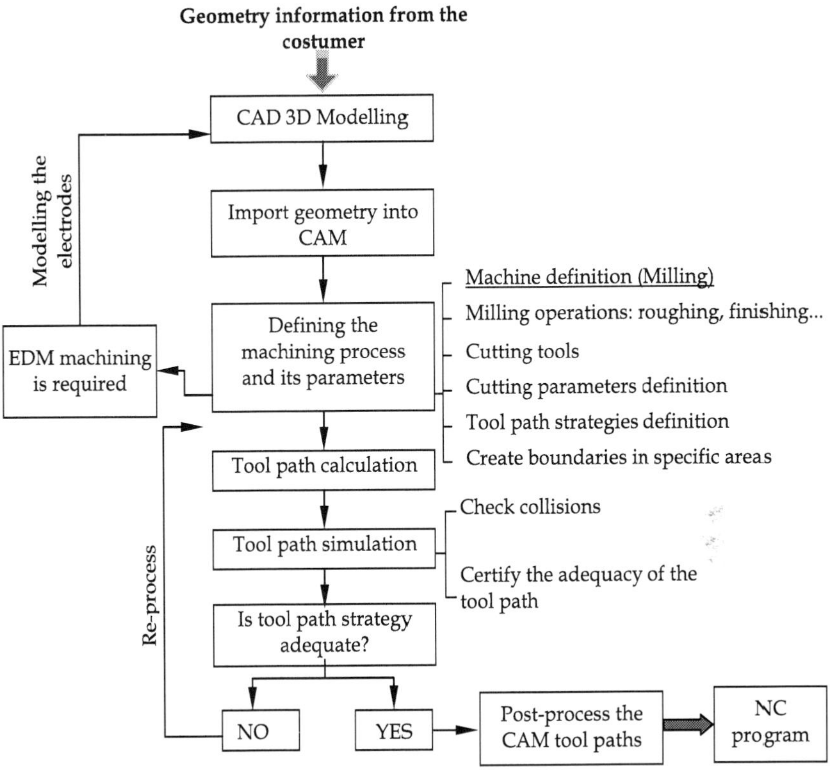

Fig. 4. Procedure to manufacturing free form geometries by CAM software

The mainly features of free form milling can be listed as following.

4.1 Roughing milling in free form geometry manufacturing

This operation can happen in two circumstances:

i. The raw material is furnished in a prismatic format. The software needs this information to calculate the extensions of the path calculated, the height, the width and the length. The geometric block, when prismatic, should be defined easily with the CAM software.

ii. The geometry of the raw material is furnished in specific format. It could be manufactured before, by foundry, or any other processes. Like this the prime material for milling should have geometry close to the final form. (Fig. 5b). In these cases the user should model the geometry form of the prime material using a CAD software. This geometry must be exported to CAM to recognize the geometry as being the block of raw material to be milled. In this way CAM calculates the path just between the block of prime material and the material being milled. This avoids unnecessary movements between the block and the tool, which economizes milling time.

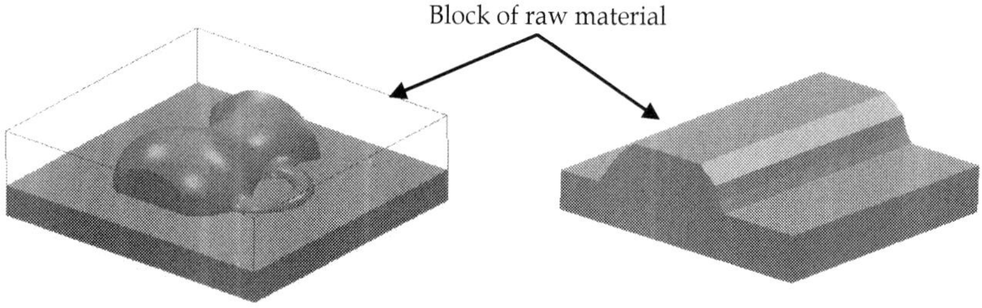

(a) Row material cubic (b) Row material near net shape

Fig. 5. Blank material options

The roughing employs movements in 2 ½ axis. The tool is positioned at a determined height in Z. After movements are executed in 2 axes for the removal of material at this level. When the removal of material at this level is finalized the tool is then positioned at a new level of Z (defined by the depth of cut), the material is removed at this plan. The process is repeated until the roughing is complete. End-mill with corner radius is the cutting tool frequently used for this operation.

4.2 Stock of material for finishing operations (stock material)

Normally some stock of material (also called thickness) is left for the finishing operation. The value of this stock of material is set in the CAM by the user, and it is an offset of the geometry to be machined. For steels machining an average of values that are employed are between 0.5 mm and 1 mm of stock that should be removed in the finishing operation. One value given to the thickness of the leftover of metal could represent a significant effort in finishing and this could prejudice the quality of the surface and also prejudice two situations: a) geometric errors caused by deflection of the tool; b) elevated surface roughness. Low stock can also prejudice the quality of the surface therefore tool can come across difficulties when carrying out properly, the material being cut. This causes passing and crushing material that should be cut. Once again each case ought to be methodically evaluated.

When milling complex shapes besides this theoretical stock, a volume of material cannot be removed from the roughing, due to topological limitations between the surface complexity and cutting tool geometry.

An extra volume of stock can be found on the degrees of material formed after roughing as illustrated in Figure 6(a). This extra volume depends of the depth of cut, the curvature of the surface and the geometry of the tool. Cutting tools with an extreme radius or spherical tools minimize but do not eliminate this condition.

To optimize roughing, tools with greater diameters are employed. In regions of corners, many times the cutting tool is limited to reach and to carry out the milling. Then again an extra volume of material is left by the roughing (Figure 6).

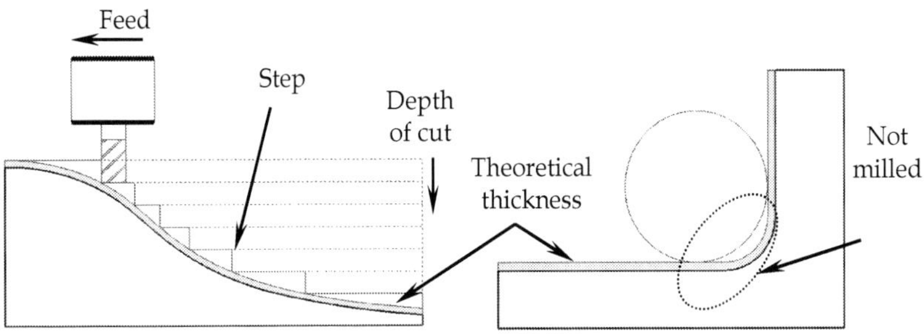

Fig. 6. Thickness to finishing

Frequently the differences of the volume of material beyond the theoretical stock is not cautiously evaluated, which has repercussions with possible problems in the milling. It is common that the machinists of the CNC machines reduce the speed during the milling process manually in some regions of the piece. To minimize this problem of semi-finishing operations have the objective to remove this excess of material and maintaining the stock as close as possible to the theory value (offset of geometry).

4.3 The tool-surface contact in free form milling

Finishing operation of machining can be considered the most important along the manufacturing free form shapes, once that implicates directly on the final product, in terms of surface integrity and geometric errors. A ball-end cutting tool is the most suitable to finish free form shapes by milling due to this geometry which permits to reach different curvature shapes. However, this tool implicates in specifics characteristics in milling free form shapes, once that the contact tool/surface alternates constantly and affects directly the milling process, in terms of cutting speed, cutting force, surface quality and so on (see also SOUZA et al, 2010).

First, the cutting speed usually cannot be calculated by nominal tool diameter. It has to be taken into account the effective diameter according to the axial cutting depth together to shape curvature. For a planar surface milled by a ball-end tool (Figure 7), the maximum effective diameter can be calculated by equation 2. But it has to be taken into account the radius of curvature of a free form geometry alternates constantly (Figure 7b). Therefore, the effective tool diameter following this oscillation and thus, the cutting speed as well. This is one of the drawbacks because the machining process became unstable.

$$R_{ef} = \sqrt{R^2 - (R - ap)^2} \qquad (2)$$

Where:

Ref = Effective radius [mm]

ap =axial depth of cutting [mm]

R = Tool radius [mm]

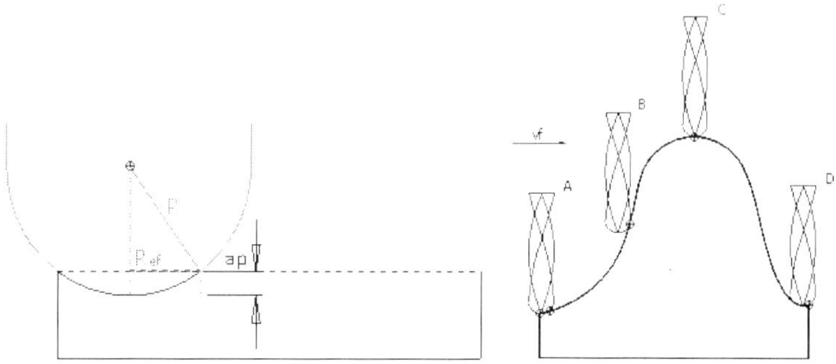

(a) effective radius (b) contact tool-surface

Fig. 7. Effective tool diameter and contact tool-surface in free form milling

$$R_{ef} = \sqrt{R^2 - (R - ap)^2} \tag{2}$$

Where:

Ref = Effective radius [mm]

ap =axial depth of cutting [mm]

R = Tool radius [mm]

Another weakness of this process is the surface roughness. First, the cutting speed is zero in the extreme centre of the cutting tool. Therefore, when the centre of the tool is in the cutting area, the tool smash and do not cut the material. This phenomenon affects negatively the surface roughness. The ball mill also leaves a material remain between the cutting passes, known as cusp high. The cusp high depends upon the axial (ap) and radial depth per cut (ae), tool diameter and surface curvature (see Figure 8).

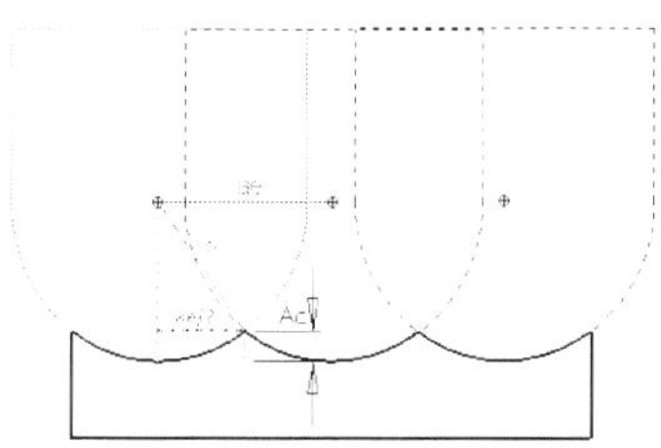

Fig. 8. Cusp high in a planar surface and contact tool-surface in free form milling

Where:

Ac= cup high [mm]

ae = radial depth of cut [mm]

R = tool radius [mm]

4.4 Machine limits

High frequency spindles, in combination with high feed rates are a more precise characterization for HSC (High Speed Cutting), especially for die and mould manufacturing. Literature mention HSC operations using feed rates as high as 20.000 mm/min. However, the feed rate varies severely along the path. These feed rate oscillations depend on two limitations: mechanical (inertia) and electronic (CNC control). In the first case, the machine does not achieve acceleration and deceleration as fast as required by a specific programmed feed rate. In the second, CNC limits the feed rate to a maximum value according to its processing capacity, considering a tool path is described by small linear segments. Feed rate oscillation happens even in the most updated CNC machines and it drastically affects machining time and surface quality. The ordinary CAM software estimates the machining time considering that the feed rate is kept constant as set in the NC program; the software does not consider the CNC and machine's limitations. Therefore, the machining time estimated by CAM becomes very inaccurate [COELHO et al, 2010].

4.4.1 Feed rate oscillation in free form milling

A free form tool path is described by an small straight line interpolation. The respective CNC program generated by a CAM software create an huge NC programs and it represents a great amount of information for the CNC machine to process in short periods of time. As a result, the CNC machine may have to reduce its feed rate to manage such large amounts of information. Therefore, the linear segment length is directly related to the feed rate during milling and, as a consequence, to the machining time. Better description about the topic can be found ahead in this chapter.

4.4.2 Machine control

A typical CNC control of a machine tool has two major responsibilities: i) the servo control loop and ii) the interpolator, responsible for coordinating the machine tool axis motions. The NC program, as well as others error compensation programs and routines, is located in a higher level of the control hierarchy. Supervisory tasks such as final part measurements and corrections can be accomplished by the CNC getting feedback from other controllers.

From the machine point of view, acceleration, deceleration and kinematics characteristics could be the feed limits in a machining process. According to ELBESTAWI (1997), the main issues in high speed servo control are feed rate planning and servo control loop laws. The users and/or the machine tool builders have to choose between path error and feed rate error. If the path error is allowed, the machine tries to run at the programmed feed rate; however, the trajectory is not guaranteed. If the path error has to be low, the machine keeps the path accuracy, but the feed rate can be lower than the programmed one.

In free form milling, the linear interpolation of straight line segments means a high volume of information to be processed by the CNC. Therefore, in order to manage the tasks to be carried out in real time during machining movements (i.e.: machine control loop, jerk limitation, interpolation, and so on), the CNC reduces the feed rate according to its capacity.

a. **Block Processing Time**. To run the machine in linear path, the points of the NC program have to be interpolated to obtain the movement in one or more machine axis. This task is done by the CNC interpolator. The interpolator has to calculate the speed of each servomotor of the machine tool in every block. The real-time interpolator converts the CL path to the motion trajectories of the orthogonal machine axes so as to coordinate their motion in three-axis surface machining. This task is known as CNC block processing time. The main CNC tasks are presented as follows:
b. **Control loop.** In addition to the interpolation, the current CNC controller also has to check feedback devices in order to maintain the machine's high precision. The machine closes a control loop using devices such as encoders and linear scales. In the CNC machine system, the practical tool location is measured by the feedback devices at a specific time, i.e., sampling interval. The sampling interval is about 0.001 seconds, depending on the CNC milling machine; this represents another task to be calculated by the CNC.
c. **Motion algorithms**. Other algorithms developed to aid machine motion and positioning, such as look-ahead, feed forward control, and jerk limitation are also implemented in present CNC machine. Therefore, all the tasks mentioned above have to be done by the CNC in a very short time.

When machining complex geometries, the machine often encounters curved surfaces, where short segments are required to closely follow the profile. When that occurs with the NC program set with high feed rates the machine slows down and sometimes oscillates imitating a vibration behaviour.

Feed rate oscillation happens even in the most updated CNC machines and depends on servo response, feedback devices, CNC characteristics and its implementation on the machine tool.

4.5 Tool paths described by polynomial spline

The polynomial Spline curve can also be used to describe complex tool paths (SOUZA et al, 2010). Therefore the NC program generated will not contain the traditional commands (G01, G02 or G03), but there is a new syntax which illustrates the tool path and the polygon control which defines the curve. The Cartesian points are used to define the polygon. There are different syntaxes for the entrance of the spline data. This variation agrees with the numerical commands used. Below can be found an example of syntax for Spline movement which is generated in accordance with nomenclature of CNC FANUC.

For the generation of CNC programs using polynomials Spline, the CAM systems permit this resource which do not generate directly. The CNC programs contain Spline functions that are generated leaving a CNC linear program in straight segments. With this initial tolerance is used to calculate the linear program adding the tolerance calculation of the Spline program. For the process to be adequate the user should scientifically plan the most efficient way to work with these tolerances. Bibliographical research done during the development of this work has found very practical analysis.

YAU e KUO (2001) developed an application that converted traditional CNC generated programs for interpolation linear in CNC programs with Spline format. This method of work has implemented sophisticated numerical commands. In this case the data entered into a CNC program containing commands G01 and internally the CNC converts the straight paths into Spline equations, proportioning better performance than the traditional method of straight segments, however this conversion must be attributed to another value of tolerance. A deeper understanding can be found in SOUZA (2004) and SOUZA & COELHO (2007)[a,b].

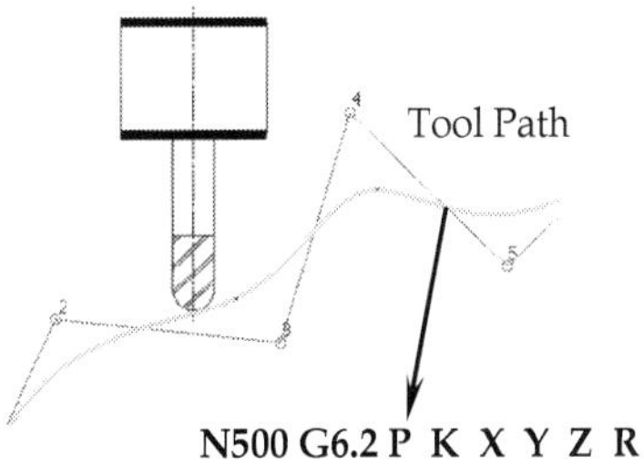

Where:

G6.2 - syntax of NURBS
[P] - degree of the polynomial equation
[K] - vector on the control point (knot)
[X,Y,Z] - coordinate of the control points
[R] - weight on the control points (weight)

Fig. 9. Spline program line for the FANUC command

5. CAM systems (Computer Aided Manufacturing)

In the milling of complex forms, due to complexity of the paths of tool, the generation of the programs for CNC machines should be deeds with aid of software, the CAM (Computer Aided Manufacturing). Although the acronym CAM expresses a generic system for support a chain of manufacture, the term CAM is commonly employed for the generation of programs for CNC machines.

The CAM calculates the paths of the tool for carry out milling, besides simulating the movement of the machine for certification, and generates the CNC programs for the manufacturing of the piece.

At present, the biggest application of CAM software is for the programming of machining by milling, in special 3 to 5 simultaneous axes. CAM software can also generates CNC commands for EDM machines, lathes, and laser or water pressure cutting machines, among other applications carried out by equipment controlled by a CNC.

The CAM systems can be independent software that works in assembly with any application CAD, in general 3D, needing this, the importing of geometries in format-standard, such as, IGES, STL, VDA-FS, etc.

A CAM system can be used to develop three main activities: i) calculate the tool paths for the machining the CNC machine; ii) carry out the simulation of the machining calculated;

iii) generate programs for specific CNC machines. Doubtless the first function is more important and can represent the main difference between the diverse CAM systems available in the present market.

The operations of milling, focusing on the making of complex forms, can be classified as: roughing, semi-finishing, relief of corners and finishing. For each one of these operations exist different techniques and peculiarities of programming with CAM.

5.1 Tool paths for roughing

Upon carrying out machining by milling of a surface, the tool can traverse different routes. The CAM systems offer different tool path options. Although any one of these paths can carry out the operation of machining, they can have repercussions in the trial, differentiating itself mainly regarding the time of machining, demands of the tool of cut, and superficial quality. The identification of these options should be carried out by the user. The CAM systems presented are not able to identify the best strategies of machining for specific cases, according to geometry it be machined, mainly in complex forms.

For the operations of roughing for milling, besides the paths of parallels passes, the CAM systems still developed other strategies as: paths in offset, paths based in the internal contour of the geometry, paths based on the external contour of the geometry.

In the roughing operations, the dimensions of the block of raw material can be utilized to limit the tool paths calculated by the CAM. In this way, this resource permits the user create machining paths in a specific region of the piece.

5.2 Programming operations of semi-finishing

Although there is not a standard definition of semi-finishing for this application, here it will be naming semi-finishing the milling operations that has for objective become constant the value of the stock of material to be milled after by the finishing operation.

For this reason, the regions with remaining material after roughing besides the theoretical stock should be machined. In general, the semi-finishing is an operation of "roughing" in 2 ½ axes, in specific regions and can be divided in two phases: a) operations for removal of the degrees and b) operations for relief of corners, as presented previously.

Programming operations for the removal of degrees

This algorithm, available in some CAM systems, identifies automatically the steps left by the operation of roughing, considering the axial depth of cut and the geometry of the tool, in function of the curvature of the machined surface.

Some systems create a model with the geometry of the piece after roughing, with all the excess of material. This geometry then is utilized like a block of raw material for the semi-finishing (Figure 10).

Programming operations for the relief of corners

This operation has the purpose of remove the material not machined in roughing, in areas not reached by the cutting tool, due to relatively big diameter tools used (tools characteristics of the operations of roughing), as presented previously.

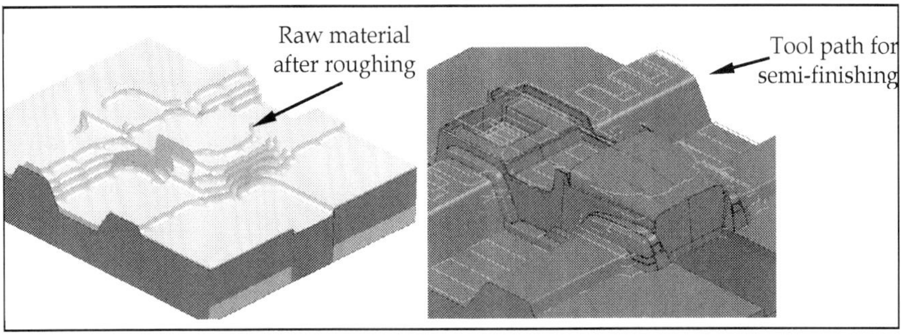

Fig. 10. A methodology to define the block of raw material for the semi-finishing

Therefore, in the corners relief operation should be employ tools with diameters smaller than the tools utilized in roughing.

The more modern CAM software presents a tool for identifying these regions automatically. The algorithm identifies the corners of the geometry of the piece where it has a radius less than the ray of the tool of thins. In the finishing of this algorithm the software requests the information of the diameter of the tool that was employed in roughing, identifies the areas and calculates the machine paths of these regions, employing a tool of smaller diameter.

5.3 Programming finishing operations

This is an operation of three-dimensional milling, utilizing 3, 4, or 5 free axes, according to arrangement of the machine-tool. This operation is going to remove all excess material of the previous operations and finalize the operations of milling.

In the finish, the axial cut depth parameter passes the tolerance value and depending on the position of the tool-piece, this depth will not be necessarily axial, as illustrated the Figure 11. The tool parameter of geometrical cut itself maintains the same concept.

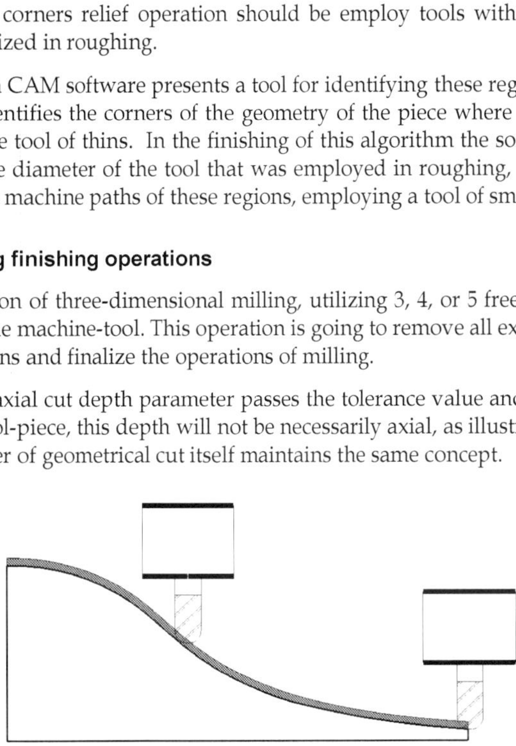

Fig. 11. Finishing depth of cut

The CAM systems present offer different algorithms to calculate machine paths for finishing. The main principles are:

a. **Parallel passes horizontally constant.** Probably this is one of the first algorithms for finish path calculation to be developed, and is widely employed. Known as lower the tolerance band, the closer will be the tool paths from the surface to be machined, however, smaller will be the followings segments of lines, CNC programs will possess greater volume of data, since more lines comprise the program and this characteristic will be able to limit the feed rate during machining, particularly in complex forms. The machine reduces the programmed feed rate, as studied by SOUZA & COELHO (2007).

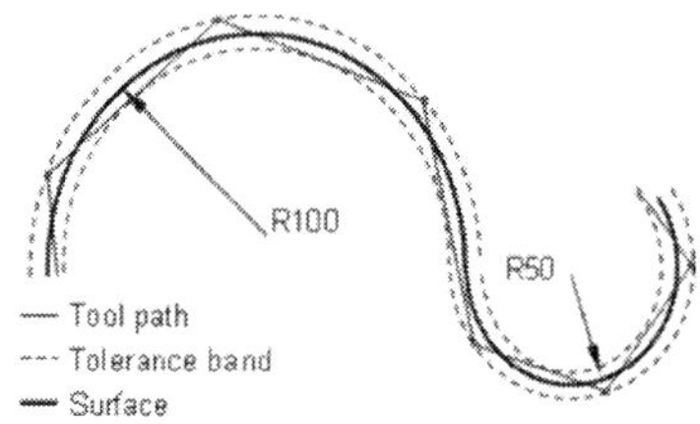

Surface radius	Segment Length
100 mm	9.0 mm
50 mm	6.3 mm

Fig. 16. Tolerance to calculate the CNC programs

The path of the tool calculated by a system CAM represents the central extremity of the tool, and by that, the path is not a simple offset of the surface it be machined. To develop the calculation, the software should identify two positions: a) Point of contact of the tool with the surface, known as CC - Cutter Contact and b) the point that represents the path of the tool (its central extremity), named of CL - Cutter Location.

The lines of the programs CNC, referring the paths of the tool, contain the Cartesian information of each points CL. It must be noticed that the Cutter Contact is the point which should be inside this band of tolerance and not the path of the tool or the Cutter Location.

Deferential ways exist to calculate both, the CC and CL. These algorithms are virtues of a CAM, therefore can have repercussions in the time of calculation, time of machining and quality of the surface machined. Initially, in a certain moment, and then it should calculate the final point the tool centre (CL) points that are linearly interpolated to generate the tool paths.

An approach for calculate the CC in polynomial surfaces is the division of the own polygon of control of that describes the curve or surface. This subdivision can identify points about the curve or surfaces, which can be the CC. Others approaches are developed for such end.

After it identified the CC, the CL must be calculated. The Figure 17 exemplifies a possibility to calculate the CL for two points of a path, considering that a previous algorithm already have found the CC. In this example, knowing to normal of the surface in the point desired (CC) and angle that this normal is found regarding the axis of the coordinates, is able to the CL be found through the ray of the tool.

5.7 Circular and linear tool path interpolation

Due to the inconveniences of the traditional linear interpolation method for machining complex forms, an alternative is to use circular interpolation associated with linear interpolation to describe a tool path. That combination can maintain the continuity of the curve C^1 and showed many advantages about the isolated utilization of segments of straight line (SOUZA, 2004). This approach, however, can only be applied to planar motion by interpolating two axes of the CNC machine, since the great majority of CAD / CAM / CNC are unable to carry out circular interpolation in three dimensions. Thus, it has aimed utilize polynomials Spline to describe complex tool paths.

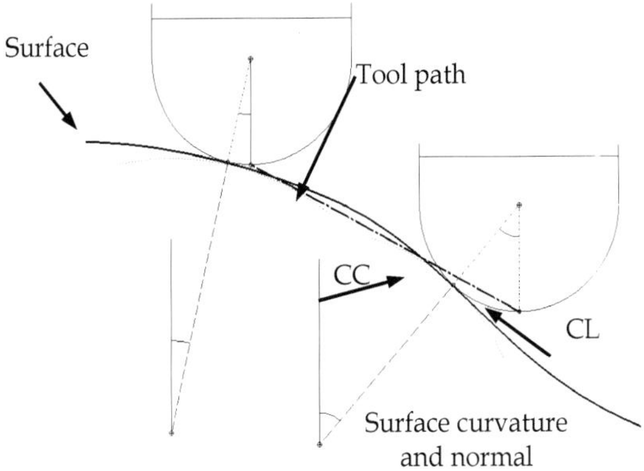

Fig. 17. Determination of CL by CC

5.8 Post-process task

After the CAM calculates and simulates the tool paths, there is the phase of post-processing, which is responsible for transforming these paths into CNC programs.

The initial calculation in the system CAM generates a generic file, without specifications. This file is known like Cutter Location File (CLF).

The post-processing has the function of encode the paths calculated by any software CAM in specific commands for a given configuration machine/CNC.

Therefore, two elements are necessary for create the program CNC:

- Tool path calculated by the CAM in computational format (CLF).
- Post-processor - computational program that reads the file CLF and converts it into a program to be performed in a given combination of machine/CNC.

For each equipment, machine/CNC should be developed a specific post-processor. The procedure for the creation of post-processors and development of programs CNC is presented in the Figure 18.

The information of the machine is related to its physical characteristics, as maxim movement of the axes, number of axes, speeds etc. Regarding the numerical control, there is a wide range of CNCs on the market today, ISO 66025 standardizes the main functions codes CNC. There are several distinctions among suppliers. The simplest cases involve, as example, the utilization of point or comma for separate decimal places; number of decimal places demanded; block end character need; command for beginning and end of program, among others.

To assist in the correct definition of a specific CAD platform for a given application, CAD systems can be classified as: i) Low-end 2D CAD systems, ii) Middle-end 3D CAD systems modelling of solid objects or surfaces, iii) High-end CAD hybrid system allows the modelling of solid objects and surfaces, among other features. A CAD solid or surface model refers to its core used for the mathematical representation of three-dimensional objects as shown below.

6.1 CAD surface modelling – Middle end

The surface model uses mathematical tools that allow creation of complex shapes with greater interactivity. The geometry created in a model of surfaces have no thickness, geometry is a "shell" and therefore no mass and centre of gravity (although it can be estimated). These are the main features of a CAD surface model. In CAD, the modelled objects may be closed. Surfaces are created using interpolation between the lateral and longitudinal representing the outline of the geometry. Although current CAD systems can identify regions of the opposite surface, this does not necessarily represent regions of an internal or external object.

The surface modelling CAD systems are able to perform the modelling of any geometric shape, regardless of their complexity. They have sophisticated features for manipulation of the geometry modelled by allowing the geometry to be changed interactively to obtain the desired product. The manipulation of the surface is given by the movement of any point of the geometry in any direction.

Besides this possibility, four are associated tangent vectors at each point on the surface. Vectors can be edited as the direction and magnitude. These devices allow changing any part of the surface. The possibility and ease of editing of any point of geometry, ensuring continuity of tangency and curvature, is an important feature that distinguishes the CAD surface model.

Figure 19 illustrates examples of issues in the geometric model. In this case, using point geometry (Figure a) there were three editions: Figure b) point geometry has moved in the vertical direction, Y axis in this case. It is observed that the geometry accompanying the extent varied. Figure c) pushed up the value of the magnitude of a vector. Figure d) has changed the angle of one of the vectors. Importantly, the changes in vector magnitude and slope can occur before and/or after the selected point.

It should be noted also that the edited point corresponds to an intersection between a lateral and longitudinal. So when it wants to change a point on the surface where there is no such intersection, the user must create lateral and longitudinal in order to obtain an intersection point. The creation of these entities is trivial in geometric surface models. However, the interactivity to identify their position is characteristic of each specific software, facilitating or hindering the work for the user.

The continuity of a surface can be considered in a simplified form as its degree of smoothness. Surfaces are called Class A high-quality surfaces in terms of continuity of tangency and curvature. This term is often used in the automotive industry (Souza & Ulbrich, 2009).

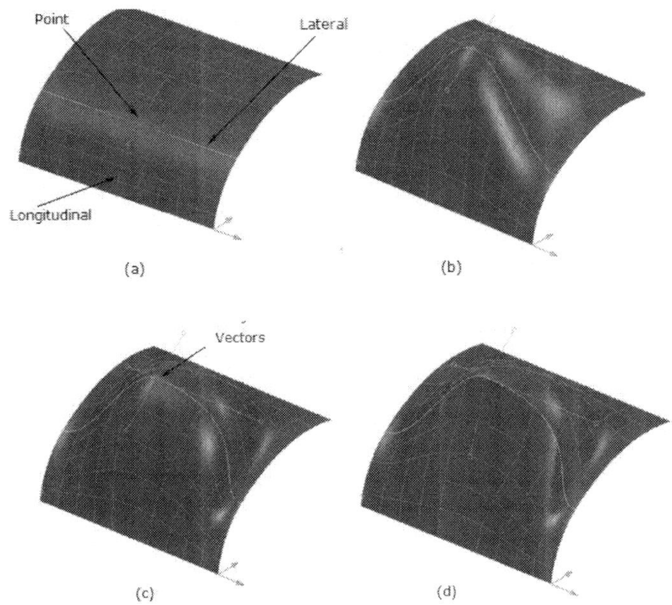

Fig. 19. Geometric edition on the surface

6.2 Mathematics for constructing curves

The quality of a curve is related to the mathematics used for this representation. Advanced CAD systems can offer different possibilities to generate a curve, in order to allow modelling of complex geometric forms, smooth and seamless.

Using parametric polynomial equations, Bezier-based methodology, the developers responsible for the development of each CAD software creates specific mathematical resources in order to allow greater control and ease of work to the user. This development of each software house is confidential and may represent the differential of the CAD software. Thus, only generic formulations of curves are discussed in the user guides for each software. Therefore, we will discuss below, some methodologies that differ from the creation of curves in different CAD systems.

Other mathematical formulations based on Bezier method have been developed for the representation of curves and surfaces in computer systems. The B-spline curve is also used for a parametric polynomial equation with modifications of the proposal Bezier, allowing among other things, to represent a curve using a polynomial of low degree, facilitating computations, also allowing local changes of the curve.

The Spline NURBS (Non-Uniform Rational B-Spline) represents the state of the art for the representation of curves and surfaces in CAD system, providing better control of the curve, allowing local issues, and allows more efficient calculation. Basically, the methodology is based on NURBS B-Spline method, adding two main functions:

Non-Uniform: The vectors (knot) that indicate which portion of the curve is affected by a single point of control are not necessarily uniform.

Rational: You can set the intensity (weight) that each control point "attracts" the curve (exemplified below). It also allows the representation of elementary geometric entities: cylinders, cones, and plans, as well as conic curves, such as circles, ellipses, parabolas and hyperbolas.

In summary, these features mean more control factors can be applied to the curve, so more complex surfaces can be represented with a smaller number of turns, maintaining the continuity of curvature and tangency. For these reasons, the NURBS method is the state of the art in the representation of complex curves and surfaces in CAD systems.

The most advanced CAD systems can offer users different methodologies to create a curve. These methods are implemented by each developer and may involve from the choice between different mathematical functions, such as Bezier curves, B-Spline curve, NURBS curve, etc. As well as to allow changing the degree of the equation used.

It should be noted that each software has individual characteristics. The options of mathematical CAD software are not the same as other CAD software. These options, as well as construction tools and quality analysis of the geometries built, may be important differences for specific applications. These differences may involve the continuity of a curve, chances of building and editing, as well as computational workload.

The construction of curves in CAD systems is divided into two main mathematical methods:

a. **Interpolation method**. interpolation is to create a curve that passes through the points provided by the user. The curve of Figure 20 shows the interpolation process to create a curve, using four points as an example: P1 (0, 0), P2 (50, 50), P3 (100, -50), P4 (150, 0). The Hermite curve is a curve described by the example of interpolation method.
b. **Approach method**. In this case, the points provided by the user are used to create the control polygon will represent the curve, as shown in Figure 20 curve b. The Bezier curve is a curve described by the example of approximation method.

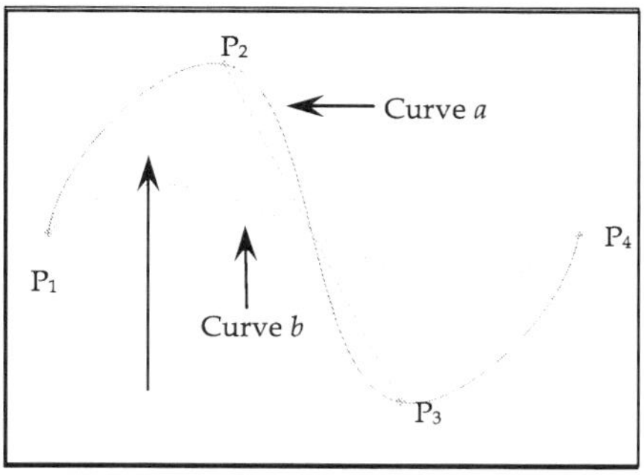

Fig. 20. Curves created by the methods of interpolation and approximation of points

Besides the method, interpolation or approximation, CAD systems can still provide other resources for modelling curves, such as:

a. **Degree of mathematical equation.** Some CAD systems allow changing the degree of the equation used to construct the curve. Using the points P1 (0, 0), P2 (50, 50), P3 (100, -50), P4 (150, 0), curves were generated by changing the degree of the polynomial, as shown in Figure 21. Note that due to the number of control points, according to the mathematical rule are allowed to generate curves of maximum degree 3 (the degree of the equation is 1 minus the number of control points). Also note that a polynomial of degree 1 generates a sequence of lines.

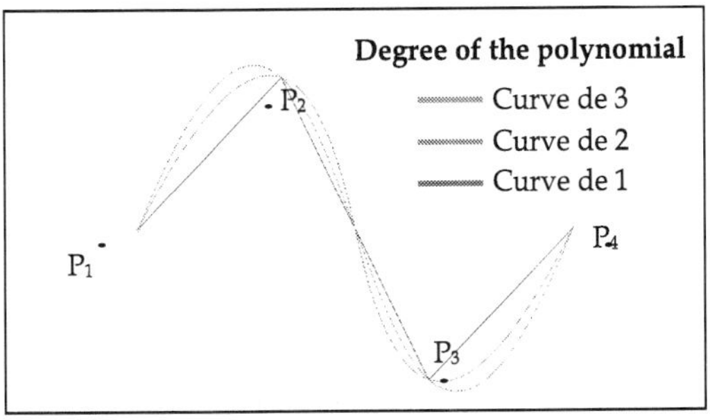

Fig. 21. Curves constructed by polynomials of different degrees

b. **Different equations.** Some CAD systems do not allow changing the degree of the polynomial used to construct the curve. However, many systems offer different equations to build a curve, as illustrated in Figure 22.

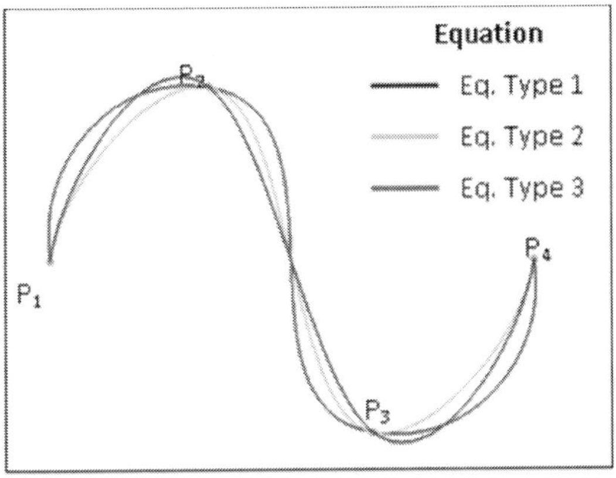

Fig. 22. Curves for different mathematical built using the same points

It is important to note that the same sequence of points can build distinguish ways in the same CAD system. Thus, it is difficult to build the same geometrical shape in two different CAD systems using the same sequence of points.

c. **Influence of the control points (weight).** Some CAD systems allow changing the degree of influence of each control point of the polygon under the curve calculated (Figure 23).

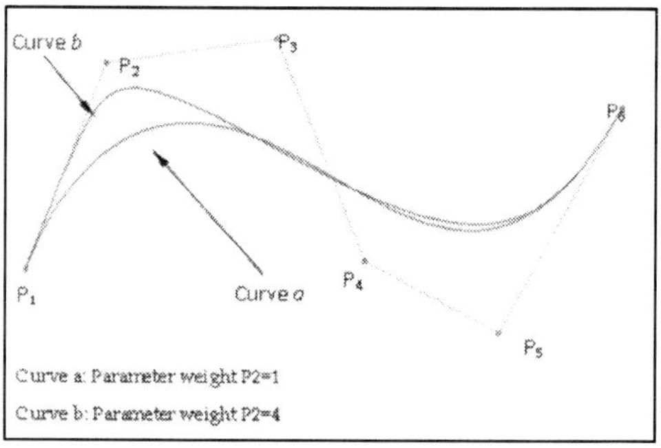

Fig. 23. Changed the influence of the point P2 on the curve

These are resources that help advanced modelling of complex geometries containing forms.

6.3 Curvature continuity

The continuity of geometry is one way to assess the quality of a geometric entity generated by CAD software. The level of continuity of geometry is related to the degree of smoothness along its length. Geometry can represent discontinuous aesthetic and functional implications in a product. The continuity of surfaces is a concept used to extrapolate the curves. Thus, it is used the concepts of continuity of curves to present and discuss this topic. A discontinuous curve may generate a discontinuous surface. The continuity of a geometry, curve or surface, can happen in two cases:

a. Going over a single entity.
b. Continuity in the connection between two distinct entities.

There are three size classes for the continuity of curves and surfaces.

1. **Continuity of position:** occurs when two segments of the same entity or two separate entities, have in common only the point of contact (Figure 24b). This is the case of less smoothness.
2. **Continuity of position and tangent:** occurs when a point between the segments analyzed, or two separate entities, have in common other than the contact tangent vectors (Figure 24c).
3. **Continuity of position, tangency and curvature:** occurs when a point between the segments analyzed, or two different geometries have in common, the contact point, tangent vector and the same degree of curvature (Figure 24d). This is the case for better continuity.

Figure 24 illustrates the case for continuity in the connection between two different curves, connected by a user or other mechanism of software construction, as in the construction of borders to make cuts at the intersection of surfaces. These boundaries are spline curves calculated by the software and have degree of continuity. The quality of this curve can influence the surface generated by it.

As mentioned, the continuity can occur along a single curve, depending on the model used for its representation.

This can partition the curve internally into segments known as patches, and it often is not noticed by the user.

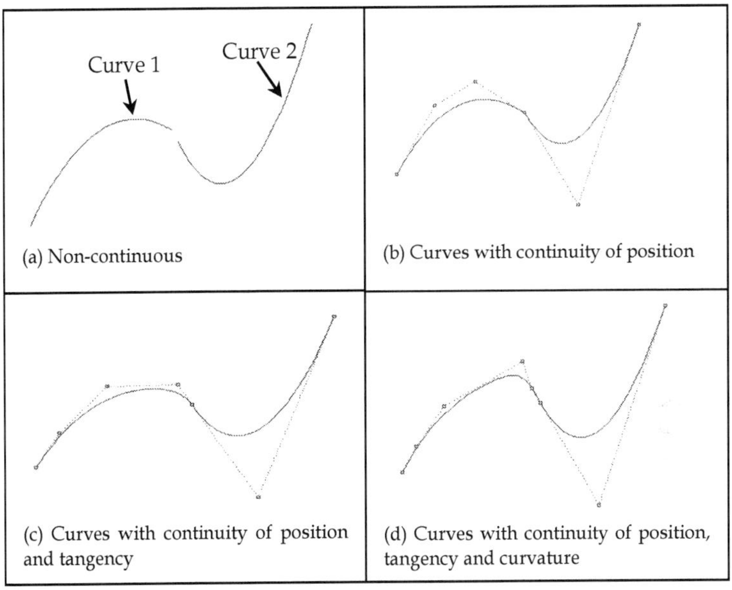

Fig. 24. Degree of continuity of continuity of a curve

6.4 Quality of surfaces generated by CAD systems

In many cases it is necessary for modelling three-dimensional boundary conditions that go beyond the requirements of design and ergonomics. Surfaces are required in these cases with a high degree of continuity in order to maintain the smoothness in the shape of the product. The quality of a complex surface generated by a CAD system is related to the model used for its creation in three circumstances:

a. The smoothness of a surface can be related to the continuity of the curve used for its creation.
b. The connection between different areas may represent regions with impaired smooth.
c. Areas of intersection with rays (Fillets) calculated "automatically" by the software.

When you want to develop surfaces for products requiring strict-looking, such as appliances, the external body of a car, etc., small discontinuities in the surfaces of the 3D CAD model can be replicated in the final product, be it cast or stamped. When creating surfaces through the interpolation of two sections and non-planar curves with different radius of curvature, and the need to maintain continuity of the tangential surface created in which the directions of the surface change smoothly, the surface must also have continuity of curvature. Applying tests of light reflection on the product such geometric discontinuities can be found.

CAD systems have advanced graphical tools to detect such discontinuities in a surface. Figure 25b illustrates a surface with discontinuities. The CAD system identifies the discrete regions through hazards. The parallel strips and gently represent regions consistent with good degree of continuity. The discrete regions are illustrated by the tracks with abrupt changes (Figure 25c). This may occur for mathematical limitations and / or the method used to construct the surface. The surface of Figure 25 (b and c) is a surface Fillet, calculated automatically by the software.

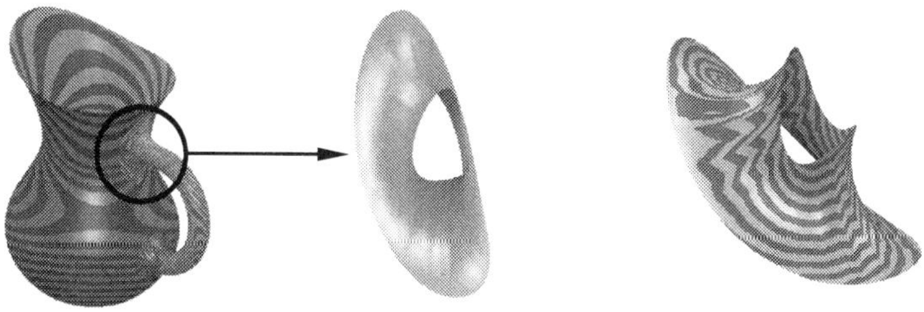

(a) Product (b) Discontinuous surface (c) Abrupt changes in gray bands

Fig. 25. Surface emphasized the problem of discontinuity

The abrupt changes of the bands indicate regions with continuity compromised. However, after identifying these regions and/or surfaces, CAD systems are still not able to help troubleshoot the problem. The reasons that caused specific discontinuities are not presented.

Thus, the user must develop empirical ability to solve or minimize the problem. In the case illustrated in Figure 25, the surface of agreement (Fillet) was calculated by the software with a large number of longitudinal and lateral, creating the problem of continuity observed. Performing a work setting "manual" to delete longitudinal and lateral, it was observed an improvement in continuity of surface.

6.5 CAD solid modelling – Middle end

The mathematics used in the solid model ensures the reality of constructive geometric entity modelled. The solid model performs mathematical methods to validate real possibility of the existence of the modelled geometry. Because of this geometric validation, among other factors, a CAD solid model is great limitations to the modelling of complex shapes, such as, for example, a computer mouse, the panel of a car, bottle packaging, among others objects. There are difficulties in parts of issues surface in the tangent of specific points and ensuring continuity. To work with geometries with these characteristics, a CAD modeller surfaces may be more appropriate.

A CAD solid model should possess the following characteristics:

- Geometry unambiguous and real.
- Generation of simple objects (compared to the models of surfaces).
- Suitable for engineering geometry.

- Cost less than the CAD surface model.
- Limited to create complex geometric shapes.
- Use relatively simple.

Different computational techniques for solid modelling. The internal representation of a solid object is commonly used to classify the modeller. The internal representation is how the software stores the model. This is different from the external representation, which presents the model on the computer screen. Two techniques of internal representation are: a) B-Rep (Boundary Representation). This representation employs models described by boundaries, proposed by Ian Braid, Cambridge University. b) CSG (Constructive Solid Geometry). This method was proposed by Voelcker and Requicha the University of Rochester.

Both methods were applied in commercial systems in the late 70's. Other methods of representation of solids were developed, such as octree, quadtree, cell breakdown, among others. However, the methods of representation of solids used in most current CAD systems are the CSG (Constructive Solid Geometry) and B-Rep (Boundary Representation).

In this method, solid objects are built with a combination of simple solid primitives such as cylinders, cubes, cones, spheres, etc. Through Boolean operations (George Boole, 1815-1864, developed the Boolean algebra) addition, subtraction and intersection between solid objects in case of CAD software. When an object is constructed with a CSG modeller, the system saves a history of creation (creating topographic tree), which includes all solid primitives with their size and positioning, and information about all the Boolean operations used to construct the object.

6.6 CAD hybrid modelling – High end

The hybrid model allows building three-dimensional geometries employed the techniques of combining modelling surfaces in solid modelling. This provides greater versatility to work for the user, can benefit from the best features of both methods.

6.7 Techniques of modelling

There are three main techniques, common in CAD systems for the construction of three-dimensional geometries: i) primitive geometries ii) intersecting geometries iii) domain-based wireframe geometries. The latter will be the most used, in this case the objects are constructed using geometric information in wireframe, and may be straight lines, curves, arcs, points or a combination thereof. The software uses this field and applies the function selected by the user, which are: operation of revolution or extrusion of a profile on a directional vector; extrusion of a profile on another profile directional; interpolation between curves; migration between cross sections. More complex surfaces, which require patches with three or five boundaries, are more complex representation. These surfaces are found in regions with corners at the intersection of surfaces, as well as the junction of three Fillets.

7. Application of CAE systems

The Computer Aided Engineering (CAE) systems were developed to simulate real applications, and enabling auxiliary steps of projects and manufacturing. There exists CAE software specifically for each application. The software SigmaSoft® and Moldflow® are examples of CAE to simulate the transformation of plastic and none metallic products. The software Ansys® and Nastran® are more specific for simulations of mechanical demands dynamic statistics.

Simulations of products and processes have been highly desired for the last decade for various applications. Secure results of these simulations can prevent accidents, reduce manufacturing costs, optimise products behind for example the reduction of the thickness, how to help in the substitution of materials, collecting materials, and making processes and products more efficient.

The CAE systems make sophisticated mathematical algorithms to perform calculations. The quality of a simulation involves a degree of approximation between the results obtained and the real situation. This relation of the mathematical algorithm employed for the generation of the finite elements mesh and for the calculations asked for. This way the information entered, such as the geometry, mechanical demands and materials.

Information about the process to be simulated must be attributed to the CAE system to make the calculations. This information is called entrance variables. Both variables are specifically defined in function of the process and/or the product studied.

To carry out the simulation calculations, CAE uses finite technical elements. The geometry originating from CAD can be used to analyse the primitive elements, such as triangles. Therefore the software generates a geometry dominated by a finite elements mesh, which represents an object or a process to be simulated. Finite elements and methods are widely employed by CAE systems.

The working procedures to carry out simulations with CAE systems can be divided in three main phases: pre-processing (generation of the sweater of finite elements and variables of entrance), processing (calculation of the demands) and post-processing (evaluation and interpretation of the answer of the software), approached with more property in the sequence.

In the pre-processing the modelling of the product is supplied by the CAD system. In case it cannot be integrated with the systems CAD/CAE, the geometry should be exported by the CAD through a standard interface (IGES, by example) and imported in the CAE system. After importing the geometry, the CAE system should calculate the finite elements mesh. This is a crucial step for the quality of simulation, because the nodes of the finite elements mesh are used as the basis for all calculations performed in CAE applications.

The quality of the finite elements mesh generated by the software is attributed mainly by the homogeneity of the elements that it composes. The best condition will be generated by software that contains the largest number of elements and dimensions equilateral homogeneous. The mesh calculated by the software can at times be unstable, having repercussions in future simulations. In these cases the user should be qualified to identify and correct such limitations of software, which can be related with the complexity of the surface, rank of refinement of the mesh, quality of the surface imported of the CAD via communication patterns, beyond the mathematical limitations made for generation of the mesh. Figure 26 (a) illustrates the CAD model, (b) example of mesh with none equilateral regions containing elements and a fault of connection between the elements, which represents errors or impossibilities of calculation.

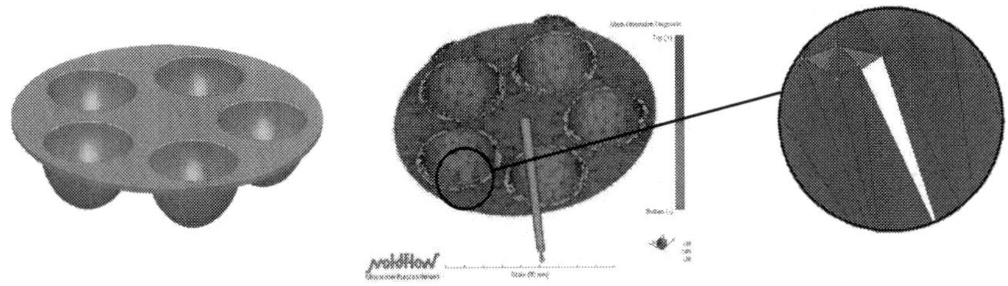

(a) (b)

Fig. 26. Finite elements mesh with problems. a) Product CAD; b) Mesh. Plans not oriented and non-equilateral triangles; open mesh

In cases of problems in the automatic construction of the finite elements mesh, the user should interact with the software in order to provide an adequate mesh. The dimensions of the elements that compose the mesh can be defined by the user and will have repercussions in the precision of the calculation. Still in the pre-processing, all the conditions of contour should be supplied, such as: mechanical stresses, strength of the product material, temperature, among other information that composed the simulation.

The results, in general, are presented in the form of multicoloured graphics, according to the analysis requested. Figure 27 illustrates the result of the simulation of the injection of a plastic product, having like variables in the temperature injection necessary, the time of injection of the product, and the pressure of injection.

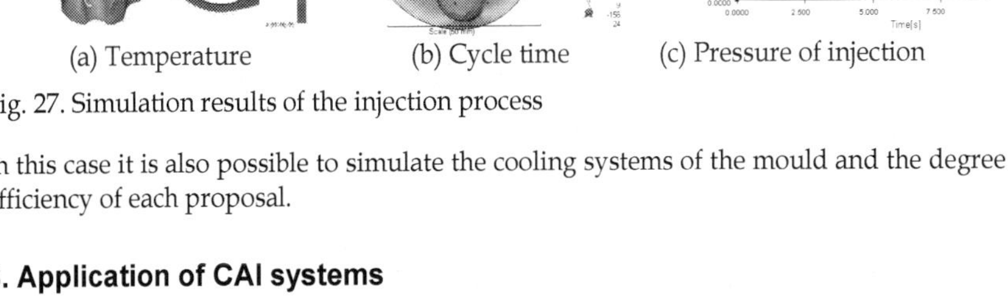

(a) Temperature (b) Cycle time (c) Pressure of injection

Fig. 27. Simulation results of the injection process

In this case it is also possible to simulate the cooling systems of the mould and the degree of efficiency of each proposal.

8. Application of CAI systems

The Computer-Aided Inspection (CAI) represents the final integration of manufacturing processes, this step is performed as a comparison between the geometric model constructed in the CAD, which represents the theoretical form of the design geometry without dimensional errors, with the product manufactured, which has so errors arising from mechanical limitations of both machine tool and manufacturing process.

At this stage, the CAI software receives the geometric information of CAD and information representing the final product, obtained by a coordinate measuring machine, through a cloud of points. CAI compares the geometric errors between the real object with the CAD design and report on the regions with the geometrical deviations, considering the tolerances specified in the project (Figure 28).

Fig. 28. Integration of CAD/CAI/MMC

9. Acknowledgments

The authors would like to thank CAPES for supporting this research project under the grant PROCAD 139/2007 and PROENGENHARIAS PE 27/2008. The authors also would like to thank the research group PROMOLDE from IST/SOCIESC, Villares Metals for supplying the raw material and Sandvik Corromant.

10. References

Coelho, R. T.; Souza, A. F.; Rodrigues, A. R. (2009). Mechanistic approach to predict real machining time for milling free form geometries applying high feed speed. International Journal of Advanced Manufacturing Technology, v. 1, p. 1433-3015-1433-3015.

Souza, A. F.; Coelho, R. T, (2007). "Experimental Investigation of Feed Rate Limitations on High Speed Milling Aimed at Industrial Applications". Int. Jou. Adv. Manufacturing Technology. DOI 10.1007/s00170-006-0445-2.

Souza, A. F.; Coelho, R. T.; Rodrigues, A. R. (2010). Manufacturing complex geometries using high speed cutting technology. VDM Verlag. v. 1. 130 p.

Souza, A. F.; Coelho, R. T. (2007). Investigation of Spline polynomial for describing a free form tool path for high speed milling. In: International Conference on High Speed Machining. San Sebastian, Spain.

Souza, A. F.; Ulbrich, C. B. L. (2009). Engenharia Integrada por Computador e Sistemas CAD/CAM/CNC. Princípios e Aplicações. 1. ed. São Paulo: Artliber. v. 1. 332p.

Souza A.F. (2004). Contribution for free form milling applying the High Speed Cutting technology. Doctored thesis. University of Sao Paulo. 186pp.

Yau, H.T.; Kuo, M.J. (2001). NURBS machining and feed rate adjustment for high-speed cutting of complex sculptured surfaces. International Journal of Production Research, London, v.39, n.1, p.21-41, Jan.

Elbestawi, M.A. (1997). High-speed milling of dies and molds in their hardened state. Annals of the CIRP, Paris, v.46, n.1, p.57-62.

Part 5

New Approaches for Mechanical Engineering Education and Organization Systems

25

Modern Methods of Education, Research and Design Used in Mechanical Engineering

Borza Sorin-Ioan, Brindasu Paul Dan and Beju Livia Dana
University "Lucian Blaga" of Sibiu, Romania

1. Introduction

Today's society is characterized by extraordinary dynamics, information systems have revolutionized society and the pressure of change from all fields is increasingly pregnant upon individuals. More free access to information and to continuous education are the main solutions for the adaptation of individuals to social pressure. In today's complex society systems influence each other and it often happens that a discovery in a field be useful in other areas. The emergence of IT technologies is an obvious aid to education, design and research, leading to modern and efficient solutions.

In the past, the teacher was the core of classical education; conversely, today's educational actions are focused on the students, everyone's role changes and educational methods have become fundamentally different. Starting from the idea that "a man is only old once he has given up learning ", experts in education seek to provide modern learning methodologies, which can be applied irrespective of time and location.

Design and research activities employ an increasing amount of information and communication technology, which is a huge step forward. However, the importance of involving the most valuable specialists in any field has not diminished. The customers have an increasingly active role in the design and research processes now, and whereas they used to take on a much less involved stance in the pass, nowadays the entire action is directed towards them.

The Internet (abbreviation of the American expression *"interconnected networks"*) is the distribution environment of information and a communication channel between stakeholders (tutors, students, administrators, engineers, beneficiary etc.). There is a multitude of applications, such as: email, instant messaging, discussion forums, the World Wide Web etc. that allow the exchange of information between users.

2. E-learning

In the technical literature, there are many definitions of the term "E-learning", all of which are associated with distance learning via the Internet.

In a broader sense, "E-learning" means "all educational situations in which ICT (Information and Communication Technology) resources are significantly used ".

In the narrow sense, the term "E-learning" is "a type of distance education, organised by an institution that provides online materials to be assimilated by the students in their own way."

2.1 E-learning models

Currently, there are a number of E-learning models (Butz, 2006) developed and used worldwide in distance education. Each university, or other institution provides such programs by adopting an appropriate model - a model that it deems to be the most effective for its intended type of training.

The first model is the ***self-directed E-learning***. This model is the simplest with respect to the number of functionalities, and is aimed at experienced learners in the process of continuous professional development, who wish to deepen specific topics. The information content consists of web pages, multimedia presentations, audio-video presentations and other materials. The content is made available to users via a hosting Web server.

The second model is called ***facilitated E-learning***. This system combines the self-directed e-learning architecture model with a range of communication facilities, such as the email or the discussion forums. Documents are often transferred through the forum, where all users have access.

The third model is the ***advanced E-learning***, which uses advanced Web technologies for the management of the entire educational process. Such a model uses audio-video real time transmission techniques via video telephony or video conferencing applications, chat or smart boards. These facilities are added on top of the previous model. Also, the educational process follows a timetable that all the students are required to meet. Such a model is built on the LMS system (Learning Management System). The information content is covered gradually, allowing the student to deepen all aspects of a topic.

2.1.1 Virtual university

A Virtual University is an exclusively virtual institution that will provide specific functions for the development of educational programs (including the design of the learning materials), such as: tutorials and guidance, production, storage and distributing of educational materials, management and accounting (financial problems), marketing, evaluation and accreditation.

Conventional education systems often attempt to create secondary distance learning programmes, however their administrative structures are often incompatible or poorly adaptable to these. Also, traditional institutions are usually sceptical towards distance education, deeming it unable to achieve satisfactory results. Therefore, they are reluctant in invest in this direction.

A virtual university, on the other hand, can be fully devoted to this educational model, therefore being more likely to meet the requirements of distance learning students.

2.1.2 Trends in E-learning system development

The most significant context in which E-learning technology is used is distance education. Used primarily as an education tool for remote locations, especially in Australia, this system became widespread as a result of the development of communication technologies and the rise of the Internet. Distance education is just an expression of the new trend for the consumers of the various training institutions. The accumulation and transfer of credits as well as the modularisation of courses are parts of the same structural transformation of the educational system.

Widespread acceptance of the various types of distance education and training, from study by correspondence to E-learning, is far from complete. The inclusion of distance courses in the common practice of training will have be preceded by multiple efforts of raising the quality standards, supported by research in the field.

2.2 The E-learning architecture

Increasing pressure faced by organizations (universities included) compel them to be organized in a better performing manner. High flexibility is another requirement, so as to enable them to quickly implement continuous improvement processes. A few reference architectures were created in order to assist the organizations in this work. They provide frameworks (templates) for certain fields. These templates typically consist of a list of modules, domains and specific functions and interactions between them. They also contain elements and features located outside the studied architecture.

Figure 1 shows an example of a reference architecture specific to E-learning educational systems. The product life cycle stages (of system development and use) specific to E-learning system are presented on an axis and the main entities and actors interacting with it, on the other. At the intersection points of these modules the main activities that characterize an E learning system are described.

We will further dwell upon certain issues that were outlined in the framework of investigations conducted at the Faculty of Engineering of Sibiu (Romania) in the Industrial Engineering team, as they have completed the construction of a learning system in the cutting mechanical processing and cutting tools. The most interesting aspects are the development of an on-line course and the issuing of the tutorials.

2.3 Development of an on-line course

The devising of an E-learning course requires a greater effort than in the case of its classic counterpart. This is because the system is lacking the possibility of direct human interaction and thusly requires a well set up series of images, animations, questions and evaluations to compensate for this.

2.3.1 Design stages of an on-line course

Designing a course goes through a set of predefined steps, as showcased below:

1. Identifying the strategic role of the new product, which should be reflected in the strategic plan of the faculty and in the curriculum. It is important to settle upon the type of the course (new or improved), its objectives, the required knowledge, the specific skills that will get at the completion of the course etc;
2. Generating the "idea", with the application of methods and techniques (including creative ones), in order to complete the new course. The teaching strategy includes: type learning

(individual or in groups), the possibility of presenting movies and animations, access to some web pages, tutorials, availability of interactive materials, bibliographical research and classification of information, application activities and design, support offered by the teacher in the event that the pupil encounters obstacles etc.

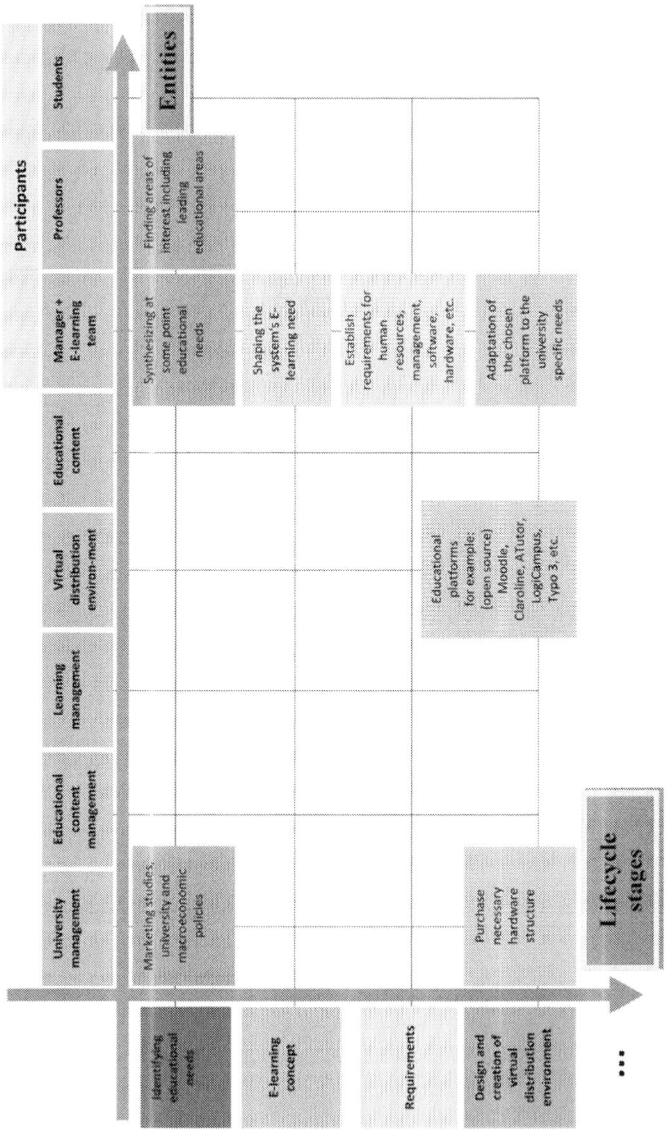

Fig. 1a. E-learning architecture

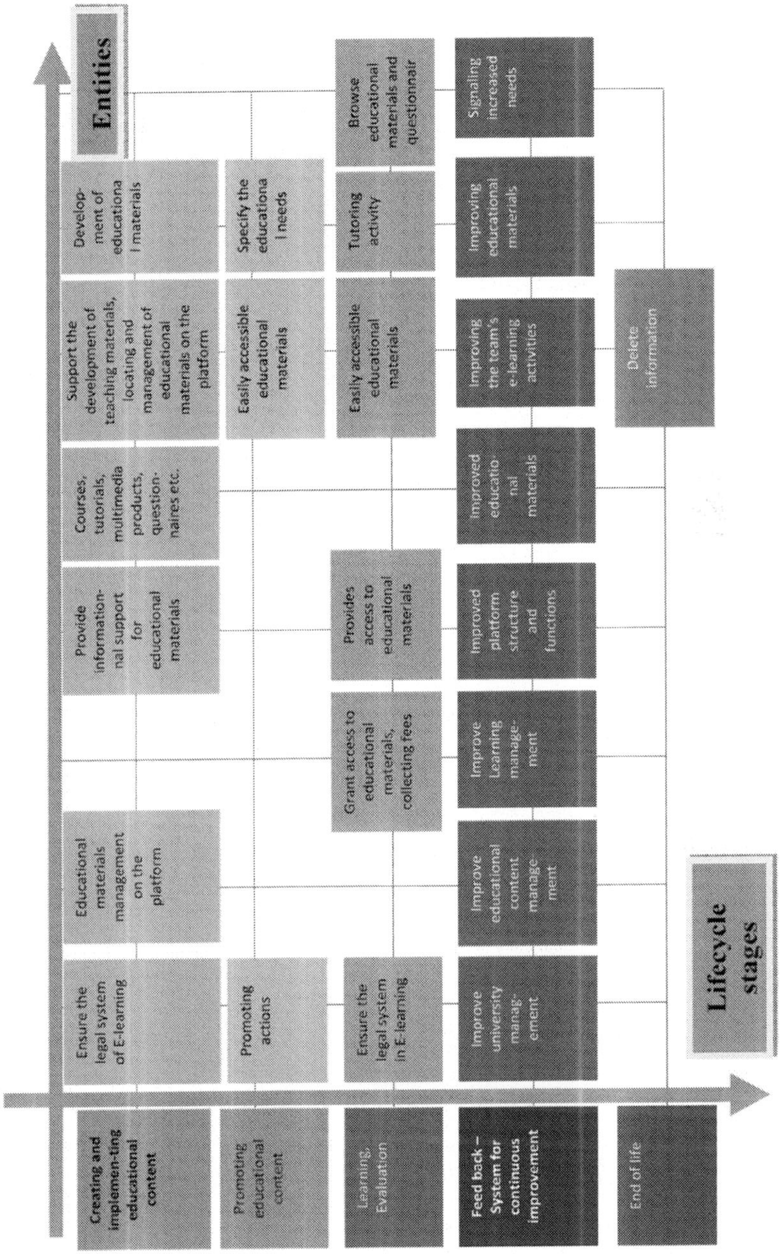

Fig. 1b. E-learning architecture

3. The positioning of the new course to the other offered courses (topic, audience and importance within the program) to predict the cost of implementation, efficiency and monitoring;
4. Completion and testing of the first versions of the course, including the ability to do the new course by the faculty team;
5. The economic analysis, identifying the economic potential of the new product, including effectiveness in learning, for students;
6. Developing and testing the new course based on specific methods and techniques, which results in a report of the chosen solution and a definition of its essential characteristics;
7. Market testing, which validates the proposed strategies and corrects errors done in the preview stage;
8. Promotion of the new product;
9. Launch of the course within the study program, which comprises its implementation and validation of the accuracy of the results obtained in the earlier stages of the "product" development (reality test).

2.3.2 Example development of an E-learning course

As an example, we will go into further detail upon an E-learning course in the area of cutting tools.

Its main aspects are:

- The objectives:
  - development of creative thinking applied to the field of cutting tools;
  - knowledge of the role of cutting tools in the technological system;
  - gaining an understanding of the design and manufacturing principles of cutting tools;
  - ability to make the right choice of tools for different machining operations;
  - identification of the key issues in the manufacturing and management of modern cutting tools;
  - getting users familiar with the software for constructive and technological design of the cutting tools;
  - creative design via the internet.
- The previous knowledge necessary for browsing this course encompasses:
  - the geometry of the elementary tools;
  - the parameters of cutting tools and the optimal choice for these parameters ;
  - surface generation;
  - cutting phenomena: chip formation, cutting forces, cutting tool wear, surface quality etc.
- The specific skills that will be gained at the completion of the course:
  - knowledge and understanding: Students will learn the role of cutting tools in the technological system; they will grasp the principles of construction and design of the cutting tools; they will know how to choose tools for various machining operations;
  - explanation and interpretation: the attaining of the technical general principles in the cutting tools area and ability to implement them; the discovery of known constructive solutions (tools), as well as the ability to develop new solutions by customizing generalized models;

- applied skill: students will identify key issues in the operation and management of modern cutting tools; they will work with software for constructive and technological design of cutting tools; they will attain an appropriate command of the technical language;
- attitude: the ability and habit of using modern means of documentation and communication; internet use, etc.

The application was focused on teaching the students creative design of cutting tools by using the Internet, and therefore the material lacks some specific elements (access to other web pages, specialty movies, group activities, work on the forum, etc.).

In figure 2 the main page of the cutting tool platform can be seen *(http://edesign.ulbsibiu.ro/moodle/).*

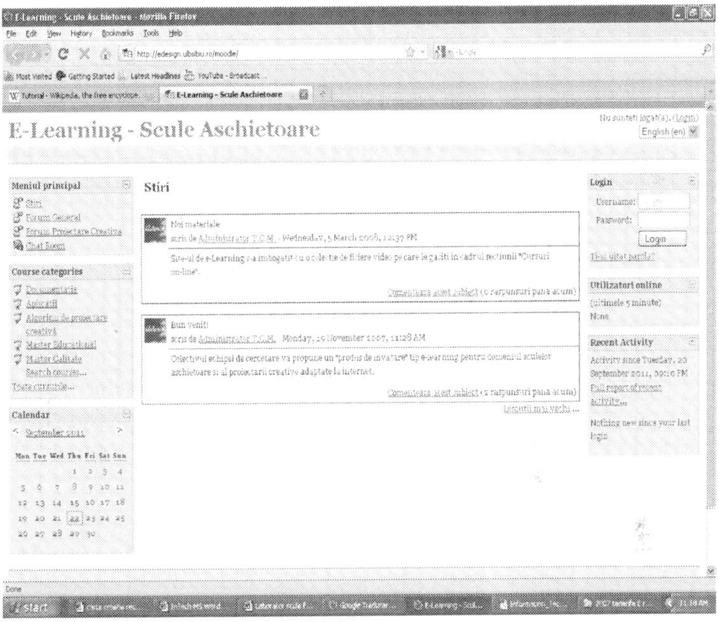

Fig. 2. Main page of cutting tool platform

The beneficiaries of such a training, are students who have a "Cutting Tools" discipline in their curriculum (Manufacturing Engineering program), as well as users from companies, who want advanced constructive solutions for cutting tools used in manufacturing processes and who co-opt specialists from several locations (consulting centres, designers, cutting tool providers, teachers, etc.).

The teaching strategy of the course is described in a written presentation of general "information packages", but also by modules (depending on the type of cutting tools), with drawings, images, animations and movies that explain the construction, the functioning, as well as further subtleties of the cutting process. In addition, references are made to various sites of the major cutting tool companies.

Afterwards, design algorithms are presented for different types of tools as well as separately, methods and creative design techniques.

An objective of the proposed on-line course is, as it has been pointed out, the comprehension of the creative design, on the Internet. Therefore, at the beginning, the designing of algorithms for each type of cutting tool is presented. This makes it possible to achieve the level of knowledge required in this area. The next steps are the study of creative techniques adapted to the Internet, ways of working in groups and, finally, team design, visualisation, on-line building of prototypes etc.

Regarding the assessment (for the presented case), complex online tests are used and gradually developed. After the bibliography and the specific materials (course) are completed and after reaching a certain level of competence, the E-learning system allows the formation of working groups with the participation of a professor, in order to design cutting tools using the Internet (e.g. on the forum).

The testing method provides a good balance between the objectives, their qualities and the time required for development. It also provides a high level of objectivity and repeatability.

A test sample used in the chapter on turning tools is given below, where knowledge is gradually targeted. The students identify at first the principles of operation and the construction of various parts of the turning tools. Only afterwards does he discover the constructive solution.

In order to develop complex tests with powerful imagery, the "Quizmaker" software was used. The program provides twenty kinds of questions with which tests can be built. These include dual-choice items, completion items, multiple choice items, etc. A picture, sound or video can be added to each question. Also, the score may vary from one question to another, and the percentage required for promotion may be changed. The default is 70%.

It is impossible to skip a question without giving an answer. It is possible to limit the time for each question or for the entire test.

You can choose to complete the test multiple times (as allowed by the creators of the test), and every time you start the answers will come in a different order. At the end of the test, the student can get a message informing him or her if he has passed. In the end he or she can see the correct and incorrect answers and print the results.

An example is presented next.

Choose the insert, the tool body and the clamping system in order to build a turning tool with changeable insert:

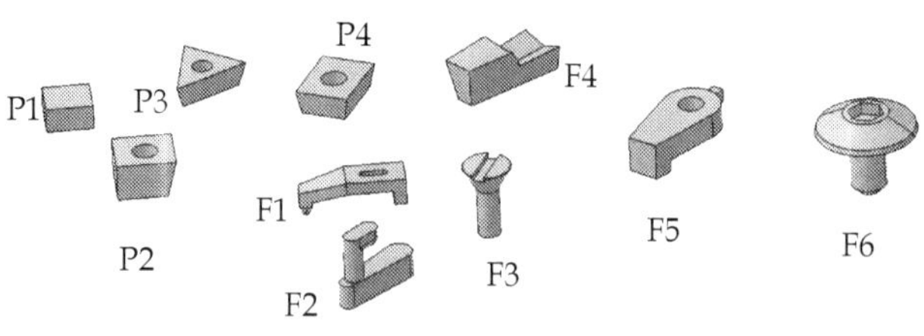

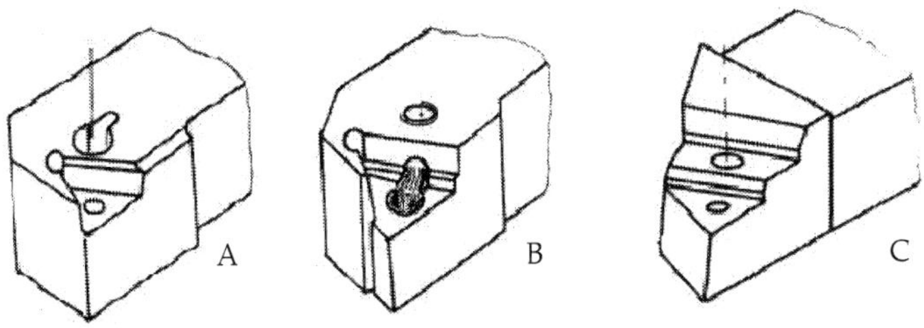

	Tool body	Insert	Clamping system
Turning tool			

The field of cutting tools is extremely dynamic and therefore continuous improvement is very important.

In our example, at some point the professors involved added a number of new features to the product: animations, videos and a streaming facility, through which users can observe the activity of the CNC Realmeca-Compa-ULBS Laboratory and of the Smart Factory Laboratory of the Faculty of Engineering (ULBS) remotely.

In order for the supervisors of the project to receive feedback, the students respond to a number questionnaires about the activity that they have performed, about the quality of the programme, about the the collection and analysis of tests etc.

2.4 Tutorials

The notion of a tutorial emerged together with the development of IT technologies.

The tutorial is a tool often used nowadays in various fields in order to present a specific software product or business activities-products of a given company, etc. In making a tutorial there are several concepts to be taken into account, such as:

- Entrain Multimedia presentations in business and education;
- Enlight concept of dynamic templates for online communication;
- Enlife web concept training and E-learning

The user has the option of creating material-support improving message and audience impact. Tutorials are used in different situations:

- as an auxiliary tool for university courses or other teaching activities;
- for supporting communication in scientific sessions, symposia, workshops, etc.;
- for communication between employees at the level of economic organizations;
- for communication between economic organizations, in the vendor-client relationship;
- for the dissemination of information via public computers (electronic information systems, electronic kiosks, info-kiosks);
- for development of tutorial-based applications, in closed systems, mixed or open (in the Internet environment).

The presentations are integrated with the Office software systems, being aligned with the principles of interaction and with the available instruments. They are geared towards solving the problems of communication and ensuring the possibility of personalization by including elements of interactivity with the receiver user. The development of several versions for each application leads to the diversification of methods, which range from a simple slideshow to rich multimedia applications.

The tutorial aims to introduce participants to the educational process in a given subject, with the final goal of assimilating a minimum level of knowledge. But there are some fundamental differences from other educational processes:

Educational methodology is rich in media content (static content, motion screenshots, animation or video instructions), which is necessary in order to exemplify the necessary steps in solving the problem.

- There is a focus on a particular topic, without details about adjacent concepts.
- A tutorial cannot replace a course, so theoretical aspects are usually overlooked in favour of practical ones. Theory is only mentioned when it becomes absolutely necessary to explain an applied situation.
- In general, if software is permanently installed, the tutorial makes reference to the way of using multimedia software;
- A minimum of IT resources and documentation (for the hardware, the software and the prerequisite knowledge), time and motivation are the only things necessary to acquire a tutorial.

2.4.1 The steps involved in making a tutorial

Usually the tutorial starts with an introductory section that informs the student about the objectives and the nature of the lesson (Borza et al., 2007). The information is then presented in a determined form. Questions that are addressed to the student receive an answer. The program considers the student's response and provides a reaction in order to strengthen the understanding and to increase student performance.

The constituent parts of the tutorial are:

- Introduction;
- Monitoring of the student on browsing the lesson;
- Motivating the student;
- Presentation of the information;
- Questions and answers;
- Answer analysis;
- Additional guidance depending on the accuracy of the responses;
- Sequencing/Segmentation of the lesson;
- Completion of the tutorial.

Usually there are several steps to be followed when creating a tutorial:

1. establishment of the subject that will be treated in the tutorial;
2. determining the target audience who should attend the tutorial. The difficulty level, as well as the estimated time of browsing must be determined. These two affect the degree of motivation for learning. It is preferable to have a short tutorial that covers a more limited topic, instead a richer one that demands for a lot more effort.

3. choice of the type of technology to be used. The elaboration of a tutorial that explains a software application requires complex IT elements.
4. compiling, structuring and placing information in the tutorial;
5. tutorial test, and correction of any conceptual and/or implementation mistakes that may arise

2.4.2 Multimedia

Multimedia is everything you can hear or see: texts, books, pictures, music, sounds, CDs, videos, DVDs, Records, Films, and more. Multimedia comes in many different formats. On the Internet you will find many of these elements embedded in web pages, and today's web browsers have support for a number of multimedia formats. The support for sounds, animations and videos is handled in different ways by different browsers. Some elements can be handled inline, some requires a plug-in and some requires an Active control. Multimedia elements (like sounds or videos) are stored in media files. The most common way to discover the media type is to look at the file extension. When a browser sees the file extensions .htm or .html, it will assume that the file is an HTML page. The .xml extension indicates an XML file and the .css extension indicates a style sheet.

Picture formats are recognized by extensions like .gif and .jpg. Multimedia elements also have their own file formats with different extensions.

2.4.3 Multimedia sound formats

The WAVE (waveform) format is developed by IBM and Microsoft. It is supported by all computers running Windows, and by all the most popular web browsers. Sounds stored in the WAVE format have the extension .wav.

MP3 files are actually MPEG files. But the MPEG format was originally developed for video by the Moving Pictures Experts Group. We can say that MP3 files are the sound part of the MPEG video format. MP3 is one of the most popular sound formats for music recording. The MP3 encoding system combines good compression (small files) with high quality. Expect all your future software systems to support it. Sounds stored in the MP3 format have the extension .mp3, or .mpga (for MPG Audio). The AIFF (Audio Interchange File Format) was developed by Apple. AIFF files are not cross-platform and the format is not supported by all web browsers. Sounds stored in the AIFF format have the extension .aif or .aiff.

2.4.4 Multimedia video formats

The AVI (Audio Video Interleave) format was developed by Microsoft. The AVI format is supported by all computers running Windows, and by all the most popular web browsers. It is a very common format on the Internet, but not always possible to play on non-Windows computers. Videos stored in the AVI format have the extension .avi.

Microsoft develops the Windows Media format. Windows Media is a common format on the Internet, but Windows Media movies cannot be played on non-Windows computer without an extra (free) component installed. Some later Windows Media movies cannot play at all on non-Windows computers because no player is available. Videos stored in the Windows Media format have the extension .wmv.

The MPEG (Moving Pictures Expert Group) format is the most popular format on the Internet. It is cross-platform, and supported by all the most popular web browsers. Videos stored in the MPEG format have the extension .mpg or .mpeg.

Apple develops the QuickTime format. QuickTime is a common format on the Internet, but QuickTime movies cannot be played on a Windows computer without an extra (free) component installed. Videos stored in the QuickTime format have the extension .mov.

To achieve multimedia tutorials one can use both Internet and INTRANET technologies based on what was mentioned above. This depends the type of software used. For image processing, one can use software like: Photoshop, Corel Draw, or open source products, like GIMP. To achieve animation we can use the FLASH product or software that captures the screen image like Camtasia. We will write some words about Camtasia in the "Video Tutorials" section of this paper.

2.4.5 Streaming applications

The increase of the accessibility of video equipment has allowed the development of online educational resources. Making interactive distance has now become possible by following a multimedia technology developed recently, namely streaming technology.

Streaming data is transmitted in the form of streams, which refers to the ability of an application to play synchronized media data streams in a continuous manner, while those streams are transmitted to the client through a data network. Streamed data is divided into packets whose size is suitable for transmission from one server to the client. Thusly, a user can view a package, while at the same time receiving and uncompressing the following parts. The great advantage here is that they don't have to wait for the end of the transmission in order to view the material.

Applications that can be built using streaming services are viewed in figure 3 and divided into:

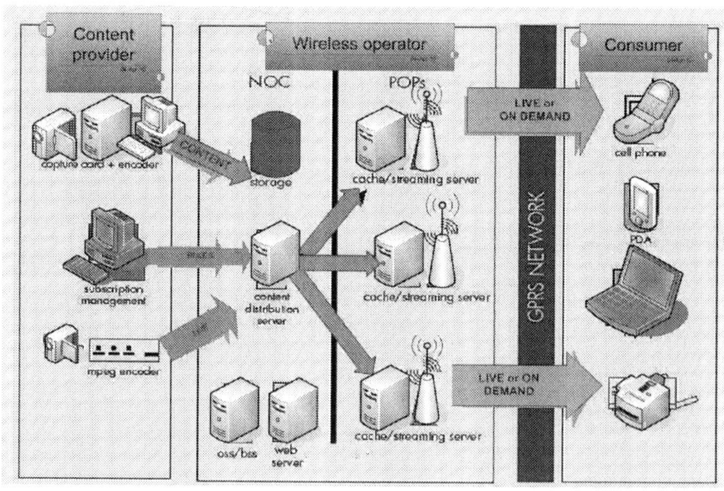

Fig. 3. Streaming Applications

- on demand applications - news, music, movies, multimedia tutorials;
- live information - radio and TV programs directly.

On-demand streaming is based on files stored on a server for a longer period. Files are available to customers as they submit their requests to the server. Live streaming videos are transmissions that occur as the performances or events depicted take place. Streaming technology has caused a profound change in education, business, media and entertainment, as it combines the richness of television content with the interactivity of the Internet.

2.4.6 Video tutorial in engineering educational process

Video tutorials are not a new concept. The concept has emerged in the late 1960s, when Bandura, first commented on the issue of self-modelling. Since then, video modelling has been described as "instructional videos" or "video tutorials". Most video tutorials are produced as part of a strategy to promote guidance in developing new skills and abilities - for example, to improve teacher training or to prepare medical students for their internship experience. Also, video tutorials have been used to assess their effectiveness in improving the learning efficacy of students.

Video tutorials are one of the most important factors in the improvement of the educational process. As we also mentioned in the conclusions of this work, the video tutorial has a strong impact on students. It can exist in two modes. The first one is as a separate entity with no text. In this case several independent tutorials treat a particular topic. The second mode is having it embedded in a website, or as part of a topic that is presented in the context of web page or slideshow. In both cases the video tutorial is recorded using specialized software.

By offering audio-visual demonstrations, the teacher can use a video tutorial to develop a particular skill or strategy. Thus, video tutorials can be used as a medium to encourage students to learn independently outside the regular classroom schedule. The materials presented in the audio explanations and the visual demonstrations become achievable goals that the learner can reach by understanding and applying a similar process to produce their own project. The video tutorial resources may used to engage learners to explore their technical skills. Exposure to software operation in the target language might empower students to become familiar with the target technical lexical items thus enabling them to discuss their difficulties with their teacher or peer. Hence, the teachers' role becomes one of facilitator or guide when a student needs further assistance, and their oral interaction has the potential to change from language learning to purposeful communication exchange whereby problem-solving is the focus.

2.4.7 Making video tutorials using the Camtasia software

Camtasia is one of the applications that can be used to make a video tutorial. TechSmith produces it. Video tutorials created with Camtasia assists in creating demonstration videos by recording computer screen activities. For example, one could video record the screen as they type. Even though Camtasia is fairly user friendly, some video editing skills are necessary.

After recording the screen activity, Camtasia stores the recording on its editing environment. During the editing stage it is possible to add narration, music, photos or other movie elements. Finally the finished video tutorial can be saved - either on a CD or DVD, or, it can be uploaded on a website for online access.

2.4.8 Example of a video tutorial for learning the G language used in mechanical engineering

To ease the study of the programming language G, an on-line tutorial has been designed, which uses multimedia elements to provide examples for the main notions involved. The tutorial contains 17 lessons (fig. 4.), which are titled as follows: About Labview, Windows,

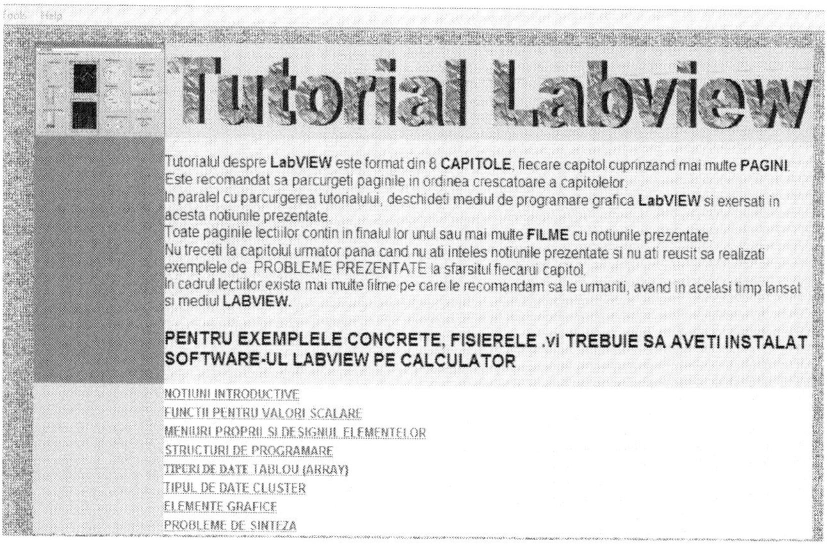

Fig. 4. The Home page of G Language Tutorial

Build VI, Data Type, SubVI, Debug, While Loop, For Loop, Array, Clusters, Case Structure, Sequence structure, String, Files, Build an Application. Each lesson contains text and pictures, in which the content is illustrated. The tutorial begins with an introduction that provides some initial background (figure 5.). "LabVIEW is a tool by National Instruments Company. It is an integrated development environment (with graphic interface), based on the G language. It is different from other development environments in that programming in LabVIEW is based on the concept of connecting icons.

When using LabVIEW there is no need to write any source code, it is enough to create a graphic representation of the desired program, using icons. Connected icons make up diagrams representing the data flow from the data sources to the program output. The inputs and outputs can be either external devices or user interfaces (on a diagram they are represented by an appropriate icon). This unified interface between LabVIEW and external devices makes it easy to set up the computer as a supervising and controlling device. LabVIEW is often used in research centres".

3. E-design, E-research

3.1 E-design, E-creative design a redefined notion

A definition of the notion of "E-design" could be "the design process that occurs on the Internet with the help of the computer". The E-design concept is starting to be better known and used, even though for the moment it is only more prominent in the design and consultation sites in the field of arts and architecture. E-technological equipment in the design is still in its infancy.

By applying creative methods in the aforementioned processes, one can achieve a more advanced e-design, namely E-creative design (Brîndasu & Beju, 2007).

This notion has appeared in the technical literature, but under a different meaning - the meaning of an original and effective platform. The team from Sibiu, however, considers that this term, namely E-creative design, is a superior stage of the Internet design for using creative methods specific to group activities.

Some benefits derived from implementation of an E-creative design in technological equipment are (Creswell, 2009):

- A decrease in design time;
- An increase in the degree of innovation (e.g. cutting tools);
- Use of the capacity of the most valuable resource (human, material) in the project;
- Generation of new tools, technologies and networks for specific applications;
- An ensuring of the compatibility requirements for the integration in the social / economic Europe.

The adaptation of the methods and techniques of creative design for the network involves the establishment of working procedures, the enabling of the design (e.g. Cutting tools) in an innovative way, as soon as possible. That should, however, not hinder the work of the designer (expert), but rather it should facilitate and enable the participation of the consumer, even to the process itself.

3.2 E-research

The concept of E-research refers to research that is carried out in virtual teams, using the Internet, and to which the participants lie in various locations.

The research methodology prescribes the following steps:

- description of the problem. This must be performed as clearly as possible, as it determines the type of research to be carried out.
- research planning using project management methodology;
- tracking of the actual conduct of research;
- design and implementation of a prototype;
- experiments on the prototype;
- final analysis - interpretation and dissemination of the results;
- audit of the project;
- lessons for the "use" and shipment the product to the customer.

3.3 Trends in design-research centres

The trend in terms of research and design centres is international cooperation and collaboration. These can be achieved by organizing virtual research and design centres.

The development of communications and computer technologies (ICT) offers the possibility of using the Internet in research design activities (E-design E-research), within virtual centres (Kincheloe, 2006).

Virtual research combines a variety of activities, as follows:

- Information retrieval is eased by the search capabilities of the platform;
- Identification of the general objectives should only be made by taking into account the world trends;
- Whenever goals are set for shorter terms, their planning can be achieved with better accuracy;
- The organization of resources is much more efficient due to the wider access to human resources. Physical attendance is unnecessary, user on-line connectivity is sufficient;
- The realization phase in the virtual research centres is performed much faster due to greater information transfer speed;
- The research results can better meet the initial needs because of on-line prototyping. Whenever the prototype development indicates a deviation from the proposed objectives, corrective action can be undertaken.
- Virtual events can take place in the virtual centre, in a chat room, or a forum;
- Advice can be provided via email, chat;
- Cooperation in solving problems or discussions in the study group can be achieved through the forum or chat, virtual communities, international scientific collaboration via Web/Internet;
- The system allows shared editing of documents.

The major objective of the future research/design centres has to be the improvement in the performance of the centre for international exchange of knowledge.

Future research-design centres can be viewed as appropriate mechanisms to establish links with world research, as well as to prevent the need for researchers to work "behind the walls". This could be a solution to stop the migration of researchers in countries who have a much higher technological development of information and communication.

A proposal for organizing a research-design activity within the virtual research-design centres was put forward in the domain of mechanical engineering and, more specifically, in the field of cutting tools, by using the E-creative design concept.

3.4 E-research – E-design methodology

Because the research activities are carried out in teams and because researchers are located in different places, there is a clear need for a precise methodology for communication research. These are supported by the TEAM CENTER software.

The stages of a research methodology are:

- clear description of the problem;

- determining the type of research that needs to be done;
- planning of the research, using the methodology of "Project Management";
- research activity and tracking;
- the design and manufacturing of a prototype (if necessary);
- experiments and analysis of the final product;
- the interpretation and dissemination of results;
- completion of the audit of the project and raising awareness upon the "lessons" to be learned;
- research delivery to the beneficiary.

The Industrial Engineering department from „Lucian Blaga" University of Sibiu did a research design platform focused on creativity. The main theoretical and practical issues of the research activity performed on this platform are presented below. For an easier understanding, an example was used.

3.4.1 Problem definition

In management and engineering, a problem can be defined as the difference between the real and the desired situation. In general, the real causes are not known and a careful analysis of the root causes is needed in order to determine the true nature of the problem. Correct and complete understanding of the issue is very important and necessary in every situation, especially in the situation of a collaborative network, in which participants exchange information in synchronous or asynchronous mode. The problem needs to be explained in a textual and/or graphical way and afterwards communicated over the network.

The main steps that must be followed in order to define a problem in the communication network are as follows:

- the formation of a virtual team for defining the problem;
- individual studies or E-sessions, performed in order to determine the fields (e.g. technical, managerial, economical, ecological, legal, etc.) and in order to identify key indicators, parameters or behaviours. Additionally potential feedback loops caused by different parameters, contradictions, as well as other risks and uncertainties have to be acknowledged;
- E-sessions, in order to determine the problem, as well as research and design topics.

In order to communicate information in network in the best possible manner, it is very effective to use the methods of the "root cause analysis". These methods are as follows: 5Whys, Ishikawa diagrams, failure mode and effects analysis, fault tree analysis, Pareto analysis, inference, TRIZ methodology etc.

In order to perform a correct and complete definition of the problem, it is useful to organize several E-session with that purpose.

The theme of the provided example was, as shown in figure 5, the designing of a boring head. This boring head was required to process at several diameters, as well as to create a chamfer at the end of the hole (that the chamfer tool could cut different dimensions depending on the boring head diameter). The material to be processed had "high" features and the manufacturing series was small.

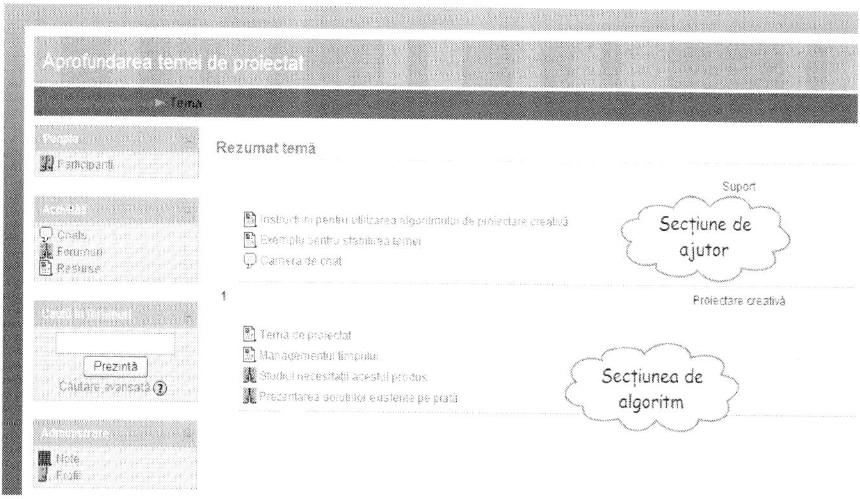

Fig. 5. E design - e research platform (problem definition, research planning, presentation of the up to date information)

3.4.2 Determining the type of research that needs to be done

After defining the precise research topic it is important to determine the types of research to be conducted. Each research concludes with a model. There are several types of models:

- Mathematical models (analytic). In this case a formal theory is developed, and the results of this theory can be compared against empirical values;
- Experimental models. These models require an experimental design observation of the phenomenon, data collection, data processing, and the validation of the model. There are three types of experimental research based on the level of understanding of the model:
  - *scientific models*: The phenomenon is carefully observed and the model proposes a theory to explain it. Measurements and analyses will be carried out to validate the model;
  - *engineering models*; Existing solutions are being studied, measurements are being undertaken and improvements of these solutions are proposed and implemented. These activities are repeated until no more optimisations are required;
  - *empirical models;* In this case there is no need for a theory or an explanation for the phenomenon. The model works as a black box with inputs and outputs. Some experiments are usually performed, together with a statistical research, analysis and validation of the model.

Experimental investigations are run through a few steps, namely:

- Preliminary design of the experiment. The variables are selected, as well as their level of investigation. It is necessary to settle the parameters to be measured, their characteristics, measurement units, format, etc.;

- Choice of the most suitable model for the desired outcome. The main experimental methods of research are: full factorial Experiment, fractional factorial Experiment, Latin Square, Taguchi Method;
- Conducting experiments and elaboration of the model;
- Verification of the model. A few questions are typically asked:
  - The model reflects the system target in an acceptable way?
  - For input data, does the simulation provide the correct output data?
  - Are there statistical methods that allow the experimental verification of the model? The mostly used method is ANOVA.

To solve this issue, the chosen option was to build a representative model that describes all the manufacturing possibilities of the drilling heads. This model consists of an imaginary body with several steps, which can do several movements and has different adjustment possibilities. It also possesses several active edges, all with different shapes and positions.

Through customization, one can get different boring heads with various functions. The need of the representative tool is not only aimed towards a particular solution, but also towards synthesising the knowledge in this area and completing the network database.

3.4.3 Research planning, use of project management methodology

After the description of the problem and the choice of the research type that needs to be done are completed, the next step is the planning of research. This is performed by using the project management methodology. The most often used software in this field is Microsoft Project. The main stages of the planning process are shown in figure 6.

Outside of research or optimization, a project can contain various modules for product design, design technology, prototype design and/or prototype technology design.

In order to make a realistic plan, it is very important to have a network database with complete information about: researchers, laboratories, knowledge in various fields, costs, resources available, etc.

3.4.4 The research activity

Proper implementation of the project must be made in accordance with the timetable of the project. Creativity in design research meetings (fig.7) may be carried out between research team members in a synchronous or asynchronous fashion. Experimental activities, the drawing up of a model and the design and manufacturing activities may be undertaken in one or more locations. The current software enables organising meetings with the participation of specialists located in different geographical areas (Beju & Brindasu 2008).

Software products such as "Teamviewer" and "Teamcenter" enable that all participants of such a remote meeting make the viewing, the modifying and the completion of the drawings. The final version of the model (drawing) can be discussed by all of the participants and accepted by consensus or by vote. In order to develop new solutions, it is necessary to establish a virtual team. The methodology for choosing the team members is shown below in the pseudocode:

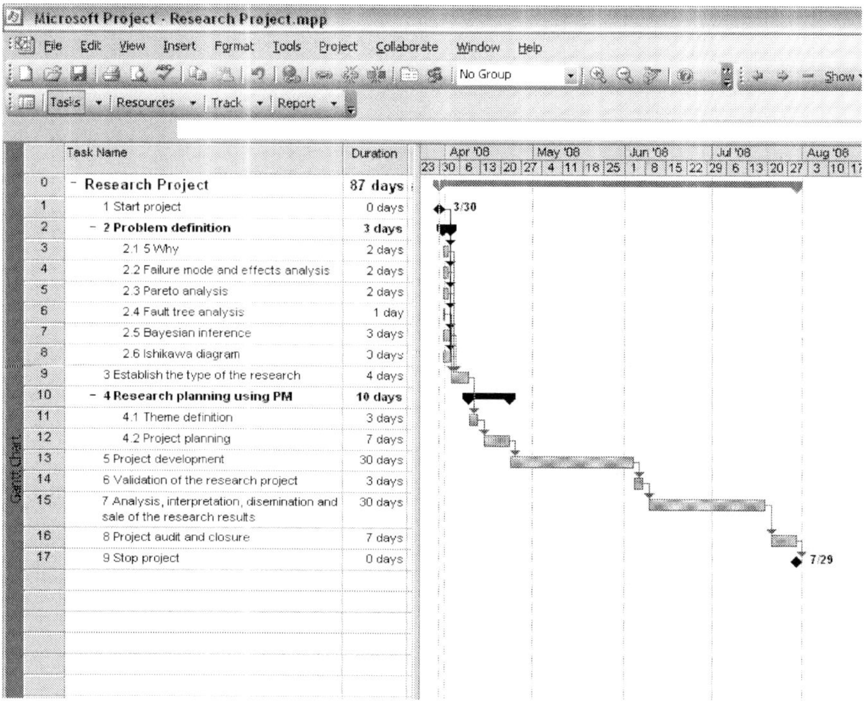

Fig. 6. Project planning

```
ESTABLISH the members of the virtual E-research Group;
DO WHILE the E-research group is complete
  CONTACT a member of the virtual E-research group;
  IF the person agrees to participate
    THEN
      EXPLAIN the working principles;
            IF the level of knowledge must be checked (students)
             THEN
                    The person solves a test;
                    IF the test results are good
                       THEN the person is accepted in group
                             ELSE select another person from the virtual group
      END IF
            ELSE
                    Accept the person in the group;
           END IF
  END IF
END DO
```

Research design sessions on the network can be done synchronously or asynchronously.

For the synchronous sessions, the stages can be:

- ESTABLISH working team members;
- INFORM research - design theme- to all group members by email;
- ORGANIZE a network meeting for establishing the analysis criteria. Each member will rate all the criteria. A specific software will determine the most important criteria as well as an indicator of their importance;
- ORGANIZE a networking session with the aim to develop new constructive solutions in connection with the proposed theme and in accordance with the criteria set out above - by using a creativity method;
- ORGANIZE a networking session with the aim of analyzing the existing constructive solutions based on the criteria set out above (consider all possible variants -the Morphological Method is very useful for this task);
- ORGANIZE a new session on the network (if necessary).

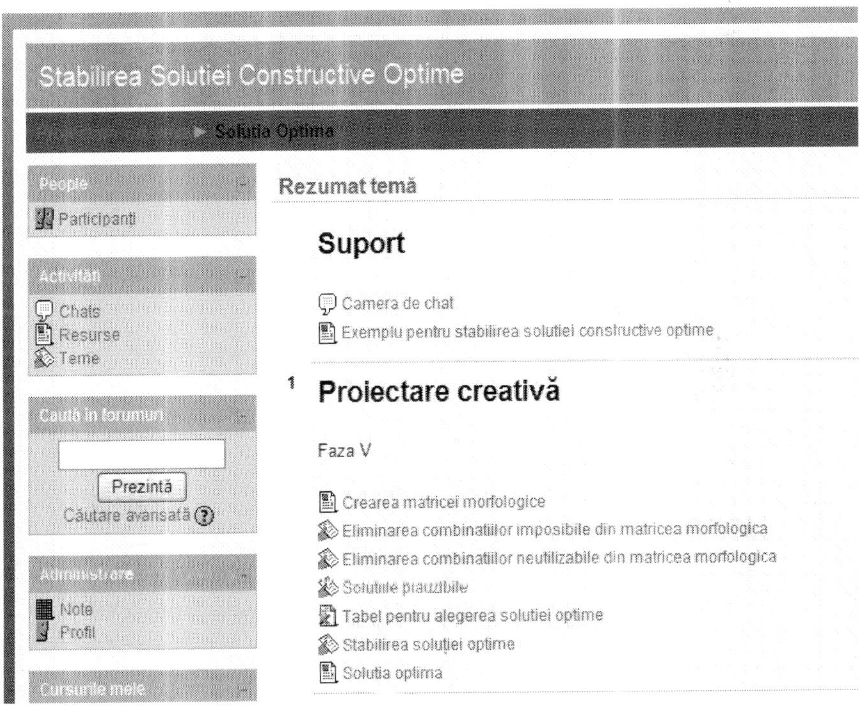

Fig. 7. E-design - E-research platform (Creative design page)

Acordati note solutiilor plauzibile in functie de criteriile de clasificare.

Notele reprezinta:

Valoarea	Seminificatie	Valoarea	Semnificatie
0	Nu este tolerabil	6	Solutie buna dar cu lipsuri
1	Insuficient	7	Solutie buna fara lipsuri
2	Solutie dificila	8	Solutie foarte buna
3	Acceptabil	9	Mai mult decat presupus
4	Satisfacator	10	Ideal
5	Foarte satisfacator		

Acordati note completand tabelul:

Criteriul de analiza	Uşurinţa în schimbarea zonei active		Preţul total al produsului		Fiabilitate		
Soluţia constructivă	Nota	Nota* Pondere(0.45)	Nota	Nota* Pondere(0.33)	Nota	Nota* Pondere(0.22)	∑Nota* Pondere=
T1P1M1	7	3.15	9	2.97	6	1.32	7.44
T2P1M1	7	3.15	9	2.97	7	1.54	7.66
T3P1M1	7	3.15	9	2.97	6	1.32	7.44
T4P1M1	7	3.15	9	2.97	7	1.54	7.66
T5P1M1	7	3.15	8	2.64	7	1.54	7.33
T1P2M1	10	4.5	10	3.3	8	1.76	9.56
T2P2M1	10	4.5	10	3.3	9	1.98	9.78
T3P2M1	10	4.5	10	3.3	9	1.98	9.78
T4P2M1	10	4.5	10	3.3	10	2.2	10

Fig. 8. E-design, E-research platform (page for establishing the best solution)

For asynchronous E-sessions, the problem definition, the research themes, as well as other necessary documents are sent to the team members, as showcased in the algorithm below:

- Each participant will receive a code (e.g. P1, P2, P3, P4, P5, P6);
- After the acknowledgement of the problem, each participant will write three proposals to solve the problems;
- The file is sent from one participant to the other in the order of their codes, until every participant has gotten hold of it
- Each participant fills in details, makes improvements, establishes details, makes comments on the previous ideas, until all the ideas are discussed by all the participants;
- One of the participants (usually the project manager) analyses, groups and compares the proposals (value analysis can be used, by marking each solution for all the proposed criteria and by all the team members);
- The project manager determines the best solution based on the marks awarded during the previous step
- A preliminary design of this very solution is performed;
- The chosen solution is analysed by all team members;

- At the end, the chosen solution is optimised (fig.8) . All attendees will once again award marks to the found solution. To adopt it, in general, five of the six team members must award the highest mark. Otherwise, the process starts again, until the best solution is obtained.

E-sessions can be organized in a classical manner or by using creative methods. The creative methods have the advantage of overcoming the psychological inertia. A number of methods can be used: consonance, extrapolation, combination, analogy, inversion, empathy, modifying, improving, INPUT-OUTPUT techniques, interrogative lists. These E-sessions are organized by topic, by the type of research, by the importance of the theme, by team skills, etc.

In the proposed case, a synchronous network meeting was organised, in order to build a representative model of boring tools. Some creativity methods were used. In order to break the psychological inertia of participants, questions were asked, such as:

- Imagine yourself in the place of the boring tool. How would you work easier?
- What material would you prefer being built of?
- From what part of you would you like cutting fluid to be provided?
- How would the edges better be placed?
- What happens if you look at the process from inside?....

The main purpose of the on-line creativity sessions was the representation of several modules with cylindrical or toroidal shapes. Based on these shapes, several inserts were placed on the cylindrical side and/or on the frontal side, in radial or tangent position.

After these steps, the final shape of the representative tool (fig.9) for hole manufacturing was obtained in asynchronous communication. The representative model (fig. 9) consists of several conical and cylindrical bodies placed on the same axis (the z-axis). The main form of the tool can be spindle or hole type. On the cylindrical, conical and frontal surfaces, inserts are placed in radial or tangent positions (Popescu et al., 2010).

Through customization, several real boring heads was obtained (fig.10).

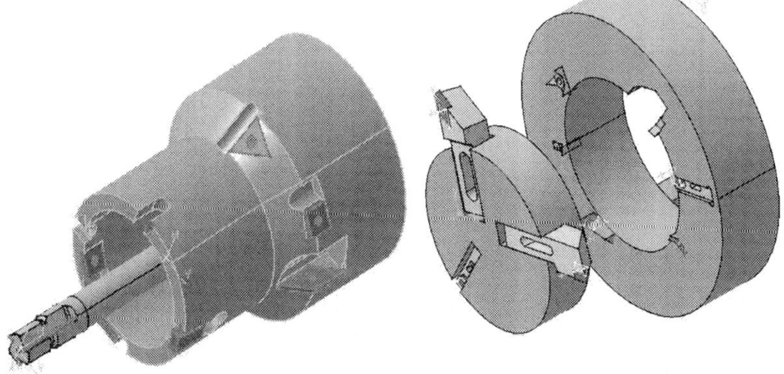

Fig. 9. Representative model of the boring tools

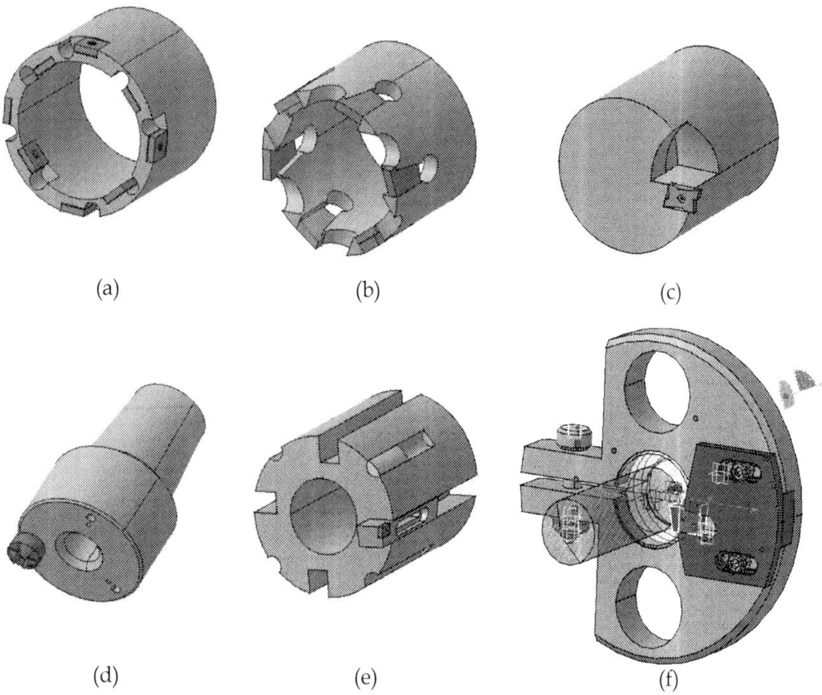

Fig. 10. Boring heads (a) with tangential inserts attached to the cylindrical part, (b) with inserts placed on the frontal side by the elasticity of the tool body, (c) with tangential insert placed on the frontal face, (d) with insert with several edges, (e) with tangential insert placed the cylindrical side of an adjustable body, (f) modular tool)

A more outstanding solution is a modular boring tool with a carbide helical drill or a drill with carbide inserts (in the case of larger diameter) and a special adjustable body for chamfer. The insert is fixed due to the elasticity of the tool body (fig. 10,f). This was the solution, which was finally chosen by the research team.

The following step represents the optimisation of the adopted solution.

Optimisation criteria were:

- small tension values, distributed as uniformly as possible;
- the type and form of the clamping system;

To this solution, there were several aspects that could be optimized: the shape of the insert, the clamping system and the tool body. Optimization began with the estimation of the forces. The study continued with a static (fig.11) and dynamic modelling process, and it finalised with a Finite Element Method analysis.

The clamping system due to the elasticity of the tool body is very economical in terms of space. It is necessary to precisely determine the shape of the elements, in order to obtain a secure and stable insert. The purpose of this phase of research was to determine the inclination of the insert and of the slot. Solving of the equation allowed the establishing of the optimal values θ_1=2…15° and θ_2=2…6°.

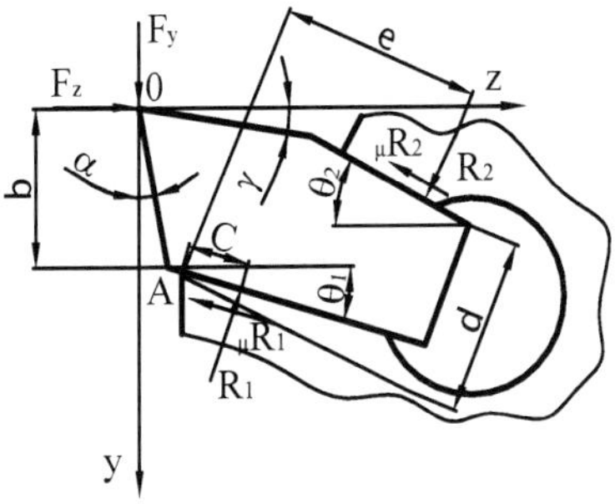

Fig. 11. Equilibrium analysis of the clamping system due to elasticity of the tool body

3.4.5 Validation of the project results

The result of the research project was discussed and validated by all the team members. Typically, other specialists and the beneficiaries are allowed to attend the synchronous meetings.

3.4.6 Analysis, interpretation, dissemination and sale of research results

The analysis of the results is performed on different processes, using the theoretical and experimental data. The results are graphics, tables and databases. These synthetic results can be presented in reports, research papers, databases, E-learning products etc.

3.4.7 Audit of the research project

The completion of the audit of the project is of great importance because it allows:

- to establish the manner in which the project was done;
- to ensure the satisfaction of the client;
- to track the flaws of the project during its development, as well as the causes of its success;
- to identify the changes that can improve it in the future.

Audit and project reports are tools of continuous improvement of the organization.

3.4.8 Delivery of the research results to the beneficiary

Accomplished research results should be delivered to the beneficiary in a useful manner.

4. Conclusions

Changes in the economic, social, political areas, as well as the availability of information determine changes in the needs and educational offers. One successful solution is E-learning. In order to respond to the dynamic needs of students, the E-learning educational systems must be modelled using the life cycle management and reference architectures. This paper presents an original model for reference architecture in the field of E-learning.

Among the most important activities that were highlighted by this reference architecture were the methodology for the realization of on-line courses and tutorials. These same activities are also the ones that raised the most important issues to the authors during their work. This paper presents the steps for elaborating an on-line course, as well as practical aspects that appeared in designing courses for mechanical engineering and cutting tools. The stages of developing a tutorial were also presented, alongside the main software products used in this field.

As for the issue of modern research and design, the team from Sibiu has redefined the notion of creative design, a process that uses creative methods in the Internet-based remote design process. Additionally, a model that describes the necessary stages of a remote research design activity was elaborated and an example of creative design carried out in the field of cutting tool was also presented.

The models presented in this paper correspond to the present-day information technology available. Future changes in technology will lead to changes in the hereby-presented technology.

5. References

Beju, L.D., Brindasu, P.D. (2008) Methodology For Organizing E-Creative Research And E-Creative Design Sessions, Proceedings of the 19th International DAAAM Symposium, Pages: 83-84, ISSN 1726-9679, ISBN 978-3-901509-68-1, Editor Branko Katalinic, October 2008, Trnava, SLOVAKIA

Borza, S; Brindasu, P.D.; Beju, L.D. (2007) E-research centre and E creative design, new trends for E activities platform, Proceedings of The 6th WSEAS International Conference on E-Activities Pages: 86-93, December 2007, Puerto de la Cruz, SPAIN

Brîndaşu, P.D., Beju, L.D. (2007) E-creative Design for Cutting Tools Area, Annals of the MTeM for 2007 and Proceedings of the 8th International Conference Modern Technologies in Manufacturing , ISBN 973-7937-07-04, Editor Cs Gyenge, October 2007, Cluj Napoca, Romania

Butz, M. (2006) Rule based evolutionary Online Learning System, Springer Verlag, ISBN 3-540-25379-3, Berlin Heidelberg.

Creswell, J. (2009) Research Design. Qualitative, Quantitative and Mixed Method Approaches, Sage Publication, ISBN 978-1-4129-6556-9, London.

Kincheloe, J. (2006) Teachers as Researchers. Qualitative Inquiry as a Path to Empowerment, Routledge, Taylor & Francis Group, ISBN 0-415-27645-4, London

Popescu, L.G., Brindasu, PD; Beju, LD (2010) The Strategic Analysis Model For A Flexible Virtual Research Centre, Proceedings of the 6th International Seminar on the Quality Management in Higher Education, vol I, Pages: 659-662, July 2010, Tulcea ROMANIA

26

Mechanical Engineering Education: Preschool to Graduate School

Emily M. Hunt[1], Pamela Lockwood-Cooke[2] and Michelle L. Pantoya[3]
[1]*Engineering and Computer Science West Texas A&M University,*
[2]*Mathematics, Physics, and Chemistry West Texas A&M University,*
[3]*Mechanical Engineering Department Texas Tech University,*
USA

1. Introduction

Google decided to re-invent television by creating Google TV: which is basically software that can access everything available on regular television channels and the vast sea of content on the Internet, all on the biggest screen in the house. One motivation was to transform their current 1 billion market share associated with computer and hand held browsers to 4 billion TV watchers. When this feat is accomplished, the current statistics that cite 70% of 4 to 6 year olds have used computers and been exposed to the Internet prior to kindergarten will likely increase to 100%. In these exciting times there is a need to integrate this multi-modal influence into engineering education on a massive scale. According to studies, this new generation of Millenials (born early 1980-2000) places significant emphasis on meaningful careers. By introducing impactful, engineering education to this generation by integrating literature, technology, and successful teaching and learning methods into their culture, there are no limits to the meaningful contributions that future engineers will make toward improving our way of life. This chapter will highlight mechanical engineering education from kindergarten to functioning society member. We will discuss what works and how it works with the new student and citizen of today.

2. Early engineering literacy

Engineering education at the youngest ages is largely predicated on hands-on activities using manipulatives such as *LEGOS™* [1]. But at the young ages of P-2nd grade, there is significant emphasis on language and literacy skills such that little time is devoted to science or engineering education in the classroom [2, 3]. Therefore, integrating engineering concepts into language and literary skills designed for young children could impact the early development of engineering thinking while simultaneously enabling more instruction and exposure to engineering concepts than currently exist. It is important to understand how purposefully prepared engineering literature presented in the format of picture book children's stories impacts learning in emergent readers. The influence of literature on children's thinking about *engineering* and the connection children make between science and engineering can be observed through illustrative data and feedback after exposure to engineering literature [4-8].

By first grade, readers have developed an understanding of the alphabet, phonological awareness, and early phonics [2, 3]. They have command of a significant number of high-frequency words and developing a much better grasp of comprehension strategies and word-attack skills. They can recognize different types of text, particularly fiction and nonfiction, and recognize that reading has a variety of purposes. Typically books for this reading level contain: increasingly more lines of print per page, more complex sentence structure, and less dependency on repetitive pattern and pictures [3]. Examining this developmental reading level will enable a link between how engineering literature is presented and how children process the information[8]. Researchers are currently working to create improved books targeting this specific developmental level. Engineering books available at this developmental level are severely limited.

The idea that engineering learning could be promoted through literature is supported by the theoretical perspectives of situated cognition and distributed cognition [9-14]. Especially from the perspective of a young child, engineering activities can be described as socio-cultural such that a person's cognition is enmeshed with a situation and activity in a community of practice [9]. In other words, *concepts* are formed by both *culture* and *activity*, and the meaningfulness of learning is constrained by all three conditions. In this way, the literature needs to present an engineering concept in the framework of a culture (i.e., characters in a story) ensnarled in an activity (i.e., venturing through the story's plot) [13].

Engineering Elephants [6] is a children's book that introduces the engineering profession as well as fundamental Science, Technology, Engineering, and Mathematics (STEM) concepts to young children. The book teaches children about relevant topics such as nanotechnology, renewable energy, and prosthetics by engaging them through an interactive journey of an elephant and his questioning of the world around him. The authors worked with early childhood literacy experts, science museums, and local school districts to strategically develop the text. The text was composed using the language of engineering (i.e., asking questions) and introduces vocabulary relevant to engineering using a narrative text structure and lyrical pattern children are familiar with as well as vibrant water color artwork that provide context clues and deeper understanding (see Fig. 1)[6].

Fig. 1. Excerpts from **Engineering Elephants** that illustrate the interactive, engaging presentation of engineering concepts tailored to young ages.

As an example of current research results in engineering literature, the following study will be discussed. A group of children is examined where half had been exposed to **Engineering Elephants** and the other half had not (the control group). After reading and discussing the story, each classroom engaged in a creative paper-and-pencil activity in which the students were asked to draw what they would design if they were an engineer and explain their picture with corresponding text. This drawing assignment was also given to the classrooms that had not been exposed to **Engineering Elephants**. Figures 2 and 3 show representative illustrations from both groups of students.

Fig. 2. Engineering illustration by student not exposed to Engineering Elephants. Text reads *"Me Driving a Train in the Daytime"*.

Science is guided by observations and builds and organizes knowledge in the form of testable explanations and predications about the world [15, 16]. Engineering can be described as part investigative scientist and part creative inventor with the goal of solving practical problems using both math and science. Engineering is not synonymous with science but uniquely distinct yet synergistically entwined with overlapping epistemologies. The key learning objective in this study is teaching children what engineering is and how it is different than science.

Fig. 3. Engineering illustration by student exposed to Engineering Elephants through class reading and discussion. Text reads "*A firework becoming a rainbow*".

The children's drawings show the advanced ways that they think about ideas [5, 7]. When asked to draw what they would design when they were an engineer, the students did not hesitate to immediately picture themselves in this role. The student who had never been exposed to **Engineering Elephants** or in any type of classroom instruction (Fig. 2) held the common belief that engineers drive trains or work on trains. Several students drew pictures of flowers or clouds and appeared to be unable to make any connection to engineering at all, which is also very typical of this grade level. The students that had read **Engineering Elephants** in class and participated in class discussion about engineering showed elevated knowledge in their drawings with direct correlations to topics covered in the book. For example, Fig. 3 shows a firework becoming a rainbow. **Engineering Elephants** uses fireworks to explain combustion. It is encouraging that the students are obviously learning through this text because their drawings show they have begun to develop concrete ideas about engineering [5].

The purpose behind the development and use of **Engineering Elephants** or children's literature in general is not mastery of all engineering concepts, but to introduce children to the idea of engineering and problem solving and encourage them to begin to imagine all of the things that they could potentially create. Results from integrating Engineering Elephants into 1st grade classrooms show that engineering literature inspires heightened levels of creativity and instilled a concrete sense for what engineers can do. These results show the need for engineering based literature that complements current scientific curriculum such that the stories can more easily be integrated into every classroom and foster early enthusiasm for engineering.

3. Integration of junior – And high-school science clubs and university engineering societies

The Technology Student Association (TSA) is one example of a national non-profit education organization dedicated to promoting engineering and helping students discover their potential for the engineering or technology-based professions [17]. A solid framework of secondary school educators, corporations, professional organizations and universities incorporate pre-college engineering programs in local communities throughout the United States. Another example is the Junior Engineering Technical Society (JETS) which employs a unique and innovative approach—explore, assess, experience—and through which thousands of diverse students are enticed to pursue engineering majors and careers each year [18]. Collegiate student sections of the American Society of Mechanical Engineers (ASME) [19] have worked with local high school professional organizations in an effort to generate future engineering talent. Through this collaboration, TSA or JETS and ASME engages students in a variety of educational programs, increasing awareness of what engineers do and showing how math and science are used to make tangible differences in the world. Students participate together in local, regional, and national engineering competitions, conduct local service projects together, and participate in several social events structured to make connections and build friendships between the students. Foundations for student impact are built upon providing career resources and experiences not often found in traditional learning environments; opening students' minds to their own career possibilities by removing social barriers and negative attitudes about engineering; and addressing major industry needs for a qualified, engineering-literate workforce. These collaborations also provide unique mentor/mentee relationships between high school students and undergraduate engineering majors that can provide the support needed for college transition.

4. Mentoring

"I am here today because I had (chose one of the following): teacher, counselor, mentor in the community, college professor, principal, who believed in me and opened their (chose one of the following): classroom after school or during lunch, research lab, workplace to me and let me see the real world of learning and science beyond the classroom."[20] Mentoring is quite simply an older student, teacher, or professional taking an interest in the life and aspirations of a younger protégé. More formally Kram defines mentoring as a relationship between an experienced individual and an understudy where the experienced individual acts as a role model, providing support and direction[21]. The quote above paraphrases the comments of successful graduates from the Academy for Math, Engineering, and Science, AMES, a Title One science, technology, engineering, and mathematics (STEM) early college high school in Salt Lake City. The graduates of this program when speaking of college and

A curricular learning community in engineering is created by requiring a cohort of first year students to dual enroll in two or more math, science or engineering courses. Some examples are the following:

- A first semester Intro to Engineering course and Precalculus
- A first semester Intro to Engineering course and Calculus I
- Calculus I, Physics I, and/or Intro to Engineering
- Calculus II, Physics I, and/or Engineering Statics

Each cluster course is taught by a member of the discipline faculty. Although research indicates a simple learning community model with no curricular adaptations will impact first year retention, this model is improved upon when faculty work to integrate the curriculum of the courses. The implementation of problem-based learning is one way to integrate the foundation disciplines of mathematics and physics into significant engineering problems that increase student engagement while improving student problem solving skills.

Key elements of a successful Engineering Learning Community model are:

- Emphasizing to the students the goals of the learning community initially and throughout the semester
- Consistent integration of the clustered course curriculum throughout the semester
- Implementing PBL projects in cluster courses that allow students to apply theoretical engineering, science and mathematics principles in the solution of significant engineering design problems
- Frequent communication between the instructors regarding the status of the clustered courses

It is difficult for students to assess the impact of a learning community experience without knowing what to expect. Emphasizing the goals of the learning community initially at advising and registration and throughout the course will allow students to assess whether or not their expectations have been met. Engineering faculty expect students to work together to solve engineering problems, much as engineers in the field work in teams. This same team structure promoted in the learning community can enable students to successfully complete the first year hurtles of Calculus I, II, Physics, and Engineering Statics, courses where frequently students determine whether or not they will remain in engineering.

It is important for students in a curricular learning community to see the interconnection between the disciplines and courses of the learning community. There are two means in which this can be accomplished. 1) Instructors of the clustered courses work to integrate the curriculum on a consistent basis throughout the semester; and 2) Assigning dual problem based learning projects whose solution requires the integration of content from all clustered courses. Integrating course content on a consistent basis can be challenging depending upon the courses involved. In a learning community linking mathematics and engineering, one method is through the introduction of new course content. Each new topic in mathematics is introduced in the context of an engineering problem or application. Similar applications can then be assigned as additional homework problems. When introducing the concept of the derivative, the following problem integrates the engineering and physics concept of position, velocity and acceleration while helping the student develop a conceptual understanding of a derivative of a function.

The velocity of a vehicle starting from rest at position x=0 is shown in the figure below:

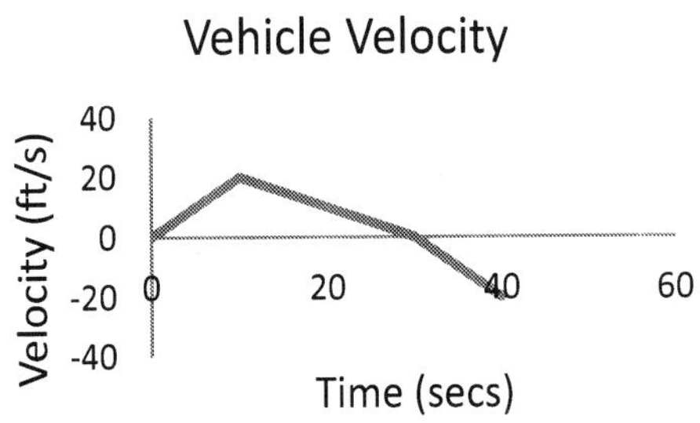

Fig. 4. Example problem for integration of course content.

Knowing that acceleration is the rate of change of velocity, sketch a graph of the acceleration curve. When later introducing integration the same problem can be used with the following question:

"Knowing that velocity is the rate of change of position *x(t)*, if the maximum position is 100 feet and the final position is 20 feet, sketch the graph of the position function *x(t)*." [46]

The introduction of a new topic can also been used as the startup of a PBL project. Introducing the project before covering the content allows students to hypothesize a solution and then build on that hypothesis as student knowledge of the content expands. Function Optimization in Calculus I may be introduced through a PBL project where students optimize the cost of laying an oil pipeline around or through a swamp. A map and scale is given indicating where the pipeline originates and must end. The costs of laying the pipeline through the swamp and on dry land are given per unit foot and student must write the equation for the cost as a function of the path chosen. Engineering faculty appreciate this problem because of its emphasis in modeling and design. No information is given to the students regarding an appropriate shape to model the swamp. Students must determine a shape that will have a mathematical solution and yet accuracy must also be considered. [47]

When linking two or more courses in a learning community model communication between the instructors of the courses is vital. Clustered courses that have common objectives are easier to link that those that do not. For example, Calculus II and Engineering Statics share common topics in applications of integration such as calculating area, volume, surface area, moments, work and pressure against a surface by a fluid. Scheduling and communication are essential when attempting to arrange for both courses to discuss these topics at the same time in the semester. When common objectives are not available it is helpful if flexibility is allowed in the cluster courses for the creation of interdisciplinary learning opportunities. If vector dot products and cross products are not part of the standard Calculus II curriculum, finding a day to discuss these topics in Calculus II will both reinforce the link between mathematics and engineering for the students as well as provide a mathematical framework for the engineering application. The techniques of integration discussed in Calculus II can be motivated by an engineering beam stress problem with a complex distributed load. Allowing for flexibility in the clustered course curriculum creates engaging opportunities for students to approach all problems from an interdisciplinary standpoint and experience where they will utilize the concepts in the future.

The results of a successful curricular learning community can be significant. It is important for institutions to develop a means to assess the impact of the learning community experience through both tracking of student enrollment data as well as student impressions of their first year experience. Focus groups conducted with learning community participants and surveys administered to all first year engineering students can be used to compare student impressions of their learning gains for those in the learning community versus students in the traditional curriculum. Some results that may be seen comparing student impressions of learning gains for learning community students with traditional curriculum students are:

- Greater intent to persist in the engineering field
- Greater student impressions of learning to work as a member of a team
- Larger gains in student ability to identifying and formulating an engineering problem
- Larger gains in student ability to apply engineering principles
- Larger gains in understanding engineering principles
- Larger gains in critical thinking skills
- Significant gains in ability to use mathematics to solve engineering problems
- Significant differences in student ability to find fellow students with whom they could study.

The long term impact of a curricular learning community experience can be assessed by the tracking of an engineering cohort. Data must be collected on retention in engineering, enrollment in subsequent science, math, and engineering courses and grades in these courses. The long term results of a successful learning community are:

- Greater retention in engineering for learning community participants
- Higher grades in key math, science, and engineering courses for learning community participants
- More consistent progression through the engineering curriculum for learning community participants
- Shorter time to degree completion.

Curricular learning communities are not difficult to implement at any size institution and are a perfect match with the engineering curriculum. It is essential for engineering students to learn early in their academic career to work as a part of a team. The learning community experience can create in the first semester, study groups that will assist students through the gateway courses in mathematics, science, and engineering; while providing opportunities to strengthen student problem solving and critical thinking skills, developing interdisciplinary problem solving strategies.

7. Informal public science education and mechanical engineering

With technology moving at such a rapid pace, it has become increasingly important for citizens to be scientifically literate. While children are growing up with these technological advances, there are still several indicators showing that US science literacy is low [48, 49] and experiences with math and science outside of the classroom is crucial to increasing technological literacy not just for children but also the general public. Local science museums and science centers can serve as these pathways to math and science education which forms the background and sparks the interest for engineering. The increased involvement of university-level researchers in science outreach has become part of the national discussion over the last few years with the White House[50, 51]. For some researchers these opportunities are straightforward, since their universities participate in engineering outreach programs to connect to the general public by volunteering at science fairs, offering K-12 teacher professional development opportunities, and by arranging classroom visits [52]. Much more common, however, for many educators such infrastructure just does not exist.

One effective model for informal engineering education and outreach is NanoDays [53] which is a project funded and sustained by the National Science Foundation. NanoDays is a nationwide festival of educational programs about nanoscale science and engineering and its potential impact on the future. Each year, NanoDays events are organized by participants in the Nanoscale Informal Science Education Network (NISE) and take place at over 200 science museums, research centers, and universities across the country from Puerto Rico to Hawaii. NanoDays engages people of all ages in learning about this emerging field of science, which holds the promise of developing revolutionary materials and technologies. The whole idea is to teach the general public about nanotechnology using an informal, hands-on approach in a comfortable, stimulating environment. These activities are scalable and transferable to any age and background. While NanoDays is a national program, it is run locally by science centers and in some cases, university faculty members which creates a successful link between the university and the public. Because the public is generally more comfortable in the science center, NanoDays is conducted in the local science center auditorium. NanoDays programs combine simple hands-on activities for young people with events exploring current research for adults [53]. NanoDays activities demonstrate different, unexpected properties of materials at the nanoscale -- sand that won't get wet even under water, water that won't spill from a teacup, and colors that depend upon particle size [53].

In this model, NanoDays involves faculty who present their research on nanotechnology in a way that is active and engaging and can connect effectively with the public [54]. Undergraduate students are also involved in the hands-on components and demonstrations. Since 2008, interactive presentations have been made by faculty on nanotechnology research such as explosives, new materials for technology, and medicine [54]. It is imperative to choose faculty members that can speak and communicate in a way that reaches the general public. Tips for selecting and training faculty members to be successful in outreach are well-described in [55] and include: use analogies to common day objects when describing scientific phenomena; limit the use of jargon or new words to five (scientific) terms; target talks to 7th graders (12 to 13 year olds); use lots of visuals and demonstrations when possible; and, describe size or scale relative to the human body. The author goes onto say that during training she poses the question to presenters, "How would you explain this to your grandparents?" Lastly and most importantly, she suggests that researchers put their presentation in a narrative or story if possible where the audience can see development from a problem, attempts to solve the problem, a climax and then a conclusion [55] because it has been found that audiences connect with the story of science as well as its facts [56].

Informal science education is an impactful method for relating the general public to current, technology-driven research. NanoDays activities bring university researchers together with science museum educators and the public which creates a unique learning/teaching experience for all and provides real connections for children and adults in engineering.

8. Conclusions

In the highly multi-modal digital age of the youngest generation, science, technology, engineering and math education and learning is confronted with new challenges that require innovative approaches exploiting our understanding of how children and adults learn engineering. Several new structural models for STEM education have been discussed that combine the best features of formal and informal learning. By introducing impactful, engineering education to this generation by integrating literature, technology, and successful teaching and learning methods into their culture, there are no limits to the meaningful contributions that future engineers will make toward improving our way of life.

9. References

[1] T. Karp, "Generation NXT: Building Young Engineers With LEGOs," *IEEE Transactions on Education,* vol. 53, pp. 80-87, 2010.

[2] D. Johnson, *Critical Issue: Addressing the Literacy Needs of Emergent and Early Readers*: North Central Regional Educational Laboratory, Editorial Offices: NCREL, 1120 E. Diehl Rd., #200, Naperville, IL 60563. Tel: 800-356-2735 (Toll Free). For full text: http://www.ncrel.org/sdrs/areas/issues/content/cntareas/reading/li100.htm., 1999.

[3] *National Science Education Standards observe, interact, change, learn*. Washington, D.C.: National Academy Press, 1996.

[4] J. A. Shymansky, "Elementary-School Teachers Beliefs About and Perceptions of Elementary-School Science, Science Reading, Science Textbooks, and Supportive Instructional Factors," *Journal Of Research In Science Teaching,* vol. 28, pp. 437-454, 1991.

[5] M. Varelas, "Exploring the Role of Intertextuality in Concept Construction: Urban Second Graders Make Sense of Evaporation, Boiling, and Condensation," *Journal of Research in Science Teaching,* vol. 43, pp. 637-666, 2006.

[6] *Engineering Elephants*: Authorhouse, 2010.

[7] M. Varelas, C. C. Pappas, and T. I. Team, *Young Children's Own Illustrated Information Books: Making Sense in Science through Words and Pictures*. Arlington, Virginia: National Science Teachers Association Press, 2006.

[8] K. Wendell, K. Connolly, C. Wright, L. Jarvin, C. Rogers, M. Barnett, and I. Marulca, "Incorporating Engineering Design into Elementary School Science Curricula," presented at the ASEE Annual Conference and Exposition, 2010.

[9] J. S. Brown, "Situated Cognition and the Culture of Learning," *Educational Researcher,* vol. 18, pp. 32-42, 1989.

[10] S. R. Goldman, "Toward a functional analysis of scientific genres: Implications for understanding and learning processes," pp. 19-50, 2002.

[11] G. R. Kress, *Reading images the grammar of visual design*. London: Routledge, 1996.

[12] J. Lave, *Situated learning legitimate peripheral participation*. Cambridge [England]: Cambridge University Press, 1991.

[13] P. Mantzicopoulos, A. Samarapungavan, and H. Patrick, ""We Learn How to Predict and be a Scientist": Early Science Experiences and Kindergarten Children's Social Meanings About Science," *Cognition & Instruction,* vol. 27, pp. 312-369, 2009.

[14] P. Mantzicopoulos, H. Patrick, and A. Samarapungavan, "Young children's motivational beliefs about learning science," *Early Childhood Research Quarterly,* vol. 23, pp. 378-394, 2008.
[15] G. Kress, "Before Writing: Rethinking the Paths to Literacy," 1997.
[16] K. Popper, *The Logic of Scientific Discovery* New York, NY, 2002.
[17] (October 24, 2011). *Technology Student Association* Available: http:www.tsaweb.org
[18] (2011, October 24, 2011). *Junior Engineering Technical Society.* Available: http://www.jets.org
[19] (2011, October 24, 2011). *American Society of Mechanical Engineering.* Available: www.asme.org
[20] A. Church, "STEM Mentoring--Aspiration to Achievement," *NCSSSMST Journal,* vol. 16, pp. 13-14, 2010.
[21] K. E. Kram and L. A. Isabella, "Mentoring Alternatives: the Role of Peer Relationships in Career Development," *Academy of Management Journal,* vol. 28, pp. 110-132, 1985.
[22] J. Abdul-alim, "Mentor Program Provides STEM Options," *Education Week,* vol. 30, pp. 1-11, 2011.
[23] J. Jones, "Survey Results Show ACE Mentor Program Is Surpassing Goals," *Civil Engineering (08857024),* vol. 80, pp. 40-43, 2010.
[24] S. C. Hockings, K. J. DeAngelis, and R. F. Frey, "Peer-Led Team Learning in General Chemistry: Implementation and Evaluation," *Journal of Chemical Education,* vol. 85, p. 990, 2008/07/01 2008.
[25] D. C. Lyon and J. J. Lagowski, "Effectiveness of Facilitating Small-Group Learning in Large Lecture Classes," *Journal of Chemical Education,* vol. 85, p. 1571, 2008/11/01 2008.
[26] S. E. Lewis, "Retention and Reform: An Evaluation of Peer-Led Team Learning," *Journal of Chemical Education,* vol. 88, pp. 703-707, 2011/06/01 2011.
[27] M. C. Loui, "Work-in-Progress - Assessment of Peer-Led Team Learning in an Engineering Course for Freshmen," *FIE: 2008 IEEE Frontiers In Education Conference, Vols 1-3,* pp. 581-582, 2008.
[28] "Learning Assistant Model for Teacher Preparation in Science and Technology (LA-TEST) " University of Colorado Boulder2010-2011.
[29] (October 20). *Peer Led Team Learning Website.* Available: www.pltl.org
[30] D. R. Woods, "Applying problem-based learning approach to teach elementary circuit analysis," *IEEE Transactions on Education,* vol. 50, pp. 41-48, 2007.
[31] G. R. Norman and H. G. Schmidt, *Academic Medicine,* vol. 67, pp. 557-565.
[32] D. H. Dolmans, W. H. Gijselaers, H. G. Schmidt, and S. B. van der Meer, *Academic Medicine,* vol. 68, pp. 207-213.
[33] J. C. Perrenet, P. A. J. Bouhuijs, and J. G. M. M. Smits, "The Suitability of Problem-based Learning for Engineering Education: theory and practice," *Teaching in Higher Education,* vol. 5, pp. 345-358, 2000.
[34] L. R. J. Costa, M. Honkala, and A. Lehtovuori, "The Suitability of Problem-based Learning for Engineering Education; theory and practice " *Teaching in Higher Education* vol. 5, pp. 345-358, 2000.
[35] C. Eugene, "How to teach at the university level through an active learning approach: Consquences for teaching basic electrical measurements," *Measurement,* vol. 39, pp. 936-946, 2006.
[36] P. Cawley, *A Problem-based Module in Mechanical Engineering* 1991.

Following on from this research, the modified internal rate of return method is the least used among the methods studied in the research, although this method avoids the deficiencies that occur with the internal rate of return method (e. g., one avoids the presumption that all payments are reinvested at the same rate of return as the internal one).

It is also interesting that the research showed that most experts simultaneously use a number of discounted cash flow methods, a technique which diminishes the possibility of false conclusions. 57.1 percent of experts with an education in mechanical engineering who evaluate investments by means of discounted cash flow methods most often use a combination of two methods, but only 35.7 percent of experts in other sciences use a combination of two methods.

Research also shows that, according to their personal opinion, project managers are not sufficiently aware of the limitations of discounted cash flow methods. In establishing the level of knowledge about flaws in discounted cash flow methods, Pšunder and Ferlan's research (2007) established that less than half (43.2 percent) of experts (in all sciences) are familiar with multiple internal rate of return, and only 16 percent know the problem of results (conflicting advice) between the internal rate of return method and the net present value method. Major differences occur between experts of different profiles (Table 1).

Flaw	Mechanical engineering	Other technical sciences	Mathematical and natural sciences	Other
Conflicting advice of net present value and internal rate of return	37.5 %	33.3 %	0.0 %	43.8 %
Multiple internal rate of return	37.5 %	6.7 %	0.0 %	18.8 %
Deformation due to presumption that all payments are reinvested according to the rate of return identical to the internal one	37.5 %	40.0 %	20.0 %	56.3 %
Other	0.0 %	0.0 %	0.0 %	12.5 %

Table 1. Knowledge of flaws of discounted cash flow methods by field of education (Pšunder and Ferlan, 2007).

2. Theoretical background

We can divide project evaluation methods into two groups: a group of non-discounting methods and the discounting one. Among the most commonly used methods from the first group is certainly the payback period method.

Non-discounting methods contain several simplifications that can lead to flaws in results. It has to be stressed that non-discounting methods do not consider the time value of money. A cash inflow to be received in the future is weighed equally with a cash inflow to be received now. Since maturity of payments plays a significant role, this simplification can cause erroneous advice.

Non-discounting methods cannot be adopted for different risk levels, which means that only projects and investments with the same risk level can be mutually compared. Neither can capital gain be included in the calculation. This means that non-discounting methods are based on the assumption that every investment will have the same capital gain or loss.

Despite the evident deficiencies of non-discounting methods, they are wide spread among project managers and other decision makers. According to Pšunder and Ferlan (2007), approximately 80 percent of project managers with an education in mechanical engineering and approximately two thirds of other project managers with an education in engineering (e.g., civil engineering or electrical engineering) are still using them.

In contrast to non-discounting methods, in the discounting ones the calculation is based on the time value of money. That means the differences in maturity of payments can be considered. With discounting methods, capital gain or loss can be included in the (last) payment of the project. Since the methods deal with a discount rate, the differences in the risk premium can be considered as well.

Discounting methods always deal with cash flow analysis of a project or an investment. There are several methods - e.g. the net present value index and modified internal rate of return method - but by far the most commonly used among the discounting methods are the net present value method and the internal rate of return method (Pšunder and Ferlan, 2007).

The main advantage of discounting methods over non-discounting ones is the consideration of the time value of money. This is particularly important in engineering projects where duration of projects is usually long, and payments can be vastly deferred. Thus, by using discounting methods for project analyses, we can overcome the time inconsistency of payments.

By using the net present value method, we compare the present value of future payments with initial investment. In this way we determine the surplus from a project or an investment in present value terms. The advice of the method is positive if the net present value of project or investment is greater than or equal to 0, which means a project or an investment will generate a surplus by a given discount rate. Since the risk premium is included in the discount rate, the surplus represents extra gain for the investor, measured in present value terms. Of course, the investor does not receive the sum immediately, but that "extra gain" represents the present value of future surpluses from the cash flow of a project or an investment. Mathematically, the net present value can be calculated by using equation 1:

$$NPV = -I_0 + \sum_{i=1}^{n} \frac{CF_i}{(1+r)^i} \qquad (1)$$

In equation 1, CF_i stands for the cash flow in the period i, n represents the number of periods and r is the discount rate.

The internal rate of return method is quite similar to the net present value method, but despite these similarities, it produces different results. With the internal rate of return method, we calculate the return rate by equalizing the net present value with 0. The result is a measure of the rate of return earned on that capital used in the project during the time that the capital is used, after allowing for the recoupment of the initial capital outlay (Holmes, 1998).

According to Puxty and Dodds (1991) the internal rate of return method is no more difficult to understand. Though the mathematics are just as easy, it is trickier because in normal circumstances the solution can only be found by trial and error. The goal of the method is to find an interest rate at which inflows exactly equal outflows. The internal rate of return is calculated from the following equation:

$$0 = -I_0 + \sum_{i=1}^{n} \frac{CF_i}{(1+IRR)^i} \tag{2}$$

In equation 2, CF_i stands for the cash flow in period i, n means the number of periods and IRR is the internal rate of return.

With investment projects in engineering we often encounter a residual value when the life-cycle of the project or investment is ended. Ling and Archer (2008) emphasise that it is necessary to take into account the cash flow from the sale of a property and not only the periodic investment inflows of cash. In such cases, it is important to include in the last projected cash flow any potential (marketable) residual value of a project or an investment. Recent research by Pšunder and Cirman (2011) states that the residual value of an investment usually appears as a positive cash flow, but in some cases it can also be a negative one: for example, if we are dealing with the removal of a completely derelict property or a property with a very low value, then the cash outflows for the removal are greater than the inflows from the liquidated property.

Both the net present value method and the internal rate of return method derive from the same time value of money formula. However, they give a different type of indication. The net present value method gives an absolute size, while the internal rate of return method gives a relative indication. For optimal decision making, both methods should be used in practice. Both are valuable pieces of information in the decision process. In most cases, both the net present value method and the internal rate of return method will give the same advice (Brozik, n.d.).

3. Flaws in use of discounted cash flow models for evaluation of engineering projects

The net present value method principally does not contain any methodological assumptions or deficiencies that could impact the result of evaluation. However, the results are harder to understand in comparison to the percentage from the internal rate of return method. It would not be very meaningful if the investor were to determine that the net present value of a project or an investment were, for example, 25,000 EUR. It certainly may be difficult to compare this result with alternative investments, like real estate investment, bonds or cash deposits. It would be certainly more understandable if the result were expressed as a percentage of internal rate of return. This is probably the main reason that the internal rate of return method is often more popular.

In contrast to the net present value method, the theoretical findings identify numerous limitations connected with the internal rate of return method. The most significant are the following:

- reinvestment assumption,
- internal rates of return are non-additive,
- possibility of multiple internal rate of return,
- possibility of conflicting advice with the net present value method.

3.1 Reinvestment assumption

Reinvestment assumption means that all cash flows from the proposed investment are reinvested at the same rate of return as the internal one. Brozik (n.d.) explains that reinvestment assumption means that, if you are really going to get, e.g., an 8 percent return on the proposed investment, each cash flow must earn 8 percent for the life of the project. The more common way to state this is that, in order to achieve an 8 percent return on the entire investment, all cash flows must be reinvested at 8 percent until maturity. No cash flows can be diverted for other purposes. No better investments can be taken should they come along. The cash is essentially tied up for the life of the project, and you must find projects that will return 8 percent for the various time horizons each cash flow faces.

It is unrealistic to expect that all the cash flows from the proposed project will be reinvested at exactly the same rate of return as the internal one, which means that the internal rate of return will return distorted results. How intense the distortion will be depends mainly on the difference between the internal rate of return of the proposed investment and the reinvestment rate of return.

The problem can be overcome by using modified internal rate of return. However, although the modified internal rate of return method clearly overcomes several problems of the internal rate of return method, according to Pšunder and Ferlan (2007), it is the least popular discounting method among engineers. Half of mechanical engineers use the modified rate of return method for project or investment evaluation, but only 13.3 percent of engineers other than mechanical engineers do so. Moreover, only 37.5 percent of mechanical engineers are aware of the possibility of flaw in results due to reinvestment assumption.

3.2 Internal rates of return are non-additive

An interesting phenomenon in connection with the internal rate of return method is that we cannot add the internal rater of return of two projects; moreover, we get very surprising results when two or more combined projects are considered together. Several authors have discussed this in the past. Treynor and Black (1976) and after them Puxty and Dodds (1991) have found that inclusion of a third project can affect the choice between the first two.

Let's suppose we have two projects (A and B) and combine them with a third one (C), as shown in Table 2.

Project	I_0	PMT_1	PMT_2	IRR
A	-100	0	125	11.8
B	-100	110	0	10.0
C	-100	130	0	30.0
A + B	-200	130	125	18.0
B + C	-200	240	0	20.0

Table 2. Example of non-additivity of internal rate of return (modified from Treynor and Black (1976) and Puxty and Dodds (1991)).

If we compare project A and B, A is the better choice, as its internal rate of return is higher. But if we include project C and there are enough funds to finance both projects, the attractiveness of A and B change: B becomes preferable (Puxty and Dodds, 1991).

3.3 Multiple internal rate of return

Multiple internal rate of return is probably the best known flaw of the internal rate of return method. The problem is often experienced when the internal rate of return equation yields two (or more) results. Of course, in that case the mathematical result isn't meaningful for investment decision making purposes.

The reason for two or more results from the equation lies in its structure. The internal rate of return equation is a polynomial. Lumby and Jones (1999) explain that the fifteenth-century mathematician Descartes proved with his "rule of sign", that there are possible solutions to polynomial equation for each change of sign. Thus, any particular investment project may have more than one internal rate of return (i.e., there may be more than one discount rate that will reduce a project's cash flow to a zero net present value), or it may not have any internal rate of return at all.

In projects, where the cash flow can change the sign more than once among different periods, the internal rate of return method yields more than one root of the equation, which means more than one internal rate of return (Figure 1)

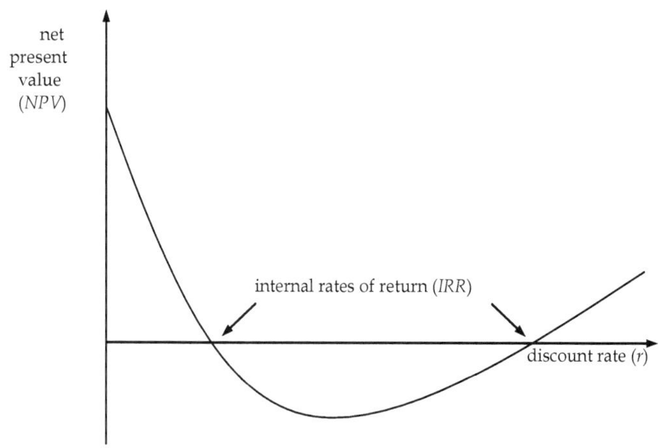

Fig. 1. Two roots of polynomial equation - multiple internal rate of return.

Several internal rates of return cannot form the basis for a justified decision, since the internal rates of return may differ several times. In cases of multiple internal rates of return, financial calculators and spreadsheet processing programs offer a choice of guessing the right internal rate of return; however, without knowing the theoretical process of avoiding multiple internal rates of return, this may not be the perfect solution either.

In theory, there are projects with conventional cash flows and projects with non-conventional cash flows. Puxty and Dodds (1991) explain that a conventional cash flow project is one where a cash outflow, or series of outflows, is followed by a cash inflow or series of inflows. A projects with a conventional cash flow has only one change in sign (+, –)

between the time periods. Non-conventional cash flows can therefore be defined as those that involve more than one change in sign. Such projects are due to modifications, reconstructions and overhauls, which require intensive investments that often cause negative cash flow, quite common in engineering projects.

The problem of more than one change in sign can be overcome with the elimination of second and further changes in sign by discounting such part of equation to the article of the equation with the same sign in cash flow as the discounted one.

A further possibility for overcoming the multiple internal rate of return problem, according Puxty and Dodds (1991), involves the net present value rule. It would have no difficulty in giving the correct advice: to reject the project because it has a negative net present value or to accept it because it has a positive one at the given discount rate.

In connection with the internal rate of return, it is worth mentioning at least one further problem: a non-existent internal rate of return. Since the problem is very unlikely to appear in practice, we will not give it any in depth attention.

According to Pšunder and Ferlan's (2007) research among Slovenian project managers, only 37.5 percent of project managers with an education in mechanical engineering know about the multiple internal rate of return problem.

3.4 Conflicting advice of the net present value and the internal rate of return

Sometimes it becomes necessary to compare two projects in practice. In such cases the net present value and the internal rate of return may give opposite advice. To understand why the results of both methods differ, it is necessary to present the lines of two projects in a diagram (Figure 2).

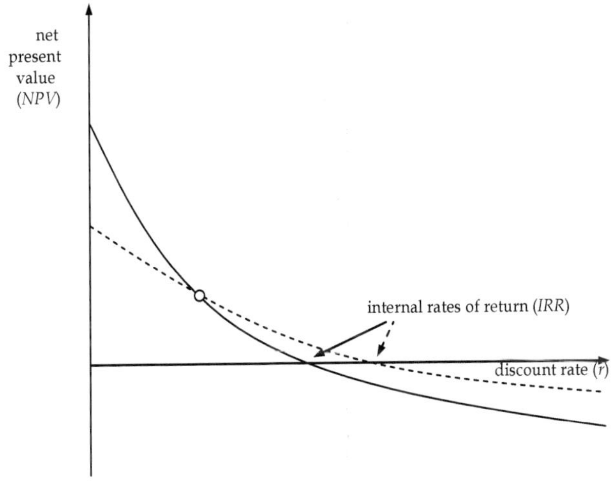

Fig. 2. Conflicting indication from net present value and internal rate of return (modified from Holmes, 1998).

The two curves belonging to two projects cross at a certain discount rate. Above this rate, the full line belongs to the superior project, below this rate, the dotted line represent the better investment. Puxty and Dodds (1991) explain the reason: the bulk of cash inflows from the project with the dotted line arrives in the later years, and hence at higher rates of interest, they are discounted more heavily.

Holmes (1998) writes that the reason for conflicting advice is the two techniques make different assumptions regarding what will happen to cash inflows from investment projects. According to his findings, in such cases the net present value rule gives the correct advice.

According to their own opinion, only 37.5 percent of Slovenian projects managers with an education in mechanical engineering, 33.3 percent of Slovenian project managers with education in other technical sciences, and 43.8 percent of project managers with other education know the problem of conflicting advice from the net present value method and the internal rate of return method (see Table 1).

4. Impact of the discount rate on engineering project evaluation

Pšunder and Cirman (2011) claim that the key factor in estimating the present value of future cash flows is the discount rate. If it were 0, the net preen value equation (equation 1) would be reduced merely to the addition and subtraction of cash flows during different periods, without considering the time value of money. The discount rate determines the required rate of return that an investor demands for a certain investment in accordance with the risk associated with the investment, and it has a profound impact on the net present value. A higher discount rate results in a reduction of the net present value, whereas a lower one results in its increase, an effect that is evident in Figure 3.

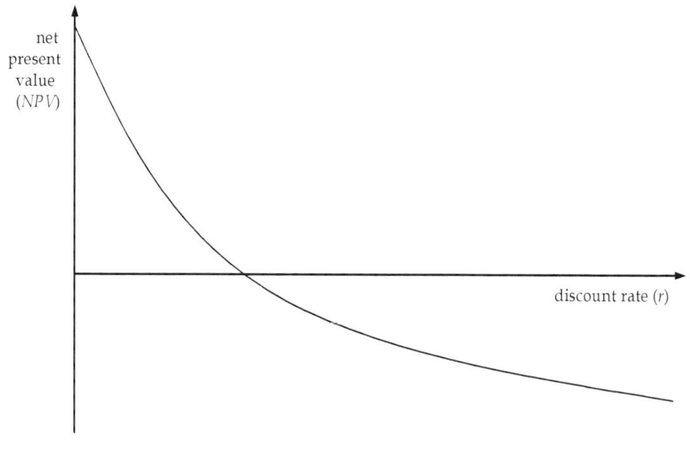

Fig. 3. Influence of the discount rate on the net present value.

From Figure 3 we can see the relationship between the net present value and the internal rate of return. When increasing the discount rate, the net present value decreases. When the net present value reaches 0, the discount rate is equal to the internal rate of return.

Pšunder and Cirman (2011) state that the discount rate is the rate at which future cash flows are converted into their present value. The differences between the discount rates have significant impact on the result of investment or project analysis. The contemporary theory of the determination of the discount rate favours a more precise definition of the discount rate. As the discount rate does not include a capital recovery premium, it can only be used for assessing an investment where we do not expect changes in the value of the investment, or where we can expect that changes in the value of the investment will be considered when selling property or at the termination of the investment (adopted from Friedman and Ordway, 1989, and The Appraisal of Real Estate, 2008). Ling and Archer (2008) emphasize that it is also necessary to take into account the cash flow from the sale of a property and not only the periodic investment inflows of cash. In such cases, it is important to include in the last projected cash flow any potential (marketable) residual value of an investment. The latter usually appears as a positive cash flow, but in some cases it can also be a negative one; for example, if we are dealing with the removal of a completely derelict property or of a property with a very low value, then the cash outflows for the removal are greater than the inflows from the liquidated asset.

By definition, the discount rate represents the rate of return that can be obtained in the financial market for a similar investment with comparable risk. What rate of return will be required for a certain investment depends on the risk associated with the specific investment and on the rate of return on investments with a comparable risk (Mramor, 1993).

According to Pšunder and Cirman (2011), different risk premiums must be taken into account for different projects, enabling us to compare two quite different projects. Certainly, machinery and equipment investments are mostly subject to deterioration and obsolescence, which are the reasons an investment loses value in the long run. The loss of value can be included in cash flow from the residual value (the last cash flow in the equation when individual cash flows can be considered).

Pšunder and Cirman (2011) write that the discount rate has a significant influence on the result of the present value method; that is why the correct choice of a discount rate is a precondition for an appropriate analysis.

The above mentioned authors state that when analyzing a certain project, the size of the initial investment, expected cash flows and estimated duration of the project are known. The key factor that influences the result of the analysis is the discount rate. The discount rate is a decisive factor when evaluating whether projects are acceptable or not. The impact of the applied discount rate significantly increases with the duration of a project.

Presuming a limited period until the end of the project and constant annual cash flow of the project, we can derive a present value of future annuities (PVFA). In the case of these presumptions, the equation 1 takes following form:

$$NPV = -I_0 + \sum_{i=1}^{n} \frac{CF}{(1+r)^i} = -I_0 + CF \cdot \sum_{i=1}^{n} \frac{1}{(1+r)^i} \tag{3}$$

In equation 3, CF stands for the constant annual cash flow, n represents the number of periods and r is the discount rate.

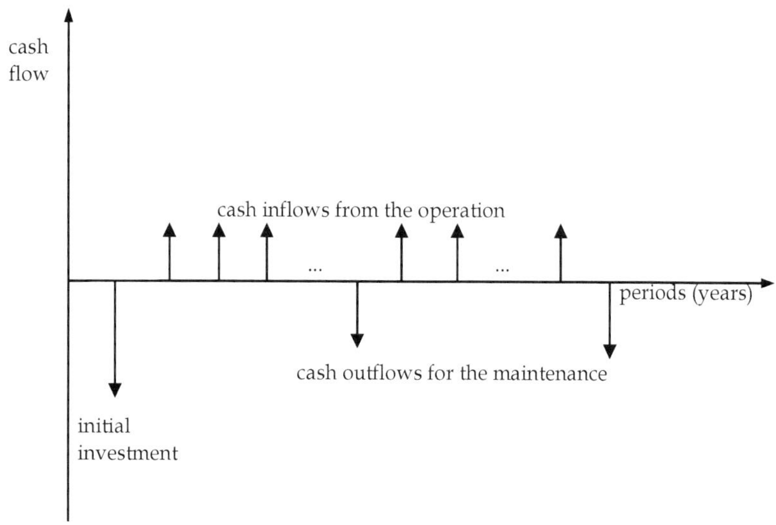

Fig. 4. Non-conventional cash flows in engineering projects, owing to maintenace and overhauls.

According to Pšunder and Ferlan (2007) only slightly more than one-third of mechanical engineers are acquainted with multiple internal rate of return, so overcoming the multiple internal rate of return can cause problems in practice. However, the problem can be overcome by discounting the article with the change of sign to the period, represented by the article of the same sign in cash flow as the discounted one. A further possibility for overcoming the multiple internal rate of return problem is, according to Puxty and Dodds (1991), using the net present value rule. It would have no difficulty in giving the correct advice.

5.2 Use of the discount rate

The net present value method has numerous advantages over the internal rate of return method. Above all, it is simpler to calculate and it isn't sensitive to the flaws arising from the multiple internal rate of return, the non-additive internal rate of return and other specific flaws that are associated with the internal rate of return method.

However, although we can avoid all the flaws of the internal rate of return method when using the net present value method, we encounter a problem of setting up the discount rate.

Let's take a case study on an investment project with a duration of 20 years. For easier calculation, a presumption of constant annual cash flow is made. A project is feasible when the net present value, calculated from equation 3, is equal to or higher than 0.

In a 20-year project, the constant cash flow at a discount rate of 10 percent must amount to at least 10.6 percent of the investment on the annual level in order for the analysis to provide a positive indication. Meanwhile, at a 15 percent discount rate, it needs to be approximately 1.5 times as high, as seen from Table 4.

Discount	Project duration (years)		
rate (r)	5	10	20
5	0.130	0.080	0.065
6	0.136	0.087	0.073
7	0.142	0.094	0.081
8	0.149	0.102	0.089
9	0.156	0.110	0.097
10	0.163	0.117	0.106
11	0.170	0.126	0.115
12	0.177	0.134	0.124
13	0.184	0.142	0.133
14	0.192	0.151	0.143
15	0.199	0.160	0.152

Table 4. Constant annual cash flows necessary (as a proportion of the initial investment outlay) for a positive indication of the present value analysis (modified after Pšunder and Cirman, 2011).

When a 20-year project generates a constant annual cash flow that amounts to 10 percent of the investment on the annual level, the net present value will be higher than 0 at a discount rate of 9 percent, but lower than 0 at a discount rate of 10 percent. The advice from the method will change, owing to changes in the discount rate. This example shows how important precise setting of the discount rate is.

The impact of the discount rate is comparatively diminished when projects are shorter and when the discount rate level is higher. However, even with 5-year projects and discount rates over 10 percent, the setting up of the discount rate could be decisive for the advice from the net present value method.

The survey conducted in February and March 2011 among appraisers in Slovenia by Pšunder and Cirman (2011) shows that experts have a quite unified opinion about the risk free rate. The average nominal risk-free rate in the survey was 4.20 percent. This average corresponded to the returns on bonds issued by the Republic of Slovenia with a maturity of seven to eight years. By using the calculated average real rate of return, they established that the respondents' average expected future inflation was 2.25 percent, which means the respondents expect slightly higher inflation in Slovenia than the target inflation set by the

European Central Bank (2.00 percent). The univariate statistics on risk-free rates of return are presented in Table 5.

Risk-free rate	Mean	Std. Dev
Nominal rate	4.20	1.25
Real rate	1.91	0.74
Expected inflation	2.25	

Table 5. Risk-free rates of return and expected inflation (Pšunder and Cirman, 2011).

The experts' opinion was not so unified considering the risk premium. The experts considered the risk premium for improvements (additions to building etc.) at 4.91 percent (mean). The standard deviation was 3.07. As the risk premium is added to the risk free rate when calculating the discount rate, the discount rate may also vary immensly. Consequently the net present value method might give very varying advice when used.

6. Discussion and conclusion

Research conducted by Pšunder and Ferlan (2007) about knowledge of faults in investment project evaluation methods shows that, regardless of the field of education, non-discounting methods are still more popular than discounting methods. The research also shows that experts who have an education in mechanical engineering and are involved in investment project evaluation are more likely than average to use discounting methods. They also possess above-average knowledge about deficiencies in investment project evaluation methods. In using the internal rate of return method, experts with an education in mechanical engineering are most familiar with multiple internal rate of return. The level of familiarity with the multiple internal rate of return is above average, most probably because big projects in the field of mechanical engineering are, owing to variable cash flow, typically subject to multiple internal rate of return.

Discounting methods are far more precise in advice about investment projects than non-discounting methods; however, the discounting methods have their flaws and drawbacks, as well. In particular, internal rate of return has numerous drawbacks that can lead to incorrect conclusions. The most frequent drawbacks involve multiple internal rates of return, reinvestment assumptions and non-additive projects.

These drawbacks of the internal rate of return method are the reason that the net present value method is the better method of the two. It is easy to calculate, but results are harder to understand in comparison to the percentage from the internal rate of return method. It certainly may be difficult to compare the result with alternative investments.

The discount rate is a key factor in proposed project analysis. The size of the discount rate can greatly influence decisions about an investment, since even small changes in the discount rate can dramatically change a decision about the acceptance or rejection of an investment. This is why its determination needs experience and should not be left to practice-based approximations. The proper application of the net present value method is based on the accuracy of the applied discount rate. Recent research shows that experts' opinions about risk free rate are quite unified, but not by the risk premium (Pšunder and Cirman, 2011).

Despite the wide use of discounting methods, there is still incomplete knowledge about them. To ensure the reliability of discounting methods, project managers must be acquainted with the deficiencies and drawbacks of the methods, particularly of the internal rate of return method. With the net present value method, it is vital to apply professionally proven methods when establishing the discount rate. Only this will ensure that the results of the analysis will be reliable and credible.

7. References

Brozik D. (n.d.). The Reinvestment Assumption. Date of access: 28.9.2011, Available from: http://webpages.marshall.edu/~brozik/reinvestmentweb.pdf

Friedman J. P., Ordway N. (1989). *Income Property Appraisal and Analysis*. Prentice Hall, ISBN 0-13-453655-X, New Jersey.

Holmes P. (1998). *Investment Appraisal*, International Thomson Business Press, ISBN 1-86152-212-6, London

Klammer T. P., Walker M. C. (1984). The continuing increase in the use of sophisticated capital budgeting techniques. *California Management Review*, Vol. 27, No. 1, pp. 137-48.

Kušar J., Bradeško L., Duhovnik J., Starbek M. (2008). Project management of product development. *Strojniški vestnik - Journal of Mechanical Engineering*, Vol. 54, No. 9, pp. 588-606, ISSN 0039-2480.

Ling D. C., Archer W. R. (2008). *Real Estate Principles* (2nd edition), McGraw-Hill/Irwin, ISBN 978-0--07-110130-1, New York.

Lumby S., Jones C. (1999). *Investment Appraisal & Financial Decisions* (6th edition), International Thomson Business Press, ISBN 1-86152-257-6, London.

Mramor D. (1993). *Uvod v poslovne finance*, Gospodarski vestnik, ISBN 86-7061-070-1, Ljubljana.

Pike R. H. (1988). An empirical study of the adoption of sophisticated capital budgeting practices and decision making effectiveness. *Accounting and Business Research*, Vol. 18, No. 72, pp. 341 - 51.

Pšunder I., Cirman A. (2011). Discount rate in methods based on discounted cash-flow for the purpose of real estate investment analysis and valuation. *Geodetski vestnik*, Vol. 55, No. 3, pp. 563 - 577, ISSN 0351-0271.

Pšunder I., Ferlan N (2007). Analysis of the knowledge and the use of investment project evaluation methods in the field of mechanical engineering. *Journal of Mechanical Engioneering*, Vol. 53, Vol. 9, pp. 569-581, ISSN 0039-2480.

Pšunder I., Ferlan N. (2008). Use of capital budgeting methods for investment project evaluation in the field of civil engineering. *American Journal of Applied Sciences*, Vol. 5, No. 1, pp. 18-21, ISSN 1546-9239.

Puxty A. G., Dodds J. C. (1991). *Financial management: method and meaning* (2nd edition), Ed. Wilson, R. M. S., Chapman & Hall, ISBN 0-412-46000-9, London. *The Appraisal of Real Estate* (13th edition) (2008). Appraisal Institute, ISBN: 978-0-922154-98-2, Chicago.

Treynor J.L., Black F. (1976). Corporate investment decisions. In: *Modern Developments in Financial Management*, Mayers S.C., Praeger, New York.

28

Configuration Logic of Standard Business Processes for Inter-Company Order Management

Carsten Schmidt and Stefan Cuber
Institute for Industrial Management (FIR) at RWTH Aachen University, Germany

1. Introduction

Today's manufacturing companies embedded in non-hierarchical production networks are facing multiple and dynamic customer-supplier-relationships. In the course of increasing complexity of products and growing needs for flexibility and product variation companies focus more on core competencies and thus more production processes are shifted to external suppliers. This complex environment leads to growing coordination-efforts and wasteful turbulences throughout the entire network. The result is a delivery reliability of usually less than 65% within the European machinery and equipment industry generating an estimated loss of efficiency of 1 billion Euros per year. Besides additional costs the missing delivery reliability entails poor customer satisfaction and increased lead times compromising the competitiveness of individual companies as well as the entire machinery and equipment industry (Gunasekaran, 2000; Reinhart, 2006).

In order to handle the given complexity, the procurement processes have to be managed according to the needs of the respective situation determined by the ordered product and the involved supplier. That means, the design, standardization and configuration of practically applicable order management processes according to the certain business context becomes a key factor. A context-aligned process configuration would lead to a tailored capacity to act and could improve performance and delivery reliability within the machinery and equipment industry.

The issue of poor delivery reliability is addressed within the publicly funded research and development project "inTime" (Funded by the 7th Framework Program of the European Commission, EU FP7-NMP, No. NMP2-SL-2009-229132). Within the project typologies are derived based on practical input from companies which distinguish and characterize certain product / service-types as well as certain business-relationship types, which have a direct influence on the design and handling of the respective order management processes between the involved companies. These types then serve as starting point for the limitation and design of relevant reference process phases of order-handling, respectively procurement-handling. The processes are described in detail and are enriched by the detailed definition of the transferred information within the inter-company interfaces.

2. Types of information exchanged in the course of the order execution

The exchange of information is a basic prerequisite for the coordination of the inter-company order execution. That is to say also for the division of labor in general. According to (Schuh & Westkämper, 2006) and (Wiendahl & Meyer, 2006), the information exchanged between the network partners in the course of the inter-company order execution can be distinguished based on the typical, time-logical sequence of tasks concerning the interaction of the involved parties. In this context, the first task of interaction usually refers to a declaration of intent, which basically initiates the customer-producer-interaction or respectively the producer-supplier-interaction in the first place. The corresponding type of information exchanged at that point is consequently called intent information. This type for instance encompasses inquiries or requests in written form. The aforementioned task is usually followed by the task of clarifying the specifications of the product or service to be exchanged. The corresponding information exchanged between the network partners in this context is thus called specification information. On the one hand, this second type of information respectively encompasses order-independent master data, such as standard drawings or parts lists. On the other hand, it furthermore includes order-specific data, which basically represent the customers influence on the product or service to be provided by the producer or respectively the producer´s influence on the product or service to be provided by a certain supplier. As far as the complex, highly individualized products or services of the addressed target group are concerned, existing master data is usually modified or extended to a considerable extend by order-specific data to comply with order-specific requirements.

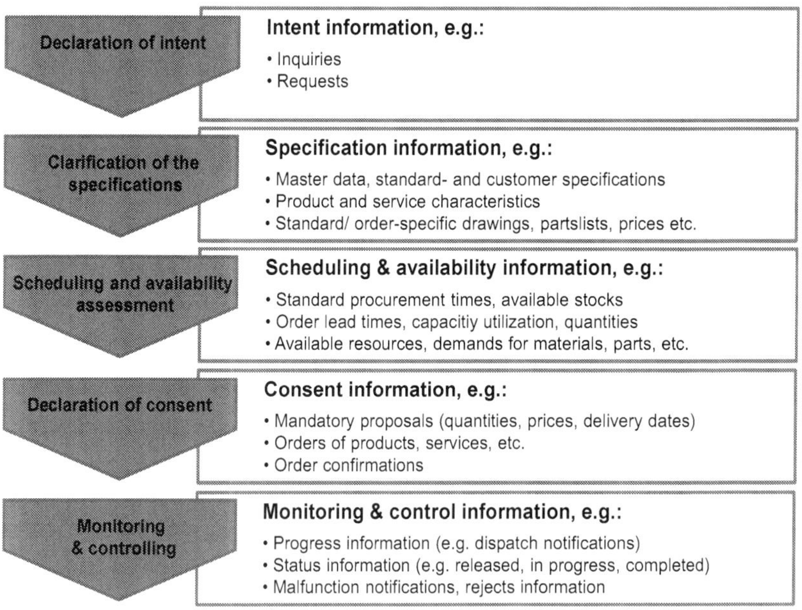

Fig. 1. Types of information exchanged in the course of the order execution

The next interaction task basically deals with the scheduling of the intended product or service exchange as well as with the assessment of availabilities. The information exchanged in this context is hence called scheduling & availability information. This type of information again includes master data, such as procurement times of standard parts, and order-specific data, such as current demands for materials or parts and required resources. In the following course of the companies´ interaction, a number of written documents typically have to be exchanged in order to provide a legally binding basis for the scope of supply to be provided. These documents, among which are e.g. mandatory proposals, reservations, orders and order confirmations, mainly stipulate the terms and conditions agreed on during the clarification of the specifications as well as the during the scheduling and the availability assessment. The type of information exchanged in this context is called consent information. The final interaction task of monitoring & controlling eventually leads to the last type of information exchanged in the course of the inter-company order execution. That is the corresponding type of monitoring & control information. Monitoring & control information are for instance internal or external progress or status reports, dispatch or release notifications as well as sporadic malfunction information.

3. The basics of typification

A typification can basically be regarded as a process of aggregation and abstraction, which describes, structures and eventually reduces the complexity of an issue to essential aspects, via the definition types. Numerous projects from research and practice have proven the method of typification to be a suitable tool for the structuring of complex organizational situations, which include a variety of different business forms (Grosse-Oetringhaus, 1974; Büdenbender, 1991; Diemer, 1992). In this context, a type represents a number of objects with a common set of features and attributes. A conceptual and structural basis for the creation of types is typically provided by morphological schemes, also called morphologies. Morphologies are usually matrices, which visualize the set of features and their respective attributes characterizing a certain type (cf. Figure 2). A fictive "type 1" could for instance be characterized by the combination of the attributes highlighted in grey (A2, B1, C2, D2) whereas a fictive "type 2" could be described by the attributes highlighted in green (A3, B2, C3, D3).

Feature	Attribute 1	Attribute 2	Attribute 3	...
A	A1	A2	A3	...
B	B1	B2	B3	...
C	C1	C2	C3	...
D	D1	D2	D3	...
...	...	...	...	...

Fig. 2. Structure of a morphological scheme

The respective choice of features within a morphology highly depends on the requirements of the desired analysis. Since the creation of types is therefore always subject to a specific purpose, no universally valid types can usually be derived. A derived type is in fact rather linked to the particular interest of the observer and has to comply with his goals and problems (Förster, 1988).

4. Type-based characterization of producer-supplier interaction

The characteristics of the producer-supplier-interaction are substantial, both to the design of order execution reference processes and to the design of the corresponding processes of interaction. That is because those characteristics influence the time-logical arrangement of the tasks and producer-supplier-interfaces as well as the respective way of interaction between the producer and the suppliers at those interfaces.

4.1 Derivation of morphologies

Basic influencing factors on the producer-supplier-interaction are such factors, that have a major effect on when and how the main producer and his 1st-tier suppliers have to interact in the course of the inter-company order execution as well as which information have to be exchanged during their interaction.

The first basic influencing factor regarded is the type of product or service exchanged between a producer and a supplier. Complex products or services for instance typically require significantly higher efforts of coordination at earlier phases of a project than less complex products or services due to a far more distinctive exchange of specification information (Schmidt, 2008). Besides the type of product or service exchanged, it is in particular the type of business relationship between the producer and the supplier, which characterizes the producer-supplier-interaction. For instance, different levels of confidentiality between the producer and his suppliers exist and thus necessitate a very differentiated and individualized design of information flows in the context of the producer-supplier-interaction (Meyer et al., 2006).

In order to derivate morphologies of typification both product- and service-related features as well as business-relationship related features and attributes have been analyzed and determined.

4.2 Product/service morphology

The feature level of individualization of a product or service provided by a supplier describes if and to which extent customer- or producer-induced modifications are necessary in the course of the order execution to eventually comply with certain order-specific requirements (Besslich & Lumbe, 1994; Scherer, 1991; Schomburg, 1980). The feature level of specification characterizes how detailed the requirements and characteristics for a product or service to be exchanged are specified in the beginning of the order-specific producer-supplier-interaction. The higher the level of individualization and the lower the level of specification, the higher is typically the necessary effort of coordination and thus the more intense is the interaction between the producer and the supplier in the course of the inter-company order execution (Hillebrandt, 2002; Schulte-Zurhausen, 2005). Furthermore suppliers of highly individualized and poorly specified products or services have to be involved in much earlier project phases than suppliers of less individualized and more specified products or services.

Product-/service-related features: Feature	Attributes				
Level of individualization	Order-specific development	Order-specific modification	Standardized		
Level of specification	Requirements set, free solution	Requirements set, solution partially set	Requirements set, solution mainly set	Requirements set, solution set	
Complexity (service)	Several, heterogenous subservices	Several, similar subservices	Few, similar subservices		
Complexity (product)	Multi-part product with complex structure	Multi-part product with simple structure	Few-part product with simple structure		
Changes to the specifications	Frequent producer-induced changes	Frequent supplier-induced changes	Occasional producer-induced changes	Occasional supplier-induced changes	Changes insignificant
Substitutability of supplier	No substitution possible	Substitution difficult	Substitution uncomplicated		
Influence on the substitution	No influence	Indirect influence (external compensation)	Direct influence (internal compensation)		

Fig. 3. Product/service morphology

The feature complexity refers to the design-based, structural characteristics of the product or service exchanged between the producer and the supplier (Scherer, 1991; Schomburg, 1980; Wildemann, 2000). The complexity of a product or service thus affects the producer-supplier-interaction in a similar way as the level of individualization and the level of specification. Moreover, complex products or services often require a rather frequent interaction of the producer and the supplier throughout the entire order execution. That is due to their typically high importance to the overall system ordered by the customer and the corresponding necessity of distinctive monitoring & controlling procedures (Rotering, 1993).

The feature changes to the specifications refers to the frequency and the initiator of the changes applied to the specifications of a product or service exchanged between the producer and a supplier. In particular frequent changes to the specifications of a product or service can, depending on the level of individualization, specification and the complexity of a product, result in considerable extra-efforts of coordination and thus in an increase of the intensity as well as the frequency of the producer-supplier-interaction. This is due to the necessity of repeating processes that have already been started or were even completed. The feature changes to the specifications therefore indicates how intense the producer-supplier-interaction is in the course of the inter-company order execution but provides little information about when this interaction is taking place.

The feature substitutability of the supplier includes if and to which extent a specific supplier can be substituted in the context of a specific project (Eberle, 2005; Scherer 1991; Schuh et al. 2006). The higher the level of individualization and the higher the complexity of a product or service, the less suppliers are typically capable of providing it (Arnold, 2004; Homburg, 1995; Schwerk, 2000). Whereas the substitution of suppliers is therefore often not possible or at least quite difficult as far as highly individualized and complex products or services are concerned, the substitution of suppliers is rather uncomplicated as far as standard products or services are concerned. The feature influence on the substitution represents the producer´s influence on the substitution of a supplier (Arnold, 2004; Eberle, 2005; Kraljic, 1988).

4.3 Business-relationship morphology

The characteristics of business relationships are influence the interaction of the main producer and his 1st-tier suppliers in the course of the inter-company order execution. Five features were derived and are set in the framework of a business-relationship morphology.

The feature level of institutionalization addresses the intensity of the business relationship between the participating parties in a business network. The level of institutionalization is closely related to the second feature information infrastructure. If a supplier is for instance both legally and economically independent from the producer, their business relationship is project-specifically institutionalized. In this case, the business partners often do not have a common information infrastructure but apply a rather conventional exchange of information via fax, email or telephone. However, the more intense a business relationship is and the more integrated the information systems of the partners are, the easier it is usually to cope with the need for information in the course of the order execution. The feature stipulation of the sourcing conditions determines to which extent the collaboration, concerning rights and duties as well as their enforceability, between the producer and the supplier is legally set prior to the actual producer-supplier-interaction. The legal framework, set by the governmental legislation, provides the basis for optional, far more detailed contracts and policies negotiated between the business partners (Wildemann, 2000).

<table>
<tr><th></th><th>Feature</th><th colspan="4">Attributes</th></tr>
<tr><td rowspan="6">Relationship-related features</td><td>Level of institutionalization</td><td>Project-specific</td><td colspan="2">Framework agreement</td><td>Capital participation</td></tr>
<tr><td>Information infrastructure</td><td>Conventional exchange of information</td><td colspan="2">Standardized exchange of information</td><td>Integrated information systems</td></tr>
<tr><td>Stipulation of the sourcing conditions</td><td>Conditions not stipulated</td><td colspan="2">Conditions partially stipulated</td><td>Conditions fully stipulated</td></tr>
<tr><td>(Specifications set for...)</td><td>Supplier and product</td><td colspan="2">Supplier, product, quantities and prices</td><td>Supplier, product, quant., prices, procurement & delivery times ...</td></tr>
<tr><td>Direction of interaction</td><td colspan="2">Bilateral</td><td colspan="2">Unilateral</td></tr>
<tr><td>Availability of planning information</td><td>On request</td><td>Direct access</td><td>Information stipulated ex ante</td><td>Information and feedback-times stipulated ex ante</td></tr>
</table>

Fig. 4. Business-relationship morphology

The last two relationship-related features, namely the direction of interaction and the availability of planning information, are both closely related to the last mentioned feature stipulation of the sourcing conditions (Wildemann, 2000). These two features basically determine the producer´s options of accessing planning-relevant information of the supplier. An unilateral interaction implies, that relevant planning information is provided by the supplier without any further interference of the producer. It necessitate an early stipulation of information and the determination of adequate feedback-times, which results in high efforts of coordination and planning during the early phases of a project but rather small efforts at later phases. The exact opposite typically prevails for bilateral interactions during which planning information is only available on request. The two features thus also provide information about when the producer and a particular supplier interact during the inter-company order execution.

4.4 Definition of product/service types

Referring to the product/service morphology presented in chapter 4.2, four different product/service types can be identified. These types are in the following termed as the bottleneck-, the long-running-, the standard-outsourced- and the diverse-source product or service.

The standard-outsourced type refers to few-part products with simple structures. This type is not regarded as the most relevant type in the context of late deliveries since production times are comparably short, there's a huge supplier base offering those products and substitution is comparably simple. The diverse-sourcing product/service type covers products or services that can be provided by external suppliers as well as by the producer himself or at least that the producer is basically able to produce these products/services in-house. Since the producer can thus directly influence the supplier-substitution by providing the product himself, the diverse-sourcing products are less critical regarding the supply risks. Therefore standard-outsourced and diverse-sourcing products or services are not described more in detail within this paper.

Concerning the criteria mentioned above the bottleneck product or service and the long-running product or service best represent the stated problems of late deliveries in the machinery and equipment industry. Therefore these two types are presented in the following. The analysis about the information requirements in the coordination-points in the following chapters will also base upon these two product/service-types.

The bottleneck product or service, is primarily characterized by low levels of specification and high levels of individualization. This often necessitate an order-specific development or at least a substantial modification of existing product or service structures. The product/service requirements are usually set, whereas its characteristics to comply with those requirements can either fully or partially be determined by the supplier. Bottleneck products are furthermore usually very complex and consist of numerous parts. Bottleneck services typically include several, heterogeneous subservices. Those products or services are frequently subject to specification changes during the order execution, which are mainly induced by the final customer or the producer. Supplier-induced changes to the specifications are also conceivable though. This is not at least a result of the typically small number of available suppliers for bottleneck products or services and their respective market power.

At the same time, this makes it difficult or even impossible to substitute a supplier of bottleneck products or services in the course of the order execution. The main producer thus has to face considerable market-induced supply risks as far as this type of products or services is concerned. Fig. 5 shows the described characteristics of bottleneck products or services - marked in purple.

Concerning the relevance for the producer-supplier-interaction, products or services of this type are first of all characterized by the necessity of involving the corresponding suppliers in the very beginning of a project. This is in particular due to the prevailing risks in supply and the necessity of determining the product/service specifications according to the customer´s requirements. The exchange of information in those early phases hence basically refers to specification information as well as scheduling and availability information. As far as the project monitoring and controlling is concerned, bottleneck products furthermore require stringent progress and malfunction procedures in order to reveal supplier delays as soon as possible.

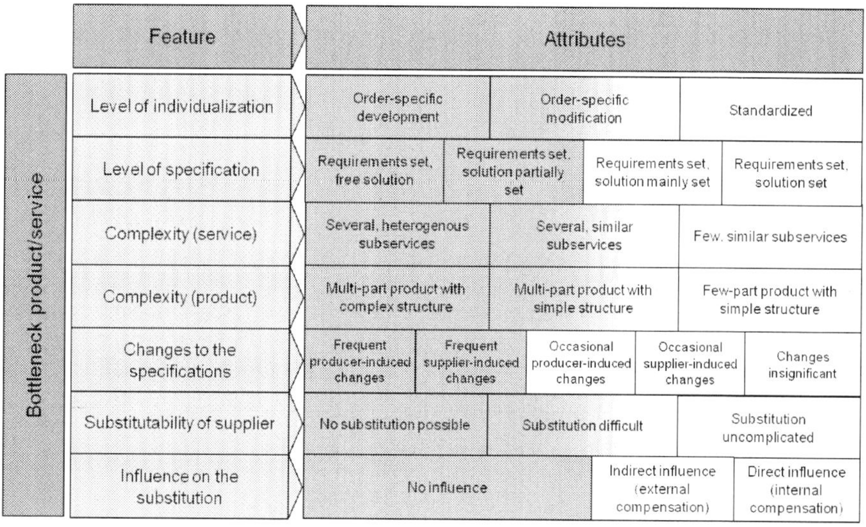

Fig. 5. Characteristics of "bottleneck" products or services

Products or services counted among the long-running type, are typically standardized or order-specifically modified (cf. Fig. 6). They are however in particular characterized by considerable long procurement times. Products or services of this type are either completely or mainly specified prior to the order-specific producer-supplier-interaction. Customer- or supplier- induced changes in specifications are usually rare. In contrast to the aforementioned bottleneck type, long-running products or services cannot be distinguished according to their complexity since basically all three complexity-attributes are possible. Even for highly standardized long-running products or services, there is however a considerable market-induced supply risk due to temporary shortages in supply. Those shortages are mainly caused by unexpected, short-term increases in demand and the resulting demand-overlap, which typically leads to decisive extensions of the anyway long procurement times. Against this background, a substitution of long-running product/service suppliers is typically not possible in the course of the order execution or at least very difficult.

Due to the typically long procurement times, the corresponding suppliers already have to be involved in the early phases of a project, that is for instance during the bid preparation process. Furthermore, the availability of long-running products or services should be continuously checked and assessed during the project planning process. The effort of coordination between the main producer and the respective supplier is, compared to bottleneck products or services, usually smaller. This is due to the typically high levels of product specification and low levels of individualization. The exchange of information in this context thus mainly refers to scheduling and availability information. However, since long-running products or services often represent critical schedule- and production-related restrictions, intense and stringent project monitoring & controlling procedures are needed.

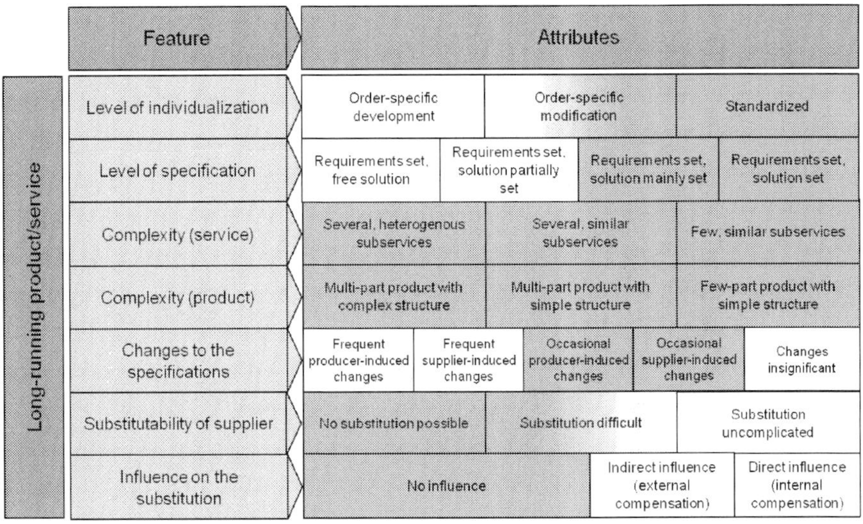

Fig. 6. Characteristics of "long-running" products or services

4.5 Definition of business relationship types

Four ideal business-relationship types can be distinguished. These are the market-autarkic type, the cooperatively-autarkic type, the cooperatively integrated type and the hierarchically integrated type.

Project-specific co-operations occur in the market-autarkic as well as in the cooperatively-autarkic type whereas in the cooperatively-integrated type the supplier is economically highly dependent on the producer and in the hierarchically-integrated type the producer and the supplier have dependent power-structures and oftentimes integrated information infrastructure. Hence, in the context of this paper focusing on the improvement of delivery- and planning reliability in non-hierarchical networks of the machinery and equipment industry, characterized by project-related cooperations, only the market-autarkic and the cooperatively-autarkic types are further analyzed.

The market-autarkic business relationship type is primarily characterized by a frame contract - free collaboration between the main producer and its supplier and shows a low intensity of binding (cf. Fig. 7). The supplier in market-autarkic business relationships is both legally and economically independent from the main producer. Their collaboration is furthermore project-specifically institutionalized. Investments in a common information infrastructure are therefore untypical and the applied communication media rather conventional (e.g. telephone, fax, email or mail). Due to the project-specific collaboration, the sourcing conditions - such as prices, quantities etc. - are usually not set prior to the actual producer-supplier-interaction in the course of the inter-company order execution. Only the product characteristics as well as the requirements for the supplier are specified in the first place. The exchange of information is usually bilateral and initiated by requests from either side.

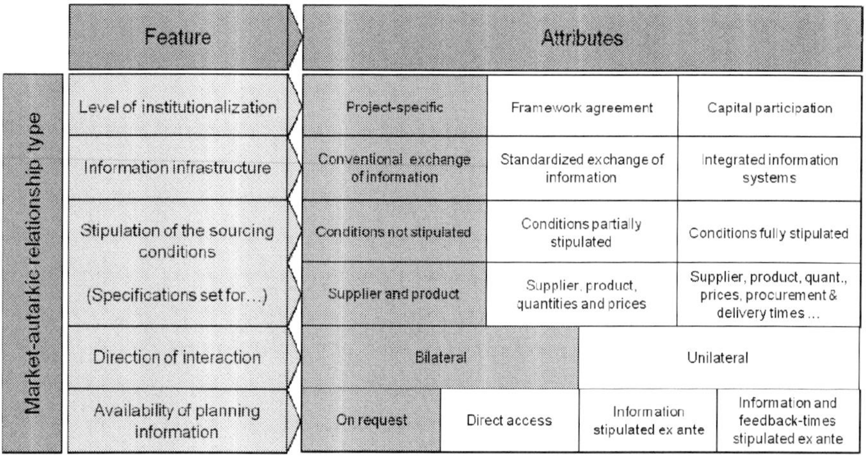

Fig. 7. Characteristics of "market-autarkic" business relationships

Concerning the relevance for the producer-supplier-interaction, market-autarkic business relationships are in particular characterized by the fact, that the main producer has got almost no options of influencing the supplier´s behavior. This matter of fact results from the project-specific cooperation and thus the absence of further legal agreements or even capital participations. Due to the rudimentary specification of the sourcing conditions, the producer furthermore needs to gather a lot of additional information during the gross and detail planning phase via time-consuming bilateral question-answer-interaction. Among those information are for instance additional specification information (e.g. prices) as well as scheduling and availability information (e.g. delivery times and dates) and monitoring and control information during later project phases. The supplier´s input therefore considerably effects the plans and schedules created by the producer and often represents a restriction for the further planning.

The cooperatively-autarkic relationship type is characterized by a cooperative but at the same time market-oriented collaboration (cf. Fig. 8). This means, that similarly to the market-autarkic relationship type, the supplier is legally and economically independent from the main producer. However, their cooperation is based on a legal framework agreement, which e.g. contains further sourcing conditions. In addition to the product characteristics and the requirements for the supplier, quantities and the respective prices of the product to be exchanged, are thus typically specified prior to the actual producer-supplier-interaction (partially set sourcing conditions). Further information, e.g. information concerning delivery times etc., are usually exchanged during the order execution via reciprocal, bilateral processes of interaction. This information usually has to be specifically requested by the producer. The communication media applied in this context is rather conventional (e.g. telephone, fax, email or mail). A partially standardized exchange of information, e.g. via internet-based sourcing platforms, may however also be found in the context of cooperatively-autarkic relationships.

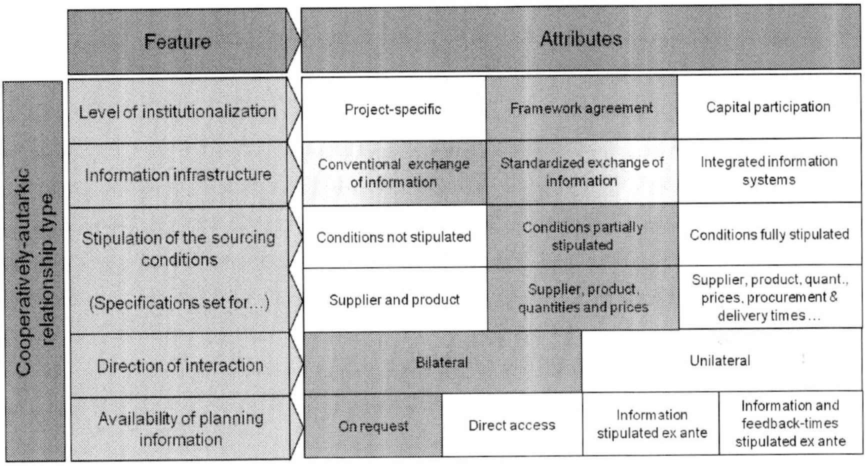

Fig. 8. Characteristics of "cooperatively-autarkic" business relationships

Compared to the market-autarkic type of business relationships, the effort of coordination is smaller during the order execution, due to the additional specifications stipulated within the framework agreement. such as quantities and quantity-related prices.

5. Order-execution reference processes and inter-company interfaces

In the course of the following chapters, the reference processes for the inter-company order execution of producers in non-hierarchical networks of the machinery and equipment industry are worked out. For this purpose, the order-execution tasks will be described in detail and put in their time-logical order within the corresponding order-execution process. In addition, the points of interaction, that is the informational inter-company interfaces, between the producer and the suppliers will be pointed out. There are points of interaction between the customer and the producer as well and these are also shown in the figures. As these points of interaction are not of primary interest in the context of the paper, they are only mentioned for the sake of completeness but are not further explained. With regard to chapter 2, the types of information exchanged at the producer-supplier interfaces will furthermore be determined.

Figure 9 shows a detailed overview of the process activities within the general order-handling process in the machinery and equipment industry deviated from the Aachener Model for Production Planning and Control (PPC) (Schuh, 2006).

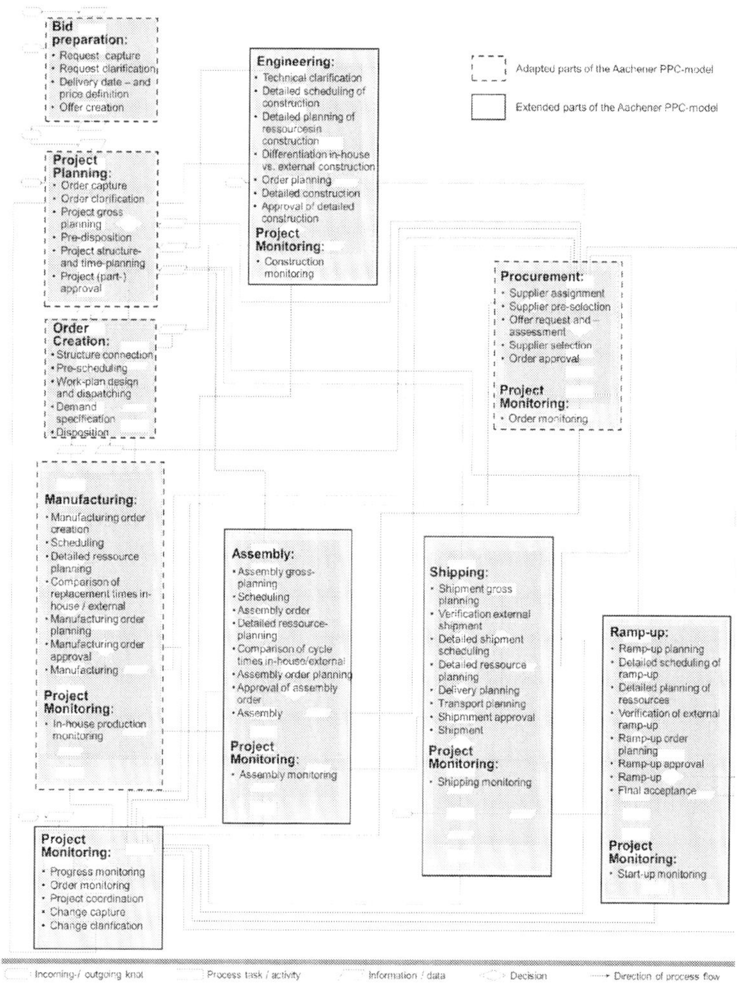

Fig. 9. Schematic overview of the detailed process-steps (Schmidt, 2008)

As stated before, the design of an optimal order-handling process is dependent on the product/service-type and the relationship type of each single transaction. The product/service-type determines the specific starting point of the order management process and its further handling. That means the product/service-type affects the consideration and order of the specific process steps performed for its accomplishment. The relationship-type determines the design and content of the coordination-points between producer and supplier. Consequently, the first step towards a configuration logic is the determination of the relevant type-specific process activities and their order on a gross level. These gross processes are visualized in Figure 10 in form of a process landscape.

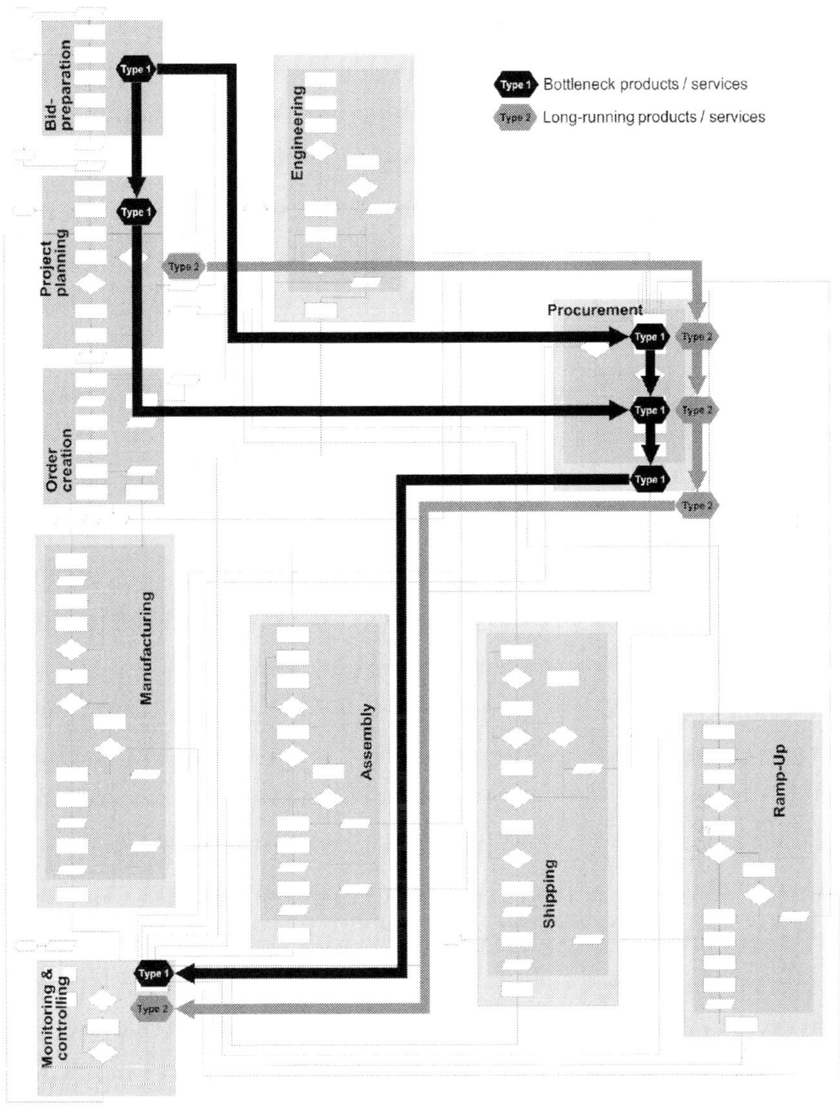

Fig. 10. Sequence of coordination points for bottleneck- and long-running products / services

The procurement of a complex bottleneck product e.g. requires the order-handling process to start already during the bid preparation phase of the producer. The procurement demand has to be allocated via the procurement department, concretized during the project planning in terms of specification and availability and afterwards to be ordered by the procurement department and supervised in form of project monitoring. The process regarding an already known long-running product or service on the other hand starts within the project planning. When the long-running product or service has been requested and ordered by procurement, the progression of the purchased parts/services is also monitored.

Figure 11 illustrates which process steps are accordingly relevant for the order-management process to be analyzed - namely the bid preparation process, the project planning process, the procurement process and the monitoring and controlling process are illuminated as these process steps incorporate the main variables and leverages influencing delivery reliability in the focus of the determined product/service types of bottleneck- and long-running-products/services.

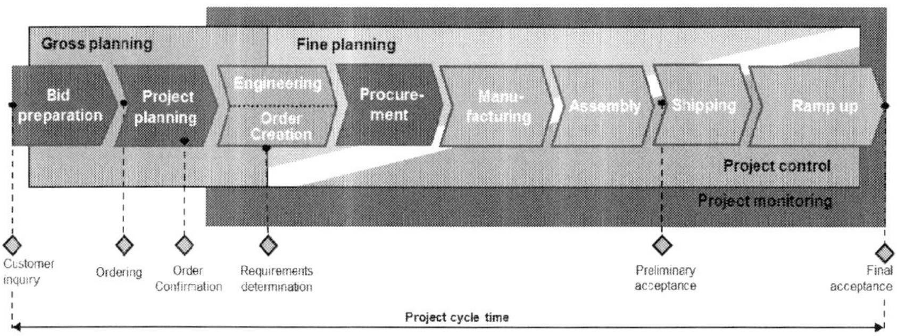

Fig. 11. Gross structure of planning and order-handling processes in the machinery and equipment industry

5.1 The bid preparation process

In the machinery and equipment industry, the process of bid preparation is typically initiated by a single customer inquiry. Accordingly, the first task within this process is usually the inquiry entry, during which the customer inquiry is registered and usually administrated in an IT-based information system (Cuber & Schmidt, 2012).

The inquiry entry is typically followed by the inquiry clarification & assessment. In the course of the inquiry clarification, it is of primary interest to transfer the more or less detailed customer requirements into adequate product specifications. The inquiry assessment serves the purpose of determining at what risks and benefits and to which extend the requested product or service and the therewith related project is realizable (Schuh & Schmidt, 2006).

Especially regarding rather complex structures, during the determination of the delivery date it is hard to obtain precise planning data. Therefore in this task at first empirical data or knowledge, such as data from comparable proposals and projects or the experience and

The next task in the course of the project planning process is the in-advance planning of long-running parts & materials, which includes the identification of those parts and materials as well as the planning of their sourcing. "Long-runners" are those parts or materials, whose standard procurement times exceed the planned lead time of the corresponding order-execution process and which therefore have to be procured in advance. To assure a high quality of planning data, technical parameters as well as master- and standard-specifications have to be updated and the standard procurement times have to be checked (cf. KL3 in Fig. 13). Depending on whether the long-runners are produced internally or externally a production- respectively order-proposal is forwarded to the manufacturing- or procurement-department.

The in-advance planning of long-runners is followed by the assessment of whether the project is feasible as planned or not. If the project is considered as feasible, its project structure and schedules are subsequently passed. In the course of the passing of the project structure and schedules, the written order confirmation is send to the customer in which the commercial conditions as well as the confirmed deadline is binding stated. The final task within the project planning process, the (sub-)project approval, eventually launches the respective subprojects according to the project schedules and initiates the corresponding detail engineering & design order. In this context, each detail engineering & design order successively concretizes the project structure in terms of a continuously growing part list. Furthermore the sub-project approval triggers the transmission of order data and time framework to the process of order creation.

5.3 The procurement process

The question of where the needed materials, components or services should be purchased - that is the assignment of suppliers - is the starting point for the procurement process (Cuber & Schmidt, 2012). While some suppliers might be known for instance due to already existing business relationships, others still need to be found and chosen. Thus, the search for suppliers is the next task within the procurement process, if an appropriate supplier of a certain material, component or service is in fact unknown. Hence, this task basically serves the purpose of identifying potential suppliers. To determine whether a supplier is generally capable of providing the required materials, parts or services, a corresponding exchange of specification information is necessary at this point. This interaction between the producer and potential suppliers is represented by a producer-supplier-interface in the course of the inter-company order execution (cf. KL4 in Fig. 14).

The suppliers are asked - in form of bid requests - to submit a respective proposal. That means intent information as well as consent information are exchanged. After the suppliers´ bids have been received, those are assessed according to company-specific criteria such as lowest price or earliest date of delivery. If it turns out in the following, that the created order plans do not comply with the overall project requirements, the created plans will have to be adjusted and the described activities need to - at least partially - be executed again. If the requirement of the producer is satisfied by the request of the supplier, the bid is classified as realizable and the respective supplier is assigned. Within the following task of order-approval, the producer first of all dispatches the orders to the involved suppliers. The corresponding exchange of consent information leads to the producer-supplier-interface KL5 in the course of the inter-company order execution (cf. Fig. 14). Each order is eventually monitored and controlled in the course of the order monitoring & controlling, during which monitoring & controlling information have to be exchanged (cf. KL6 in Fig. 14).

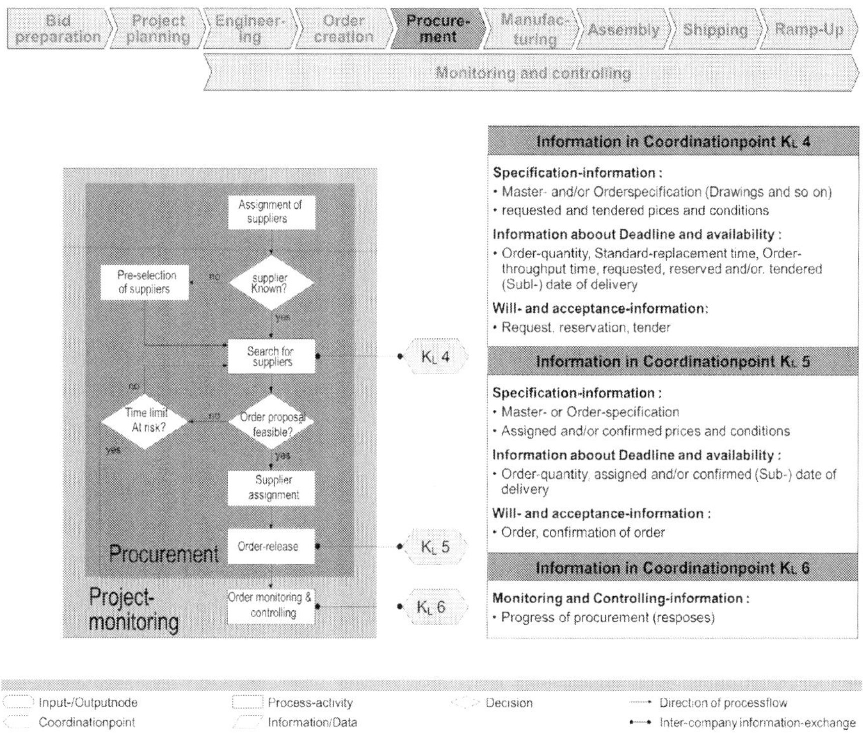

Fig. 14. Detailed structure of the procurement process (Schmidt, 2008)

5.4 The monitoring and controlling process

The monitoring & controlling process is initiated upon completion of the project gross planning and timely ranges from this point on to the end of the order execution. This process thus basically accompanies the order-execution processes of engineering, order-creation, procurement, manufacturing, assembly, shipping and ramp-up. The created detailed plans provide the basis for the monitoring of progress. The monitoring of progress basically serves to track the overall project at all times and to thereby provide an up-to-date overview of the project situation, concerning costs-, quality- and schedule-related aspects, based on monitoring information provided by the involved in-house departments and suppliers. It is in particular subject to the monitoring of the progress in this context to identify divergences between the primary project plans (to-be performance) and the actual project performance (as-is performance). The monitoring of suppliers furthermore necessitates a regular exchange of monitoring information in the course of order-monitoring, which is represented by a further producer-supplier-interface (cf. KL7 in Fig. 15). Another important task within the process of monitoring & controlling is the Entry of Change Requests. Those changes mainly refer to specification- or schedule-related aspects. During the following clarification of Change Requests, the feasibility of the requested changes is assessed together with the customer in due consideration of the current project situation. Since unexpected events such as the mentioned changes or delays due to malfunctions or labor slacks are naturally unpredictable, the primarily set project plans have to be adjusted continuously according to the information created in the course of the monitoring of progress and the clarification of Change Requests. That is in fact the main content of the project coordination. Upon the coordination of the project plans, it is subject to the project coordination to anticipate the occurred changes, as well as to compensate and counter their effects. In order to still realize the guaranteed delivery dates, the modified project schedules and schemes have to be forwarded to the respectively affected in-house departments or suppliers, where the detailed plans should be adjusted accordingly. If the delivery date is no longer realizable, the process-activity of project planning and order coordination of the project planning process has to be carried out again (Cuber & Schmidt, 2012).

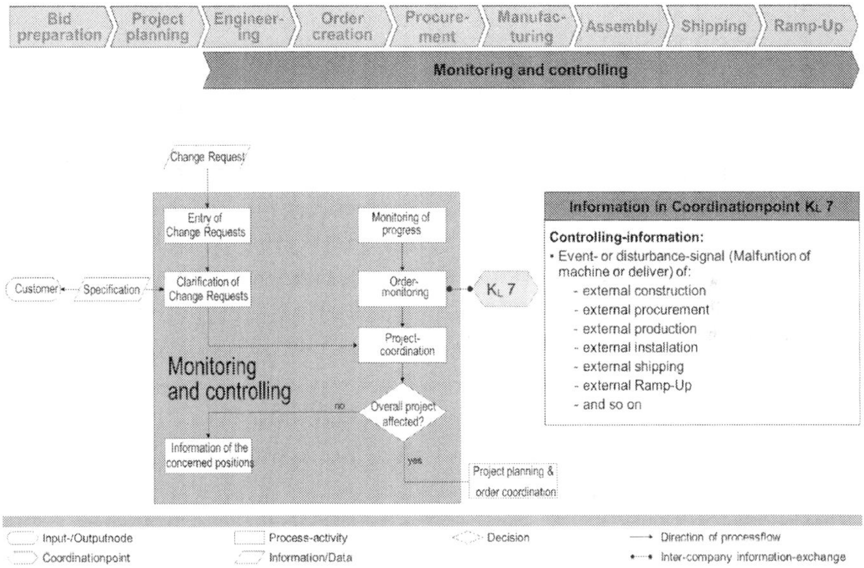

Fig. 15. Detailed structure of the monitoring and controlling process (Schmidt, 2008)

6. Information requirements in the coordination points

The product/service-type-related sequence of order-handling-processes is in combination with the different product/service-types as well as the different business-relationship types the basis for a mapping of the concrete contents to the specific interaction processes. Regarding the relevant contents it is necessary to distinguish whether information from the producer's point of view serves as an input or as an output for his coordination-processes. If an information is both Input and Output, it is a bilateral update of existing planning information. The discussion of the business-relationship-type specific classification of the contents that are relevant for the coordination of the processes of order-handling, interaction and planning will be held in the following using again the example of the bottleneck and the long-running products or services.

The bilateral clarification of the specification between the producer and his supplier within the "Determination of date of delivery" (KL 1) as well as within the "Project planning and order coordination" (KL 2) is a fundamental part of the coordination process for a complex bottleneck product or service. First of all the technical realizability of the product or service has to be verified independent of the relationship-type. In the course of the specification clarification specific function parameters of the specification (output) of the customer specification and the technical restrictions (input) are verified. In addition in market-autarkic relationship- types target prices (input) and other procurement conditions are agreed between the supplier and the producer based on the individual order specification (cf. Figure 16). The examination of the availability with regard to dates is as well carried out in the coordination points KL 1 and KL 2. In both relationship-types the required order quantity (output) and the desired date of delivery (output) are requested by the producer. The supplier replies with a first estimation of a possible delivery date (input).

The clarification of the specification and the delivery date, is followed by the declaration of intent. This happens within the coordination-points "search for suppliers" (KL 4) and "order approval" (KL 5). First of all the technical specification of the product or service (e.g. figures, specification sheet), the preferred date of delivery and the target price of the producer are merged in a inquiry, reservation or order (output). Afterwards this will be sent to the supplier. The producer thereupon receives a proposal respectively a confirmation of reservation and an order confirmation from its supplier (input).

Because of the very high risk of date for a bottleneck product or service the progress- and incident massages in the course of "Order monitoring & controlling" (KL 6) and "Order monitoring" (KL 7) are of particular importance when considering bottleneck products- or services. In both relationship-types the producer generates a request of progress (output) and receives thereupon a corresponding message of progress (input) from its supplier. In case of unexpected events the supplier proactively delivers a message about a possible delivery-failure to the producer.

In case of long-running products/services, the "Determination of date of delivery" is replaced by the "In-advance planning of long-runners" (cf. KL 3 in Figure 17). In this case the master specification and the determination of standard prices is clarified between the producer and his supplier in a bilateral process. Considering long-running products/services in contrast to bottleneck products or services, in the coordination-point KL 4 instead of sending an inquiry or reservation, there is always a reservation send. In the opposite direction of information-flow there is no proposal but directly a confirmation of reservation send. The coordination-point of "Projectplanning and order-coordination", which is relevant considering bottleneck-products/services (cf. KL 2 in Figure 17) is not relevant in case of long-running products/services.

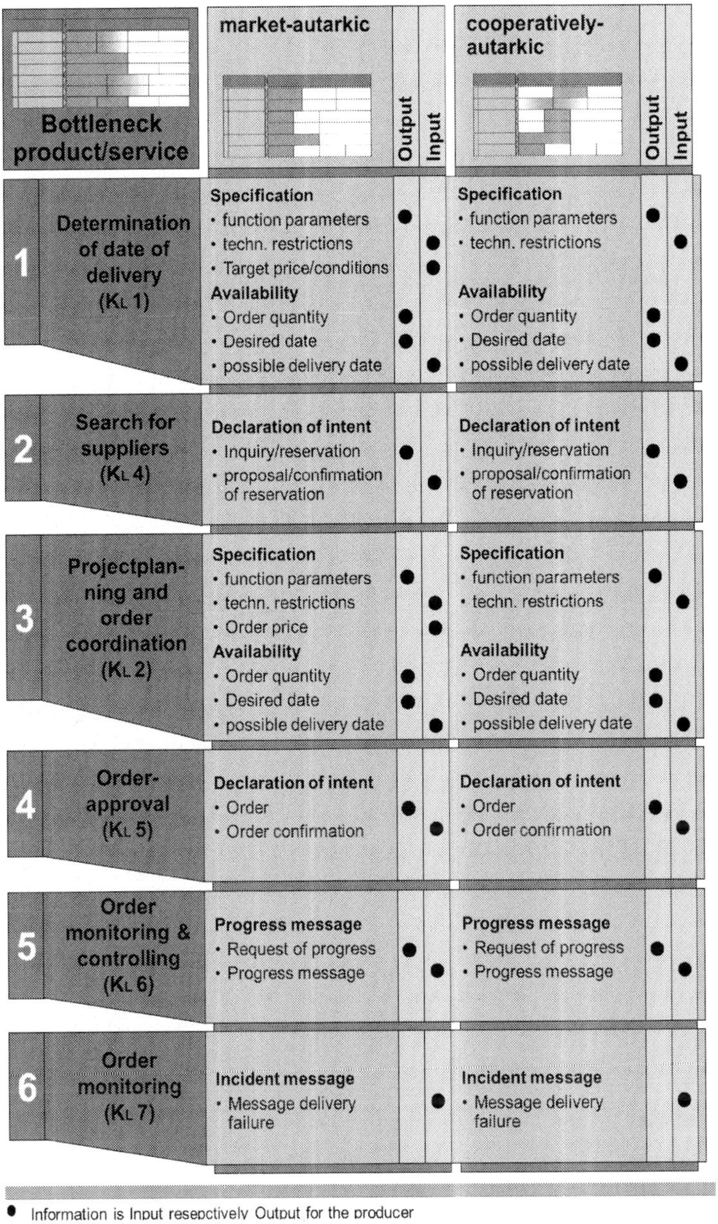

Fig. 16. Exchanged information in the context of bottleneck products or services (Schmidt, 2008)

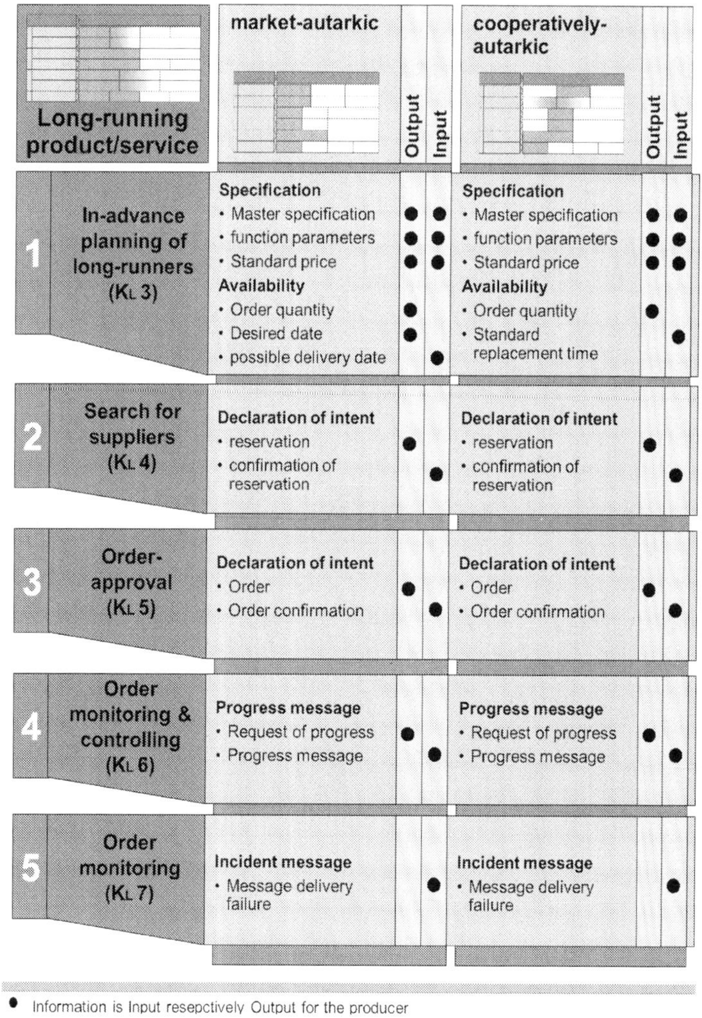

Fig. 17. Exchanged information in the context of long-running products or services (Schmidt, 2008)

7. Conclusion

A considerable part of the problem of late deliveries has its origin in unsound process structures and little process control internally but even more regarding inter-company coordination. However, dealing with a transparent process structure and being able to manage each process according to its requirements and status in the overall project, can already prevent a good portion of late deliveries or at least of uncertainty in the process.

When the target process is structured, it can be managed like a workflow with checklists at the coordination points. By doing so, the transfer of incorrect or incomplete data and information transfer can be reduced. Even though a part may still be late, a delay can earlier be anticipated by a consistent process control. An adjustment of project plans can earlier take place which may provide further options of reaction. Hence, as more transparent and controlled a process is managed, as faster a company is able to realize deviations and as faster it can react to those disturbance variables.

8. References

Arnold, B. (2004): Strategische Lieferantenintegration. Dissertation Technische Universität Berlin. Deutscher Universitäts-Verlag, Wiesbaden 2004

Besslich, J., Lumbe, H. J. (1994): Erster Schritt: Bestandsaufnahme der Material- und Lieferantenstruktur. In: Beschaffung aktuell o. J. (944) 10, p. 22-25

Büdenbender, W. (1991): Ganzheitliche Produktionsplanung und -steuerung. Konzepte für Produktionsunternehmen mit kombinierter kundenanonymer und kundenbezogener Auftragsabwicklung. Dissertation RWTH Aachen, Springer Verlag, Berlin, Heidelberg 1991

Cuber, S., Schmidt, C. (2012) : Advances in Production Management Systems. New Challenges, New Approaches. Springer, Berlin, Heidelberg 2012

Diemer, H. (1992): Grundtypen industrieller Warenverteilung und Möglichkeiten ihrer Gestaltung. Dissertation Universität Würzburg, Eigenverlag, Würzburg 1992

Eberle, A. O. (2005): Risikomanagement in der Beschaffungslogistik - Gestaltungsempfehlungen für ein System, Dissertation Hochschule für Wirtschafts- Rechts- und Sozialwissenschaften (HSG), St. Gallen 2005

Förster, H.-U. (1988): Integration von flexiblen Fertigungszellen in die PPS. In: Hackstein, R. (Hrsg.): FIR+IAW Forschung für die Praxis, Band 19. Springer-Verlag, Berlin 1988

Grosse- Oetringhaus, W. F. (1974): Fertigungstypologie unter dem Gesichtspunkt der Fertigungsablaufplanung. Drucker & Humboldt Verlag, Berlin, 1974

Gunasekaran, A.; Forker, L. & Kobu B. (2000). Improving operations performance in small company. International Journal of Operations & Production Management, 20(3), 316-335

Hillebrandt, V. (2002): Gestaltung und Auswahl von Koordinationsschwerpunkten zwischen Produzent und Logistikdienstleister. Dissertation RWTH Aachen, Shaker Verlag, Aachen 2002

Homburg, C. (1995): Kundennähe von Industriegüterunternehmen. Konzeption, Erfolgsauswirkungen, Determinanten. Gabler-Verlag, Wiesbaden 1995

Kraljic, P. (1988): Zukunftsorientierte Beschaffungs- und Versorgungsstrategie als Element der Unternehmensstrategie. In: Henzler, H. (Hrsg): Handbuch strategische Führung. Gabler Verlag, Wiesbaden 1988, p. 477-497

Meyer, M., Walber, B., Schmidt, C. (2006): Produktionsplanung und –steuerung (PPS) in temporären Produktionsnetzwerken des Maschinen- und Anlagenbaus. In: Schuh, G. (Hrsg.): Produktionsplanung und -steuerung. Grundlagen, Gestaltung und Konzepte. 3. Edition, Springer Verlag, Berlin, Heidelberg 2006, pp. 511-541

Reinhart, G. & Bredow, M. (2006). Kooperationsgestaltung in Produktionsnetzwerken. Vernetzt planen und produzieren: Vol. 5 (pp. 241-244). Chemnitz

Rotering, J. (1993): Zwischenbetriebliche Kooperation als alternative Organisationsform. Ein transaktionskostentheoretischer Erklärungsansatz. Schäffer-Poeschel Verlag, Stuttgart 1993

Scherer, J. (1991): Zur Entwicklung und zum Einsatz von Objektmerkmalen als Entscheidungskriterien in der Beschaffung, Dissertation Universität Köln, Fördergesellschaft Produkt-Marketing, Köln 1991

Schmidt, C. (2008): Konfiguration überbetrieblicher Koordinationsprozesse in der Auftragsabwicklung des Maschinen und Anlagenbaus. Shaker Verlag, Aachen 2008

Schomburg, E. (1980): Entwicklung eines betriebstypologischen Instrumentariums zur systematischen Ermittlung der Anforderungen an EDV-gestützte Produktionsplanung und -steuerungssysteme im Maschinenbau. Dissertation RWTH Aachen. Shaker Verlag Aachen 1980

Schuh, G. (2006): Produktionsplanung und -steuerung. Grundlagen, Gestaltung und Konzepte. 3., völlig neu bearbeitete Auflage, Springer, Berlin, Heidelberg 2006

Schuh, G., Gierth, A., Schiegg, P. (2006): Prozessarchitektur. In: Schuh, G. (Hrsg.): Produktionsplanung und -steuerung. Grundlagen, Gestaltung und Konzepte. 3. Edition, Springer Verlag, Berlin, Heidelberg 2006, p.11-27

Schuh, G., Schmidt, C. (2006): Prozesse. In: Schuh, G. (Hrsg.): Produktionsplanung und -steuerung. Grundlagen, Gestaltung und Konzepte. 3., völlig neu bearbeitete Auflage, Springer Verlag, Berlin, Heidelberg 2006, p. 108-194

Schuh, G., Westkämper, E. (2006): Liefertreue im Maschinen- und Anlagenbau. Stand - Potenziale - Trends. Aachen, Stuttgart 2006

Schulte-Zurhausen, M. (2005): Organisation. 4., überarbeitete und erweiterte Auflage, Verlag Franz Vahlen, München 2005

Schwerk, A. (2000): Dynamik von Unternehmenskooperationen. Dissertation Freie Universität Berlin, Duncker und Humblot Verlag, Berlin 2000

Wiendahl, H.-H., Meyer, M. (2006): Methodische Grundlagen. In: Schuh, G., Westkämper, E. (Hrsg.): Liefertreue im Maschinen- und Anlagenbau. Stand - Potenziale - Trends. Aachen, Stuttgart, p.11-18

Wildemann, H. (2000): Einkaufspotenzialanalyse: Programme zur partnerschaftlichen Erschließung von Rationalisierungspotenzialen. TCW Transfer-Centrum-Verlag, München 2000